Hot Hadronic Matter
Theory and Experiment

NATO ASI Series

Advanced Science Institutes Series

A series presenting the results of activities sponsored by the NATO Science Committee, which aims at the dissemination of advanced scientific and technological knowledge, with a view to strengthening links between scientific communities.

The series is published by an international board of publishers in conjunction with the NATO Scientific Affairs Division

A	**Life Sciences**	Plenum Publishing Corporation
B	**Physics**	New York and London
C	**Mathematical**	Kluwer Academic Publishers
	and Physical Sciences	Dordrecht, Boston, and London
D	**Behavioral and Social Sciences**	
E	**Applied Sciences**	
F	**Computer and Systems Sciences**	Springer-Verlag
G	**Ecological Sciences**	Berlin, Heidelberg, New York, London,
H	**Cell Biology**	Paris, Tokyo, Hong Kong, and Barcelona
I	**Global Environmental Change**	

PARTNERSHIP SUB-SERIES

1. **Disarmament Technologies**	Kluwer Academic Publishers
2. **Environment**	Springer-Verlag
3. **High Technology**	Kluwer Academic Publishers
4. **Science and Technology Policy**	Kluwer Academic Publishers
5. **Computer Networking**	Kluwer Academic Publishers

The Partnership Sub-Series incorporates activities undertaken in collaboration with NATO's Cooperation Partners, the countries of the CIS and Central and Eastern Europe, in Priority Areas of concern to those countries.

Recent Volumes in this Series:

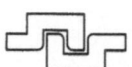

Series B: Physics

Hot Hadronic Matter
Theory and Experiment

Edited by

Jean Letessier

LPTHE
Université Paris 7
Paris, France

Hans H. Gutbrod

Gesellschaft für Schwerionenforschung
Darmstadt, Germany and
CERN
Genèva, Switzerland

and

Johann Rafelski

University of Arizona
Tucson, Arizona

Springer Science+Business Media, LLC

Proceedings of a NATO Advanced Research Workshop on
Hot Hadronic Matter: Theory and Experiment,
held June 27–July 1, 1994,
in Divonne, France

NATO-PCO-DATA BASE

The electronic index to the NATO ASI Series provides full bibliographical references (with keywords and/or abstracts) to about 50,000 contributions from international scientists published in all sections of the NATO ASI Series. Access to the NATO-PCO-DATA BASE is possible in two ways:

—via online FILE 128 (NATO-PCO-DATA BASE) hosted by ESRIN, Via Galileo Galilei, I-00044 Frascati, Italy

—via CD-ROM "NATO Science and Technology Disk" with user-friendly retrieval software in English, French, and German (©WTV GmbH and DATAWARE Technologies, Inc. 1989). The CD-ROM also contains the AGARD Aerospace Database.

The CD-ROM can be ordered through any member of the Board of Publishers or through NATO-PCO, Overijse, Belgium.

Library of Congress Cataloging-in-Publication Data

On file

ISBN 978-1-4613-5798-8   ISBN 978-1-4615-1945-4 (eBook)

DOI 10.1007/978-1-4615-1945-4

© 1995 Springer Science+Business Media New York

Originally published by Plenum Press, New York in 1995

Softcover reprint of the hardcover 1st edition 1995

10 9 8 7 6 5 4 3 2 1

PREFACE

The past decade has seen the development of the operational understanding of fundamental interactions within the standard model. This has detoured our attention from the great enigmas posed by the dynamics and collective behavior of strongly interacting particles. Discovered more than 30 years ago, the thermal nature of the hadronic particle spectra has stimulated considerable theoretical effort, which so far has failed to 'confirm' on the basis of microscopic interactions the origins of this phenomenon. However, a highly successful Statistical Bootstrap Model was developed by Rolf Hagedorn at CERN about 30 years ago, which has led us to consider the 'boiling hadronic matter' as a transient state in the transformation of hadronic particles into their melted form which we call Quark-Gluon-Plasma (QGP).

Today, we return to seek detailed understanding of the thermalization processes of hadronic matter, equipped on the theoretical side with the knowledge of the fundamental strong interaction theory, the quantum chromo-dynamics (QCD), and recognizing the important role of the complex QCD-vacuum structure. On the other side, we have developed new experimental tools in the form of nuclear relativistic beams, which allow to create rather extended regions in space-time of **Hot Hadronic Matter**. The confluence of these new and recent developments in theory and experiment led us to gather together from June 27 to July 1, 1994, at the Grand Hotel in Divonne-les-Bains, France, to discuss and expose the open questions and issues in our field.

This volume presents a complete account of the presentations made at the workshop. We begin with a broad, historical introduction to the field, which is largely theoretical in nature. Subsequently, we present the topical highlights, scheduled in the form of mini-workshops at the meeting. These include in particular the search for mechanisms of thermalization and entropy production in fundamental interactions, the particle correlation and multi-particle studies of the underlying dynamics, and the experimental and theoretical efforts to develop observables of the new phase of matter, the QGP, the presumed source of boiling hadronic matter. The last section of this volume is devoted to the future and is mostly experimental in nature. We thank all the contributors for their willingness to submit in written form, and in a very timely fashion, the accounts of their ongoing research.

Our special appreciation goes to NATO-Scientific Affairs Division for the help and support given, which was essential to get this program going. It is with considerable concern that the scientific community considers the current suspension of the fundamental workshop activity, as the priorities at NATO shift towards the development of cultural links with the former Soviet Union nations. The vacuum left behind as NATO withdraws its basic science workshop support is considerable and we do not see an easy remedy. As our own meeting has shown, NATO-Scientific Affairs Division has over the years acquired a very instrumental and highly successful role in facilitating basic research amongst the researchers from member states. We sincerely hope that this highly important long-term cultural activity will soon be resumed in the area of basic sciences.

The meeting was held in close cooperation with CERN, the European Nuclear Physics laboratory, where much of the European activity in our research field is centered. We received considerable 'in kind' support from the CERN-PPE, -TH and -DG divisions for which we are very thankful. Further financial support to the meeting was provided by the Soros Foundation-International Science Foundation, which has supported the participation at the meeting of a number of key former Soviet scientists. Many participants were funded in full or in part by their home laboratories and we thank all for this very important contribution to the success of the meeting.

The last vote of thanks goes to all those who have helped us in making this gathering possible: the staff of the Grand Hotel and in particular Laurette Belleville, for their assistance, understanding and kind cooperation; Rosemarie Audria (CERN) for her invaluable organizational services, before, during and after the meeting. We would like to acknowledge kind support and assistance from colleagues who helped us with the organization of the meeting; we particularly thank Hans Bethe, Andrzej Bialas, Pierre Darriulat, Steven Frautchi, Rolf Hagedorn, and Helmut Satz.

It is fortuitous that the year 1994 has also seen the 75th birthday of **Rolf Hagedorn** to whom we dedicate this volume, expressing our great appreciation for his seminal role in the development of the foundations of our today well established and broad research field. For this reason we open the volume with some personal reminiscences about the times of Rolf Hagedorn and the beginning of **Hot Hadronic Matter**.

Jean Letessier, Hans H. Gutbrod and Johann Rafelski
Paris, Geneva, Darmstadt and Tucson

November 15, 1994

CONTENTS

EXPONENTIAL MASS SPECTRUM — HISTORY

PHASES OF HADRONIC MATTER

THERMALIZATION — ENTROPY

MULTIPARTICLE PRODUCTION: CORRELATIONS AND INTERMITTENCY

PHOTONS AND DILEPTONS

PARTICLE SPECTRA AND STRANGENESS

APPROACHING FUTURE

TRIBUTE TO ROLF HAGEDORN

Torleif Ericson,[1] Maurice Jacob,[1] Johann Rafelski[2,3] and Helmut Satz[1,4]

[1]CERN, 1211 Geneva 23, Switzerland
[2]Department of Physics, University of Arizona
Tucson, AZ 85721, USA
[3]Laboratoire de Physique Théorique et Hautes Energies *
Université Paris 7, Tour 24, 5ᵉ étage
2 Place Jussieu, F-75251 Paris Cedex 05, France
[4]Fakultät für Physik der Universität, D-33615 Bielefeld, Germany

TORLEIF ERICSON

Dear Rolf, it is now nearly 35 years since we have been friends and office neighbors at CERN. I am glad and honored to have this opportunity to address you on the occasion of your 75th birthday celebration. It gives me a special pleasure to do so in the scientific setting of this meeting with tributes to the seminal influence you have had on the scientific community. Of course this room looks in no way like the offices and seminar rooms at CERN in which most of your scientific discussions took place in the past, but still it is very much a scientific and lively meeting. This is a greater tribute to you than any address and a most appropriate one.

I do not want in this address to dwell so much on your great scientific achievements. This has already been covered comprehensively during the sessions at this meeting. But I must say that I appreciated greatly your very nice lectures about the background of the statistical bootstrap model and related areas of strong interaction physics. Instead I want to recall the atmosphere at CERN when I first met you in 1960 and some episodes that I remember from our time together at CERN. I will also say a few words about the miraculous events that brought you into physics in the first place, and to CERN.

Rolf brings to my mind a particular word: fortuitous. The word 'fortuitous' has several meanings. First of all fortuitous means something that happens by chance, it's statistical, and statistical of course brings Rolf to mind immediately. But fortuitous has also the overtone of something of good fortune, good luck, and that again is very much what I associate with Rolf's impact on all of us. It also applies very much to his path into science and choice of research area, which has also been associated with a couple of steps of chance and good fortune.

---

*Unité associée au CNRS: D 0280

I first met Rolf, when I arrived at CERN as a postdoc in 1960. CERN was a quite different place 35 years ago, not at all like CERN today. It was considerably smaller, although Rolf probably would say it had already become a very large organization during the time he was there before. But in 1960, when I came to CERN, the Theory Division had only some 30 people or so and now we are 170. So everybody knew everybody and we discussed our research with everybody else. I had the good fortune to be assigned an office next door to Rolf's, and luckily, fortuitously, even in later years we have always had offices side by side, so we have been in very close contact. I had been working myself on statistical reactions in nuclear physics, and in particular on compound nuclei and related topics, and it was of course interesting to see what this could mean in particle physics. I knew before I arrived of Rolf's work, at least vaguely. Once at CERN I quickly wanted to know what Rolf was up to. I immediately found out that he was a true strong interaction man — I recall that when I first went to his office, I did not look at his door. I just knocked and stepped in — ignorant of a huge sign on the door saying 'Ne pas déranger' —'Do not disturb'! I did not know that Rolf had the habit of taking a nap for an hour or so at noon, and I just barged in there. As you can imagine, there was a pretty strong interaction! I was a newly arrived, fresh postdoc, while he was a staff member fast asleep. But ever since I only had very pleasant interactions with Rolf.

Rolf Hagedorn, June 30, 1994 *(PHOTO-CERN)*

What was the Theory Division at this time? Let us look at the people who were around. In fact some of these people are here with us. There was one important fixture of the Theory Division through all the years, our beloved Tatiana Fabergé, who is here with us today. She was already then running the TH-division, and has done it ever since, independent of whoever happened to be the division leader. The late Léon Van Hove came to take up the leadership of the Theory Division on nearly the same day I arrived and he was already a major physics figure. He had been on and off at CERN before, but that was about the time he started to take care of us. He is unfortunately no longer with us, but his wife Jenny is here. I am very happy about, that since Léon was a very old and dear friend of Rolf's. Jaques Prentki was member of the division from the early days. John and Mary Bell had arrived at CERN about a year before — Mary was not part of the Theory Division, but was in close contact with it. Since Rolf also was interested in accelerators, they were also in close contact with him. André Martin had shown up the year before and was like me a Fellow in the Theory Division as was Daniele Amati. Both became staff members a couple of years later. Roland Omnès had left just before I arrived, and it is a pleasure to see him here. Frans Cerulus had also left just before, and he had of course interacted very strongly with Rolf. André Petermann was busy with calculating QED corrections. Another staff member was the late Vladimir Glaser, who was a very brilliant abstract theoretician, who had been a fellow postdoc with Rolf in Germany. He was a staff member for many years and a great specialist in finding ingenious counter-examples to theoretical conjectures. That covers the staff members, and most of the remainder were Fellows, with a few visitors from the USA. It was really quite a small group, very different from the 170 postdocs and senior people who are in the group today.

In these early years I had a lot of discussions with Rolf about statistical hadronic physics and its basis, but I never got quite into the field. Rolf explained to me repeatedly the background of the Fermi statistical model and its assumptions. I always thought I understood it, when he explained it, but afterwards I never quite understood it. I do not know today if I understood it or did not; it was on and off. I speculated if the kind of statistical fluctuations effects I worked on in low energy nuclear physics could be applied to these high energy situations. Steve Frautschi developed some of these ideas later on. I instead turned to intermediate energy physics, which took off just then. While I did not quite personally get involved into Rolf's research line, I followed it very closely.

One day, which I remember very vividly, some time in 1964, I ran into you, Rolf. You were just bubbling over to a degree I have not seen you ever. Your eyes were quite bright, and you described to me all these fireballs: fireballs going into fireballs living on fireballs forever and all in a logically very consistent way. This must have been only a few days after you had invented the statistical bootstrap picture. I really had the impression of a man who had just found the famous stone of the philosophers, and that must have been exactly how you felt about it. Clearly, you recognized at once the importance of the novel idea you introduced there. It was very interesting to observe, Rolf, how deeply you felt about it yourself from the very beginning.

Let me now tell you a few words about the Rolf's career — how he got into physics in the first place. This is a very interesting story, I have collected from Rolf in various forms at various times throughout the years. Until rather late in his life there was little that indicated that Rolf would make an exceptional career as a scientist. Rolf's life as a young man was deeply marked by the upheavals of the war years in Europe, of the greater evils that took place at that time. He graduated from high school just before the war, and was immediately drafted into the army, and shipped off into North Africa as an officer in the Rommel Afrika Korps. He has told me in fact how much he enjoyed the vast spaces and the quiet, pitch dark nights with brilliant stars in the desert. As you know the Afrika Korps was captured after the Allied invasion of North Africa in 1943 and Rolf spent the rest of the war in an officer prison camp in the United States. They were all very young people in the camp and there was not much to do, so they set up their own 'university'. There were of course no senior professors, so they taught each other what they happened to know. Presumably this was similar to Viki Weisskopf's saying: 'Physicists are persons who explain to each other what they do not know'. So maybe that was when Rolf became a physicist. Rolf got training not so much in physics, but in other subjects such as mathematics. He ran into an assistant of Hilbert's, who taught him a lot of mathematics. When he came back to Germany a year or more after the war, German universities were destroyed. In fact the students usually had to start at the universities literally carrying bricks to build up the institutes. Rolf in his mid-twenties was not a very young student. Because of his training in the prison camp he succeeded, after some non-trivial effort, to be accepted as a fourth semester student at the university of Göttingen — one of the few undestroyed ones and at the same time with a great tradition in physics and mathematics. After having completed his studies with the usual diploma and doctorate (with a thesis you would not quite expect: on solid-state physics) he was accepted at the Max Planck Institute for Physics as a postdoc. This was still at Göttingen at the time, and not as now in Munich, and the director was Werner Heisenberg. He was there in an exceptional group and I think some names might interest you. There was Bruno Zumino, Harry Lehmann, Wolfhart Zimmermann, Kurt Symanzik, Gerhard Lüders, Reinhard Oehme, Vladimir Glaser, Carl Friedrich von Weizäcker and a couple more. I suppose at the time Rolf thought that this was a pretty normal group, but there was something very special at Göttingen at that time, for all of these have made important marks on physics.

Rolf Hagedorn, after lecture, in discussion with T. Ericson
(also shows G. Feinberg and J. Rafelski) *(PHOTO-CERN)*

In 1954 he got the opportunity to go for some months to CERN, this very new place coming up in Geneva. It was not known much, and not even formally and totally approved yet. He went to CERN to help with some accelerator design problems, to calculate some non-linear oscillations in particle orbits. The pioneering work on linear orbit theory had just been completed by Gerhard Lüders, who wished to go back to Göttingen. He asked Heisenberg to send somebody to replace him, and Heisenberg asked Rolf if he was interested for a couple of months. Being paid only 300 DM per month at the Max Planck Institute (not much for a family of three) he jumped the occasion. Important events in life happen like that: you come for a couple of months and you end up for the rest of your life, and that's exactly what happened to Rolf.

During this early period Rolf calculated non-linear orbital oscillations of the CERN–PS with some clever novel approximation techniques extended from one dimension to two dimensions. When the CERN theory group came to Geneva from Copenhagen, where it had been located at first, Rolf joined it, and it was where I met him 35 years ago. In the small CERN environment of the time it was very easy to move between very varied activities and he had of course been a theoretical physicist all the time. I want to emphasize that it is this unusual background that has marked Rolf very deeply in his scientific evolution; without it he would have had a very different and probably less original impact on physics. We all know that Rolf is not a physicist who shoots from the hip when you bring up a problem with him. He takes his time, he works it over, he wants to penetrate it. Maybe this reflects the thoughts of long nights in the desert and the prison camp, in which time of reflection is a great virtue. I am sure that we see here the mark of this background. You were brought up in an environment with plenty of time for informal discussions, with no professor just breathing down your neck and pushing you to produce something very quickly. You prefer to penetrate things slowly, and you do it in a very deep fashion.

Maybe that is why, in a strange way, Rolf differs from a large number of other prominent physicists I met during my life. Most of the physicists are recognized rapidly after their contribution. After that quick recognition, their impact as time goes on gets disseminated and integrated, and people often notice it less and less. With Rolf, the opposite happened. It is like the best of wines. It is not so palatable in the early years, but as time goes on it just gets more and more remarkable. Statistical Bootstrap was looked upon with considerable skepticism in the beginning, at least inside the CERN Theory Division, and as time has gone on, it has taken on bigger and bigger dimensions and has become more and more important. This is the sign of truly original work, of something that really had influence on our thinking. It is a real tribute from the physics community to the leadership that you have provided. We are all very thankful to you for what you have done. Thank you much for all these years together. I wish you many happy anniversaries to come.

MAURICE JACOB

When Jan Rafelski asked me to say a few words about Rolf, I strongly felt the honor thus bestowed on me but, at the same time, I had to express my worries. I was afraid not to do justice to Rolf's achievements in physics in front of such a distinguished assembly of specialists which counts many members who could do that far better than me. But it is a great pleasure for me to praise Rolf, and I took this duty willingly, since I am expected to speak about Rolf as a man as well as about Rolf's achievements as a physicist. Rolf has made very important contributions to our field of research. The Hagedorn Temperature and the Statistical Bootstrap Model are concepts which are here to stay. They have stimulated much further research and this conference bears magnificently witness to that! But Rolf is also a wonderful person and, saying that, does not require a specialist.

I first met Rolf in Madras in the fall of 1963. I had seen him only briefly earlier and knew about him as a member of the CERN theory group, but in Madras, where we spent three full months together, that I really got to know him. We had independently responded to the call of Alladi Ramakrishnan and were both giving a one term lecture course at Matscience.

If I remember well, Rolf was teaching on relativistic mechanics after a book which he had recently written on the subject.

We spent a lot of time together, sharing this fantastic Indian experience, and I much appreciated the friendship which he extended to me despite the difference in age. I strongly felt the great sensitivity with which he was reacting to everything. I was impressed by his thoughtful kindness and his benevolent understanding in front of many features of Indian life which were bringing a complicated mixture of great admiration and some of the times revulsive feelings to our western eyes. His long war years had shaped up his compassion for humankind.

I never saw Rolf loose his temper. His only strong reaction which I can recall was during a night which we spent in a guest house in an Indian wild life reserve at Bandipur. This was during a wonderful four day trip which Lise and I shared with him through the magnificent Mysore area. During the middle of the night the guest house, where we had been alone in the evening, was invaded by a very noisy group of visitors who stormed it past midnight as if it was their own. I still remember Rolf shouting "Since it is a guest house you should have realized that there could be guests inside".

Ch. Llewelyn-Smith is querried after his lecture
July 1, 1994 *(PHOTO-CERN)*

Rolf was a magnificent traveling companion and the great pleasure which he had in discovering these magnificent temples of southern India was communicative. I remember him summarizing his visits saying "Now I have seen something beautiful. I am happy". He was a zealous photographer and he actually collected a magnificent series of photographs which I could admire later. He had clever ways to take close up pictures of people without embarrassing them. I admired his skills. I also recall his appreciation for the magnificent Indian music which we could hear in Madras. We listen together to "monuments" such as M.S. Subulakshmi and Ravi Shankar. The Carnatic music of Subulakshmi is very different from our classical western music but Rolf would say "Music can take many forms but one can always recognize and fully appreciate great music".

If I may at this point venture an hypothesis, I would say that music may have had even a role in Rolf joining the CERN Theory group. Rolf had come to CERN as an accelerator physicist, working with Schorr on accelerator theory. This was in the middle fifties. But his office was next to that of Jacques Prentki and Bernard D'Espagnat and Jacques and Rolf could often share their love for music during many discussion and also talk physics. Jacques told me that he remembers Rolf talking in a seminar about "an ensemble of accelerators". We already notice the influence of thermodynamical concepts! But, back to music still, I was told that they were surprised to discover a common and great appreciation for Heinrich Schutz. A student of Monteverdi, Schutz is considered by the experts as the initiator of Baroque music. He may not compare to many to the giants who followed him and in particular to Bach but Rolf would say "Bach is like the Alps when Schutz is rather like the Jura, and... I like the Jura". This we know he does, living in the countryside where the slope of the Jura starts to rise sharply, and sharing with his wife this pleasant and quite country life with cats, horses and lots of music.

Torleif Ericson, for many years his neighbor in the TH Division has told us about Rolf place at CERN. I will turn now to the very difficult task of trying to summarize in plain words Rolf's contributions to our field of research.

First with generalities, one may say that it is in the line with the brilliant germanic tradition in statistical Thermodynamics and Rolf may find a well deserved pride in having his name associated with a temperature. It is also in line with the phenomenological approach whereby one tries to understand and predict according to models. It is finally in line with the desire of any theorist to achieve a powerful synthetic view providing a rationale for the observed phenomena. The Statistical Bootstrap Model, which is Rolf's brain child, fits perfectly the latter point.

Particle production processes are now so well known that they are taken for granted. Nevertheless the fact that very high energy collisions generally result in the production of many secondaries first came as a surprise to many. Having admitted that this is the case, the idea to try to apply the wide body of knowledge of statistical thermodynamics to such production processes may naturally come to mind. However, difficulties quickly speak for themselves. Great minds like Fermi and Landau indeed made clever attempts but with unsatisfactory results. Particle physics and statistical physics were long much separated. This is no longer the case! In particular, we now know of the great successes which were later met at the interface of statistical physics and field theory.

The contribution of Rolf bears on the application of statistical physics to the phenomenology of hadronic interactions, something in which Leon Van Hove was also much interested at CERN and which he described as "A meeting ground between particle and statistical physics, a dialogue between theory and experiment".

Rolf's work started much before I came to CERN. My understanding is that Bruno Ferretti, who was head of the Theory Division when he joined it, asked him to try to predict particle yields in the accelerator high energy collisions of the time. This he started with Frans Cerulus. There were little clues to begin with but they made the best of the fireball concept which was then supported by Cosmic Ray studies and used it to make predictions about particle yields and therefore the secondary beams to be expected from the machine beam directed at a target. Many key ingredients brought soon afterward by experiment helped refine the approach. Among them one should quote the limited transverse momentum with which the overwhelming majority of the secondary particles happen to be produced. They show an exponential drop with respect to the transverse mass. One should also quote the exponential drop of elastic scattering at wide angle as a function of incident energy. Such exponential behaviors strongly suggest a thermal distribution for whatever eventually comes out of the reaction and it is the great merit of Rolf to have clung to this thermal interpretation and use it to built production models which turned out to be remarkably accurate at predicting yields for the many different types of secondaries which originate from high energy collisions.

It is his great merit since many objections were raised at the time. What could actually be "thermalized" in the collisions? Applying straightforward statistical mechanics to the produced pions was indeed giving the wrong results. But, even if there was a thermalized system at all, why was the temperature apparently constant? Shouldn't one have expected it to rise with incident energy or with the mass of the excited fireball, according to the Boltzmann law?

It was Rolf's great merit to interpret the apparently limiting temperature which could be associated with the transverse mass distribution of the secondaries as resulting from an exponential spectrum for the many resonant states in which hadrons can be excited into before these resonance would fragment into less massive ones to eventually give, at the end of the line, the observed secondary particles.

The rise of the temperature is associated with the population of higher and higher energy levels by the elements of a system. If there is an exponentially increasing number of level offering themselves to be filled, the temperature saturates. It is the entropy which eventually increases linearly with the collision energy but the temperature gets then stuck to a limiting value. This is the Hagedorn temperature. It is of the order of 150 MeV, close to the pion mass.

The impressive number of states which have now to be considered at the same time leads to a new writing of equations based on statistical physics. The factorial n factor, which was plaguing statistical calculation focusing on pions only and which was introduced to rightfully avoid multiple counting in phase space integrals, had now to be dropped since each one of the many states was unlikely to have a population exceeding one. Agreement between experiment and statistical calculations prevailed at long last.

In his recent popular book, "The quark and the Jaguar", Murray Gell Mann explains how progress in physics often results from the dropping of a condition which was long considered as mandatory and which had not been properly challenged. This applies very well to this factorial n factor which as Rolf concluded had not to be there after all.

Despite the great success of the Hagedorn approach at predicting particle yields, we may still have reservations at speaking about a temperature in collisions among elementary particles but, as we shall see later, this applies to heavy ion collisions which are attracting an increasing interest and attention. But now comes Rolf's great achievement in pioneering the development of the Statistical Bootstrap Model. Rolf has beautifully described its genesis in his talk at this conference. His paper of '65 on the subject has become a citation classics.

To put it in a nutshell , as I should tonight, one may say that each of the many resonant state in which hadrons can be excited through a collision is itself a constituent of a still heavier one while being also composed of lighter ones. What Rolf showed is that when one puts logic and hard work into the idea one cannot escape an exponential spectrum of resonant states. The temperature of such a system is then limited from above. This limit is the Hagedorn temperature. If one takes a more global view talking about "fireballs" (in the old language) or "clusters" (in the more modern vernacular) rather than of resonances, the conclusion is that the temperature of such objects is independent of their mass. One can then also understand why the limiting temperature is of the same order as the mass of the smallest mass state, the pion.

The concept of an exponential spectrum is now part of our understanding of hadron phenomena. It has been reached through different approaches such as that offered by dual models. It fits beautifully the hadronic level counting which can now be followed up to over 4000 catalogued resonances Rolf was first at pinning it down through his Statistical Bootstrap Model. The Statistical Bootstrap Model has been at the origin of much further works which have refined it. Rolf was thus at the origin of an important and very fruitful line of research.

Can one go beyond the limiting temperature set by Hagedorn? The answer is yes but one has to consider a phase transition whereby one leaves the hadronic phase to reach a new phase where the hadron constituents, the quarks and the gluons, are no longer confined. The limiting temperature becomes a phase transition temperature which can be calculated through lattice gauge theory calculations.

This conference bears witness to all the fascinating activities which now go applying statistical and thermodynamical concepts to heavy ion collisions. Rolf can take a fully justified pride in having pioneered and followed this line of approach for a long time and this despite of the many stumbling blocks which he had to overcome. We can rejoice with him that many of his views have been vindicated by recent and promising developments.

We can also rejoice with him that his "pro forma" retirement, already 10 years ago, has kept him close to CERN and we are very happy to have his frequent visits. We are happy to see him following closely and participating in present research. His talk at this conference bears magnificent witness to that.

In connection with Rolf "pro forma" retirement, I should conclude by saying that I was most happy to have Rolf as a test case when pushing through the Management Board, together with Adolf Minten and Allan Wetherell, a special status for "retired scientists willing to continue research". This of course met with difficulties but they could be overcome and the new status was finally approved in the summer of '84. Rolf was the first person at CERN to be granted the new status. Great was my joy when I could leave with him the letter from the DG written to that effect which had followed by a mere few days the approval of the new scheme. Ten years later we can see how good this was.

Many happy returns, Rolf !

Let me try to summarize topics much discussed at this conference through a modest limerick:

There are many hadrons with strong interaction

Which behave thermally in their curious motion

Rolf Hagedorn shown long ago

How S-B-M can make it go

Exponential spectrum is the explanation.

JOHANN RAFELSKI

This meeting on HOT HADRONIC MATTER probably would not be possible, had Rolf Hagedorn not pioneered this field some 30 years ago, his work being driven by experimental data. Even today we still do not understand what is going on when we speak about temperature. The profound difficulty of the subject matter derives from our incomprehension of temperature concept in the context of an elementary collision: temperature requires many degrees of freedom, and equilibration of energy among them. It is possible that a full answer lies beyond our current understanding of the fundamental laws of Nature, since this behavior is what we see in experiments, as Hagedorn has shown us.

I first met with Rolf in Winter 1975/76, when I attended his Colloquium on the Statistical Bootstrap Model. At that time I knew nothing about the subject, only little about Relativistic Statistical Mechanics, and hardly anything about the experimental data he worked on. After the lecture I asked him privately a few quite simple questions and was astonished to see that he took every matter very seriously, did not dismiss any remark and followed with very clear explanations what could be answered on the spot. One thing was clear to me right away, I could learn something from Rolf, and I realized that I was interested in finding out what he knew. I asked if I could visit him at CERN and he suggested I consider a short term position application. After one short visit I arrived as fellow at CERN in September 1977.

Rolf was a wonderful teacher for the following two years of our very close collaboration and subsequent occasional cooperation that lasted till his retirement in 1985. I think our very different approaches to Physics, and our very different carrier paths matched in a very special way. While we wrote just a few papers together, I can say with certainty that Rolf has critically influenced my development as a Physicist.

Our collaboration was in the beginning very unusual: Rolf did not mind to describe to me his ideas and theoretical work slowly, repeating many details till I understood everything. Sometimes we set in his office for hours, from the morning till evening, and for Rolf it did not matter. I occasionally was very worried if I did not waist to much of his time, and tried out other collaborations. But I always returned to Rolf, driven not only by the subject but perhaps equally by his personality which I found most fascinating.

Rolf Hagedorn, with Helmut Satz, Johann Rafelski and Tatjana Faberge
June 30, 1994 *(PHOTO-CERN)*

One day, we had begun to implement a few ideas into publications. You would imagine that this was the moment a younger collaborator goes to the computer to work out the ideas of his professor: WRONG – Rolf Hagedorn did not operate this way at all: he enjoyed showing me how simple it was to make these computations with *Sigma*, the user oriented computer language he had helped to develop (see the paper by J. Reinfelds in this volume). He was able to complete all the calculations very rapidly, to obtain graphic representations of our results; indeed he even helped me on other projects.

Rolf Hagedorn always wants to help those that are in need. I recall for example that when his colleague and collaborator arrived from Hungary, then under communist regime, and told him that he has decided not to return home, Hagedorn spend long days looking for a place that would take him, he was prepared to make any effort to reach this goal. Similarly, the fate of the prominent Soviet dissidents Andrei Sakharov, Youri Orlov, preoccupied him much of the time, and he left no stone upturned to obtain their release and provide them with scientific information and other support. When the State of War in Poland made life there miserable in the early eighties, and one of our friends and collaborators, was imprisoned, Rolf would work in all his free time to ease the burden on his friends in Poland. Certainly the Soviet empire was not brought down by Rolf alone, but positively he has been an important force which has helped our colleagues in the East fight for their freedom, knowing that people like Rolf where there to stand behind them in bad times. Today the cards are dealt differently, and Rolf is fighting for the rights of those that the revolution in the East has suddenly left in limbo, as they may be looked upon with scorn by those who think of them as collaborators. Hagedorn knows better, and has the moral privilege to have stood in other days for freedom and truth. His cause is now, as before right one, and I am convinced that he will prevail.

Another deep impression, which I would like to share concerns Rolf Hagedorn taking his retirement. In the early eighties Hagedorn told me that he was attending a course on 'How to Retire' and was persuaded that he must follow one of their advises: he aught to start to reduce his work load gradually, perhaps even before reaching the age of 65, such that when he reaches 70 he would approach near zero scientific activity. He was persuaded that the worse thing for him to do would be to work full steam till the age of 65, and than to drop the pen. He thought this was very unhealthy and for most scientists impossible, as it is impossible for aged men to work full steam. As we all know the menace of retirement makes many work at unhealthy level and others to effectively retire without plan and often too early. Hagedorn is in this regards definitively an example to study and follow. Maurice Jacob (see above) has said a few words about the ensuing negotiations that have lead to the recognition of the Emeritus status at CERN. I can only add that in one way Hagedorn was wrong; he is bouncing back from the 10% level of activity to over 100% at this meeting!

HELMUT SATZ

HOT HADRONIC MATTER

In days of old
a tale was told
of hadrons ever fatter.
Behold, my friends, said Hagedorn,
the ultimate of matter.

Then Muster Mark
called in the quarks,
to hadrons they were mated.
Of colors three, and never free,
all to confinement fated.

But in dense matter,
their bonds can shatter
and they freely move around.
Above T_H, their colors shine
as the QGP is found.

THE LONG WAY TO THE
STATISTICAL BOOTSTRAP MODEL

Rolf Hagedorn

Theoretical Physics Division, CERN
CH – 1211 Geneva 23

"It is the nature of a hypothesis when once a man has conceived it, that it assimilates everything to itself, as proper nourishment; and, from the first moment of your begetting it, it generally grows the stronger by everything you see, hear, read or understand. This is of great use."[1]

INTRODUCTION

The statistical bootstrap model (SBM) is a statistical model of strong interactions based on the observation that hadrons not only form bound and resonance states but also decay statistically into such states if they are heavy enough. This leads to the concept of a possibly unlimited sequence of heavier and heavier bound and resonance states, each being a possible constituent of a still heavier resonance, while at the same time being itself composed of lighter ones. We call these states clusters (in the older literature heavier clusters are called fireballs; the pion is the lightest "one-particle-cluster") and label them by their masses. Let $\rho(m)dm$ be the number of such states in the mass interval $\{m, dm\}$; we call $\rho(m)$ the "SBM mass spectrum". Bound and resonance states *are due* to strong interactions; if introduced as new, independent particles in a statistical model, they also *simulate* the strong interactions to which they owe their existence. To simulate all *attractive* strong interactions we need all of them (including the not yet discovered ones), that is: we need the complete mass spectrum $\rho(m)$. To simulate *repulsive* forces we may use proper cluster volumes à la Van der Waals. In order to obtain the full mass spectrum, we require that the above picture, namely that a cluster is composed of clusters, be self-consistent. This leads to the "bootstrap condition and/or bootstrap equation" for the mass spectrum $\rho(m)$. The bootstrap equation (BE) is an integral equation embracing all hadrons of all masses. It can be solved analytically with the result that the mass spectrum $\rho(m)$ has to grow exponentially. Consequently any thermodynamics employing this mass spectrum has a singular temperature T_0 generated by the asymptotic mass spectrum: $\rho(m) \sim \exp(m/T_0)$. Today this singular temperature is interpreted as the temperature where (for baryon chemical potential zero) the phase transition (hadron gas) $\Leftrightarrow$ (quark-gluon plasma) occurs.

The main power of the SBM derives from the postulate that the strong interaction – as far as needed in statistical thermodynamical models – is completely simulated by the presence of clusters with an exponential mass spectrum and with mass-proportional proper volumes; this postulate implies that

> *in SBM the strongly interacting hadron gas is formally replaced by a non-interacting (i.e., ideal) infinite-component cluster gas with Van der Waals volume corrections and exponential mass spectrum,*

which can be handled analytically without recourse to perturbative methods.

The story of how this model was first conceived in the language of the grand canonical ensemble, reached maturity in the language of the microcanonical ensemble (i.e., phase space) and was finally equipped with finite particle volumes in order to become applicable to heavy-ion collisions and to the question of the phase transition has been told elsewhere.[2]

In this lecture I wish to describe the long way from the first theoretical ideas about multiple particle production up to the situation in which constructing SBM seemed natural. The story starts in 1936, and in my record I omit everything that did not lie on or near the way leading to SBM. What I wish to show is that SBM did not suddenly appear in 1965 as a *deus ex machina*, but was rather the logical consequence of a history of almost 30 years. Thus, of a large network of observations and theoretical ideas I pick only a few lines, chosen for their common end point: SBM. A complete and impartial picture of this history up to 1972 is presented by E.L. Feinberg in his exhaustive and instructive report,[3] which is an indispensable complement to the present biased lecture.

I will try to be as non-technical as possible. Formulae are meant merely as illustrations (often oversimplified); for hard information the reader should consult the quoted literature.

Units are $\hbar = c = k$ (Boltzmann) = 1; energy in MeV or GeV.

FROM 1936 TO 1965

We list here experimental facts and theoretical concepts which were important and instrumental to the construction of SBM.

Fireballs

How did we come to believe that "fireballs" – the things called "clusters" in SBM – exist?

Multiple production: Heisenberg (1936)

Before Yukawa's paper postulating the pion,[4] one tended to believe that the particles produced in cosmic ray events were electron-positron pairs. The only field theory then known, quantum electrodynamics (as yet without a consistent renormalization scheme), suggested that events with many secondaries should have vanishingly small cross-sections (proportional to $(e^2)^n$). This led many theoreticians to the interpretation that such events must be the result of many interactions with different nucleons in the same heavy nucleus, each single interaction producing just one pair, a point of view[5-7] persisting even after the advent of meson theory and in spite of growing experimental evidence in favour of multiple production. Heisenberg – still unaware of Yukawa's paper – was the first to claim that in a single elementary interaction many secondaries might be produced,[8] which at that time was a heretical idea – the pion was discovered 11 years later! Heisenberg followed this idea through many years (until ~ 1955) and devised different theoretical approaches to it, all invoking strong non-linearities and/or diverging field theories. The final, irrevocable, decision between his views and his opponents' only came with the first hydrogen bubble chamber pictures: Heisenberg's revolutionary idea had been right*. We summarize this line of thought in "Lesson 1":

> In a single elementary hadron–hadron collision, many secondaries can be produced.   (L1)

Today this is so obvious that stating a "Lesson" seems ridiculous, but seen in a historical perspective it challenged a strong prejudice.

Dulles–Walker variables (1954)

Assume a source of particles (a "fireball") moving with velocity β [Lorentz factor $\gamma = (1 - \beta^2)^{-1/2}$] as seen from the lab and assume further that this source emits particles with velocities β_i' isotropically in its own rest frame. We put the z-axis in the direction of motion and call θ_i the polar angle under which particle i is emitted. Quantities in the fireball's rest frame are primed, those of the lab frame are not. In any book on relativistic kinematics one finds the formula for the angle transformation:

$$\tan\theta_i = \frac{1}{\gamma}\frac{\sin\theta_i'}{\cos\theta_i' + \beta\beta_i'} \approx \frac{1}{\gamma}\tan\frac{\theta_i'}{2}\,, \tag{1}$$

where the last $\approx$ is true when β and β_i' are both near 1, which will be assumed from now on.

The fraction F of particles emitted inside a cone of polar angle θ' is, from elementary geometry, in the fireball's frame:

$$F = \frac{1}{2}(1 - \cos\theta') = \sin^2\frac{\theta'}{2}\,, \tag{2}$$

while in the lab the same particles – and the same fraction F – will be found inside the cone of angle [Eq. (1)]

$$\tan\theta \cong \frac{1}{\gamma}\tan(\theta'/2) \tag{3}$$

so that in the assumed approximation

$$\gamma^2\tan^2\theta \cong \frac{F}{1-F}\,; \tag{4}$$

hence

$$\log\frac{F}{1-F} \cong 2\log\gamma + 2\log\tan\theta\,. \tag{5}$$

We note in passing that with $F = 1/2$ we find the angle $\theta_{1/2}$ of the cone into which half of the particles fall:

$$\tan\theta_{1/2} = \frac{1}{\gamma}\,. \tag{6}$$

Now define for each particle i the fraction F_i of particles falling inside the cone of polar angle θ_i (i.e., those having an angle smaller than or equal to that of particle i) and plot the points $y_i = \log F_i/(1-F_i)$ versus $x_i = \log\tan\theta_i$. These points will – *under the supposed conditions :* $\beta_1\beta_i' \approx 1$ *and isotropy* – scatter about the straight line given by Eq. (5) with slope 2 and intercepts

$$x(y = 0) \;=\; -\log\gamma \tag{7}$$
$$y(x = 0) \;=\; 2\log\gamma$$

as depicted in Fig. 1.

---

*Although the various theoretical models he constructed, and which he himself considered as preliminary, did not give final answers to the why's and how's.

The discovery of these variables by Dulles and Walker[9] proved of great importance for the analysis of cosmic ray events: if the points are plotted according to the above rule, then if anything similar to a straight line would emerge, an isotropically emitting centre had to be conjectured and its Lorentz factor γ could be read off. Although things were not that simple, the method revealed a lot of information, as we shall soon see.

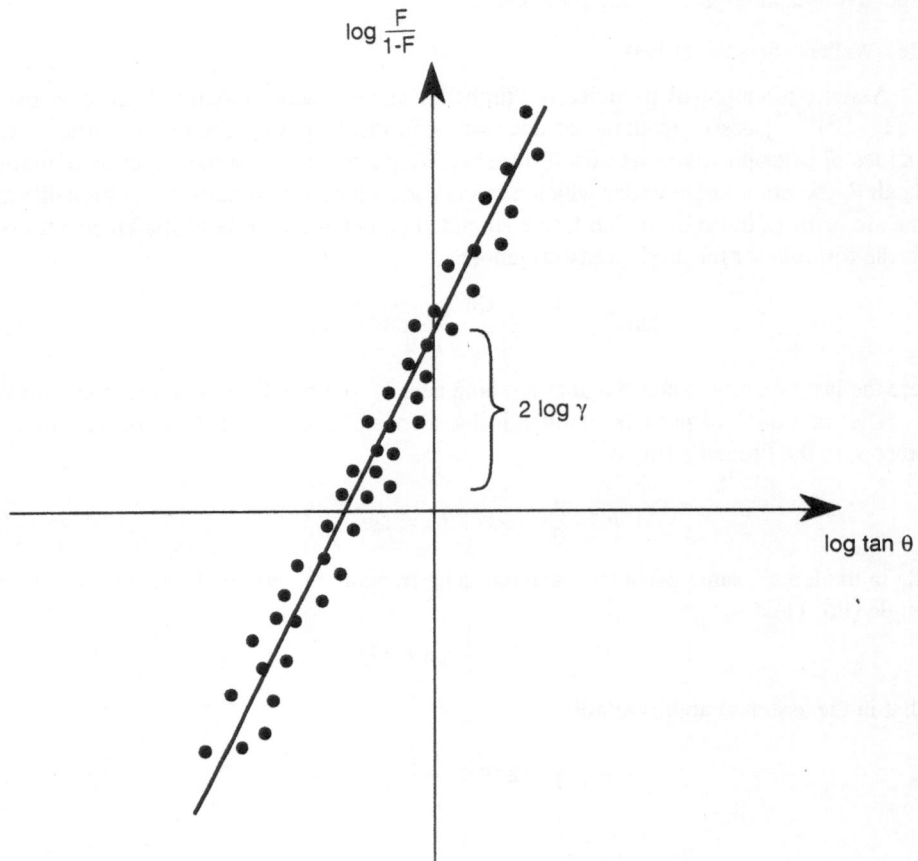

Figure 1. Relativistic particles ($\beta_i' \approx 1$) emitted isotropically in the "fireball" frame, which itself moves with $\beta \approx 1$ seen from lab., will in a plot with Dulles–Walker variables scatter about a straight line with slope 2 and y–intercept $2 \log \gamma$.

"Constant" mean transverse momentum (1956)

The invariance of the transverse momenta (of the produced particles) under a Lorentz boost in the z-direction made them interesting from the beginning. The amazement was therefore great when it gradually turned out that their average $\langle p_\perp \rangle$ seemed to be practically independent of the primary energy of the collision from which they emerged. This was reported in so many papers over so many years that I cannot quote all of them. Probably J. Nishimura was the first to have pointed it out.[10] The result was by 1958 rather well confirmed[11] and remained so until the "large transverse momenta" were discovered in 1973,[12] which – important as they were – corrected this result only slightly. We write down "Lesson 2":

(L2)

The two-centre model (1958)

The most prominent qualitative feature of the particle tracks in emulsions and/or cloud chambers was that they were arranged in two cones: a wide one and, inside it, a narrow one. No measurements were needed to see this and to guess a simple mechanism that would produce it: two "centres", one moving slowly and another one moving fast[†] along the collision line, both emitting particles isotropically and with rather small, energy-independent momenta [Lesson 2] in their respective rest frames. I do not know whether one could say that a particular physicist had this idea first (it might have been Takagi,[13] but I am not sure): it must have appeared obvious to anyone who saw pictures of these events. It was another thing to analyse such pictures quantitatively. The pioneers were the Cracow–Czech collaboration[14] and G. Cocconi.[11] They exploited the powers of the Dulles–Walker representation.

The story went like this: one applied the Dulles–Walker plot to the available events[11,14] and, instead of finding the points representing the tracks scattering about a straight line – as expected for a single "fireball" – one found something very different. The result is show in Fig. 2 and Fig. 3 which I copy from Cocconi's paper.[11] The spirit of Cocconi's paper is so well concentrated in a few original passages that I repeat them here. Cocconi says:

> "It is evident from an examination of Fig. 2 (our Fig. 2) that in most cases the relativistic secondaries are separated into two groups as if they were emitted, in the CM centre of the collision, not by a single centre but by two bodies, as described in Section II(d)[‡]. The evidence is so striking that we are going to analyse these events in a slightly different manner, more adjusted to the model.
>
> Instead of considering all the relativistic particles produced in the collision together, let us divide them into two groups: the forward group, b_1, and the backward group, b_2 (the narrow and wide cones).
>
> Let n_1 and n_2 be the number of particles falling in each group and let us analyse them in terms of $\log \tan \theta$ versus $\log[F_1/(1 - F_1)]$ and versus $\log[F_2/(1 - F_2)]$. The results are plotted in Fig. 3 (our Fig. 3)."

The figures 2 and 3 and Cocconi's remarks need no further comment. It should be noted, however, that he is aware of the possibility of other interpretations, in which not individual "fireballs" but a two-jet structure produces much the same effect.

The two-centre model was popular for a long time, as witnessed by the review paper written by Gierula[15] in 1963, five years later, and based on more than 100 events with $E_{lab} \gtrsim 10^3$ GeV. If I remember well from those years, the model did not always work, sometimes three or more fireballs had to be invoked, but on the whole it was rather successful. That it seems never to have been disproved came perhaps from the shift of interest to other questions arising from working with accelerators, where single events were analysed mostly in the hope of discovering new particles but not to prove or disprove a two-centre model. The famous "flat rapidity plateau" was, of course, no argument against a two- (or few-) centre model, as it arose from averaging over the impact parameter in many collisions contributing to the measured inclusive distributions. We thus draw "Lesson 3":

[†]In the lab; in the CM frame: one forward, one backward.

[‡]Cocconi proposes a two-centre model in Section II(d) of his paper (with two "leading nucleons" not contained in the "fireballs").

17

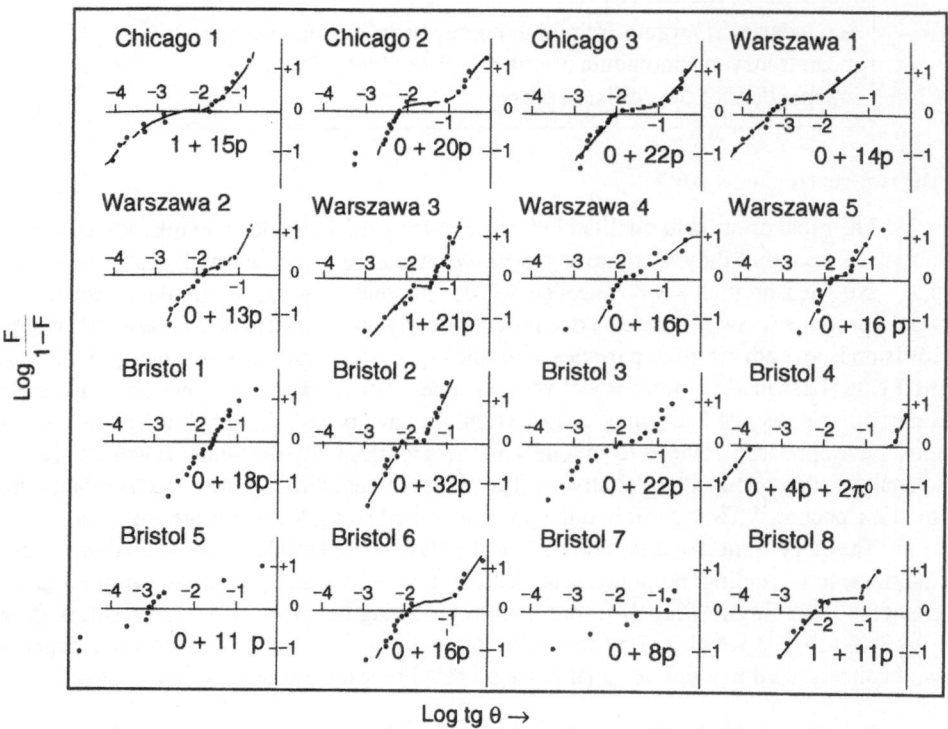

Figure 2. Experimental data plotted in Dulles–Walker variables.

> Secondaries produced in elementary hadron collisions seem
> to be emitted from few (≈ 2) "fireballs" rather isotropically   (L3)
> with small momenta in the fireballs rest frame.

Conclusion: Fireballs with limited $\langle p \rangle$ exist

We conclude that "fireballs", decaying with limited momenta, do exist. In other words: *lumps of highly excited hadronic matter keep together for a very short time before they decay in a very specific and – on this level – not yet understood way.*

Statistical and Thermodynamical Methods

Having collected, in the previous section, the arguments in favour of the existence of "fireballs", we now turn to their description.

The methods and the models used eventually for this description were developed long before the existence of their final objects was established. In fact, the story goes back to two early theoretical ideas:

- the compound nucleus of Bohr in 1936;

- the incorporation of interaction in statistical thermodynamics via scattering phase shifts by Beth and Uhlenbeck in 1937.

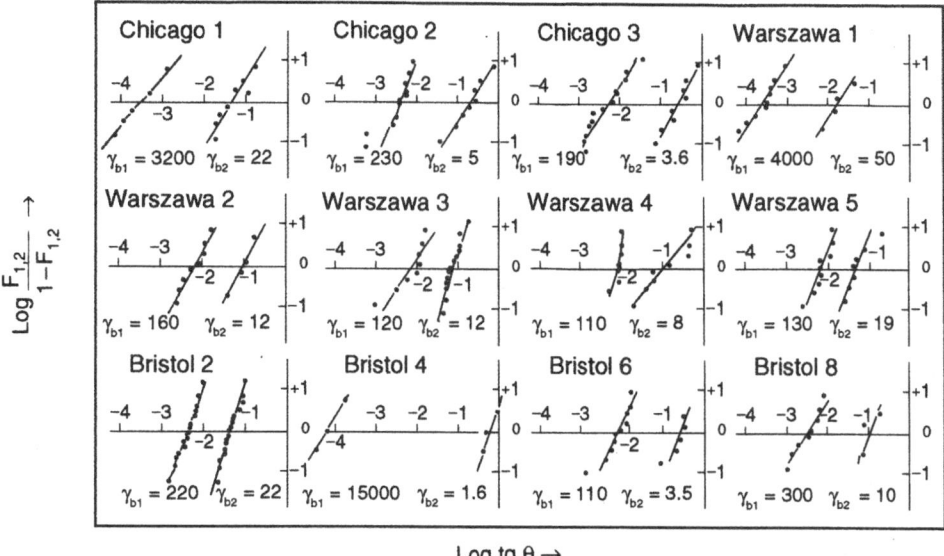

Figure 3. Some of the events of Fig. 2. Forward and backward (CM) particles plotted separately.

Bohr's compound nucleus (1936)

Bohr[16] proposed the following picture for a certain class of nuclear reactions: if a heavy nucleus is hit by a nuclear particle, then the strong interaction among the constituents and with the projectile can often lead to a complete dissipation of the available energy, so that no single nucleon gets enough of it to escape at once. This excited "compound nucleus" will then live a rather long time before it decays by emitting such nucleons, which accidentally obtained sufficient kinetic energy to overcome the binding force. This picture, of course, cries for a statistical description.

The Weisskopf evaporation model (1937)

One had not a long time to wait for it. Weisskopf[17] writes, in his famous paper on nuclear evaporation:

"... Qualitative statistical conclusions about the energy exchange between the nuclear constituents in the compound state have led to simple explanations of many characteristic features of nuclear reactions. In particular the use of thermodynamical analogies has proved very convenient for describing the general trend of nuclear processes. The energy stored in the compound nucleus can in fact be compared with the heat energy of a solid body or a liquid, and, as first emphasized by Frenkel[18] §, the subsequent expulsion of particles is analogous to an evaporation process."

Weisskopf is cautious: he does not claim right away that thermodynamics is applicable to nuclei; he rather derives first from elementary quantum mechanics a formula for the emission of a neutron by the excited nucleus A, leaving another excited nucleus B behind. For this he uses the principle of detailed balance by considering the inverse reaction $B + n \to A$ of which the cross-section is supposed to be known. From this the emission probability can be calculated; it is a very simple expression containing the above cross-section, the level densities $\omega_A(E_A)$ and $\omega_B(E_B)$, of nuclei A and B, at their respective energies and the phase

---

§Our list of references.

19

space available for a neutron of given kinetic energy ϵ. And then he introduces quite formally an "entropy"

$$S(E) = \ln \omega(E) \tag{8}$$

and a "temperature"

$$T(E) = (dS/dE)^{-1} . \tag{9}$$

In these variables the derived formula for the emission probability assumes the usual form of an evaporation probability with a Boltzmann spectrum:

$$W(\epsilon)d\epsilon \sim \epsilon \exp(-\epsilon/T)d\epsilon . \tag{10}$$

The rest of the paper discusses when the formula is valid and what corrections are necessary. What interests us here is that this is (to my knowledge) the first time that it was shown quantitatively that thermodynamics might be applied to such a tiny system as a nucleus. *The reason is the enormous level density* of heavy nuclei at high excitation energy. Note also that the formula was derived for the emission of a single neutron with only a few degrees of freedom (phase space). We conclude with "Lesson 4":

> Thermodynamics and/or statistics might be (cautiously!) applied to very small systems, provided these have a very large level density (whatever that means).　　(L4)

When Weisskopf wrote his paper, not much was known about the level densities of nuclei and he proposed to learn about them from the observed emission spectra, taking his formulae for granted.

Koppe's attempt and the Fermi statistical model (1948/1950)

Although traditionally all credit for the invention of a statistical model for particle production goes to Fermi (see below), it was actually H. Koppe who proposed, already two years earlier, the essence of such a model. He wrote[19]

"In a recent paper[20]¶ a simple method has been given for the calculation of the yield of mesons produced by the interaction of light nuclei. It was based on the assumption of strong interaction between mesons and nucleons which should make it possible to treat a nucleus as a 'black body' with regard to meson radiation and to calculate the probability for emission of a meson by statistical methods."

At that time, available energies were not high (Berkeley: α-particles of ~ 380 MeV) and consequently the temperatures remained small (~ 15 MeV), much below the pion rest mass. Yet the model worked not so badly. Note that (for him) the high level density justifying the treatment was located not in the meson field but in the interacting nuclei ("black body").

Fermi[21] does then the important step to consider the pion field itself as the thermal (or better: statistical) system without need of a background "black body" à la Koppe. Thus he claims that, e.g., a proton–proton collision could be treated statistically; he writes:[21]

"When two nucleons collide with very great energy in their CM system, this energy will be suddenly released in a small volume surrounding the two nucleons. We may think pictorially of the event as of a collision in which the nucleons with their surrounding retinue of pions hit against each other so that all the portion of space occupied by the nucleons and by their surrounding pion field will be suddenly loaded with a very great amount of energy. Since the interactions of the pion field

---

¶Our reference; the paper is written in German.

20

are strong, we may expect that rapidly this energy will be distributed among the various degrees of freedom present in this volume according to statistical laws. One can then compute statistically the probability that in this volume a certain number of pions will be created with a given energy distribution. It is then assumed that the concentration of energy will rapidly dissolve and that the particles into which the energy has been converted will fly out in all directions."

After some further discussion he writes down his basic formula for the production of n pions (in modern notation):

$$ S(n) = \frac{1}{n!} \left[\frac{V_0}{(2\pi)^3} \right]^n \int \delta\left(E - \sum_{i=1}^{n} E_i \right) \prod_{i=1}^{n} 4\pi p_i^2 dp_i \; ; \quad E_i = \sqrt{p_i^2 + m^2} \, , \qquad (11) $$

where V_0 is the "interaction volume" [order $(4\pi/3)m_\pi^{-3}$ and Lorentz-contracted or not, according to taste], E the total CM energy and m the pion mass (or that of another species if considered).

The rest of his paper discusses applications at low, medium and very high energies; in the latter case a thermodynamic formulation is proposed, where the temperature is proportional to $E^{1/4}$ – that is: a Stefan-Boltzmann gas of (massless) pions is assumed. The way to this is already prepared when he discusses medium energies, where a good number of pions are produced: in order to use the only existing analytical expressions for n-body momentum space (namely for $m = 0$ and/or for $m \to \infty$) he treats pions as massless and nucleons as non-relativistic. At this time (1950) these assumptions were reasonable: the discovery of "limited transverse momenta",[10] which of course would invalidate them, was to come only six years later. He also mentions angular momentum conservation, but only to argue that it is unimportant; he soon comes back to this question in an attempt to explain the observed anisotropy in CM,[22] which failed. We pass over these details.

What is important for us is that Fermi actually tries to describe the disintegration of what we called above "fireballs" – eight years before they were discovered experimentally[11,14].

While the model fails quantitatively (Heisenberg[23] quotes a measured event with an estimated primary energy of 40 GeV, where actually about 27 pions were produced, in contrast to 2.7 predicted by Fermi in the thermodynamic version) it is nevertheless the starting point of the development leading to the SBM.

Looking back we see a line of thought that leads from the Bohr compound nucleus directly to the *theoretical* concept of a hadronic fireball and its statistical (thermodynamical) description.

Beth–Uhlenbeck, Belenkij (1937/1956)

At the beginning of this Section two main theoretical ideas were said to be essential for a statistical description of fireballs; one was the Bohr compound nucleus, leading directly to the Fermi model.

The second is found in a paper by Beth and Uhlenbeck.[24] The authors incorporate interaction in statistical thermodynamics quantum mechanically via scattering phase shifts. We sketch the idea only; details may be found in[25] and.[26]

Suppose you have an ideal gas of N non-interacting particles with masses $m_1, m_2, \ldots, m_N$ at total energy E enclosed in a volume V. Let the level density of this gas be $\sigma_N(E, V, m_1, \ldots, m_N)$. If a force acts between particles Nos. 1 and 2, they may form a bound state m_{12} and (if nothing else happens) the level density of this new system becomes $\sigma_{N-1}(E, V, m_{12}, m_3, \ldots, m_N)$. The interaction has changed the level density and the system with interaction would be described as a mixture of two ideal gases with and without bound states.

21

Beth and Uhlenbeck extend this argument to the case where the interaction leads not to a bound state but only to scattering.

Single out from our gas two particles that are to scatter on each other, and take as normalization volume a sphere of radius R centred at the point of impact of the two particles. The density of states of this two-particle subsystem will be affected by the scattering process in that the ℓth partial wave of the common wave function of our two particles will be, asymptotically:

$$\psi_\ell(r,p) \sim \frac{1}{pr}\ \sin\ \left[pr - \frac{\ell\pi}{2} + \eta_\ell(p)\right],\qquad(12)$$

where p is their relative momentum, r their distance, and $\eta_\ell(p)$ the scattering phase shift. The wave function should vanish at $r = R$:

$$pR - \frac{\ell\pi}{2} + \eta_\ell(p) = n\pi\ ;\quad n = 0, 1, 2, \ldots \qquad(13)$$

Thus n labels the allowed (discrete in R) two-body momentum states $\{p_0, p_1, \ldots\}$. For a fixed p' there are $n(p')$ states below p'; thus the density of states near p' is

$$\frac{dn}{dp'} = \frac{R}{\pi} + \frac{1}{\pi}\ \frac{d}{dp'}\ \eta_\ell(p')\ .\qquad(14)$$

Without interaction, $\eta_\ell(p) \equiv 0$; hence

the interaction changes the two-particle density of states by $(1/\pi)d\eta_\ell/dp$.

Of course this argument has to be repeated for all partial waves and all particle pairs with the final result that the sum over ℓ gives a contribution to the partition function containing the derivative of the scattering amplitude[25,27]. The *formal* extension of this method to include all interaction is due to Bernstein, Dashen and Ma.[28]

For the following argument of Belenkij[29] the simple equation (14) is most illustrative: let the two-body subsystem have a sharp resonance at relative momentum p^*; then there the phase shift rises by π within a short interval, so that $(1/\pi)d\eta_\ell(p)/dp \approx \delta(p - p^*)$. Such a δ-function appearing in the density of states is equivalent to introducing an additional particle with mass $m^* = m(m_1, m_2, p^*)$ into the system, very much like a bound state would be introduced. The actual proof is somewhat complicated due to the switching between different sets of momenta. Belenkij does this in detail. We state "Lesson 5":

> If in a statistical–thermodynamical system two-body bound and resonance states occur, then they should be treated as new, independent particles. Thereby a corresponding part of the interaction is taken into account.   (L5)

Note that after doing so, the system is still formally an "ideal gas", but now with some additional species of particles (simulating part of the interaction).

Belenkij's motivation for his work had been the known fact that Fermi's model gave wrong multiplicities: "This discrepancy may be as high as 20 times". He had hoped that his new remedy [expressed by (L5)] would cure the disease of the Fermi model; it did so only partially for reasons to become clear soon.

When we adopted Belenkij's argument for including resonances, we did so because it was intuitively obvious that resonances should be included even when the formal derivation could not be directly invoked as for instance in a process $A + B \to$ resonance $\to n$ particles, where a phase shift increasing by π is not defined. Incorporating resonances quite generally was later justified by Sertorio via the S-matrix approach to statistical bootstrap, in an important paper.[30]

The CERN statistical model (1958–1962)

When in 1957 the CERN-PS was near completion, planning of secondary beam installations required estimates of particle production yields and momentum spectra. Bruno Ferretti, our division leader at that time, asked Frans Cerulus and me to do some calculations with the Fermi model ("just a fortnight of easy work . . ." he said), not surmising that by that request he triggered a new development. In fact we soon found out that the Fermi model, as it was, could not be used:

1 - neither (in the fireball rest frame) were pions ultrarelativistic nor nucleons non-relativistic; indeed Lesson 2 (limited transverse momenta) excluded this; thus there were no analytic formulae available to calculate momentum space integrals [Eq. (11)];

2 - interaction between the produced particles might be important; the ideal gas approximation could lead to large errors.

For the second problem Belenkij had already given the solution: include all known particles and resonances (Lesson 5). For the rest, we were confident: fireballs seemed to exist (Lesson 3 was known to us by hearsay) and their statistical description in principle possible (Lesson 4). We earnestly hoped that an improved Fermi model would do. Problem 2 being trivial (thanks to Belenkij), once problem 1 was overcome, we concentrated on the calculation of momentum space integrals. Cerulus had the idea to use the Monte Carlo method and we worked it out. At that time this was a new method, not familiar to physicists; moreover the first CERN computer was only to come in a year or so. So we had tried our new methods[31] with the help of the Institute of Applied Mathematics at Darmstadt, where an IBM computer was available (not very powerful in 1958!) and we had found that momentum space integrals with up to ~ 15 particles could be computed in reasonable time and reliably with *prescribable* error (5-10%). I then took it over to write a program (my first!) for the expected CERN computer, a Ferranti Mercury. It was an adventure: I had to learn to program a non-existing machine, still under development; no possibility to check written parts of the program; everything was to be expressed in machine code – a simple addition required four lines of code and all store addresses were absolute; one had to keep book of where each number (including intermediary results) was stored – and there were thousands of them (momentum spectra of some 15 species of particles). After half a year I had finished the program (so I thought) and went to Saclay in France to test it on the first delivered Mercury machine. It failed beyond any expectation. Correcting was even more tricky than to write the program (which consisted of thousands of lines of numbers – no letters, no symbols; find the error!). It even required manual skill: input and output was via punched paper tape and one had to find the erroneous part (reading the tape by eye had to be learned, too), cut it out and replace it carefully by the corrected piece, glue everything properly together, unless the teletape reader would refuse or tear it to pieces. In short: it was a mess, but finally it came to work and in hundreds of hours we produced kilometres of tape with our precious results: the first accurate evaluations of the Fermi model including some interaction (all known particles plus some resonances) at several primary energies from 2 to 30 GeV (lab) and for pp and πp collisions. Cerulus[32] had used a very elegant group theoretical method to solve the problem of charge distribution, and then employed the same method to implement angular momentum conservation in phase space[33] in the hope to reproduce the known anisotropy. It failed (because requiring more computing than possible and also) because the process was not so statistical as we had hoped: angular momentum conservation could not produce the pronounced anisotropy found in cosmic ray events and well accounted for by two-centre models[11,14] – a fact strongly suggesting that phase relations between partial waves survived the statistical mixing assumed in the Fermi model. In principle we had the tools to build and correctly evaluate a two–centre model but it would have required at least ten times more computing (summing over impact parameters with varying fireball energies), which was impossible (we had already spent some years to do all the computing for single fireballs). Thus the angular distribution could not be described adequately.

But we had a more important success:[34] from the calculated momentum spectra it followed that the mean kinetic energy of all particles was practically independent of the primary lab energy (6 to 30 GeV). Thus the model produced about correctly the empirical fact stated in Lesson 2 (limited transverse momenta); however, this was only a numerical result, due to the large number of species of particles entering our calculations: counting spin, isospin and antiparticles of the included ones ($\pi, \rho, \omega, K, N, \Delta, \Lambda, \Sigma, \Xi$) we came to 83 different particle states, equivalent to 83 species. This proved important in the following development, where it was the key opening the door to the statistical bootstrap model, when it was tried to understand this mechanism analytically.

We state "Lesson 6":

> A properly evaluated Fermi model with some interaction (resonances; L5) produces, in a limited interval of primary energy, practically constant mean kinetic energies of second–aries and reasonable multiplicities.   (L6)

A review of our work is given in.[34] See also.[35]

The Decisive Turn of the Screw: Large–Angle Elastic Scattering

Early in 1964 evidence was growing that the elastic pp cross-section around 90° (CM) was decreasing exponentially with the total CM energy, at least in the then known region $10 \leq E \leq 30$ GeV of primary energies (lab). Many experiments contributed to this, which we cannot list here. The situation – theoretical and experimental – is well described in a paper by G. Cocconi,[36] where references to the original experiments are given.

One can include angles a little below and above 90° by using transverse momentum $p_\perp = p \sin \theta$. J. Orear[37] obtained in this way an impressive fit to large-angle elastic scattering (in what follows, E is always the total CM energy):

$$E^2 \frac{d\sigma}{d\omega}\bigg|_{el} = \text{const.exp} \left[-p_\perp/0.158\right], \qquad (15)$$

which is shown in Fig. 4; the cross-section follows this empirical formula over nine orders of magnitude in the interval $1.7 \leq p_0 \leq 31.8$ GeV/c (primary lab momentum). Moreover, the reaction $p + p \rightarrow \pi + d$ as well as πp elastic scattering showed the same behaviour; in particular the exponential decrease had the same slope as in pp elastic scattering.[37]

It is tempting to interpret (15) as a thermal Boltzmann spectrum. In that case 0.158 GeV would be the "temperature" at which the two nucleons were emitted*.

It thus seemed that there was something statistical, a suspicion strangely corroborated by the observation that the same law (with slightly different "temperature") was obeyed by secondaries in inelastic processes.[36]

Almost two years before the Orear plot was published, L.W. Jones proposed already a statistical interpretation: the two colliding particles would form sometimes a fireball – in

---

*If E is used at 90° instead of $p_\perp \cong E/2$, the exponent becomes $-E/0.31$, which was sometimes misinterpreted as $T \approx 0.3$ GeV.

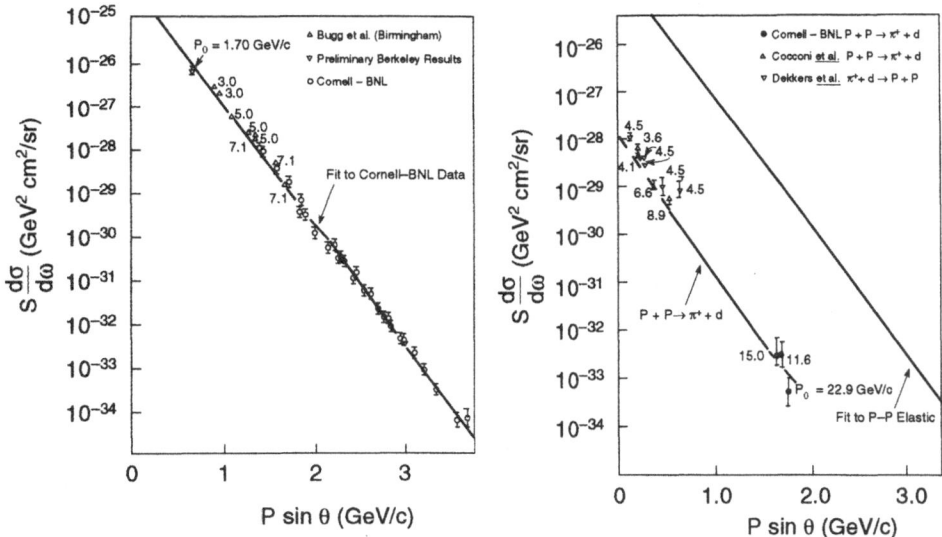

Figure 4. The two original figures of Orear[37] for large-angle pp scattering and for $p + p \to \pi + d$, showing the exponential decrease with the same slope.

analogy to the compound nucleus – which would decay into many channels, among them the two-body channel containing the original particles. This two-body decay could only be observed far outside the diffraction peak, that is, at large angles. He asked me whether such a picture could be quantitatively described by our statistical model.

Statistical model description of large-angle elastic scattering

For some obscure reasons we had archived all results obtained since 1958[†] and even included the two-body channel. It was simple to analyze them again and to find the amazing result

$$E^2 \frac{d\sigma_{el}}{d\omega}\bigg|_{90°} \sim E^2 \frac{P_0}{\sum_b P_b} \cong \text{const.} \ E \ \exp[-3.17E] \, , \qquad (16)$$

where P_0 is the probability of the two-body channel and ΣP_b the sum over the probabilities of all channels; P_0 and all the P_b were numerical results from hundreds of phase-space calculations as described above. When we established Eq. (16), no free parameters were available, everything was in our archived data. This result[38] agreed reasonably well with early experimental data,[36] but when the Orear fit[37] was published, the agreement became perfect: the number 3.17 in the exponent of (16) corresponds to a temperature $T = 0.158$ if $E = 2p_\perp$ (at 90°) is inserted; then Eq. (15) results (the factor E in front of the exponential is negligible).

Thus L.W. Jones' proposal was immensely successful. An independent confirmation by the observation of Ericson-fluctuations[39] would have been desirable, but I do not know if it was ever tried. It probably was too difficult.

Thermal description

Our results were so convincing to me (unfortunately not to most others; among the few exceptions was G. Cocconi) that I firmly believed that in (16) we really had an exponential function and not something approximately exponential. This belief (which was directly

---

[†]Several people had contributed to the accumulation of statistical model results: J. von Behr, F. Cerulus, H. Faissner, G. Fast, K.H. Michel, J. Soln and myself.

leading to the statistical bootstrap model) had to be justified better than by Eq. (16), which was merely a numerical result established in a rather small range of energy $\sim 2.4 \le E \le 6.8$ GeV.

The belief was, however, seriously challenged by A. Białas and V.F. Weisskopf[40] who had given a thermodynamical description based on assumptions that I considered unsuitable, but that nevertheless also gave a good fit to the data.

Which were these assumptions? Mainly these:

- the compound system is a hot gas;

- as constituents only pions are considered; K mesons and resonances are assumed to be unimportant;

- pions are taken as massless.

These were exactly the assumptions that Fermi[21] had already made and that had led to wrong multiplicity estimates[23,29].

From the above assumptions it follows immediately that the gas is at the black-body temperature (Stefan–Boltzmann law)

$$T(E) = \text{const.}\,E^{1/4} \, . \tag{17}$$

Therefore a Boltzmann spectrum for elastic scattering at 90° would be of the form

$$\exp[-p_\perp/T(E)] = \exp[-\text{const.}\,E^{3/4}] \tag{18}$$

instead of our result $\sim \exp[-\text{const.}\,E]$ as given in (16). The authors derive for $(d\sigma/d\omega)_{90°}$ an expression that contains (18) as the essential part, the rest being algebraic factors.

The difference is in principle fundamental, but it is numerically insignificant in the range of energies then available. Apparently the Orear plot,[37] which might have pleaded in favour of a pure exponential, was not yet available to the authors (as seen from the dates of reception of the two papers).

Exponential or not?

This question was so important that I wish to formulate it in two different ways.

i) Our result was [see Eq. (16)] that ΣP_b grows exponentially with E (the other factors being algebraic). Now, a given phase-space integral for b particles is the density of states of the b-particle system at energy E; thus ΣP_b is the total density of states of the "fireball" at energy E

If our result [Eq. (16)] was to be true, it would mean that the density of states of hadronic fireballs would grow exponentially with their mass ($= E$) up to at least $m = 8$ GeV.

ii) The second formulation is a consequence of the first. The entropy is the logarithm of the density of states, hence the entropy of a fireball would be

$$S(E) = \text{const.}\,E \tag{19}$$

and therefore its "internal temperature" would be

$$T = (dS/dE)^{-1} = \text{const.} \tag{20}$$

In words:

> *If our result [Eq. (16)] was to be true, it would mean that the internal temperature of hadronic fireballs would be independent of their mass ($M \equiv E$).*

This would also explain (in a thermodynamic language) why our phase-space calculations had given "constant" mean kinetic energies [Lesson 6]: particles were emitted with a Boltzmann spectrum at an energy-independent temperature; we had suspected that this behaviour was due to our including interaction by admitting all relevant species of particles and resonances known to us, but that had remained a speculation so far.

Cocconi had clearly seen what was going on; he writes:[36]

> "If the dependence of S on E is of the form $S = aE^n$ it follows that $d\sigma/d\omega = $ const. $\exp[-aE^n]$ and that the temperature of the compound system is $T = (naE^{n-1})^{-1}$. The value of n characterizes the "gas" of the compound system ...; $n = 1$ corresponds to the case of a system in which, *for E increasing, the number of possible kinds of particles increases so as to keep the energy per particle, and hence the temperature, constant*"[‡]

and, commenting on our phase-space results:[34]

> "This model produces an essentially "constant temperature" because, in the compound system, *beside the nucleons and mesons, also the known excited states are counted separately*"[‡].

All this can conveniently be summarized in Lesson 7:

> The exponential decrease of the elastic pp cross-section at large angles up to a CM energy of about 8 GeV had empirically established the existence of "fireballs" (clusters; compound states) up to at least $m = 8$ GeV. Moreover: their density of states had to grow exponentially as a function of their mass up to at least $m = 8$ GeV, which means that if the level density is interpreted as a mass spectrum, there was an unexpectedly large number of resonance-like states above those few then explicitly known.　　(L7)

The question now was

> *Could a reasonable analytical model for fireballs be constructed, which would lead to an exponentially growing density of states and, consequently, to an energy-independent temperature?*

Asympotics of momentum space

The question just formulated was in the mind of several people who therefore investigated the asymptotic behaviour of momentum space integrals for $E \to \infty$[41–43]. They all consider essentially a pion gas and show that for $E \to \infty$ the masses become negligible and that there the asymptotics can be evaluated. All authors agree that (in general) the density of states for $E \to \infty$ then grows like $\exp[\text{const}.E^{3/4}]$, just as for a gas of particles with $m = 0$. Vandermeulen as well as Auberson and Escoubès consider also the pathologic case where the usual factor $1/n!$ in front of the phase-space integral is omitted; they discover the amazing fact:

[‡]The italics are mine.

If the factor $1/n!$ in front of the phase-space integral is omitted, then the density of states for $E \to \infty$ grows like $\exp[\text{const}.E]$.

I give here a simple derivation of this result, taking all masses $m = 0$ from the outset and passing over subtleties such as the difference between $\langle E \rangle$ and E in thermodynamics.

For zero masses the particle's energies equal their momenta and the n-body phase-space integral ($= n$-body density of states at energy E) with spatial volume V becomes:

$$\sigma_n(E,V) = \frac{1}{n!} \left[\frac{V}{(2\pi)^3} \right]^n \int \delta\left(E - \sum_{i=1}^n p_i\right) \prod_{i=1}^n 4\pi p_i^2 dp_i \; ; \tag{21}$$

the n-body partition function is then the Laplace transform of σ:

$$Z_n(T,V) = \int_0^\infty \sigma_n(E,V) e^{-\beta E} dE = \frac{1}{n!} \left[\frac{V}{8\pi^3} \right]^n \left[\int_0^\infty e^{-\beta p} 4\pi p^2 dp \right]^n \; , \tag{22}$$

where $\beta = 1/T$ (we use β and T for convenience). The last integral equals $8\pi T^3$, so that

$$Z_n(T,V) = \frac{1}{n!} \left(\frac{VT^3}{\pi^2} \right)^n = \frac{1}{n!} Z_1(T,V)^n \; . \tag{23}$$

Summing over n gives the partition function for our gas with particle number not fixed:

$$Z(T,V) = \Sigma Z_n(T,V) = \exp[Z_1(T,V)]$$
$$\ln Z(T,V) = Z_1(T,V) = \frac{VT^3}{\pi^2} = -\frac{F}{T} = -\frac{1}{T}(E - TS) = S - \beta E \; , \tag{24}$$

where F is the free energy and S the entropy. It follows that

$$E = -\frac{d}{d\beta} \ln Z = \frac{3VT^4}{\pi^2} \; , \tag{25}$$

which is the Stefan–Boltzmann law for Boltzmann statistics. Further

$$S = \beta E + \ln Z = \frac{4VT^3}{\pi^2} \; . \tag{26}$$

If S is expressed as a function of E (as it should be), we have

$$S = \left(\frac{256V}{27\pi^2} \right)^{1/4} E^{3/4} \tag{27}$$

and the density of states becomes, as derived more rigorously in the above papers:

$$\sigma(E,V) = e^S = \exp[\text{const}.E^{3/4}] \; . \tag{28}$$

The situation changes drastically if the factor $1/n!$ is omitted.

Go back to Eqs. (23) and (24) and drop $1/n!$ The sum now gives

$$\left. \begin{array}{c} Z(T,V) = \frac{1}{1 - Z_1} = \frac{1}{1 - \frac{VT^3}{\pi^2}} = \dfrac{\frac{T_0^3}{T_0^3 - T^3}}{} \\[2mm] T_0 = \left(\frac{\pi^2}{V} \right)^{1/3} \end{array} \right\} \tag{29}$$

For $T \to T_0$ the partition function diverges. Hence T_0 is a singular temperature for this gas.

Now the miracle happens: assume V to be the usual "interaction volume" of strong interactions[21] (without Lorentz contraction):

$$V \approx \frac{4\pi}{3} m_\pi^{-3} \quad \text{gives} \quad T_0 \cong \left(\frac{3\pi}{4}\right)^{1/3} \cdot m_\pi \cong 0.184 \text{ GeV} . \tag{30}$$

This is almost the mysterious "constant" temperature so often encountered in this report.

Following the standard procedure, we calculate the energy and entropy; both become simple for $T \to T_0$:

$$\begin{aligned} E &\approx 3T_0^4/(T_0^3 - T^3) \\ S &\approx \frac{E}{T_0} + \ln \frac{E}{3T_0} \end{aligned} \tag{31}$$

The energy diverges for $T \to T_0$ (therefore T_0 is the *maximum* temperature for this gas). For the level density we obtain

$$\sigma(E, V) = e^S \cong \frac{E}{3T_0} \cdot \exp(E/T_0) , \tag{32}$$

that is: omitting the factor $1/n!$ leads to a maximum temperature and to an exponentially growing density of states. Equation (31) implies: for $E \geq 10T_0$ one always finds a temperature $0.9T_0 \leq T < T_0$, hence practically constant.

This result brought me – *me*, but nobody else – in a state of obsession. Did it not explain one of the most intriguing features of strong interaction processes? And was it not obviously wrong because of its unrealistic assumptions? Yet, there was an interpretation that opened the way to a better model.

Interpretation: distinguishable particles; Pomeranchuk's ansatz

The factor $1/n!$ in front of the phase-space integral (21) serves to compensate "double" counting: given a set of fixed momenta $\{\bar{p}_1 \bar{p}_2 \ldots \bar{p}_n\}$, all $n!$ permutations of this set occur during the integration over $p_1, p_2, \ldots, p_n$. If the particles are indistinguishable, one has therefore to divide by $n!$. If all n particles are different from each other, one should not divide. This was exactly the point:

in our statistical model calculations we had used ~ 80 different particle states and had therefore to replace

$$\frac{1}{n!} \to \frac{1}{\Pi n_k!} ; \quad k = 1 \ldots 80 \tag{33}$$

in front of the phase-space integrals.

As, however, the number of produced secondaries remained far below 80, the values n_k remained, *for all essentially contributing phase-space integrals,* either 0 or 1; hence practically all $n_k! = 1$; thus $1/n!$ was *effectively* replaced by 1.

If *this situation was to be simulated* analytically by a solvable model [namely all masses $= 0$ (for $E \to \infty$)], then in order to come near to reality, the factor $n!$ should be dropped, as if the particles were distinguishable.

This argument had led Auberson and Escoubès to look at the case where $1/n!$ is dropped; they also consider a scenario corresponding to Eq. (33), namely where there are r different species of particles, while inside each species particles are indistinguishable. They are cautious in the interpretation of their results; they write:[41]

"If it is probable that the discernibility hypothesis is the most realistic at low energies, one cannot very well locate the energy at which this hypothesis must be abandoned (if at all)"

and later:

"Clearly r could be larger than 3, to take into account the resonances at high energies [§]. [If, however, in reality the strongly interacting particles should have an infinity of excited states ... we fall back essentially on discernible particles.]"

They leave the question open.

In the present context a paper by Pomeranchuk[44] must be mentioned: he proposes to improve the Fermi model by admitting that real pions are not point-like. Therefore n pions would not find place in a volume V_0 ($\approx \frac{4\pi}{3}m_\pi^{-3}$) but need at least a volume $n \cdot V_0$. Thus the space volume factor in front of the integral (21) would be $[nV_0/8\pi^3]^n$ instead of that written in Eq. (21). As, however, for large n

$$n! \approx \sqrt{2\pi n}\, n^n e^{-n} , \qquad (34)$$

the factor n^n arising from the corrected volume will essentially cancel the factor $1/n!$ and one thus arrives effectively at a model with "distinguishable" particles in a volume $e \cdot V_0$. Thus Pomeranchuk also obtains a maximal temperature of the order of m_π, that is, a practically constant temperature. His paper came about thirteen years too early – or at least five, since the constant mean transverse momenta became popular only after 1956[10] (the decisive large-angle scattering took shape around 1964).

While the model of distinguishable particles was useful because it produced the surprise that motivated the investigations described in the following section, it was clear that all further efforts had to be made on the realistic basis of massive particles with Bose and/or Fermi statistics. The principal lesson to be kept in mind was that there should be many, many different species of particles.

THE STATISTICAL BOOTSTRAP MODEL (SBM)

Up to here we have collected everything that helped to motivate the construction of SBM. We now describe this construction. For what follows, a few formulae need be recalled; although everybody knows them, it is necessary to have them ready at hand in order not to interrupt the argument. We use Boltzmann statistics for simplicity (the first paper on SBM used correct statistics[45]).

[§]r is the above number of different species of particles, ≈ 80 in our old phase-space calculations.

A Few Well-Known Formulae

We go back to Eq. (21), rewrite it for relativistic massive particles and follow the same derivations as there. The density of states of n particles of mass m enclosed in a volume V at energy E is then

$$\sigma_n(E, V, m) = \frac{1}{n!} \left[\frac{V}{(2\pi)^3}\right]^n \int \delta\left(E - \sum_{i=1}^n E_i\right) \prod_{i=1}^n 4\pi p_i^2 dp_i \; ; \quad E_i = \sqrt{p_i^2 + m^2} \,. \tag{35}$$

Its Laplace transform is the n-particle partition function:

$$Z_n(T, V, m) = \frac{1}{n!} \left[\frac{V}{(2\pi)^3}\right]^n \left[4\pi \int e^{-\beta\sqrt{p^2+m^2}} p^2 dp\right]^n \; ; \tag{36}$$

the integral is

$$\int e^{-\beta\sqrt{p^2+m^2}} p^2 dp = m^2 T K_2(m/T) \,, \tag{37}$$

where K_2 is the second modified Hankel function [which for $m \gg T$ goes like $(\pi T/2m)^{1/2} \exp(-m/T)$ and for $m \ll T$ like $(T/m)^2$].

We obtain

$$Z_n(T, V, m) = \frac{1}{n!} \left[\frac{V}{(2\pi)} m^2 K_2(m/T)\right]^n = Z_1(T, V, m)^n/n! \tag{38}$$

by which the "one-particle partition function" Z_1 is defined. Summing over n gives the (grand canonical) partition function for an unfixed particle number:

$$Z(T, V, m) = \sum_{n=0}^{\infty} \frac{1}{n!} Z_1^n = \exp\left[\frac{VT}{2\pi^2} m^2 K_2\left(\frac{m}{T}\right)\right] \,. \tag{39}$$

For a mixture of two gases with particles of masses m_1 and m_2 respectively we have $Z(T, V, m_1, m_2) = Z(T, V, m_1) \cdot Z(T, V, m_2)$. We generalize this to a mixture of gases of many different sorts of particles by introducing the (yet unknown) mass spectrum $\rho(m)$:

$$\rho(m)dm = \text{number of different species of particles in} \{m, dm\} \tag{40}$$

and obtain

$$Z^{(\rho)}(T, V) = \exp\left[\frac{VT}{2\pi^2} \int_0^{\infty} m^2 K_2\left(\frac{m}{T}\right) \rho(m)dm\right] \,. \tag{41}$$

On the other hand, any $Z(T, V)$ can be written as

$$Z(T, V) = \int_0^{\infty} \sigma(E, V) \exp\left(-\frac{E}{T}\right) dE \,. \tag{42}$$

Given $Z(T, V)$, the energy spectrum (density of states) $\sigma(E, V)$ can be calculated, and vice versa.

Doing the same steps: summing over n and introducing a mass spectrum without, however, executing the Laplace transformations, yields the phase-space analogue of Eqs. (41), (42):

$$\left. \begin{aligned} \sigma^{(\rho)}(M, V) &= \sum_{n=0}^{\infty} \frac{1}{n!} \left[\frac{V}{(2\pi)^3}\right]^n \int \delta(M - \sum_{i=1}^n E_i) \prod_{i=1}^n 4\pi p_i^2 \rho(m_i) dp_i dm_i \\ \text{with } E_i &= (p_i^2 + m_i^2)^{1/2} \end{aligned} \right\} \tag{43}$$

Note the enormous difference between the two densities of states, $\rho(m)$ and $\sigma(M, V)$: suppose a single species with mass m_0; then $\rho(m) = \delta(m - m_0)$ is zero everywhere except at $m = m_0$, while $\sigma(M, V)$ grows (for $M \gg m_0$) like $\exp[\text{const.} M^{3/4}]$ as shown in Eq. (28).

Introducing the Bootstrap Hypothesis

From an article that does not otherwise concern our subject, R. Carreras[46] picked up a *bon mot* which may serve very well as a motto for this section:

"... all of these arguments can be questioned, even when they are based on facts that are not controversial."

Here are these arguments:

- if anything deserves the name "fireball", then it is the lump of hadronic matter in the state just before it decays isotropically into a two-body final state, as observed in large-angle elastic $[p + p \rightarrow p + p \quad ; \quad \pi + p(n) \rightarrow \pi + p(n)]$ or two-body inelastic scattering $[p + p \rightarrow \pi^+ + d)]$;

- this fireball answers, within experimental accuracy, to the description by an improved Fermi statistical model, as witnessed by the agreement of our phase-space results with the Orear plot (Fig. 4);

- we therefore postulate that fireballs describable by statistical models do exist, provided that in such models interaction is taken into account by including known particles and resonances (Lessons 5,6,7);

- while *practically* a limited number of sorts of particles and resonances was already sufficient to describe, within experimental accuracy, fireballs up to a mass of 8 GeV, we should *in principle* include all of them with the help of a – yet unknown – mass spectrum $\rho(m)$;

- recalling Eqs. (41), (42) and (43) we observe that there are two mass spectra appearing in the statistical description:

    $\sigma(M, V_0)dM$ is the number of states (of species) of *fireballs* (volume V_0) in the mass interval $\{M, dM\}$,
    $\rho(m)dm$ is the number of species of possible *constituents* (of such fireballs) having a mass in $\{m, dm\}$;

- a glance at the "Review of Particle Properties"[47] informs us, under the headings "Partial Decay Modes" that heavy resonances [to be counted in $\rho(m)$] have many decay channels, some of them containing again resonances. Thus: heavy resonances "consist" (statistically) of particles and lighter resonances – just as fireballs do;

- therefore *there is no principal difference between resonances and fireballs*: the states counted in $\sigma(M, V_0)$ should also be admitted as possible constituents in fireballs of larger mass – that is: they should be counted in $\rho(m)$;

- we conclude that $\rho(m)$ and $\sigma(m, V_0)$ count essentially the same set of hadronic masses and that therefore they must be – up to details – the "same" function;

- they cannot be exactly equal, because $\rho(m)$ starts with a number of δ-functions (π, K, $p, \ldots$) while $\sigma(m, V_0)$ is continuous above $2m_\pi$;

- leaving the door open for such and other differences, we postulate only that the corresponding entropies should become asymptotically equal:

$$\frac{\log \sigma(m, V_0)}{\log \rho(m)} \xrightarrow[m \to \infty]{} 1 . \tag{44}$$

We call this the "bootstrap condition",[45] which is a very strong requirement in view of the great difference between ρ and σ in "ordinary" thermodynamics (remark at the end of the last section).

The Solution

The rest is mathematics (and the above motto no longer applies). It could be shown[45] that ρ and σ have to grow asymptotically like const$m^{-\alpha}\exp(m/T_0)$, while possible solutions growing faster than exponentially are inadmissible in statistical thermodynamics. Nahm[48] proved that by adding certain refinements, the condition (44) could be sharpened and that the power of the prefactor is then $\alpha = 3$; he also derived sum rules, which allowed to estimate T_0 to lie in the region of 140 to 160 MeV, results which agreed with Frautschi's (and collaborators) numerical results.[49] Thus the question put after Lesson 7 had been answered in the affirmative: SBM was born.

It is a self-consistent scheme in which the "particles" – i.e., clusters or resonances or fireballs: call them as you like – are at the same time

- *object of the description*

- *constituents of this object*

- *generating the interaction which keeps the object together.*

Thus it is a "statistical bootstrap"[49] embracing all hadrons.

Further Developments

All that happened to SBM after its birth was reported in some detail, and with all references known to me, in a review talk,[2] which I shall not try to sum up here.

Even so, a few more important steps must be mentioned.

- The thermodynamic description of fireballs is so simple that it can be combined with collective motions (as in two-centre models[11]) and summed over impact parameters; leading particles and conservation laws are easily taken care of. In this way J. Ranft and myself constructed the so-called "thermodynamical model", which proved useful for predicting particle momentum spectra.[50]

- S. Frautschi, in a most important paper,[49] had written down and solved the first phase-space formulation of SBM; his work triggered an avalanche of further papers reviewed in,[2] leading to a new development. His "bootstrap equation" (BE) is much stronger and more elegant than our above "bootstrap condition" (44). Later it was put in a manifestly Lorentz invariant form and analytically solved by Yellin.[51] This formulation has become standard. The Laplace-transformed BE is a functional equation for the Laplace transform of the mass spectrum; this equation was already known in 1870[52] ¶ and independently rediscovered by Yellin. All this was so important that I cannot resist to illustrate it with the help of a simple toy model, in which clusters are composed of clusters with vanishing kinetic energy. In this limit the Frautschi–Yellin BE reads

$$\rho(m) = \delta(m - m_0) + \sum_{n=2}^{\infty} \frac{1}{n!} \int \delta\left(m - \sum_{i=1}^{n} m_i\right) \prod_{i=1}^{n} \rho(m_i)dm_i \,. \qquad (45)$$

In words: the cluster with mass m is either the "input particle" with mass m_0 or else it is composed of any number of clusters of any masses m_i such that $\Sigma m_i = m$.

¶In another context – as one might guess.

33

We Laplace-transform Eq. (45):

$$\int \rho(m)\exp[-\beta m]dm = \exp[-\beta m_0] + \sum_{n=2}^{\infty} \frac{1}{n!} \prod_{i=1}^{n} \int \exp[-\beta m_i]\rho(m_i)dm_i \quad (46)$$

Define

$$
\begin{aligned}
z(\beta): &= \exp[-\beta m_0] \\
G(z): &= \int \exp[-\beta m]\rho(m)dm \ .
\end{aligned}
\quad (47)
$$

Thus Eq. (46) becomes $G(z) = z + \exp[G(z)] - G(z) - 1$ or

$$z = 2G(z) - \exp[G(z)] + 1 \ , \quad (48)$$

which is the above-mentioned functional equation for the function $G(z)$, the Laplace transform of the mass spectrum. This function proved most important in all further development; for instance: the coefficients of its power expansion in z are directly related to the multiplicity distribution of the final particles in the decay of a fireball.[53]

It is most remarkable that the "Laplace-transformed BE", Eq. (48), is "universal" in the sense that it is not restricted to the above toy model but turns out to be the same in all (non-cut-off) realistic SBM cases[49,51]; moreover, it is independent of

- the number of space-time dimensions,[54]
- the number of "input particles" (z becomes a sum over modified Hankel functions of input masses),
- Abelian or non-Abelian symmetry constraints.[55]

What is wanted is of course $G(z)$, given implicitly by Eq. (48); solutions are reviewed in.[2] Most simple is its graphic solution: we draw $z(G)$ according to Eq. (48) and exchange the axes (Fig. 5).

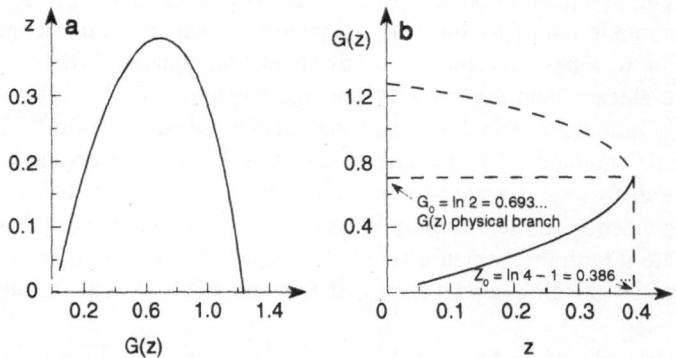

Figure 5. (a) $z(G)$ according to Eq. (48): (b) $G(z)$, the graphical solution of Eq. (48).

One sees immediately that (universally!)

$$z_{\max}(G) =: z_0 = \ln 4 - 1 = 0.3863\ldots$$
$$G_0 = G(z_0) = \ln 2 .$$

The parabola-like maximum of $z(G)$ implies a square root singularity of $G(z)$ at z_0 – first remarked by Nahm[48] – leading, upon inverse Laplace transformation, to $\rho(m) \sim m^{-3} \exp[\beta_0 m]$ where – in our present case (not universally!):

$$\beta_0 = -\left(\frac{1}{m_0}\right) \ln z_0 = \frac{0.95}{m_0} \qquad \text{[see Eq. (47)]} . \tag{49}$$

Putting $m_0 = m_\pi$ we find a reasonable value for $T_0 = \beta_0^{-1}$:

$$T_0(\text{toy model}) = 0.145 \, \text{GeV} . \tag{50}$$

Thermodynamics of a gas of the above clusters (45) has T_0 as a singular temperature.

Thus, the simple toy model already yields all essential features of SBM. For a very short representation of the more realistic case of pion clustering in full relativistic momentum space, see Section 2 of.[53]

- Much work was done by groups in Bielefeld, Kiev, Leipzig, Paris and Turin to clear up the relation of SBM to other trends in strong interaction physics (Regge, Veneziano, ...) and to the theory of phase transitions; this is reviewed in.[2]

- In the mid-seventies J. Rafelski arrived at CERN and immediately began pushing me: SBM should be polished up to become applicable to heavy-ion collisions. There were two problems: baryon-number (and, eventually, strangeness) conservation and proper particle volumes – point-like heavy ions would be nonsense. Thus we introduced baryon (strangeness) chemical potential and – less trivially – proper particle volumes, first in the BE,[56] then in the ensuing thermodynamics.[57] We found that particle volumes had to be proportional to particle masses with a universal proportionality constant. The argument was the following.

Let a cluster (fireball) of mass m and volume V be composed of constituents with masses m_i and volumes V_i. Different from usual assumptions in thermodynamics, the cluster is not confined to an externally imposed volume; it rather carries its volume with it (as already stressed by Nahm[48]); and so does each of its constituents. Let anyone of them have four-momentum p_i^μ, then its volume moves with four-velocity p_i^μ/m_i. With Touschek[58] we define a "four-volume"

$$V_i^\mu = \frac{V_i}{m_i} p_i^\mu . \tag{51}$$

The constituents' volumes have to add up to the total cluster volume and their momenta to the total momentum

$$\frac{V}{m} p^\mu = \sum_{i=1}^{n} \frac{V_i}{m_i} p_i^\mu$$
$$p^\mu = \sum_{i=1}^{n} p_i^\mu . \tag{52}$$

This is possible for arbitrary n and p_i^μ if and only if

$$\frac{V}{m} = \frac{V_i}{m_i} = \text{const} = 4B , \tag{53}$$

where the proportionality constant is written $4B$ in order to emphasize the similarity to MIT bags,[59] which have the same mass-volume relation. As, moreover, the energy spectrum of SBM clusters and MIT bags is the same – even in details[60,61] – one is led to consider these two objects to be the same, at least in the sense that statistical thermodynamics of MIT bags is identical to that of SBM clusters; this "identity" is interesting, because MIT bags "consist of" quarks and gluons, SBM clusters of hadrons: it suggests the phase transition from the one to the other.

- Thermodynamics of clusters with proper volumes still had a singularity at T_0, but a weaker one: while in the old (point-particle) version of SBM the energy density diverged at T_0 (thus making T_0 an "ultimate" temperature), now the energy density was finite at T_0, making a phase transition (already anticipated by Cabibbo and Parisi[62]) to a quark-gluon plasma possible[57,63,64];

- Technical problems in handling explicitly the particle volumes could be elegantly solved[65-69] by using the "pressure (or isobaric) partition function" invented by Guggenheim.[70] This technique allows one to treat the thermodynamics of bags with an exponential mass spectrum (pioneered by Baacke[71]) in all beauty: Letessier and Tounsi[72-74] succeeded, following the methods of the Kiev group[65-69], in describing with a *single* partition function the hadron gas, the quark-gluon plasma and the phase transition between the two in a realistic case. This opens the way to solve a number of problems connected with proving (or disproving?) the actual presence of a quark-gluon phase in the first stage of relativistic heavy-ion collisions.

SOME DISORDERED REMARKS

The Difficulty to Kill an Exponential Spectrum

The most prominent feature of SBM is its exponentially increasing mass spectrum. Many objections to it were put forward: symmetry constraints would forbid a number of the states counted in it; correct Bose-Einstein and Fermi-Dirac statistics would also reduce the number of states; in composing clusters of clusters ... one should take into account the Pauli principle, which again might eliminate many states; the original argument for including resonances was based on the Beth-Uhlenbeck method [Eq. (14); Lesson 5], which invokes phase shifts and their sudden rise by π when going through a resonance. But then the phase shifts should go to zero for infinite momentum and the Levinson theorem states that $\delta_\ell(0) - \delta_\ell(\infty) = N_\ell \pi$ with N_ℓ = number of bound states with angular momentum ℓ. Therefore phase shifts cannot go on increasing by π for each of our (exponentially growing number of) resonances – they must decrease again: that is, each and every one of the masses added somewhere to the mass spectrum must be (smoothly) subtracted later on. How then can an exponential mass spectrum survive?

All these objections turn out to leave the mass spectrum intact, because the exponential function is extremely resistant to manipulations: multiplication by a polynomial of any (positive or negative) order, squaring, differentiating or integrating it will not do much harm (consider the leading term of its logarithm!) – at most change its exponential parameter $(T_0 \rightarrow T_0')$.

Therefore, once our self-consistency requirement – crude as it may be – has led to this particular mass spectrum, it is difficult to get rid of it. Incidentally, in the first paper on SBM[45] correct Bose-Einstein/Fermi-Dirac statistics was employed (easy in the grand canonical formulation, most awful in phase space[75-79]) and the result was that the mass spectrum $\rho(m) = \rho_{\text{Bose}}(m) + \rho_{\text{Fermi}}(m)$ had to grow exponentially. The role of conservation laws has been dealt with in the literature (see references in[2]). All these explicit attacks did not kill the leading exponential part of the mass spectrum.

It remains, as an illustration of the resistance of $\rho(m)$, to assume that, for whatever reason, every mass once added to it, has to be eliminated again (I do not know of any serious argument which would require this; the Levinson theorem derived in non-relativistic potential scattering[80] cannot be invoked in a situation where all kinds of reactions between constituents take place – but assume it had to be so). Then, arriving at mass m we subtract everything that had been added at $m - \Delta m$: thus

$$\rho(m) \rightarrow \rho(m) - \rho(m - \Delta m) \approx e^{\beta_0 m}[1 - e^{-\beta_0 \Delta m}]$$

and the leading exponential remains untouched (kind of differentiation).

What is the Value of T_0?

The most fundamental constant of SBM, T_0, escapes precise determination. There are several ways to try to fix T_0:

a) Theoretically

    i) inside SBM

    ii) from lattice QCD

b) Empirically

    i) from the mass spectrum

    ii) from the transverse momentum distribution

    iii) from production rates of heavy antiparticles $(\overline{\text{He}^3}, \bar{d})$

    iv) from the phase transition to the quark-gluon phase

We obtain

ai) T_0 from inside SBM :

    - the crude model Eq. (45) yields with pions only: $T_0 \approx 0.145$ GeV
      if K and N were added to the input: $T_0 \approx 0.135$ GeV

    - the unrealistic model of distinguishable, massless particles
      as described by Eq. (31) gives $T_0 \approx 0.184$ GeV

    - a more realistic model (pions + invariant phase space)
      yields[81] $T_0 \approx 0.152$ GeV

    - Hamer and Frautschi[82] solve their BE by numerical
      iteration and read off $T_0 \approx 0.140$ GeV

    - Nahm derives a sum rule from which he found, under different
      assumptions (which particles are admitted?):[48] $T_0 \approx 0.154$ GeV
                                                                  or 0.142 GeV

aii) T_0 from lattice QCD

The determination of T_0 from lattice QCD is still hampered with difficulties. First estimates using pure gauge gave rather high T_0 while the introduction of quarks pose their own principal problems. Nevertheless compromises have been devised, which circumvent these problems and allow dealing with quarks (by paying a prize depending on what one is after). I quote here a table from a review article by F. Karsch,[83] where the methods are described and references to original work is given (Table 2 from;[83] see also[84])

n_f	T_0 from $\sqrt{\sigma}$	T_0 from m_ρ	T_c from m_N
0	239 ± 13	239 ± 23	225 ± 30
2	–	145 ± 7	113 ± 9
4	160 ± 22	130 ± 7	105 ± 9

$T_0 \approx 0.145$ GeV or 0.130 GeV

$\sqrt{\sigma}$ is the string tension $= 0.42 \pm 0.02$ GeV and n_f the number of light flavours, $n_f = 2$ or 4 being the most physical choice. It is believed that the value "T_0 from m_ρ" is the most realistic one. It agrees rather well with the above-listed estimates of T_0 from inside SBM (except that for distinguishable massless particles which – accidentally? – lies nearer to the pure gauge ($n_f = 0$) value).

bi) from the mass spectrum

Here the difficulty is that approximate completeness of the empirical mass spectrum ends somewhere around 1.5 GeV, because the density of mass states increases and the production rate decreases (both exponentially, as predicted by SBM) and the identification of all masses rapidly becomes impossible. On the other hand we know only the asymptotic form of the mass spectrum $\sim m^{-3} \exp[m/T_0]$ and have to guess an extrapolation towards lower masses, which does not diverge for $m \to 0$. Various attempts (after 1970):

- Hagedorn and Ranft[85] obtain, with large uncertainties $\qquad T_0 \cong 0.148$ GeV

- Letessier and Tounsi[86] find $\qquad T_0 \cong 0.155$ GeV

- See also contribution of A. Tounsi to this workshop (I wish to thank J. Letessier and A. Tounsi for communicating this result to me before the workshop.) $\qquad T_0 \cong 0.158$ GeV

bii) from $p_\perp$ spectrum

- Large-angle elastic scattering; Orear[37] finds an apparent temperature $T = 0.158$ GeV, which should lie near to T_0; hence $\qquad T_0 \gtrsim 0.158$ GeV

- Folklore has it that the $p_\perp$ distribution in the soft region is $\exp[-6p_\perp]$. The exact formula for the distribution is[87] quite different from $\exp(-p_\perp/T)$, but for $p_\perp \gg m_\pi$ one might generously accept $\exp[-6p_\perp]$; then $\qquad T \approx T_0 \cong 0.167$ GeV

- A serious attempt to fix T_0 from the $p_\perp$ distribution is reported in.[88] In a region where $p_\parallel$ is very small (no integration over $p_\parallel$) the authors fit the $p_\perp$ distribution to a Bose-Einstein formula and obtain the surprisingly low result: $\qquad T_0 > T \approx 0.117$ GeV

They give reasons why they identify this with T_0, although the primary momentum is only 28.5 GeV (Brookhaven), so that T_0 could still lie somewhat higher (as I believe).

Remark

The determination of T from $p_\perp$ suffers from a number of perturbing effects, which have been discussed in detail in:[87] resonance $(\rho, \Delta \ldots)$ decay, leakage of "large $p_\perp$" down to the soft region, etc. It seems that all these effects do not influence the two-body large-angle scattering, so that the value found by Orear[37] might be more trustable than the values obtained by fitting the soft $p_\perp$ distribution by various formulae (partly not well justified).

biii) From production ratios of heavy antiparticles

Production rates of anti-^{3}He, antitritium $(\bar{t})$, antideuteron $(\bar{d})$ and of many other particles have been measured[89] *. By taking ratios we avoid (at least in part) problems coming from the pion production rate, not well known theoretically, and from (not much) different momenta and target material. From the quoted paper[89] we take ratios $\overline{^3\text{He}}/\bar{d}$ and $\bar{t}/\bar{d}$ and average the values [all around $(0.8$ to $2) \times 10^{-4}$]; we find

$$\left\langle \frac{\overline{^3\text{He} \text{ or } \bar{t}}}{\bar{d}} \right\rangle = (1.4 \pm 0.7) \times 10^{-4} \ . \tag{54}$$

From SBM – taking into account that for each produced antibaryon another baryon must be produced along with it – one easily works out (spin, etc., factors included)

$$\left\langle \frac{\overline{^3\text{He} \text{ or } \bar{t}}}{\bar{d}} \right\rangle = V \cdot \left(\frac{3}{2} \right)^{3/2} \left(\frac{m_N T}{2\pi} \right)^{3/2} \cdot \frac{8}{3} \exp[-2m_N/T] \ . \tag{55}$$

The exponential is easily understood: $\overline{^3\text{He}}$ or $\bar{t}$ require production of six nucleon masses, $\bar{d}$ requires 4, in the ratio four of the six cancel out.

Putting in numbers, one finds that the experimental values (54) are obtained with a temperature between 0.15 and 0.16 GeV when for V we assume the usual $4\pi/(3m_\pi^3)$. Hence (at a primary momentum of ~ 200 GeV/c): $\qquad T_0 \gtrsim 0.155$ GeV

biv) From the phase transition to the quark − gluon phase

As the existence of the quark-gluon phase is still hypothetical, no direct measurement is available. A theoretical estimate was proposed by Letessier and Tounsi:[90] they require that the curves $P = 0$ for an SBM hadron gas and for a quark-gluon phase coincide "as well as can be achieved". They find:

$$T_0 \approx 0.170 \text{ GeV}$$

Remark 1

The collective motions expected in the expansion and decay of the "fireballs" produced in relativistic heavy-ion collisions will "Doppler-shift" the temperatures read off from transverse momentum distributions. Too high temperatures will result if the collective transverse motion is not corrected for. Thus in our early work on heavy-ion collisions[91] we have (erroneously?) assumed a value $T_0 \approx 0.19$ GeV, which does not seem, in view of all the other estimates, to be realistic.

---

*I am grateful to P. Sonderegger (CERN) for making me aware of this work.

Remark 2

While no precise value can yet be assigned to T_0, it is satisfying that so many different methods yield values which differ typically by less than 20%. An average over all values listed above gives[†]:

$$T_0 = 0.150 \pm 0.011 \text{ GeV}$$

Where is Landau and Where are the Californian Bootstrappers?

History as told above makes it evident that Landau's model[92] is orthogonal to our approach and did not influence its development. There was, however, one moment after the formulation of SBM, namely when we tried to combine it with collective motions to obtain momentum spectra of produced particles, when we considered a combination of Landau's hydrodynamical approach with SBM – only to discard it almost immediately: Landau dealt with "prematter" expanding after a central collision, we needed the evolution of hadron matter after collisions averaged over impact parameter; we had to take into account various sorts of final particles (π, K, N, hyperons and antiparticles) obeying conservation laws (baryon number and strangeness); we had to care for leading particles, etc. All that forced us to the semi-empirical "thermodynamical model"[50] whose aim was not theoretical understanding but practical predictions for use in the laboratory.

Moreover there was a psychological obstacle which I never overcame: namely the Lorentz-contracted volume from which everything was supposed to start. True: when two nucleons hit head-on, then just before the impact they are Lorentz-contracted (seen from the CM); then they collide, heat up and come to rest. When they start to expand, they are at rest and hence no longer Lorentz-contracted. On the other hand, one can conceive that the mechanical shock has indeed compressed them; but into what state would be again a complicated hydro-thermodynamical problem in which their viscosity, compressibility, specific heat and what not would enter[‡]. Why should this state of compression, which constitutes the initial condition for the following expansion, be exactly equal to the Lorentz "compression" before the shock? As among the people working on this model nobody shared my uneasiness, I guessed it was my fault – but it somehow prevented me to ever become enthusiastic about the Landau model.

This is the place to mention another Russian physicist whose work would have inspired us, had we been aware of it: in 1960, five years before SBM, Yu.B. Rumer[93] wrote an article with the title "Negative and Limiting Temperatures" in which he states the necessary and sufficient conditions for the existence of a limiting temperature – namely an exponential spectrum – and gives an example: an ideal gas in an external logarithmic potential. Unfortunately he remained essentially on the formal side of the problem and did not connect it with particle production in strong interactions. Otherwise ...

Now the Californian Bootstrappers:

Even a most modest account of what has been done in the heroic effort of a great number of theoreticians on the program of "Hadron Bootstrap" or "Analytic S-Matrix" would fill a whole book. For me the question is: Did it in any way help to the conception of SBM in 1964 – and the answer is no. One only has to remember the above-reported history up to 1964 and hold it against the best non-technical expositions of the basic ideas and the general philosophy of "Hadron Bootstrap": namely the two articles by G.F. Chew:

"Bootstrap: A scientific idea?"[94]

[†]The two extreme values 0.117 and 0.184 were omitted; the error quoted is the mean standard error arising from the listed values (without their individual errors taken into account).

[‡]These are problems occupying now theoreticians working on relativistic heavy-ion collisions – could these problems be trivial? See, however, the Post Scriptum at the end of this paper.

"Hadron bootstrap: triumph or frustration?"[95]
to realize that. After 1964, however, the influence was enormous, although not technically. But it was of great value for all those who worked on SBM to see their philosophical basis shared with others. Most influential, of course, was that S. Frautschi, one of the leading Californian Bootstrappers, joined our efforts and not only lent his prestige but indeed gave SBM a turn that upgraded it (he also coined its name: "statistical bootstrap") and made it acceptable to particle physicists: the phase-space formulation and its numerous consequences.

The Californian Bootstrappers credo was the analytic S-matrix: Poincaré-invariant, unitary, maximally analytic (with crossing, pole-particle correspondence), which was believed – or hoped – to be uniquely fixed by these requirements. Another aspect was that it should generate the whole hadron spectrum, where each hadron *plays three different roles: it may be a 'constituent', it may be 'exchanged' between constituents and thereby constitute part of the force holding the structure together, it may* **be** *itself the entire composite*".[94]

I know of only two realizations of this aspect: the $\pi - \rho$ system[96,97] and SBM – both remaining infinitely behind the ambitious bootstrap program. In spite of a large number of other achievements obtained in and around the program (dispersion relations, Regge poles, Veneziano model – even string theories) one sees today a strong resurrection of Lagrangian field theories, which now have taken the lead in the run to a "theory of everything". I believe that the bootstrap philosophy and the Lagrangian fundamentalism must complement each other. Each one alone will never obtain complete success.

CONCLUSION

Had I been asked to speak only five minutes, my lecture might have been much better. Here it is.

On the long way to SBM we stopped at a few milestones:

- the realization that in a *single* hadron-hadron collision many secondaries can be produced (1936);

- the discovery of limited $\langle p_\perp \rangle$ (1956);

- the discovery that fireballs exist and that a typical collision seems to produce just two of them (1954-1958);

- the concept of the compound nucleus and its thermal behaviour (1936-1937);

- the construction of simple statistical/thermodynamical models for particle production in analogy to compound nuclei (1948-1950);

- the introduction of interaction into such models via phase shifts at resonance (1937, 1956);

- the discovery that large-angle elastic cross-sections decrease exponentially with CM energy (1963), and

- the discovery that a parameter-free and numerically correct description of this exponential decrease existed already, buried in archived Monte Carlo phase-space results (1963).

The birth of SBM in 1964 was but the logical consequence of all this. Between 1971 and 1973 the child SBM became a promising youngster, when it was reformulated and solved in phase space. It became adult in 1978-1980, when it acquired finite particle volumes. Today, another fourteen years later, it shows signs of age and is ready for retirement: in not long a time, all of its detailed results will have been derived from QCD – maybe from "statistical QCD".

So then – was that all? I believe that something remains: SBM has opened a (one-sided but) intuitive view on strong interactions, revealing that

- their "thermal behaviour" and thus

- their accessibility to statistical thermodynamical descriptions*;

- the existence of clusters (fireballs);

- the production rates of particles and

- their typical multiplicity distributions;

- large-angle elastic scattering;

- the "universal" soft-$p_\perp$ distribution (plotted against $\sqrt{p_\perp^2 + m^2}$, please!);

- the existence of a singular temperature T_0 where a phase transition takes place:

that all these are nothing else than obvious (and calculable) manifestations of one single, fundamental property of strong interactions, namely that

$$\boxed{\text{they possess an exponential mass spectrum } \rho(m) \sim \exp(m/T_0).}$$

And most remarkably: the constant T_0 is roughly equal to the *lowest* hadron mass, m_π. I believe the simple interpretation to be that strong interactions are as strong as they can possibly be, before becoming *too* strong. Too strong – that is: if they would cause the mass spectrum to grow only a little faster than exponentially[†], say: $\rho(m) \sim \exp[(m/m_0)^{1+\alpha}], \alpha > 0$, the "entropy" of its clusters would be $S_c \sim \ln \rho \sim (m/m_0)^{1+\alpha}$; then clusters would swallow each other (if in reach) to become giant superclusters – a sort of hadronic black holes – which could not live in thermal equilibrium with each other while they would be cold inside: $T_{\text{interior}}(m) = (dS/dm)^{-1} = \text{const}(m_0/m)^\alpha \underset{m \to \infty}{\to} 0$. Maybe, even, the joined forces of gravitation and superstrong interactions might have stopped the expansion of the Universe at some early state. Anyway – we can save the effort of working out all consequences of superstrong interactions: the state of the world seems not to favour such a hypothesis.

Don't ask me why strong interactions are actually as strong as permissible – this will have to be answered by some future unified theory (maybe the only possible one?) not yet known (to me).

I would like to know the reaction of a physicist who, in 50 years, comes accidentally over this lecture and takes the trouble to read it. He might quote[98]

"How finely we argue upon mistaken facts!"

---

*Many-body systems can often be described statistically; final states emerging from single two-body collisions only in strong interactions.

[†]In view of the "stability" of the exponential function (Section 4.1), this might require a superstrong interaction which could be *much* stronger than the actual one.

ACKNOWLEDGEMENTS

J. Rafelski has initiated and largely organized this workshop; I wish to thank him for this occasion to deliver a paper which otherwise would not have been written and I wish to thank Marie-Noëlle Fontaine for the beautiful typescript and her infinite patience with it (and with me).

Post Scriptum

As kindly pointed out to me by E. L. Feinberg, my uneasiness about Lorentz contracted interaction volumes mentioned under the leading "Where is Landau and..." is indeed due to my fault — or better: to my ignorance about some fundamental ingredients of Landau's model. The relevant points are put forward in E.L. Feinberg's beautiful paper "Can the relativistic change in the scales of length and time be considered the result of action of certain forces?"[99] I wish to thank him for bringing it to my attention.

REFERENCES

1. Laurence Sterne, "The life and Opinions of Tristram Shandy Gentleman" (London, 1760), Book II, Ch. 19.

2. R. Hagedorn, in "Quark Matter '84", Helsinki Symposium, ed. K. Kajantie, published in "Lecture Notes in Physics" **221** (Springer, Heidelberg, 1985).

3. E.L. Feinberg, *Physics Reports* **5** (1972) 237.

4. H. Yukawa, *Proc.Phys.Math.Soc.Japan* **17** (1935) 48.

5. W. Heitler, *Proc.Cambr.Phil.Soc.* **37** (1941) 291.

6. W. Heitler and H.W. Peng, *Proc.Cambr.Phil.Soc.* **38** (1942) 296.

7. L. Janossy, *Phys.Rev.* **64** (1943) 345.

8. W. Heisenberg, *Z.Phys.* **101** (1936) 533.

9. N.M. Dulles and W.D. Walker, *Phys.Rev.* **93** (1954) 215.

10. J. Nishimura, *Soryushiron Kenkyu* **12** (1956) 24.

11. G. Cocconi, *Phys.Rev.* **111** (1958) 1699.

12. F.W. Büsser et al. (CERN-Columbia-Rockefeller Coll.), *Phys.Lett.* **46B** (1973) 471, and a wealth of further papers in conference reports and review articles.

13. S. Takagi, *Progr.Theor.Phys.* **7** (1952) 123.

14. P. Ciok, T. Coghen, J. Gierula, R. Holyński, A. Jurak, M. Miesowicz, T. Saniewska, O. Stanisz and J. Pernegr, *Nuovo Cimento* **8** (1958) 166 and **10** (1958) 741 [the second without O. Stanisz].

15. J. Gierula, *Fortschr.d.Physik* **11** (1963) 109.

16. N. Bohr, *Nature* **137** (1936) 344.

17. V.F. Weisskopf, *Phys.Rev.* **52** (1937) 295.

18. I. Frenkel, *Soviet Phys.* **9** (1936) 533.

19. H. Koppe, *Phys.Rev.* **76** (1949) 688.

20. H. Koppe, *Zs.f.Naturforschung* **3a** (1948) 251.

21. E. Fermi, *Progr.Theor.Phys.* **5** (1950) 570.

22. E. Fermi, *Phys.Rev.* **81** (1951) 683.

23. W. Heisenberg, *Naturwissenschaften* **39** (1952) 69.

24. E. Beth and C.E. Uhlenbeck, *Physica* **4** (1937) 915.

25. L. Landau and E. Lifshitz, "Statistical Physics" (MIR, Moscow, 1967).

26. R. Hagedorn, in Cargèse Lectures, Vol. 6 (1973), p. 643, ed. E. Schatzmann (Gordon and Breach, New York, 1973).

27. K. Huang, "Statistical Mechanics" (John Wiley and Sons, New York, 1963).

28. J. Bernstein, R. Dashen and S. Ma, *Phys.Rev.* **187** (1969) 1.

29. S.Z. Belenkij, *Nucl.Phys.* **2** (1956) 259.

30. L. Sertorio, *Rivista del Nuovo Cimento* **2** (1979) No 2.

31. F. Cerulus and R. Hagedorn, *Suppl.Nuovo Cimento* **9** (1958) 646, 659.

32. F. Cerulus, *Nuovo Cimento* **19** (1961) 528.

33. F. Cerulus, *Nuovo Cimento* **22** (1961) 958.

34. R. Hagedorn, *Nuovo Cimento* **15** (1960) 434.

35. S.Z. Belenkij, V.M. Maksimenko, A.I. Nikisov and I.L. Rozental,
    a) *Usp.Fiz.Nauk* **62** (1957) 1, in Russian;
    b) *Fortschr.d.Physik* **6** (1958) 524, same in German.

36. G. Cocconi, *Nuovo Cimento* **33** (1964) 643.

37. J. Orear, *Phys.Lett.* **13** (1964) 190.

38. G. Fast, R. Hagedorn and L.W. Jones, *Nuovo Cimento* **27** (1963) 856;
    G. Fast and R. Hagedorn, *Nuovo Cimento* **27** (1963) 208.

39. T.E.O. Ericson, *Phys.Lett.* **4** (1963) 258.

40. A. Białas and V.F. Weisskopf, *Nuovo Cimento* **35** (1965) 1211.

41. G. Auberson and B. Escoubès, *Nuovo Cimento* **36** (1965) 628.

42. H. Satz, *Nuovo Cimento* **37** (1965) 1407.

43. J. Vandermeulen, *Bull.Soc.Roy.Sci.Liège (Belgium)* **34** (1965) Nos 1-2.

44. I. Pomeranchuk, *Dokl.Akad.Nauk SSSR* **78** (1951) 889.

45. R. Hagedorn, *Suppl. Nuovo Cimento* **3** (1965) 147.

46. R. Carreras, "Picked up for you this week", No 26 (CERN, Geneva, 1993).

47. Particle Data Group, *Phys.Rev.* **D50** (1994) No 3, Part I.

48. W. Nahm, *Nucl.Phys.* **B45** (1972) 525.

49. S. Frautschi, *Phys.Rev.* **D3** (1971) 2821.

50. R. Hagedorn and J. Ranft, *Suppl. Nuovo Cimento* **6** (1968) 169;
    H. Grote, R. Hagedorn and J. Ranft, "Particle Spectra" (CERN, Geneva, 1970).

51. J. Yellin, *Nucl.Phys.* **B52** (1973) 583.

52. E. Schröder, *Z.Math.Phys.* **15** (1870) 361.

53. G.J.H. Burgers, C. Fuglesang, R. Hagedorn and V. Kuvshinov, *Z.Phys.* **C46** (1990) 465.

54. H. Satz, *Phys.Rev.* **D20** (1979) 582.

55. K. Redlich and L. Turko, *Z.Phys.* **C5** (1980) 201.

56. R. Hagedorn, I. Montvay and J. Rafelski, Proc. Workshop on Hadronic Matter at Extreme Energy Density, Erice, 1978, p. 49 eds. N. Cabibbo and L. Sertorio (Plenum Press, New York, 1980).

57. R. Hagedorn and J. Rafelski, *Phys.Lett.* **97B** (1980) 136.

58. B. Touschek, *Nuovo Cimento* **B58** (1968) 295.

59. A. Chodos, R.L. Jaffe, K. Johnson, C.B. Thorn and V.F. Weisskopf, *Phys.Rev.* **D19** (1974) 3471.

60. J.I. Kapusta, *Nucl.Phys.* **B196** (1982) 1.

61. K. Redlich, *Z.Phys.* **C21** (1983) 69.

62. N. Cabibbo and G. Parisi, *Phys.Lett.* **59B** (1975) 67.

63. R. Hagedorn and J. Rafelski, Bielefeld Symposium on "Statistical Mechanics of Quarks and Hadrons", p. 237, ed. H. Satz (North Holland, Amsterdam, 1981).

64. J. Rafelski and R. Hagedorn, as in Ref. (63), p. 253.

65. M.I. Gorenstein, *Yad.Fiz.* **31** (1980) 1630.

66. M.I. Gorenstein, V.K. Petrov and G.M. Zinovjev, *Phys.Lett.* **106B** (1981) 327.

67. M.I. Gorenstein, G.M. Zinovjev, V.K. Petrov and V.P. Shelest, *Teor.Mat.Fiz.* **52** (1982) 346.

68. R. Hagedorn, *Z.Phys.* **C17** (1983) 265.

69. M.I. Gorenstein, S.I. Lipskikh and G.M. Zinovjev, *Z.Phys.* **C22** (1984) 189.

70. E.A. Guggenheim, *J.Chem.Phys.* **7** (1939) 103.

71. J. Baacke, *Acta Phys.Pol.* **B8** (1977) 625.

72. J. Letessier and A. Tounsi, *Nuovo Cimento* **99A** (1988) 521.

73. J. Letessier and A. Tounsi, *Phys.Rev.* **D40** (1989) 2914.

74. J. Letessier and A. Tounsi, Preprint PAR/LPTHE 90-17, Univ. Paris VII (1990) presented at IInd Internat. Bodrum Physics School, 1989, University of Istanbul, Turkey.

75. K. Fabricius and U. Wambach, *Nucl.Phys.* **B62** (1973) 212.

76. M. Chaichian, R. Hagedorn and M. Hayashi, *Nucl.Phys.* **B92** (1975) 445.

77. J. Engels, K. Fabricius and K. Schilling, *Phys.Lett.* **59B** (1975) 477.

78. J. Kripfganz, *Nucl.Phys.* **B100** (1975) 302.

79. M. Chaichian and M. Hayashi, *Phys.Rev.* **D15** (1977) 402.

80. M. Levinson, *Kgl.Danske Vid.Selsk.Mat.Fys.Med.* **25** (1949) No 9.

81. R. Hagedorn and I. Montvay, *Nucl.Phys.* **B59** (1973) 45.

82. G.J. Hamer and S. Frautschi, *Phys.Rev.* **D4** (1971) 2125.

83. F. Karsch, in "QCD 20 Years Later", Aachen Symposium, 1992, Vol. 2, p. 717, eds. P.M. Zerwas and H.A. Kastrup, (World Scientific, Singapore, 1993).

84. J. Fingberg, U. Heller and F. Karsch, *Nucl.Phys.* **B392** (1993) 493.

85. R. Hagedorn and J. Ranft, *Nucl.Phys.* **B48** (1972) 157.

86. J. Letessier and A. Tounsi, *Nuovo Cimento* **13A** (1973) 557.

87. R. Hagedorn, *Rivista del Nuovo Cimento* **6** (1983) No 10.

88. A.T. Laasanen, C. Ezell, L.J. Gutay, W.N. Schreiner, P. Sch'ublin, L. von Lindern and F. Turkot, *Phys.Rev.Lett.* **38** (1977) 1.

89. A. Bussière et al., *Nucl.Phys.* **B174** (1980) 1.

90. J. Letessier and A. Tounsi, *Nuovo Cimento* **99A** (1988) 521.

91. R. Hagedorn and J. Rafelski, *Phys.Lett.* **97B** (1980) 136.

92. L.D. Landau, "Collected Papers", ed. D. Ter Haar (Gordon and Breach, New York, 1965).

93. Yu.B. Rumer, *Soviet Phys. (JETP)* **11** (1960) 1365 [Translation from original: JETP **38** (1960) 1899].

94. G.F. Chew, *Science* **161** (1968) 762.

95. G.F. Chew, *Physics Today* **23** (1970) 23.

96. G.F. Chew and S. Mandelstam, *Nuovo Cimento* **19** (1961) 752.

97. F. Zachariasen, *Phys.Rev.Lett.* **7** (1961) 112.

98. Laurence Sterne, "The Life and Opinions of Tristram Shandy Gentleman" (London, 1761), Book IV, Ch. 27.

99. E.L. Feinberg, *Sov. Phys. Usp* **18** (1976) 624 [Translation from original: *Usp. Fiz. Nauk* **116** (1975) 709].

ENTROPY FOR HADRONS

Luigi Sertorio

Dipartimento di Fisica Teorica
Università di Torino

WHY HADRON THERMODYNAMICS

The formalism of statistical thermodynamics is generally used in situations where the confinement of the bulk piece of matter is either obtained in the laboratory or is offered by Nature. Examples:
– Newtonian gas of atoms. By mentioning atoms we imply temperatures below the ionization threshold namely $T \approx 10^3 K$. The sample exists in nature, e.g. the atmosphere, or can be obtained in the laboratory.
– Photon gas, massless relativistic gas. Temperatures in a large range; for very low T the container is for instance the non massive part of the Universe itself, fossil black-body, or, in the laboratory, any material cavity. Above $10^5 K$ the container can be the surface of a star where gravity confines the plasma which in turn radiates. For extremely high temperature, or energy density, the photon gas may couple with mass (not the simple scattering within the blackbody cavity), and the concept of photon gas disappears.
– Phonon gas. It belongs to phenomena at low temperature and the container is a crystal.
– Plasma. It requires a non equilibrium description, and the confinement is given by gravity.
– Crystals and quasi crystals. The temperatures are in the medium range, that in which the molecular wave functions with their complicated spatial distributions provide both the lattice siting and the lattice interactions.

We should add to this list an exceptional case: the hot hadron matter. It is exceptional because the sample is non existing in the solar system, therefore is not accessible to thermodynamical experiments [1]. This fact poses the remarkable problems that characterize this kind of research. These problems are profound and touch inner feelings about the object of fundamental research. These inner feelings in turn pull philosophical strings: eternal reversible laws versus evolutionary laws, of which thermodynamics is the prototype. I shall try to summarize these problems, limiting myself to two remarks.

The first remark is that among high energy physicists there is a tendency to polarize the general cultural interest to the programme of unified field theories, the only ambition that deserves to be called true and fundamental. From the final theory everything derives. This is the domain of free mathematical creativity which is rewarded by the eventual obedience of Nature to the ultimate abstract system. This attitude comes together with a weak interest for thermodynamics, the plumber sequitur of the truth. Another way of arguing is that, in the first place, without thermodynamics and the possibility of conceiving the arrow of time, modern cosmology could not even exist. We would be stuck to the seventeenth century Lagrangian initial-condition-explanation of the entire universe. Now, it is exactly in cosmology that high energy physics finds its domain of validity. Testing the theory in cosmology and then saying that this knowledge is knowledge of the building blocks of Nature (reductionism) is not true, even in principle, because unlike the cases of the non relativistic limit of relativistic expressions, or the classical limit of quantum operator expectation values, there is no confluence of hot matter whether described in bulk or microscopically, into cold matter. More precisely we are not going to rebuild nuclear physics in terms of quarks; where quarks exist nuclei don't, and vice-versa. The presence of quarks in nuclei is nominal, it is like knowing that the bricks of my house are made of clay: the dynamics of clay, mixing and cooking, comes prior to my house, clay manipulation and brick manufacture are disjoint from architecture. There are indeed limits to architecture given by the properties of bricks; other shapes and sizes may come from concrete and steel utilization, and so on. With respect to the clay furnace and the steel foundry the house is the fissile, not the dynamical consequence. In modern language we say that nuclei are fossils of the cosmological evolution, from eras in which other equations were valid, other constituents were dynamically in action. Reductionism has a long history, where sometimes it was linked to a great progress but at other times was an alibi.

Therefore rather than polarizing interest to the Hamiltonian formalism we should greatly prise the developments of high energy thermodynamics, as that is the main pillar of cosmology which is, let us repeat it again, the true domain of the whole of high energy physics.

The second remark is somehow philosophical. It is about the history of the dualism between reversible and irreversible. Note that these words are rather modern; for instance in Maxwell's work there is no mention of time reversal for his equations. This dualism can be considered to have been born, in its present connotations, with Boltzmann, where it was restricted to the confrontation of the classic Newtonian laws (considered at that time more fundamental than electromagnetism) and entropy. But this profound dualism, under different spelling and articulation with respect to the evolving cultural reference frames, goes back to the early Greek philosophers. In the modern version there is the persistent dogma that irreversible must be deduced from reversible. On the other side there is the attitude that reversible and irreversible laws are complementary in the description of the ever evading unknown territories of Nature [2].

There is a lot to read on all this. An interesting reference is the book "The Entropy Law and the Economic Process" of Georgescu-Roegen [3]. This author does not favor the ancillary role of irreversibility.

On the former side we have the following curious entry, from no less author than Galileo.

There is a debate between the Jesuit scientist father Grassi and Galileo. Grassi, writing in the "Libra Astronomica ac Philosophica" and quoting the Byzantine lexicographer Suida who lived about the year 1000, says "...twirling eggs in a sling...the Babylonians cooked them with this motion". Galileo replied in the Saggiatore: "if we do not get an effect which others managed to get at another time then there must be something missing in what we are doing, namely, that which caused the effect to take place and, since only one thing is missing then this single thing must be the real cause. Now, we do not lack eggs, slings or sturdy men to twirl them, but still they do not cook. Therefore, since we do not lack anything except being from Babylon, this being from Babylon is the cause of the eggs hardening". For Galileo the study of viscous motion was not a part of real science [4]. What the preceding two remarks have to do with hadron thermodynamics?

Concerning the first remark, pay attention to the fact that in the hypothetical hadron era the thermodynamical situation might have been that of a phase transition between hadron resonances and quarks. In some way the arrow of time must have been pointing somewhere. The sharpest non agnostic hypotheses on those events have been put forward by the theoretical laboratory of the bootstrap thermodynamic school.

Concerning the second remark, we may notice that the philosophical debates evolve according to the developments in the sophistication of knowledge in the physical sciences, the life sciences, the human sciences. The development of bootstrap thermodynamics has these connotations of sophistication; it has stretched the formalism of statistical mechanics to its extreme limits, in the relativistic domain. This is not a small fact. The bootstrap school may not have achieved agreement with a crucial experiment, but advance in science is not only discovery of facts, it is an interplay of facts and an extension of language, it is in fact when there is harmony among these two elements that a piece of scientific work is beautiful.

After about thirty years of development we may say that the bootstrap statistical thermodynamics can be considered to be one of the four main branches of statistical physics. In statistical mechanics the best known calculations are those with the ideal gases, either Boltzmann, or Bose-Einstein, or Fermi-Dirac. Next we have the non relativistic perturbative expansions either classical or quantum; heavy formalism that enjoyed limited success. Third we have the lattice approach, invented by Heisenberg, first developed by Ising, and proliferated successively to a great extent. Fourth we have the Hagedorn bootstrap approach.

Let us list some theoretical acquisitions of the hot hadron matter research.

A: Invention of a method that is not perturbative in the interaction for the calculation of the partition function.

B: Extension of the S-matrix approach to statistical mechanics, the Dashen and Ma formalism, to the high energy non perturbative domain.

C: Exploration of a limiting domain of applicability of statistical counting, namely the domain of the maximum possible clusterization of states, which is what the bootstrap is, and the discovery of the relative equations of state, etc.

D: Finally, the revisitation of the scholarly saying that all statistical ensembles must be equivalent; of course they should, if the state equation is to be uniquely determined. Try to do the easy cases and then trust. The bootstrap statistical thermodynamics has implied all possible acrobatics from the grand-microcanonical to the grand-pressure ensembles. Correspondingly the bootstrap statistical dynamics has taken various forms:

– A functional equation.
– A differential equation.
– A recursive equation.
– A graph-cluster combinatorial equation.
– An integral equation.

After these general comments we are back to the start with the fact that bulk hot hadronic matter does not exist in the laboratory so that the typical experiments of thermodynamics, namely the measurements of the response functions are not feasible.

NO SAMPLE OF HOT HADRONIC MATTER

This is the most serious problem of hadron thermodynamics, which creates conceptual ambiguities; we shall try to review briefly these ambiguities.

In the accelerator experiments we have informations on the matrix elements $T_{2,n}$ but not on the matrix elements of the kind $T_{m,n}$. Remember that the partition function is given by

$$Z = \text{Trace} \left[e^{-\beta H} \right] = \int e^{-\beta H} \text{Trace } \delta(E - H) dE,$$

if the field theoretical Hamiltonian is known. The above formula implies calculations where all multiparticle matrix elements of H appear.

If instead we work with the S-matrix alone the partition function is given by

$$Z = \frac{1}{4\pi i} \int e^{-\beta E} \text{Trace} \left[S^{-1} \frac{\partial S}{\partial E} - \frac{\partial S^{-1}}{\partial E} S \right] dE,$$

The operator in square brackets contains products with all multiparticle scattering states. In the Hamiltonian case the calculations are feasible in the perturbative approximation; the hadron S-matrix case is particularly difficult because the experimental information covers only the subset of $2, n$ matrix elements.

In the absence of a sample we have only one chance to keep our research on hadron thermodynamics alive and that is to assume that in production scattering there is an intermediate transient in which at least a partial thermalization occurs. This idea has been first proposed by Fermi and soon after refined by Landau; the approach of Hagedorn, ten years after that of Landau, is revolutionary with respect to the hadron equation of state, but is similar with respect to the fluid motion.

– Fermi assumed that in the $p - \bar{p}$ scattering the annihilation energy decays into as many pions as possible, namely that the T matrix elements are constant for any multiplicity n, and this implies obviously that the scattering cross sections are pure phase space, which is like saying that the intermediate state is a grand canonical of free pions. In other words the interaction enhances or decreases the phase space only at low pion-pion relative energies.

– Landau assumed again that the multiparticle pion state is a free gas state but the thermalization occurs not at rest but during a fluid expansion along the collision axis. This fluid motion contains no additional hadronic information exactly as in Fermi. This is a point worth clarifying. Landau uses the relativistic version of the Newtonian fluid dynamic equations that he himself had formulated in those years. I write here the non relativistic version which contains the same ingredients with half the number of indices.

$$\frac{\partial \rho}{\partial t} = -\vec{\nabla} \rho \vec{v}, \qquad \frac{\partial}{\partial t}(\rho v_i) = -\frac{\partial}{\partial x_k} \Pi_{ik}, \qquad \frac{\partial e}{\partial t} = -\vec{\nabla} \vec{e},$$

where

$$e = \rho \epsilon(T) + \frac{1}{2} \rho v^2, \qquad e_i = -k \frac{\partial T}{\partial x_i} + v_k P_{ik} + \rho v_i e,$$

$$\Pi_{ik} = P_{ik} + \rho v_i v_k, \qquad P_{ik} = P(T)\delta_{ik} - \sigma_{ik},$$

and where

$\epsilon = \epsilon(T)$ is the energy equation of state,
$P = P(T)$ is the pressure equation of state,
σ_{ik} is the viscous tensor,
k is the coefficient of thermal expansion.

Landau takes ρ constant, neglects σ_{ik} and assumes for ϵ and P the ideal gas equations of state. Therefore the derivative of the momentum tensor, namely the fluid force, is given simply by the ideal gas pressure gradient, and this implies that the longitudinal motion is not carrier of any residual interaction. The model is elegant and self consistent, but the agreement with experiment cannot be considered satisfactory.

– Hagedorn assumes also that there is a fluid motion along the collision axis but leaves the velocity distribution function undetermined. The moving, locally thermalized, fluid volume element is not a parcel of ideal gas but is instead a cluster gas, a brand new concept invented by Hagedorn. This new concept of thermal state is brilliant:

a) It abandons the description in terms of elementary components and their interactions.

b) It introduces the concept of level counting of interacting states, or statistical degeneracy of "cluster states". These concepts were consistent in the sixties with the bootstrap S-matrix theory of Chew, Frautschi, Stapp, and show the best example of complementarity between reversible and irreversible physics [5]. In the domain of statistical thermodynamics that was a revolution, as I have emphasized briefly in the preceding section.

Let us return to scattering. The partition function is derived in an abstract way from a minimum amount of assumptions. The velocity distribution function remains instead phenomenological. Therefore if the scattering model agrees with experiment the agreement proves that: "the scattering model is good & the partition function is correct"

The statement "the partition function is correct" comes conjugated to the correctness of the velocity distribution function. Hagedorn and Ranft have shown that the distribution function is independent of the final state masses and quantum numbers; but nobody, to my knowledge, was able to derive such distribution from first principles, or simply with an economy of assumptions comparable to the economy of the statistical bootstrap itself, which contains only one parameter, the cluster volume.

Therefore we may say that after the contributions of Fermi, Landau, and Hagedorn the indirect information on hadron matter obtained by necessity from scattering, still remains a conceptual problem. We are fossils of the dynamics and thermodynamics operating at high energy. Note that in the Newtonian description of the universe there are no energy scales and that the inverse development from rarefied matter to dense matter meets no obstacles. It is instructive to remark that the quantum relativistic theory, which has opened the door to modern cosmology, also indicates a conceptual barrier of observability. Looking in retrospect to the classical dynamic approach to cosmology, and to the previous static descriptions of the pre-galilean philosophers, the above mentioned conceptual problem appears to be a very important scientific achievement [6]. The development of scientific knowledge is not only a way to augment the set of proven statements about Nature, but also an extension of the domain of the remaining unknown part of Nature.

HADRON THERMODYNAMICS AND THE EXTREME BOUNDARIES OF RELATIVISTIC STATISTICAL MECHANICS

In general we may begin by considering the following Boltzmann grand microcanonical relativistic series:
$$\tau(M) = \Sigma_n \tau_n(M),$$
where τ_n is the n-particle microcanonical volume. The case of several particle species and quantum statistics can be added with heavier formalism. We may put

$$\tau_n = f_n(M)\omega_n(M),$$

which becomes for the ideal Boltzmann gas

$$\tau^0(M) = \Sigma_n \tau_n^0(M) = f_n^0 \omega_n(M),$$

$$f_n^0 = \frac{1}{n!}, \qquad \omega_n(M) = \int \delta(P - \Sigma p_i) \Pi^n \Omega p_i d^4 p_i \delta_0\left(p_i^2 - m^2\right),$$

$$\omega_n(M) \propto \frac{V^n M^{3n-2}}{\Gamma\left(3n - \frac{1}{2}\right)}.$$

Ω is the four-volume, with fourth component V in the rest frame of the total momentum P. $f_n(M)$ contains the dynamics; only if the n particles are free (ideal gas), we have $f_n^0 = \frac{1}{n!}$. In general, the functions $f_n(M)$ are given by:

$$f_n(M) = \frac{1}{4\pi i} \text{Trace}_{(n)} \left[S^{-1} \frac{\partial S}{\partial E} - S \frac{\partial S^{-1}}{\partial E} \right].$$

Notice that the trace washes away the detail of the S-matrix; this is why a model can reproduce the essence of the microscopic theory. In this spirit let us adopt for the $f_n(M)$ a parameterization in terms of a discrete index in the following way,

$$f_n(M) \rightarrow f_n^k = \frac{1}{(n!)^k}.$$

It is interesting to define additionally

$$d_n^k = \frac{f_n^k}{f_n^0} = \frac{1}{(n!)^{k-1}}.$$

For the ideal gas:

$$f_n^0 = \frac{1}{n!}, \qquad d_n^k = 1, \qquad k = 1.$$

The numbers d_n^k have the meaning of:
degeneracy of the n-state, if $d_n^k > 1$,
order of the n-state, if $d_n^k < 1$.

The index k can be called order parameter. In terms of k we have then this classification:
– Weakly filled microcanonical space

$$d_n^k < 1, \qquad \infty > k > 1.$$

$k > 1$ means that there is subcounting, or some order in the n-space, or some information among the momenta p_i $(i = 1, 2, ..n)$.
– Strongly filled microcanonical space

$$d_n^k > 1, \qquad 0 < k < 1.$$

$k < 1$ means that there is additional degeneracy in the n-particle space, or supercounting, or clusterization.

We cannot consider countings which imply

$$\bar{M} \neq \hat{M},$$

where

$$\bar{M} = \frac{\int_0^\infty M e^{-\frac{M}{T}} \tau(M) dM}{\int_0^\infty e^{-\frac{M}{T}} \tau(M) dM},$$

and where $\hat{M}$ is the peak value of $\tau(M) e^{-\frac{M}{T}}$ because the violation of the equality between peak energy and average energy corresponds to the violation of the equivalence between canonical and microcanonical. Given the grand microcanonical volume we can calculate the entropy, also as a function of k, and from $\frac{1}{T} = \frac{\partial S}{\partial E}$ we can calculate $T = T(M, k)$ namely the energy equation of state as a function of k. The explicit calculation, starting from the logarithm of $\tau(M)$, gives

$$T = \left(\frac{\partial S}{\partial M}\right)^{-1} = \left[\frac{d}{dM} C M^{\frac{3}{3+k}}\right]^{-1}.$$

In conclusion we can draw in the plane T, M the domains of all possible equations of state according to the k-labelling (see fig. 1). The domain of acceptable partition functions has two boundaries, the curve (1) which is reached for $k \to \infty$, limit curve which is forbidden because at that limit $\bar{M} \neq \hat{M}$.

The second boundary is curve (3), $k = 0$, which is allowed, but we cannot cross it because below that curve we have an unphysical entropy, S decreasing with increasing M. The equation of state on the curve (2) corresponds to

$$\tau(M) = e^{(\frac{M}{m})^{\frac{3}{4}}},$$

and its equation is

$$T = M^{\frac{1}{4}},$$

the well known Stefan-Boltzmann equation of state.

Finally, the curve $k = 0$, which is allowed, corresponds to

$$d_n^k = n! \qquad \tau(M) = e^{\frac{M}{m}}, \qquad M \propto \frac{1}{\sqrt{T_0 - T}}.$$

This is the limiting case of statistical counting. This case is realized by the Hagedorn model [7]. Hagedorn was aware since the beginning that the exponential mass spectrum, or $d_n^k = n!$, was the indication of the existence of a purely combinatorial law. The statistical bootstrap equation was written as a cluster combinatorial equation by Fre and Sertorio. Later Hagedorn found that the bootstrap combinatorial equation could be reduced to a problem of pure topology that was formulated and solved long ago.

Note that the jet model of production scattering is consistent with

$$< n > \propto \log \frac{M}{m},$$

$$\sigma_{tot}^{inel} \propto \tau(M) \propto c M^h.$$

In this case the energy $\hat{M}$ turns out to be

$$\hat{M} = hT,$$

while the average $\bar{M}$ turns out to be $\bar{M} = (h + 1)T$. This is the case $k = \infty$.

We can say that in the limit $k \to \infty$ the collective degrees of freedom that are not excited tend to infinity, we are no longer in true statistical counting, the series of $\tau(M)$ represents no thermodynamics. If we accept that production scattering is well represented by a jet-like model, there cannot be a thermalization in a single volume with a single temperature. This is the argument that refutes the original simple Fermi hypothesis.

THE CONSISTENCY TABLE

Concerning this consistency table, which I leave to the reader for his meditation on the fortunes and misfortunes of theoretical physics, I wish to add only a short remark.

Stat. Thermodynamics	Microscopic Theory
Fermi model equilibrium 1950 Landau model, non-equilibrium, 1953 Hagedorn bootstrap partition function, in fixed Fermi volume 1964	Old field theory: Bethe de Hoffmann 1955 S matrix theory: Chew 1961, 1966 Frautschi 1963 Eden et al. 1966
Thermodynamic multiperipherism: Hagedorn and Ranft 1966, 1968	Peripherism: Ferrari and Selleri 1961, 1963 Multiperipherism: Fubini et al. 1965 Zachariasen review, 1971
Bootstrap statistical model, micrograndcanonical in fixed Fermi volume: Frautschi 1971 Cosmological fluctuations 1971 Bahcall, Frautschi S matrix formulation of statistical mechanics: Bernstein, Dashen, Ma 1969	Dual S matrix models: Rubinstein, Veneziano, Virasoro 1968 Dual S matrix mass spectrum: Fubini et al 1969, Veneziano 1971 Dual S matrix review: Alessandrini, Amati, Olive 1973 End of S matrix theory — — — — — — — — — — — — — — — — — —
Relativistic extension: Sertorio, Toller 1972 Sertorio, Bassetto 1973. Sertorio, Fiore: Volume criticism, 1977, 1978 Volume correction: Hagedorn, Montvay, Rafelski 1978: Phase transition hadron $\rightleftharpoons$ quarks Pressure ensemble, new Hagedorn model 1980, 1985, 1990	Enters QCD: Fritsch, Gell Mann 1972 Gross, Wilczek, Politzer 1973

The paper of Fubini and Veneziano showing that the dual S-matrix model mass spectrum behaves exponentially in agreement with Hagedorn, uses the orthodox Dashen-Ma formalism and makes use of all $S_{m,n}$ matrix elements. For the short period of time in which the dual S matrix model was considered acceptable, the consistency between microscopic theory and the Hagedorn thermodynamics was perfect. But while the non inclusion of the hadron resonance lifetimes in the statistical counting is formally acceptable, in the S-matrix theory the use of delta functions is an insuperable formal violation of unitarity. If, on the other hand, one starts with Bassetto and Sertorio, with a true unitary and analytic S matrix, the discussion of the statistical bootstrap becomes conceptually and formally very difficult. In fact Bassetto and Sertorio have met serious difficulties with the connectedness structure of the S-matrix and these difficulties indicated (their conclusions were not a proof, but only an indication) that the concept of a fixed bootstrap volume could not be held. This criticism was also supported on different grounds by the Russian school. Several papers were written on the volume problem, until the issue was settled by Hagedorn and Rafelski with a brilliant hypothesis on the relativistic covolume, and a sophisticated use of the pressure ensemble, resurrected from the labyrinthine depositories of statistical mechanics.

Now, thirty years after Hagedorn's first paper , it is reasonable to consider the analogy between nuclear physics and bootstrap physics; formalisms which are appropriate for a certain energy range. We human beings are coeval of the nuclei, atoms, molecules; we are the fossils instead of the fireball. It is for this mismatch of eras that nuclei are observable, fireballs are not, or perhaps not, or perhaps yes.

The final difference is that we are not going to have hadron engineering, hadron weapons, hadron powerplants.

Or is that what somebody was hoping for?

NOTES AND REFERENCES

[1] In fact the experimental evaluation of the response functions implies the determination of energy exchanges ΔQ and also the determination of values of P, T, V.

[2] Three concepts are touched here, irreversibility versus reversibility, reductionism, and fossil. They are shorthand connotations of issues which are interesting if considered in a dynamic context.

Take a fossil; a species which is extinct with respect to the species dynamics and remains as a specimen of a longer lasting form of matter. So, no gain in calling the skeleton of the emperor Augustus a fossil, since the human species is still in operation including the historical legacy of Augustus himself. It appears more appropriate to concentrate on the understanding of the meaning of joint dynamics, for instance among species in mutual interaction, and conversely of disjoint dynamics.

This brings us immediately to the concept of reduction to components of a complex system. If components and compound are joint dynamically, the reduction is interesting; that is the case of a nervous system composed of neurons. Disjoint dynamics, instead, is the quark-gluon plasma, which, after the nucleo-synthesis, becomes extinct, and therefore dynamically disjoint from the events of the present era.

Finally the concept of component plays a role in the reversible-irreversible issue. In a rather narrow range of temperatures we may ask if the macroscopic gas behaviour is consistent with a reduction to mass points moving according to the Newtonian reversible dynamics (molecular dynamics). It is comforting to find that the two descriptions, thermodynamic and microscopic, are consistent. However, when experiment indicated that the ratio $\gamma = \frac{C_P}{C_V}$ increases for decreasing temperature (discrepancy between the Hamiltonian theory and thermodynamics), it was the molecular model which was abandoned, not in favour of another classical model with different interactions, different walls, etc.; but rather the Newtonian description itself was abandoned. Quantum mechanics re-establishes the coexistence with thermodynamics, and we discover how fertile this philosophy of coexistence is.

This is not to deny the importance of the issue of the uniqueness of reality, which may or may not be identified with uniqueness of knowledge. But this is much beyond the limits of this note.

[3] The book was first published by Harvard University Press in 1971.

Why an economist rather than a man from within the business? With Galileo, physics becomes hard science and thereafter keeps developing with experimental tools and languages that are more and more autonomous. In fact, despite the personal defeat of the abjure, Galileo succeeded in disjoining physics from the revealed truth; and unleashing the language of science. This independence of language and the fast development of physical science result in a bias when physicists talk of philosophy, in the sense that they tend to extrapolate from their successful world to other areas of knowledge. The science of economics never enjoyed such success nor did it achieve such a maturity of language. It is perhaps for these reasons that the book of Georgescu Roegen has remarkably unbiased philosophical foundations which make it unusual and stimulating.

[4] Obviously Galileo erred and missed the thermodynamics of the sling. But this reference must be understood in a very wide and deep framework. The polemic vein characterized most of the life and writing of Galileo. His major epistemological effort was a quantification of the knowledge of Nature in order to stipulate a univocal fundamental role to the confrontation with experiment. In that line he was pitiless with the scientists who extracted their arguments from books and not from the direct observation of Nature. His fight with books generating books, citations generating citations, remains his sacred legacy, given the recurrent propensity of human beings for replacing truth with consensus.

[5] Note that the hadron statistical bootstrap exponential mass spectrum came five years before the result obtained with the dual S-matrix.

[6] The concept of "cosmological fossil" is used with epistemological correctness in the powerful, concise, (and difficult) paper of S. Frautschi: *Entropy in an expanding universe*. Science 213 (13 August 1982).

[7] The square root singularity of the energy equation of state is the outcome of the fixed volume bootstrap. The second version of the Hagedorn bootstrap shows a non singular equation of state. See the Hagedorn paper at this Meeting.

Acknowledgments

I wish to thank Prof. J. Rafelski for his kind invitation to contribute to the NATO ARW "Hot Hadronic Matter, Theory and Experiment" - Divonne les Bains, June 27 - July 3, 1994.

STATISTICAL STUDIES OF HADRONS

Steven Frautschi

California Institute of Technology
Pasadena, CA 91125 USA

In recent years my activities have shifted towards interesting beginning students in physics, helping to produce a 26-hour video called "The Mechanical Universe" on freshman physics, and the like. I am not currently active in studies of Hot Hadronic Matter. Therefore, this talk will be historical and reminiscent in nature. In brief, I'll describe how I started with few-body issues, how the physics then led me to many-body and statistical studies, and how finally I learned that it could all be viewed as a manifestation of string theory.

REGGE POLES

In 1960, I and other members of Chew's group at Berkeley were studying two-body reactions

$$A + B \rightarrow C + D \quad \quad (1)$$

within the framework of the analytic S-matrix and bootstrap philosophy. Seemingly we were far from the statistical description of many bodies! Two main considerations led us to introduce complex angular momentum into the theory:

i. There was a deep problem with hadrons of spin $J > 1$. On the one hand they were known to exist experimentally, as in the famous $J^P = \frac{3}{2}^+$ resonance of πN scattering. On the other hand, interactions of such particles are not renormalizable in quantum field theory. In particular, when a particle with $J > 1$ is exchanged, say between pion and nucleon, it produces cross sections that increase without limit at high energies, in an unacceptable manner.

ii. There was an elegant analysis of the spin dependence of bound states and resonances in *non-relativistic* quantum mechanics by Regge, which introduced the Regge pole as an interpolating function between different physical spin states.

Chew and I[1] conjectured that Regge poles would also exist in relativistic quantum mechanics, allowing the spin J to take on values > 1 where needed to describe high spin particle states, but to interpolate down to $J < 1$ in other kinematic regions as needed to avoid inadmissible infinities in calculations. As regards the detailed nature of the interpolation we simply drew a straight line (Figure 1), since a straight line is the simplest interpolation between two points, and two experimental points was the most we had at that time.

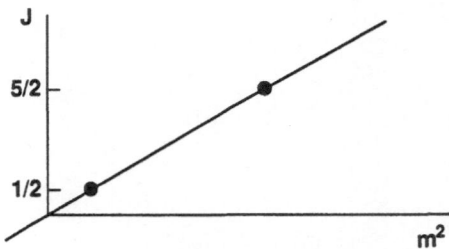

Figure 1. A straight line Regge trajectory

There followed studies by many people of various non-relativistic potential models where Regge poles could be calculated. Typically the result was something like Figure 2, and the speaker would say "See the straight line!" to general laughter. If hadron physics

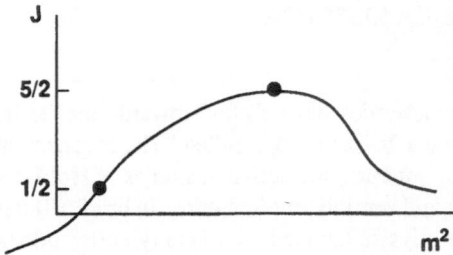

Figure 2. Regge trajectory for a typical non-relativistic potential.

had actually behaved like such models, Regge poles would not have implied a large hadron spectrum.

But experiments, first on ρ exchange[2] in

$$\pi^- + p \rightarrow \pi^0 + n, \tag{2}$$

then on N and Δ exchange in *backward* πN scattering,[3] and eventually on many reactions, pointed to straight line trajectories. And straight line trajectories *do* imply a large hadron spectrum.

There is one model that gives a straight line in a plot of J versus m^2: the relativistic string! So with time the straight line trajectory, originally just a crude approximation, came to be seen as a guiding principle and eventually as a hallmark of the string model. Furthermore, string models also imply parallel daughter trajectories (Figure 3) and a huge number of states.

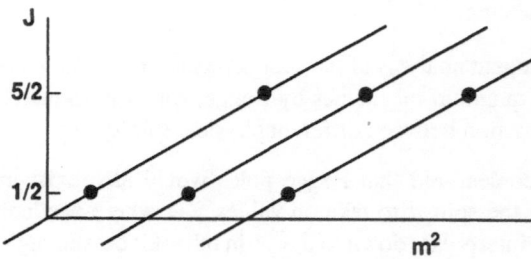

Figure 3. Regge trajectories in a string model.

So *we had been led from the study of a few modes to the study of many by the physics.*

STATISTICAL TREATMENT OF HIGH ENERGY SPECTRUM AND REACTIONS

When many states are involved, statistical considerations are likely to be important. As early as 1950, Fermi[4] had introduced a statistical model of hadron reactions at high energies, estimating the relative production of different multiparticle states in terms of phase space.

An application where Fermi's simple statistical considerations had some success was the annihilation of $\bar{p}$'s at rest:

$$p + \bar{p} \to n\pi. \tag{3}$$

Fermi's model predicted the shape of the pion momentum distributions quite well, and could fit the distribution in number n of pions although an unrealistically large interaction volume was required. Belenky[5] (1956) showed that attractive interactions between the pions could pack more quantum levels with given total energy into a given volume. To a good approximation, resonances could be counted as extra particles. When pion resonances such as

$$\rho \to \pi + \pi , \qquad \omega \to \pi + \pi + \pi , \tag{4}$$

were discovered, including them in the statistical model as prescribed by Belenky led to a good fit of the number distribution of pions with a reduced, physically reasonable interaction volume.

Hagedorn[6] (1965) carried Belenky's idea much further. He treated pions, nucleons, and all their excited states and resonances on the same footing, and set up a consistent *bootstrap condition* linking the hadron spectrum to the states of hadrons in a box of fixed hadronic volume. Notable results were the famous exponential increase of level density

$$\rho(m) \xrightarrow{m \to \infty} c\, m^a \exp(m/T_H) \tag{5}$$

and the equally famous Hagedorn limiting temperature

$$T \leq T_H \tag{6}$$

beyond which the partition function diverged as the exponential growth in states overwhelmed the Boltzmann exponential.

By 1968 Hagedorn and Ranft[7] were fitting a wide variety of high energy multiparticle production data with a model that combined longitudinal flow of hot hadronic matter with thermally controlled transverse momentum and particle type distributions. Their work implied that hot hadronic matter thermalizes and that its temperature saturates at $T \simeq T_H$ with

$$T_H \simeq 160\text{MeV}. \tag{7}$$

In 1971 I introduced a minor technical modification of Hagedorn's bootstrap condition.[8] The modification consisted of working directly with phase space, enforcing momentum as well as energy conservation, rather than working with the partition function. In other words, I considered the *microcanonical ensemble* instead of the *canonical ensemble*. This change permitted a tightening of the consistency relation; now the power factor m^a as well as the exponential $\exp(m/T_H)$ could be made consistent. Among the results were:

i. The power a, previously estimated at $a = -5/2$, was now determined to be -3.

ii. The seemingly minor change in power had a remarkable effect on the predicted decay products of a resonance of mass m. With $a = -5/2$, decay into a number $n = c' \ln m$ of secondary particles, each of similar mass, had been statistically favored. With the new power $a = -3$, we found that *resonances decay predominantly into one heavy*

59

secondary particle that gets almost all the mass, *plus one or two light particles*, with probabilities

$$\text{Resonance} \quad \rightarrow \quad \begin{array}{l} \text{2 bodies (69\%)} \\ \text{3 bodies (24\%),} \quad \text{etc.} \end{array} \tag{8}$$

iii. A subsequent numerical study by Hamer and Frautschi[9] showed that the level density approaches its asymptotic limit rapidly. They were able to *calculate* $T_H \simeq m_\pi$ in rough agreement with the earlier experimental fits.

RELATION TO DUALITY AND STRING THEORIES

The prediction that two-body channels dominate was highly compatible with *dual resonance* and *Veneziano* models. In fact, Hagedorn's bootstrap condition can be viewed as linking the number of resonances to the number of channels, and is thus closely related in spirit (though not in technique) to models that saturate channels with resonances. Indeed the $\exp(m/T_H)$ growth in level density was deduced from duality by Krzywicki[10] and Brout[11] in 1969, and from the Veneziano model that same year by Fubini and Veneziano,[12] Bardakci and Mandelstam,[13] and Olesen.[14] Further application of the Veneziano model allowed Huang and Weinberg[15] to specify the range of possible powers as $a = -5/2, -3, ...$, and Brower[16] to pinpoint $a = -3$. Subsequently, the Kiev group of Miransky, Shelest, Struminsky, and Zinovjev[17] displayed even more detailed correspondences: for example, that the Veneziano model leads to secondary pion energy distributions of Boltzmann form as predicted by Hagedorn. Thus, the analytic S-matrix, when generalized to a suitably global perspective, was shown to embody the insights of the statistical bootstrap model.

Another approach developed around this time was string theory. In the familiar nonrelativistic case, excitations of the transverse modes of a string satisfy an equal spacing rule with frequencies $\omega, 2\omega, 3\omega, ...$. The associated quantum mechanical energy E_N can be achieved by various combinations of modes:

$$\begin{aligned} E_N - E_0 = N\hbar\omega, \quad & \text{or} \quad (N-1)\hbar\omega + \hbar\omega, \\ & \text{or} \quad (N-2)\hbar\omega + 2\hbar\omega, \\ & \text{or} \quad (N-2)\hbar\omega + \hbar\omega + \hbar\omega, \quad \text{etc.} \end{aligned} \tag{9}$$

In other words, the number of states is just the number of ways an integer N can be composed of positive integers. The solution to this integer counting problem is known to be of form

$$\rho(N) \sim \exp\left(2\pi\sqrt{\frac{N}{6}}\right). \tag{10}$$

In the case of a *relativistic* string, it is m^2 that increases in equally spaced integer steps,

$$N = \alpha_0 + \alpha' m_N^2. \tag{11}$$

For Regge poles this gives the linear increase of J with m^2 described earlier; for the level density of strings that can vibrate in D transverse dimensions it gives

$$\rho(m) \underset{m \to \infty}{\sim} \exp\left(2\pi m\sqrt{\frac{D\alpha'}{6}}\right), \tag{12}$$

in remarkable agreement with Hagedorn's approach.

To summarize the position reached by the early 1970's, just as Regge poles had led from a few modes to infinite sequences of them, the statistical approach had led from a few modes (Fermi) to an infinite proliferation of them (Hagedorn). In both cases, the results could be viewed as a manifestation of string theory.

LATER DEVELOPMENTS

After the early 1970's, it became universally accepted that hadrons are described by the QCD theory of quarks and gluons. The Hagedorn temperature $T_H \simeq 160$ MeV is now viewed, not as a limiting temperature, but as the temperature of a phase transition from hadrons to quark-gluon plasma. Hagedorn and Rafelski[18] have provided an insightful model of the phase transition. For the hadronic phase, a concrete *phenomenological* model of hadronic strings as gluonic flux tubes connecting the valence quarks has been developed. Meanwhile the *fundamental* theory of strings has been transformed to include all interactions, but with a completely different energy scale based on the Planck length rather than hadronic interactions. Though the excited levels of these fundamental strings no longer represent hadrons, the Hagedorn spectrum still applies on this expanded scale and may describe cosmic strings as well as events on the very early Planck time scale.

While these momentous theoretical developments were taking place, I was making modest attempts to enlarge the *experimental* evidence for Hagedorn's hadron spectrum. Such attempts were necessary because the rate of direct experimental discovery of new hadron resonances, whose fast pace had helped fuel the development of the Hagedorn statistical model in the 1960's, slowed thereafter. Partly this reflected a shift of interest towards other experimental areas where the fundamental QCD and electroweak theories could be tested more directly. But it is probably also a consequence of the very richness of the Hagedorn spectrum. When the resonances in a channel with given baryon number, electric charge, strangeness, J^P, etc., are more closely spaced than the typical resonance width of m_π or more, as is predicted to happen at a quite low energy if $\rho(m) \sim \exp(m/T_H)$, then the overlapping resonances no longer stand out as distinct peaks, and even detailed phase shift analysis would be hard put to disentangle them.

In *nuclear* reactions, overlapping nuclear resonances leave a trace, the *Ericson fluctuations*,[19] even when they are individually undetectable. The reason is that when a two-body channel couples to many overlapping resonances, their individual spacings, and their coupling strengths to the channel, are expected to be statistically distributed. Fluctuations occur in the cross section as the energy is charged by a typical resonance width.

I conjectured[20] that similar fluctuations should occur in two-body hadronic reactions, in those non-exotic channels that have many resonances (e.g., $\pi + N \to \pi + N$, but not $N + N \to N + N$). There are indeed fluctuations, e.g., in an extensive series of πN data[21] taken at fixed momentum transfer as the energy was varied in small steps, but it has not been possible to show conclusively that they are statistical because (unlike the nuclear case), the scale of hadron resonance widths ($\Gamma \simeq m_\pi$) is as large as the scale on which other dynamical effects change. Thus, my attempt to follow the Hagedorn spectrum into the region of overlapping resonances was not really successful.

A FINAL EXAMPLE: POLYMERS

I'd like to conclude this report with an application that I just learned of recently. It illustrates that not only are the ideas of Hagedorn alive, you don't need to go to expensive accelerators or remote cosmological places and times to study them.

Polymers are strings, and (over a finite energy range) they have an exponentially rising level density. One of our physical chemists at Caltech, Jack Beauchamp, studies biopolymers. He does beautiful experiments in which intact biological macromolecules (10^5 to 10^6 nucleons) are extracted from their normal wet environment into the gas phase, and electrically charged so they can be manipulated like Millikan oil drops or trapped in ion traps. While trapped, their structures, properties, and dynamics can be probed by spectroscopy, irradiation, etc.

When the biomolecules are heated gently enough, it is found[22] that usually just a single protein breaks off the end – the biomolecules do *not* break near the middle as you might expect. Why not? Not because of a particular affinity for water or fat: the biomolecule has been extracted from its usual watery or fatty environment. Not because only the end is sticking out accessible on the surface: though the ground state of these molecules is tightly folded, they are pretty well unfolded and spread out at the temperatures where fission occurs (in the free energy $F = E - TS$, there is an energy penalty for each unfolding, but there is a counterbalancing entropy gain as unfolding makes more length available to oscillate).

Beauchamp thinks the reason for the highly asymmetric breakup is entropic: the biomolecule is a polymer, polymers are strings, unfolded strings have transverse vibrations, and breaking into a long string and a short string is statistically favored (allows the most vibrational states). In other words, the polymer behaves just like massive resonances in the statistical bootstrap model which break, as we've discussed, into a slightly less massive resonance plus a pion or two!

REFERENCES

1. G. Chew and S. Frautschi, *Phys. Rev. Lett.* 7:394 (1961); 8:41 (1962).

2. I. Mannelli et al., *Phys. Rev. Lett.* 14:408 (1965); P. Sonderegger et al., *Phys. Rev. Lett.* 14:763 (1965); ibid *Phys. Lett.* 20:75 (1966).

3. V. Barger and D. Cline, *Phys. Rev. Lett.* 16:913 (1966).

4. E. Fermi, *Progr. Theor. Phys.* (Kyoto) 5:570 (1950).

5. S.Z. Belenky, *Nucl. Phys.* 2:259 (1956).

6. R. Hagedorn, *Nuovo Cimento Suppl.* 3:147 (1965).

7. R. Hagedorn and J. Ranft, *Nuovo Cimento Suppl.* 6:169 (1968).

8. S. Frautschi, *Phys. Rev. D* 3:2821 (1971).

9. C.J. Hamer and S. Frautschi, *Phys. Rev. D* 4:2125 (1971).

10. A. Krzywicki, *Phys. Rev.* 187:1964 (1969).

11. R. Brout (unpublished).

12. S. Fubini and G. Veneziano, *Nuovo Cimento* 64A:811 (1969).

13. K. Bardakci and S. Mandelstam, *Phys. Rev.* 184:1640 (1969).

14. P. Olesen, *Nucl. Phys.* B18:459 (1970); B19:589 (1970).

15. K. Huang and S. Weinberg, *Phys. Rev. Lett.* 25:895 (1970).

16. R.C. Brower, *Phys. Rev. D* 6:1655 (1972).

17. V.A. Miransky, V.P. Shelest, B.V. Struminsky, and G.M. Zinovjev, *Phys. Lett.* 43B:73 (1973).

18. R. Hagedorn and J. Rafelski, *Phys. Lett.* 97B:136 (1980).

19. T.E.O. Ericson and T. Mayer-Kuckuk, *Ann. Rev. Nucl. Sci.* 16:183 (1966).

20. S. Frautschi, *Nuovo Cimento* 12A:133 (1972).

21. K.A. Jenkins, et al., *Phys. Rev. Lett.* 40:425 (1978); 40:429 (1978).

22. J. Beauchamp, private communication.

HAGEDORN'S TEMPERATURE
AND THE DUAL RESONANCE MODEL:
A 25 YEAR OLD LOVE AFFAIR

Gabriele Veneziano

Theory Division, CERN
1211 Geneva 23, Switzerland

A SURPRISE THAT SHOULD NOT HAVE BEEN ONE

There is a simple (a posteriori!) physical argument for the necessity of a Hagedorn-like spectrum of excited states in any model that satisfies duality. Let me remind you that, by definition of (Dolen-Horn-Schmit) duality, the asymptotic behaviour of (the imaginary part of) any scattering amplitude should be correctly described **either** by Regge pole exchange in the t-channel **or** by resonance formation and decay in the s-channel.

The Regge-pole description gives a power-like behaviour, hence, if we work to exponential accuracy, a constant amplitude at high energy. The resonance description yields instead (say for an elastic 2-body collision):

$$Im A_{el} \simeq \frac{1}{E} \sum_{Res} \Gamma_{2b}^R = \sum_{Res} \frac{\Gamma_T^R}{E} B_{2b}^R \leq N(E) \cdot \bar{B}_{2b} , \qquad (1)$$

i.e. gives a bound on $Im A_{el}$ in terms of the number of states $N(E)$ at energy E and of their average branching ratio $\bar{B}_{2b}$ into the particular two-body channel under consideration. By asking for consistency of the two descriptions we thus find, to exponential accuracy:

$$N(E) \geq \bar{B}_{2b}^{-1} \qquad (2)$$

Making now the (reasonable) assumption that $\bar{B}_{2b}$ behaves like the ratio of two-body phase space to total phase space, and using the fact that the latter becomes exponentially small at high energy, we arrive at the "prediction" of an exponentially growing spectrum for duality-fulfilling resonances. The (crucial) power of E appearing in the exponent is not fixed, however, by this argument.

We may ask why this conclusion was not immediately reached in the very early days of duality, some 26 years ago. The reason was, I believe, that one was accustomed to associate each resonance with a separate pole in the scattering amplitude; now, the number of poles occurring in the dual-model amplitudes was **not** growing with increasing energy. All one could see was the fact that the Nth pole contained resonances of spin up to N, but that fact in itself could only account for a power-like degeneracy, not for an exponential one.

Of course we know very well now (and since 25 years) the answer to this apparent paradox. The exponential growth of the number of states in the dual resonance model is hidden behind an exponential degeneracy! This degeneracy, which is neither predicted by Hagedorn's arguments, nor implied by the duality-based reasoning given above, is the truly new ingredient brought in by the dual resonance model. As I shall now argue this degeneracy is directly related to an underlying string picture for the resonances appearing in dual models.

FROM T_H TO THE STRING

When, in the summer of 1968, I first told Sergio Fubini in Torino about my new ansatz for the scattering amplitude his reaction was: "very nice, but....what about negative norm states?". Two months later, I had barely landed in Boston/MIT that we started counting and labelling states, clearly a necessary preliminary step before we could compute their norm.

By early 1969 we had learned how to count (it took longer to answer Sergio's original question about the norm, but eventually people proved, under certain restrictions, the celebrated no-ghost theorem). The rather unexpected result that Sergio and I (and independently Bardakci and Mandelstam) found was that the individual states were labeled by a set of integers $\{N_1, N_2, \ldots\}$ with the mass of the state given by the simple formula:

$$\frac{\alpha' M^2}{\hbar} = N = \sum_{n=1}^{\infty} n \cdot N_n \tag{3}$$

where α', the universal Regge-slope parameter, sets the energy scale of the theory at $\Lambda \equiv \sqrt{\hbar/\alpha'}$.

The Hagedorn spectrum then simply comes from the observation that a given (allowed) mass $M = \sqrt{N/\alpha'}$ can be obtained via eq. (2.1) in as many ways as the number of ways in which the integer N can be written as a sum of integers. This "partitio numerorum" number is known to grow like $exp(c\sqrt{N})$ (with c a known constant) hence like $exp(c\sqrt{\alpha'/\hbar E})$. This gives immediately a Hagedorn temperature $T_H = c^{-1}\Lambda$.

In the operator reformulation of the dual resonance model the mass-square operator can be written as

$$\frac{\alpha' M^2}{\hbar} = \sum_{n=1}^{\infty} n a_n^\dagger a_n \tag{4}$$

where $a_n^\dagger, a_n$ are an infinite set of ordinary harmonic oscillator creation and destruction operators satisfying the usual commutation relations:

$$[a_n, a_m^\dagger] = \delta_{n,m} \tag{5}$$

The high degree of degeneracy of the spectrum obviously comes from the presence of the higher harmonics ($n = 2, 3 \ldots$ in eq.(2.2)), but this is just what characterizes a vibrating string!

We conclude that the combination of the Hagedorn spectrum and of degeneracy leads straight into strings. Conversely, if a string picture is assumed for hadrons, one immediately predicts

i) duality as implied by drawing (duality) diagrams in which string splitting and joining are the basic processes underlying hadronic reactions.

ii) a linear relation between mass and entropy (another way of defining Hagedorn's spectrum) coming from the fact that the energy stored in the string is proportional to its length ($l = \alpha' \cdot E$), while there is a unit of entropy per bit of length.

The unit (bit) of length, λ_s, is a quantum object and is related α' by:

$$\lambda_s = \sqrt{\alpha'\hbar} \tag{6}$$

For the hadronic string λ_s is of the order of $10^{-13}cm$: there is about one bit of information for every fermi of string length.

CRISIS, REINTERPRETATIONS

One of the main motivations (successes) behind Hagedorn's model was the exponential fall off of the transverse spectrum of produced particles in high energy collisions:

$$d\sigma/dp_T \simeq exp(-p_T/T_H) \tag{7}$$

This holds well in a sizeable region of p_T. Not surprisingly, a similar behaviour was found to occur in the dual model (in string theory).

The discovery of hard constituents inside the hadron, revealing themselves, e.g., through the power-like drop of jet (or exclusive) cross sections, came as a serious blow to both Hagedorn's model and to the hadronic string.

Amusingly, they both survived, with some reinterpretation, in the emerging new theory of strong interactions, QCD. The Hagedorn temperature got reinterpreted as a deconfining phase transition (rather than as an ultimate) temperature, while strings become an effective description of hadrons as composite systems of quarks which are kept together by a thin tube of chromoelectric field.

Around the time that QCD took over, Joel Scherk and John Schwarz came up with the daring proposal that fundamental strings should be relevant for describing all fundamental interactions (including gravity) at much shorter scales than $10^{-13}cm$. The natural scale for the new string is simply the Planck length:

$$\lambda_P = \sqrt{G_N\hbar} \tag{8}$$

where G_N is Newton's constant. Numerically, $\lambda_P \sim 10^{-33}$ cm.

Obviously, also the new string has a Hagedorn temperature: simply it is shifted upward by some 18 orders of magnitude to about $10^{17} - 10^{18}$ Gev. In the last part of this talk I shall give one example of some new uses of Hagedorn's temperature in this new context: I shall argue that, in analogy with the reinterpretation of the old T_H as deconfining temperature, the new T_H will play the role of a limiting temperature for Black holes, a kind of gravitational-deconfinement temperature, if you like.

25 YEARS LATER...

There is a (so far undisproved) conjecture by J. Bekenstein that the entropy S of any physical system of energy E and size R cannot be arbitrarily large, i.e.

$$S \leq S_{BB} \equiv \frac{E \cdot R}{\hbar} \tag{9}$$

This is called the Bekenstein bound (BB). Let us consider now a black hole, i.e. a system of energy E contained in a spherical region of space of radius $R < G_N \cdot E$ (this is just the definition of a collapsed state in gravity).

Let us now compare the entropy of the black hole,

$$S_{BH} = \frac{G_N E^2}{\hbar} , \tag{10}$$

with S_{BB} in eq. (4.3), allowing the maximal size for R, $R = G_N \cdot E$. We see that the bound is precisely saturated, something suggesting that a black hole maximizes the entropy of a system of given energy and spatial extension.

We may now ask if a string of energy E satisfies the BB. As said before

$$S_{string} = \frac{\alpha' \cdot E}{\lambda_s} \tag{11}$$

which satisfies the BB (4.3) only if the size of the string is larger than λ_s, a conclusion that can be reached also by independent considerations.

I have told you that λ_s is of the same order as λ_P, but the more precise relation is actually

$$\lambda_P = \alpha_{gut}^{1/2} \lambda_s , \tag{12}$$

expressing the physical fact that, in string theory, gravitational and gauge interactions become identical at the (distance) scale λ_s.

We can finally compute the ratio between the string and black hole entropies and find

$$\frac{S_{BH}}{S_{String}} = \frac{G_N E \lambda_s}{\alpha' \hbar} = \frac{E}{M_P} \cdot \frac{\lambda_P}{\lambda_s} = \frac{\alpha_{gut}^{1/2} \cdot E}{M_P} \tag{13}$$

This ratio becomes 1 at $E = \alpha_{gut}^{-1/2} M_P$. At this energy both entropies are of order α_{gut}^{-1}, thus probably between 10 and 100. Above this energy, entropy considerations favour the black hole while below a string state is favoured. It is easy to see that such a state is not collapsed at all since its physical size, as we argued above, has to be larger than λ_s which, in turn, is larger than the gravitational radius $G_N \cdot E$ of the system.

As Hawking has shown, black holes evaporate, loosing mass while increasing their temperature. In ordinary gravity this process would continue, until a curvature singularity (and an infinite temperature) is reached. The arguments given above (and due to several people) suggest that, in string theory, black hole evaporation should stop at a certain point, leaving behind a non-collapsed string system of entropy $O(\alpha_{gut}^{-1})$.

The Hawking temperature of the black hole at this point is just the Hagedorn temperature of the string theory under consideration. We can thus say that an interpretation of T_H in the new incarnation of string theory is that of a "decollapse" temperature if you allow me to use such a word for the gravitational analogue of deconfinement. On the other hand, the analogue of the quark-gluon-plasma phase of QCD in quantum string gravity is still clouded with mystery (is space-time itself "melting"?).

There could be also a cosmological analogue of the limiting temperature for black holes, as it is known that there is a (Hawking) temperature associated with the event horizon of inflationary cosmologies and easily expressible in terms of the Hubble parameter during inflation. This could lead to new insights in the way string-Hagedorn models deal with the very beginning of the Universe.

CONCLUSION

The love affair between dual and Hagedorn models is still well and alive after 25 years: let's hope it will continue like that for still 25 more!

Acknowledgement

I am grateful to Johann Rafelski for his kind but firm insistence without which this written version of my talk would have probably never existed.

MASS SPECTRUM OF q-DEFORMED DUAL STRING THEORY

M. Chaichian,[1,2] J.F. Gomes,[3] and R. Gonzalez Felipe[1]

[1]Laboratory of High Energy Physics , Department of Physics
P.O. Box 9 (Siltavuorenpenger 20C) FIN-00014
University of Helsinki, Finland
[2]Research Institute for High Energy Physics
P.O. Box 9 (Siltavuorenpenger 20C) FIN-00014
University of Helsinki, Finland
[3]Instituto de Física Teórica-UNESP
Rua Pamplona 145, 01405-900 São Paulo, Brazil

The relation between the spin and the mass of an infinite number of particles in a q-deformed dual string theory is studied. For the deformation parameter q a root of unity, in addition to the relation of such values of q with the rational conformal field theory, the Fock space of each oscillator mode in the Fubini-Veneziano operator formulation becomes truncated. Thus, based on general physical grounds, the resulting spin-(mass)^2 relation is expected to be below the usual linear trajectory. For such specific values of q, we find that the linear Regge trajectory turns into a square-root trajectory as the mass increases.

As is well known string theory provides a promising approach to describe the different forces and particles observed in nature. The development of this theory is, more or less, directly related to the Veneziano's discovery[1] of a four-point crossing-symmetric scattering amplitude with linear Regge trajectories. Afterwards Fubini, Gordon and Veneziano[2] provided an elegant operator formalism for the dual amplitudes, which further led to the interpretation of the Veneziano model as a theory of interacting strings in physical space-time.

In recent years the mathematical structure of the quantum groups (see, e.g.[3]) has been extensively explored in connection with several important aspects of physical phenomena. Among them, the classification of two dimensional conformally invariant field theories plays an important role in understanding the structure of string theory (see, e.g.[4]). In particular, for rational conformal field theories, described by a finite number of primary fields, the fundamental properties of fusing and braiding of conformal blocks can be casted within the product of representations of q-deformed algebras for the specific values of the deformation parameter q being a root of unity.[5] For such values of q quantum algebras exhibit rich representation behaviours and are important in two-dimensional conformal field theories and in statistical mechanics models.[6]

In Ref.[7] the operator formalism proposed in[2] was followed to present a q-deformed dual string model which possesses crossing symmetry in exchanging the s- and t-channels and factorization of the dual amplitudes in such a way that they can be constructed as a field theory with Feynman-like diagrams out of products of vertices and propagators as the building blocks.

Here we propose an alternative q-deformed dual string amplitude, which has not only the required properties of crossing symmetry and factorization, the correct pole structure and a suitable asymptotic behaviour, but also leads to a spin-(mass)^2 relation, which in our opinion possesses all the acceptable features from the physical point of view. As in[7] we shall introduce an infinite number of q-oscillators which build up a Fock space with well-known properties. However, for the specific values of the deformation parameter q being a root of unity, only a finite number of oscillators for each (harmonic) mode possesses non-vanishing norm. This is a fundamental property which is responsible for a drastic change in the high energy behaviour of the amplitudes and, therefore, in the mass spectrum of the physical particles. Earlier attempts by introducing a logarithmic behaviour for the Regge trajectories with only a finite number of oscillators[8] or by replacing the ordinary gamma functions by their q-analogs[9] have been proposed. However, they cannot bring to a field theoretical formulation of the dual string model.

Let us consider first the usual Veneziano[1] 4-point dual amplitude given by

$$A_4 = \int_0^1 z^{-\alpha(t)-1}(1-z)^{-\alpha(s)-1}dz, \tag{1}$$

where $\alpha(s) = \alpha's + \alpha_0$ is the linear Regge trajectory, with α' and α_0 the Regge slope and intercept,respectively. We will take our units so that $\alpha' = 1$. The amplitude (1) can be factorized by introducing an infinite set of oscillators[2] satisfying

$$[a_m^\mu, a_n^{\nu+}] = \delta_{mn}g_{\mu\nu};$$

μ and ν are Lorentz indices; $n, m = 1, 2, \ldots, \infty$ correspond to the different oscillator modes. We have then the identity

$$(1-z)^{-A\bar{A}} = \prod_{n=1}^\infty \langle 0|e^{\frac{Aa_n}{\sqrt{n}}} z^{na_n^+ a_n} e^{\frac{\bar{A}a_n^+}{\sqrt{n}}} |0\rangle , \tag{2}$$

where the contraction of the Lorentz indices is understood in all the scalar products. The four-vectors A_μ and $\bar{A}_\mu$ correspond to incoming and outgoing momenta respectively. In what follows we denote

$$a = A\bar{A} \equiv \alpha(s) + 1 , b \equiv \alpha(t) + 1$$

and omit the Lorentz indices.

To perform the q-deformation within the operator formalism, it seems most natural to use the q-deformed oscillators[10] instead of the usual ones in the factorization procedure, as this was done in,[7] since q-oscillators have many features in common with the usual harmonic oscillators and from the quantum group point of view are the straightforward generalization of the latter. Therefore, we will replace the identity (2) by the q-deformed expression

$$F(a, z) = \prod_{n=1}^\infty {}_q\langle 0|e_q^{\frac{Aa_n}{\sqrt{n}}} z^{nN_n} e_q^{\frac{\bar{A}a_n^+}{\sqrt{n}}} |0\rangle_q = \prod_{n=1}^\infty \sum_{\ell=0}^\infty \left(\frac{a}{n}\right)^\ell \frac{z^{n\ell}}{[\ell]!} = \prod_{n=1}^\infty e_q^{\frac{a}{n}z^n}, \tag{3}$$

where

$$[\ell] = (q^\ell - q^{-\ell})/(q - q^{-1}), [\ell]! = [1][2]\cdots[\ell], [0]! = 1$$

and

$$e_q^x = \sum_{\ell=0}^\infty \frac{x^\ell}{[\ell]!}$$

is the q-exponential function. The q-oscillators entering in (3) satisfy[10]

$$a_m a_m^+ - q a_m^+ a_m = q^{-N_m} ,$$

$$[N_m, a_m^+] = a_m^+ , \quad [N_m, a_m] = -a_m , \tag{4}$$

with all the other *commutation relations* corresponding to different indices of oscillators, vanishing. The operator N_m is the number operator corresponding to the mode m.

The main difference between the expression (3) and the corresponding one proposed in[7] (cf. Eq. (8) in[7]) is that we have assumed the total energy operator of the genuinely free system to be

$$H = \sum_{n=1}^{\infty} n N_n$$

instead of $H = \sum_{n=1}^{\infty} n a_n^+ a_n$ and therefore, we have replaced $a_n^+ a_n$ by the number operator N_n in the exponent of z. As a consequence of this assumption, the amplitudes defined in the present model will exhibit poles in s when $\alpha(s)$ is a positive integer, or poles in t when $\alpha(t)$ is a positive integer as in the undeformed case. For the amplitudes defined in[7] the poles are in general located at non-integer values.

In order to preserve duality, expressed by the symmetry $z \rightarrow 1 - z$ in (1), we define then the q-deformed 4-point amplitude as

$$A_4^q = \int_0^1 F(a,z) F(b,1-z) dz = \prod_{n=1}^{\infty} {}_q\langle 0| e_q^{\frac{A a_n}{\sqrt{n}}} \int_0^1 z^{n N_n} F(b,1-z) dz \, e_q^{\frac{A a_n^+}{\sqrt{n}}} |0\rangle_q . \tag{5}$$

Inserting a complete set of orthonormal states

$$1 = \sum_{\lambda} |\lambda\rangle_q \, {}_q\langle\lambda|,$$

with

$$|\lambda\rangle_q = \prod_i \frac{(a_i^+)^{\lambda_i}}{\sqrt{[\lambda_i]!}} |0\rangle_q$$

we find A_4^q in its factorized form

$$A_4^q = \sum_{\lambda} V(A,\lambda) D(\lambda,b) V^+(\bar{A},\lambda) , \tag{6}$$

where

$$V(A,\lambda) = {}_q\langle 0| \prod_{n=1}^{\infty} e_q^{\frac{A a_n}{\sqrt{n}}} |\lambda\rangle_q \tag{7}$$

is the vertex operator (with only one leg off shell) and

$$D(\lambda,b) = {}_q\langle\lambda| \int_0^1 dz \, z^{\sum_{n=1}^{\infty} n N_n} F(b,1-z) |\lambda\rangle_q \tag{8}$$

is the propagator.

The other vertex operator Γ with two legs off shell , which is needed for higher n-point functions ($n \geq 5$), is given now by

$$\Gamma(\lambda, \lambda', p, \bar{p}) = {}_q\langle\lambda| \prod_{n=1}^{\infty} e_q^{\frac{pa_n^+}{\sqrt{n}}} e_q^{\frac{pa_n}{\sqrt{n}}} |\lambda'\rangle_q ,\tag{9}$$

where $p = \bar{p}$ is the momentum of the unexcited leg in the vertex. The 5-point q-deformed amplitude, e.g., now can be written as a Feynman-like diagram in terms of products of vertices and propagators in the tree approximation:

$$A_5^q = \sum_{\lambda, \lambda'} V(A, \lambda)D(\lambda, b_1)\Gamma(\lambda, \lambda', p, \bar{p})D(\lambda', b_2)V^+(A, \lambda'),\tag{10}$$

where b_1, b_2 are the Mandelstam variables of the corresponding channels.

Notice that the vertices (7) and (9) and the propagator (8) differ from those proposed in[7] since there, two infinite sets of independent unphysical oscillators ($b_i, b_i^+, i = 1, 2$), which were just auxiliary and not necessary, were introduced. The corresponding q-deformed n-point amplitudes, however, are identical whether one uses these auxiliary oscillators or not. Notice also that (7), (8) and (9) lead to the usual undeformed expressions for the vertices and the propagators in the limit $q \to 1$. The latter feature was not present in the vertices and propagators defined in.[7]

Let us first examine the singularities of A_4^q as a function of s and t. Since the integral in (5) is symmetric in s and t, we need only to analyse the singularities in one of the two variables, e.g. in the variable t. We first observe that near the singular point $z = 0$ the integrand $F(b, 1 - z)$ in (5) can be approximated as

$$F(b, 1 - z) \sim z^{-b},\tag{11}$$

and therefore the integral (5) is not defined when $\alpha(t) \geq 0$ (or $\alpha(s) \geq 0$) . Following the standard procedure[11] we can define it by analytic continuation to show that the amplitude A_4^q exhibits poles in t at any integer. The residue at the n-th pole in the t-channel is a polynomial in s of degree $J \leq n$, which when decomposed into spherical harmonics describes the exchange of a set of particles of spins $\leq J$.

Let us consider now the effects of the deformation in the high energy behaviour of A_4^q. In the undeformed case the behaviour of the 4-point amplitude (1) for a large can be easily obtained (see,e.g.,[12]) from the contour integral,

$$A_4 \sim \oint_{z=0} \frac{dz}{2\pi i}(1 + az + \cdots + \frac{a(a + 1)\cdots(a + n - 1)}{n!} z^n + \cdots)z^{-b},\tag{12}$$

leading to

$$A_4 \sim \frac{a(a + 1)\cdots(a + b - 2)}{\Gamma(b)} \sim \frac{a^{b-1}}{\Gamma(b)},\tag{13}$$

where Γ is the gamma function.

In order to discuss the high energy behaviour of the q-deformed 4-point amplitude (5), we will follow a procedure similar to the one used for the undeformed case , namely, we write A_4^q as the contour integral

$$A_4^q \sim \oint_{z=0} \frac{dz}{2\pi i}F(a, z)z^{-b} .\tag{14}$$

Consider now a generic term in the product (3) which can be written as

$$C_{n_1,...,n_m} a^{n_1+n_2+\cdots+n_m} z^{n_1+2n_2+\cdots+mn_m}, \tag{15}$$

where $m = 1, 2, \ldots, \infty$ and $C_{n_1,...,n_m}$ are coefficients which depend on q. Since we are interested in the high energy behaviour of (14), i.e. in the case when a is large, we should pick in $F(a, z)$ the highest power of a which gives nonvanishing contribution to the contour integral (14). We are therefore led to solve the algebraic equation

$$n_1 + 2n_2 + \cdots + mn_m = b - 1, \tag{16}$$

with the condition that the spin

$$J = n_1 + n_2 + \cdots + n_m, \tag{17}$$

which is the power of a in (15), takes its maximum value.

In what follows we shall consider two cases:

(i) Assume that q is real. Since in this case the n_i in (16) range from 0 to ∞, we obtain that the maximum value of J in (17) will be given by $J = b - 1$ when $n_1 = b - 1, n_i = 0$ for $i \geq 2$. Thus we obtain from (15) that for high energies

$$A_4^q \sim a^{b-1}. \tag{18}$$

We notice that the high s-behaviour of the 4-point q-deformed amplitude is proportional to s^t, which leads to a mass spectrum with a linearly-rising Regge trajectory as in the undeformed case.

(ii) Let us consider now the case when the deformation parameter q is a K-th root of unity, i.e. $q = \exp(2i\pi/K)$. The q-analogs $[\ell]$ thus become

$$[\ell] = \sin(2\pi\ell/K)/\sin(2\pi/K).$$

We should point out here the main difference with respect to the undeformed case: due to the truncation of the Fock space for each oscillator mode in the Fubini-Veneziano operator formulation, each term in the product (3) consists of a finite series ending at $\ell = \tilde{K}$, where

$$\tilde{K} = \begin{cases} K - 1, & \text{for } K \text{ odd} \\ K/2 - 1, & \text{for } K \text{ even}. \end{cases} \tag{19}$$

In this case $0 \leq n_i \leq \tilde{K}$ and we have two possibilities:

If $b \leq \tilde{K} + 1$, then it is always possible to find a solution of (16) such that $n_1 = b - 1 \leq \tilde{K}$ and $n_i = 0, i \geq 2$, which gives the maximum value of J in (17). Then as before the high energy behaviour will be given by eq. (18) and, thus, *the trajectories will be linear*.

If $b > \tilde{K} + 1$, it is easy to see that for a generic m the highest power of a in (15) is $J = m\tilde{K}$ and is obtained when all the n_i are equal to their maximum value $\tilde{K}$. Then according to (16) J will satisfy the second-order algebraic equation

$$\frac{J(J + \tilde{K})}{2\tilde{K}} = b - 1, \tag{20}$$

which has the positive solution

$$J = \frac{\tilde{K}}{2}\left\{-1 + \sqrt{1 + \frac{8(b-1)}{\tilde{K}}}\right\}. \tag{21}$$

Thus finally we obtain that for large values of a and for $b > \tilde{K} + 1$

$$A_4^q \sim a^J, \tag{22}$$

where J is given by (21).

We notice now from (21) and (22) that the high s-behaviour of the q-deformed amplitudes is proportional to $s^{\sqrt{t}}$ which leads to the mass spectrum with a *square-root Regge trajectory*

$$\alpha(t) = \sqrt{t} + const.$$

It appears, as was expected by physical arguments, that the crucial change in the high energy behaviour of A_4^q occurs due to the truncation in the series in (3) and as mentioned before, is a direct consequence of q being a root of unity. This change can be understood by the finite character of the Fock space as a consequence of the assumed values of q.

It is worth mentioning that Eq. (21) gives the position of the poles in the leading Regge trajectory for the values of the spin $J = m\tilde{K}$. In general, for an arbitrary value of the spin

$$J = m\tilde{K} + r \ , r = 0, 1, \ldots, \tilde{K} - 1 \ ,$$

it is possible to show that the poles of the amplitude are located at the points given by the solution of the equation

$$\frac{(J+r)(J+\tilde{K}-r)}{2\tilde{K}} = \frac{(2J - m\tilde{K})(m+1)}{2} = b - 1 \ . \tag{23}$$

In Fig.1 the leading Regge trajectories for both cases, when q is real and $q = \exp(2i\pi/K)$, are shown . The dots denote the position of the poles and the linearly-rising Regge trajectory corresponds to the case when the deformation parameter q is real. We observe that when q is a K-th root of unity, the trajectory is linear for $t \leq \tilde{K}$ and turns into a square-root trajectory for $t > \tilde{K}$, with $\tilde{K}$ given by Eq.(19). We also notice that for smaller values of the parameter K (i.e. when the truncation of the series in (3) occurs earlier), the effect of the deformation is enhanced. As $K \to \infty$, then $q \to 1$ and we recover the usual linear trajectory. Since the amplitude is symmetric in s and t, the resonances in the s-channel also lie on the same Regge trajectory.

Thus quantum groups and the q-deformation provide with a new phenomenon of a linear Regge trajectory turning into a square-root trajectory for higher masses in the case when q is a root of unity. The latter case is of upmost physical interest and in particular, appears in rational conformal field theories. It will be of interest to consider, for comparison, the q-deformed dual amplitude proposed in[7] and study its high energy behaviour in order to obtain the corresponding Regge trajectories. It is our belief, however, that the features of the latter amplitude will make the model proposed in[7] less appealing from the physical point of view. This question will be discussed in detail in a forthcoming publication.

It will be interesting to find out whether the proposed dual string model still has an exponentially increasing mass spectrum,[2] and its implication on the Hagedorn statistical bootstrap model exponential spectrum.[13]

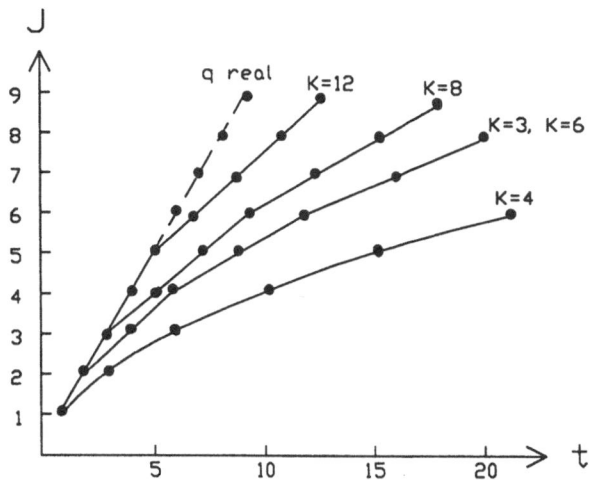

Figure 1. The behaviour of the Regge trajectories for the q-deformed dual string model proposed here. The dots denote the position of the poles and the straight (dashed) line corresponds to the case when q is real, with linearly-rising Regge trajectory. When q- is a K-th root of unity, $q = \exp(2i\pi/K)$, the trajectory is linear for $t \leq \tilde{K}$ and turns into a square-root trajectory for $t \geq \tilde{K}$, with $\tilde{K} = K - 1$ for K odd and $\tilde{K} = K/2 - 1$ for K even (see Eq. (19)). The curves are shown for the particular values of $K = 3, 4, 6, 8, 12$.

REFERENCES

1. G. Veneziano, Il Nuovo Cimento **57A** (1968) 190.

2. S. Fubini, D. Gordon and G. Veneziano, Phys. Lett. **29B** (1969) 679; S. Fubini and G. Veneziano, Nuovo Cimento A **64** (1969) 811; Ann. Phys. **63** (1971) 12.

3. L.D. Faddeev, N.Yu. Reshetikhin and L.A. Takhajan Algebra Anal. **1** (1989) 178.

4. M.B. Green, J.H. Schwarz and E. Witten, *Superstring Theory* (Cambridge Univ. Press, Cambridge ,1987).

5. L. Alvarez-Gaumé, C. Gomez and G. Sierra, Phys. Lett. B **220** (1989) 142; Nucl. Phys. **B319** (1989), 155; Nucl. Phys. **B330** (1990) 347.

6. H.J. de Vega, Nucl. Phys. **B374** (1992) 692.

7. M. Chaichian, J.F. Gomes and P. Kulish, Phys. Lett. B **311** (1993) 93.

8. D. Coon, S. Yu and M. Baker, Phys. Rev. D **5** (1972) 1429.

9. L.J. Romans, in Proc. Conf. High Energy Physics and Cosmology (Trieste, 1989), eds. J.C. Pati *et al.* (World Scientific, Singapore, 1990).

10. A.J. Macfarlane, J. Phys. A: Math. Gen. **22** (1989) 4581; L.C. Biedenharn, J. Phys. A: Math. Gen. **22** (1989) L873; M. Chaichian and P. Kulish, Phys. Lett. B **234** (1990) 72.

11. S. Mandelstam, Phys. Rep. **13** (1974) 259.

12. M. Chaichian, Phys. Rev. D **5** (1972) 425.

13. R. Hagedorn, Suppl. Nuovo Cimento **3** (1965) 147; S. Frautschi, Phys. Rev. D **3** (1971) 2821; J. Yellin, Nucl. Phys. **B52** (1973) 583.

HAGEDORN'S REINCARNATION IN STRING THEORY

John Ellis

Theory Division, CERN
CH-1211 Geneva 23, Switzerland

ABSTRACT

After a review of Hagedorn's first incarnation in the quark-hadron phase transition, universality properties of the state level density in string theory are discussed, as well as the possible transition to a new phase beyond the string Hagedorn temperature. Studies of string black holes suggest that the concepts of space and time are likely to disappear in this reincarnation of Hagedorn.

FIRST INCARNATION

The Hagedorn spectrum first emerged[1] from the statistical bootstrap model of hadrons, which predicted a level number of density of the form

$$n(m) \simeq m^{-a} \exp(bm) \tag{1}$$

where the exponential parameter $b = \mathcal{O}(1/m_\pi)$, or $\mathcal{O}(\sqrt{\alpha'})$ in Regge language. An exponential level density implies special behavior of thermodynamic quantities at the Hagedorn temperature $T_H = 1/b$, since the integrals

$$\rho, p, \ldots \simeq \int dm\, m^{p_i} \exp\left(b - \frac{1}{T}\right) m \tag{2}$$

over masses are ill-defined for $T > T_H$. The behaviours of thermodynamic quantities[2] as $T \to T_H - \epsilon$ depend on the pre-exponential factor a in the level density (1). The density ρ and pressure density p will approach finite values at $T \to T_H - \epsilon$ if

$$a > \frac{1}{2}(D + 3) \tag{3}$$

where D is the dimensionality of space-time. This is the situation which is now believed to obtain in QCD: the hadronic Hagedorn temperature T_H is related to the critical temperature T_C for the transition from the hadronic phase to the quark-gluon phase of QCD, with $T_H = T_C = \mathcal{O}(\Lambda_{\mathrm{QCD}}) \simeq 150\ MeV$.

MODELLING THE QUARK-HADRON PHASE TRANSITION

At low energies, hadrons can be described by an effective field theory that incorporates the symmetries of QCD. Foremost among these is chiral symmetry, which is spontaneously broken by quark condensation in the vacuum:

$$\langle 0 \,|\, \bar{q}q \,|\, 0 \rangle = c \neq 0 \; : \; m_\pi^2 = c \frac{m_q}{f_\pi^2} \tag{4}$$

with pions (and other pseudoscalar mesons) interpreted as pseudo-Goldstone bosons. Their interactions can be described by an effective Lagrangian of nonlinear σ–model form:[3]

$$\mathcal{L}_c = \frac{f_\pi^2}{16} \mathrm{Tr}\left[\partial_\mu U \partial^\mu U^\dagger\right] - c\,\mathrm{Tr}\left[m_q(U + U^\dagger)\right] + \cdots \tag{5}$$

where

$$U(x) = \exp\left(i \sum_i \frac{\lambda_i \pi_i(x)}{f_\pi}\right), \qquad \langle 0 \,|\, \bar{q}q \,|\, 0 \rangle = c.\langle 0 \,|\, (U + U^\dagger) \,|\, 0 \rangle \tag{6}$$

and the dots represent higher-order terms. Within this framework, baryons appear as topological solitons,[4] *i.e.*, "lumps" in the chiral field $U(x)$, whose stability is guaranteed by a topological quantum number identified with baryon number.

Finite-temperature corrections to this description have been calculated,[5,6] and yield a reduction in the quark condensate as the temperature increases:

$$\langle 0 \,|\, \bar{q}q \,|\, 0 \rangle_T = \langle 0 \,|\, \bar{q}q \,|\, 0 \rangle_0 \left(1 - \frac{8}{3}\frac{T^2}{f_\pi^2} + \cdots\right) \tag{7}$$

if one neglects the light quark masses. The lightest baryon mass $m_B \to 0$ as $\langle 0 \,|\, \bar{q}q \,|\, 0 \rangle \to 0$, and its radius $R_B \to \infty$. These results are very suggestive that quarks become deconfined when quark condensation vanishes, as happens when the temperature is raised.[7] However, it is not clear that this joint transition can be described adequately by the chiral Lagrangian (5) above.

One feature of QCD that it does not reflect is that of scale symmetry: most of the QCD Lagrangian ($F_{\mu\nu}F^{\mu\nu}, \bar{\chi}\not{D}\chi$) has scale dimension 4, whilst the quark mass term $\bar{\chi}\chi$ has scale dimension 3. These properties can be incorporated most simply in the effective Lagrangian approach by introducing a single scalar field χ of scalar dimension 1,[8] with a vacuum expectation value $\langle 0 \,|\, \chi \,|\, 0 \rangle \equiv \chi_0$ that is related to gluon condensation

$$\mathcal{L}_{\mathrm{eff}} = \frac{f_\pi^2}{16}\left(\frac{\chi}{\chi_0}\right)^2 \mathrm{Tr}\left[\partial_\mu U \partial^\mu U^\dagger\right] - c\left(\frac{\chi}{\chi_0}\right)^3 \mathrm{Tr}\left[m_q(U + U^\dagger)\right] + \cdots$$
$$+ \frac{1}{2}\partial_\mu\chi\partial^\mu\chi - B\left[\chi^4 \ell n\left(\frac{\chi}{e^{1/4}\chi_0}\right) + \frac{1}{4}\right]. \tag{8}$$

As is well known, scale invariance is broken by renormalization, which induces a trace anomaly $(\beta/2g)\,F_{\mu\nu}F^{\mu\nu}$ whose effects are modelled by the form of the effective χ potential in (8).[9]

Because of the Coleman-Weinberg form of this effective potential, any χ transition must be first-order in $\langle 0 \,|\, \chi \,|\, 0 \rangle$.[7] Since the quark condensate is now represented by

$$\langle 0 \,|\, \bar{q}q \,|\, 0 \rangle = c\left(\frac{\chi}{\chi_0}\right)^3 \langle 0 \,|\, (U + U^\dagger) \,|\, 0 \rangle \tag{9}$$

the quark transition may either be first-order, if it is driven by $\langle 0 \,|\, \chi \,|\, 0 \rangle \to 0$, or second-order, if the chiral dynamics dominates as in (7). In the former case, the quark and gluon transitions would be simultaneous, whereas the quark transition would occur before the gluon transition in the latter case. Even if the transition is first order, it is likely to relatively weakly so: one estimate of the latent heat L and the bubble surface tension suggest[7] that

$$L = B \sim 4 \times 10^{-4}\, GeV^4\,, \quad \sigma \sim (70\, MeV)^3\,. \tag{10}$$

This would have resulted in relatively small separations between bubble nucleation sites in the early Universe, which is also consistent with constraints on inhomogeneties in the early Universe derived from primordial light element abundances.[10]

The evaluation of the finite-temperature corrections to the effective χ potential requires[7] the inclusion of all hadronic degrees of freedom, in principle, since all their masses are proportional to $\langle 0\,|\,\chi\,|\,0\rangle$ in the chiral limit:

$$\mathcal{L}_{\text{eff}} \supset -\frac{m_B^2}{2}\, B^2\chi^2\,, \quad -m_f \bar{f}f\chi \tag{11}$$

up to chiral-symmetry-breaking corrections, and $T_C = \mathcal{O}(\Lambda_{\text{QCD}})$ is not expected to be small compared to a typical hadronic mass scale. Including a Hagedorn mass spectrum of the form (1), we find[7] an apparent singularity in the pressure at $T_H = 1/b$:

$$P(\chi, T) \propto \frac{T^{5/2}}{(2\pi)^{3/2}} \left(\frac{TT_H}{T_H - T}\right)^{(3/2)-a} \Gamma\left(\frac{5}{2} - a\,, \frac{y}{2}\right) : y \equiv 12\chi\left(\frac{T_H - T}{TT_H}\right) \tag{12}$$

with a critical temperature $T_C < T_H$ where the free-energy curves for the hadronic and quark-gluon phases cross. Closer study[7] of the finite-temperature effective potential reveals the likelihood that the local minimum at $\langle 0\,|\,\chi\,|\,0\rangle \neq 0$ disappears at a temperature $T_\chi : T_C < T_\chi < T_H$. Superheating in a metastable hadronic state is possible in the temperature range $T_C < T_\chi < T_H$, and supercooling in a metastable quark-gluon phase may be possible for a range of $T < T_C$.

REINCARNATION OF THE HAGEDORN SPECTRUM IN STRING THEORY

As you know, string theory originated as a theory of the strong interactions from attempts to model the apparently infinite spectrum of hadrons with squared masses

$$m^2 \simeq m_0^2 + \frac{n}{\alpha'} : \alpha' \sim 1\, GeV^2\,. \tag{13}$$

The original Veneziano[11] model for the 2-to-2 scattering amplitude

$$A = \frac{\Gamma(1 - \alpha's)\Gamma(1 - \alpha't)}{\Gamma(1 - \alpha's - \alpha't)} \tag{14}$$

can be regarded as an implementation of the original bootstrap idea, since direct-channel states are seen to be built up out of an infinite set of crossed-channel states, and vice versa. The discovery (described here by Gabriele Veneziano) of the exponential spectrum followed,[12] and then it was realized[13] that the Veneziano model was a string theory, which would require a string tension

$$\frac{E}{L} = (\alpha')^{-1} \sim 1\, GeV^2\,. \tag{15}$$

This version of string theory was superseded by QCD as a theory of the strong interactions, but was soon resurrected[14] as a possible Theory of Everything (TOE) to describe all particle interactions, as a result of the observation that the spectrum of a closed string theory includes a massless spin-2 particle that can be interpreted as a graviton. However, this idea remained a minority interest for several years until string theory was shown to be consistent at the quantum level, *i.e.*, free of anomalies.[15] This little historical review indicates that string theory was discovered early, apparently by chance, but with valuable help from Hagedorn via the bootstrap[1] idea.

The biggest problem of all in theoretical physics is to reconcile Quantum Mechanics with General Relativity. This has both perturbative aspects—how to make perturbative quantum loop corrections calculable—and nonperturbative—what happens to quantum coherence in space-time backgrounds with event horizons, such as black holes? Conventional point-like quantum field theories are unable to solve these problems, and my expectation is that radically new geometric ideas are needed, just as new geometrical ideas reconciled Maxwell's equations with Newtonian mechanics—the space-time continuum of Special Relativity—and Special Relativity with gravity—the Riemannian geometry of General Relativity. This is what I find the strongest motivation for thinking that the TOE is based on string.

A generic closed string theory may be written in terms of degrees of freedom moving in opposite directions around the string loop. In the early days, the left- and right-movers were assumed to be identical, either bosonic or supersymmetric. In the bosonic string case, the total amount of "stuff," as counted by the conformal anomaly c, should be

$$c = D + \text{(internal)} = 26 \tag{16}$$

where D is the number of space-time dimensions, which may be less than 26 if there are extra internal degrees of freedom, whereas in the supersymmetric case

$$\frac{2}{3}c = D + \text{(internal)} = 10 . \tag{17}$$

Subsequently, it was realized that one could construct heterotic string models in which the left- and right-moving degrees of freedom are different. For example, one could have the left-moving sector bosonic with $c_L = 26$, and the right-moving sector supersymmetric with $\frac{2}{3}c_R = 10$.[16] Such heterotic string theories can even be formulated directly in $D = 4$ dimensions, with the extra left-moving degrees of freedom providing internal quantum numbers for gauge and matter fields.[17]

The density of states in a general flat space-time background has two remarkable universality properties.[18,19] The first is that the exponential growth parameter b in the general formula (1) is given by

$$b = \sqrt{6\alpha'}.\pi.\left(\hat{b}_L + \hat{b}_R\right) \tag{18}$$

where

$$\hat{b} = \sqrt{c - 2} \tag{19}$$

for a bosonic sector, and

$$\hat{b} = \sqrt{c - 3} \tag{20}$$

for a supersymmetric sector. This means that

$$b = 4\pi\sqrt{\alpha'}, \quad (2 + \sqrt{2})\pi\sqrt{\alpha'}, \quad 2\sqrt{2}\pi\alpha' \tag{21}$$

for bosonic, heterotic and supersymmetric strings, respectively, independently of any other details of the model. The second universal feature is that the pre-exponential factor a depends only on the dimensionality of the flat space-time:

$$a = D \tag{22}$$

also independently of the details of the string model.

As in the hadronic case discussed earlier, $T_H = 1/b$ is a candidate for a limiting temperature, or perhaps a critical temperature for a transition to a new phase, depending whether the density and pressure p, ρ have singularities as $T \to T_H - \epsilon$. Comparing Eqs. (3) and (22), we see that $D = 4$ is the lowest dimension for which $p, \rho \to$ constants as $T \to T_H - \epsilon$, and a transition to a new phase is to be expected.

MODELLING THE STRING PHASE TRANSITION

Here we adopt a strategy similar to that used for QCD in Section 2. Particle physics at energies and temperatures $\ll 10^{19} \, GeV$ is described by an effective field theory analogous to the chiral Lagrangian of QCD. Particles are excitations of closed string loops, analogous to glueballs, whereas pions could be regarded as the lowest excitations of open QCD strings with a quark and an anti-quark at either end:

$$\pi(x) \leftrightarrow \bar{q}(x)\gamma_5 P \exp\left(i \int_x^y A_\mu(x)\, dx^\mu\right) q(y) \,. \tag{23}$$

When a QCD system is heated, quarks and gluons become deconfined: what happens if the Standard Model is heated to $10^{19} \, GeV$? what is the nature of the high-T phase? is something deconfined? is there a topological field theory, as some[20] have argued? what are the order parameters analogous to $\langle 0 \,|\, \bar{q}q \,|\, 0 \rangle$ and $\langle 0 \,|\, F_{\mu\nu} F^{\mu\nu} \,|\, 0 \rangle$?

We will not discuss all the details here, which have in any case not been fully worked out, but provide qualitative answers to some of these questions. The effective potential at zero temperature, analogous to the χ potential in Eq. (8), is provided by some generalized no-scale supergravity model,[21] which guarantees $V_{\text{eff}} \geq 0$. The order parameters include the moduli of string compactification, which are model-dependent, but there is one that is universal, namely the dilaton S whose v.e.v. fixes the gauge and other string couplings:

$$\frac{1}{g^2} = \langle 0 \,|\, S \,|\, 0 \rangle \,. \tag{24}$$

As the temperature T is increased, it appears[22] that $\langle 0 \,|\, S \,|\, 0 \rangle \to 0$, just as $\langle 0 \,|\, \chi \,|\, 0 \rangle \to 0$ in the QCD case. As in QCD, there are order parameters that appear at high temperature. In the QCD case, there are time-like Polyakov loops $L(x)$, whose effective potential may resemble

$$\mu^2 \,|\, L \,|^2 + \frac{1}{2} A \left(L^3 + L^{*3} \right) + \lambda \,|\, L \,|^4 \tag{25}$$

where μ^2 becomes tachyonic at the Hagedorn temperature:

$$\mu^2 = \mu_0^2 \left(\frac{1}{T^2} - \frac{1}{T_H^2} \right) \,. \tag{26}$$

Analogously, a tachyon appears at the Hagedorn temperature in string theory, associated with a condensate of world-sheet vortices.[23] This causes the effective potential to become negative in some domain of parameter space, and since $V_{\text{eff}} \propto g^2$, simple energetics favour $g^2 \to \infty$, or $S \to 0$.

BEYOND HAGEDORN—BLACK HOLES

In order to gain some insight into the nature of the new string phase beyond the Hagedorn temperature, we first consider some relevant aspects of black holes and their description in string theory, so as to gain some insight into what might happen if a smooth space-time background is not assumed. It is well known that the entropy of a black hole is[24]

$$S = A = \frac{1}{4}\, m_{BH}^2 \tag{27}$$

where A is the area of the event horizon, for a black hole without electric charge or angular momentum. Correspondingly, the number of density of states is

$$n(m_{BH}) \sim \exp A = \exp\left(\frac{1}{4}\, m_{BH}^2\right) \tag{28}$$

and there is a non-zero effective temperature, found by Hawking:[25]

$$T_n = \frac{1}{8\pi\, m_{BH}}\,. \tag{29}$$

In fact, there are problems in defining a thermodynamic ensemble, since integrals over masses with a Boltzmann distribution are not well defined:

$$\rho, p, \ldots \propto \int dm\, \exp\left(\frac{m^2}{4} - \frac{m}{T}\right) \tag{30}$$

which is formally divergent at any non-zero temperature T. This reflects the well-known instability of a black hole/particle system in a box.

As discussed in Section 3, the asymptotic number density of states in a generic bosonic string theory is of the generic form

$$\ell n\, n(m_{BH}) \sim \sqrt{c - 2}\, m_{BH}\,. \tag{31}$$

A two-dimensional (spherically symmetric four-dimensional) string black hole solution is known,[26] whose mass is related to the central charge:

$$m_{BH} \propto \sqrt{c - 2}\,. \tag{32}$$

Thus the asymptotic density of black hole states in string theory is[27]

$$n(m_{BH}) \sim \exp\left(\hat{\hat{b}} m_{BH}^2\right) \tag{33}$$

which is of the same form as (26). However, every string black hole states is believed to be distinguishable in principle, since string theory has an infinite set of gauge symmetries.[28] Associated with these are an infinite set of conserved charges available to label string states. Specifically, in the two-dimensional string black hole case there is a W_∞ symmetry, whose classical algebra is

$$\left[W_j^m,\, W_{j'}^{m'}\right] = [jm' - j'm]\, W_{j+j'}^{m+m'}\,. \tag{34}$$

This possesses an infinite-dimensional Cartan subalgebra, the W_j^j, which label the states and are in principle measurable.[29] Thus the entropy of a string black hole can be reduced to zero. Moreover, there is a well-defined S matrix for scattering light particles of a fixed black-hole background,[26] and black-hole decay can be regarded as a quantum loop effect,[27] with a rate

$$\tau_{BH}^{-1} = \Gamma_{BH} \propto \mathrm{Im}\,(m_{BH}) \tag{35}$$

where the dominant contribution comes from the absorptive part of a two-point correlation function on a world-sheet torus.

The infinite stringy symmetries are spontaneously broken in general, for example by the v.e.v.

$$\langle 0 \,|\, g_{\mu\nu} \,|\, 0 \rangle = \eta_{\mu\nu} \neq 0 \qquad (36)$$

and they relate string states with different masses $\Delta m^2 \propto 1/\alpha'$. So what is the nature of the unbroken phase, which presumably appears above the string Hagedorn temperature?

Some insight into this may be gleaned[30] by looking at the core of a string black hole, close to what would be a singularity in point-like field theory. Physics close to this "singularity" is described by a topological field theory with action[26,31]

$$S = \int d^2 z \left[\omega \epsilon_{ij} F^{ij} + Da.\bar{D}b + \ldots \right] \qquad (37)$$

where F^{ij} is a $U(1)$ gauge field strength, D and $\bar{D}$ are covariant derivatives, ω parametrizes the "singularity" and (a, b) are "matter" fields that describe the space-time coordinates. The appearance of a v.e.v. $\eta_{\mu\nu} = \langle 0 \,|\, g_{\mu\nu} \,|\, 0 \rangle \neq 0$ is equivalent to the formation of a matter condensate analogous to $\langle 0 \,|\, \bar{q}q \,|\, 0 \rangle \neq 0$ in QCD.

In the absence of such a condensate, in the high-temperature phase of string theory, there is no metric, and the concepts of space and time do not exist. In this phase, one can say poetically that the influence of Hagedorn is both ubiquitous and timeless, which seems like a good time and place to stop!

REFERENCES

1. R. Hagedorn - Nuov. Cim. Supp. 3 (1965) 147; CERN preprint TH.7190/94 (1994), talk presented at this meeting.

2. K. Huang and S. Weinberg, Phys. Rev. Lett. 25 (1970) 895.

3. J. Gasser and H. Leutwyler, Nucl. Phys. B250 (1985) 465.

4. T.H.R. Skyrme, Nucl. Phys. 31 (1962) 556.

5. P. Binetruy and M.K. Gaillard, Phys. Rev. D32 (1985) 931.

6. J. Gasser and H. Leutwyler, Phys. Lett. B184 (1987) 83 and B188 (1987) 477.

7. B.A. Campbell, J. Ellis and K.A. Olive, Nucl. Phys. B345 (1990) 57.

8. J. Ellis, Nucl. Phys. B22 (1970) 478.

9. J. Schechter, Phys. Rev. D21 (1981) 3393.

10. D. Thomas, D.N. Schramm, K.A. Olive and B.D. Fields, Ap. J. 406 (1993) 569.

11. G. Veneziano, Nuov. Cim. 57A (1968) 190.

12. S. Fubini and G. Veneziano, Nuov. Cim. 64A (1969) 811.

13. P. Goddard, J. Goldstone, C. Rebbi and C. Thorn, Nucl. Phys. B56 (1973) 109.

14. J. Scherk and J.H. Schwarz, Nucl. Phys. B81 (1974) 118.

15. M.B. Green and J.H. Schwarz, Phys. Lett. 149B (1984) 117.

16. D.J. Gross, J.A. Harvey, E. Martinec and R. Rohm, Phys. Rev. Lett. 54 (1985) 502.

17. K.S. Narain, Phys. Lett. 169B (1987) 41.

18. I. Antoniadis, J. Ellis and D.V. Nanopoulos, Phys. Lett. B199 (1987) 402.

19. M. Axenides, S.D. Ellis and C. Kounnas, Phys. Rev. D37 (1988) 2964.

20. J. Atick and E. Witten, Nucl. Phys. B310 (1988) 291.

21. E. Cremmer, S Ferrara, C. Kounnas and D.V. Nanopoulos, Phys. Lett. B133 (1983) 61; J. Ellis, C. Kounnas and D.V. Nanopoulos, Nucl. Phys. B241 (1984) 406 and B247 (1984) 373.

22. B.A. Campbell, J. Ellis, S. Kalara, D.V. Nanopoulos and K.A. Olive, Phys. Lett.B255 (1991) 420.

23. Y.I. Kogan, J.E.T.P. Lett. 45 (1987) 709; B. Sathiapalan, Phys. Rev. D35 (1987) 3277.

24. J. Bekenstein, Phys. Rev. D12 (1975) 3077.

25. S.W. Hawking, Comm. Math. Phys. 43 (1975) 199.

26. E. Witten, Phys. Rev. D44 (1991) 314.

27. J. Ellis, N.E. Mavromatos and D.V. Nanopoulos, Phys. Lett. B278 (1992) 246.

28. G. Veneziano, Phys. Lett. 167B (1985) 388; J. Maharana and G. Veneziano, Nucl. Phys. B283 (1987) 126.

29. J. Ellis, N.E. Mavromatos and D.V. Nanopoulos, Phys. Lett. B284 (1992) 33, 43.

30. J. Ellis, N.E. Mavromatos and D.V. Nanopoulos, Phys. Lett. B288 (1992) 23.

31. T. Eguchi, Mod. Phys. Lett. A7 (1992) 85.

INTERACTIVE COMPUTER LANGUAGES:
PAST AND FUTURE OF SIGMA

Juris Reinfelds

Department of Computer Science
New Mexico State University, Las Cruces, NM 88003, USA

HISTORY

Physicists take great pride in their ability to perform back-of-an-envelope-calculations to obtain order of magnitude estimates of otherwise elaborate and time-consuming problem solutions. Interactive programming systems and languages were born out of the desire to use computers to enlarge that envelope by several orders of magnitude.

The first published interactive system appeared in 1961. It was MADCAP by Mark Wells, then at Los Alamos,[1] later at New Mexico State University. Mathematician Glen Culler[3] and physicist Bert Fried,[2] developed one of the first interactive systems that became widely used and in retrospect turned out to be better than most of its successors. The list of important research papers that had been completed with the help of the Culler-Fried system by 1968 is truly impressive.[2]

The Culler-Fried System

Programming languages use single numbers or characters as basic data units. Culler introduced a fixed-size vector of 124 components as the basic data unit of his system. The user had the option to use real numbers or complex numbers as components. All the usual single-number operators such as arithmetic and relational operators or trigonometric or transcendental functions were available and their operations were defined as component-wise independent. For example, for arrays X and Y, the expression

$$X + Y \tag{1}$$

performed 124 additions

$$X_i + Y_i \quad for \quad 1 \le i \le 124 \tag{2}$$

The Culler-Fried system behaved like a sophisticated desk calculator for 124-number aggregates. Any sequence of calculational steps could be stored for later reuse. A rich selection of mathematical functions was available. A plot function was provided to display the values of one array (y-coordinates) against the corresponding values of another array (x-coordinates).

Culler also provided a number of operators that affected the vector as a whole. For example, the DIFF operator was defined to generate the forward difference for each element (with a special case for the last element which has no successor)

$$(DIFF(Y))_i = y_{i+1} - y_i, \; for \; i = 1, 2, \ldots (n-1) \tag{3}$$

$$(DIFF(Y))_n = 2y_n - 3y_{n-1} + y_{n-2} \tag{4}$$

Operators like DIFF provide the problem solving power of the Culler-Fried system. DIFF could be used to generate numerical derivatives or to separate low and high frequency components from a mixed frequency signal. RightShift and LeftShift operators, with extrapolation at the missing endpoint, permitted the construction of various numerical integration or data smoothing operators. The original and powerful speech waveform analysis described by Culler[3] would take more effort and more programming on most of our systems in 1994 than it took with Culler's system in 1967.

In today's language, the Culler-Fried system provided a SIMD (Single Instruction Multiple Data) desk calculator, augmented by powerful cross-element operators.

The AMTRAN System

At NASA, Marshall Space Flight Center, Seitz, Reinfelds and Woods[4] built on the ideas of Culler-Fried and developed the AMTRAN (Automatic Mathematical TRANslation) system on an IBM 1620 computer. AMTRAN added syntax analysis to the system and provided BASIC-like programming language capabilities, so that users could write expressions, assignment statements, conditional statements and loops.

In 1968 - 1970, with the help of research grants from NASA and NSF, the AMTRAN Team at the University of Georgia (Reinfelds, Bodenseher, Engelbert, Eskelson, Jedlicka, Kopetz and Kratky), designed and implemented an AMTRAN[5] system for the IBM 1130 computer. The most remarkable achievement of this effort was, that with a small modification of the IBM 1130 hardware, the system could accommodate two time-shared users with the whole system running in heavily overlayed 16k bytes of memory which was all the memory that was available on the 1130 at that time.

The 1130 AMTRAN language was an extension of Huntsville AMTRAN. The basic data structure was still a one-index vector, as in Culler-Fried, but AMTRAN vectors could have arbitrary length, as long as they, the program and any intermediate storage requirements did not exceed the 4,980 byte data and program storage area of the system. The Georgia implementation was clean, robust and efficient. At its peak it was used by some 100 sites in North America, Europe and Australia.

The GAMMA System

In 1968 at CERN, Vandoni and Hagedorn designed and implemented GAMMA (Graphically Aided Mathematical MAchine). GAMMA was a Culler-Fried inspired interactive system for one-dimensional array computation. It ran at CERN on CDC 3000 Series computers. GAMMA's main contribution was the provision of excellent display capabilities through the integration of the facilities of the CERN FORTRAN graphics package into the GAMMA system.

SIGMA

In 1971 Reinfelds and Vandoni decided to join forces and together with physicists Hagedorn and Van Hove, a more general multi-index array-based language was designed and implemented at CERN over the period 1971 - 1975. The new language was called SIGMA[6] (System for Interactive Graphical and Mathematical Applications).[7] It combined the best features of GAMMA and AMTRAN and provided a significant extension with its multi-indexed array manipulation capabilities. Vandoni was the chief implementor and Helga Rafelski, Mosczinski and Barta made major contributions to the implementation.

A special effort was made to design a syntactically simple, yet versatile and powerful, DISPLAY command so that graphs of single curves or multiple curves or families of curves or wire-frame displays of three dimensional surfaces, could be displayed with a single command with reasonable, automatically provided scaling and axes-labeling defaults. These defaults could be overridden by explicit requests from the user, but in most cases that was not necessary. For example, the following SIGMA program

$$X \ = \ ARRAY(300, 0 \# 8) \tag{5}$$

$$G \ = \ COSH(X) + SIN(1/(0.1 + X * X)) \tag{6}$$

$$DISPLAY \ G : X \tag{7}$$

will display the function

$$g(x) = coshx + sin(\frac{1}{0.1 + x^2}) \tag{8}$$

using 300 points over the range $0 \leq x \leq 8$, as in Fig.1. Automatic scaling chooses the

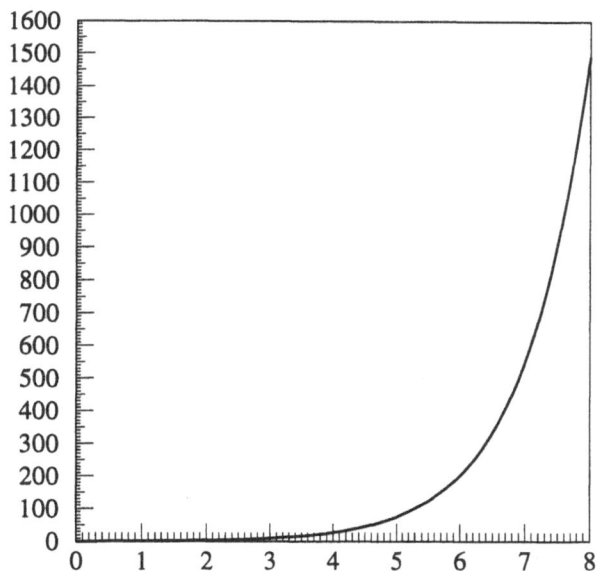

Figure 1. Default DISPLAY scales automatically to show all points

y-coordinate range so that no points are lost and as a consequence, the sine oscillations are not visible. If we specify a smaller range explicitly, a subset of the previously calculated points may be displayed, as shown in Fig.2, with the command

$$DISPLAY(0 \# 5 : 0 \# 3) \ G : X, \ [+]G : X \tag{9}$$

In this display we have requested fixed ranges $0 \leq y \leq 5$ and $0 \leq x \leq 3$ so that we will only see those points that fall within the specified limits. We requested that the points be displayed twice: once as a curve (G:X) and once as explicit points denoted by the character '+' as in

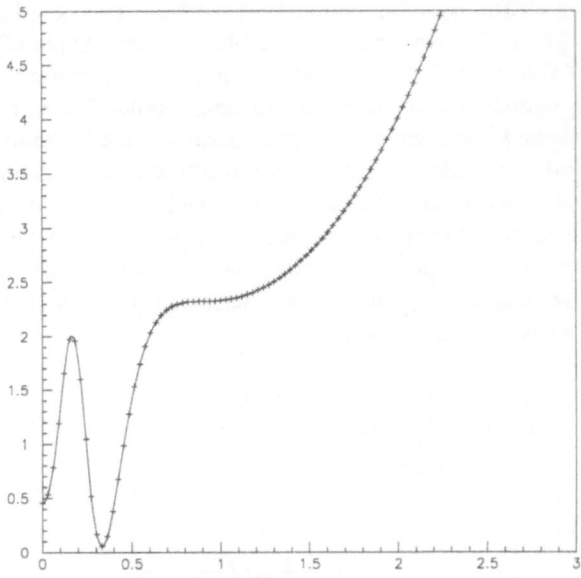

Figure 2. Explicit scaling shows any subregion of interest

([+]G:X). Fig.2 shows that the oscillations are now clearly visible and the point density is sufficient for a faithful rendering of the oscillations.

The display command incorporates the most commonly used curve drawing options into a simple and easy to remember syntactic structure. Even today, this structure seems in many ways preferable to the smorgasbord style option selection menus of today's graphic systems. There is room for new research to find simple yet powerful display commands that will extend notational simplicity to today's wide variety of available display capabilities.

In 1974, Richard Miller, at the University of Saskatchewan, implemented SIGMA[8] on IBM 370 computers. Miller chose to rewrite only the essential graphics components while CERN SIGMA used the whole graphics package. As a consequence, Miller's SIGMA reduced the 25,000 FORTRAN lines of CERN SIGMA to about 2,500 FORTRAN lines for the Saskatchewan SIGMA with practically no loss of functionality. In 1976 Reinfelds and Miller compiled the Saskatchewan SIGMA on Univac 1106 computers. This SIGMA was used at the University of Wollongong for a number of years.

When the last CDC 6000 Series computers were retired at CERN, SIGMA was transferred to become a part of the Physics Analysis Workstation (PAW) software, where it is still in use today. In computer science it is remarkable and unusual that a language design and implementation is still interesting and useful after 20 years of continuous use without any significant upgrades or modifications.

COMPATIBLE ARRAYS

The Culler-Fried system demonstrated the problem solving power of a desk calculator for fixed size vectors as basic data elements. AMTRAN extended the language to handle one-index arrays of arbitrary size and gave the language a syntax comparable in scope to the BASIC programming language. GAMMA extended the graphic display facilities and SIGMA added multidimensional array capabilities.

Especially due to the insight and insistence of Hagedorn at CERN, SIGMA developed the array concept and array handling mechanisms much further than its one-index predecessors. Today we can say with hindsight, that SIGMA developed an array algebra mechanism that has not been matched by any of its successors.

The basic data type of SIGMA is an n-index rectangular array (Riemannian manifold). Each index i_k spans its own index range

$$1 \leq i_k \leq m_k \tag{10}$$

The shape of an array is defined by the **shape vector** of the array that contains the m_k for each index (all indices start with 1). This vectors determines the number of components in each dimension, hence each index is often called a dimension) and SIGMA refers to it as the NCO (Number of COmponents) vector. For example, the array with three rows and five columns

$$
\begin{array}{llllll}
Z & = & 11 & 12 & 13 & 14 & 15 \\
  &   & 21 & 22 & 23 & 24 & 25 \\
  &   & 31 & 32 & 33 & 34 & 35
\end{array} \tag{11}
$$

has two indices so that its NCO vector is

$$NCO(Z) = 3\,5 \tag{12}$$

Single numbers (scalars) are represented by a one-index, one-component array. The array name without subscripts, by convention, represents the first element of the array so that if the array contains only one element, we can still write

$$Pi = 3.14159 \tag{13}$$

to define what we consider as scalar values in the usual way. For two arrays of the same shape, arithmetical or conditional or logical operations are performed independently on corresponding element pairs as in Culler-Fried, AMTRAN or GAMMA, but, in addition, SIGMA introduces the notion of **compatible arrays**:

Two arrays are compatible, if their shapes are made identical by duplication of maximal subarrays of one or both arrays.

For example, if we generate Y as the five component vector

$$
\begin{array}{lll}
Y & = & ARRAY(5, 1\#5) \\
\end{array} \tag{14}
$$
$$
\begin{array}{lll}
Y & = & 1\,2\,3\,4\,5
\end{array} \tag{15}
$$

and X as the 3 by 1 array

$$
\begin{array}{lll}
X & = & ARRAY(3\&1, 1\#3)
\end{array} \tag{16}
$$
$$
\begin{array}{lll}
X & = & 1
\end{array} \tag{17}
$$
$$
2 \tag{18}
$$
$$
3 \tag{19}
$$

then we can generate the previously given 3 by 5 array Z with the statement

$$Z = 10 * X + Y \tag{20}$$

where the one-element array (10) is expanded by subarray duplication to a column of three 10's to perform component by component multiplication with X. Then the 3 by 1 column is replicated five times and the 1 by 5 row is replicated three times to produce two 3 by 5 arrays which are added componentwise to produce a 3 by 5 result.

Indexing

Sigma permits the use of one-dimensional arrays as indices, so that array shape can be reorganized in a few operations and mostly without explicit loops. Subsets of array values can be chosen and reshaped into new arrays. For example, if we have a three index array A, the expression

$$B = A(1\&3, 2, 1\&2\&1) \tag{21}$$

generates a new array B with the shape and values

$$\begin{matrix} A_{121} & A_{122} & A_{121} \\ A_{321} & A_{322} & A_{321} \end{matrix} \tag{22}$$

The Row Principle

If we assume that each index of an array describes an independent dimension of some problem, then the vector operations defined by Culler-Fried (such as DIFF or smoothing or numerical integration) should apply to only one dimension at a time. SIGMA chose the rightmost index (the 'row-index') as the default dimension to which these operators apply. For example, we can define a family of parabolas

$$f(a, x) = ax^2 \tag{23}$$

by generating a 100 component equally spaced row array for X and a 10 component column for A

$$X = ARRAY(100, -1\#1) \tag{24}$$
$$A = ARRAY(10\&1, 1\#10) \tag{25}$$

so that the array

$$Parabs = A * X * X \tag{26}$$

represents a family of ten parabolas as shown in Fig.3. The row-index represents the independent variable X and the column-index represents the parameter A.

Topological Combination of Arrays

Any pair of arrays can be expanded by subarray duplication to a common size that at worst has an NCO vector whose elements are products of the corresponding NCO vector element pairs of the two arrays. Even in problems of modest size such expansions can become very large. In many problems we are interested in specific subsets of the components of the expansion only. For example, matrix multiplication can be viewed as a contraction of the four-dimensional topological direct product of two two-dimensional arrays. Suppose that we have matrices A and B such that

$$\begin{aligned} NCO(A) &= 10 \quad 20 \\ NCO(B) &= 20 \quad 30 \end{aligned} \tag{27}$$

then the four-index topological direct product D of A and B will have 120,000 elements and a shape vector

$$NCO(D) = 10 \quad 20 \quad 20 \quad 30 \tag{28}$$

However, the matrix product C

$$C_{ik} = \sum_{j=1}^{m_j} A_{ij} B_{jk} \tag{29}$$

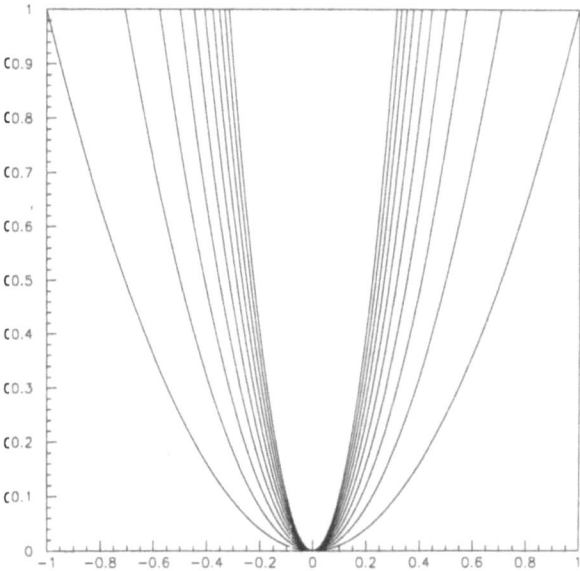

Figure 3. A simple way to draw families of curves

needs only those components for which the second and third index of D have equal values (6,000 instead of 120,000 components). The array algebra introduced by SIGMA defines array combination operations so that component explosion can be avoided and unneeded components are not generated. To construct the matrix product (which we can provide as a built-in operation for speed and convenience) we reshape the array A to give AA with unit row-dimension and with the same components as A

$$AA = ARRAY(NCO(A)\&1, \ A) \tag{30}$$

so that we have (any shape vector can be extended with leading ones)

$$\begin{aligned} NCO(AA) &= 10 \ \ 20 \ \ \ 1 \\ NCO(B) &= \ \ 1 \ \ 20 \ \ 30 \end{aligned} \tag{31}$$

If we multiply AA by B, compatibility expansion will generate D3 such that

$$\begin{aligned} D3 &= AA * B \tag{32} \\ NCO(D3) &= 10\,20\,30 \tag{33} \end{aligned}$$

The matrix product is obtained by taking the sum and contracting over the middle index. Such an operation is called a TRACE in SIGMA

$$C = TRACE(D3, 2) \tag{34}$$

We can combine the above steps to write matrix multiplication as a single statement

$$C = TRACE(((ARRAY(NCO(A)\&1, \ A)) * B), 2) \tag{35}$$

In this way, user control over the topological direct product expansion permits effective solutions to many problems that would be time or space-wise impossible with full direct products.

This insight to use the compatible array concept to control the direct product expansions is Hagedorn's most significant contribution to SIGMA. With today's large memories and single-instruction-multiple-data parallel computers, it has become a most promising approach to the efficient processing of large numerical array structures. Even today, 20 years after the introduction of SIGMA, the problem solving power of compatible arrays has not been fully developed because, as we will see in subsequent sections, array research took a different direction under the influence of APL and the recursive list structures championed by LISP.

OTHER INTERACTIVE SYSTEMS

Over the last twenty years many interesting interactive systems have come and gone without leaving much of a ripple in the history of computer science. Interactive system designers and developers seem to prefer a more isolated and self-centered existence than most computer scientists. A glance at the reference lists of their papers shows their strong belief in their own version of the holy grail as well as their reluctance to compare their ideas and results with those of others. Under such circumstances it would take an extraordinary effort to do historical justice to all developments and we do not make any such attempt in this paper. We will restrict our discussion to those systems that influenced the development of SIGMA and to the array models that have influenced the interactive systems that are most popular today.

APL

In 1962 Iverson[9] published a new notation that was intended as a bridge between conventional mathematical notation and conventional programming language notation. The title of Iverson's book was "A Programming Language", so the notation became known as APL. In 1966 Breed, Lathwell and Falkoff[10] implemented APL on IBM System/360 computers. Many other implementations and an enthusiastic and fiercely loyal user community developed around APL. APL is still in use today.

APL introduced n-dimensional rectangular arrays, but without the array algebra and compatibility notions of SIGMA. Instead, APL introduced higher order functions that today's functional programmers call "fold" and "map" and APL generalized the inner and outer product concept of vector analysis and provided powerful shape restructuring operations. APL inspired the development of a theory of arrays by More[11] who considered the array as a recursive data structure so that array elements were allowed to be arrays of any shape or size to any depth of recursion. This array model does not have an obvious mapping to a parallel computer architecture and it does not fare well with the notions of strong typing. It is similar to the typeless list structures of Lisp or Prolog.

More's array theory inspired Mullin's[12] "Mathematics of Arrays" which is a mathematical reasoning system for algorithm research involving flat arrays of numbers. Mullin's MOA comes closest to SIGMA arrays in structure but it does not exploit the advantages generated by the conformable array concept.

Mathematica

Perhaps the most popular interactive system of today is Wolfram's[13] Mathematica. Mathematica combines symbolic expression manipulation capabilities with numerical capabilities. Arbitrary precision integer arithmetic and complex number handling are built-in. Arrays are implemented as lists. List handling concepts dominate. An array is a second class citizen, regarded as a list with an indexing function. Mathematica provides an extensive and powerful collection of mathematical problem solving methods, but its array handling capabilities are restricted by its basic list orientation. Mathematica has a path to C-programs, so an interesting research problem would be to investigate how to extend Mathematica with SIGMA style arrays.

Maple

Maple[14] from Waterloo, Canada is a powerful and reliable interactive system for doing mathematics on computers. Maple is particularly strong on symbolic calculations such as indefinite integrals or linear equation solutions. Its numerical arrays are listbased as in Mathematica.

Matlab

... in tens of thousands of industrial, government and academic settings MATLAB has become the premier software package for interactive, numeric computation, data analysis and graphics ...

This is how Matlab 'modestly' introduces itself on page (xi)[15]. Arrays in Matlab are restricted to a maximum of two dimensions. Vectors of index values may be used as subscripts so that new matrices may be extracted from existing matrices without explicit loops. The model of computation of Matlab is strongly influenced by the mathematics of matrices and column and row-vectors.

Graphic Capabilities

Mathematica, Maple and Matlab provide versatile and powerful graphic display capabilities, but the user has to find the right combination of options and alternatives by hunting and pecking through a large offering of menu driven possibilities.

CONCLUSIONS

Over the last twenty years, several array management models and notations have been developed and implemented. The conformable array concept of SIGMA was the basis of one of the first such implementations. Its close correspondence to the Single Instruction Multiple Data concept of parallel computer architectures, together with its powerful, yet simple, curve display notation, should awaken new interest in research on SIGMA style array algebras as well as on their implementation on SIMD parallel computers.

The array algebra of SIGMA provides the user with sensitive control over topological direct product expansions as well as over component-wise independent and row-wise independent operations. As Culler-Fried showed, many ingenious problem solutions can be built without too much tedious work from such building blocks. Today's most popular interactive systems favor large libraries of packaged solutions to much larger problem components or to whole problems from which it is sometimes impossible to build solutions for much larger problems. Perhaps we should agree with mathematician and philosopher A. N. Whitehead[16] who said

"It is a profoundly erroneous truism, repeated by all copy books and by eminent people when they are making speeches, that we should cultivate the habit of thinking of what we are doing. The precise opposite is the case. Civilization advances by extending the number of important operations which we can perform without thinking about them."

On the other hand, perhaps we should increase our research efforts to find array algebras that permit versatile yet powerful formulation of parallel solutions to interesting problems.

Acknowledgments

The author acknowledges the contributions of the pioneers of interactive computing on whose ideas we built AMTRAN and SIGMA. Many thanks to Bob Seitz and the Huntsville AMTRAN Team, the Georgia AMTRAN Team who met in 1994 for the 25th-year reunion of that work, Carlo Vandoni and the CERN SIGMA Team, especially the guidance and contributions of Rolf Hagedorn and Leon Van Hove in the development of the array algebra that has obtained new significance with the rise in importance of parallel computing on SIMD computer architectures. Special thanks to Richard Miller for the Saskatchewan implementation. A big thank you to the many users, installers, modifiers, improvers and maintainers of SIGMA over the last twenty years.

REFERENCES

1. M. B. Wells, *MADCAP: A Scientific Compiler for a Displayed Formula Textbook Language*, *CommACM* Vol.4, No.1, pp.31-36 (1961).

2. B.D.Fried, *On the User's Point of View*, pp.11-21, *Interactive Systems for Experimental Applied Mathematics*, Eds. M.Klerer and J.Reinfelds, Academic Press, NY (1968).

3. G.J.Culler, *Mathematical Laboratories: A new Power for the Physical Sciences*, pp.355-384, *Interactive Systems for Experimental Applied Mathematics*, Eds. M.Klerer and J.Reinfelds, Academic Press, NY (1968).

4. L.H.Woods, J.Reinfelds, R.N.Seitz, P.L.Clem, *The AMTRAN System*, *Datamation* Vol.12, No.10, pp.22 (1966).

5. J.Reinfelds, *AMTRAN 70*, pp.370-375, Proceedings IFIP'71, Ljubljana, Yugoslavia, North Holland, Amsterdam (1972).

6. R.Hagedorn, J. Reinfelds, C.E.Vandoni, L.Van Hove, *SIGMA, A New Language for Interactive Array-Oriented Computing*, CERN 78-12, Geneva, Switzerland (1978).

7. J.Reinfelds, C.E.Vandoni, *SIGMA 76*, pp.963-968, *Information Processing 77*, Ed. B.Gilchrist, North Holland, Amsterdam (1977).

8. R.Miller, *SIGMA Implementation Report*, Mathematics Department, University of Saskatchewan, Saskatoon, Canada (1974).

9. K.E.Iverson, *A Programming Language*, Wiley, NY (1962).

10. L.M.Breed, R.H.Lathwell, *Implementation of APL/360*, pp.390-399, *Interactive Systems for Experimental Applied Mathematics*, Eds. M.Klerer and J.Reinfelds, Academic Press, NY (1968).

11. T.More Jr., *Axioms and Theorems for a Theory of Arrays*, *IBM J.Res.Develop.*, Vol.17, No.2, pp135-175 (1973).

12. M.A.Jenkins, L.M.Mullin, *A Comparison of Array Theory and A Mathematics of Arrays*, pp.237-268, *Arrays, Functional Languages and Parallel Systems*, Eds. L.M.R.Mullin et al., Kluver Academic Publishers, Boston (1991).

13. S. Wolfram, *Mathematica, A System for Doing Mathematics by Computer*, Addison Weley, Reading (1988).

14. M.L.Abell, J.P.Braselton, *The Maple V Handbook*, Academic Press, NY (1994).

15. Mathworks, *The Student Edition of MatLab*, Prentice Hall, Englewood Cliffs (1992).

16. K.W.Smillie, *APL and the Teaching of Probability*, unpublished Lecture Notes, UNSW, Australia (1979).

DECONFINEMENT: CONCEPT, THEORY, TEST

Helmut Satz

Theory Division
CERN
CH-1211 Geneva 23, Switzerland
and
Fakultät für Physik
Universität Bielefeld
D-33501 Bielefeld, Germany

INTRODUCTION

The study of strongly interacting matter is a fascinating subject for a variety of reasons. It is needed to describe the first few microseconds in the evolution of our universe, and it provides the equation of state for stellar matter at high density. It gives us the statistical closure of whatever strong interaction dynamics we consider, from pion exchange to QCD. As such, it was and remains a laboratory to investigate the critical behaviour arising from new forms of dynamics. Hagedorn's ultimate temperature of hadronic matter[1] is one particularly striking result obtained in this laboratory – and one which did much to further our understanding of dense matter. In the past decade, two new developments made strong interaction thermodynamics even more attractive.

On the theoretical side, large-scale computer simulations allow for the first time in physics an *ab initio* calculation of the thermodynamic behaviour of an actual physical system. From QCD as input dynamics, we can today evaluate directly, i.e., without intermediary models, the states of matter and the critical behaviour of strongly interacting systems. The most important result which has so far come out of these studies is a quantitative confirmation of the idea that matter at sufficiently high density will undergo a transition to a state of deconfined quarks and gluons.

On the experimental side, an extensive program of high energy nuclear collision studies was initiated, with the specific aim of investigating the critical behaviour and the new deconfined state of strongly interacting matter under controllable laboratory conditions. The collision of two heavy nuclei will, as we hope, provide droplets of such matter sufficiently large, hot and thermalised to form the quark-gluon plasma predicted by statistical QCD. The experiments carried out so far, using beams of rather light nuclei on heavy targets, have shown that the analysis of very high multiplicity final states is feasible, and they have led to a number of results which seem to indicate that we are indeed on the road to strong interaction thermodynamics.

In this survey of deconfinement, we will first have a look at the physical concepts which led to the phenomenon, and then turn to statistical QCD and to the results obtained from it by computer simulation on the lattice. Finally we will indicate how one can test if the matter produced in high energy nucleus-nucleus collisions indeed consists of deconfined quarks and gluons.

COLOUR SCREENING IN DENSE MATTER

An increase in the energy density of a pion gas leads not only to the production of more pions, but also to abundant resonance production. One way to accommodate this is to consider the thermodynamics of an ideal resonance gas.[2] Its partition function is

$$\ln Z_R(T, V) = \frac{VT^3}{\pi^2} \int dm\, \rho(m)\, e^{-m/T},$$ (1)

where $\rho(m)$ describes the resonance mass spectrum; for $\rho(m) = \delta(m - m_\pi)$, we get back the ideal pion gas. To fix it in general, we need an input from hadron dynamics. Hagedorn[1] used a composition law, the statistical bootstrap model, requiring resonances to be composed of resonances in a universal way. It yields

$$\rho(m) \sim m^a e^{bm},$$ (2)

with the constant a dependent on the details of the model, while b is fixed by the lowest hadron mass, m_π, giving $b \simeq 150$ MeV. The most detailed model for hadronic interactions, the dual resonance model,[3] assumes the interaction amplitude to be built up by resonances in all channels; it leads to the same form (2). In this case, b is determined by the Regge slope governing the spin degeneracy of resonances; the resulting value agrees remarkably well with that from the statistical bootstrap.[4] A similar form had previously also been obtained through geometric arguments;[5] here b was determined through the size of the basic hadrons, leading again to a value near 150 MeV.

Inserting the spectral form (2) into the partition function (1), we note that the integral diverges for $T > T_H \equiv 1/b$. What is the significance of this temperature? Hagedorn had initially proposed it to be the ultimate temperature of matter, since for the values of a he had obtained, $T \to T_H$ was possible only for an energy density $\epsilon \to \infty$. However, it was noted soon afterwards[6] that for slightly smaller values of a, ϵ remained finite at $T = T_H$; only higher derivatives of the partition function diverged, suggesting a transition to a new phase of matter.

Whatever interpretation we choose, we are left with the conclusion that there is an intrinsic limit to hadron thermodynamics; appearently something must happen at the Hagedorn temperature $T_H \simeq 150$ MeV. Today we consider hadrons as bound states of quarks, and it therefore seems natural to associate T_H to a transition from hadronic to quark matter.

This raises some questions. What happens to colour confinement in such a transition? How can the binding of quarks be dissolved, when the confinement potential $V_{conf}(r) \sim r$ diverges as $r \to \infty$?

The mechanism responsible for colour deconfinement is charge screening in a dense medium. The Coulomb potential between two electric charges becomes Debye-screened in a dense ionised gas,

$$\frac{e^2}{r} \to \left(\frac{e^2}{r}\right) e^{-\mu_D r},$$ (3)

with a screening radius $r_D = 1/\mu_D$. For the same reason, the colour potential will experience screening in a medium of colour charges,

$$\sigma\, r \;\rightarrow\; \left(\frac{1 - e^{-\mu r}}{\mu\, r}\right)\sigma\, r, \tag{4}$$

with $r = 1/\mu$ for the colour screening radius. The effect of this screening is seen in Fig. 1. With increasing colour charge density, μ increases (just as μ_D does for an increasing density of electric charges), and so eventually any bound state is dissolved. At low density, we thus have a *colour insulator* : hadronic matter, consisting of colour neutral bound quark states. At high density, this turns into a *colour conductor* : the quark-gluon plasma, whose constituents are unbound coloured quarks and gluons. Colour deconfinement thus seems to be the QCD

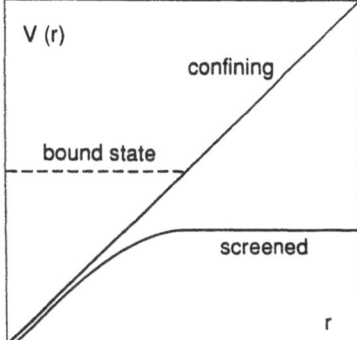

Figure 1. Effect of screening on a confining potential.

analogue of the insulator-conductor transition in QED. And if we assume the transition to occur when the screening radius has reached the typical hadronic scale of 1 fm, we find[7] a transition temperature of about 150 MeV.

What happens to the mass of the constituents in this transition? We know that the effective mass of the electron in a metal is not its vacuum mass – it becomes modified by medium effects (lattice vibrations, electron cloud). For strong interactions, with "bare" quark masses $m_q \simeq 0$ in the QCD Lagrangian, we have in the confined phase constituent quarks with an "effective" mass $m_q^{\text{eff}} \simeq m_p/3 \simeq m_\rho/2$. In this confined state, the chiral symmetry of the Lagrangian with massless fermions must therefore be spontaneously broken. At sufficiently high density, we expect the gluon dressing to come off, turning massive constituent quarks into massless current quarks and thereby restoring chiral symmetry.

For a Lagrangian with $m_q=0$, we thus expect a shift in the effective quark mass leading to chiral symmetry restoration, as a further phase transition in strongly interacting matter. Deconfinement and chiral symmetry restoration can, but need not coincide. In atomic physics, we can have an insulator turning with inceasing density into a superconductor, which for still higher density becomes a conductor. The first transition, decoupling electron-ion binding, corresponds to deconfinement; however, the liberated electrons don't remain free, but couple to Cooper pairs. In the second transition, the Cooper pairs are broken up and the electrons obtain their "metallic" effective mass. In QCD, we could correspondingly have hadronic matter, constituent quark matter, and the quark-gluon plasma. Since constituent quarks, just like Cooper pairs, are probably easy to break up by thermal motion, such a constituent quark state may well exist only at low temperatures and high baryon number density. At zero baryon number density, deconfinement and chiral symmetry restoration occur at the same temperature, as lattice calculations (see the next section) show.

The schematic phase diagram of strongly interacting matter, as it emerges from these considerations, is shown in Fig. 2.

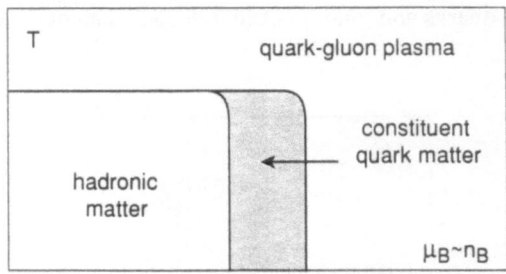

Figure 2. Schematic phase diagram for strongly interacting matter.

STATISTICAL QCD AT FINITE TEMPERATURE

Statistical QCD allows a *direct* determination of the phase structure of strongly interacting matter, starting from the QCD Lagrangian as dynamical input,

$$\mathcal{L} = -\frac{1}{4}(\partial_\mu A_\nu^a - \partial_\nu A_\mu^a - g f_{bc}^a A_\mu^b A_\nu^c)^2 - \sum_f \bar{\psi}_\alpha^f (i\gamma^\mu \partial_\mu + m_f - g\gamma^\mu A_\mu)^{\alpha\beta} \psi_\beta^f, \quad (5)$$

with m_f for the masses of N_f current quark species. The basis of thermodynamics is the QCD partition function, most conveniently written in Euclidean form as functional integral over gluon (A) and quark (ψ) fields,

$$Z(T,V) = \int DA \, D\psi \, D\bar{\psi} \, e^{-S(A,\psi,\bar{\psi})}, \quad (6)$$

where the action

$$S(A,\psi,\bar{\psi}) = \int_0^{1/T} d\tau \int_V d^3x \, \mathcal{L}(\tau = ix_0, \vec{x}) \quad (7)$$

is calculated directly from the Lagrangian $\mathcal{L}$. Derivatives of $Z(T,V)$ give us the thermodynamic quantities of interest, such as the energy density ϵ,

$$\epsilon(T) = \left(\frac{T^2}{V}\right)\left(\frac{\partial \ln Z}{\partial T}\right)_V \quad (8)$$

or the pressure P,

$$P(T) = T\left(\frac{\partial \ln Z}{\partial V}\right)_T. \quad (9)$$

Equations (5) - (7) thus constitute the complete basis of statistical QCD. We are left "only" with the question of how to carry out the calculation of the partition function (6/7).

So far, the computer simulation of the lattice formulation[8] is the only generally viable method for this, applicable in particular also in the critical region, where perturbative methods break down. Lattice QCD is obtained through three steps. First, we replace the space-time continuum $\{\tau = ix_0, \vec{x}\}$ in eq. (7) by a lattice of $N_\tau \times N_\sigma$ discrete points, with $\{\tau = N_\tau a;\ x_i = N_\sigma a,\ i = 1, 2, 3\}$ and an isotropic lattice spacing a; hence $T = 1/(N_\tau a)$, $V = (N_\sigma a)^3$. Next, the quark and antiquark field integrations are carried out, which is possible since they enter the integral in the form $\exp(-\bar{\psi} Q_F \psi)$; this leads to a factor $(det\ Q_F)$ in the remaining integral over the gluon fields A.[9] In the third step, these integrations are changed to the (compact) variables $U = \exp(ixA)$, where U denotes a matrix of the SU(3) colour gauge group of QCD; such variables U are associated to the links between any two points on the lattice. After these steps, the partition function (6) has the structure

$$Z(T, V) = \int \prod_{\text{links}} dU\ e^{-S(U)} \tag{10}$$

of a "generalized spin problem", something like an Ising model partition function, with SU(3) matrices U instead of spins, and with these on lattice links instead of lattice sites. The QCD action $S(U) = S_G(U) + S_F(U)$ consists of the gluon part $S_G(U) \sim (1 - UUUU)$ and the quark part $S_F(U) \sim \log(det\ Q_F)$. The product of four "spins" U on a closed loop of four links ("plaquette") in the gluon action assures local gauge invariance,[10] which the Ising action with its product of two adjacent spins does not have. This completes the lattice formulation. The results which we want to obtain from it are supposed to hold in the continuum limit, in which the number of lattice points becomes infinite while the lattice spacing simultaneously goes to zero, keeping T and V constant.

Now we want to study the critical behaviour arising in statistical QCD. Since phase transitions are in statistical physics quite generally associated with changes in symmetry, we look for the symmetries of the QCD Lagrangian $\mathcal{L}$ or of the QCD action $S(U)$ in eq. (10).

Consider the limit of very heavy quarks: for $m_f \to \infty$, $S_F \to 0$, and we get the thermodynamics of pure SU(3) gauge theory. This pure gluon system already contains many of the crucial features of QCD, quite in contrast to QED, where the pure photon system forms an ideal boson gas. The reason for the difference is the term $gA_\mu A_\nu$ in eq. (6), which allows gluons, unlike photons, to interact directly, without intermediate fermions.

If we now carry out the "spin flip"

$$U(n_0, \vec{n}) \to U'(n_0, \vec{n}) = zU(n_0, \vec{n}) \quad \forall\ \vec{n}, \tag{11}$$

where $z = \exp(i\pi r/3)$, $r = 0, 1, 2$, is an element of the center Z_3 of the SU(3) gauge group, then this leaves the action $S_G(U)$ invariant. Such an operation is just the analog of flipping all spins $s_i \to -s_i$ in the Ising model, where it also leaves the Ising action invariant. However, the "generalized spin"

$$L(U) \equiv \prod_{n_0=1}^{N_\tau} U(n_0, \vec{n}) \tag{12}$$

is "flipped" by this operation: $L(U') = zL(U)$. Its average, $\langle L \rangle$, therefore measures a possible spontaneous breaking of the Z_3 symmetry of the state of matter in which the system is, just as the average spin $\langle s \rangle$ in the Ising model indicates if the Ising system is in an ordered state with spontaneous magnitization or in a disordered state with $\langle s \rangle = 0$.

The transition from unbroken to broken Z_3 symmetry corresponds in fact to deconfinement. The order parameter $\langle L \rangle$ can be related to the potential $V(r)$ between a static quark and antiquark in the limit of infinite separation,

$$\langle L \rangle \sim \lim_{r \to \infty} e^{-V(r)/T}. \tag{13}$$

For a confined state, $V(r)$ diverges for $r \to \infty$, so that $\langle L \rangle = 0$; therefore Z_N-symmetry signals confinement. In a deconfined state, $V(r)$ remains finite as $r \to \infty$, since colour screening cuts out the diverging (and hence confining) long distance part of the potential; thus the spontaneous breaking of the Z_N symmetry indicates deconfinement. Therefore the "generalized spin" $\langle L \rangle$ (in lattice gauge theory usually referred to as Polyakov loop) constitutes the theoretical probe for confinement or deconfinement. As noted, this is strictly true only in the limit of infinite quark mass, i.e., in pure $SU(N)$ gauge theory. For finite quark mass, thermal pair production leads to string breaking and hence to a finite $\langle L \rangle$ for all $T \neq 0$. This is quite similar to the actual insulator-conductor transition, where thermal ionisation also prevents the conductivity from vanishing in the insulator phase for $T \neq 0$. In both cases, however, the order parameter remains exponentiallly small in the "symmetric" phase: $\langle L \rangle \sim \exp(-m_h/T)$, with $m_h \sim m_\rho$ as typical hadron mass, corresponds to the conductivity $\sigma \sim \exp(-\Delta E/T)$, where ΔE denotes the ionisation energy.

We now turn to the already mentioned shift in the mass of the constituents in connection with deconfinement and consider the Lagrangian in the limit of vanishing quark mass, $m_f = 0$. The four-spinor fields for the quarks then decompose into a direct sum of a left-handed and a right-handed two-spinor, $\psi^{(4)} \to \psi_L^{(2)} + \psi_R^{(2)}$. The chiral "rotation" between these,

$$\psi \to \psi' = e^{i\alpha\gamma_5}\psi, \tag{14}$$

leaves invariant the $m_f = 0$ Lagrangian, since $\bar{\psi}'\gamma_\mu\psi' = \bar{\psi}\gamma_\mu\psi$. However, the mass term in the original $m_f \neq 0$ Lagrangian would not remain invariant, since $\bar{\psi}'\psi' = \bar{\psi}e^{2i\alpha\gamma_5}\psi \neq \bar{\psi}\psi$. Hence $\langle\bar{\psi}\psi\rangle$ measures if the system is in a chirally symmetric state or not. If $\langle\bar{\psi}\psi\rangle \neq 0$, the chiral symmetry of the state is spontaneously broken, the quarks have acquired a non-vanishing effective mass. For $\langle\bar{\psi}\psi\rangle = 0$, the chiral symmetry is restored and the constituents of the systems are quarks with a vanishing effective mass, i.e., the current quarks of the massless Lagrangian. In the real world, the quarks in the Lagrangian cannot be massles, since the pion mass $m_\pi \sim m_q^2$ is not zero, so that we have approximate chiral symmetry only.

Statistical QCD thus leads to two critical phenomena, deconfinement and chiral symmetry restoration. Deconfinement is associated to a global Z_3 symmetry of the Lagrangian for $m_f \to \infty$, with $\langle L \rangle$ as order parameter; the chiral symmetry of the Lagrangian for $m_f \to 0$ is measured for a given state of the system by the order parameter $\langle\bar{\psi}\psi\rangle$. Although in the real world both these symmetries are only approximate, we nevertheless believe that their remnant effects will show up in the transition from low to high density behaviour of strongly interacting matter.

So the basic quantities to be studied in statistical QCD are the deconfinement measure $\langle L \rangle$, the chiral symmetry measure $\langle\bar{\psi}\psi\rangle$, the energy density ϵ, the pressure P, and thermo-dynamic quantities derived from these, such as entropy, specific heat, and others. The actual calculation of these "observables" is highly non-trivial, since we are studying a relativistic field theory for an interacting system near a critical point, where perturbative methods are not applicable. As already mentioned, the only viable evaluation method in this region is the computer simulation[11] of the lattice formulation of the problem:[8] one creates on a sufficiently large and fast computer a "world according to QCD" and then "measures" in this world the quantities of interest; for a recent review, see.[12] One draw-back of this method is that it can be applied up to now only to systems of zero over-all baryon number density, i.e., only to hot matter at $n_B = 0$, not to dense matter at $T = 0$; another is that its precision is limited by computer size and speed. Nevertheless, it provides us with the unique chance to calculate the critical behaviour directly from the underlying fundamental theory, and the mentioned shortcomings will hopefully be removed or reduced in the future.

The main results from the computer simulation of lattice QCD at finite temperature

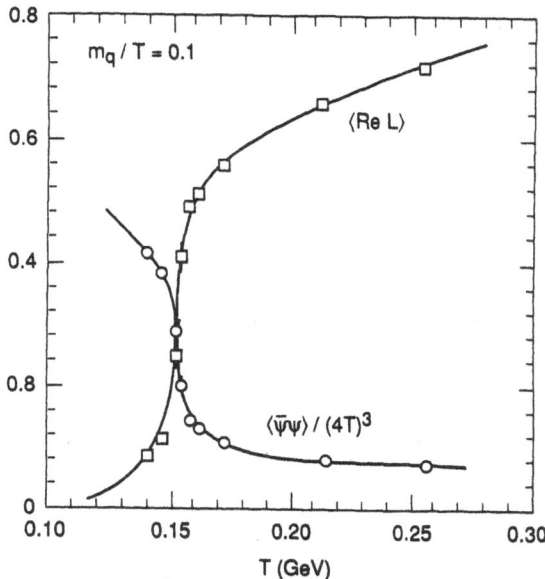

Figure 3. Temperature dependence of deconfinement measure $\langle Re\ L \rangle$ and chiral symmetry measure $\langle \bar{\psi}\psi \rangle$, from.[13]

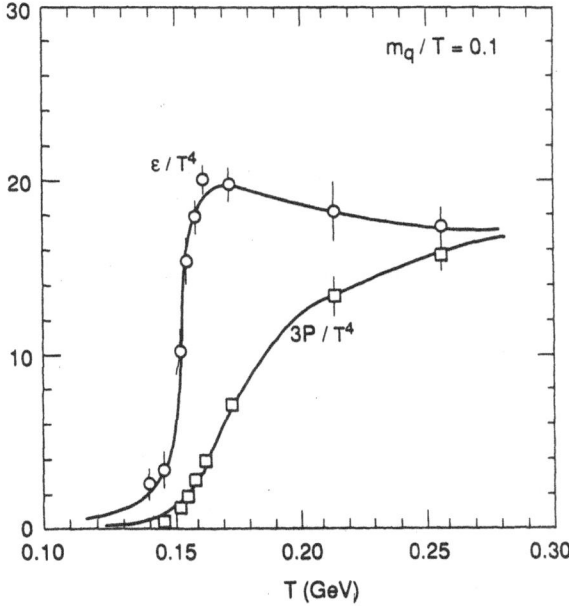

Figure 4. Temperature dependence of energy density ϵ and pressure P, from.[13]

are summarized in Figs. 3 - 4, which were obtained in recent calculations for two light (staggered) quark species on $8^3 \times 4$ and $16^3 \times 4$ lattices.[13] In fig. 3, we see abrupt changes in the deconfinement measure $\langle L \rangle$ and the chiral measure $\langle \bar{\psi}\psi \rangle$ at the same temperature $T = T_c \simeq 150$ MeV; hence for $n_B = 0$, the two critical phenomena coincide. We also note that neither order parameter is ever really zero; for $\langle L \rangle$ this is due to finite lattice size and finite quark mass, for $\langle \bar{\psi}\psi \rangle$ due to the non-zero quark mass. In spite of this, both quantities show clear transition signals and thus allow the definition of a critical temperature T_c. In Fig. 4, we see that at this temperature, the dimensionless energy density measure ϵ/T^4 increases abruptly from a value near that of an ideal pion gas ($\simeq 1$) to one near that of an ideal quark-gluon plasma ($\simeq 20$). This increase in the energy density $\Delta\epsilon$ represents something like the "latent heat of deconfinement". We also observe that the pressure grows with temperature at a much slower rate, and the ideal gas relation $\epsilon = 3P$ appears to become fulfilled only for $T \geq T_c$. Hence in the region $T_c \leq T \leq 2T_c$, there are definite interaction effects in the plasma. In spite of a number of model considerations (see e.g.[14] and further references there), their origin still constitutes at present one of the most interesting open problems in finite temperature lattice QCD.

We should note here that the value of the critical temperature in physical units ($T_c \simeq 150$ MeV) is obtained by calculating T_c in units of the ρ-mass and then using the experimental value of this mass.

Such a value of T_c means that an energy density $\epsilon_c \simeq 1.0$ GeV/fm^3 is needed in order to reach deconfinement. Energy densities of this magnitude existed in the early universe for the first 10 - 20 microseconds after the big bang, and it is hoped to obtain them in small volumes and for short times in high energy nuclear collisions. Abundant multiple scattering among the incident nucleons and the many secondaries produced in the collisions is expected to provide a rapid increase in entropy sufficient to reach thermal equilibrium. And for high enough incident energies (15 -20 GeV cms energy per nucleon-nucleon collision), the initial energy density will be above the deconfinement threshold of 1 GeV/fm^3.

CONFINEMENT/DECONFINEMENT IN NUCLEAR COLLISIONS

The ultimate aim of high energy nuclear collisions is thus to produce and study the predicted deconfined state of strongly interacting matter. In whatever state matter is produced in such collisions, it will evidently consist of quarks and gluons – the crucial feature to check is whether they are deconfined. This can be done indirectly, using the equation of state of the quark-gluon plasma together with a specific evolution pattern as input and then looking for the consequences this leads to for the hadrons emitted at the end of the collision process. On the other hand, we can try to check directly if the quarks and gluons in the primordial state are confined or not. This, however, requires hard probes, capable of resolving the short-distance structure of the quark-gluon plasma; it cannot be achieved by studying the usual large hadrons formed at freeze-out. In this section we want to use J/ψ production to show how confined and deconfined matter can be distinguished by hard probes.[15]

For matter in a confined state, the constituents will be hadrons; for simplicity, consider an ideal gas of pions. Their momentum distribution will be thermal, i.e., for temperatures not too low it will be given by $\exp(-E_\pi/T) \simeq \exp(-p_\pi/T)$. Hence the average momentum of a pion in this medium is $\langle p_\pi \rangle = 3\,T$. The distribution of quarks and gluons within a pion is known from structure function studies; the gluon density is $g(x) \simeq 0.5(1-x)^3$ for large $x = p_g/p_\pi$.* As a consequence, the average momentum of a gluon in confined matter is given

---

*For very small x, recent results from HERA indicate a steeper increase; however, this does not effect our considerations here.

100

by

$$\langle p_g \rangle_{\text{conf}} = \frac{1}{5}\langle p_\pi \rangle = \frac{3}{5}T. \tag{15}$$

Hence in a medium of temperature $T \simeq 0.2$ GeV, the average gluon momentum is around 0.12 GeV.

In contrast, the distribution of gluons in a deconfined medium is directly thermal, i.e., $\exp(-p_g)$, so that

$$\langle p_g \rangle_{\text{deconf}} = 3T. \tag{16}$$

Hence the average momentum of a gluon in a deconfined medium is five times higher than in a confined medium[†]; for $T = 0.2$, it becomes 0.6 GeV. We thus have to find a way to look for such a hardening of the gluon distribution in deconfined matter.

The lowest $c\bar{c}$ vector state, the J/ψ, provides an ideal probe for this. It is very small, with a radius $r_\psi \simeq 0.2$ fm $<< \Lambda_{QCD}^{-1}$, so that J/ψ interactions with the conventional light quark hadrons probe the short distance features, the parton infra-structure, of the latter. It is very strongly bound, with a binding energy $\epsilon_\psi \simeq 0.65$ GeV $>> \Lambda_{QCD}$; hence it can be broken up only by hard partons. Since it shares no quarks or antiquarks with pions or nucleons, the dominant perturbative interaction for such a break-up is the exchange of a hard gluon.

We thus see qualitatively how a deconfinement test can be carried out. If we put a J/ψ into matter of temperature $T = 0.2$ GeV, then

- if the matter is confined, $\langle p_g \rangle_{\text{conf}} \simeq 0.12$ GeV, which is much too soft to resolve the J/ψ as $c\bar{c}$ as a bound $c\bar{c}$ state and much less than the binding energy ϵ_ψ, so that the J/ψ survives;

- if the matter is deconfined, $\langle p_g \rangle_{\text{decon}} \simeq 0.6$ GeV, which with some momentum spread is hard enough to resolve the J/ψ and to break the binding, so that the J/ψ will disappear.

Our arguments thus provide a dynamical basis for J/ψ suppression by colour screening,[16] and they indicate in fact that J/ψ suppression in dense matter will occur if and only if there is deconfinement.

To put these arguments on a firm theoretical basis, we need the cross section for the inelastic interaction of a J/ψ with a light hadron h, $\sigma_{\psi h}$, from which we can then calculate if a J/ψ can be broken up on its passage through hadronic matter. Because of the small radius and large binding energy of the J/ψ, $\sigma_{\psi h}$ can be calculated in the short-distance analysis of QCD.[17,18] The crucial feature for this calculation is the fact that heavy and tightly bound quarkonium states can be broken up by scattering on usual light hadrons only through the exchange of hard gluons. The momentum distribution of gluons within such a light hadron incident on a J/ψ, $g(x)$, with $x = p_g/p_h$, is a universal non-perturbative input, determined e.g. from parton counting rules or from deep inelastic processes. It has for large x in general the form $g(x) \simeq (1-x)^k$, with $k \simeq 3$ for mesons, $k \simeq 4$ for nucleons. The resulting $J/\psi - h$ cross section then becomes[15]

$$\sigma_{\psi h}(s) \simeq 3 \text{ mb} \, (1 - \lambda_o/E_h)^{k+5/2}, \tag{17}$$

with $E_h = (m_h^2 + p_h^2)^{1/2}$ and $\lambda_o \simeq (\epsilon_\psi + m_h)$. For low collision energies $\sqrt{s}$, the cross section is thus determined by the behaviour of the gluon distribution at large x, as already noted above, and this leads to a very slow growth from threshold towards the asymptotic value of 3 mb. The functional form of this behaviour is shown in Fig. 5.

Only in the limit of large quark mass can the quarkonium-hadron cross section can be calculated rigorously in short-distance QCD. We thus have to ask if the charm quark mass is

---

[†]We could equally well assume matter at a fixed energy density, instead of temperature. This would lead to gluons which in case of deconfinement are approximately three times harder than for confinement.

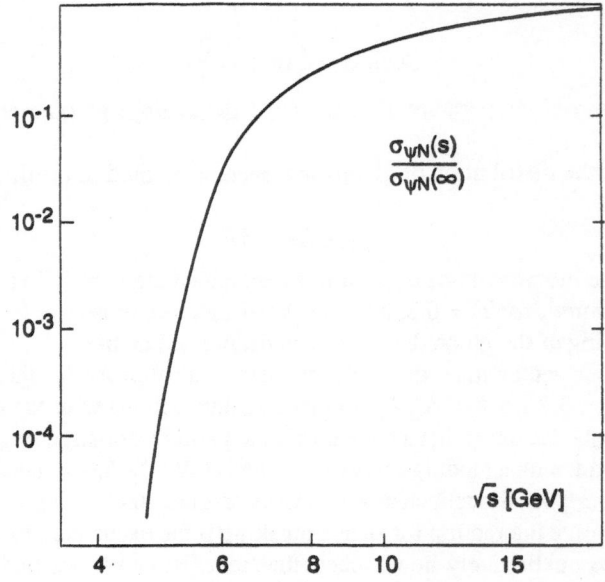

Figure 5. Energy dependence of the J/ψ-nucleon cross-section, from.[15]

sufficiently large to apply the results of heavy quark theory. For large M_Q, the quarkonium binding becomes completely Coulombic, and for Coulomb bound states, radius and binding energy are related by

$$\epsilon_{(Q\bar{Q})} = \frac{1}{M_Q r^2_{(Q\bar{Q})}}. \tag{18}$$

For the J/ψ parameters $\epsilon_\psi \simeq 0.65$ GeV, $r_\psi \simeq 0.2$ GeV and $M_c \simeq 1.5$ GeV, relation (18) is in fact quite well satisfied. For an empirical test, one can use the same approach to calculate the cross section for the photoproduction of open charm, $\sigma_{\gamma h \rightarrow c\bar{c}}(s)$. The result is

$$\sigma_{\gamma h \rightarrow c\bar{c}}(s) \simeq 1.2 \ \mu\text{b} \ (1 - \nu_o/\nu)^k, \tag{19}$$

with $2\nu = s - m_h^2$ and $2\nu_o = M_\psi^2 + 2M_\psi m_h$. For this process, there are data,[19] and in Fig. 6 it is seen that they agree well with the heavy quark theory result (19).

A direct test of eq. 17 will in fact also soon become possible. The ideal way to study the fate of J/ψ's in confined matter is in principle provided by hadron-nucleus collisions,

$$h \ A \ \rightarrow J/\psi \ + ..., \tag{20}$$

where the J/ψ is then observed by measuring its decay dimuons, $J/\psi \rightarrow \mu^+\mu^-$. In practice, however, the muons are identified by passing through an absorber, and for this they have to be sufficiently energetic. As a consequence, experiments in which a hadron beam is incident on a nuclear target lead to J/ψ's which are very fast ($x_F \geq 0$) in the rest frame of the nucleus and hence leave the nuclear medium long before ever becoming a fully formed physical resonances. Such experiments thus mainly study the behaviour of a coloured $c\bar{c}$ pair in the medium, and the behaviour of this system is quite different from that of a colour-singlet J/ψ. To test the interaction of the latter in the nuclear medium, we need J/ψ's in the momentum range $-1 \leq x_F \lesssim -0.3$, and as noted, these lead to dimuons too slow to be observable in the usual pA experiments. However, with the advent of the Pb-beam at the CERN-SPS, it will become possible to study J/ψ-production from a nuclear beam incident on a hydrogen target.

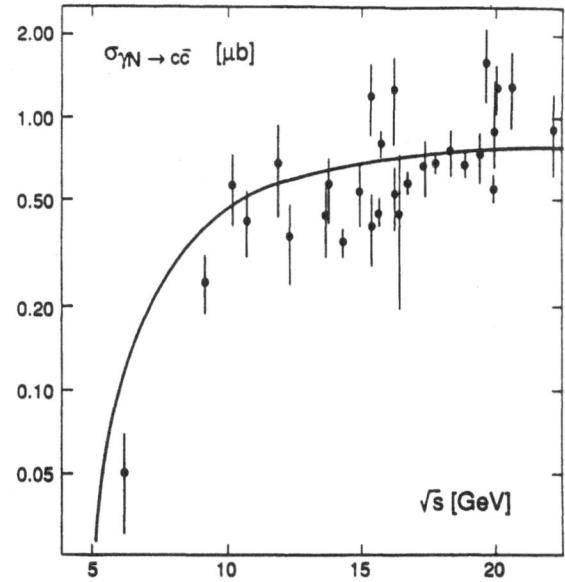

Figure 6. Energy dependence of open charm photoproduction, from.[15]

Fast J/ψ's in the lab system will then be slow in the nuclear rest frame and thus pass through the confined nuclear medium as fully formed physical resonances. The short-distance QCD analysis presented here predicts essentially no absorption for this passage, in contrast to a suppression of 25% or more if the asymptotic break-up cross-section is used instead of eq. (17).

To estimate the survival chances of J/ψ's in confined hadronic matter of higher than standard nuclear density, we consider the survival probability

$$S = e^{-nL\sigma_{\psi h}}. \tag{21}$$

For hadronic matter of nuclear dimension ($L \simeq 5$ fm) but of 10 times higher density ($n \simeq 10\ n_o$), we still find essentially no suppression up to temperatures of 500 MeV. Eq. (21) does not take into account the expansion of the system, which would reduce even further any possible chances for a break-up. We thus conclude that there will not be any noticeable J/ψ absorption in confined hadronic matter of the volume and density obtainable in nuclear collisions, and that only deconfined matter can lead to J/ψ suppression. J/ψ production thus provides us indeed with a clear-cut probe to test if a given sample of strongly interacting matter is confined or deconfined.

In closing, we comment briefly on the interpretation of the J/ψ suppression observed in $O - U$ and $S - U$ collisions at CERN.[20] At this time, there remain at least two open questions which prevent us from concluding that this suppression is due to deconfinement. Our present considerations exclude J/ψ absorption in confined *matter*. The temporal sequence of J/ψ formation and equilibration in nuclear collision is not yet clear, however, and so energetic pre-equilibrium hadrons could have broken up the suppressed J/ψ's, either as fully formed resonances or in a nascent state. Such break-up processes can be studied theoretically, and it should be possible to determine their effect. Another uncertainty arises from J/ψ production through χ decay, which leads to a sizeable fraction of the overall production rate (30 - 40 %). Since the χ's are much less tightly bound, their break-up into $D\bar{D}$ pairs is easier, and hence much of the observed supression could in fact be χ suppression. In $p - Li$ and

$\pi^{\pm} - Li$ interactions at 300 GeV beam momentum,[21] the production rates of the different hidden charm final states (J/ψ, χ_1, χ_2, ψ') were measured, and if this would be done also in nuclear collisions, the effect of χ suppression could be determined. Thus both theorists and experimentalists have some work left to do before the relation between the observed J/ψ suppression and colour deconfinement is fully clarified.

Acknowledgments

I thank L. Kärkkäinen for providing me with the results shown in Figures 3 and 4 prior to publication. This work was supported in part by contracts with the German Ministry of Science and Technology (BMFT 06 BI 721) and with the European Community (CHCRX-CT92-0051).

REFERENCES

1. R. Hagedorn, Nuovo Cim. Suppl. 3 (1965) 147.

2. E. Beth and G. E. Uhlenbeck, Physica 4 (1937) 915.

3. G. Veneziano, Nuovo Cim. 57A (1968) 190.

4. S. Fubini and G. Veneziano, Nuovo Cim. 64A (1969) 811;
   K. Bardakci and S. Mandelstam, Phys. Rev. 184 (1969) 1640.

5. I. Pomeranchuk, Doklad. Akad. Nauk SSSR 78 (1951) 889.

6. N. Cabibbo and G. Parisi, Phys. Lett. 59 B (1975) 67.

7. H. Satz, Nucl. Phys. A418 (1984) 447c.

8. K. Wilson, Phys. Rev. D10 (1974) 2445.

9. T. Mathews and A. Salam, Nuovo Cim. 12 (1954) 563.

10. F. J. Wegener, J. Math. Phys. 10 (1971) 2259.

11. M. Creutz, Phys. Rev. D21 (1990) 2308.

12. F. Karsch and E. Laermann, Rep. Prog. Phys. 56 (1993) 1347.

13. T. Blum, L. Kärkkäinen, D. Toussaint and S. Gottlieb, "The Equation of State for Two Flavor QCD", contribution to the XIIth International Symposium on Lattice Field Theory (Lattice '94), Bielefeld 1994.

14. V. Goloviznin and H. Satz, Z. Phys. C 57 (1993) 671.

15. D. Kharzeev and H. Satz, Phys. Lett. B 334 (1994) 155.

16. T. Matsui and H. Satz, Phys. Lett. B 178 (1986) 416.

17. M. E. Peskin, Nucl. Phys. B 156 (1979) 365;
    G. Bhanot and M. E. Peskin, Nucl. Phys. B 156 (1979) 391.

18. M. A. Shifman, A. I. Vainshtein and V. I. Zakharov, Phys. Lett. 65B (1976) 255.
    V. A. Novikov, M. A. Shifman, A. I. Vainshtein and V. I. Zakharov, Nucl. Phys. B136 (1978) 125.

19. S. D. Holmes, W. Lee and J. E. Wiss, Ann. Rev. Nucl. Sci. 35 (1985) 397;
    T. Nash, in Proceedings of the International Lepton/Photon Symposium, Cornell 1983, p. 329;
    L. Rossi, "Heavy Quark Photoproduction", Genova Preprint INFN/AE-91-16;
    A. Ali, "Heavy Quark Physics in Photo- and Leptoproduction Processes at HERA and Lower Energies", DESY Preprint 93-105, 1993.

20. C. Baglin et al., Phys. Lett. B220 (1989) 471; B251 (1990) 465, 472; B225 (1991) 459.

21. L. Antoniazzi et al. (E705), Phys. Rev. D46 (1992) 4828; Phys. Rev. Lett. 70 (1993) 383.

HADRONIC MATTER EQUATION OF STATE
AND
THE HADRON MASS SPECTRUM

Ahmed Tounsi[1], Jean Letessier[1] and Johann Rafelski[1,2]

[1]Laboratoire de Physique Théorique et Hautes Energies *
Université Paris 7, Tour 24, 5e étage
2 Place Jussieu, F-75251 Paris Cedex 05, France
[2]Department of Physics, University of Arizona, Tucson, AZ 85721, USA

INTRODUCTION

The Statistical Bootstrap Model[1-3] (SBM) based on the hypothesis that hadrons are made of hadrons, with constituent and compound hadrons being treated on the same footing, implies a hadronic mass spectrum of the asymptotic form

$$\rho(m) \approx c\, m^{-a} \exp(m/T_H). \tag{1}$$

Details of the theoretical description lead to the preferred values 5/2 or 3 of the mass power a. The parameter T_H was interpreted as a limiting temperature reached at infinite energy density. The partition function of the system is singular at $T = T_H$, even when considering a finite system volume. This is due to the assumption that hadrons are point-like. For $T > T_H$ there is no partition function and of course no equation of state (EoS), in SBM. However, for $a \geq 7/2$, T_H is attained at finite energy density and given the singularity of the partition function, it must be interpreted as a phase transition temperature.[4,5]

However, the SBM partition function of a *pointlike* hadron system cannot be easily continued to temperature greater than T_H to describe the new, high temperature phase, presumably a Quark-Gluon Plasma (QGP). In many phenomenological models one uses two different partition functions, or equations of state, one for the hadron phase and the other for the QGP phase, connected arbitrarily at some rather subjectively selected transition temperature. More relevant theoretically would be a single, even thought complex expression for the partition function, and hence equation of state (EoS), describing both phases at the same time. This has been the primary objective of lattice calculation studies of QCD[6] during the past 15 years. This technique works today in a satisfactory way for the case of zero baryon number density, but runs into intricate and yet unsolved difficulties for the today relevant case

---

*Unité associée au CNRS: D 0280

Hot Hadronic Matter: Theory and Experiment
Edited by J. Letessier *et al.*, Plenum Press, New York, 1995

of finite baryon number density. Another approach follows formally the line of argument of the SBM approach with two crucial modifications:

1. hadrons are viewed as composed objects, akin to the MIT bags of quarks and gluons and therefore their proper volume[7,8] is taken into account when establishing the partition function;

2. only $SU_C(3)$ singlet states may be counted[9-12] among the possible statistical combinations in the partition function of a system of quarks and gluons, with total energy m and volume v. This is achieved by means of group theory technique of projection operators;[13-15] furthermore, with a view to the heavy ion collision data interpretation, also $U(1)$ conservation groups of baryon number[10,16] and strangeness have to be implemented as well. As an aside we note that for these conservation laws one can use either the projection operator technique or one can introduce chemical potentials corresponding to the conserved quantum numbers.

We derive here the density $\tau(m, v)$ of such bag (hadron) states, the analogue of the spectrum given by Eq. (1). Within the pressure ensemble we then explore the singularity structure of the generalized SBM and obtain the temperature and chemical potentials along the phase transition boundary, and more generally the form of the equations of state, below *and* above the critical temperature. At this time we can complete these calculations only when neglecting the residual $\mathcal{O}(\alpha_s/\pi)$ interactions between the constituents of hadronic bound states. We explore in detail how the experimentally determined mass spectrum of hadrons agrees with the SBM theoretical shape and we find a remarkable agreement between earlier studies based on much smaller sample of hadronic sates and our current work — exact agreement is obtained for the bag constant $B^{\frac{1}{4}} = 174$ MeV. We believe that this is an important result, in particular since it has been shown in an empirical least error fit to hadronic masses,[17] that this value of B is also the preferred value.

THE BAG DENSITY OF STATES

Consider a colorless bag of massless free quarks and gluons, which transform respectively according the fundamental and the adjoint representations of $SU_C(3)$. Using the projection technique[13-15] its grand canonical partition function can be derived from the function

$$\tilde{Z}_Q(\beta, v, \mu_q, \mu_S) = \int_{SU_C(3)} d\mu(g) \int_v \frac{d^3 R}{v} \left\{ Tr_G \widehat{U}_G(g) e^{-\beta H_G + i \vec{P}_G \cdot \vec{R}} \right\} \tag{2}$$

$$\left\{ Tr_q \widehat{U}_q(g) e^{-\beta H_q + i \vec{P}_q \cdot \vec{R} - i \mu_q \widehat{N}_q} \right\} \left\{ Tr_{\bar{q}} \widehat{U}_{\bar{q}}(g) e^{-\beta H_{\bar{q}} + i \vec{P}_{\bar{q}} \cdot \vec{R} + i \mu_q \widehat{N}_{\bar{q}}} \right\}$$

$$\left\{ Tr_s \widehat{U}_s(g) e^{-\beta H_s + i \vec{P}_s \cdot \vec{R} - i \mu_S \widehat{N}_s} \right\} \left\{ Tr_{\bar{s}} \widehat{U}_{\bar{s}}(g) e^{-\beta H_s + i \vec{P}_s \cdot \vec{R} + i \mu_S \widehat{N}_s} \right\},$$

where G stands for gluons, q for the quarks u and d and s for the strange quark. The integral $\int_{SU_C(3)} d\mu(g) \widehat{U}(g)$ represents the projectors on the subspace of all the states that transform under the singlet representation of $SU_C(3)$. Indeed, for any compact Lie group $\mathcal{G}$ corresponding to a symmetry of the system the projector $\mathcal{P}_\alpha$ on the subspace of all states that transform under the irreducible representation α, is[13,15]

$$\widehat{\mathcal{P}}_\alpha = d_\alpha \int_{\mathcal{G}} d\mu(g) \chi_\alpha^*(g) \widehat{U}(g), \tag{3}$$

where $\widehat{U}(g)$ is a unitary representation of $\mathcal{G}$, $d\mu(g)$ is the normalized Haar measure on $\mathcal{G}$ and where d_α and $\chi_\alpha^*(g)$ are respectively the dimension and the character of the irreducible representation α. In Eq. (2), $d_\alpha = 1$ and $\chi_\alpha^*(g) = 1$, corresponding to the singlet representation. Similarly $\int_v \frac{d^3R}{v} e^{i\vec{P}\cdot\vec{R}}$ is the projector on the states of zero momentum for a system enclosed in a volume v: this allows to exclude the motion of the whole bag when the density of states is calculated. Conservation of baryon number and strangeness is ensured by the introduction of the quark chemical potentials μ_q and $\mu_s \equiv \mu_q - \mu_S$, where μ_S is the strangeness chemical potential. The physical partition function $Z_Q(\beta, v, \mu_q, \mu_S)$ will be obtained by considering[13] the Wick rotation $\mu \rightarrow -i\beta\mu$

Standard calculations show[14,15] that $\tilde{Z}_Q(\beta, v, \mu_q, \mu_S)$ can be written in the form:

$$\tilde{Z}_Q(\beta, v, \mu_q, \mu_S) = \int_{SU_C(3)} d\mu(g) \int_v \frac{d^3R}{v} e^\Theta, \tag{4}$$

where

$$\Theta = \frac{\beta v}{\pi^2 (R^2 + \beta^2)^2} [\, d_G \mathcal{U}(g) + d_q \mathcal{V}(g, \mu_q) + d_s \mathcal{V}(g, \mu_S)], \tag{5}$$

with

$$\mathcal{U}(g) = \sum_{k=0}^\infty \frac{1}{k^4} \chi_{adj}(g^k), \tag{6}$$

and

$$\mathcal{V}(g, \mu) = Re \sum_{k=0}^\infty \frac{(-1)^{k-1}}{k^4} e^{ik\mu} \chi_{fund}(g^k), \tag{7}$$

where $\chi_{fund}(g) = \sum_{i=1}^{i=3} e^{i\theta_i}$ is the character of the fundamental representation whose eigenvalues $e^{i\theta_i}$ depend on the class parameters θ_i ($i=1, 2, 3$) restricted by $\sum_{i=1}^{i=3} \theta_i = 0$, (mod 2π). The character of the adjoint representation under which the gluons transform is given by

$$\chi_{adj}(g) = |\chi_{fund}(g)|^2 - 1 = 2 + 2\sum_{i<j}^3 \cos(\theta_i - \theta_j). \tag{8}$$

Then

$$\mathcal{U}(g) = \frac{\pi^4}{45} + 2\sum_{i<j}^3 u(\theta_i - \theta_j), \tag{9}$$

and

$$\mathcal{V}(g, \mu) = \sum_{i=0}^{i=3} v(\theta_i + \mu), \tag{10}$$

where[18]

$$u(\theta) = \sum_{k=0}^\infty \frac{1}{k^4} \cos(k\theta) = \frac{\pi^4}{90} - \frac{\pi^2}{12}\theta^2 \left(1 - \frac{|\theta|}{2\pi}\right)^2 \qquad |\theta| < 2\pi, \tag{11}$$

and

$$v(\theta) = \sum_{k=0}^\infty \frac{(-1)^{(k-1)}}{k^4} \cos(k\theta) = \frac{7\pi^4}{720} - \frac{\pi^2}{24}\theta^2 \left(1 - \frac{\theta^2}{2\pi^2}\right). \tag{12}$$

For large v the integration in Eq. (4) is done by the saddle point method. The maximum of Θ is reached for $R = 0$, and $\theta_i = 0$. Expanding Θ near this point to second order in R and θ_i we find

$$\Theta = \frac{1}{3}\frac{v}{\beta^3} U(\mu_q, \mu_S)(1 - \frac{2R^2}{\beta^2}) - \frac{v}{2\beta^3} C(\mu_q, \mu_S) \sum_{i=0}^{i=3} \theta_i^2, \tag{13}$$

with

$$U = \frac{\pi^2}{30} \left[8d_G + \frac{21}{4}(d_q + d_s) - \frac{45}{2}d_q \left(\frac{\mu_q}{\pi} \right)^2 \left(1 - \frac{\mu_q^2}{2\pi^2} \right) - \frac{45}{2}d_s \left(\frac{\mu_S}{\pi} \right)^2 \left(1 - \frac{\mu_S^2}{2\pi^2} \right) \right], \quad (14)$$

where $d_G = 2$, $d_q = 4$ and $d_s = \simeq 2$ are the number of degrees of freedom of gluons and quarks, not counting the color multiplicity considered here explicitly, and remembering the here relevant finite mass of strange quarks was neglected. Furthermore

$$C(\mu_q, \mu_S) = \left[d_G + \frac{d_q + ds}{6} - \frac{d_q}{2} \left(\frac{\mu_q}{\pi} \right)^2 - \frac{d_s}{2} \left(\frac{\mu_S}{\pi} \right)^2 \right]. \quad (15)$$

Integration over R and θ_i gives

$$\tilde{Z}_Q(\mu_q, \mu_S) = \frac{9}{2^3} \sqrt{2\pi} C^{-4}(\mu_q, \mu_S) U^{-\frac{3}{2}} \left(\frac{\beta^3}{v} \right)^{\frac{13}{2}} \exp \left[\frac{1}{3} U v \right]. \quad (16)$$

Wick rotation transforms this $\tilde{Z}_Q((\mu_q, \mu_S)$ into the physical grand canonical partition function $Z_Q(\mu_q, \mu_S)$. Finally the density of states of a bag of volume v, at energy W, is derived from the partition function $Z_Q(\mu_q, \mu_S)$ through inverse Laplace transform:

$$\sigma(W, v, \mu_q, \mu_S) = \frac{1}{2\pi i} \int_{\beta_0 - i\infty}^{\beta_0 + i\infty} d\beta \exp(\beta W) Z_Q(\beta, \mu_q, \mu_S). \quad (17)$$

Using once more the saddle point method for large W one obtains

$$\sigma(W, v, \mu_q, \mu_S) = \frac{9}{16} C^{-4} U^{7/2} v^{-3/2} W^{-11/2} \exp \left[4/3 \, U^{1/4} v^{1/4} W^{3/4} \right]. \quad (18)$$

In the following we need to express $\sigma(W, v, \mu_q, \mu_S)$ in terms of the baryon number and strange quark fugacities, $\lambda = e^{3\mu_q/T}$ and $\lambda_s = e^{\mu_s/T}$. Then the function $U(\lambda_q, \lambda_s)$ and $C(\lambda_q, \lambda_s)$ take the form:

$$U(\lambda, \lambda_s) = \frac{\pi^2}{30} \left\{ 8d_G + \frac{21}{4}(d_q + d_s) + \frac{5d_q}{2\pi^2} \left(\ln^2(\lambda) + \frac{1}{18\pi^2} \ln^4(\lambda) \right) \right.$$
$$\left. + \frac{45}{2\pi^2} d_s \left[\left(\frac{1}{3} \ln \lambda - \ln \lambda_s \right)^2 + \frac{1}{2\pi^2} \left(\frac{1}{3} \ln \lambda - \ln \lambda_s \right)^4 \right] \right\}, \quad (19)$$

and

$$C(\lambda, \lambda_s) = d_G + \frac{1}{6}(d_q + d_s) + \frac{(d_q + d_s)}{18\pi^2} \ln^2(\lambda)$$
$$+ \frac{d_s}{2\pi^2} \left(\ln^2(\lambda_s) - \frac{2}{3} \ln(\lambda_s) \ln(\lambda) \right). \quad (20)$$

If the colorlessness condition is omitted (no integration on the $SU_C(3)$ parameters) then the partition function (16) becomes

$$Z_c = \left(\frac{3}{2} \right)^{\frac{3}{2}} \pi^{\frac{3}{2}} U^{-\frac{3}{2}} \left(\frac{\beta^3}{v} \right)^{\frac{5}{2}} \exp \left[\frac{1}{3} U v \right]. \quad (21)$$

In this case one recovers the bag density of states obtained[19] using the microcanonical ensemble:

$$\sigma_c(W, v, \mu_q, \mu_S) = \frac{3\sqrt{3}}{8} \pi \, U^{1/2} v^{-1/2} W^{-5/2} \exp \left[4/3 \, U^{1/4} v^{1/4} W^{3/4} \right]. \quad (22)$$

108

TOWARDS QCD BASED EQUATION OF STATE

The partition function of a system of hadrons (bags) with mass m_i and proper volume v_i, enclosed in a volume V, can be written[8,10]

$$Z_H(T, V, \lambda, \lambda_s) = \sum_{N=0}^{\infty} \frac{1}{N!} \int \prod_{i=1}^{N} \left[\frac{d^3 p_i}{(2\pi)^3} dm_i \, dv_i \left(V - \sum_{j=1}^{N} v_j \right) \right.$$
$$\left. \times \, e^{-\beta\sqrt{p_i^2 + m_i^2}} \, \tau(m_i, v_i, \lambda, \lambda_s) \right] \theta \left(V - \sum_{j=1}^{N} v_j \right), \quad (23)$$

where $\tau(m, v, \lambda, \lambda_s) \, dm \, dv$ is the number of bag states in the mass interval $(m, m + dm)$ and volume interval $(v, v + dv)$. As is well known[20] phase transitions appear as singularities of the partition function in the thermodynamic limit $V \to \infty$. In order to investigate this problem an isobaric partition function $\Pi(T, s, \lambda, \lambda_s)$ is introduced[10,21] It is the Laplace transform of $Z_H(T, V, \lambda, \lambda_s)$ with respect to the volume V:

$$\Pi(T, s, \lambda, \lambda_s) \equiv \int_0^{\infty} dV \, e^{-sV} Z_H(T, V, \lambda, \lambda_s). \quad (24)$$

Substituting (23) in (24) one obtains

$$\Pi(T, s, \lambda, \lambda_s) = \frac{1}{s - f(\beta, s, \lambda, \lambda_s)}, \quad (25)$$

where

$$f(\beta, s, \lambda, \lambda_s) = \int \frac{d^3 p}{(2\pi)^3} dm \, dv \, \tau(m, v, \lambda, \lambda_s) e^{-sv - \beta\sqrt{p^2 + m^2}}. \quad (26)$$

The singularities of $\Pi(T, s, \lambda, \lambda_s)$ can be a pole s_0 given by

$$s_0(\beta, \lambda, \lambda_s) = f(\beta, s_0(\beta, \lambda, \lambda_s), \lambda, \lambda_s), \quad (27)$$

and/or a singularity $s_c(\beta, \lambda, \lambda_s)$ induced by the singular behavior of the function $f(\beta, s, \lambda, \lambda_s)$ in itself. Because of the analytic properties of the inverse Laplace transform the asymptotic behavior of the partition function $Z_H(T, V, \lambda, \lambda_s)$ in the thermodynamic limit is governed by the furthest to the right singularity of $\Pi(T, s, \lambda, \lambda_s)$. The existence and the relative position of s_0 and s_c can be established when the spectrum $\tau(m, v, \lambda, \lambda_s)$ in Eq. (26) is known. It is obtained[9] through the derivative of $\sigma(W, v, \lambda, \lambda_s)$ (see Eq. (18)) with respect to the volume:

$$\tau(m, v, \lambda, \lambda_s) = \frac{\partial}{\partial v} \sigma(W, v, \lambda) \bigg|_{W = m - Bv}, \quad (28)$$

which gives

$$\tau_{as}(m, v, \lambda, \lambda_s) = \frac{3}{16} C^{-4} U^{15/4} v^{-9/4} (m - Bv)^{-19/4} \exp\left[\frac{4}{3} U^{1/4} v^{1/4} (m - Bv)^{3/4} \right]. \quad (29)$$

Expression (29) is in fact the asymptotic part of $\tau(m, v, \lambda, \lambda_s)$. For small m and v one introduces a phenomenological function

$$\tau_0(m, v, \lambda, \lambda_s) = \sum_i d_i(\lambda, \lambda_s) \delta(m - m_i) \delta(v - v_i), \quad (30)$$

which describes the low part of the hadronic spectrum, $d_i(\lambda, \lambda_s)$ being the multiplicity of hadron i. Introducing (29) and (30) in (26) one can show[10–12] that

$$f(\beta, s, \lambda, \lambda_s) = f_0(\beta, s, \lambda, \lambda_s) + \varphi(\beta, \lambda, \lambda_s) \Gamma\left[-4, V_0(s - s_c(\beta, \lambda, \lambda_s)) \right]$$
$$\times \left[s - s_c(\beta, \lambda, \lambda_s) \right]^{-4}, \quad (31)$$

where

$$s_c(\beta, \lambda, \lambda_s) = \frac{1}{3} U(\lambda, \lambda_s) \beta^{-3} - B\beta, \tag{32}$$

and where f_0 is the contribution to f coming from τ_0. Eq. (27) entails that the transition temperature $T_C(\lambda, \lambda_s) = \beta_C^{-1}(\lambda, \lambda_s)$ is given by the equation

$$s_c(\beta_c(\lambda, \lambda_s), \lambda, \lambda_s) = s_0(\beta_c(\lambda, \lambda_s), \lambda, \lambda_s) = f(\beta_c(\lambda, \lambda_s), s_c(\beta_c(\lambda, \lambda_s), \lambda, \lambda_s), \lambda, \lambda_s), \tag{33}$$

which expresses the matching of the pole s_0, see Eq. (27), with the singularity s_c at $T_C(\lambda, \lambda_s)$.

For $T < T_C(\lambda, \lambda_s)$, $s_0 > s_c$ and the behavior of the system is governed by s_0 and all thermodynamic quantities are expressed in terms of s_0, the furthest to the right singularity. For example the energy density is given by

$$\varepsilon(T, \lambda, \lambda_s) \equiv - \lim_{V \to \infty} \frac{\partial}{\partial \beta} \ln Z_H(T, V, \lambda) = - \frac{\partial}{\partial \beta} s_0(\beta, \lambda, \lambda_s). \tag{34}$$

For $T > T_C(\lambda, \lambda_s)$ the leading singularity is s_c and in view of Eq. (32)

$$\varepsilon(T, \lambda, \lambda_s) \equiv - \frac{\partial}{\partial \beta} s_c(\beta, \lambda, \lambda_s) = U(\lambda, \lambda_s) T^4 + B. \tag{35}$$

Equation (35) is the EoS of a quark-gluon plasma. Thus in this approach the same partition function describes a hadron phase for $T \leq T_C$ and a QGP for $T > T_C$. Fig. 1 shows the energy density versus temperature. The solid curves correspond to the equation of state for $\lambda = \lambda_s = 1$ and for $B^{1/4} = 200, 225$ MeV. The dashed dotted curve shows the behavior of hadronic phase, when the value of $B^{1/4}$ is so large that the transition occurs outside of the temperature range of the figure. The dashed lower curve corresponds to an ideal massless gas. It is displayed for comparison.

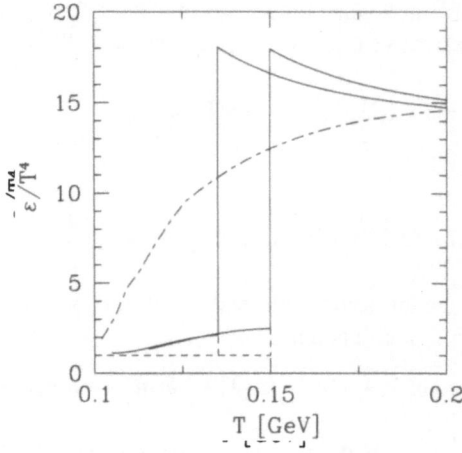

Figure 1. Energy density versus temperature. The solid curves give the EoS in the case $\lambda = \lambda_s = 1$, for $B^{1/4} = 200, 225$ MeV corresponding to $T_C = 135, 150$ MeV. The lower dashed curve corresponds to an ideal pion gas.

CRITICAL TEMPERATURE

In the bag model $v = \dfrac{m}{4B}$ and $W = \dfrac{3}{4}m$ and thus the expression of τ_{as} in Eq. (29) becomes

$$\tau_{as}(m) = \frac{\text{const.}}{m^7} C^{-4} U^{15/4} B^{9/4} \exp\left[\left(\frac{U}{3B}\right)^{1/4} m\right]. \tag{36}$$

Formally in the present model, the analogue of Hagedorn temperature is $T_0 = (U/3B)^{-1/4}$. However this T_0 is neither a limiting nor a critical temperature, but is related to these cases by simple algebraic relations we shall discuss below. In view of Eq. (32) it is actually the temperature at which a singular point s_c appears in $\Pi(T, s, \lambda, \lambda_s)$ at the origin of the real axis of the s-plane. We note here that the bag constant B can in principle be determined by fitting the spectrum (36) to the hadronic experimental mass spectrum.

We have fitted the spectra (1) and (36) put into the form

$$\rho(m) \approx c \, (m_0^2 + m^2)^{-a/2} \exp(m/T_0) \tag{37}$$

onto the experimental mass spectrum known today. Each particle is counted with a statistical weight $z = (2J + 1)(2I + 1)$ multiplied by 2, if the particle is different from its particle, and then represented by a Gauss function normalized to z. In each fit the value of a is fixed and c, m_0 and T_0 are let free. The quality of the fits is very good as can be seen in the two cases of particular interest here: Figs. 2 and 3 show the smoothed mass spectrum as it developed from 1967 to 1994 and the function (37) for $a = 3$ and $a = 7$. In both cases we obtain a good fit although the underlying physics is different. The different results are summarized in table 1 and Fig. 4. Note that T_0 is a decreasing function of a while m_0 is slowly increasing with a.

Table 1. Fitted parameters for given a.

a	c	m_0	T_0
2.5	0.83479	0.6346	0.16536
3.	0.69885	0.66068	0.15760
3.5	0.58627	0.68006	0.15055
4.	0.49266	0.69512	0.14411
5.	0.34968	0.71738	0.13279
6.	0.24601	0.73668	0.12341
7.	0.17978	0.74585	0.11489

Given the value of T_0 we can identify the appropriate value of the model parameter $B^{\frac{1}{4}}$. Furthermore, we can determine the critical temperature T_C when transition occurs — recall that $a = 7$, when we carry out the color projection, and $a = 3$ when we do not project on color singlets. As it can be seen in Eq. (31) the singular part of $f(\beta, s, \lambda, \lambda_s)$ vanishes when $s_c \to s_0$, that is, when $T \to T_C$. Thus the transition temperature is determined by the equation

$$f_0(\beta_c, s_0 = s_c, \lambda, \lambda_s) = \frac{1}{3} U(\lambda, \lambda_s) \, \beta_c^{-3} - B\beta_c. \tag{38}$$

The function f_0, the contribution of τ_0 (see Eqs. (30) and (26)) can be written

$$f_0(\beta, s_0, \lambda, \lambda_s) = \frac{1}{2\pi^2\beta} \sum_i d_i(\lambda, \lambda_s) \, m_i^2 \sum_{k=1}^{\infty} \frac{(-\gamma_i)^{k-1}}{k^2} K_2(k \, m_i \beta_c) \exp\left[-s_0 \frac{m_i}{4B}\right], \tag{39}$$

where the summation over k takes into account quantum statistics, with $\gamma_i = -1$ for bosons and $\gamma_i = 1$ for fermions. In the actual calculations we have included in (39) non strange and

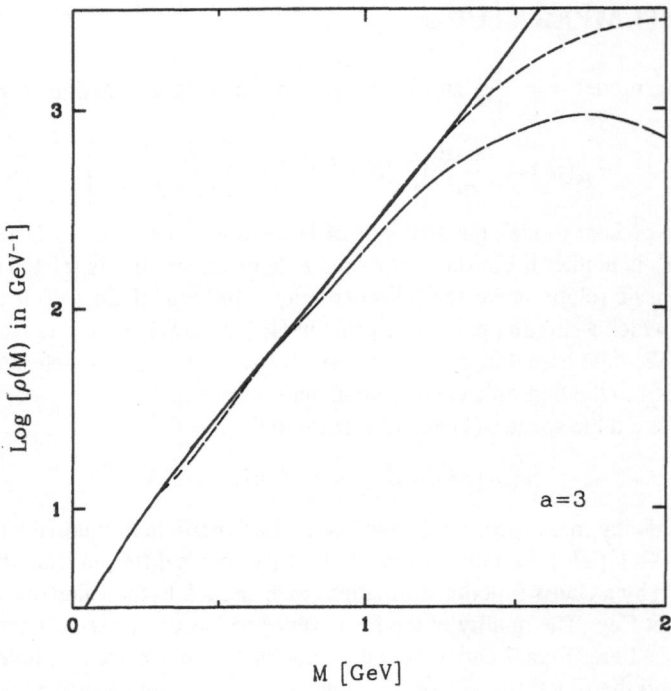

Figure 2. The smoothed hadronic mass spectrum. Long-dashed line: 1411 states (1967). Dashed line: 4558 states (1994). The solid line represents the function (37) with $a = 3$, $m_0 = 0.66$ GeV and $T_0 = 0.158$ GeV.

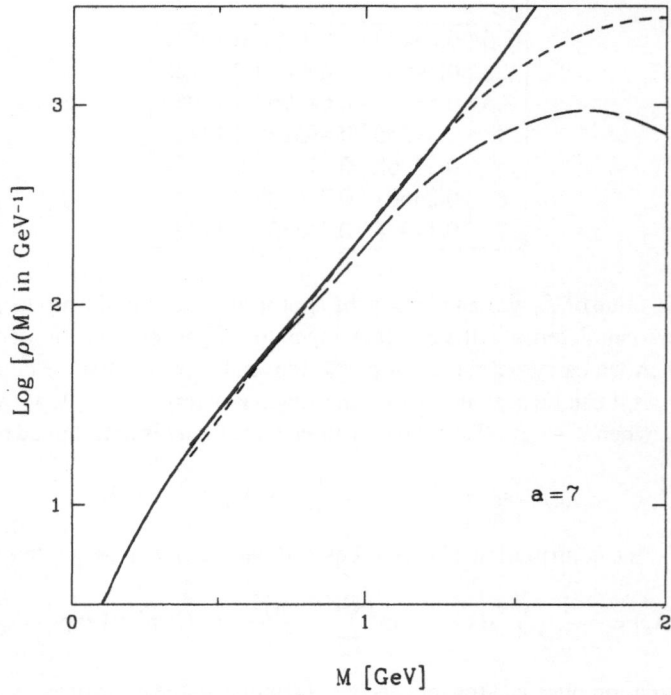

Figure 3. Same as Fig. (2). But the solid line represents the function (37) with $a = 7$, $m_0 = 0.746$ GeV and $T_0 = 0.115$ GeV.

112

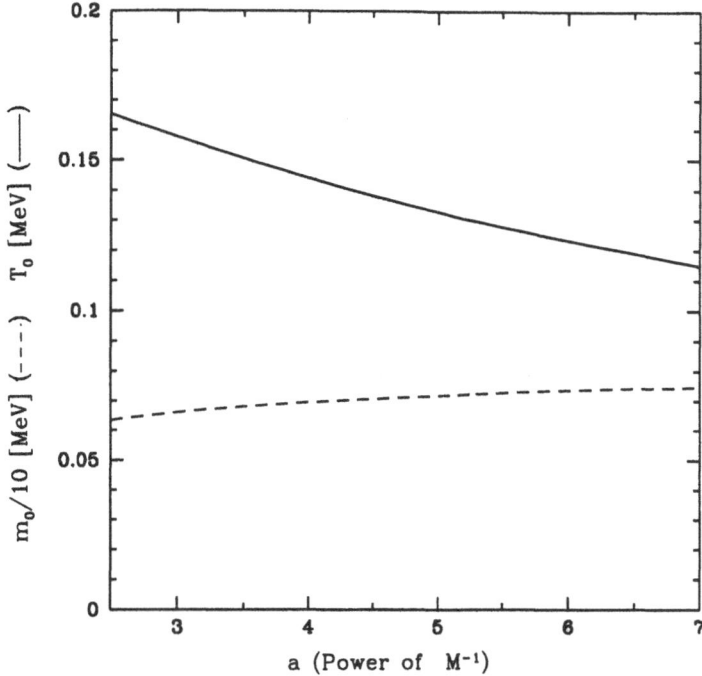

Figure 4. Variations of m_0 and T_0 in (37) with the mass power a.

strange hadrons with masses up to $M_0 = 1672$ MeV with the corresponding $V_0 = M_0/4B$ (see Eq. (31)), above which the asymptotic expression (29) begins to hold. However numerical calculations show that the transition temperature $T_C(\lambda, \lambda_s)$ depends very little on quantum statistics and is essentially determined, for a given value of the bag constant $B^{1/4}$ smaller than 300 MeV, by the pion mass. This can be understood by looking at Eq. (39). Not only the Hankel function $K_2(k\, m_i\beta_c)$ decreases with increasing k and m_i but also the exponential factor is strongly decreasing for with m_i. Thus in the first member of the Eq. (38) defining T_C only the pion contribution is relevant. However for higher values of $B^{1/4}$, T_C becomes sensitive to the inclusion of higher masses. The variation of T_C on B is displayed in Fig. (5). It deviates from the linear behavior found in the case where hadronic matter is reduced to a massless pion gas.

Using the value of T_0 found for $a = 7$ we get using the implied definition in Eq. (36): $B^{\frac{1}{4}} = 1.51 T_0 = 174$ MeV corresponding to $T_C = 117$ MeV. Both the value of B and T_C are remarkably consistent with the empirical studies of the phase transition properties. As noted before, the value of $B^{\frac{1}{4}} = 172$ MeV has been previously shown to give a an excellent empirical fit to the hadronic mass spectrum.[17] A phase transition at so low temperature would occur below the freeze-out limit for hadrons, which thus needed not to reequilibrate. This would naturally explain why under certain conditions hadronic final state strange particles retain information about particular QGP properties, such as $\lambda_s = 1$ as was found in empirical analysis of the 200 GeV A data.[22] On the other hand for some studies, such as QCD on the lattice, this is about 20% below the favored transition temperature range around 150 MeV. This difference to our finding based on the shape of the mass spectrum can be explained in magnitude and sign considering perturbative QCD-QGP effect on the relation between $B^{\frac{1}{4}}$ and T_C, as the effect of the residual QCD interactions among the constituents of hadronic bound states, neglected so far in our study.

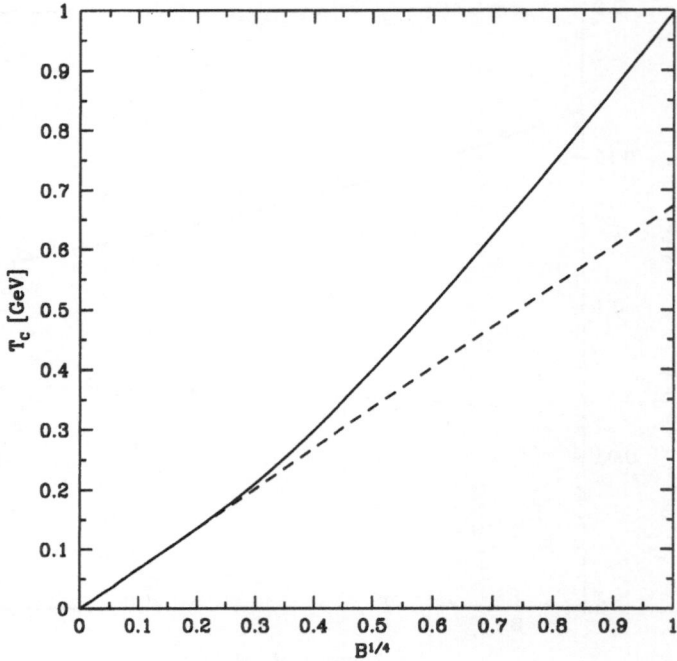

Figure 5. Transition temperature as a function of $B^{\frac{1}{4}}$, for $\lambda = \lambda_s = 1$. Solid line: all hadrons. Dashed line: only massless pions.

The general variations of T_C with λ and λ_s for a given B are explicitly obtained. We give in table 2 the results for three sets of particularly interesting chemical conditions.

Table 2. Critical temperature T_C in MeV, for given chemical conditions, **a)** for $\lambda = 1$ and $\lambda_s=1$; **b)** for $\lambda = 3.25$ and $\lambda_s = 1$ determined[22] for the collision system S–W at 200 GeV A (WA85, SPS-CERN); **c)** for $\lambda = 47$ and $\lambda_s = 1.72$ determined[23] for Si–Au collision systems at 14.6 GeV A (E802/E859, AGS-BNL).

$B^{\frac{1}{4}}$ [MeV]	T_C [MeV] (a)	T_C [MeV] (b)	T_C [MeV] (c)
175	117.5	116	107.5
200	135	133	123

CONCLUSION

Using group theory projection technique we have obtained the hadronic mass spectrum for colorless states with baryon number and strangeness conservation. From the same partition function embodying this spectrum, we have derived an equation of state which takes naturally two different forms corresponding to two different branches appearing in the solution at low and high temperatures, one for the hadron phase and one for the QGP phase. We have considered here as input the structure of hadrons based on a model of bags of *free* massless quarks and gluons, in which case the projection technique applies easily because of the trace factorization in Eq. (2). Although the neglect of quark masses and of residual QCD interactions seems to be reasonable at high temperature, we think that in the future, in order to obtain a more accurate equation of state, one aught allow for finite strange quark mass and incorporate quark-gluon interactions in the derivation of the bag colorless partition function.

Our quantitative study has shown that the current theoretical knowledge about the hadronic mass spectrum is sufficient to determine the value of transition temperature: we derived within the finite size hadron model the pre-exponential mass spectrum factor and in a fit of the exponentially rising shape we find a bag constant $B^{1/4} = 174$ MeV, which in turn implies a critical temperature of the magnitude $T_C \simeq 120$ MeV (keeping in mind that the explicit value of $T_C(B)$ is also at 10–20% level a function of α_s — inclusion of α_s would lead to such an increase of T_C at given B). Within our approach one can see clearly how the critical temperature is in fact primarily determined by the magnitude of the pion mass.

Acknowledgments

A. Tounsi wishes to thank G. Auberson for useful discussions on group theory projection technique. J. Rafelski acknowledges partial support by US DOE grant DE-FG02-92ER40733 and thanks his colleagues for their kind hospitality in Paris.

REFERENCES

1. R. Hagedorn, *Suppl. Nuovo Cimento* **3** (1965) 147.
   R. Hagedorn and J. Ranft, *Suppl. Nuovo Cimento* **6** (1968) 169.
   R. Hagedorn, *The Long Way to the Statistical Model*, in this volume.

2. S. Frautschi, *Phys. Rev.* **D3** (1971) 2821.

3. A. Tounsi, *Statistical Bootstrap and Thermodynamical Model of High Energy Strong Interactions*, in: Phenomenology of Particles at High Energy, Academic Press, London, 1974. 14th Scottish Universities Summer School in Physics, Middleton Hall, 1973.

4. N. Cabibbo and G. Parisi, *Phys. Lett.* **B59** (1975) 67.

5. J. Letessier and A. Tounsi, *Bootstrap statistique et transition de phase dans la matière hadronique*, Orsay preprint, IPNO/TH 76-14.

6. H. Satz, *Ann. Rev. Nucl. Part. Sci.* **35** (1985) 245 and references therein.
   J. Cleymans, R. Gavai and E. Suhonen, *Phys. Rep.* **C130** (1986) 218.

7. J. Baacke, *Acta Phys. Pol.* **B8** (1977) 625.

8. R. Hagedorn, I. Montvay and J. Rafelski, *Thermodynamics of Nuclear Matter from the Statistical Bootstrap Model* in: Hadronic Matter at Extreme Energy Density, N. Cabibbo and L. Sertorio, eds., Plenum Press, New York (1980);
   R. Hagedorn and J. Rafelski, *Phys. Lett.* **B97** (1980) 136.

9. M. I. Gorenstein, G. M. Zinovjev, V. K. Petrov and V. P. Shelest, translated from *Teor. Mat. Fiz* **52** (1982) 346.

10. M. I. Gorenstein, S. I. Lipskikh and G. M. Zinovjev, *Z. Phys.* **C22** (1984) 189.

11. J. Letessier and A. Tounsi, *Phys. Rev.* **D40** (1989) 2914.

12. M. Kataja, J. Letessier, P.V. Ruuskanen and A. Tounsi, *Z. Phys.* **C55** (1992) 153.

13. K. Redlich and L. Turko, *Z. Phys.* **C5** (1980) 201.

14. M. I. Gorenstein, S. I. Lipskikh, V. K. Petrov and G. M. Zinovjev, *Phys. Lett.* **B123** (1982) 437.

15. G. Auberson, L. Epele, G. Mahoux and F. R. A. Simão, *J. Math. Phys.* **27** (1986) 1688.

16. H. T. Elze, W. Greiner, J. Rafelski, *Z. Phys.* **C24** (1984) 361.

17. A. Aerts and J. Rafelski, *Phys. Lett* **B148** (1984) 337.

18. L. S. Gradshtein, and S. I. Ryzhic, *Table of Integals, Series and Products*, Academic Press (1965).

19. J. I. Kapusta, *Phys. Rev.* **D23** (1981) 2444.

20. See, e.g., K. Huang, *Statistical Mechanics*, J. Wiley Inc., New-York (1963).

21. R. Hagedorn, *Z. Phys.* **C17** (1983) 265.

22. J. Letessier, J. Rafelski, A. Tounsi, *Phys. Lett.* **B292** (1992) 417;
    Strangeness Conservation in Hot Fireballs J. Letessier, A. Tounsi, U. Heinz, J. Sollfrank, J. Rafelski. Paris LPTHE/93-27. Submitted to *Phys. Rev. D*.

23. M. Danos and J. Rafelski, *Phys. Rev.* **C50** (1994) 1686.

DECONFINEMENT OF CONSTITUENT QUARKS AND THE HAGEDORN TEMPERATURE

O.D. Chernavskaya and E.L. Feinberg

P.N. Lebedev Physical Institute of the Russian Academy of Sciences
Russia, 117924 Moscow, Leninsky prospect 53

INTRODUCTION

The overwhelming majority of theoretical papers on phase transition of hadronic matter H to quark gluon plasma ($H \leftrightarrow QGP$) ignores constituent quarks Q which we shall call below briefly *valons* (the term proposed by R.Hwa). The present paper is fundamentally based on the conception of those valons as real entities having the same quantum numbers as current quarks and the mass which was more than once calculated theoretically as some 300 MeV and used for explanation of experiments (e.g., see[1]). The possibility of existence of a special state of strong interacting matter, that of *deconfined valon gas with still broken chiral symmetry*, is discussed here.

Such a possibility, i.e. appearance of the third, Q phase intermediate between H and QGP had been discussed already in papers[2-5] (we apologize for possibly missing some other papers). The idea may be traced back to E.Shuryak who pointed out that deconfinement and restoration of chiral symmetry might not coincide[6]. However in the papers mentioned above the result turned out pessimistic. E.g., on the $\mu - T$ diagram (μ - chemical potential, T - temperature) possible Q phase (allowing also for admixture of pions necessary first of all as Goldstone particles) occupies but a very small area at large T and rather small μ (see e.g. Fig.1 taken from[5]).The "constituent quark phase is strongly suppressed"[2].

Meanwhile the intuitive physical pattern[7,8] based on the bag model ideology, although being extremely rough, points to the quite a different possibility schematically shown in Fig.2. In fact, a nucleus consists of nucleons N which in turn are bags containing valons, which are nothing else than a massive clouds of strong interacting current quarks q, antiquarks $\bar{q}$ and gluons g with quantum numbers of a single quark. If we compress the nucleus (Fig.2a) until nucleons become close packed and fill up the entire space (Fig.2b) they merge into a single bag with valons deconfined (still with chiral symmetry broken) and freely moving within the common bag (Fig.2c). Further compression leads to a new situation, that of close packed and merging valons (Fig.2d). Now $q, \bar{q}$ and g which earlier were confined within valons can freely propagate within a common bag (Fig.2e). Here they are deconfined and massless, chiral symmetry is automatically restored and this very state is QGP.

Thus we expect two phase transitions to take place: the one, $H \leftrightarrow Q$, and (at higher density and/or temperature) another, $Q \leftrightarrow QGP$.

Hot Hadronic Matter: Theory and Experiment
Edited by J. Letessier *et al.*, Plenum Press, New York, 1995

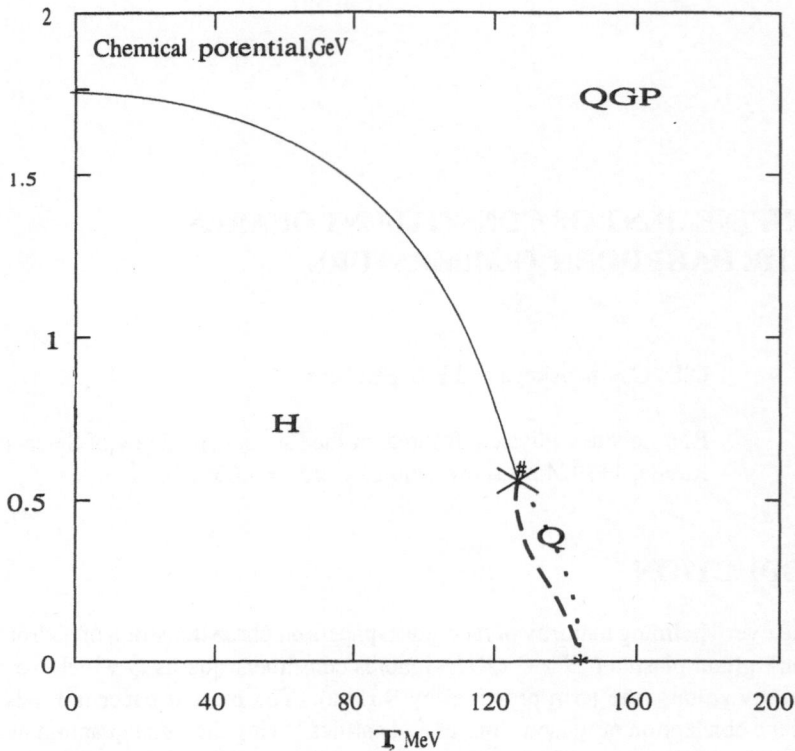

Figure 1. Phase transition curves in the $T - \mu$ plane(from[5]): Q-phase region is between dashed and dotted lines.

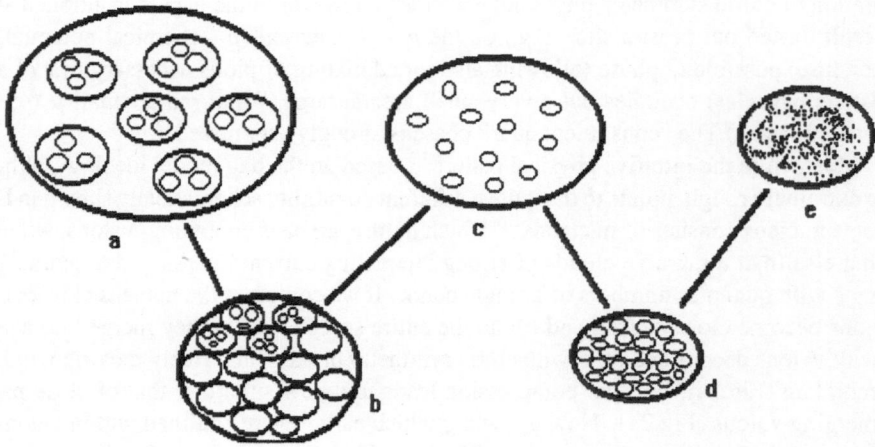

Figure 2. Schematic picture of nuclear matter transformation under compression.

In this paper we are going to show, in distinction to[2-5], that with reasonable values of bag parameters B_Q for the valon phase and B_q for QGP there actually appears possibility for a pattern shown in Fig.2, i.e. the possibility of the overall $H \leftrightarrow QGP$ transition to proceed only via the intermediate Q phase.

The first $(H \leftrightarrow Q)$ transition establishes limitations on possibility of existence of normal hadrons. This is the reason for interpreting it as the Hagedorn limit and to consider the temperature of transition, i.e. of valon deconfinement, T_d, as the Hagedorn temperature T_H. The second transition restores chiral symmetry and is characterized by the temperature T_{ch}. Thus the Hagedorn temperature appears to be not merely a very rough approximation to chiral symmetry restoration/ breaking temperature T_{ch} as is frequently believed. It has its own physical meaning of temperature of constituent quark ($\equiv$valon) deconfinement: $T_H \equiv T_d$.

THERMODYNAMICAL FORMALISM

We are treating phase transitions following traditional manner [2-5], fundamentals of which were given by Hagedorn[9]. As the first step we consider the partition function $Z^0(T, \mu, V)$ depending on T, μ, and system volume V for each phase in the first approximation (which means: for pointlike particles):

$$\ln Z_j^0(T, \mu, V) = \frac{V}{T} \sum_i \left\{ \frac{g_i^B}{6\pi^2} \int \frac{k^4 dk}{\sqrt{k^2 + m_i^2}} \frac{1}{exp\left(\frac{\sqrt{k^2+m_i^2}}{T}\right) - 1} + \right.$$

$$\left. \frac{g_i^F}{6\pi^2} \int \frac{k^4 dk}{\sqrt{k^2 + m_i^2}} \left[\frac{1}{\exp\left(\frac{\sqrt{k^2+m_i^2}-\mu_i}{T}\right) + 1} + \frac{1}{\exp\left(\frac{\sqrt{k^2+m_i^2}+\mu_i}{T}\right) + 1} \right] \right\} + \ln Z_{vac}^0 \tag{1}$$

where $g_i^{B,F}$ and m_i are degeneracy factors and masses for Bose and Fermi i-th type particles respectively for each j-th phase (j designates hadronic H, valonic Q, and current (QGP), q phases respectively). $\ln Z_j^0(T, \mu, V)$ is a result of summation over particles which play considerable role (under T and μ in question) in the given phase. In particular, for hadronic phase H we have taken into account nucleons (protons and neutrons, $M = 940$ MeV), pions ($m_\pi = 140$ MeV) and Λ hyperons ($m_\Lambda \simeq 1150$ MeV). The constituent phase Q contains light valons $\equiv$ constituent quarks ($m_u \simeq m_d \simeq 320$ MeV), some admixture of strange valon quarks ($m_s \simeq 512$ MeV), and again pions. The term $\ln Z_{vac}^0$ stands here for Q and QGP phases and reflects effective interaction of quarks and gluons with the QCD-vacuum. This interaction in Q and QGP phases is assumed as

$$\ln Z_{vac,Q}^0(T, \mu, V) = -\frac{V}{T} B_Q , \tag{2}$$

$$\ln Z_{vac,q}^0(T, \mu, V) = -\frac{V}{T} B_q \tag{3}$$

where B_Q, B_q are some constant bag parameters differing from each other (to be discussed below).

Then by usual differentiations of $Z_j^0(T, \mu, V)$ we obtain for each phase energy density $\epsilon_j^{(0)}$, pressure $p_j^{(0)}$, and particle density $n_j^{(0)}$ as functions of internal parameters T, μ, V.

Having this as a basis and following the procedure proposed[2-5] (more exactly in[2]) we introduce mutual interaction of particles by ascribing them hard core radii in the Van der Waals manner. Namely, we substitute $n^{(0)}$ with real particle density $n^{(r)}$ and then obtain "real" energy density $\epsilon^{(r)}$ and pressure $p^{(r)}$ for each of three phases ($j = H, Q, QGP$):

$$n_j^r(T, \mu) = \frac{n^0(T, \mu)}{1 + n^0(T, \mu) * v_j} \,, \tag{4}$$

$$\epsilon_j^r(T, \mu) = \frac{\epsilon^0(T, \mu)}{1 + n^0(T, \mu) * v_j} \,, \tag{5}$$

$$p_j^r(T, \mu) = \frac{p^0(T, \mu)}{1 + n^0(T, \mu) * v_j} \,, \tag{6}$$

where each of the parameters v_j is hard core volume characterizing j-th phase particles:

$$v_N = \frac{4}{3}\pi r_N^3 \,, \tag{7}$$

$$v_Q = \frac{4}{3}\pi r_Q^3 \,, \tag{8}$$

current quarks and gluons are of course pointlike ($v_q = 0$).

At given T and V of the system the most preferable (i.e. stable) phase is the one having the largest pressure p_j and smallest μ_j. At equilibrium of each two phases their pressures and effective chemical potentials are to be equal. Three types of phase equilibrium curves are to be analyzed: the one corresponding to the **valon deconfinement** transition ($H \leftrightarrow Q$)

$$\mu_H = 3\mu_Q; \qquad p_H(T_d, \mu_H) = p_Q(T_d, \mu_H/3) \,, \tag{9}$$

direct transition ($H \leftrightarrow QGP$)

$$\mu_H = 3\mu_q; \qquad p_H(T_c, \mu_H) = p_q(T_c, \mu_H/3) \,, \tag{10}$$

and **chiral** transition ($Q \leftrightarrow QGP$)

$$\mu_Q = \mu_q; \qquad p_Q(T_{ch}, \mu_Q) = p_q(T_{ch}, \mu_H/3) \,. \tag{11}$$

The point of **coexistence of all three phases** (i.e. the *triple point*) is defined by the following conditions:

$$p_H(T^\#, \mu^\#) = p_Q(T^\#, \mu^\#) = p_q(T^\#, \mu^\#) \tag{12}$$

Before turning to study of equilibrium conditions and phase transitions let us discuss the chosen set of parameters.

BASIC PARAMETERS

There are two pairs of decisively important parameters: firstly, particle sizes, e.g. average root square radii of nucleons N, r_N, understood as $(\bar{r^2}_N)^{1/2}$, and valons, r_Q, which designates $(\bar{r^2}_Q)^{1/2}$. Secondly bag pressures, B_Q for the Q phase and B_q for QGP.

Let us discuss sizes. They are used[2-4] to describe interaction of particles in a simplified form by considering them as black balls forming a Van der Waals gas. This means that we approximate their interaction potential $U(r)$ as $U = \infty$ for $r < r_c$ and $U = 0$ for $r > r_c$, r_c being the core radius (in other calculations we have used also some continuous potential $U(r)$). In[5] the nucleon interaction was instead described by two versions of the continuous $U(r)$ satisfactorily describing nuclear properties. It seems sufficient in this paper to consider

N and Q simply as having cores[2,3]. For nucleon radius and volume r_N, v_N we usually assume, as most reasonable, the values:

$$r_N \simeq 0.8\text{fm}; \qquad v_N \simeq 2.13\text{fm}^3 \qquad (13)$$

(however sometimes we shall consider also $r_N = 0.7$ fm). Since the nucleus has the radius $R_A = r_0 * A^{1/3}$ (A - atomic weight) with $r_0 \simeq 1.10 - 1.15$fm, the fraction of the nucleus volume occupied by nucleons is

$$(r_N/r_A)^3 \sim 0.26 \div 0.38 . \qquad (14)$$

Thus it seems sufficient to compress a nucleus only 3-4 times to come to close packing of nucleons.

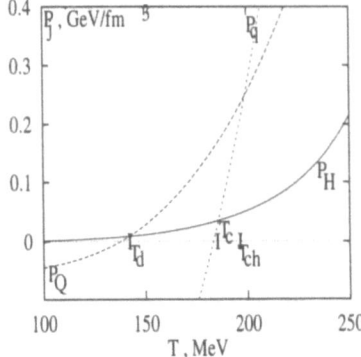

Figure 3. Phase diagram for $\mu = 0$ case at the $P - T$ plane for standard parameter set(17): the solid, dashed and dotted lines correspond to the the pressures of hadronic (H), valonic (Q) and current $q(QGP)$ phases respectively.

The size of valons has been estimated[10] from analysis of properties of various hadrons within the Additive Quark Model (AQM); when expressed by the square root radius r_Q(for $r_N = 0.8$ fm) this estimate sounds as:

$$0.20 \le r_Q \le 0.36\text{fm}. \qquad (15)$$

We shall concentrate on some middle reasonable value, $r_Q = 0.3$fm, although other values within limitation (15) will be also used (let us remark that estimating r_Q within black ball valon model and AQM from QQ collision cross section $\sigma_{QQ} = \pi(2r_Q)^2 \simeq (1/9)\sigma_{NN}$ where σ_{NN} is the NN collision cross section taken, at not too high energies, as 40 mb, we get $r_Q = 0.2$ fm. For tevatron energies it is much larger).

Now, about B_Q and B_q. The choice of their values is much more disputable, being at the same time very essential. Previous authors[2-5] were (seemimgly) guided by the results of lattice calculations in which confinement/deconfinement (c/d) and chiral symmetry breaking/restoration (b/r) transitions had been found to coincide at $\mu = 0$. Moreover "deconfinement" was (and still is) usually understood as the straightforward $H \leftrightarrow QGP$ transition. Accordingly, e.g. a condition was superimposed[5] on pressures p_H and p_q of H and QGP phases at the transition point (for $\mu = 0$) : $p_H(T_c, 0) = p_q(T_c, 0)$. Herefrom the ratio $\beta = B_q/B_Q$ was obtained thus leaving only one of these two parameters undefined. In other works similar coincidence of (c/d) and (b/r) at $\mu = 0$ was also taken as a basis. Herefrom, for a freely chosen B_Q, followed the value of B_q. All authors accordingly used mainly the values $\beta = B_q/B_Q \sim 3 \div 4$ with B_Q more or less close to the MIT bag value, $B_{MIT} \simeq 56 \,\text{MeV/fm}^3$ ($B_Q = 56 \div 150 \,\text{MeV/fm}^3$, $B_q = 200 \div 500 \,\text{MeV/fm}^3$, keeping in all cases[2-5] $\beta \sim 3 \div 4$).

Assumption concerning $B_Q \simeq B_{MIT}$ seems to be reasonable at least in the vicinity of the $H \leftrightarrow Q$ transition. In fact, here valon density in Q phase, at least in the beginning of the

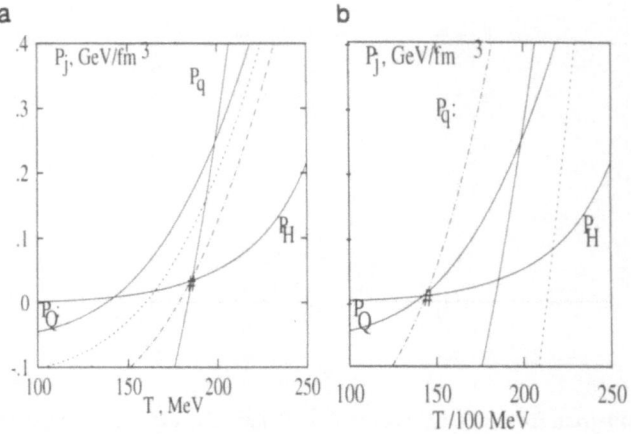

Figure 4. The same plot as in Fig.3 for standard parameter set (solid lines) and varying B's: **(a)** - $B_Q \simeq 100$ (dashed line), 160 MeV/fm^3(dotted line); **(b)** -$B_q \simeq 1000$ (dashed line), 200 MeV/fm^3 (dotted line);"#" marks the triple points.

valon deconfinement process, is the same as in a nucleon (see Fig.2b). Accordingly we shall use the assumption:

$$50 \leq B_Q \simeq B_{MIT} \leq 100 \,\text{MeV/fm}^3. \tag{16}$$

On the contrary, argumentation which have lead to the above mentioned estimate of $\beta = B_q/B_Q \sim 3$ is not admissible for us since *difference* between confinement/deconfinement (c/d) of valons (which takes place with broken chiral symmetry) and chiral symmetry breaking/restoration (b/r) is *exactly the phenomenon we are looking for*. Accordingly this difference should not be excluded at the start of the analysis as has been, as a matter of fact, done e.g. in[5]. Thus B_q is a parameter which needs its own physical foundation. Since it refers to QGP, it should be tightly connected with the vacuum pressure $B_{vac} \simeq 500 \div 1000 \,\text{MeV/fm}^3$ and be close to it. Accordingly we are going to use larger values of $\beta = B_q/B_Q$ putting it equal to $\beta \sim 10$. Below we shall use the following set of reasonable parameters as the *standard*

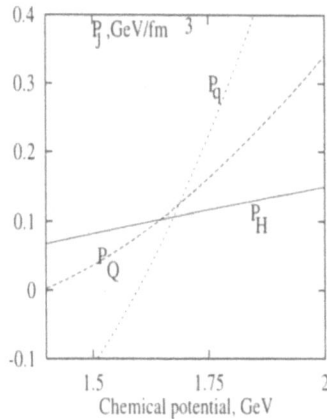

Figure 5. Phase diagram in the $p - \mu$ plane for T=0 and standard parameter set; designations are the same as in Fig.3.

$$B_q = 500 \, \mathrm{MeV/fm^3}; \quad B_Q = 50 \, \mathrm{MeV/fm^3}; \quad r_N = 0.8 \mathrm{fm}; \quad r_Q = 0.3 \mathrm{fm}. \qquad (17)$$

The influence of B_Q and B_q as well as r_j variations are partially investigated.

MAIN RESULTS

Let us start with extreme cases: $\mu = 0, T \neq 0$(Figs.3,4) and $T = 0, \mu \neq 0$(Fig.5).

Fig.3 clearly demonstrates characteristic pattern of double phase transition at the $p - T$ plane for chosen standard parameter set (17). It is evident that two phase transitions are well separated, Q is the most stable one (i.e. its pressure p_Q is higher than that of other phases) within the temperature interval of $\Delta T = T_{ch} - T_d \sim 50 \, \mathrm{MeV}$, with absolute values of transition temperatures being of the order of the direct phase transition $(H \leftrightarrow QGP)$ temperature T_c: $T_c \simeq 180 MeV$, $T_d \simeq 145 Mev$, $T_{ch} \simeq 195 Mev$. The value of this *"temperature corridor"* of Q phase depends on B_q (see Fig.4(b)) so that it becomes larger (up to 90 MeV) for larger values of B_q (when $B_q = 1 \, \mathrm{GeV/fm^3}, \beta \sim 20, T_{ch} \sim 230$ MeV) and shrinks to a *triple point* (which are marked by the symbol "#" within all the figures) $T^{\#} \sim 140$ MeV for most frequently chosen $B_q \simeq 200 \, \mathrm{MeV/fm^3}$ $\beta \sim 4$, close to that in[5]. The same effect stays for B_Q dependence: the closer it is to B_q, the narrower is T-region allowing for Q phase existence, with the biggest value (corresponding to the triple point $T^{\#} \sim 180$ MeV) $B_Q \sim 160$ MeV (i.e. $\beta \sim 4.5$). For all combinations of B's the chiral symmetry restores at T_{ch} which is always larger than the critical temperature of direct transition $T_c \equiv T_{H \leftrightarrow QGP}$ obtained in calculations when ignoring existence of valons. Note that we do not show the influence of r parameters because it is negligible in this $T - \mu$ region.

The similar pattern is presented in Fig.5 for $T = 0, \mu \neq 0$. for the standard parameter set(17). Here again there is a possibility for appearance of Q phase region but now it is greatly suppressed: if $\beta = 10$ while B_Q has its minimal reasonable value $50 \, \text{MeV}/\text{fm}^3$, the area covered by Q phase on $\{p, \mu\}$ diagram is rather small and it shrinks into a triple point already at $B_Q \simeq 80 \, \text{MeV}/\text{fm}^3$; this region becomes of course broader for larger β values. However it is essential that valons tell even at $T = 0$. For more details concerning $T = 0$ see [11].

Now we go over to the general case of $T, \mu \neq 0$. The $\{\mu, T\}$ diagram for the standard parameter set (17) is shown in Fig.6. It is clear that there exists a wide corridor around the curve for direct $H \leftrightarrow QGP$ transition which remains as well for various combinations of parameters $(B_q, B_Q, r_N, \text{and } r_Q)$ (see Figs.7). It shrinks and even disappears with μ growth only at rather definite conditions, namely for small β(the shape of Q phase region for $\beta < 3$ shown in Fig.7(a), as well as in Fig.1, reminds a thin banana) and for extremely large r_Q

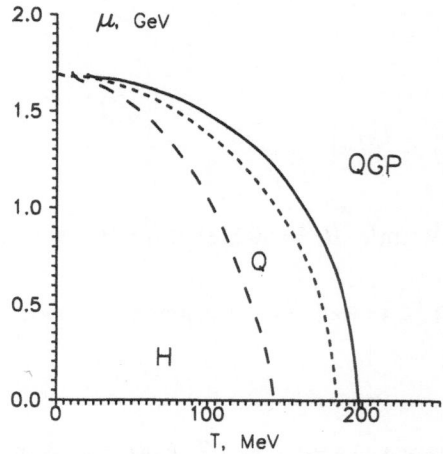

Figure 6. Phase transition curves in the $T - \mu$ plane: valon deconfinement (dashed line), direct transition ignoring valons (solid line), chiral transition (short dashed).

(Fig.7(b)). Otherwise transition from H phase to QGP proceeds only via valon phase (for details see [12]).

Note that under not too large $\mu \leq 1$ GeV the master parameter is $\beta = B_q/B_Q$ while influence of particle interaction (i.e. choice of values of r_N and r_Q) is quite negligible. Symbol line in Fig.7(b) corresponds to the case of all particles taken point like. It differs considerably from the real particle diagram only at $\mu \sim 1$ GeV, and covers the area having the shape of some exotic fruit ("pineapple").

Critical values ϵ^{cr} for energy densities (that is the ones on phase equilibrium curves) are presented in Figs.8. Let us point out (see Fig.8(a)) that the *latent heat*, i.e. the jump $\Delta \epsilon^d = (\epsilon_Q^d - \epsilon_H^d)$ at the deconfinement transition is much smaller than that at the chiral one, $\Delta \epsilon_{ch} = (\epsilon_q^{ch} - \epsilon_Q^{ch})$, and much smaller than that for direct transition ignoring valons (Fig.8(b)). The deconfinement transition requires rather "soft" experimental conditions.

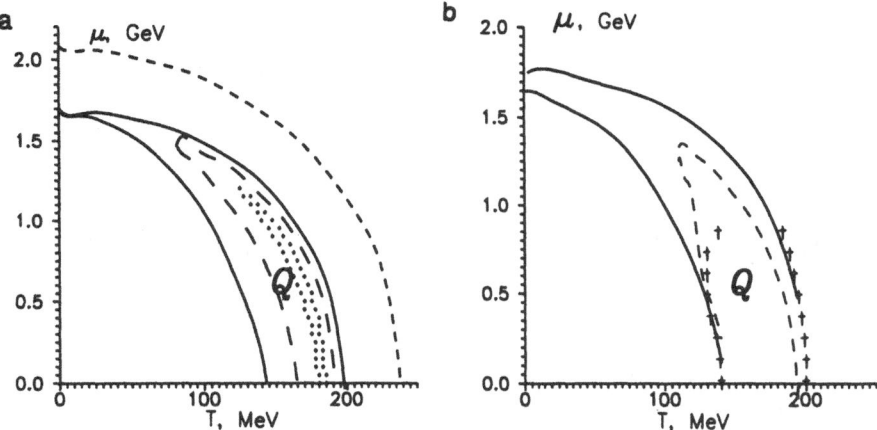

Figure 7. Transition curves bounding the Q-phase region: **(a)** - for standard r_j and B_q, $B_Q = 50$ MeV/fm³ (solid lines), 100 MeV/fm³ (long-dashed lines) and 150 MeV/fm³ (dotted lines); short-dashed line corresponds to $B_q = 1$ GeV/fm³; **(b)** - for standard B's and varying $r_Q = 0.2$ fm (solid lines), 0.4 fm (dashed lines) and $r_j = 0$ († lines).

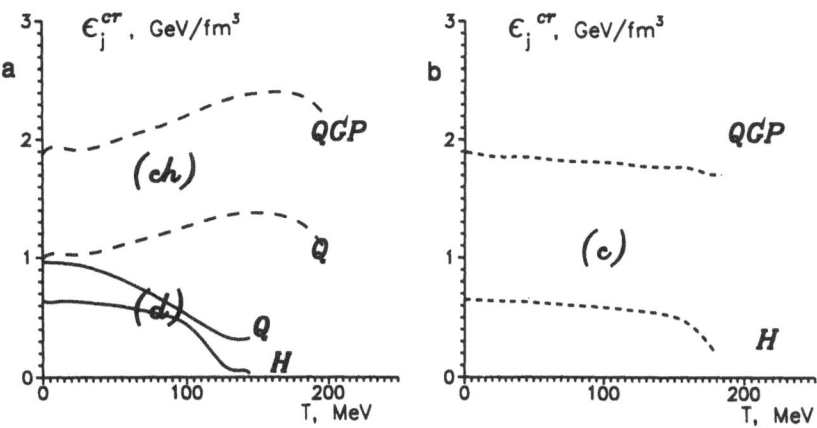

Figure 8. Energy densities ϵ_j^{cr} curves for: **(a)** H and Q matter at deconfinement transition (solid lines), Q and QGP matter at chiral transition (dashed lines); **(b)** H and QGP matter at direct transition ignoring valons (dotted lines).

DISCUSSION AND SUMMARY

Of course, obtained numerical values, in particular those of T_d and T_{ch}, should not be taken literary. E.g., they essentially depend on B_Q and B_q values chosen in a rather rough manner. According to e.g. Shuryak [13], for hadronic substance at $T = 0$, B_{eff} (efficient B_{eff} describing the summarized influence of interaction of particles between themselves and the vacuum) is not reducible to two constants but changes continuously increasing with baryon number density n_B from some lower value of the order of B_{MIT} up to $B_{eff} \sim B_{vac}$ for QGP.

However, our calculations [12] show that baryon number density in Q phase, n_Q, is much closer to that in the H phase than to the one in the QGP. Thus the use of only two values - B_Q and B_q - seems to be not too bad.

Further, our calculations show that particle density of H phase in the vicinity of transition curve to Q phase at relatively large $\mu > 1$ GeV is very close to its limiting value $\sim v_N^{-1}$ (close packing) and use of the Van der Waals gas approximation for it is very suspicious. But the same weak point is met in the papers which lead to pessimistic conclusions concerning possibility of the Q-phase[2-4]. Repulsive potential for nucleons considered in [5] seems of course to be more realistic; similar potential will be discussed in our paper [12]. This does not change the final result qualitatively.

Nevertheless qualitative and even semiquantitative results seem to deserve attention. Comparing Figs.6,7 with Fig.1 we see that for sufficiently large values of B_q/B_Q the pattern declared in the beginning of this paper (Fig.2) is almost fully supported. We may conclude that there actually exists possibility of hadronic phase to QGP transfer (at seemingly reasonable choice of parameters) to proceed only via an intermediate valon, i.e. constituent quark plus pions, phase. Therefore, we expect existence of two phase transitions. For $\mu = 0$ transition temperatures differ by some 50 MeV. One of them, deconfinement of valons, proceeds at the *Hagedorn temperature* $T_d = T_H$, above which, in the deconfined valon phase, hadrons exist but as a small thermal admixture. The second phase transition at $T_{ch} \sim 200$ MeV corresponds to deconfinement/confinement of current quarks and gluons and, simultaneously, restoration/breaking of chiral symmetry. Thus difference between T_d and T_{ch} is physically meaningful.

The Q phase with deconfined constituent quarks (valons) having mass ~ 320 MeV, due to $m_Q/T > 1$, is nonrelativistic, its state equation is that of nonrelativistic Van der Waals gas (since r_Q is small). It is only in the vicinity of the phase transition to QGP that valon size begins to tell. However, this last transition at not too large $\beta = B_q/B_Q$ takes place long before close packing of valons is attained[12]. This result is rather interesting and has more physical sense than had been put into the model. In fact the model of hard black balls, being quite sufficient for nucleons due to their high stability, is hardly valid for constituent quarks which may become partially transparent and larger. Overlapping of valon tails securing exchange of current quarks and gluons between valons may happen at distances exceeding the assumed black ball radius. Thus the transition to QGP phase does not necessarily require for close packing of valons. This is not so for nucleons at $H \leftrightarrow Q$ transition where the transition happens long after the considerable close packing is attained; this may be ascribed to necessity of overcoming the surface tension of nucleons. However for more detailed analysis it is desirable to consider more realistic repulsive potential for interaction of valons than simple hard core model (e.g. at the same manner as it has been done in[14]).

The obvious first application of these results of course is to be to the evolution of matter formed at collision of highly relativistic heavy nuclei. Let us stress two most important items.

First, as it have been shown above (Figs.8) the latent heat value for $H \leftrightarrow Q$ transition is much smaller than that for the "upper" transition $Q \leftrightarrow QGP$ (and much smaller than people had obtained for direct, i.e. $H \leftrightarrow QGP$ transition). The same is true[12] for transition pressure and n_B values. This means that first deconfinement transition may proceed under much more "soft" conditions than it was expected for direct transition $H \leftrightarrow QGP$: it is sufficient to compress the nucleus only 3-4 times applying not very high pressure. Thus this transition may happen at not too high energies of colliding nuclei, i.e. *even at Bevalac or in Dubna*[7,8] where appearance of pure QGP is hardly possible. Accordingly, e.g. in the Dubna experiments one may expect to observe some effects impossible for simple pp collisions. In particular, our estimates show that admixture of Λ particles may attain in central rapidity region some 10 percents of nucleons and this reminds us Okonov group Dubna experiment [15] in which $\Lambda's$ were especially studied.

Secondly, transition pattern described above should influence essentially the character of hydrodynamical expansion and cooling of initially hot matter both if it takes place either in Q or in QGP phase. For QGP initial state the following picture seems to be natural. The system experiences the phase transition to QGP phase. After cooling down to chiral transition temperature $T_{ch} \sim 200 MeV$, it would further expand and cool down within a large temperature interval according to nonrelativistic state equation. Since at present (and later when RHIC will be operating) the attainable initial temperature may exceed T_{ch} but slightly, the time spent for getting T_{ch}, as well as duration of cooling down in H phase from $T_d \sim 150 MeV$ to freezing temperature, $T_f \sim 140$ MeV, may be smaller than the time of existence of Q phase with its nonrelativistic state equation. Therefore the existence of Q phase should tell itself essentially. Here detailed calculations for resulting rapidity distribution, etc. are necessary to compare with experiment (at present they are in progress).

The results obtained seem to contradict to the statement based at lattice calculations that at $\mu = 0$ two phase transitions are to coincide. The situation here is however disputable. Although existence of Q phase was seemimgly never especially looked for in lattice calculations there are some works in which existence of "heavy modes" was assumed in order to get rid of some results looking physically unsatisfactory[16] (the QGP phase does not have properties of an ideal relativistic gas). Lattice specialists should answer the question whether the double phase transition with Q mass at 320 MeV could be noticed in lattice calculations performed until now and how it should show itself.

Acknowledgments

The authors wish to express their gratitude for critical remarks by I.V. Andreev, I.M. Dremin and I.I. Royzen. We are also indebted to G.M. Zinoviev and M. Polikarpov for fruitful discussions. The work was supported partly by the Russian Foundation for Fundamental Researches, grants Nos. 94-02-15558 and 94-02-3815.

REFERENCES

1. B.L. Ioffe, V.A. Khoze, L.N. Lipatov, "Hard Processes". Volume 1. "Phenomenology, Quark-Parton Model", North Holland, Amsterdam, 1984

2. J. Cleymans, K. Redlich, H. Satz, E. Suhonen, Z.Phys.C 33(1986), 151

3. S. Sohlo, E. Suchonen, J. Phys. G. Nucl. Phys. 13(1987), 1487

4. H. Kuono and F. Takagi, Z. Phys. C 42 (1989), 209

5. D.V. Anchishkin, K.A. Bugaev, M.I. Gorenshtein, E. Suhonen, Z. Phys. C45 (1990) 687

6. E.V. Shuryak Phys. Lett. 107B (1981) 103; Nucl. Phys. B203 (1982) 140.

7. E. Feinberg, On Deconfinement of Constituent and Current Quarks in Nucleus-Nucleus Collisions, P.N. Lebedev Institute Preprint, 1989, N 177.

8. E. Feinberg in Relativistic Heavy Ion Collision, ed. by L.P. Czernai and D.D. Strottman, World Scientific, 1991 (see Chapter 5, Section 7)

9. R. Hagedorn and J. Rafelsky, in: Thermodynamics of quarks and hadrons. H.Satz (Ed.). Amsterdam: North Holland 1981;
   R. Hagedorn: Z. Phys. C - Particles and Fields 17 (1983) 265.

10. V.V. Anisovich, M.N. Kobrinsky, Y. Niri and Y.M.Shabelski, Sov. Phys. Uspekhi, 4 (1984) 553 (see section 1).

11. O.D. Chernavskaya, "Double phase transition at zero temperature" (to be published)

12. O.D. Chernavskaya and E.L. Feinberg, "The possibility of double phase transition via the deconfined constituent quark phase" (to be published).

13. E.V.Shuryak, The QCD Vacuum, Hadrons and the Superdense Matter, World Scientific, 1988, see fig.(9.2).

14. Ch. Barter, D. Blaschke, H. Voss, Phys. Lett. B 293 (1992) 423.

15. M. Anikina et.al. Z.Phys. C25 (1984) 1; Phys.Rev. C 23 (1986) 895.
    Okonov E.O. in: Proceed. of Intern. Symposium on Modern Developments in Nuclear Physics, Novosibirsk, 1987, p.166, World Scientific.

16. F. Karsch Prep. CERN TH/13/94.

CRYSTALLINE QUARK-HADRON PHASE
IN NEUTRON STARS *

Norman K. Glendenning

Nuclear Science Division
Lawrence Berkeley Laboratory
Berkeley, California 94720

The textbook example of a first order phase transition is for what we refer to as a simple system, one that has a single conserved charge or independent component. The constancy of the pressure throughout the mixed phase independent of the proportion of the two phases is its most special feature, and one that has noticeable consequences under many circumstances. For example, if a neutron star underwent such a constant pressure phase transition, the mixed phase would be totally absent, since the pressure in a gravitating body is monotonic. But such constant pressure phase transitions are a special case and not at all general. For a general substance having more than one conserved charge we prove two theorems: (1) The pressure and the *nature* of the phases in equilibrium change continuously as the proportion of the phases varies from one pure phase to the other. (2) If one of the conserved charges is the Coulomb force, an intermediate-range order will be created by the competition between Coulomb and surface interface energy. Their sum is minimized when the coexistence phase assumes a Coulomb lattice of one phase immersed in the other. The geometry will vary continuously as the proportion of phases. We illustrate the theorems for a simple description of the hadron to quark phase transition in neutron stars and find a crystalline phase many kilometers thick.

OUTLINE

The only place where quark matter may exist in nature is in the cores of neutron stars. Since the pioneering work of Baym and Chin[1,2] in the mid-seventies it has been debated whether the density is high enough to create this phase. Curiously throughout the original paper and all succeeding ones the true nature of the phase transition was never questioned. It was always described in complete analogy with the text-book example of the water-vapor transition. This contains only one independent component or conserved charge, and is a very

---

*The right is reserved to the author and his sponsor, DOE, to separately publish parts of or an expansion of this conference paper in other conference proceedings or in a book.

special example of a first order phase transition, and not at all general. We call it a 'simple' substance.

In the traditional way of looking at it, the core of the star is purely a quark gas, and the exterior regions a neutron matter liquid, with the inevitable thin solid crust of rather ordinary metals at the surface. The mixed phase is absent because pressure is monotonically changing with depth in a star, while the above description leads to a constant pressure mixed phase, independent of the proportion.

I will show that when full account is taken (1) of the physics of a first order phase transition in a 'complex' substance, such as neutron star matter, which has more than one conserved charge (baryon and electric), and (2) the physical constraint of charge neutrality of stars in its unrestricted sense, that is to say, as a *global* and not a *local* constraint, the situation is much different and much richer.[3-6] From a consideration of the bulk energy, there should exist in the star a broad region that is in the coexistence phase, and taking account of the forces at play, most especially the isospin restoring force in nuclear matter, the Coulomb force, and the surface interface energy, this region should be a Coulomb lattice of quark matter spheres of negative charge immersed in a background of positively charged hadronic matter with size and spacing of sites that minimize the Coulomb and surface interface energies. At greater depth within the star, the roles of quark and confined hadronic matter are interchanged. Between, the geometric structure will vary through other geometries, – rods and slabs. The thickness of the solid crystalline region of the mixed quark-hadron phase in compact stars of somewhat different mass should vary considerably, lending individuality to them. We estimate the thickness of the solid region to be a few kilometers, extending from the center to several kilometers from the edge of the lighter ones while in the heavier ones the central portion is expected to be pure quark matter with the solid mixed phase region outside it and somewhat thinner.

Many observable properties of compact stars will be effected by this thick solid dense interior region, including transport properties, both thermal and electrical, and pulsar glitches are likely to be influenced by if not actually centered in this solid. Indeed, the phenomenology of a compact star is likely to be so strongly coupled to this broad solid region, that I venture to suggest that in time the most convincing evidence of the existence of the quark-hadron phase transition will come from observed phenomena of compact stars and not from laboratory experiments.

DEGREE(S) OF FREEDOM

All research up until 1990, and some later, had made one or other of two idealizations which had the side-effect of freezing out a degree of freedom. Nature on the other hand exploits every degree of freedom to find the lowest energy state. Therefore all work of the last quarter century on this phase transition in neutron stars described an excited, unrealizable state of the star. It is easy to state precisely what the degree of freedom is in the case of a neutron star, and as easy to see the generalization to arbitrary systems. Stars must be electrically neutral. This is because they are bound by the gravitational force, and any net charge would reduce the binding. A charged star would simply acquire from the interstellar medium such charge as needed to bring it to its ground state. *However* charge neutrality is a *global* property. It does not at all imply that the charge *density* is identically zero. There are many examples. The neutron itself has finite charge distributions; – they simply integrate to zero. There is another more relevant example. At the surface of a star the pressure is zero. Therefore the iron on the surface of a neutron star is much like the atomic iron found here on earth at one atmosphere of pressure. But a little way into the interior, the pressure rises, bringing the nuclei closer together. The atomic structure is destroyed. The atoms

130

become fully ionized. Nevertheless the short-range nuclear force holds the nuclei together at sub-nuclear density, rather than allowing them to fill space uniformly at low density. So we have positively charged nuclei in an almost uniform background of electrons. The lowest energy state is a Coulomb crystal.[7] The matter is neutral, but not of vanishing charge density. The system finds it energetically favorable to maintain neutrality globally but not locally, in accord with the nature of the internal force(s). Formerly, charge neutrality was *imposed* locally, either (1) by assuming identically neutral configurations, like pure neutron matter and quark matter with twice as many d-quarks as u-quarks,[2] or (2) by explicitly enforcing local neutrality on a system with charged particles and in beta equilibrium.[1] This had the effect of making the charge chemical potential discontinuous at the boundary between the two phases.

The effect of freezing out the degree of freedom associated with a non-uniform distribution of charge in a two component system is equivalent to reducing it to a one component system, such as water. We recall that giving the system access to a degree of freedom that has been frozen out, is either neutral or *lowers* the energy of the system!

FIRST ORDER PHASE TRANSITIONS IN SIMPLE SUBSTANCES

To better understand the more interesting nature of the mixed phase in a 'complex' substance, we recall the characteristic features in a 'simple' substance. The water-vapor transition is a familiar example of a first order phase transition in a simple substance; the single independent component is H_2O. The remarkable and well known features of the mixed or coexistence phase are: (1) The pressure remains constant on an isotherm for all proportions of the two phases. (2) The nature of each phase in equilibrium is identical for all proportions.

We illustrate the well known pressure relation of a one-component substance in Fig. 1. The coexistence region is H,Q. All points between these two simply represent different

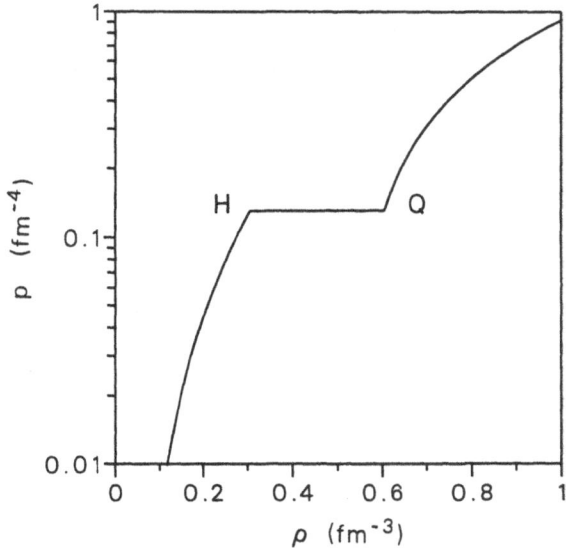

Figure 1. Behavior of the pressure vs. density in the vicinity of a first order phase transition in a simple substance having one chemical potential corresponding to conserved baryon number. Points labeled H and Q mark the end of phase 1 and beginning of phase 2.

volume proportions of the two phases in the states (their corresponding densities, and energies) denoted by H and Q. The total energy is therefore a volume *linear* combination of two *constants*, and this is consistent with the constancy of the pressure since it is the negative of the volume derivative of the energy.

If this were the correct description of the hadron-quark phase transition in a neutron star it would have quite peculiar consequences. The equations of hydrostatic equilibrium, either classical or general relativistic, assure that the pressure is a monotonic decreasing function of distance from the center of the star, just as the pressure in our atmosphere is monotonic. Consequently the end points of the mixed phase, H and Q of Fig. 1, and all between, are mapped onto a single radial point in the star, as in Fig. 2. The densities of the two phases at H and Q are however different. So the mass-energy profile of a star, in a constant pressure transition is discontinuous as in Fig. 2.

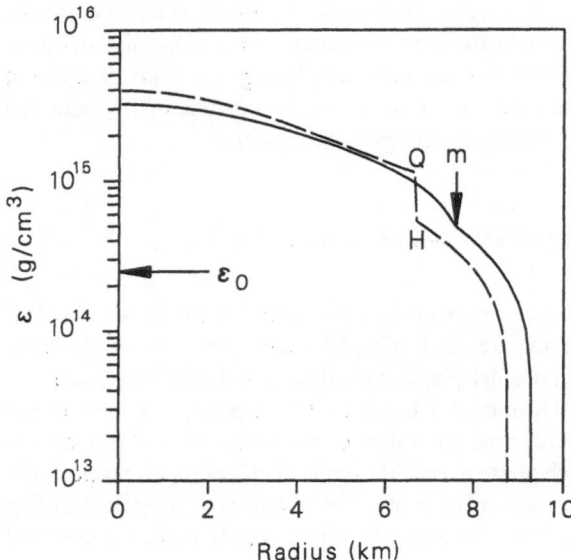

Figure 2. The mass-energy profile of a star showing the mapping of the points H,Q of Fig. 1 in case the transition is treated as a simple one. The solid curve shows the profile when the transition is treated as the 'complex' one it is. The point 'm' marks the transition from mixed to pure hadronic phase.

This is completely analogous with water in equilibrium with its vapor in a heat bath in the presence of the earth's gravitational field, where we would plot h, the height above the bottom of the container instead of r. This is rigorously the only possibility for a *simple* system, one with one component! This analogy of the quark-hadron phase transition with the water-vapor transition has gone unchallenged until very recently.[3,4]

NEUTRON STARS ARE NOT 'SIMPLE' SYSTEMS

There are profound differences between a first order phase transition in a substance which has only one conserved charge as compared to two or more. (Conserved 'charge' means *additive independent attribute* like baryon number, electric charge, strangeness, number of H_2O, etc, etc.)

By one idealization or the other in earlier work, the transition in stars was inadvertently rendered as that of a simple system by one of the restrictions listed below, with the physically unacceptable consequence listed in parenthesis:

(1) Star is *purely* neutron, and converts to quark matter with the same types of quarks, $n_d = 2n_u$. (Both phases are β-unstable).

(2) Star is in β-equilibrium but is required to be everywhere *locally* charge neutral. (This leads to a discontinuous electron chemical potential at the interface of the phases and the constraint is inconsistent with Gibb's criteria!).

Both idealizations freeze out a degree of freedom available to a complex system which allows, even demands, that the pressure vary as the proportion of phases! I will explain this shortly. Moreover, both restrictions rigorously exclude the mixed phase from a star, as we have seen, whereas, with the restrictions lifted, the mixed phase is present and has a most intricate crystalline structure.

By the circumstances of its birth a neutron star has a fixed *baryon* charge, which being conserved, it keeps. Because of beta-equilibrium it also has *charged* particles even though the *net* charge is zero, as explained earlier. Therefore a neutron star has *two* conserved charges and two chemical potentials corresponding to them. We will now see what a difference this makes.

FIRST ORDER TRANSITION IN 'COMPLEX' SYSTEMS

Consider a substance composed of two conserved 'charges', $-Q$ of one kind, B of the other. In the case of a star, these could denote the net electric charge number (in units of e) and baryon charge number. Let the substance be closed and in a heat bath. (This corresponds to the the loss of any energy produced in reaching the ground state of the star by neutrino and photon radiation.) Define their ratio (concentration),

$$r = Q/B. \tag{1}$$

Is this ratio fixed? One would certainly think so since Q and B are fixed. But the ratio is fixed *only* as long as the system remains in one pure phase or the other! When in the mixed phase the ratio in each of the regions of one phase or the other may be different and they are restricted only by the conservation on the total numbers,

$$r' = Q'/B', \quad r'' = Q''/B'', \quad (Q' + Q'' = Q, \quad B' + B'' = B). \tag{2}$$

If the internal forces can lower the energy of the system by rearranging the ratio, it will be done.

In fact it is not possible to insist on a prescribed ratio and at the same time satisfy Gibbs' criteria. There are too many conditions for the number of variables.[4]

The above observations allow us to prove the first theorem stated in the abstract. Consider the system at the density or pressure where the neutron star matter has just begun to condense some quark matter. There is little scope for the internal forces to optimize the concentrations, r', r'', in the two phases, since the small quantity of quark matter can neither receive nor donate much of either charge. However, at higher density or pressure, the proportion of the two phases will become more comparable, and the internal forces now have more scope to optimize the concentrations in the two phases, always consistent with overall conservation of the two charges. From this observation, we learn: (1) The nature of each phase in equilibrium changes with the proportion of the phases. Consequently since the total energy is now the volume proportion of the energy densities of the two phases, each of which *varies* with the proportion, the derivative with respect to volume is no longer a constant. The pressure varies as the proportion of phases!

In contrast to a simple system, the mixed phase of a complex one is *not squeezed out by gravity*. It can occupy a finite radial extent in a star corresponding to the pressure spanned by the mixed phase! We find this crystalline region to be many kilometers thick.

ROLE OF INTERNAL FORCE(S)

We have spoken of the internal force(s) as driving the mixed phase to optimize the concentrations of the conserved charges in the two phases in equilibrium. We now discuss this explicitly, again in terms of a neutron star.

In nuclear matter there is an isospin symmetry restoring force. It arises in part because of the Fermi energies, which if equal for a neutron-proton system minimizes the sum of eigen-energies, and in part because any 3-component of isospin couples to the rho meson, and costs the system the energy of the rho condensate (proportional to the square of the isospin 3-component of the mean rho field strength). From the valley of beta stability we know the preference for symmetry in N (neutron) and Z (proton) numbers. Suppose a system was prepared with N>>Z or, as in a neutron star, is required to obey this condition so that the Coulomb force does not disrupt the star. Then the part of the star that is in the pure hadronic phase is sitting high above the minimum in energy as far as the nuclear force is concerned because of its high isospin. However at some depth within the star where quark matter first condenses, the isospin symmetric nuclear force will be able to rearrange charges to make the hadronic phase more symmetric to the extent permitted by the conditions of overall conservation of charges and the minimal energy. The hadronic matter will become electrically more positively charged, and the quark matter will become negatively charged by a rearrangement of the quark Fermi surfaces, so as to achieve overall electric charge neutrality. Electrons may also permeate the mixed phase region but since they do not contribute to the baryon charge that the star possess by the accident of its birth, and since they simply cost energy if the system can neutralize itself among the baryon carrying charged fermions, they will tend to be quenched.

We note here that the constraint of *local* charge neutrality, which goes beyond what is required for Coulomb stability of a star, freezes out this degree of freedom.

INEVITABILITY OF COULOMB LATTICE

We saw above the role of the internal force(s) in rearranging charges between phases in the coexistence phase so as to minimize the energy consistent with overall conservation of the charges of the system. When one of these corresponds to the long-range Coulomb force, as it does in a neutron star, an intermediate-range ordering of regions of the two phases will minimize the energy as we show.

Charged regions cannot become too extensive because of the repulsive Coulomb self-energy. So they will tend to divide into small regions. The surface interface energy will resist this. *So at any proportion of the phases there is an optimum size and shape of the regions that will arrange themselves into a lattice so as to minimize the surface and Coulomb energies.*

The above considerations prove the second theorem.

The geometric structure of the mixed phase and its variation with proportion of the two phases is easy to compute, in principle. The surface energy per unit volume of a quark drop of radius r in a nuclear background of radius R, chosen so that there is zero net charge in R (Wigner-Seitz cell) is

$$E_S/V = [4\pi r^2 \sigma]/[(4\pi/3)R^3] = (3\sigma\chi)/r \equiv S(\chi)/r, \qquad (3)$$

where for droplets, $\chi = (r/R)^3$ is the volume proportion of the quark phase. Likewise, while more involved to prove,[7] the Coulomb energy per unit volume has the form,

$$E_C/V = C(\chi)r^2 \,. \tag{4}$$

Their sum is a minimum when $E_S = 2E_C$. These equations lead at once to

$$r^3 = S(\chi)/(2C(\chi)) \,. \tag{5}$$

Thus at each proportion χ a definite size of quark drops immersed in the nuclear matter is specified and through $\chi(= (r/R)^3$ for droplets) a definite separation between sites. We note that the long-range of the Coulomb force is screened by the formation of the lattice.

The functions C and S and the proportion χ, expressed in terms of the geometry of the one phase immersed in the other, have quite definite forms for each geometry, for droplets, which merge to form rods, which merge to form slabs, in analogy with the sub-nuclear crystal structure of nuclei immersed in an electron gas, which is believed to form the crust of a neutron star.[7]

LOCAL VS GLOBAL CHARGE NEUTRALITY

Since a star is bound by gravity and any net charge would reduce its binding, a star is charge neutral. This is a *global* condition, not a *local* one. But it can be satisfied by either imposing the conservation of charge in the mixed phase in the *local* sense by demanding that both phases in equilibrium be *separately* charge neutral

$$q_H(\mu_b, \mu_e) = 0, \quad q_Q(\mu_b, \mu_e) = 0 \,, \tag{6}$$

or in the *global* sense,

$$(1 - \chi)q_H(\mu_b, \mu_e) + \chi q_Q(\mu_b, \mu_e) = 0 \,. \tag{7}$$

Since each of these procedures satisfies the constraint of charge neutrality of the star, one may consider the energetics to decide between them. As already explicit in our whole discussion, the second way of calculationally enforcing the physical constraint of charge neutrality allows the internal forces, – the isospin-symmetric nuclear force in the case of a star, – to redistribute charge so as to lower the energy. A system on which stronger constraints are imposed lies higher in energy. So the second way of expressing charge neutrality, (7) is the physical one. One can go further than this. We show that it is not consistent with Gibbs' criteria for phase equilibrium to specify the concentrations r', r'' for the two phases in equilibrium. Thermal, chemical and mechanical equilibrium demand that,

$$p_H(\mu_b, \mu_e, T) = p_Q(\mu_b, \mu_e, T) \,, \tag{8}$$

and baryon number conservation might be expressed in the proportional form (or any other individual form for each phase that conserves baryon number)

$$\rho_H(\mu_b, \mu_e) = (1 - \chi)B/V, \quad \rho_Q(\mu_b, \mu_e) = \chi B/V \,, \tag{9}$$

where ρ_B, ρ_Q are the baryon number densities in the two phases and V is the local volume, (not to be confused with the volume of the star. Rather it is the volume surrounding a sample spacetime point in the star within which the equivalence principle assures that spacetime is flat to arbitrary accuracy. The volume V contains B baryons, and is smaller toward the center of the star. In this way we see that the conventional discussion of a sample substance contained in a laboratory vessel of volume V with a plunger that allows one to vary the pressure, and a surrounding heat bath to keep the substance at constant temperature holds for such local volumes in the star. The pressure in this case is chosen according to the depth in the star and the volume by the requirement that it contain B baryons.)

Alternately, baryon number in the mixed phase can be conserved overall,

$$(1 - \chi)\rho_H(\mu_b, \mu_e) + \chi\rho_Q(\mu_b, \mu_e) = B/V. \qquad (10)$$

Now at any specified proportion of phases, χ the alternative way of global conservation has three unknowns, μ_b, μ_e, V in just that number of equations (7,8,10). The values of the unknowns obviously depend on the proportion of the phases in equilibrium, χ, which formally proves our verbal proof given above. It also shows that the common pressure (8) in the mixed phase varies as the proportion of the phases. This is shown in a sample calculation, Fig. 3. The particle populations are shown in Fig. 4, and in this case the pure quark phase is not achieved in the star. The central region of many kilometers of thickness is in the mixed phase and therefore is of varying crystalline structure.

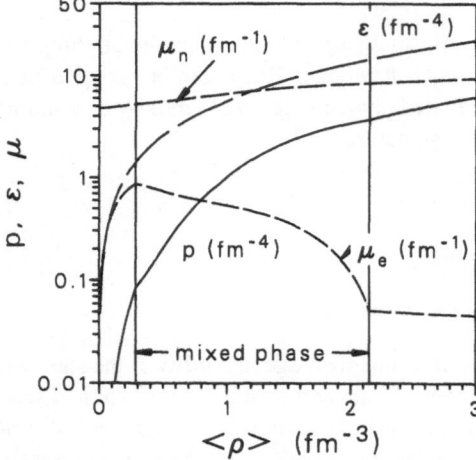

Figure 3. Showing the variation of the pressure in the mixed phase of a complex substance, in particular neutron star matter in the hadronic and quark matter phases.[4]

However, in the local way of attempting to satisfy the conservation laws and phase equilibrium, one sees that there are 5 equations, (6,8,9) that must be satisfied with the same 3 variables as above. Hence Gibb's criteria for phase equilibrium in a first order phase transition of a complex substance *cannot* be satisfied by applying the conservation laws in the *local* sense. The charges cannot be specified arbitrarily in each phase in equilibrium with the other, but rather the internal forces in the substance must be allowed to determine how they are distributed, ie how much of each of the phases contains how much of the each of the conserved charges!

BULK PROPERTIES SET RADIAL SCALE IN STAR

The total energy is, $E_{\text{Total}} = E_{\text{Bulk}} + E_{\text{Surf.+Coul.}}$. The first term dominates and sets the radial scale of the mixed phase in the star. The Surface-Coulomb term alone is a fine variation

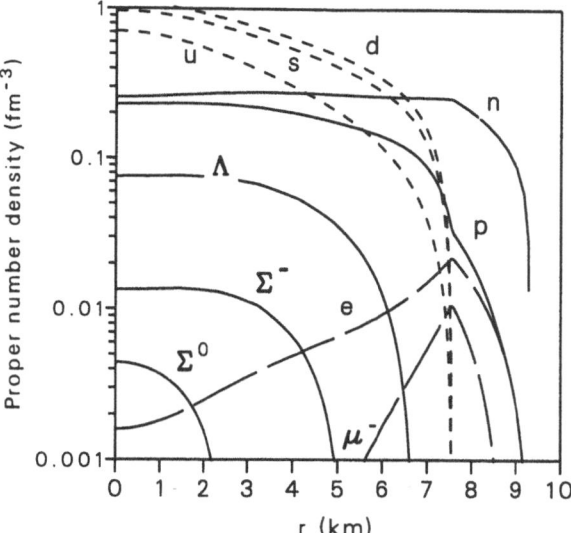

Figure 4. Particle populations as a function of radial distance in a neutron star in which the mixed crystalline phase occupies most of the star from the center to about 7.5 km.[4]

about the first. The general equilibrium condition $E_{\text{Surf}} = 2E_{\text{Coul}}$ assures this. So for assessing the radial *extension* of the solid mixed phase in the star, the bulk properties dominate. The surface and Coulomb energies are important for studying the transport properties and strength of the solid.

CALCULATIONAL EXPEDIENT: BULK AND SURFACE

Ideally one would like to compute the energy of a star allowing full freedom to find the lowest energy state of the Lagrangian or Hamiltonian used to describe its matter. Usually this is a very difficult problem especially when two phases may coexist and the internal force may rearrange the conserved charges optimally in the coexistence phase with a resultant formation of geometrical structure. The approximation of treating large systems in the bulk approximation, in which surface effects are at first ignored, and then later incorporated though a consideration of the surface and Coulomb energy of pieces of the bulk matter, perhaps in the Wigner-Seitz approximation, is well known and generally accurate.

As a matter of principle, we note that the less constrained way of imposing charge neutrality discussed in this paper, as a global condition and *not* as a local one must in general either be neutral or lead to a lower energy. Therefore there is no question as to what range of surface energies will give the ordered phase a lower energy and what range a higher energy, a question raised in ref.[5] The surface energy ought to be computed consistently with the bulk energies of the phases, and ought never to lead to a higher energy for the ordered crystal phase. In this sense we disagree with Heiselberg at al. who otherwise made a start on the study of the geometric structure[5] but use as arbitrary input, the surface interface energy coefficient, unrelated to the theories of the phases or the bulk energies of the phases and their difference.

We note that three authors using the presently described theory of a first order phase transition in a compact stars all agree that the mixed hadronic-quark phase begins to appear at densities between 2-3 of normal nuclear matter density.[4-6] The original estimates of Baym and Chin placed the transition around 10 nuclear matter density. As explained in ref.,[4] this was because highly excited states of both phases were used in the estimate, a purely neutron star with a quark matter core of twice as many d quarks as u, and no strange quarks.

ACKNOWLEDGEMENTS

This work was supported by the Director, Office of Energy Research, Office of High Energy and Nuclear Physics, Division of Nuclear Physics, of the U.S. Department of Energy under Contract DE-AC03-76SF00098.

REFERENCES

1. G. Baym and S. A. Chin, Phys. Lett. **62B** (1976) 241;
   B. D. Keister and L. S. Kisslinger, Phys. Lett. **64B** (1976) 117;
   G. Chapline and M. Nauenberg, Phys. Rev. D **16**(1977) 450;
   W. B. Fechner and P. C. Joss, Nature **274** (1978) 347;
   H. A. Bethe, G. E. Brown and J. Cooperstein, Nucl. Phys. **A462** (1987) 791;
   B. D. Serot and H. Uechi, Ann. Phys. (N. Y.) **179** (1987) 272.

2. N. K. Glendenning, in *Proceedings of the 1989 International Nuclear Physics Conference*, August 1989, Sao Paulo, Brasil, Vol. 2, p. 711, ed. by M. S. Hussein et al., (World Scientific, Singapore, 1990);
   N. K. Glendenning, in *Relativistic Aspects of Nuclear Physics*, August 1989, p. 241, ed. by T. Kodama et al., (World Scientific, Singapore, 1990);
   J. Ellis, J. Kapusta and K. A. Olive, Nucl. Phys. **B348** (1991) 345;
   A. Rosenhauer, E. F. Staubo, L. P. Csernai, T. Overgard, and E. Ostgaard, Nucl. Phys. **540** (1992) 630.

3. N. K. Glendenning, Nuclear Physics B (Proc. Suppl.) **24B** (1991) 110.

4. N. K. Glendenning, Phys. Rev. D, **46** (1992) 1274.

5. H. Heiselberg, C. J. Pethick, and E. F. Staubo, Phys. Rev. Lett. **70** (1993) 1355.

6. V. R. Panharipande and E. F. Staubo, in *International Conference on Astrophysics*, Calcutta, 1993, ed. by B. Sinha, (World Scientific, Singapore).

7. D. G. Ravenhall, C. J. Pethick and J. R. Wilson, Phys. Rev. Lett. **50** (1983) 2066; D. G. Ravenhall, C. J. Pethick and J. M. Lattimer, Nucl. Phys. **A407** (1983) 571, and references cited.

A NEW EFFECTIVE MODEL OF THE QUARK-GLUON PLASMA WITH THERMAL PARTON MASSES

A. Peshier[1], B. Kämpfer[1,2], O. P. Pavlenko[2,3], and G. Soff[1]

[1]Institut für Theoretische Physik, Technische Universität Dresden
 Mommsenstr. 13, 01062 Dresden, Germany
[2]Research Center Rossendorf Inc., PF 510119, 01314 Dresden, Germany
[3]Institute for Theoretical Physics, Kiev, Ukraine

INTRODUCTION

Recent "numerical experiments" provide thermodynamical quantities of pure SU(3) gauge theory in lattice approximation.[1] These data, corrected at least partially for finite lattice effects, are thermodynamically selfconsistent,[2] i.e., energy density and pressure are related via $e = T \partial p / \partial T - p$, and exhibit a striking difference from the popular bag model parametrization $p = e/3 - 4B/3$ (which would imply for the energy density and the pressure a temperature dependence $e, p \propto T^4$).

In Ref.[2] the data[1] were interpreted within an effective model with low momentum cut-off employing bag constant and first order perturbative corrections. Here we follow a different interpretation[3,4] of the data[1] by a non-interacting quasi-particle description with only thermal masses $m(T)$. Such temperature dependent parton masses have been previously extracted[5] from SU(2) lattice gauge data,[6] while in Ref.[7] one describes the lattice data by a bag constant and finite but constant quark masses. We show that the extracted thermal masses indeed can be fitted by a suitably regularized running coupling constant $g(T)$, i.e., $m(T) = \text{const}\, g(T)\, T$. We then extrapolate our approach to the quark-gluon plasma and derive a model equation of state which has a smaller latent heat at confinement in comparison with the bag model. Our model describes available lattice data which incorporate also quarks, despite the fact that they do not seem to reflect the same reliability as the corresponding data for a pure gluon gas.

Finite effective quark and gluon masses have distinct observable consequences, such as the modification of the strangeness production rates[7,8] or the violation of the M_T scaling of dilepton rates.[9]

THE GLUON GAS

Our first aim is to reproduce the gluon lattice data[1] by the ideal gas formula for the pressure (no bag constant is used!)

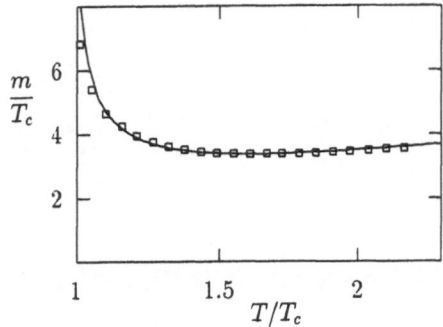

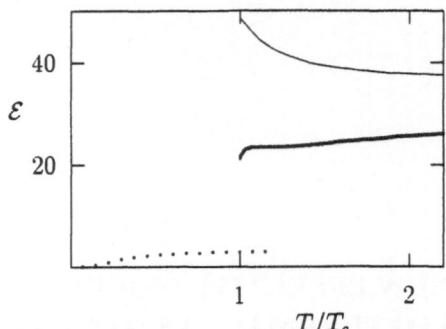

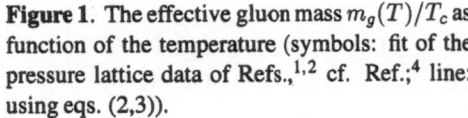

Figure 1. The effective gluon mass $m_g(T)/T_c$ as function of the temperature (symbols: fit of the pressure lattice data of Refs.,[1,2] cf. Ref.;[4] line: using eqs. (2,3)).

Figure 2. The energy density $\epsilon \equiv e/(T^4 \frac{\pi^2}{30})$ as function of the temperature for the parameters described in text. Also displayed are the pion gas (dotted line) and the bag model equation of state (thin line).

$$p(T) = \frac{d_g}{6\pi^2} \frac{T^4}{(\hbar c)^3} \int dx \frac{x^4}{A} \frac{1}{\exp(A) - 1}, \qquad A = \sqrt{\frac{m_g^2(T)}{T^2} + x^2} \qquad (1)$$

with the gluon degeneracy $d_g = 16$ and the temperature dependent effective mass $m(T)$, which is displayed in Fig. 1. Remarkably, the temperature dependence of $m_g(T)$, found in Ref.[5] for the pure SU(2) gauge theory, is very similar to the present SU(3) case: The mass displays a minimum near $1.6\,T_c$, and rises at larger temperatures to achieve at very large T the perturbative regime. Going down in temperature from the mass minimum, the mass increases steeply in approaching the confinement temperature T_c.

Our next aim is to describe the masses $m_g(T)$ as thermal masses with the ansatz

$$m_g^2(T) = \frac{1}{\Gamma_g} g_g^2 T^2 \qquad (2)$$

with a suitable coupling constant g_g. The first-order perturbative running coupling $g^2 = 16\pi^2[4\pi + (11 - \frac{2}{3}N_f) \ln(T/T_c)^2]^{-1}$, ($N_f$ is the quark flavor number put to zero here for the pure gluon gas),[10] is not appropriate since it is too less varying in the considered temperature interval. However, we find that

$$g_g^2 = \frac{16\pi^2}{11 - \frac{2}{3}N_f} \frac{1}{\ln\left(\frac{T+T_s}{T_c}\right)^2} \qquad (3)$$

reproduces the thermal masses quite well, see Fig. 1, for $\Gamma_g = 3.3$ and $T_s/T_c = 0.023$. Eqs. (1 - 3) describe the lattice data[1,2] of the pressure *and* the energy density rather perfectly.[4] The number Γ_g is close to the predictions of the high-temperature perturbative thermal gluon mass, $\Gamma_g = 3$.[10] Our form of g_g(3) can be regarded as phenomenologically regularized coupling constant with shift parameter T_s. (The SU(2) gluon masses[5] are reproduced with $\Gamma_g^{SU(2)} = 9.5$ and $T_s^{SU(2)} = 0.025$.)

THE QUARK-GLUON PLASMA

While the model (1 - 3) might be directly applied to Shuryak's hot glue scenario (i.e., a gluon gas with only a few quarks admixed) we would generalize it to the standard reference model of a quark-gluon plasma in chemical equilibrium. In doing so we employ eqs. (1 - 3) also for quarks and antiquarks (suitably modified for Fermi statistics of fermions) and use $m_q^2 = \Gamma_q^{-1} g_q^2 T^2$, and write $p_{qg} = p_g + p_q$. In expression (3) the number of accessible quark flavors N_f is now assumed to be 2, i.e., $d_q = 24$. This ansatz suggests rather large thermal quark masses > 500 MeV, therefore, one could also include more quark flavors as in Ref.[8] and consider, e.g., $N_f = 3$ or 4. Here we focus on the illustrative example with $N_f = 2$. Since the gluon self-energies are modified by quark loops, the parameters Γ_g and T_s of the pure gluon gas need not necessarily apply for the quark-gluon plasma. Lacking of high-precision lattice data which include quarks reliably we consider here the following parametrization: $\Gamma_g = 9/4$[10] and $\Gamma_q = 6$[11] and adjust T_s to get $T_c = 170$ MeV.[12] A particular choice is $g_g = g_q$. T_c is determined as usual by $p_\pi(T_c) = p_{qg}(T_c)$, where we assume, according to Ref.,[13] that the confined strongly interacting matter (i.e., the hadron gas) is well approximated by a free pion gas up to temperatures $T \sim m_\pi$ and slightly above.

Fig. 2 displays the energy density for the chosen parameters compared with the bag model equation of state. The latent heat depends on $\Gamma_{q,g}$ and T_s, but generally it is less than in the bag model; the jump in the entropy density is reduced in a similar way, too. The available lattice results with quarks[12] give $e/e_{SB} \sim 0.8$ (the subscript SB refers to the Stefan Boltzmann limit) for $T/T_c = 1.2 - 2.3$, while our parameters give $e/e_{SB} \sim 0.75$ at $T \sim 300$ MeV with e/e_{SB} slightly increasing with increasing temperature. A similar agreement with the entropy lattice data[12] is found, where e/e_{SB} slightly above 0.8 can be deduced. Somewhat different parameters are discussed in Ref.[4] The common outcome of such different parameter choices is that at $T \sim 400$ MeV the scaled energy density $e/(T^4 \frac{\pi^2}{30})$, which reflects the number of degrees of freedom, is about 27, while the bag model is near to 37 for two light quarks.

It is worth noticing that present lattice data do not constrain the parameters enough. Besides the values of $\Gamma_{q,g}$ one might also use different parametrizations of g_q and g_g.

APPLICATIONS

One of the most important application of the equation of state of strongly interacting matter is connected with the thermo-hydrodynamical description of the evolution of the quark-gluon plasma in high-energy heavy-ion collisions. We have studied the cooling behavior of the quark-gluon plasma in the idealized Bjorken scenario where $de/d\tau + (e+p)/\tau = 0$, with $\tau = \sqrt{t^2 - x^2}$ as proper time of a plasma piece at longitudinal coordinate x and time t in c.m.s. For a fixed initial entropy density the cooling of pure deconfined matter in our model is slower compared with the bag model, especially near to the transition temperature, while the duration of the mixed phase is shorter due to the reduced latent heat. One can even expect a stronger effect of the delayed phase transition due to the higher viscosity of massive partons as shown in Ref.[14]

In Ref.[4] we applied our model of thermal masses to the dilepton emission from a quark-gluon plasma. The analysis shows the presence of a kinematical threshold in the emission rate caused by the finite thermal masses. E.g., at a given transverse dilepton mass $M_T = (M^2 + Q_T^2)^{1/2}$ the minimum invariant mass M is given by twice the quark mass in case of the dominating quark fusion processes $q\bar{q} \to l\bar{l}$; therefore at large transverse dilepton momentum Q_T the rate has to approach zero due to kinematics. If the thermal quark mass is large enough it might lead to the violation of the M_T scaling of the thermal dilepton spectra as pointed out in Ref.[9] For a selfconsistent calculation the in-medium vertex modifications need to be included, e.g., in line with the Braaten Pisarski approach of resumming the hard thermal loops.

The main conclusion is that the value of the threshold at larger Q_T is mainly related to the effective quark masses at temperature being near T_c and appears to be almost independent of the initial state of the plasma and its hydrodynamical evolution.

At the same time the dilepton emission reflects another important consequence of our quark-gluon plasma model. It is related to the change of the latent heat of the deconfinement transition and slower cooling of the plasma during the hydrodynamical expansion in comparison with the standard bag model. Actually it leads to a higher initial temperature of the matter at fixed entropy density (or rapidity density of final pions), and consequently to an additional dilepton yield compared to bag model estimates in the invariant mass region $M \sim$ 2 - 3 GeV. This is in accordance with the recently observed dilepton continuum in central S - W collisions at CERN - SPS energies.[15] Although the invariant dilepton mass region and the rapidity density of secondary pions achieved in these collisions seem to correspond to the formation of quark - hadron mixed phase, the observed yield cannot be explained in the framework of a mixed phase picture which is based on the bag model.[16] Our model in principle opens a natural way for an explanation.

Acknowledgments

Helpful discussions with J. Engels, M.I. Gorenstein, U. Heinz, B. Müller, D. Rischke, H. Satz and G.M. Zinovjev are gratefully acknowledged. The work is supported in part by BMFT under 06 DR 107, and grant No. U4D000 by International Scientific Foundation.

REFERENCES

1. J. Engels, J. Fingberg, F. Karsch, D. Miller, M. Weber, *Phys. Lett.* B252: 625 (1990).

2. D. Rischke, M. Gorenstein, A. Schäfer, H. Stöcker, W. Greiner, *Phys. Lett.* B278: 19 (1992); *Z. Phys.* C56: 325 (1992).

3. B. Kämpfer, O.P. Pavlenko, FZR-93-16, *Nucl. Phys.* A566: 351c (1994).

4. A. Peshier B. Kämpfer, O.P. Pavlenko, G. Soff, preprint FZR-94-49, *Phys. Lett.* in print.

5. V. Goloviznin, H. Satz, *Z. Phys.* C57: 671 (1993).

6. J. Engels, F. Karsch, K. Redlich, preprint BI-TP 94/30

7. T.S. Biro, P. Levai, B. Müller, *Phys. Rev.* D42: 3078 (1990).

8. J. Letessier, J. Rafelski, A. Tounsi, *Phys. Lett.* B323: 393 (1994).

9. B. Kämpfer, O.P. Pavlenko, *Phys. Rev.* C49: 2716 (1994).

10. V.V. Klimov, *Sov. Phys. JETP* 55: 199 (1982).
    J.I. Kapusta, "Finite-Temperature Field Theory", Cambridge Uni. Press, Cambridge (1989).

11. H.A. Weldon, *Phys. Rev.* D26: 1394 (1982).

12. A. Ukawa, *Nucl. Phys.* A498: 227c (1989).
    B. Petersson, *Nucl. Phys.* A525: 237c (1991).
    S. Gottlieb et al., *Phys. Rev.* D47: 3619 (1993).

13. U.G. Meissner, *Nucl. Phys.* A566: 141c (1994), and further Refs. therein.
    G.G. Bunatian, B. Kämpfer, preprint FZR-93-28.

14. J. Cleymans, S.V. Ilyin, S.A. Smolyansky, G.M. Zinovjev, *Z. Phys.* C62: 75 (1994).

15. M.A. Massoni (HELIOS/3 collaboration), *Nucl. Phys.* A566: 95c (1994).

16. E.V. Shuryak, L. Xiong, preprint SUNY-NTG-94-14.

HADRONIC MATTER WITH INTERNAL SYMMETRIES
AND ITS CONSEQUENCES:
AN EXPANDING HADRONIC GAS

Ludwik Turko

Institute of Theoretical Physics, University of Wrocław
Pl.Maksa Borna 9, 50-204 Wrocław, Poland

INTRODUCTION

Let us consider a heavy ion collision which goes through the intermediate state of an hadronic matter being in the thermal and chemical equilibrium. This matter can be considered as a gas which expands longitudinally according to a hydrodynamical evolution. In the simplest case the hadronic matter is supposed to be made of non - interacting massless pions.[1] However an assumption of a thermodynamical equilibrium leads to the conclusions that a system under consideration should consist of all species of hadrons allowed by conditions of the equilibrium. We are going to consider an ideal gas consisting of all species of hadrons up to Ω^- baryon. These realistic hadrons need that conservation laws due to an internal symmetries should be taken into account. We take into account strangeness and baryon number conservation. Electric charge conservation has no influence on the solutions to the equation of state because of almost vanishing isospin for medium size nuclei.

The aim of this talk is to present some of pecularities of this more realistic hadronic gas as compared to the simple pion gas. Some of these comparison were published before to study J/Ψ suppression patterns.[2]

EQUATION OF STATE

For an ideal hadron gas in thermal and chemical equilibrium, which consists of l species of particles, energy density ϵ, baryon number density n_B, strangeness density n_S and entropy density s read ($\hbar = c = 1$ always)

$$\epsilon = \frac{1}{2\pi^2} \sum_{i=1}^{l} (2s_i + 1) \int_0^\infty \frac{dp\, p^2 E_i}{\exp\left\{\frac{E_i - \mu_i}{T}\right\} + g_i} \,, \tag{1a}$$

$$n_B = \frac{1}{2\pi^2} \sum_{i=1}^{l} (2s_i + 1) \int_0^\infty \frac{dp\, p^2 B_i}{\exp\left\{\frac{E_i - \mu_i}{T}\right\} + g_i} \,, \tag{1b}$$

$$n_S = \frac{1}{2\pi^2} \sum_{i=1}^{l} (2s_i + 1) \int_0^\infty \frac{dp\, p^2\, S_i}{\exp\left\{\frac{E_i - \mu_i}{T}\right\} + g_i} , \qquad (1c)$$

$$s = \frac{1}{6\pi^2 T^2} \sum_{i=1}^{l} (2s_i + 1) \int_0^\infty \frac{dp\, p^4}{E_i} \frac{(E_i - \mu_i) \exp\left\{\frac{E_i - \mu_i}{T}\right\}}{\left(\exp\left\{\frac{E_i - \mu_i}{T}\right\} + g_i\right)^2} , \qquad (1d)$$

where $E_i = (m_i^2 + p^2)^{1/2}$ and m_i, B_i, S_i, μ_i, s_i and g_i are the mass, baryon number, strangeness, chemical potential, spin and a statistical factor of specie i respectively (we treat an antiparticle as a different specie). And $\mu_i = B_i \mu_B + S_i \mu_S$, where μ_B and μ_S are overall baryon number and strangeness chemical potentials respectively.

In general, all densities of the left sides of eqs.1a-d are functions of time during the cooling of the hadron gas. This means that gas parameters such as temperature and chemical potentials should be functions of time. We would like to obtain these functions explicite. This can be done by solving numerically the system of equations for entropy density s, baryon number density n_B and strangeness density n_S, with s, n_B and n_S given as time dependent quantities. From the Bjorken model we have the following solution for the longitudinal expansion[1,5]

$$s(t) = \frac{s_0 t_0}{t} , \qquad n_B(t) = \frac{n_B^0 t_0}{t} , \qquad (2)$$

where s_0 and n_B^0 are initial densities of the entropy and the baryon number respectively. The overall strangeness is equal to zero during all the evolution. So to solve (1b,c,d) with s and n_B given by (2) and $n_S = 0$, we need to know initial values s_0 and n_B^0. To estimate initial baryon number density n_B^0 we use experimental results of Ref.[3].

These results are for S-S collisions, but since there are no data on baryon multiplicities for heavier nuclei we have to evaluate them in some way. We assume that the baryon multiplicity per unit rapidity in the CRR is proportional to the number of participating nucleons. For a sulphur-sulphur collision we have $dN_B/dy \cong 6$ and 64 participating nucleons.[3] For an O-U collision we can roughly estimate the number of participating nucleons at $16 + 58 = 74$. The second factor of the sum has been obtained by the following assumption: since an oxygen nucleon is much smaller than an uranium one, we can approximate the part of the uranium, through which the oxygen passes, by the cylinder of the volume equal to $\pi R_O^2 \cdot 2R_U$. The same procedure can be applied to the S-U case. Here we obtain $32 + 93 = 125$ participants. Therefore, we have $dN_B/dy \cong 7$ and $dN_B/dy \cong 11.7$ for O-U and S-U collisions respectively. Having taken the initial volume in the CRR equal to $\pi R_A^2 \cdot 1$ fm, we arrive at $n_B^0 \cong 0.25$ fm^{-3} for both cases.

To find s_0, first we have to solve (1a,b,c) with respect to T, μ_S and μ_B, where we put $\epsilon = \epsilon_0$ and so on. For ϵ_0 we have taken estimates given in Ref.[4]

RESULTS

As a result, we have obtained $T_0 \cong 212$ MeV, $\mu_S^0 \cong 34.6$ MeV and $\mu_B^0 \cong 133$ MeV for the O-U collision ($\epsilon_0 = 2.5$ GeV/fm^3) and $T_0 \cong 209.1$ MeV, $\mu_S^0 \cong 36.7$ MeV and $\mu_B^0 \cong 143.3$ MeV for the S-U collision ($\epsilon_0 = 2.3$ GeV/fm^3). Then, from (1d) we have $s_0 \cong 13.68$ fm^{-3} for O-U and $s_0 \cong 12.74$ fm^{-3} for S-U. Now, having put (2) and $n_S = 0$ into (1b,c,d), we can solve them numerically to obtain T, μ_S and μ_B as functions of time. Our results are presented in fig.1, where solid, long-dashed and dashed lines mean the temperature, the strangeness chemical potential and the baryon chemical potential respectively. Fig.1 shows results for $S - S$ collision initial conditions. The time scale is chosen in a way which enables the temperature to reach the freez-out at 140 MeV. This corresponds to the freez-out time equal to $t_{f.o.} \cong 10.4$ fm .

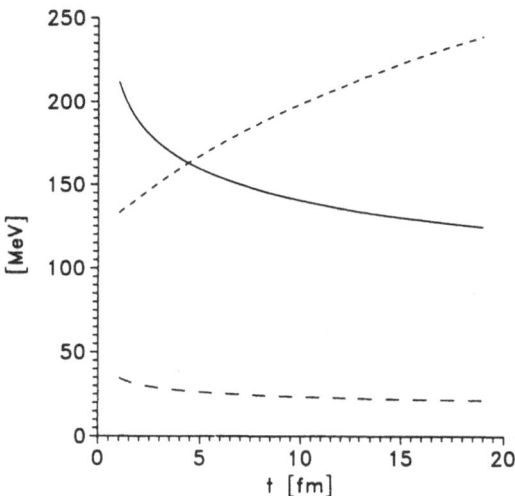

Figure 1. Dependence of chemical potentials and temperature on time for S-S collision initial condition. Solid, long-dashed and dashed lines mean the temperature, the strangeness chemical potential and the baryon chemical potential respectively.

The solution for the temperature function can be approximated by an expression of the form

$$T(t) = T_0 \cdot \left(\frac{1}{t}\right)^a , \qquad (3)$$

where
$T_0 = 212.4 MeV, a = 0.178$ for $O - U$ system
and
$T_0 = 209.7 MeV, a = 0.179$ for $S - S$ system.

We can see that all above expressions have the form known from the solution for the longitudinal expansion of a baryonless gas with the sound velocity constant, namely[5]

$$T(t) = T_0 \cdot \left(\frac{1}{t}\right)^{c_s^2} , \qquad (4)$$

where c_s is the sound velocity and we put the initial time t_0 equal to 1 fm. We have checked that for $n_B = 0$ results for the temperature function are very similar to those in eq.3. And the following approximations of the temperature function hold

$$T(t) = 215.2 \cdot \left(\frac{1}{t}\right)^{0.180} \qquad (4a)$$

$$T(t) = 208.9 \cdot \left(\frac{1}{t}\right)^{0.172} \qquad (4b)$$

We can see that the power in eqs.3 and 4 is around 0.18. Therefore a question arises: is this value connected in any way with the sound velocity of the hadron gas? For the baryonless case the answer is positive. We have checked this computing straightforward $c_s^2 = s/(T\frac{\partial s}{\partial T})$, which is the value of the sound velocity squared for a baryonless gas.[5] The results are presented in Fig. 2. For comparision, the square of the sound velocity of a pion gas is also depicted (dashed lines). We can see that in the range of the temperature 200-140 MeV the square of the speed of sound equals 0.17-0.18 indeed. We have also found a region of the temperature where the sound velocity decreases as the temperature increases. Nevertheless, the condition for the stability of the expansion,[6] $d/dT(sc_s/T) > 0$ is still valid because sc_s/T is an increasing function of time.

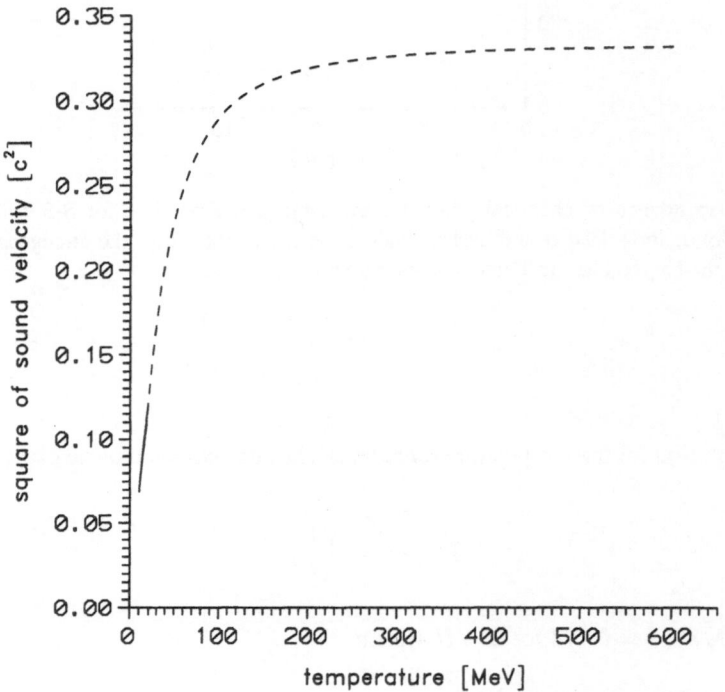

Figure 2. Time-dependence of the square of the sound velocity for the expanding hadron gas. Long-dashed and solid lines mean the square of the sound velocity for the pion gas and for the baryonless hadron case respectively.

Acknowledgments

It is pleasure to thank Prof. J.Rafelski for his kind invitation to contribute to this workshop and for his warm hospitality.

REFERENCES

1. J.D.Bjorken, Phys.Rev.D 27 (1983) 140

2. D.Prorok and L.Turko, Z.Phys.C - Particles and Fields 61 (1994) 109

3. H.Ströbele (NA35 Collab.), Nucl.Phys.A 525 (1991) 59c

4. M.C.Abreu et al. (NA38 Collab.), Nucl.Phys.A 544 (1992) 209c

5. J.Cleymans, R.V.Gavai and E.Suhonen, Phys.Rep.130 (1986) 217

6. G.Baym et al., Nucl.Phys.A 407 (1983) 541

ON THE STATISTICS-CHANGING PHASE TRANSITION
IN GAUGE THEORIES

Stefan Mashkevich and Gennady Zinovjev

Bogolyubov Institute for Theoretical Physics
National Academy of Sciences of Ukraine
252143 Kiev, Ukraine

THE CHERN–SIMONS TERM

In (2+1)-dimensional gauge theories the fermion mass term breaks P- and T-symmetry. Taking a Lagrangian which describes a gauge field coupled to a fermion field,

$$\mathcal{L} = \bar{\psi}\left(i\gamma^{\mu}\partial_{\mu} + \gamma^{\mu}A_{\mu} - M + i\gamma^{0}\mu\right)\psi \tag{1}$$

and integrating out ψ, one obtains the effective action for A_{μ} which contains the *Chern–Simons term*

$$\mathcal{L}_{\text{CS}} = \frac{\alpha}{2}\varepsilon^{\mu\nu\lambda}A_{\mu}\partial_{\nu}A_{\lambda} \tag{2}$$

where the coefficient is given by[1]

$$\alpha = -\frac{M}{|M|}\frac{1}{4\pi}\frac{\sinh(\beta|M|)}{\cosh(\beta|M|) + \cosh(\beta\mu)}, \tag{3}$$

β being the inverse temperature and μ the chemical potential. The Chern–Simons term is sometimes called topological term due to its behavior under gauge transformations: It is invariant under a topologically trivial transformation (the one which can be continuously deformed into the identity transformation) but adds a constant under a topologically nontrivial transformation, proportional to the winding number of that transformation.

Apart from the "straightforward" way of obtaining the Chern–Simons term in an initially (2+1)-dimensional theory, it may appear as a result of dimensional reduction at finite temperatures and densities, starting from a (3+1)-dimensional theory.[2] The mechanism of this phenomena is not yet completely clarified (see also[3]) but it is anticipated that it could play an essential role in determining the structure of finite temperature gauge theories.

We will consider models of matter (conserved current j_{μ}) coupled to a Chern–Simons gauge field (A_{μ}):

$$\mathcal{L} = \mathcal{L}_{\text{m}}(j_{\mu}) + \mathcal{L}_{0}(A_{\mu}) + \mathcal{L}_{\text{CS}} + j^{\mu}A_{\mu}. \tag{4}$$

ANYONS

The simplest case is $\mathcal{L}_0 = 0$, i.e. a pure Chern–Simons gauge field. The gauge field equations obtained from (4) take the form

$$\frac{\alpha}{2}\varepsilon^{\mu\nu\lambda}F_{\nu\lambda} = j^{\mu}. \tag{5}$$

Thus, the gauge field has no independent dynamics but is completely determined by the current distribution. Namely, the magnetic field $B = -F_{12}$ is proportional to the charge density j_0. In particular, a point charge $j_0 = e\delta^2(\mathbf{r})$ produces a δ-shape magnetic field

$$B = -\frac{e}{\alpha}\delta^2(\mathbf{r}). \tag{6}$$

In other words, a point charge is at the same time a magnetic flux tube with the flux $\Phi = -\frac{2\pi\Delta}{e}$ where

$$\Delta = \frac{e^2}{2\pi\alpha}, \tag{7}$$

that is, becomes a *charge-flux composite*. (Note that the gauge field should not necessarily be of electromagnetic nature; we use electromagnetic terms mainly for clarity.) Interchanging two such composites will lead to an additional phase acquired by their wave function due to the Aharonov–Bohm effect. Thus, the transformation of the wave function under such interchange is

$$\psi \rightarrow \psi \cdot \exp[i\pi\Delta] \tag{8}$$

if the particles in question "themselves" are bosons, an additional minus sign is to be added if they are fermions. As α changes continuously, the phase is in general fractional. In other words, the particles undergo an effective change of statistics, i.e. become *anyons*[4,5] whose properties continuously interpolate between those of bosons and fermions as Δ, which is usually referred to as the statistical parameter, changes from an even to an odd number. As is common for the Aharonov–Bohm effect, the vector potential corresponding to (6) is evidently a pure gauge everywhere except the origin, therefore the classical interaction force vanishes, but the wave function cannot be represented in terms of one-particle ones.

DISTANCE DEPENDENT STATISTICS

Now let $\mathcal{L}_0 \neq 0$, i.e., consider a gauge field the Lagrangian which is the Chern–Simons term plus some term of non-topological nature. The field equations will then take the form

$$\frac{\partial \mathcal{L}_0}{\partial A_\mu} - \partial_\nu\left[\frac{\partial \mathcal{L}_0}{\partial(\partial_\nu A_\mu)}\right] + \frac{\alpha}{2}\varepsilon^{\mu\nu\lambda}F_{\nu\lambda} = j^{\mu}. \tag{9}$$

In general, this means that a charge will now produce both an electric field and a magnetic field. The electric field leads to a Coulomb-like charge–charge interaction, in general screened, which is not our main concern. Now, the interchange phase acquired by a wave function of two charges is path dependent (it is proportional to the magnetic flux through the area spanned by the interchange path). If, for example, the path is a circle of radius r, then

$$\psi \rightarrow \psi \cdot \exp[i\pi\Delta(r)] \tag{10}$$

where the form of the function $\Delta(r)$ depends of course on $\mathcal{L}_0$, being determined from the solution of (9). In general, the function $\Delta(r)$ is characterized by a distance scale d and two asymptotic values, Δ_0 for $r \ll d$ and Δ_∞ for $r \gg d$. Consequently, the particles behave as anyons with statistical parameter Δ_0 and Δ_∞, being close together or well apart, respectively. It may be said that they obey *distance dependent statistics*[6]—their statistics effectively changes depending on distance between them.

An important particular case of this phenomena is the Maxwell–Chern–Simons theory,[7,8] with $\mathcal{L}_0 = -\frac{1}{4}F_{\mu\nu}F^{\mu\nu}$, one has

$$\Delta(r) = \frac{e^2}{2\pi\alpha}[1 - \alpha r K_1(\alpha r)]. \tag{11}$$

Here, $d \sim \frac{1}{\alpha}$, $\Delta_0 = 0$, $\Delta_\infty = \frac{e^2}{2\pi\alpha}$. Thus, at large distances the Maxwell term has no effect, and the particles become noninteracting anyons (in the above sense); on the contrary, at small distances the Chern–Simons term has no effect, so that there is no change of statistics but only the usual Coulomb interaction between charges. This might be anticipated from a general consideration, since the Chern–Simons term and the Maxwell term are proportional to the first and the second power of momentum and therefore dominate at small and large momenta, respectively.

PHASE TRANSITION

Finally, let $\mathcal{L}_0 = 0$ but $\mathcal{L}_m$ allow a spontaneous symmetry breaking, e.g.

$$\mathcal{L} = \frac{1}{2}(D_\mu\phi)^2 - V(\phi) - \frac{\alpha}{2}\varepsilon^{\mu\nu\lambda}A_\mu\partial_\nu A_\lambda, \tag{12}$$

with $V(\phi) = \frac{m}{2}\phi^2 + \frac{\lambda}{4}\phi^4$, $m < 0$. If there is no gauge field, it is well known that a phase transition occurs: Below a critical temperature T_c the symmetry is spontaneously broken, above T_c it is restored. What is the influence of the Chern–Simons term on this picture?

In the symmetric phase one has the general picture for the matter coupled to a Chern–Simons gauge field: Particles become anyons with the statistical parameter determined by the value of α as above. However, in the broken phase the picture changes. Acting in the standard way, we define the new fields χ and X_μ by the equalities

$$\phi = (\phi_0 + \chi)\exp[i\eta], \tag{13}$$
$$A_\mu = X_\mu - \partial_\mu\eta, \tag{14}$$

ϕ_0 being the vacuum expectation of ϕ. The Lagrangian (12) is then rewritten as

$$\mathcal{L} = \frac{1}{2}(\partial_\mu\chi)^2 - \frac{1}{2}m_\chi^2\chi^2 + \frac{1}{2}(\phi_0 + \chi)^2 X_\mu X^\mu - -\frac{\alpha}{2}\varepsilon^{\mu\nu\lambda}X_\mu\partial_\nu X_\lambda. \tag{15}$$

As it should, it describes a matter field coupled to a now massive gauge field. The mass term of the gauge field plays the role of $\mathcal{L}_0$ in (4). This situation corresponds to distance dependent statistics. Note that the structure of the two terms in the Lagrangian dictates a picture in a sense *opposite* to the one in the Maxwell–Chern–Simons theory. Since $X_\mu X^\mu$ is proportional to the zeroth power of momentum, the influence of the Chern–Simons term is suppressed at long distances and the influence of the mass term—at short ones. Therefore at extremely high densities there is no effective change of statistics, while at extremely low densities there is an abrupt change from bosons (fermions) to anyons. This change is superimposed on the continuous temperature dependence of statistics determined by the dependence of the coefficient in front of the Chern–Simons term, Eq. (3).

We conclude that in (2+1)-dimensional field theories, whenever there appears a Chern–Simons term *and* a spontaneous symmetry breaking, a "statistics-changing phase transition"[9] is possible. Still, this is a general picture involving in fact more questions than answers; in particular, it is unclear which quantity is to play the role of order parameter, and of course the actual calculation of the temperature of the transition is absent, as well as the complete understanding of the structure of the broken phase. We hope to address these questions in future publications.

REFERENCES

1. E.R. Poppitz, *Phys. Lett.* B252:417 (1990).

2. A.N. Redlich, L.C.R. Wijewardhana, *Phys. Rev. Lett.* 54:970 (1985).

3. O.A. Borisenko, V.K. Petrov, G.M. Zinovjev, Bogolyubov Institute for Theoretical Physics preprint ITP-92-8E (to appear in *Nucl. Phys.* B).

4. J.M. Leinaas, J. Myrheim, *Nuovo Cimento* 37B:1 (1977).

5. F. Wilczek, *Phys. Rev. Lett.* 48:957 (1982).

6. S.V. Mashkevich, *Phys. Rev.* D48:5953 (1993).

7. K. Shizuya, H. Tamura, *Phys. Lett.* B252:412 (1990).

8. G.M. Zinovjev, S.V. Mashkevich, H. Sato, *JETP* 78:105 (1994). [Russian original: *Zh. Eksp. Teor. Fiz.* 105:198 (1994).]

9. X.G. Wen, A. Zee, *J. Phys. France* 50:1623 (1989).

FLUCTUATION CORRECTIONS TO BUBBLE NUCLEATION

Jürgen Baacke

Institut für Physik, Universität Dortmund
D - 44221 Dortmund , Germany

ABSTRACT

I discuss the numerical computation of fluctuation corrections to instanton and sphaleron transition rates in quantum field theory. As an example of such a computation I will consider the bubble nucleation rate in the context of the electroweak phase transition. Some of the material presented here is based on work done in collaboration with V. G. Kiselev.

INTRODUCTION

Phase transitions in high energy physics are at present a subject of intensive study. While the main subject of this workshop is the transition from the quark-gluon plasma to the hadronic phase[1] and its possible observation, my present work is related to the electroweak phase transition. It is actually a technical subject to which I will address myself here, the physics of the phase transition serving as a background scenario. The technical problem occurs within the semiclassical or saddle point approximation. It is usually simple, even in field theory, to obtain the exponential factors describing the tunneling through barriers (instanton type configurations) or the Boltzmann factors describing the thermal transition over barriers (sphaleron type configurations). Matters become involved once one tries to go beyond the leading approximation and aims at determining the pre-exponential factors. These are, in the Gaussian approximation to the path integral or to the partition function, given by the determinant of the differential operator describing the fluctuations around the classical background field. Usually even the classical field configurations are known only numerically, so exact computations of the fluctuation determinant have in general to be done numerically. It is this point which I am going to discuss.

Why should we be interested in the pre-exponential factors for a transition rate? Of course, for the semiclassical or saddle point approximation to make sense, these factors should be of order 1, their logarithm should be much smaller than the leading classical action. But this precisely might not be the case; in some models , e.g. the electroweak theory, we have a large number of fluctuating fields and there is a rapid change of scale with temperature due to the phase transition; large mass scales (e.g. the top quark mass) may also lead to large corrections; so checking the validity of the semiclassical approximation is a first reason for such computations. While presumably the details of the electroweak phase transition may not be studied quantitatively by cosmological observations, such transition rates may possibly be determined to sufficient accuracy in lattice simulations.[2] Computing them with a similar accuracy may allow to check the inherent assumption that the classical or mean field configurations do really dominate the path integral or partition function and therefore describe the relevant physics.

Of course, in order to conform with the subject of this workshop, I will choose an appropriate scenario for such a computation: bubble nucleation in a first order phase transition. I will choose the electroweak phase transition, though the physics of this phase transition is by no means understood at present;[3] despite of the unclear situation there is a kind of standard scenario[4,5] which is based on the Coleman-Weinberg[6] effective potential adapted to the electroweak model. I will choose this scenario for my computations.

In the next section I will introduce the model and set up the basic relations for the bubble nucleation rate. In section 3 I will discuss the fluctuation operator. The computation of its determinant, based on a very useful theorem, will be described in section 4. In the final section I will present some results and my conclusions.

THE ELECTROWEAK PHASE TRANSITION

In the following I will use a "standard scenario" for the electroweak phase transition in the formulation of Dine et al.[5] It is described by the high temperature action

$$S_{ht} = \frac{1}{g_3(T)^2} \int d^3x \left[\frac{1}{4} F_{ij} F_{ij} + (D_i \Phi)^\dagger (D_i \Phi) + V_{ht}(\Phi^\dagger \Phi) \right] . \tag{1}$$

Here the coordinates and fields have been rescaled as[7]

$$\vec{x} \to \frac{\vec{x}}{gv(T)}, \ \Phi \to v(T)\Phi, \ A \to v(T)A . \tag{2}$$

The vacuum expectation value $v(T)$ is defined as

$$v^2(T) = \frac{2D}{\lambda_T}(T_0^2 - T^2) . \tag{3}$$

T_0 is the temperature at which the high temperature potential V_{ht} changes its curvature at $\Phi = 0$ from a minimum at $T > T_0$ to a maximum at $T < T_0$. The temperature dependent coupling of the three-dimensional theory is defined as

$$g_3(T)^2 = \frac{gT}{v(T)} . \tag{4}$$

In terms of the zero temperature parameters we have $m_W = gv_0/2$, $m_H = \sqrt{2\lambda}v_0$ with $v_0 = 246$ and we use the definitions of Dine et al.[5] modified by setting $\Theta_W = 0$ and therefore $m_W = m_Z$:

$$D = (3m_W^2 + 2m_t^2)/8v_0^2$$

$$E = 3g^3/32\pi$$

$$B = 3(3m_W^4 - 4m_t^4)/64\pi^2 v_0^4$$

$$T_0^2 = (m_H^4 - 8v_0^2 B)/4D \tag{5}$$

$$\lambda_T = \lambda - 3(3m_W^4 \ln \frac{m_w^2}{a_B T^2} - 4m_t^4 \ln \frac{m_t^2}{a_F T^2})/16\pi^2 v_0^4 . \tag{6}$$

In terms of these parameters the high-temperature potential is given by

$$V_{ht}(\Phi^\dagger\Phi) = \frac{\lambda_T}{g^2} \left((\Phi^\dagger\Phi)^2 - 2\Phi^\dagger\Phi - \frac{4E}{\lambda_T v(T)}(\Phi^\dagger\Phi)^{3/2} \right) . \tag{7}$$

The rescaling Eq. (2) with the scale $v(T)$ makes sense only for $T < T_0$. On the other hand the high temperature potential has, before rescaling, a secondary minimum at $|\Phi| = \tilde{v}(T)$ with

$$\tilde{v}(T) = \frac{3ET}{2\lambda} + \sqrt{\left(\frac{3ET}{2\lambda}\right)^2 + v^2(T)} . \tag{8}$$

This minimum is degenerate with the one at $\Phi = 0$ at a temperature defined implicitly by

$$T_C = T_0/\sqrt{1 - E^2/\lambda_{T_C}} . \tag{9}$$

T_C marks the onset of bubble formation by thermal barrier transition. In the work of Hellmund et al.[8] and Kripfganz et al.[9] the vacuum expectation value of the broken symmetry phase $\tilde{v}(T)$ is chosen for the rescaling of the fields, i.e. in Eq. (2) $v(T)$ is replaced by $\tilde{v}(T)$ and the high temperature coupling constant Eq.(4) is redefined analoguously and denoted[*] as $\tilde{g}_3(T)$. By this change of scale the high-temperature potential changes as well; it becomes (we do not introduce a tilde for the rescaled fields)

$$V_{ht}(\Phi^\dagger\Phi) = \frac{\lambda_T}{4g^2} \left((\Phi^\dagger\Phi)^2 - \epsilon(T)(\Phi^\dagger\Phi)^{3/2} + (\frac{3}{2}\epsilon(T) - 2)\Phi^\dagger\Phi \right) \tag{10}$$

with

$$\epsilon(T) = \frac{4}{3} \left(1 - \frac{v(T)^2}{\tilde{v}(T)^2} \right) \tag{11}$$

If one puts numbers into the relations presented above one finds that, e.g. for $m_H = m_W = 80.2 GeV$ and $m_t = 170$ GeV (to conform with computations in Ref.[9]), the whole process of bubble nucleation occurs at a temperature of 115 GeV within a temperature range of less than 1 GeV. So the cosmological expansion during the time of the phase transition will be negligible and we can use a time-independent metric. The typical bubble size is also (initially) smaller than the curvature scale and we may perform our computation in flat space.

The process of bubble nucleation is — within the approach of Langer,[10] followed by the work of Affleck,[11] Linde[12] and others — described by the rate

$$\Gamma/V = \frac{\omega_-}{2\pi} \left(\frac{\tilde{S}}{2\pi} \right)^{3/2} \exp(-\tilde{S}) \, \mathcal{J}^{-1/2} . \tag{12}$$

Here $\tilde{S}$ is the high temperature action, Eq. (1), with the new rescaling, minimized by a classical minimal bubble configuration (see below), $\mathcal{J}$ is the fluctuation determinant which describes the next-to-leading part of the semiclassical approach and which will be defined below; its logarithm is the 1-loop effective action. Finally ω_- is the absolute value of the unstable mode frequency.

---

[*]Our notation differs from the one of Refs.[8,9]

The classical bubble configuration is described by a vanishing gauge field and a real spherically symmetric Higgs field $\Phi(r) = |\Phi|(r)$ which is a solution of the Euler-Lagrange equation

$$- \Phi''(r) - \frac{2}{r}\Phi'(r) + \frac{dV_{ht}}{d\Phi(r)} = 0 \qquad (13)$$

with the boundary conditions $\lim_{r \to \infty} \Phi(r) = 0$ and $\Phi'(0) = 0$.

It is of course a simple numerical exercise to find the classical bubble configuration and therefore $\tilde{S}$. The computation of the fluctuation determinant on the other hand constitutes a serious problem. Various authors have developed heat kernel and/or gradient expansions, but these can certainly not replace exact computations, especially if — as for narrow bubble walls — the gradients become large. A comparison between exact and approximate computations has been given e.g. in Ref.[13] A further problem for such expansions arises in the case (relevant here) of massless fields. Then the heat kernel is not damped exponentially at large values of the heat kernel time and some intermediate infrared regularization has to be used.

Among the methods for numerical computations there are those based on the heat kernel definition of the determinant,[14,15] those based on the Euclidean Green function[16,17] and a recent one[18] based on a theorem on functional determinants.[19] The latter one is certainly the fastest available method; it is certainly also the most accurate one since the computer codes are very short and therefore under perfect control.[18,13] Before I am going to present it I will have to develop the fluctuation analysis in some detail.

FLUCTUATION ANALYSIS

The analysis of fluctuations of a spherically symmetric object can be done in a suitable partial wave basis. Such an analysis has been performed recently for the electroweak sphaleron in Ref.[16] One can take over the partial wave equations with two modifications:
— the high temperature effective potential has to be modified from the one in Eq. (7) to the one in Eq. (10);
— the classical part of the gauge field vanishes and the classical Higgs field of the sphaleron is replaced by the one of the bubble, i.e. a solution of Eq. (13). I will denote the solution as H_0 in the following.

While for the sphaleron a K spin $(\vec{K} = \vec{J} + \vec{I})$ basis has been used, one may - alternatively - use here simply an analysis based on ordinary spin. This means that the Higgs field, the Fadeev Popov field and the time components of the gauge fields are expanded with respect to spherical harmonics and the space components of the gauge fields with respect to vector spherical harmonics $\hat{x}Y_l^m$, $r\nabla Y_l^m$ and $\vec{L}Y_l^m$. The time components of the gauge fields, the Fadeev Popov fields and the magnetic components of the vector potentials all satisfy the same equation

$$\mathbf{M}_l \psi = \omega^2 \psi \qquad (14)$$

where $\mathbf{M}_l$ is the partial wave fluctuation operator. It consists of a free massless partial wave Klein Gordon operator

$$\mathbf{M}_l^0 = -\frac{d^2}{dr^2} - \frac{2}{r}\frac{d}{dr} + \frac{l(l+1)}{r^2} \qquad (15)$$

and a potential

$$V_{55} = \frac{H_0^2}{4} \qquad (16)$$

which vanishes exponentially as $r \to \infty$. There is no $l = 0$ component of the magnetic vector potential since the vector spherical harmonic $\vec{L}Y_l^m$ vanishes. In the fluctuation determinant all of these contributions cancel, only the negative s-wave Fadeev-Popov contribution survives, due to the lack of its magnetic counterpart. It is triply degenerate due to isospin.

The fluctuation operator for the scalar part of the Higgs field is given by

$$\mathbf{M}_l = -\frac{d^2}{dr^2} + -\frac{2}{r}\frac{d}{dr} + \frac{l(l+1)}{r^2} + m_H^2 + V_{44}(r)$$

$$V_{44} = \frac{\lambda_T}{4g^2}(2H_0^2 - \epsilon H_0) \tag{17}$$

$$m_H^2 = \frac{\lambda_T}{4g^2}(3\epsilon - 4) .$$

The electric components of the gauge field and the isovector (would-be Goldstone) components of the Higgs field form a coupled (3×3) system. The fluctuation operator can again be written in the form $\mathbf{M}_l = \mathbf{M}_l^0 + \mathbf{V}_l$. The free operator $\mathbf{M}_l^0$ is diagonal. It consists of free partial-wave Klein-Gordon operators with masses $(0, 0, M_H)$ and centrifugal barriers corresponding to angular momenta $(l + 1, l - 1, l)$. The nonvanishing components of the potential are given by

$$V_{11} = V_{22} = H_0^2/4$$

$$V_{33} = H_0^2/4 + (\lambda_T/4g^2)(4H_0^2 - 3\epsilon H_0)$$

$$V_{13} = V_{31} = -\sqrt{\frac{l+1}{2l+1}}\frac{dH_0}{dr} \tag{18}$$

$$V_{23} = V_{32} = \sqrt{\frac{l}{2l+1}}\frac{dH_0}{dr} . \tag{19}$$

For $l = 0$ the second component is absent due to the vanishing of the vector spherical harmonic $r\nabla Y_0^0$. These amplitudes have a triple degeneracy due to isospin.

The fluctuation determinant $\mathcal{J}$ is defined as the ratio of determinants of the fluctuation operator with background field $\mathbf{M}$ and its free counterpart $\mathcal{M}'$, i.e.

$$\mathcal{J} = \frac{\det'' \mathbf{M}}{\det \mathbf{M}^0} , \tag{20}$$

where the double prime in the numerator indicates that the zero modes (due to translation invariance here) are to be removed and the imaginary frequency of the unstable mode has to be replaced by its absolute value. The partial-wave decomposition of the fluctuation operator decomposes also its determinant, so that

$$S_{eff}^{1-l} = \ln \mathcal{J} = \sum(2l + 1) \ln \mathcal{J}_l . \tag{21}$$

We now need a method for computing numerically the determinants of the partial wave fluctuation operators. Such a method has been developed recently by V. G. Kiselev and myself[18] and will be presented in the following section.

A MOST USEFUL THEOREM

The theorem on which our computation is based has been (re-)discovered and proven several times. A very elegant proof has been presented by Coleman,[19] who also refers to earlier work; I am going to recall his argument in slightly generalized form. Consider a typical $N \times N$ partial-wave fluctuation operator $(\mathbf{M}_l)_{mn}$ with $m, n = 1, ..., N$. Introduce a space boundary at $r = R$, to be removed to infinity at the end. The eigenvalues ω_k of this operator are characterized by the existence of solution vectors $\psi_n^\alpha(\omega, r)$ with $\alpha = 1, .. N$ of

$$(\mathbf{M}_l)_{mn} \psi_n^\alpha(\omega_k, r) = \omega_k^2 \psi_m^\alpha(\omega_k, r) \tag{22}$$

which form a fundamental system regular at $r = 0$ and satisfying the bound state boundary condition $\det\{\psi_n^\alpha(\omega_k, R)\} = 0$. Consider now the equation

$$\frac{\det(\mathbf{M}_l - \omega^2)}{\det(\mathbf{M}_l^0 - \omega^2)} = \frac{\det\{\psi_n^\alpha(\omega, R)\}}{\det\{\psi_n^{0\alpha}(\omega, R)\}} \tag{23}$$

Obviously the left hand side vanishes if ω^2 equals one of the eigenvalues of $\mathbf{M}_l$; the same holds for the right hand side, its numerator being given by the bound state condition. The same relationship holds for the denominators, vanishing at the eigenvalues of the free operator $\mathbf{M}_l^0$. Left and right hand side are therefore meromorphic functions with identical poles and zeros. They therefore are identical up to an entire function. If furthermore both sides tend to unity as $|\omega| \to \infty$ in the complex plane, this entire function must be unity. This condition is essentially identical with the requirement that the S matrix becomes trivial at infinite energy, a condition satisfied by suitable potentials. For a sphaleron or instanton background field this condition is not satisfied, a shortcome to be cured possibly by some regularization. For the electroweak bubble, on the other hand, the theorem can be applied. Of course the real fluctuation operator is defined in unbounded space and one has to perform the limit $R \to \infty$ on the right hand side. The fluctuation determinant is now given by the left hand side, taken at $\omega = 0$,

$$\mathcal{J}_l = \frac{\det \mathbf{M}_l}{\det \mathbf{M}_l^0} = \lim_{R \to \infty} \frac{\det\{\psi_n^\alpha(0, R)\}}{\det\{\psi_n^{0\alpha}(0, R)\}} . \tag{24}$$

The theorem is not only beautiful but also very useful for practial numerical computations. The right hand side can be computed by one simple Runge-Kutta integration, one does not have to find the eigenvalues of $\mathbf{M}_l$ or to compute Euclidean Green functions.

A very useful further development[20] consists in writing the regular solution at $\omega = 0$ as

$$\psi_n^\alpha = (\delta_n^\alpha + h_n^\alpha)\psi_n^0 \tag{25}$$

The free solutions $\psi_n^0(R)$ are generally given by spherical Bessel functions $i_{l_n}(m_n r)$ where the angular momentum l_n and the mass m_n depend on the channnel and the component. One obtains then an inhomogenuous differential equation for h_n^α of the form[20]

$$\tilde{\mathbf{M}}_l \mathbf{h} = \mathbf{V}(1 + \mathbf{h}) \tag{26}$$

which can be integrated numerically to obtain the exact functions h_n^α and can as well be used to expand these functions perturbatively, as it is necessary for renormalization.

The determinant of the fluctuation operator is finally given simply by

$$\mathcal{J}_l = \lim_{R \to \infty} \det\{\delta_n^\alpha + h_n^\alpha(R)\} \tag{27}$$

The method is also very suited for performing the renormalization and for removing zero modes. These technical points are discussed in previous publications[13,18] to which I refer for details.

THE FLUCTUATION DETERMINANT OF THE
ELECTROWEAK BUBBLE

The application of the theorem to the case of the electroweak bubble is now straight-forward. Some points to be considered are
— the subtraction of the divergent tadpole graphs
— removing the translation zero mode
— removing a particular gauge zero mode. I will discuss these briefly.

Tadpole Diagrams

The high temperature three-dimensional theory has only linear divergences of the form of tadpole diagrams which renormalize the mass term of the Higgs field. They have to be subtracted in the numerical computation to obtain finite results. Of course they have to be regularized, renormalized and their finite parts have to added back to the numerical results. Part of these diagrams have already been taken into account in the renormalization of the four dimensional theory and in giving the vacuum expectation value of the Higgs field a quadratic temperature dependence. Some terms linear in the temperature survive however and contribute[21,14,16] (after dividing by the temperature) to the 1-loop effective action, i.e. the logarithm of the fluctuation determinant. If the mass of the field in the loop is m_i and its coupling to the external field is described by the potential V_i their contribution to the effective action is given by $-m_i/8\pi \int d^3x V_i(r)$. The fluctuating gauge fields have vanishing mass and do not contribute. However we receive contributions from the fluctuating Higgs fields. The mass circulating in the loop is then m_H which is - including the temperature dependence and rescaling - given by equation 18. The potentials are V_{33} and V_{44}. So in total we have to restitute a term

$$S_{tad}^{1-l} = \frac{m_H}{8} \int dr r^2 \left(-(1 + 24\frac{\lambda_T}{g^2})H_0^2 + 6\epsilon \frac{\lambda_T}{g^2} H_0 \right) \tag{28}$$

into the 1-loop effective action.

Translation Zero Mode

Translation invariance is broken by the classical solution, so a zero mode appears. It occurs in the $l = 1$ partial wave of the fluctuation operator of the isoscalar part of the Higgs field. It is easily removed using the prescription given in.[18] Removing three eigenvalues $\omega^2 = 0$ gives the fluctuation determinant $\mathcal{J}$ the dimension $(energy)^{-6}$. The numerical computation is based on energy units $g\tilde{v}(T)$.

Gauge Zero Mode

Though we have imposed a gauge condition there is one residual gauge degree of freedom which corresponds to the constant gauge function for the free theory. Indeed a constant gauge potential $\Lambda(\vec{x}) = g_0$ does not contribute to the vector potential and is therefore not eliminated by the gauge condition $\partial_\mu a^\mu = 0$. In the case of a background field there is a similar but nontrivial mode which satisfies the background gauge condition and is therefore not eliminated by it. It manifests itself as a zero mode in the electric system for $l = 0$ and is given by a gauge function $g(r)$ which satisfies the same differential equation as the electric and Fadeev Popov modes, Eq. (14), and becomes constant as $r \to \infty$, in analogy to the free case. This zero mode has to be removed in the same way as the translation zero mode.

RESULTS AND CONCLUSIONS

Before considering the results I would like to recapitulate the essential steps of this work. We have started with the electroweak Lagrangean (at $\Theta_W = 0$). Taking the high-temperature limit one arrives at a three-dimensional theory. If the 1-loop contributions are taken into account one arrives at the form (1). Here the vacuum expectation value is temperature-dependent, and the Higgs potential is modified by the 1-loop effective potential, i.e. the 1-loop action for constant fields. The new Higgs potential (7) contains a new term, proportional to $\Phi^3 = (\Phi^\dagger \Phi)^{3/2}$. It is this term that makes the phase transition first order and to which the bubbles owe their existence. The bubble nucleation rate (12) contains a classical (or rather mean field) part which is described by the action $\tilde{S}$. The 1-loop correction is described by the exact 1-loop action $S^{1-l} = (1/2) \log \mathcal{J}$ where $\mathcal{J}$ is the fluctuation determinant, whose computation was the subject of this presentation. As we have just remarked we have already included the 1-loop corrections for constant fields. If this was a good approximation our results should behave roughly as the Φ^3-term in the Higgs potential, computed with bubble profile $\Phi = H_0(r)$. The true correction is then ΔS^{1-l}, the difference between the the 1-loop action and the Φ^3-term.

Table 1. The results of the computation of the fluctuation determinant. The different quantities are defined in the same way as in the text. The temperatur T and the vacuum expectation value $\tilde{v}$ are given in units GeV. The unstable mode frequency ω_- is given in units $g\tilde{v}$. The effective action is based on the fluctuation determinant computed in units of $(g)^{-6}$.

T	ϵ	$\tilde{v}(T)$	$\tilde{S}$	$\omega_- \times 10^2$	S^{1-l}_{num}	S^{1-l}_{tad}	Φ^3	ΔS^{1-l}
115.725	1.934	39.74	607	1.70	-46720	-10482	-56961	-241
115.702	1.866	41.18	149	3.46	-4299	-911	-5399	187
115.675	1.800	42.67	65.95	5.22	-989.1	-186.1	-1270	95.4
115.643	1.735	44.25	35.87	7.04	-318.6	-46.8	-421.2	55.9
115.600	1.663	46.14	20.08	9.13	-107.6	-6.33	-149.7	35.71
115.555	1.600	47.91	12.33	10.83	-43.11	4.54	-66.02	27.45
115.498	1.534	49.94	7.041	12.10	-12.96	8.68	-27.98	23.69
115.428	1.466	52.19	3.450	12.20	2.729	10.11	-10.56	23.40
115.345	1.400	54.59	1.145	10.25	13.39	10.47	-2.80	26.66

The numerical results are given in Table 1. The computation is based on the values $m_H = m_W = 80.2$ GeV, $m_{top} = 170$ GeV, $v_0 = 246$ GeV and $g = .6516$. Bubble nucleation sets in at $T_C = 115.7435$ GeV and ends at the temperature $T_0 = 115.2435$ GeV where the Higgs potential becomes unstable at $\Phi = 0$. The values for the temperature chosen correspond to 10 equidistant steps of the quantity ϵ (this is equivalent to the choice of Kripfganz et al.[9]).

The 1-loop effective action is presented in two parts, S^{1l}_{num} being the 1-loop action without the divergent tadpole terms and S^{1-l}_{tad} the restored finite part of the tadpole contribution. The sum of these two should behave as the Φ^3-term given in the next column; this is indeed the case. The correction to be applied to the transition rate is $\exp(-\Delta S^{1-l})$ where ΔS^{1-l} is given in the last column. The error for this latter quantity is to be estimated on the basis of a $1 - 2\%$

error of S_{num}^{1-l}, which means that it is quite large, especially for the two highest temperatures (The first value for ΔS^{1-l} is of course obsolete, but the cancellation is impressive). Our exact results are of the same order as the analytic estimates of Kripfganz et al.[9] and show a similar temperature dependence; a detailed comparison will be made elsewhere.

In principle ΔS^{1-l} should be smaller than $\bar{S}$, it should be a *correction* to the leading approximation; this is not the case, however. It is easily seen why this was to be expected. The Lagrangean which has been used to compute the bubble profile *contains already* a 1-loop part. In particular action $\bar{S}$ contains the Φ^3 term and nevertheless can be seen to be smaller by one or more orders of magnitude. This means that the bubble owes its existence to delicate *compensations* between classical and 1-loop parts. Even if the further 1-loop corrections amount now only to a small fraction of the Φ^3-term they will nevertheless be comparable to the classical action $\tilde{S}$.

My computation shows on the one hand that the 1-loop *potential* approximates the 1-loop *action* reasonably well, and incorporating it into the high temperature action is certainly justified in the sense that the action thus obtained gives a reasonable representation of 1-loop effects. On the other hand matters are not so simple. Since the bubble profile already contains 1-loop effects, the 1-loop action contains some 2-loop effects (e.g. the masses in the loop contain 1-loop corrections), but not in a systematic way. This should be analysed more carefully. Furthermore the remaining corrections are still appreciable and therefore one can expect that a selfconsistent solution minimizing the *sum* of classical and 1-loop actions may lead to a lower value for this sum.

Certainly these results do not close the discussion on the bubble nucleation rate, but the new quantitative information should improve the basis for quantitative considerations.

REFERENCES

1. see e.g. R. Hagedorn, 'The Long Way to the Statistical Bootstrap Model', these proceedings.

2. see e.g. A. I. Bochkarev, *Phys. Lett.* B254:165 (1991) and references therein.

3. see 'Electroweak physics and the Early Universe' Eds. F. Freire and J. Romão, Plenum Publ. Corp., N. Y., in press, for a recent account of the state of art.

4. M. E. Shaposhnikov, *Nucl. Phys.* B287:757 (1987) .

5. M. Dine, R. G. Leigh, P. Huet, A. Linde and D. Linde, *Phys. Rev.* D46:550 (1992).

6. S. Coleman and E. Weinberg, *Phys. Rev.* D7:1888(1972)

7. L. Carson and L. McLerran, *Phys. Rev.* D41:647 (1990) .

8. M. Hellmund, J. Kripfganz and M. G. Schmidt, Heidelberg preprint HD-THEP-93-23 .

9. J. Kripfganz, A. Laser and M. G. Schmidt, Heidelberg preprint HD-THEP-94-13 (June 1994).

10. J. S. Langer, *Ann. Phys. (N. Y.)* 41:108 (1967); *ibid.* 54:258 (1969).

11. I. Affleck, *Phys. Rev. Lett.* 46:388 (1981).

12. A. D. Linde, *Nucl. Phys.* 216:421 (1983).

13. J. Baacke, H. So and A. Suerig, *Z. Phys.* C63:689 (1994).

14. L. Carson, X. Li, L. McLerran and R.-T. Wang, *Phys. Rev.* D42:2127 (1990).

15. D. Dyakonov, M. Polyakov, P. Sieber, J. Schaldach and K. Goeke, *Phys. Rev.* D49:6864(1994).

16. J. Baacke and S. Junker, *Phys. Rev.* **D49**, 2055 (1994); *ibid.* D50:4227 (1994).

17. J. Baacke and T. Daiber, Dortmund preprint DO-TH 94/15, to be published in *Phys. Rev.*D.

18. J. Baacke and V. G. Kiselev, *Phys. Rev.* D48:5648 (1993).

19. see e.g. S. Coleman, *The Uses of Instantons*, in *The Aspects of Symmetry*, Cambridge University Press (1985).

20. J. Baacke, *Acta Phys. Pol.* B22:127 (1991) and references therein.

21. P. Arnold and L. McLerran, *Phys. Rev.* D36:581 (1987); *ibid.* D37:1020 (1988).

A FINITE TEMPERATURE STRING PHASE TRANSITION a la VOLUME EXCLUSION

A. M. Öztaş, Y. Gündüç and T. Çelik

Department of Physics Engineering, Hacettepe University
Beytepe, 06532 Ankara, Turkey

INTRODUCTION

The density of string states between mass m and $m + dm$ has the following form

$$\rho(m)dm = Cm^a e^{bm} dm \tag{1}$$

which is characteristic of continuously extended objects. The leading contribution to $\ln Z(\beta)$ in d-dimension, apart from the numerical factors, becomes

$$\ln Z(\beta) \sim \int_{m_0}^{\infty} dm \ m^{a+d/2} \ e^{bm} \ K_{d/2}(m\beta) \tag{2}$$

where $K_{d/2}(m\beta)$ is the bessel function of second kind. For temperatures $\beta^{-1} > b^{-1}$, the integral is divergent and it simply becomes impossible to have strings in thermal equilibrium. For if one had a string gas at this temperature range, one would encounter a negative specific heat due to having one string possessing most of the energy.[1] Similar difficulties also arise in hadronic models where the spectrum is of the form (1) given above. There the appearence of an ultimate temperature is avoided by considering the coustituents as extended objects rather than pointlike particles.[2]

Here we would like to show that the extended nature of strings, when taken into account in string thermodynamics, prevents the occurrence of the singularity associated with the exponential density of string states.

EXTENDED STRINGS MODEL

In the studies of cosmic strings[3] and in the effective string picture of lattice gauge theories,[4] strings are considered as self-avoiding random walks. Scherrer and Frieman showed that a string network has the statistical properties of a set of Brownian, rather than self-avoiding, random walks.[5] The root-mean-square end-to-end distance R for a Brownian random walk scales with the length l measured along the walk as $R \sim l^{1/2}$.

On the other hand, at high temperatures, the excluded volume effects become more important and the vortex configurations due the wrapping of strings represent a phase of topologically non-trivial strings. In this dense phase, the statistical properties of the system of strings are expected to be similar to that of collapsed polymers in dense phase below the Flory temperature.[6] This conjecture has been exploited in some recent works.[4,7] The scaling[8] of R with the length for a collapsed polymer is given as $R \sim l^\nu$ where ν is the inverse of the space dimension. To maintain the simplicity, we will consider strings as topological objects extended into a volume of R^{D+1} in $D + 1$ spatial dimensions. One can choose here any fraction of this volume to allow the strings interpenetrate. We parametrize the volume of a string of mass m (defined by $m = \mu l$ where l is the proper length and μ is the string tension) as

$$v_{L/H}(m) = [m/m_0]^{\gamma_{L/H}} v_{L/H}(m_0) \tag{3}$$

and the volume taken by N strings as

$$v_{L/H}(m_1, ..., m_N) = \Sigma_{i=1}^{N} v_{L/H}(m_i) \tag{4}$$

where the subscripts L/H denote low- and high-temperature phases. In $D + 1$ spatial dimension, we have $\gamma_L = (D + 1)/2$ for the Brownian strings and $\gamma_H = 1$ for the strings in the dense phase. m_0 is the lowest mass cutoff which is introduced to ignore the tachyon and to keep the quantum-classical correspondence good. The excluded volume effects will enter into our thermodynamic treatment through the nonzero intrinsic volume $v_{L/H}(m_0)$. The total coordinate space volume $V_{L/H}(N)$ available to N strings is then obtained at low densities by the reduction of volume due to the presence of the constituents and at high densities by use of the "free-volume" approach.[9] The two limiting expressions are then

$$V_L(N) = [V - \frac{v_L(m_0)}{m_0^{\gamma_L}} \Sigma_{i=1}^{N} m_i^{\gamma_L}]^N \tag{5}$$

$$V_H(N) = N! \, [(V/N)^{1/(D+1)} - (\frac{v_H(m_0)}{Nm_0^{\gamma_H}} \Sigma_{i=1}^{N} m_i^{\gamma_H})^{1/(D+1)}]^{(D+1)N}. \tag{6}$$

We write the partition function of a box of N strings at temperature β^{-1} as

$$Z(\beta, V) = \Sigma_{N=1}^{C_{L/H}} \frac{1}{N! \, (2\pi\beta)^{N\gamma}} \{ \int_{m_0}^{M_{L/H}} dm \; V_{L/H}^{1/N}(N) \; e^{(b-\beta)m} \; m^{-(1+\gamma)} \}^N \tag{7}$$

where $C_{L/H} = V/v_{L/H}(m_0)$, $\gamma = (D + 1)/2$ and $M_{L/H} = C_{L/H}^{1/\gamma_{L/H}} m_0$ denote the maximum mass a single string can attain. We proceed in $D + 1 = 3$ spatial dimensions and put $\gamma_L = \gamma = 3/2$.

Once the partition function of (7) is evaluated, the Helmholtz free energy

$$\beta A(\beta, V) = -\ln Z(\beta, V) \tag{8}$$

and the pressure

$$\beta P_{L/H}(\beta) = \lim_{V \to \infty} [-\beta(\partial A(\beta, V)/\partial V)_\beta] = \frac{1}{v_{L/H}(m_0)} \frac{\bar{N}}{C_{L/H} - \bar{N}} \tag{9}$$

are easily obtained, where $\bar{N}(\beta)$ is determined by the saddle point method

$$\frac{\partial}{\partial N} \ln Z(\beta, V) \mid_{N=\bar{N}} = 0 \tag{10}$$

which again exist for all temperatures.

We show in Fig.1, for both the low and the high density regimes, the pressure and the number of strings versus temperature in log-log plots.

162

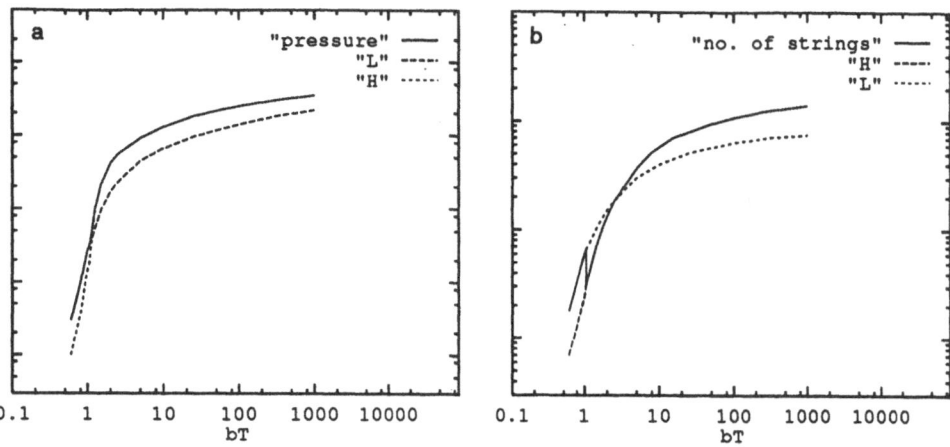

Figure 1. Log-log plots of a) the pressure and b) the number of strings versus temperature. Ordinate scales are arbitrary up to the choice of $C_{L/H}$ and $v_{L/H}(m_0)$. The corresponding curves of low- and high-temperature regimes are denoted by L and H, respectively .

In Fig.1a, the transition point from one regime to the other is determined by the requirement of minimum free energy. We have found $T_c = 1.05 \times b^{-1}$ which is slightly above the Hagedorn temperature $T_H = b^{-1}$. This crossover at $T = T_c$ corresponds to a first order phase transition, which is due only to the extended nature of strings taken into account in our thermodynamical treatment. The numerical value of T_c depends on our choice of parametrization, which turns out quite reasonable. It is interesting that our value for T_c coincides with the critical temperature obtained by Kapusta[10] for the thermodynamics of the Abelian bag model. Here the system behaves at low density and temperature like a relativistic ideal gas with unrestricted constituent mobility and energy conversion into additional string creation. On the other hand at high density and temperature the existence of many neighbours restricts the mobility and ceases the free production of strings due to the dense packing of filled volume. The number of strings versus temperature depicted in Fig.1b. For $T < T_c$ we have an ideal gas of ever increasing number of Brownian strings. About $T \sim T_c$, we encounter one long string with exponentially enhanced probability due to Eq.(1) and number of smaller strings. This mixed phase is associated with an apparent drop in the number of strings at T_H. Above T_H, however, the long string can not evolve into an infinite string simply because of the state of maximum number of collapsed strings of smaller intrinsic volumes wins over the configuration space when compared to that of one giant string.

CONCLUSIONS

In this work, we made use of the similarities between polymers and strings by taking into account the excluded volume effects in the thermodynamical treatment of strings and showed that the singularity at the Hagedorn temperature is replaced by a first order phase transition. Macroscopic fluctuations of the energy density and the negative specific heat associated with T_H are also avoided. In this two-phase picture, the low temperature phase of Brownian strings evolves through a first order phase transition into a high temperature phase of collapsed strings (or strings with a finite thickness) whose infinite temperature limit is a uniform gas of smallest size strings.

REFERENCES

1. D. Mitchell and N. Turok, Phys. Rev. Lett. **58** (1987) 1577; Nucl. Phys. **B294** (1987) 1138.

2. S. D. Odintsov, Rivista del Nuovo Cim. **15** (1992) 1.

3. T. Vachaspati and A. Vilenkin, Phys. Rev. **D30** (1984) 2036.

4. M. Caselle and F. Gliozzi, Phys. Lett. **B277** (1992) 481.

5. R. J. Scherrer and J. A. Frieman, Phys. Rev. **D33** (1986) 3556.

6. P. G. de Gennes, J. Phys. Lett. (Paris) **36** (1975) 55.

7. A. M. Allega, L. A. Fernandez and A. Tarancon, Phys. Lett. **B227** (1989) 347.

8. P. G. de Gennes, "Scaling Concepts in Polymer Physics". Cornell University Press, Ithaca (1979).

9. F. Karsch and H. Satz, Phys. Rev. **D21** (1980) 1168; V. V. Dixit, F. Karsch and H. Satz, Phys. Lett. **B101** (1981) 412.

10. J. I. Kapusta, Phys. Rev. **D23** (1981) 2444.

ENTROPY AND DECOHERENCE

Roland Omnès

Laboratoire de Physique Théorique et Hautes Energies
(Laboratoire associe au CNRS)
Université de Paris-Sud, 91405 Orsay, France

INTRODUCTION

The problem of deriving statistical models from basic principles in high energy collisions is compared with similar better known problems in elementary quantum mechanics. It is shown that several distinct problems should be expected, one of them having to do with the definition of relevant collective observables and another with the occurrence of the decoherence effect. It is furthermore indicated that, in the models one uses most frequently, energy dissipation and entropy production may become distinct questions.

Statistical models for high-energy collisions rely upon a very simple and attractive idea: an intermediate state is produced, in a thermal equilibrium with a high entropy, and the final particles are generated from this state by rapid freezing. Many experimental data support this idea and several models, more or less elaborate, have been proposed for its quantitative formulation and its derivation. Some rather delicate questions arise however when one tries to justify the approach from a fundamental standpoint and my purpose will be to call attention to a few of them. The present contribution to this meeting is not concerned however with the actual models of high-energy collisions, but with some underlying notions that are, in some sense, preliminary to their formulation. As a matter of fact, these questions become less obscure when one compares the problem at hand with similar ones that have been solved, partly or completely, in elementary quantum mechanics and, accordingly, our main topic will be concerned with this kind of analogy

The main problem one encounters is well known and most textbooks on entropy mention it right from the beginning. Let us suppose that some initial quantum state has been produced, its density operator r being either a pure state or a mixed one. The system can consist for instance of two particles directed against each other in their center of mass system. The initial entropy is given by von Neumann's expression

$$S = -\mathrm{Tr}(r \log r). \tag{1}$$

It is well known however that Schrödinger's dynamics implies that this entropy must remain the same before, during and after collision. It is a constant of motion. There must be therefore some subtlety involved in the definition of the entropy one is talking about in a statistical model and this point should better be clarified.

In most practical applications, one proceeds by using *ad hoc* simplifications, such as coarse graining of phase space or random phase approximation. We shall however consider another approach, more fundamental perhaps, which may shed a different light on the nature of our problem.

AN ANALOGY

These questions have been mostly considered in the framework of ordinary non- relativistic quantum mechanics and, moreover, in the case of a macroscopic object.[1-7] Consider for instance a pendulum starting from rest (i.e. with zero velocity) though not from its equilibrium position. Let x_1 denote the starting point. We must take the standpoint that the pendulum should be described by quantum mechanics, otherwise there would be no point in considering it as an analogue of the high energy problem with which we are really concerned. Let us therefore assume that the initial wave function of the pendulum is a narrow Gaussian function centered at x_1. It will also be convenient to assimilate the motion of the pendulum to a simple harmonic oscillator. It is then easy to write down its Schrödinger equation of motion, with still the same result, namely that the entropy remains a constant.

In some sense, we understand why. We know that entropy is generated through friction or by energy exchange with a thermostat. We shall leave aside the second possibility, because there is no analogue of a thermostat in the case of a collision between two particles. What then about friction? In the case of a real pendulum, one might think of the most obvious kind of friction, which is due to a viscous force exerted by the surrounding atmosphere, i. e. by the physical environment of the pendulum. There is no such external environment however for two high-energy particles. We must think therefore of something more intrinsic to the pendulum itself that is also responsible for friction, and this is as we shall see its own matter. As a matter of fact, this is still called an environment in the literature on the subject, an internal environment so to say. The notion warrants some comments:

As physicists, we can see what is going on: The pendulum wire is deformed because of its motion, mostly near its anchoring point. This is a continuously changing elastic deformation or, from a quantum point of view, a production of phonons. All the possible elastic deformations, either in the wire or inside the ball of the pendulum, can interact with each other by anharmonic effects, i. e. by an interaction between phonons. At the microscopic level, the existence of a phonon implies a relative motion between neighbouring atoms, and this can in turn generate some excitations of the electrons. All this means that we must take the internal state of the pendulum matter into account, also in a quantum way. In keeping with our search for an analogy with high-energy collisions, we shall not assume that some hidden energy (internal energy) is already present in the initial state and we shall therefore assume that the matter of the pendulum, including the wire, is initially in its ground state, which means that it is at zero temperature. There is no doubt however that the pendulum should not be considered any more as a one-dimensional dynamical system, but as a system involving many degrees of freedom, including those of phonons and electrons. It is then convenient to formally divide the degrees of freedom among two categories, one of them containing only the position x of the ball center whereas all the other (internal or microscopic) degrees of freedom belong to the second category. The formal dynamical system having for its coordinates these other degrees of freedom is what one calls the *environment*.

One can then split the total Hamiltonian into three parts:

$$H = H_c + H_e + H_{int} , \qquad (2)$$

where the first term is a harmonic oscillator Hamiltonian for the motion of the ball center. We may call the coordinate x of this center a *collective* observable. The second term is the environment Hamiltonian, whose average value is the internal energy as defined in thermodynamics and the last term represents the coupling between collective motion and internal environment. It is responsible for their energy exchanges, i.e., for dissipation.

In the case of a high-energy collision, one may wonder whether there is anything like this distinction between collective and microscopic environment observables. Quantum fields, with their infinite number of degrees of freedom, could certainly play the role of an environment, revealing their existence during the collision. But then, if one looks at the final state after a high-energy collision, one finds another kind of puzzle: what are the collective observables? We might say that they are the parameters one is measuring in the experiment but, when comparing the initial state with the final one, one realises that the distinction between collective and microscopic observables is far from trivial, from a fundamental standpoint. There is no definite algorithm allowing us to construct the right or the best collective observables directly from the fundamental principles. In nuclear physics for instance, the problem of identifying and constructing collective observables (which may describe deformations or other features of nuclei) has not yet been completely solved. This means that the analogy between a pendulum and high-energy collisions may be not so bad after all: both may involve a distinction between collective observables and environment but, maybe to our surprise, this separation is less obvious for collisions than for a pendulum. Clearly, the models one may construct may depend upon the choice of collective observables to be made and, also, upon the specific form of He and Hint.

DECOHERENCE, A PRELIMINARY TO DISSIPATION

Having satisfied ourselves that the analogy is not empty, we may go further by looking at the results that have been obtained in the case of the pendulum.

The number of degrees of freedom in the environment is certainly of importance. If there are only a few of them, we are confronting a typical quantum problem with no special interest. This would be the case for instance if one were to consider vibration damping in a molecule consisting of a few atoms rather than a macroscopic pendulum, or a medium energy collision between two particles rather than a very high energy collision. These finite problems belong to another realm of theory and we shall therefore consider only an environment involving many degrees of freedom. As a matter of fact, the most convenient models assume an infinite number of degrees of freedom.

Considering again the previous pendulum, starting from rest away from its equilibrium position with its matter (environment) at zero temperature (in its ground state), one expects of course some dissipation, the pendulum collective motion being damped and the internal energy increasing. One would like to understand how these features follow from quantum mechanics, even before attempting to understand what is entropy and how it is generated.

Perhaps the most important discovery resulting from this programme was to find that dissipation is not the only important effect. Another effect, unknown to classical physics and still unknown to too many physicists, occurs long before any sizeable manifestation of dissipation. This is the so-called *decoherence effect*. Perhaps the simplest way of describing it consists in comparing two pendulum motions with practically the same initial conditions, except that the starting points are respectively at two neighbouring points x_1 and x_2, everything else being the same. As its name indicates, the decoherence effect is a loss of relative phase: after a rather short "decoherence" time, the two wave functions of the environment that are associated respectively with the two slightly different collective motions become orthogonal. We shall consider this important preliminary effect from several standpoints to see why it occurs, how it can be expressed quantitatively and what its consequences are.

Why is there decoherence? Essentially, because of the very high energy level density of an environment involving many degrees of freedom (remember that this quantity increases typically like an exponential when the number of degrees of freedom increases). The simplest argument to show this connection goes as follows: If one tries to compare by perturbation theory the two slightly different motions of the whole pendulum together with the evolution of its internal matter, the perturbation terms will involve denominators containing differences $E_i - E_k$ between the eigenvalues of H_e. Some of these terms will be very large and this will make the complete wave function of the environment Ψ_e very sensitive to any slight change in the collective motion. When comparing the respective expansions of Ψ_e along the energy basis for the two cases

$$\Psi_e = c_k^{(t)}|E_k>, \tag{3}$$

the respective coefficients $\{c_k\}$ for the two different cases will rapidly lose any relative phase coherence, even if the two sums (3) were exactly the same at time zero (one starts from the ground state). The sum

$$\sum_k c_k^{(1)} * c_k^{(2)} \tag{4}$$

will then become zero, i.e. the two states of the environment will become orthogonal. This intuitive argument suggests that the effect is universal, at least when Hint is not zero, i.e. when there can be dissipation. This is not however a proof and one can even expect that a full proof will be difficult to obtain since it would involve some rather detailed knowledge of phases for a system with many degrees of freedom, and this is a very difficult problem. Of course, the argument is also somewhat circular since it also means that perturbation theory cannot be applied.

More precise and more convincing results can be obtained by using a specific model for the environment, namely a collection of harmonic oscillators (mimicking non- interacting phonons), with a simple form of Hint we shall discuss. It is also convenient to introduce the so-called collective or *reduced* density operator ρ_c, which is obtained as the trace of the full density operator r over the environment degrees of freedom. From the standpoint of interpretation, this is a density operator allowing to define completely the system as far as only measurements on the position or momentum of the pendulum are concerned. What we said about the nature of the decoherence effect, *viz.* an increasing degree of orthogonality for the environment wave functions associated with different positions, implies that, after a trace over the environment is taken, one should obtain an *increasingly diagonal reduced density operator*.[3]

Let us consider for definiteness an initial collective state involving a quantum superposition $a\phi_1(x) + b\phi_2(x)$ of two Gaussian wave functions respectively centered at two points x_1 and x_2 at time zero. Then one finds that the non-diagonal terms of the reduced density operator decrease exponentially like

$$a * b \exp\left\{-\frac{m\omega(x_1 - x_2)^2}{4\hbar}\frac{t}{\tau}\right\}, \tag{5}$$

where m and $\omega/2p$ are respectively the mass and frequency of the pendulum, and τ is its damping time. This last quantity appears because one finds that the collective position coordinate at time t is strongly concentrated along either one of two classical trajectories, one starting from x_1 and the other from x_2, both of them being described by the classical equation of motion

$$\frac{d^2x}{dt^2} + \frac{1}{t}\frac{dx}{dt} + \omega^2 x = 0. \tag{6}$$

The decoherence effect is usually identified with this spontaneous diagonalization of the reduced density operator. It is a very strong effect: taking for instance 1g for the value of m, 1 sec-1 for $\omega/2p$, $x_1 - x_2$ of the order of a few microns, t of the order of a few minutes and $t = 1$ nanosecond, one finds that the non-diagonal matrix element (4) is of the order of $\exp(-10^5)$.

The most remarkable fact from our standpoint is that dissipation is seen to come out from a quantum calculation, as shown by the occurrence of the quantity τ in Eq.(5). This damping time can be computed from a knowledge of Hint and it represents dissipation. Does this imply that we know how to compute entropy generation from the first principles? This is obviously the question one should now ask.

ENTROPY AND ITS INCREASE

Entropy is defined, in thermodynamics, in terms of the internal density matrix, i.e. the density matrix for the environment alone, while the global situation, the thermodynamical parameters, are expressed by giving the values of the collective observables. One might therefore think of the following pattern for defining this entropy: Wait until decoherence has occurred, so that the global situation is free from any possible macroscopic quantum interferences. Anything that can be observed from outside is then described by an essentially diagonal density matrix ρ_c. One is now however interested in the state of the environment for given external parameters, i.e. in an effective density operator ρ_{eff}, which is obtained from the true density operator by projecting it upon a given collective position, i.e. $\rho_{\text{eff}} = <x|\rho|x>$, the entropy would then be given by $S = \text{Tr}\{\rho_{\text{eff}} \log \rho_{\text{eff}}\}$. This procedure is a bit awkward and other ones are conceivable, but we shall not try to be more specific.

The problem we discussed, where a big oscillator is interacting (bilinearly) with a collection of little ones, can be solved by several methods, the most instructive one using a solution given by coherent states.[2] The Schrödinger equation can then be reduced essentially to a classical problem and the answer has some rather surprising features: There is decoherence and dissipation, in the sense that energy is lost by the big oscillator and in favour of the environment, but nevertheless there is no generation of entropy, if defined according to Eq.(7). This apparently paradoxical situation is strongly reminiscent of the spin echo effect, because the loss of initial coherence is incomplete and some phase information remains hidden among the environment oscillators. It can reappear, after a Poincaré cycle, in a giant fluctuation re-establishing the pendulum in its initial state.

The subtlety of the questions we are asking appears then fully. As a matter of fact, the choice of a bath of oscillators for a model of environment was rather constraining. Even if one cannot hope to obtain an analytic solution in terms of coherent states, as was possible when the collective motion was purely oscillatory, the theories of decoherence nevertheless always refer to the same oscillator model for the environment. The reason is essentially technical: One has to eliminate explicitly the environment degrees of freedom and this is done by an explicit Feynman path integration but, since the only explicitly known Feynman path integrals are Gaussian, the environment Hamiltonian must be a sum of oscillator Hamiltonian. Furthermore, when performing the trace over the environment, one must also require that the environment is initially in thermal equilibrium (in our example, at temperature zero), because the trace is again a Gaussian integral. The theory is therefore strongly constrained by the limitations of our mathematical tools.[1,5]

The theory of decoherence and entropy generation is then squeezed between what theory can reach and what physics seems to suggest: The only theory we can build requires an oscillator model for the environment. If the system is initially at zero temperature, there is dissipation (energy exchange) but no entropy. Entropy is produced if the system is initially at non-zero temperature, but this situation has no similarity with the high-energy collision problem we are interested in. As was stressed long ago by Van Hove,[8] anharmonic effects are essential in the establishment of a thermal equilibrium and the generation of entropy. But anharmonic effects are precisely what forbids us to build up a reliable theory, at least up to now. It might very well be that a full answer to these questions rely upon the fact that the full density operator, or the full wave function, has a maximal algorithmic complexity, i.e. is not computable nor perhaps even accessible to a specific theory.[9]

CONCLUSIONS

The conclusions one can reach from an analogy of high-energy collisions with an apparently simpler problem do not certainly solve the main questions, but at least they can give a better evaluation of the difficulties one is facing, and they can also be used as some sort of a warning.

1. Distinguishing between the collective parameters and the microscopic or statistical ones is a very tricky problem from the standpoint of principles. It is particularly difficult in the case of high-energy collisions where, contrary to simple macroscopic objects, the relevant collective parameters are changing during the process. Of course, one can choose them more or less arbitrarily as what is accessible to measurement in a specific experiment, but this is still far from the level of principles.

2. Explicit theories will certainly use some sort of oscillator model for the microscopic degrees of freedom. This can be a free field representing a bath of particles or the excitations of a string. In any case, one cannot be too optimistic in predicting very reliably the entropy production, except if it is put "by hand" at the beginning by assuming some sort of thermal equilibrium.

3. In the most elementary case, there are three scales of time: a very short one for decoherence, a much larger one during which energy is lost and given to the internal environment, and finally a still larger time for establishing thermal equilibrium. There is no apparent reason why these three scales of time should still be significantly different in the case of high-energy collisions, which would mean that all the problems are less clear- cut and a basic theory accordingly more difficult.

I would like to conclude this not too optimistic keynote by quoting a series of problems Arthur Wightman gave at the end of some of his lectures: **Problem 1**: Solve Problem 2. **Problem 2**: Solve Problem 3; and so on. Hint: If you cannot do it for n, you cannot do it for $n + 1$.

REFERENCES

1. R. P. Feynman and F. L. Vernon, Jr., Ann. Phys. (N. Y.) 24, 118 (1963).

2. K. Hepp and H. E. Lieb, Helv. Phys. Acta 46, 573 (1973).

3. W. H. Zurek, Phys. Rev. D 26, 1862 (1982).

4. W. H. Zurek, Physics Today, 44, 36 (1991).

5. A. O. Caldeira and A. J. Leggett, Physica A 121, 587 (1983).

6. E. Joos and H. D. Zeh, Z. Phys. B 59, 223 (1985).

7. R. Omnès, The Interpretation of Quantum Mechanics , Chapter 7, Princeton University Press, Princeton (1994).

8. L. Van Hove, Physica 18, 145 (1952).

9. C. M. Caves in Physical Origins of Time Asymmetry, J. J. Halliwell, J. Perez-Mercader and W. H. Zurek, eds, Cambridge University Press (1994).

COLORED CHAOS

Berndt Müller

Department of Physics
Duke University, Durham, NC 27708-0305

INTRODUCTION

I review recent progress in our understanding of the basis of statistical models for hadronic reactions and of the mechanisms of thermalization in nonabelian gauge theories.

Almost to the date 30 years ago, Rolf Hagedorn[1] proposed that multiparticle production and other phenomena of what today is called "soft" hadronic interactions could be explained on the basis of two assumptions:

1. The mass spectrum of hadronic states grows as $\rho(m) \sim m^{-a} e^{bm}$.

2. The available states are statistically occupied during a (soft) hadronic interaction.

We now understand that the first assumption, an exponentially growing mass spectrum, is a consequence of quark confinement, and the famous Hagedorn temperature $T_H = 1/b$ is related to the QCD string tension and the temperature T_c associated with the deconfining, chiral symmetry restoring phase transition of QCD.

The second assumption has remained more mysterious. Why is the assumption of a random statistical distribution of final states warranted in high energy reactions that last not much longer than 1 fm/c (or 3×10^{-24}s), and why are the final states not dominated by coherent quantum states or collective excitations of a small subset of the available hadronic degrees of freedom? Maybe it is good to recall that similar questions posed themselves in the context of Bohr's statistical model of compound nucleus reactions. In this case, the conceptual difficulties were eventually resolved by the insight that the highly excited compound nucleus is a chaotic quantum system[2] exhibiting rapid exchange of energy between the accessible degrees of freedom.* Here I want to show that the same mechanism is responsible for the apparent thermalization in high-energy hadron-hadron interactions: nonabelian gauge theories are strongly chaotic.

---

*Experimental evidence for the chaotic nature of the compound nucleus is mainly derived from the energy level statistics of highly excited nuclei showing the Wigner distribution characteristic of chaotic quantum systems.[3]

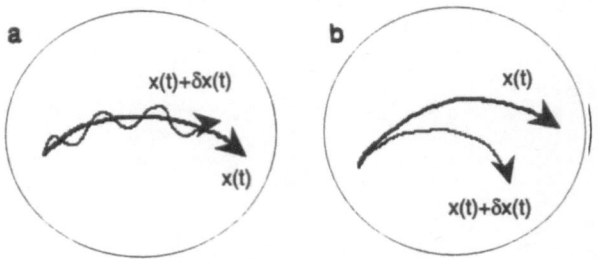

Figure 1. (a) Stable phase space trajectory; (b) unstable trajectory.

CHAOS AND ERGODICITY

A dynamical system exhibits *ergodic* behavior, if the time average of an observable A can be replaced by the phase space average

$$\lim_{T \to \infty} \frac{1}{T} \int_0^T dt\, A(t) = \langle A \rangle \equiv \frac{1}{Z_E} \int d\Gamma_E A(\Gamma_E). \tag{1}$$

$\langle A \rangle$ here denotes the *microcanonical* average, Z_E is the microcanonical partition function, and $d\Gamma_E$ is the phase space measure at constant energy. For systems with very many degrees of freedom it is equivalent to take the *canonical* average

$$\langle A \rangle = \frac{1}{Z(\beta)} \int d\Gamma A(\Gamma) \exp[-\beta E(\Gamma)], \tag{2}$$

where $Z(\beta) = \int d\Gamma \exp[-\beta E(\Gamma)]$ is the canonical partition function and the inverse temperature $\beta = 1/T$ is determined by the condition $E = -\partial(\ln Z)/\partial \beta$.

For practical applications it is crucial to know the time scale on which ergodicity is attained. It can be shown that this time scale is related to the rate h of exponential divergence of neighboring trajectories in phase space; this rate is called the (maximal) Lyapunov exponent.[4] The complete spectrum of Lyapunov exponents is defined as follows. Consider a given trajectory in phase space, $x_\alpha(t)$, where $\alpha = 1, \ldots \nu$ enumerates the degrees of freedom. $x_\alpha(t)$ is a solution of the classical equations of motion of the system. Now take a set of neighboring trajectories (see Figure 1):

$$\tilde{x}_\alpha'^{(i)}(t) = x_\alpha(t) + \delta x_\alpha^{(i)}(t), \quad \alpha = 1, \ldots, \nu. \tag{3}$$

For infinitesimal δx_α they are solutions of a second-order linear differential equation of the form

$$D[x_\alpha(t)]\delta x_\alpha^{(i)}(t) = 0. \tag{4}$$

One can then obtain a complete orthogonal set of solutions of this equation; they define ν Lyapunov exponents λ_i according to

$$\lambda_i = \lim_{t \to \infty} \frac{1}{t} \ln \frac{\|\delta x_\alpha^{(i)}(t)\|}{\|\delta x_\alpha^{(i)}(0)\|}, \quad i = 1, \ldots, \nu. \tag{5}$$

In other words, for long times one has the norm of $\delta x_\alpha^{(i)}$ growing (or shrinking) as $\|\delta x_\alpha^{(i)}(t)\| \propto \exp(\lambda_i t)$. One usually assumes the Lyapunov exponents to the ordered in size:

$$\lambda_1 \equiv h \geq \lambda_2 \geq \ldots \geq \lambda_\nu. \tag{6}$$

For conservative (Hamiltonian) systems the Lyapunov exponents occur in pairs of equal size, but opposite sign: $\lambda_i = -\lambda_{N+1-i}$. This is in accordance with Liouville's theorem which states that the volumes in phase space filled by an ensemble remains unchanged with time, implying that there must be a direction of contraction for every direction in which the phase space volume expands. For each conservation law there occur two vanishing Lyapunov exponents; the conservation of energy always ensures the existence of one such pair for a Hamiltonian system. Since the extent of the ensemble rapidly shrinks below any practially achievable resolution in the exponentially contracting directions, the *observable* volume in phase space grows as

$$\overline{\Gamma}(t) \propto \exp\left(t \sum_i \lambda_i \theta(\lambda_i) \right) \equiv \exp(\dot{S}_{\mathrm{KS}}\, t) \tag{7}$$

where the sum only includes the positive Lyapunov exponents. The exponential growth rate of the observable phase space volume implies a *linear* rate of growth of the observable "coarse-grained" entropy associated with the ensemble. This rate, $\dot{S}_{\mathrm{KS}} = \sum_i \lambda_i \theta(\lambda_i)$, is called the Kolmogorov-Sinai entropy, or short, KS-entropy. Dynamical systems that have a positive KS-entropy everywhere in phase space are called K-systems; they exhibit all the properties required for a statistical description on time scales that are long compared with the ratio between the equilibrium entropy S_{eq} and the KS-entropy, i.e. for times

$$t \gg \tau_s = S_{\mathrm{eq}}/\dot{S}_{\mathrm{KS}}\, . \tag{8}$$

An illustration of these properties is provided by the simple dynamical system

$$H(x, y; p_x p_y) = \tfrac{1}{2}\left(p_x^2 + p_y^2 + x^2 y^2 \right), \tag{9}$$

which occurs as part of the extreme infrared limit of Yang-Mills fields.[5] The system described by the Hamiltonian (9) has a positive Lyapunov exponent $\lambda \approx 0.4$. Almost all its trajectories are unstable against small perturbations[6] and the analogous quantum system has been shown to exhibit a Wigner distribution of its level spacings.[7] The remarkable ability of this system to randomize an initially localized phase space distribution is shown in Figure 2. After a rather limited time the phase space distribution is indistinguishable from a microcanonical ensemble.

CHAOS IN NONABELIAN GAUGE THEORIES

If we want to apply these concepts to nonabelian gauge theories, we must consider these as classical Hamiltonian systems with many degrees of freedom, and we need a gauge invariant distance measure in the space of field configurations. The first part is easy; the Hamiltonian formulation of lattice gauge theory by Kogut and Susskind can form the basis for a study for the gauge group SU(n) of nonabelian gauge theories as dynamical systems. The lattice Hamiltonian is expressed as (a denotes the lattice spacing)

$$H = \frac{g^2}{2a} \sum_\ell \mathrm{tr}(E_\ell E_\ell^\dagger) + \frac{2}{g^2 a} \sum_p \left(n - \mathrm{Re}\,\mathrm{tr}\,U_p \right), \tag{10}$$

where electric field strength E_ℓ and the link variables U_ℓ are defined on the lattice links, and U_p denotes the ordered product of the U_ℓ around an elementary plaquette $p : U_p = U_4^\dagger U_3^\dagger U_2 U_1$. The U_ℓ are elements of the gauge group (SU(3) in the case of QCD) and the E_ℓ are elements

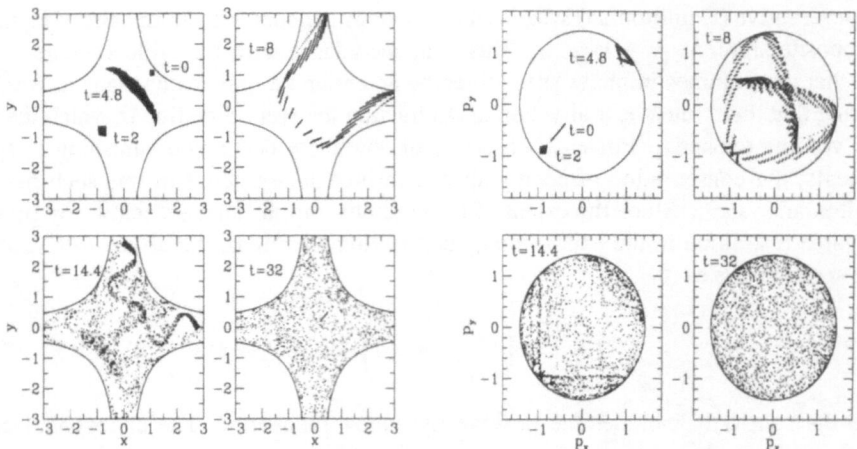

Figure 2. Evolution of the phase space distribution for the model Hamiltonian (9). All points have energy $E = 1$. The apparent volume of the phase space distribution grows rapidly, until it covers the accessible phase space (contained within the hyperbolic boundaries) homogeneously at time $t = 32$.

of the associated Lie algebra. In the classical limit, the link variables U_ℓ are functions of time, and the electric field variables are given by

$$E_\ell = \frac{a}{ig^2}\, \dot{U}_\ell\, U_\ell^\dagger. \tag{11}$$

The Hamiltonian equations then provide a set of coupled equations for the time evolution of $U_\ell(t)$ and $E_\ell(t)$, which can be integrated numerically. We have taken great care to ensure a numerically exact solution. The energy and Gauss' law remain conserved to better than 10^{-6} over the whole course of the numerical integration.

An appropriate measure for the distance between two field configurations is[8]

$$D[U_\ell, E_\ell; U_\ell', E_\ell'] = \frac{1}{2N_p}\left(a^2\sum_\ell\left|\mathrm{tr}(E_\ell E_\ell^\dagger) - \mathrm{tr}(E_\ell' E_\ell'^\dagger)\right| + \sum_p\left|\mathrm{tr}\, U_p - \mathrm{tr}\, U_p'\right|\right). \tag{12}$$

It is gauge invariant, gives a vanishing distance between gauge equivalent field configurations, and goes over into

$$D[A, E; A', E'] \propto \frac{1}{2V}\int d^3x\left(|E^2 - E'^2| + |B^2 - B'^2|\right) \tag{13}$$

in the continuum limit, measuring the local differences in the electric and magnetic field energy.[†]

If one starts from two randomly chosen neighboring field configurations and integrates these in time, one finds that the distance quickly grows exponentially, until it saturates due to the compactness of the space of gauge fields. The growth rate h quickly reaches a constant limit as function of the lattice size N^3, if the energy density is kept fixed by choosing the same average energy per plaquette E_p in each case. This is demonstrated in Figure 3 for lattices of size 2^3 up to 28^3 and the gauge group SU(2).

It is easy to see[9] that the Hamiltonian (10) exhibits a scaling behavior such that the Lyapunov exponents, if they are universal functions of the average energy density, as expressed

---

[†]If one only wants to determine the largest Lyapunov exponent, it is sufficient to consider either the electric or the magnetic contribution to $D[U, U']$.

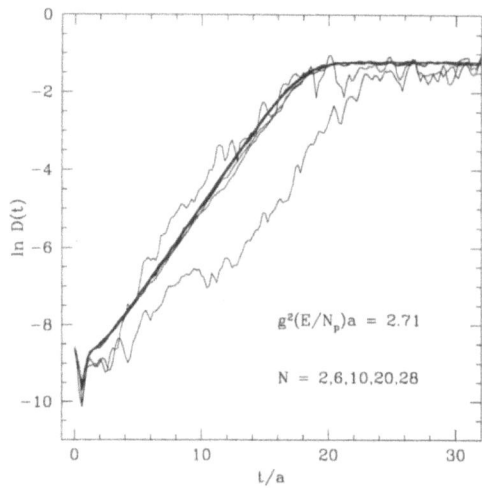

$g^2(E/N_p)a = 2.71$

$N = 2,6,10,20,28$

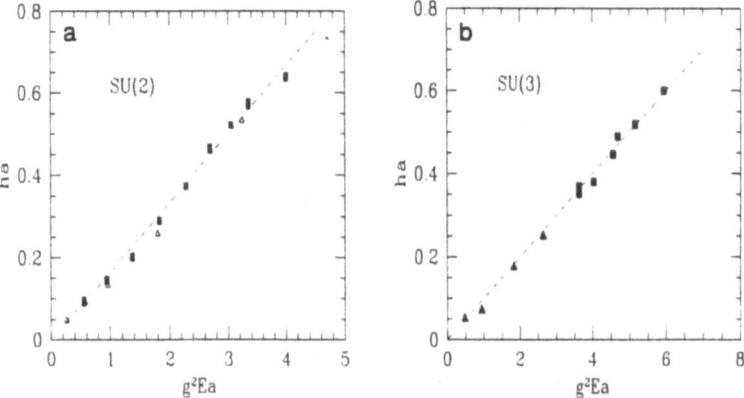

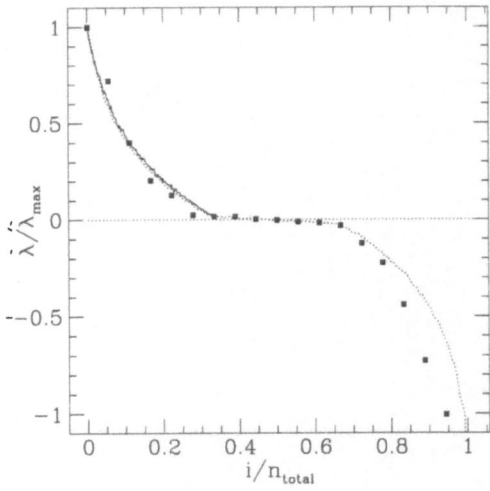

Figure 5. Spectrum of Lyapunov exponents for SU(2) lattice gauge theory. The black dots are for a 1^3, the dotted line is for a 2^3, and the solid line is for a 3^3 lattice. The $18N^3$ exponents are plotted on the fixed interval [0,1] to exhibit the scaling with N.

and $N = 3$ are indistinguishable, and there is only a small difference between $N = 1$ and $N = 2$ (see Figure 5).

The Lyapunov spectrum for SU(2) shows three separate components: there are $(6N^3 - 1)$ positive and negative exponents each, and there are $(6N^3 + 2)$ exponents that converge to zero in the limit $t \to \infty$. Their vanishing reflects the existence of $(3N^3 + 1)$ conservation laws: Gauss' law at every lattice point and the overall energy conservation.[‡] Since the density of points on the line over the fixed interval [0,1] grows as N^3, this implies that the sum over positive Lyapunov exponents increases like the volume of the lattice, yielding a constant *KS-entropy density* $\dot{\sigma}_{KS}$ in the thermodynamic limit. Gong finds

$$\dot{S}_{KS} = \sum_i \lambda_i \theta(\lambda_i) = c_2 \lambda_1 N^3, \tag{15}$$

which together with (14) and the lattice volume $V = (Na)^3$ yields

$$\dot{\sigma}_{KS} = \frac{1}{V} \dot{S}_{KS} = \frac{1}{3} b_2 c_2 g^2 \varepsilon, \tag{16}$$

where $\varepsilon = 3E_p/a^3$ is the average energy density on the lattice. (Note that there are $3N^3$ plaquettes.) No one has yet calculated the complete Lyapunov spectrum for SU(3), but I expect a similar relationship as (15) to hold in that case, too. The coefficient c_2 is not completely independent of the scaling variable $g^2 E_p a$, but has a value around 2. We will return below to the question how the physically relevant value of $g^2 E_p a$ can be chosen.

[‡]The Lyapunov exponents associated with Gauss' law obviously correspond to unphysical degrees of freedom and only show up here because we did not fix the gauge explicitly in the rescaling procedure used to determine the Lyapunov spectrum. The fact that they, indeed, vanish in the long-time limit provides support for the numerical techniques employed in the calculation of the Lyapunov spectrum.

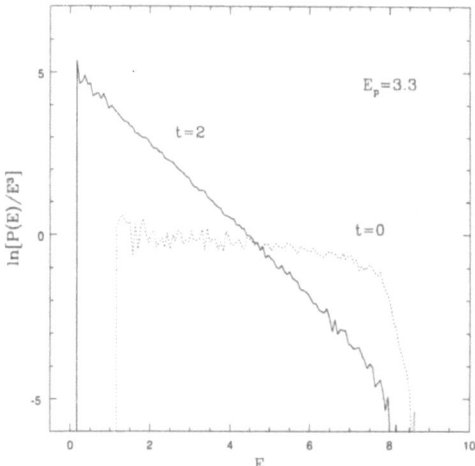

Figure 6. Logarithmic plot of the evolution of the magnetic energy density distribution on the lattice for SU(3) gauge theory. The distribution appears "thermalized" (exponential) after a time $t \approx 2$.

PHYSICS PERSPECTIVES:
THERMALIZATION TIME, GLUON DAMPING RATE

The instability of all degrees of freedom of the nonabelian gauge field (in the classical limit) leads to a very rapid "thermalization" of the energy density on the lattice. This is illustrated in Figure 6 showing the distribution of magnetic energy on the lattice plaquettes.[12] The initial state was chosen according to a random (not thermal) distribution of lattice link variables with vanishing electric field everywhere. Within two lattice units ($t/a = 2$) the energy has been equilibrated between electric and magnetic fields and, as the exponentially falling distribution shows, has assumed the form of a Gibbs distribution.[§] The time scale for this "thermalization" is in good agreement from the time scale estimated from the inverse of the maximal Lyapunov exponent which is $h \approx 0.6$ in lattice units in SU(3) at this energy density.

The fact that the energy density thermalizes on a time scale much shorter than that required for the numerical determination of the Lyapunov exponents (typically $t/a = 1000$) allows us to relate the Lyapunov exponents to quantities in the presence of a thermal environment. First, we can replace the average energy per plaquette E_p in (14) by the "temperature" T, because $E_p = \frac{2}{3}(n^2 - 1)T$ for the classical, equilibrated SU(n) gauge field. This implies that

$$h \approx \begin{cases} \frac{1}{3}g^2 T \approx 0.33 g^2 T & \text{for SU(2),} \\ \frac{8}{15}g^2 T \approx 0.53 g^2 T & \text{for SU(3).} \end{cases} \tag{17}$$

We can use this result to obtain a model independent, nonperturbative estimate of the thermalization time solely due to gauge field dynamics in QCD. To compensate for the lack of "running" of the gauge coupling constant in the context of our classical gauge field calculation, we may use the one-loop result for $g(T)^2$ in (17) to evaluate $\tau_h = h^{-1}$, as shown in Figure 7. Clearly, this time is much smaller than 0.5 fm/c for all relevant temperatures,

---

[§]I emphasize that this "thermalization" is caused by the evolution of the gauge field under its own Hamiltonian dynamics and not by some artifical coupling to a heat bath as in the standard techniques applied in Monte-Carlo simulations of lattice gauge theory. There the Monte-Carlo "time steps" have no physical meaning; here the time step is physical. The only approximation is that the lattice gauge field is treated classically.

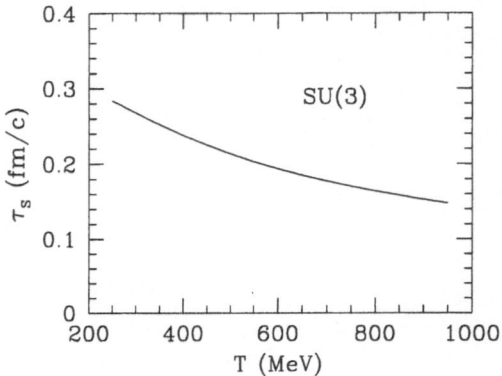

Figure 7. "Thermalization" time scale in SU(3) gauge theory as defined by the inverse of the maximal Lyapunov exponent (17), using $g(T)^2 = 16\pi^2/[11 \ln(\pi T/\Lambda)^2]$ with $\Lambda = 200$ MeV.

indicating a very rapid thermalization of the available energy. One should note that the Lyapunov exponents usually approach their asymptotic values from *above*, i.e. the dynamical instabilities are actually greater before the energy has been completely thermalized. This indicates that thermalization of field configurations far away from equilibrium may proceed even more rapidly.

It first appeared as a remarkable coincidence that the maximal Lyapunov exponents for SU(2) and SU(3) agree within numerical errors with the analytically calculated damping rate of a nonabelian plasmon at rest:[13]

$$\Gamma = 2\gamma(0) = 6.635\frac{n}{24\pi}g^2T \approx \begin{cases} 0.35g^2T & \text{for SU(2)}, \\ 0.53g^2T & \text{for SU(3)}. \end{cases} \tag{18}$$

[The reason for the factor 2 is that the plasmon pole is usually parametrized as $\omega_p(k) = \omega^*(k) - i\gamma(k)$, so that the energy density of the soft plasmon mode falls off as $\exp(-2\gamma(0)t)$.] The observation that the Lyapunov exponent is numerically evaluated in the vicinity of a "thermalized" field configuration over time scales that are much longer than those of thermal fluctuations allows us to establish a connection between these two quantities. According to (5) the Lyapunov exponents are determined from the long time behavior of solutions of the linearized equation for the fluctuation $a_\mu(x,t)$ around an exact solution $A_\mu(x,t)$ of the Yang-Mills equations. In the continuum limit this equation reads:

$$\left(g_{\mu\nu}D(A)^2 - D_\mu(A)D_\nu(A)\right)a^\nu(x,t) - 2i[F_{\mu\nu}, a^\nu(x,t)] = 0 \tag{19}$$

where $D_\mu(A)$ denotes the gauge covariant derivative and $[F_{\mu\nu}, a^\nu]$ denotes the Lie algebra commutator. Equation (19) is the usual starting point for quantization in a background field, but we will *not* impose here the background gauge constraint $D_\nu a^\nu = 0$. The initial value problem for (19) can be solved by means of the retarded propagator in the background field:

$$a_\mu(x,t) = \int d^3x' \Delta_{\text{ret}}^{\mu\nu}(x,t;x',0|A)a_\nu(x',0). \tag{20}$$

$\Delta_{\text{ret}}^{\mu\nu}$ has the formal representation as difference between the causal and the anticausal propagator:

$$\Delta_{\text{ret}}^{\mu\nu}(A) = i\theta(t)\left[D_{\text{F}}^{\mu\nu}(A) - D_{\text{F}}^{\mu\nu*}(A)\right] \tag{21}$$

Now recall that the maximal Lyapunov exponent is obtained from the long time average of the growth rate of $a^\nu(x,t)$. Assuming ergodicity, we may therefore replace the long-time average of the propagators by the canonical, i.e. thermal, average:

$$\overline{\Delta}^{\mu\nu}_{\text{ret}} = i\theta(t)\left[D^{\mu\nu}_{\text{F},T} - D^{\mu\nu*}_{\text{F},T}\right] \qquad (22)$$

where $D^{\mu\nu}_{\text{F},T}$ denotes the *exact* finite temperature Feynman propagator of the gauge field. Note that the causal Feynman propagator describes damped fluctuations, the anti-causal propagator $D^{\mu\nu}_{\text{F}}$ describes exponentially growing perturbations. The only remaining obstacle before establishing the equivalence of h and $2\gamma(0)$ is that the thermal average in (22) should be calculated for a *classical* emsemble of gauge fields, whereas the usual perturbative approach to $D^{\mu\nu}_{\text{F},T}$ is based on the quantum mechanical ensemble. However, this difference turns out to be irrelevant for the plasmon damping rate $\gamma(0)$, although it has a large effect on the effective plasmon mass $\omega^*(0)$. This is not entirely fortuitous, because the damping rate is given by tree diagrams, such as Compton scattering and bremsstrahlung, that have an exact low-energy classical limit.[12,13] We can therefore identify the maximal Lyapunov exponent with (twice) the damping rate of the most unstable mode in a thermal nonabelian plasma, which turns out to be a plasmon at rest.[14]

CONCLUSIONS AND OUTLOOK

The short thermalization time scales of less than 1 fm/c found in our studies of the time evolution of classical nonabelian gauge fields show why Hagedorn was right thirty years ago with his assumption that final states in "soft" strong interaction physics are populated statistically. The reason for the success of these classical studies is that the dynamical instabilities in thermal gauge theories are of order g^2T, which is a classical inverse length or time scale that does not involve $\hbar$.

It would be interesting to see whether other quantities of order g^2T, such as the thermal magnetic screening mass on the "spatial" string tension, can also be calculated in the framework of classical Yang-Mills theory. This consideration leads into the problem of deriving an effective quasi-classical theory for thermal Yang-Mills theories at the length scale $(g^2T)^{-1}$ that consistently incorporates quantum effects from shorter distances in the form of transport coefficients. Presumably such an effective theory will contain a gauge invariant mass term of order gT (as in the Taylor-Wong action[15]) and a Langevin noise term describing the fluctuations due to interactions with hard thermal modes.

Another interesting problem concerns the application of real-time evolution of gauge fields to processes far off equilibrium as they occur in the earliest stage of hadron-hadron or nucleus-nucleus interactions. We have recently studied the instability of the superposition of two counter-propagating plane waves, i.e. of a standing abelian plane wave, in SU(2) Yang-Mills theory.[16] Here one finds that the Lyapunov exponent is proportional to the *amplitude* of the wave, not to the energy as it is the case in random fields. Once the coherent wave is only slightly perturbed it decays rapidly, exciting modes of all wavelengths, and quickly generates a thermal energy spectrum (see Figure 8). The evolution of more realistic initial configurations, such as the interaction between nonabelian wave packets, is presently under investigation.

ACKNOWLEDGEMENTS

I thank T. S. Biró, C. Gong, S. G. Matinyan and A. Trayanov for their enthusiastic help in unraveling the intricacies of chaotic dynamics in gauge theories. I would also like to thank D. Egolf, H. B. Nielsen, S. E. Rugh, and G. K. Savvidy for illuminating discussions. This work was supported in part by the U. S. Department of Energy (grant DE-FG05-90ER40592) and the North Carolina Supercomputing Program.

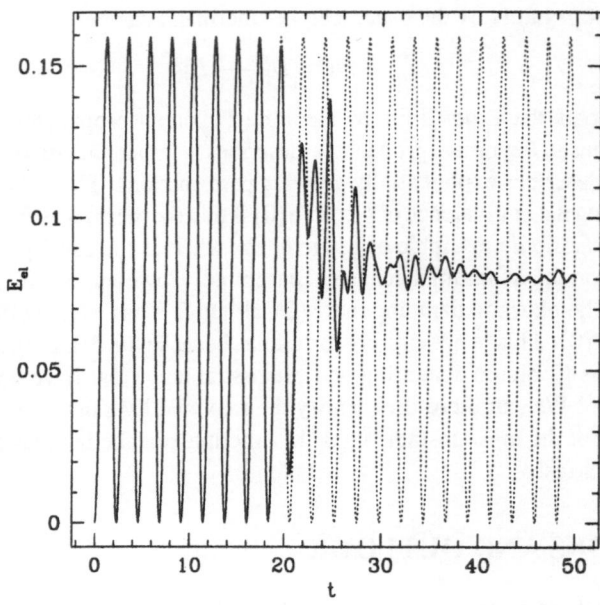

Figure 8. (a) Total electric energy for a SU(2) standing plane wave as function of time. The dotted curve shows the stability of an initially abelian wave, the solid line shows the instability in the presence of small nonabelian perturbations. In that case the wave eventually decays into a quasithermal frequency spectrum (b).

REFERENCES

1. R. Hagedorn, *Suppl. Nuovo Cimento* **3**, 147 (1965).

2. J. J. M. Verbaarschot, H. A. Weidenmüller, and M. R. Zirnbauer, *Phys. Rep.* **129**, 367 (1985); see also: C. Mahaux and H. A. Weidenmüller, *Ann. Rev. Nucl. Part. Science* **29**, 1 (1979).

3. G. E. Mitchell, E. G. Bilpuch, P. M. Endt, J. F. Shriner, and T. von Egidy, *Nucl. Instr. Meth.* **B46/47**, 446 (1991).

4. see e.g. A. J. Lichtenberg and M. A. Liebermann, *Regular and Stochastic Motion*, (Springer-Verlag, New York, 1983).

5. S. G. Matinyan, G. K. Savvidy, and N. G. Ter-Arutyunyan-Savvidy, *Zh. Eksp. Teor. Fiz.* **80**, 830 (1981) [*Sov. Phys. JETP* **53**, 421 (1981)].

6. P. Dahlquist and G. Russberg, *Phys. Rev. Lett.* **65**, 2837 (1990).

7. E. Haller, M. Köppel, and L. Cederbaum, *Phys. Rev. Lett.* **52**, 1665 (1984).

8. B. Müller and A. Trayanov, *Phys. Rev. Lett.* **68**, 3387 (1992).

9. T. S. Biró, C. Gong, B. Müller, and A. Trayanov, *Int. J. Mod. Phys.* **C5**, 113 (1994).

10. C. Gong, *Phys. Lett.* **B298**, 257 (1993).

11. C. Gong, *Phys. Rev.* **D49**, 2642 (1994).

12. C. Gong, Dissertation, Duke University, 1994 (unpublished).

13. E. Braaten and R. D. Pisarski, *Phys. Rev.* **D42**, 2156 (1990).

14. T. S. Biró, C. Gong, and B. Müller, to be published.

15. J. C. Taylor and S. Wong, *Nucl. Phys.* **B346**, 115 (1990).

16. C. Gong, S. G. Matinyan, B. Müller, and A. Trayanov, *Phys. Rev.* **D49**, R607 (1994).

STATISTICAL PROPERTIES OF RELATIVISTIC QUASIPARTICLES

Rémi Hakim and Horacio D. Sivak

Département d'Astrophysique Relativiste et de Cosmologie
Observatoire - 92190 Meudon (France)

INTRODUCTION

The concept of *quasiparticle* does appear in a large number of non relativistic physical situations where it brings both substantial technical simplifications and also far reaching clarifications of the problems under consideration. Quasiparticles are widely used in the framework of solid state physics[1] and, consequently, have been studied in details. Another case where quasiparticles, namely plasmons, have systematically been considered is plasma physics.[2]

In a relativistic context though, in spite of numerous studies on QED plasmas[3,4] and of a growing interest for the quark-gluon plasma expected to be observed in heavy-ion collisions,[5] no systematic considerations on *relativistic* quasiparticles have been undertaken. It should also be mentioned that such theoretical objects are of much interest in the context of nuclear matter whether relativistic[6] or not and in the study of Bose condensates[7] occurring in dense matter. Very simple relativistic quasiparticles [i.e. "free" quasiparticles endowed with an effective mass] are also considered in a toy model constituted by the $\lambda\Phi^4$-theory in gaussian (or other non perturbative) approximations[8,9] or in the case of the relativistic "scalar plasma".[10]

Apart from the early attempts by Jauch and Watson,[4] the most advanced study of relativistic quasiparticles seems to be the one by Migdal[7] in connection with considerations on boson condensates and, in what follows, his results are found anew in a more general context and also they are extended and specified more precisely.

The reasons why there does not exist systematic studies are multiple: (i) many peoples seem to consider that a relativistic extension of the concept of quasiparticle is trivial (it will be realized, below, that this is certainly not the case); (ii) the need for relativistic quasiparticles was not really urgent because of the relatively recent emergence (about a dozen years) of statistical considerations for relativistic quantum dense matter. We believe that, apart from an interest of their own, relativistic quasiparticles can be a source of new progresses when dealing with dense matter and, in particular, with the QCD plasma: nevertheless, it should be clear that only new physical inputs (ideas, approximations, etc) can bring new developments and, in this respect, quasiparticles can only represent a technical instrument that may express such an input.

It is, of course, not possible to deal with all cases and hence we limit ourselves to a sufficiently general one as to accommodate most situations: it is assumed that the polarization tensor (for bosons) or the mass operator (for fermions) are given not only as a function of the 4-momentum but also of *macroscopic* quantities (m.q.) such as the average 4-velocity of the system, its temperature, its density, etc. Accordingly, and at least as a first step, only free quasiparticles are dealt with. Whether the system under consideration is correctly described by an assembly of (free or interacting) quasiparticles or not is another problem. So is also the case as to the approximation method that leads to the polarization tensor (or the mass operator) at hand. Therefore, our first assumption is concerned with the general form of the equations of motion obeyed by the quasi-particle fields. For quasibosons, it reads

$$\Box\Phi(x) - \int d^4x\, \Pi(x, x'\,; \text{m.q.})\, \Phi(x') = 0\,, \tag{1}$$

while for quasifermions, one has

$$i\gamma\cdot\partial\Psi(x) - \int d^4x \sum(x, x'\,; \text{m.q.})\, \Psi(x') = 0\,. \tag{2}$$

In the following, we rather deal with the spacetime translational invariant case, where $\Pi(x, x') = \Pi(x-x', 0) \equiv \Pi(x-x')$ for bosons, and where $\Sigma(x, x') = \Sigma(x-x', 0) \equiv \Sigma(x-x')$ for fermions; and general results are given without proofs since the latter are simple extensions of the former. Therefore we treat mainly the case of Eqs. (1) (2) which are rewritten in Fourier space as

$$\{k^2 - \Pi(k)\}\Phi(k) = 0 \tag{3}$$

for quasibosons, which equations also signifies that they obey the dispersion equation

$$D(k) \equiv \{k^2 - \Pi(k)\} = 0\,. \tag{4}$$

From this "dynamical" starting point, the usual quantum field theory path is followed. First, the equations (3) for the "classical" field Φ (or Ψ) are cast into a lagrangian form; this allows the derivation of the usual conservation laws, *whenever valid* and hence explicit expressions for the four-current, energy-momentum tensor, angular-momentum, etc. Not only these expressions are needed for the macroscopic description of the system at hand but also for the quantization of the "classical" quasiparticle field. Indeed, explicit normalization of those quasiparticle plane waves, in which the field is expanded, is a necessity since a covariant expression for the scalar product of two free wave packets is needed to that end. Furthermore, from the energy-momentum tensor the quasiparticle hamiltonian is obtained. Once quantization is achieved, statistical mechanics (both for equilibrium and non-equilibrium situations) is almost a matter of routine. Next applications can follow.

CLASSICAL FIELDS

In this section, we limit ourselves to the case of complex scalar fields, the extension to other possibilities being either straightforward or explicitly dealt with. These fields obey the equations

$$\Box\Phi(x) - \int d^4\Pi(y)\Phi(x - y) = 0\,, \tag{5}$$

$$\Box\Phi^*(x) - \int d^4\Pi(y)\Phi^*(x - y) = 0\,. \tag{6}$$

with $\Pi^*(y) = \Pi(-y)$ in order that Eqs. (5) and (6) be consistent.

A possible lagrangian for these equations is

$$L = \partial_\mu\Phi^*\partial^\mu\Phi - \int d^4y\, \Pi(y)\, \Phi^*\left(x + \frac{1}{2}y\right)\Phi\left(x - \frac{1}{2}y\right)\,, \tag{7}$$

as can be checked by minimizing the action integral

$$\frac{\delta}{\delta \Phi^*(x)} \int d^4x' L[\{\Phi^*(x')\}, \{\Phi(x)\}] = 0 , \tag{8}$$

$$\frac{\delta}{\delta \Phi(x)} \int d^4x' L[\{\Phi^*(x')\}, \{\Phi(x')\}] = 0 . \tag{9}$$

However, such a derivation of the equations of motion (5)-(6) from the lagrangian (7) possesses the inconvenient of hiding some conditions to be satisfied by Φ and Φ^* and, furthermore, it needs a separate deduction of the various currents associated with symmetries whether spacetime or internal ones. Accordingly, the *nonlocal* lagrangian (7) is first transformed into a *"local"* one that involves, however, derivatives of arbitrary orders, using the series expansion of Φ (and Φ^*)

$$\Phi(x \pm \frac{1}{2}y) = \sum_{n=0}^{\infty} \frac{(\pm 1)^n}{n! 2^n} y^{\mu(n)} \cdot \partial_{\mu(n)} \Phi(x) , \tag{10}$$

where use has been made of Barut's notations[11]

$$y^{\mu(n)} \equiv y^{\mu_1} y^{\mu_2} \cdots y^{\mu_n}$$
$$\partial_{\mu(n)} \equiv \partial_{\mu_1} \partial_{\mu_2} \cdots \partial_{\mu_n} \tag{11}$$

The variation of the action with respect to the fields Φ and Φ^*

$$\int d^4x \delta\Phi(x) \left\{ -\Box\Phi^*(x) + \sum_{n=0}^{\infty} \int d^4y\, \Pi(y) \frac{(\pm 1)^n}{n! 2^n} y^{\mu(n)} \cdot \partial_{\mu(n)} \Phi^*\left(x + \frac{1}{2}y\right) \right\}$$
$$+ \int d^4x\, \delta_\mu \left\{ \partial^\mu \Phi^*(x) \delta\Phi(x) + \sum_{n,l=0}^{\infty} \int d^4y\, \Pi(y) \right.$$
$$\times \left[\partial_{\lambda(l)} \Phi^*\left(x + \frac{1}{2}y\right) \partial_{\nu(n)} \delta\Phi(x) \right] y^{\mu\nu(n)\lambda(l)}$$
$$\left. \times \frac{(-1)^n}{2^{n+l+1}(n+l+1)!} \right\} + \text{h.c.} \tag{12}$$

where, as usual, partial integrations have been performed.

This last expression contains two terms: while the first one, proportional to $\delta\Phi$, yields the equation of motion (6) [and also (5) by considering the term $\delta\Phi^*$ implicitly contained into the hermitian conjugated part], the second one - involving $\partial_\mu\{...^\mu\}$ - gives rise to a vanishing contribution after the 4-volume integral has been transformed into a surface integral *and* demanding that, not only $\delta\Phi$ but also all its derivatives $\partial_{\mu(n)}\delta\Phi(n = 1, 2, ...)$, vanish at the boundary of the spacetime domain where the action integral is evaluated.

When the system, or the lagrangian, is invariant under a given symmetry group, $\delta\Phi$ and $\delta\Phi^*$ do possess a specific form and this gives rise to a 4-current which reads

$$J^\mu[\Phi, \Phi^*, \delta\Phi, \delta\Phi^*] = \partial^\mu \Phi^* \delta\Phi + \delta\Phi^* \partial^\mu \Phi$$
$$+ \int_0^{1/2} ds \int d^4y \Pi(y) y^\mu \Phi^*\left[x + y\left(s + \frac{1}{2}\right)\right] \delta\Phi\left[x + y\left(s + \frac{1}{2}\right)\right]$$
$$- \int_0^{1/2} ds \int d^4y \Pi(y) y^\mu \delta\Phi^*\left[x - y\left(s - \frac{1}{2}\right)\right] \Phi\left[x - y\left(s + \frac{1}{2}\right)\right] \tag{13}$$

This expression constitutes the general form of a four-current for scalar particles as it arises from a general symmetry of the lagrangian. Its first term is the usual one that occurs from the

Klein-Gordon lagrangian while the second (integral) parts come from the non local part of the lagrangian (7). For a U(1)-symmetry, where $\delta\Phi = i\theta\Phi$ and $\delta\Phi^* = -i\theta\Phi^*$, the "charge" current then reads

$$J_Q^\mu = i\theta \left[\Phi^*(x)\partial^\mu\Phi(x) \right.$$
$$\left. + \int d^4y \int_{-1/2}^{1/2} ds \Pi(y) y^\mu \Phi^*\left[x + y\left(s + \frac{1}{2}\right)\right] \cdot \Phi\left[x + y\left(s - \frac{1}{2}\right)\right] \right] \quad (14)$$

which, obviously, reduces to the usual Klein-Gordon four-current when the polarization kernel $\Pi(y)$ is taken as $\Pi(y) = \mu^2 \delta^{(4)}(y)$.

The conservation laws for spacetime symmetries obviously depend on the transformation properties of the polarization tensor $\Pi(x)$ under the Poincaré group. Clearly, for equations of motion such as (3) one expects the conservation of the energy-momentum tensor: Π depends on the difference $x - x'$ and is thus invariant under spacetime translations. However, it is not so for the general equations of motion (1)-(2). Moreover, eventhough the energy-momentum tensor of the quasiparticles is conserved, this is not necessarily the case for the general angular momentum tensor.

θ^μ denoting infinitesimal spacetime translations, the corresponding variations of the field are given by $\delta\Phi = \theta^\mu\partial_\mu\Phi$ with similar equations for Φ^*. Using now the general equation (13), the equations of motion (5) (6), with the use of conventional techniques,[11-13] one obtains the following energy-momentum tensor

$$T^{\mu\nu} = \partial^{(\mu}\Phi^*\partial^{\nu)}\Phi - \eta^{\mu\nu}L[\{\Phi^*\},\{\Phi\}] + \int_0^{1/2} ds \int d^4y \Pi(y) y^\mu$$
$$\times \left\{ \Phi^*\left[x + y\left(s + \frac{1}{2}\right)\right] \partial^\nu\Phi\left[x + y\left(s - \frac{1}{2}\right)\right] \right.$$
$$\left. - \partial^\nu\Phi^*\left[x + y\left(s - \frac{1}{2}\right)\right] \cdot \Phi\left[x - y\left(s + \frac{1}{2}\right)\right] \right\} \quad (15)$$

With the use of the equations of motion (5)-(6) it can easily be checked that this tensor is conservative as expected for a system invariant under spacetime translations. It should be borne in mind that this conservation property holds as far as the *first* index is concerned: $\partial_\mu T^{\mu\nu} = 0$, and generally one has not $\partial_\nu T^{\mu\nu} = 0$. On the other hand, the energy-momentum tensor is not even symmetric, as is often the case for the *canonical* one.

FORMAL QUANTIZATION

In the preceding section the "classical" quasiparticle fields, obeying Eqs. (5)-(6), have been studied and their conservation laws derived: the latter are now used. Before quantizing these fields, it must be recalled that, in actual practice, one starts with a given physical system obeying an involved quantum field theory at finite density and/or temperature; thence approximations of various types (perturbation, one-loop, random phase, etc.) must be performed and finally lead to equations of motion for the excitations propagating within the system. It is the quantization of these excitations, of these *approximate* "free" fields which we want to consider here.

In this section, the quantization of the "classical" quasiparticle fields is first performed in a *formal* manner and next discussed in connection with specific problems such as the existence and interpretation of antiquasiparticles, etc. Such a quantization is not very complicated the more so since only free quasiparticle fields are dealt with here.

First we assume that the Φ-vacuum is normal, in the sense that $\langle \mathrm{vac}|\Phi|\mathrm{vac}\rangle = 0$, so that the field Φ can be expanded into *normalized* plane waves. Whenever $\langle \mathrm{vac}|\Phi|\mathrm{vac}\rangle$ different from zero, the average value of Φ has just to be substracted out.

The vector space of the solutions of Eqs. (5) (6) can be attributed the following hermitian product

$$
\begin{aligned}
\langle \Phi_1|\Phi_2\rangle \;=\;& \int_\Sigma d\Sigma_\mu \left[\Phi_1^*(x)\,\partial^\mu\Phi_2(x) - \partial_\mu\Phi_1^*(x)\Phi_2(x)\right] \\
&+ \int d^4y \int_{-1/2}^{1/2} ds \Pi(y)y^\mu \Phi_1^*\!\left[x + y\left(s + \frac{1}{2}\right)\right]\Phi_2\!\left[x + y\left(s - \frac{1}{2}\right)\right], \quad (16)
\end{aligned}
$$

where Σ is an arbitrary spacelike three-surface and where $\Phi_i(x)(i=1,2)$ are two solutions of Eqs. (5)-(6). It is not difficult to realize that the integrand of the expression (16) is divergent-free and hence that this hermitian product does not actually depend on Σ [use the equations of motion]. Eq. (16) allows the normalization of plane waves solutions of Eqs. (5)-(6), $< \Phi_{\mathbf{k}}|\Phi_{\mathbf{k}'} > = \delta^{(3)}(\mathbf{k}-\mathbf{k}')$ and one finally obtains

$$
\Phi_k(x) = \frac{1}{(2\pi)^{1/2}\left[2\omega_{\mathbf{k}} - \dfrac{\partial\Pi(k)}{\partial\omega_{\mathbf{k}}}\right]^{1/2}} \exp[-ik.x], \quad (17)
$$

where $\omega_{\mathbf{k}} \equiv k^0$ is a solution of the dispersion equation (3). Note that the normalization (17) was already obtained in Ref. 7 in another way.

In order to be specific, let us consider the case of a real scalar field, the extension to other fields (complex, with internal degrees of freedom, fermion, etc.) being straightforward. It is expanded into normalized plane waves as

$$
\Phi(x) = \sum_l \frac{1}{(2\pi)^{3/2}} \int \frac{d^3k}{\left.\dfrac{\partial D(k)}{\partial\omega}\right|_{\omega=\omega_l(\mathbf{k})}} \{a_{(l)}(\mathbf{k})\exp.[ik.x] + a_{(l)}^+(\mathbf{k}\exp.[+ik.x]\} \quad (18)
$$

where $D(k) \equiv k^2 - \Pi(k)$ and where the sum over l refers to a sum over all possible modes, solutions of $D(k)=0$; $a_{(l)}(\mathbf{k})$ and $a_{(l)}^+(\mathbf{k})$ are the annihilation and creation operators, respectively, of the quasi-particles under consideration and must obey the conventional commutation relations $[a_{(l)}(\mathbf{k}), a_{(l)}^+(\mathbf{k})] = \delta^{(3)}(\mathbf{k}-\mathbf{k}')$. The hamiltonian is then given by

$$
\begin{aligned}
H \;=\;& \int_{t=\mathrm{const.}} d^3x\, T^{00} \\
=\;& \sum_l \int_{t=\mathrm{const.}} d^3k\,\omega_l(\mathbf{k})\left[a_{(l)}^+(\mathbf{k})a_{(l)}(\mathbf{k}) + \frac{1}{2}\right] \quad (19)
\end{aligned}
$$

as it should be in usual cases (see below, however).

Unfortunately, actual physical situations are not so simple and they lead to difficult problems which we now mention very briefly on the case of a simple example. The following equation of motion is first chosen

$$
(\Box + m_1^2)(\Box + m_2^2)\Phi(x) = 0 \quad (20)
$$

where $\Phi(x)$ is still a real scalar field. Equations of that form, or more general ones, have been discussed a few decades ago in connection with the hope to solve the question of the divergences of quantum electrodynamics.[15-17]

Eq.(20) corresponds to a "polarization tensor" $\Pi(k)$ given by

$$
\Pi(k) = \frac{k^4 + m_1^2 m_2^2}{m_1^2 + m_2^2}, \quad (21)
$$

so that the quantity

$$2\omega_{\mathbf{k}} - \frac{\partial \Pi(k)}{\partial \omega_{\mathbf{k}}}\Big|_{\omega=\omega_l(\mathbf{k})} = 2\omega \left[\frac{m_1^2 + m_2^2 - 2k^2}{m_1^2 + m_2^2}\right]_{k^2=m_1^2 \text{ or } m_2^2} \tag{22}$$

is positive for one root of Eq.(20) and negative for the other one. It follows immediately that the plane-wave normalization (3.2) is no longer possible and must be replaced by the following one $< \Phi_{\mathbf{k}}|\Phi_{\mathbf{k}'}> = \pm\delta^{(3)}(\mathbf{k} - \mathbf{k}')$; in other words, the Hilbert space of the solutions of Eq.(3.17) is now endowed with an *indefinite metric*. In fact the $\pm$ sign refers to the sign of

$$2\omega(\mathbf{k}) - \frac{\partial \Phi(k)}{\partial \omega_{\mathbf{k}}}\Big|_{\omega=\omega_l(\mathbf{k})} \equiv \frac{\partial}{\partial \omega}[\omega^2 - \Pi(k)]|_{\omega=\omega_l(\mathbf{k})} = \frac{\partial D}{\partial \omega}\Big|_{\omega=\omega_1(\mathbf{k})} \tag{23}$$

With such an indefinite metric, the hamiltonian operator now reads

$$H = \sum_l \int d^3 k\, \varepsilon \left[2\omega_{\mathbf{k}} - \frac{\partial \Pi(k)}{\partial \omega_{\mathbf{k}}}\Big|_{\omega=\omega_l(\mathbf{k})}\right] \omega_1(\mathbf{k}) a_{(l)}^+(\mathbf{k}) a_{(l)}(\mathbf{k}) \tag{24}$$

where the creation/destruction operators now obey the following commutation relations

$$[a_{(l)}(\mathbf{k}), a_{(l)}^+(\mathbf{k})] = \varepsilon \left[2\omega_{\mathbf{k}} - \frac{\partial \Pi(k)}{\partial \omega_{\mathbf{k}}}\Big|_{\omega=\omega_l(\mathbf{k})}\right] \delta_{ll'}\delta^{(3)}(\mathbf{k} - \mathbf{k}') \tag{25}$$

and where ε is the sign function. Eq.(24) exhibits clearly that the hamiltonian operator is not positive definite with the disastrous consequence that there does no longer exist a vacuum state defined as being a minimum energy (or free energy) state.

Many solutions to these drawbacks have been proposed and discussed in the literature: transformation of the real field into a purely imaginary one; introduction of supplementary conditions; projection on positive energy states regarded as sole physical states; etc. Unfortunately, none of them can be considered as being really satisfactory.

Moreover, one has to distinguish between theories supposed to be fundamental and phenomenological ones, describing partly a given physical system in a definite class of state. While in the first case, known properties, such as causality or Lorentz invariance, have to be imperatively obeyed, in the second one, these basic properties might be relaxed at given scales.

However, in the case of dense matter, the problem of so-called "ghost" states is connected with the stability of the thermal equilibrium, i.e. of the vacuum (a more detailed discussion can be found in Ref. 18). A hint to this question can be found in a similar situation arising in plasma physics. When one considers plasma modes of oscillation, whether quantized as quasiparticles (plasmons) or not, it is essential to take the energy (and possibly quantum numbers) exchanges with the basic system into account; in other words, exchanges of energy between the vacuum and the modes. When this done, as in plasma physics, the so-called negative energy modes appear to be as physical as the positive ones.

EQUILIBRIUM AND TRANSPORT OF QUASIPARTICLES

In this section a stable "vacuum" or equilibrium state is assumed and hence no negative energy modes can be excited within the medium. When these conditions are realized, it is not very difficult to obtain the basic statistical properties of those quasiparticles propagating in the system, whether in thermal equilibrium or not. Here we still limit ourselves to the case of a complex scalar field without any internal symmetry, the extension to this latter case (or to other Bose fields) being straightforward.

From the knowledge of the quasiparticle hamiltonian and of the "charge" the equilibrium density operator retains its usual form and hence it leads to quite similar expressions for the physical quantities such as the average charge density, energy-momentum tensor (or energy density ρ and pressure P), entropy four-current, or average occupation number. These latter quantities respectively write

$$n(k) = \frac{1}{\exp(u^\mu k_\mu - \mu) - 1} , \qquad (26)$$

for the average occupation number,

$$J^\mu = \int d^4 k \delta[D(k)] \left[\frac{\partial}{\partial k_\mu} D(k) \right] n(k) , \qquad (27)$$

for the "charge" four-current, and

$$S = - \int d^4 k \delta[D(k)] \left[\frac{\partial}{\partial k_\mu} D(k) \right] \{ n(k) \log n(k) - [n(k) + 1] \log[n(k) + 1] \} , \qquad (28)$$

for the entropy four-current. When the system does not involve any other 4-vector than u^μ, then $\langle T^{\mu\nu} \rangle$ has the general form

$$\langle T^{\mu\nu} \rangle = (\rho + P) u^\mu u^\nu - P \eta^{\mu\nu} , \qquad (29)$$

with

$$\rho = \int d^4 k \delta[D(k)] k_\nu \left[\frac{\partial}{\partial k^\mu} D(k) \right] u^\mu u^\nu n(k) , \qquad (30)$$

(energy density)

$$P = - \frac{1}{3} \int d^4 k \delta[D(k)] k_\nu \left[\frac{\partial}{\partial k^\mu} D(k) \right] \Delta^{\mu\nu}(u) n(k) , \qquad (31)$$

(pressure)

In more usual notations [i.e. in a Lorentz frame where $u^\mu = (1, 0)$], one has

$$n_{eq} \equiv \frac{\langle Q \rangle}{V} = J^0 = \sum_{i=l,\pm} \int d^3 k \frac{\pm 1}{\exp.[\beta(\omega_l(\mathbf{k}) \mp \mu)] - 1} , \qquad (32)$$

$$\rho = \sum_{l,\pm} \int d^3 k \omega_l(\mathbf{k}) \frac{1}{\exp.[\beta(\omega_l(\mathbf{k}) \pm \mu)] - 1} , \qquad (33)$$

$$P = \frac{1}{3} \sum_{i,l,\pm} \int \frac{d^3 k}{\frac{\partial D(k)}{\partial \omega}|_{\omega = \omega_l(\mathbf{k})}} k^i \frac{\partial D(k)}{\partial \omega} \Big|_{\omega = \omega_l(\mathbf{k})} \frac{1}{\exp.[\beta(\omega_l(\mathbf{k}) \pm \mu)] - 1} \qquad (34)$$

The only differences with the usual case (see e.g. 19 and 20) are (i) the summation over the various modes, denoted by l, and over the "anti" quasiparticles (summation over $\pm$) and (ii) the expression of the pressure [Eq. (34)] which, in the case where $k^2 = m^2$, reduces to the known expression.

Eqs. (27)-(31) indicate that the role of the 4-velocity of the quasiparticles is played by the quantity

$$v^\mu(k) = \frac{\partial D(k)}{\partial k_\mu} , \qquad (35)$$

while their four-momentum is k^μ.

When quasiparticles do interact, or when they are in an off-equilibrium state, or when a number of problems such as the calculation of the fluctuations of various physical quantities is dealt with, the use of a covariant Wigner function presents a certain interest.[21] It is defined as

$$f(x,p) = \frac{1}{(2\pi)^4} \int d^4 R \exp.\left[-ip.R\right] K\left[\phi^*\left(x + \frac{1}{2}R\right), \phi\left(x + \frac{1}{2}R\right)\right], \qquad (36)$$

where $K[\phi^*(y), \phi(z)]$ is the correlation function of the field ϕ at points y and z, i.e.

$$K[\phi^*(y), \phi(z)] \equiv \langle \phi^*(y)\phi(z)\rangle - \langle \phi^*(y)\rangle\langle\phi(z)\rangle . \qquad (37)$$

This definition should generally be preferred in contrast with those given by other authors.[22,23] In terms of the Fourier transform of the Wigner function the average 4-current $J^\mu(k)$ is easily found to be

$$J^\mu(k) = 2\int d^4 p\, p^\mu f(k,p) - \int_{-1/2}^{+1/2} ds \int d^4 p f(k,p) \frac{\partial}{\partial p_\mu}\Pi(p + ks) \qquad (38)$$

and a similar, but more involved expression for $T^{\mu\nu}(k)$.

Using now the equations of motion obeyed by the fields ϕ^* and ϕ and the same kind of calculation as can be found in Refs. 21-23, the above definition (36) leads in a straightforward manner to the following transport equation

$$p.\partial f(x,p) + \int d^4\Pi(y) \sin[p.y] f\left(x - \frac{1}{2}y, p\right) = 0 , \qquad (39)$$

valid for *free* quasiparticles only. This last equation is the analogue of the relativistic Liouville equation[20,24] for a classical relativistic system of free particles. In fact, there also exists another equation obeyed by $f(x,p)$, which plays the role of a mass-shell equation. Eq.(36) is a basic equation when dealing with transport properties of the system; in such a case it must be supplemented by a collision term $C(f)$ which might be of the Uhlenbeck-Uehling[25] form or of any other phenomenological nature. As a matter of fact, the most reasonable one, as analysed elsewhere,[26] is probably a relativistic relaxation time approximation.[27,28] Such a relaxation time collision term is given by

$$C(f) = -u^\mu[p_\mu \nabla_\mu \Pi(p)]\frac{f - f_{eq}}{\tau} , \qquad (40)$$

where τ is the relaxation time and where the factor $u^\mu[p_\mu - \nabla_\mu\Pi(p)]$ has been chosen so that the so-called Landau-Lifschitz matching conditions[29]

$$u_\mu J^\mu_{\text{off}} = 0 , \quad u_\mu T^{\mu\nu}_{\text{off}} = 0 , \qquad (41)$$

are satisfied. This property can be checked by integrating both sides of the kinetic equation for the "small" off-equilibrium part of the Wigner function.

In many physical situations the equilibrium fluctuations of some observables are required. Here the equilibrium expression for the correlations of the Wigner function operator, namely $K[f(x,p), f(y,p')] \equiv K[f(x - y, p), f(0, p')] \equiv F(x - y; p, p')$; or, more precisely, the Fourier transform of F is evaluated below, say $F(k; p, p')$. From this quantity one can easily calculate the fluctuations of all one-quasiparticle observables.

REFERENCES

1. See e.g. C.Kittel, *Quantum Theory of Solids* [J.Wiley, New-York (1963)].

2. T. Kihara, O. Aono, T. Dodo, Nuclear Fusion **2**, 66 (1962); A. I. Alekseev, Yu. P.Nikitin, Soviet Phys. J.E.T.P. **23**, 608 (1966); E.G. Harris, Adv. Plasma Phys. **3**, 157 (1969).

3. N. P. Landsman, CH. G. van Weert, Phys. Reports **145**, 141 (1987).

4. One should also include the "phenomenological QED" of J.M.Watson, Phys. Rev. **74**, 950 (1948); *ibid.* **74**, 1485 (1948); *ibid.* **75**, 1249 (1949).

5. See e.g. this volume.

6. B. D. Serot, J. D. Walecka, *"The Relativistic Nuclear Many-Body Problem"*, in Advances in Nuclear Physics vol. **16** [J.W.Negele, E.Vogt eds.; Plenum Press; New-York (1986)].

7. A.B.Migdal, Rev. Mod. Phys. **50**, 107 (1978).

8. Among numerous articles on the subject, let us only quote the following ones: G. Baym, G. Grinstein, Phys. Rev. **D15**, 2897 (1977); T. Barnes, G. I. Ghandour, Phys. Rev. **D22**, 924 (1980); W. A. Bardeen, M. Moshe, Phys. Rev. **D28**, 1372 (1983); P. M. Stevenson, Phys. Rev. **D33**, 2305 (1985); M. Ciancitto, Nucl. Phys. **B254**, 653 (1985); etc. See Ref. 9.

9. F. Grassi, R. Hakim, H. Sivak, Int. J. Mod. Phys. **A 6**, 4579 (1991).

10. G. Kalman, Phys. Rev. **D9**, 1656 (1974).

11. A. O. Barut, *"Electrodynamics and Classical Theory of Fields and Particles"*, p. 122 ff. [Mac Millan; New-York (1965)].

12. N. N. Bogolubov and D. V. Shirkov, *"Introduction to the Theory of Quantized Fields"* [Interscience Publ.; New-York (1959)].

13. C. Itzykson, J. B. Zuber, *"Quantum Field Theory"* [Mac Graw Hill; New-York (1980)].

14. The quantization of a non local free (linear) field theory presents an interest of its own and has sometimes been considered. See e.g. R. Marnelius, Phys. Rev. **D8**, 2472 (1973).

15. See e.g. B. T. Darling, Phys. Rev. **92**, 1547 (1953); P. T. Matthews, Proc. Camb. Phil. Soc. **45**, 441 (1949); W. Pauli, F. Villars, Rev. Mod. Phys. **21**, 434 (1949); W. Pauli, Rev. Mod. Phys. **15**, 175 (1943); R. J. N. Phillips, Nuovo Cimento 1, 823 (1955); L. K. Pandit, Sup. Nuovo Cimento **11**, 157 (1959); K. L. Nagy, Sup. Nuovo Cimento **17**, 92 (1960); T. W. B. Kibble, J. C. Polkinghorne, Nuovo Cimento **8**, 74 (1958); H. M. Fried, J. Plebanski, Nuovo Cimento **18**, 884 (1960); A. O. Barut, G. H. Mullen, Ann. Phys. (N.Y.) **20**, 184 (1962); *ibid.* **20**, 203 (1962).

16. A. Pais, G. E. Uhlenbeck, Phys. Rev. **79**, 145 (1950).

17. E. C. G.Sudarshan, Phys. Rev. **123**, 2183 (1961).

18. R. Hakim, H. D. Sivak, submitted to Phys. Rev. D.

19. K.Huang, *Statistical Mechanics* [Wiley; New York (1963)].

20. J.Ehlers, *General Relativity and Kinetic Theory,* in *General Relativity and Cosmology* [Proceedings of the 1969 Varenna summer school; R. K. Sachs ed.; Academic Press, New York (1971)].

21. R.Hakim, Riv. Nuovo Cim. **1**, 1 (1978).

22. P. Carruthers, F. Zachariasen, Phys. Rev. **D13**, 950 (1976); F. Cooper, D. H. Sharp, Phys. Rev. **D12**, 1123 (1975); F. Cooper, M. Feigenbaum, Phys. Rev. **D14**, 583 (1976).

23. P. Carruthers, F. Zachariasen, Rev. Mod. Phys. **55**, 245 (1983).

24. G. E. Tauber, J. W. Weinberg, Phys. Rev.; W. Israël, J. Math. Phys. **4**, 1163 (1963); R. Hakim, J. Math. Phys. **8**, 1315 (1967); etc.

25. E. A. Uehling, G.E. Uhlenbeck, Phys. Rev. 43, 552 (1933).

26. R. Hakim, L. Mornas, P. Peter, H. Sivak, Phys. Rev. **D46**, 4603 (1992).

27. R. Hakim, L. Mornas, Phys. Rev. **C47**, 2846 (1993).

28. J. L. Anderson, H. R. Witting, Physica **74**, 466 (1974); *ibid.* **74**, 489 (1974).

29. L. Landau, E. Lifschitz, *Fluid Mechanics* [Pergamon Press; London (1959)].

QUANTUM DECOHERENCE AND ENTROPY
IN HIGH-ENERGY INTERACTIONS

Hans-Thomas Elze

CERN-Theory
CH-1211 Geneva 23, Switzerland

ABSTRACT

Or: "How to generate an ensemble in a single event?"— Following recent work on entropy in strong interactions, I explain the concept of environment-induced quantum decoherence in elementary quantum mechanics. The classically chaotic inverted oscillator becomes partially decoherent already in the environment of a single other oscillator performing only vacuum fluctuations. One finds exponential entropy growth in the subsystem with a Lyapunov exponent, which approaches the classical one for weak coupling.

INTRODUCTION

Recently the long-standing "entropy puzzle" of high-multiplicity events in strong interactions at ultra-relativistic energies has been analysed from a new point of view.[1] This is related to the concepts of an *open quantum system* and *environment-induced quantum decoherence*. The problem dates back to Fermi and Landau and is intimately connected to understanding the rapid thermalization of high energy density ($\gg 1$ GeV/fm^3) matter.[2] Why do thermal models work so well in reproducing global features of hadronic multiparticle final states? Why do they work at all?

Or, Why does high-energy scattering of pure initial states lend itself to a statistical description characterized by a large apparent *entropy* from a mixed-state density matrix describing intermediate stages in a space-time picture of parton evolution? Effectively, *unitary time evolution* of the observable part of the system breaks down in the transition from a quantum mechanically pure initial state to a highly impure (more or less thermal) high-multiplicity final state. Note that the unitary time evolution operator, $\exp(-i\hat{H}t)$, always transforms a pure state into a pure state, according to the Schrödinger equation, which *cannot* produce entropy under any circumstances (cf. below). This was discussed in detail in Refs.,[1] where more references concerning formal aspects of this work can be found. Based on analogies with studies of the quantum measurement process ("collapse of the wave function")[3] and motivated by related problems in quantum cosmology and by non-unitary non-equilibrium evolution resulting in string theory,[4] I argued that **environment-induced quantum decoherence solves the entropy puzzle** of strong interactions.

A complex pure-state quantum system can show a quasi-classical behaviour, i.e. an impure density (sub)matrix together with decoherence of the associated pointer states in an observable subsystem.[1,3,4] I will demonstrate in the following that the decoherence process is uniquely correlated with entropy production. Considering strong interactions, in particular, there is a natural *Momentum Space Mode Separation* due to confinement, which is defined in the frame of initial conditions for the time evolution and for the physical (gauge) field degrees of freedom. Thus, almost constant QCD field configurations form an *unobservable environment*, which interacts with the *observable subsystem* composed of partons. The environment modes are unobservable, since they can neither hadronize nor initiate hard scattering among themselves, whereas the partons are observable in the sense of parton-hadron duality or deep-inelastic scattering; equivalently, low-energy coloured vacuum fluctuations cannot propagate into asymptotic states.

Previously, I studied the induced quantum decoherence and entropy production in a non-relativistic single-particle model resembling an electron coupled to the quantized electromagnetic field, however, with a deliberately enhanced oscillator spectral density in the infrared. The Feynman-Vernon influence functional technique for quantum Brownian motion provided the remarkable result that in the *short-time strong-coupling limit* the model parton behaves like a *classical particle*:[1] Gaussian parton wave packets experience *friction* and *localization*, i.e. no quantum mechanical spreading, and their coherent *superpositions decohere*. The decoherence process has been shown to lead to entropy production in this oversimplified parton model.

Summarizing, my point of view is that partons feel an unobservable (gluonic) environment, which manifests its strong non-perturbative interactions on a short time scale ($\ll 1$ fm/c) through decoherence of suitable partonic pointer states [†], their quasi-classical behaviour, and entropy production. If confirmed in QCD, this will have important consequences for parton-model applications to complex hadronic or nuclear reactions. The emergence of *structure functions* from initial-state wave functions can and will be further studied in this approach.

It seems somewhat more realistic to consider two coupled scalar fields representing partons and their non-perturbative environment, respectively. In the functional Schrödinger picture employing Dirac's time-dependent variational principle, i.e. a non-perturbative method, I derived a Cornwall-Jackiw-Tomboulis (CJT) type effective action and the equations of motion for renormalizable interactions.[1] Thus, analysis of the entropy puzzle in strong interactions leads to study an observable field (open subsystem) interacting with a dynamically hidden one (unobservable environment), i.e. *quantum field Brownian motion*.

ENTROPY IN A PURE QUANTUM STATE?

Instead of representing the formalism and more technical results from Refs.[1], I want to demonstrate here in simple quantum mechanical examples the basic Why and How of the solution to the entropy puzzle.

Consider a system that can be described in terms of two normalized discrete basisstates, $|1\rangle$ and $|2\rangle$. Forming a *pure state*, $|\psi\rangle \equiv a_1|1\rangle + a_2|2\rangle$, by a coherent superposition with

---

[†]In general, these are not single-particle states but rather coherent (Gaussian) wave functionals, as constructed in the second of Refs.[1]

amplitudes $a_1 \equiv p^{1/2}$ and $a_2 \equiv (1-p)^{1/2}$, the corresponding density matrix, $\hat{\rho} \equiv |\psi\rangle\langle\psi|$, is

$$\rho_{ij} = \begin{pmatrix} a_1^2 & a_1 a_2 \\ a_1 a_2 & a_2^2 \end{pmatrix} \longrightarrow \rho_{ij}^D = \begin{pmatrix} 1 & 0 \\ 0 & 0 \end{pmatrix} , \tag{1}$$

where $\hat{\rho}^D$ is obtained by diagonalization. Note the off-diagonal *interference terms* in ρ_{ij}. Furthermore, observe that $\hat{\rho}^D$ has only one non-vanishing eigenvalue. Introducing the **von Neumann** or **statistical entropy**,

$$S[\hat{\rho}] \equiv - \operatorname{Tr} \hat{\rho} \ln \hat{\rho} , \tag{2}$$

we find $S[\hat{\rho}] = S[\hat{\rho}^D] = -[1 \ln 1 + 0 \ln 0] = 0$, i.e. *no entropy in a pure state*. Secondly, forming a *mixed state* (ensemble) such that the system is in state $|1\rangle$ with probability p and in state $|2\rangle$ with probability $1 - p$, the density matrix becomes $\hat{\rho}' = |1\rangle p \langle 1| + |2\rangle (1-p) \langle 2|$, i.e. a decoherent superposition. Hence, we obtain

$$\rho'_{ij} = \begin{pmatrix} p & 0 \\ 0 & 1-p \end{pmatrix} . \tag{3}$$

The density matrix (3) shows *no interference terms* and is diagonal per se.[‡] Then, $S[\hat{\rho}'] \equiv S(p) = -[p \ln p + (1-p) \ln(1-p)] \neq 0$, generally. In fact, $0 \leq S(p) \leq S(1/2) = \ln 2$. Total ignorance about the state of the system ($p = 1/2$) corresponds to $\ln 2$ units of entropy in a two-state system, i.e. *1 bit* of information is lost compared to certainty about its state ($p = 0, 1$).

One concludes that *entropy production can only occur, if the interference terms of the density matrix representing a more or less pure state of the observed system decay dynamically.*[§] In a *closed system* evolving unitarily in time, however, there is no way to transform, for example, $\hat{\rho}$ into $\hat{\rho}'$, see Eqs. (1), (3). Only the interaction of the system with an *environment*,[1,3,4] which may even be unobservable, can have such an effect. The presence (and integrating out) of the environment degrees of freedom essentially changes the dynamics of the observed system. This can lead to the decay of the interference terms in its density matrix, i.e. *environment-induced quantum decoherence*, which is necessary to increase its impurity and, thus, to produce entropy.

THE ANHARMONIC OSCILLATOR EXAMPLE

Next, consider a non-relativistic particle moving in a one-dimensional double-well potential presenting the observable subsystem. Let it be coupled translationally invariant to an environment consisting of a single high-frequency oscillator. The classical action is

$$S = \int dt \left\{ \tfrac{1}{2} M \dot{x}^2 + \tfrac{1}{2} m \dot{y}^2 - \tfrac{1}{2} m \omega^2 (y - x)^2 + \tfrac{1}{2} M \Omega^2 x^2 - \tfrac{1}{4!} \lambda M \Omega^4 x^4 \right\} . \tag{4}$$

For simplicity, let $M = m = \Omega$. Then, properly rescaling by Ω, one obtains

$$S = \int dt \left\{ \tfrac{1}{2} \dot{x}^2 + \tfrac{1}{2} \dot{y}^2 - \tfrac{1}{2} \omega^2 (y - x)^2 + \tfrac{1}{2} x^2 - \tfrac{1}{4!} \lambda x^4 \right\} , \tag{5}$$

[‡]Note that $\operatorname{Tr} \hat{\rho} = \operatorname{Tr} \hat{\rho}' = 1$: the system is in *some* state with total probability 1.

[§]The argument does not depend on particular physical characteristics of the system; it holds for the two-state system as well as for an interacting quantum field.

in terms of dimensionless quantities and two coupling constants, $\omega^2, \lambda \geq 0$. For $\omega = 0$ the minima of the doublewell lie at $x_\pm = \pm(\lambda/3!)^{-1/2}$, at a depth of $-3/2\lambda$ (the local maximum is zero at $x = 0$). Presently, I want to study the case that the excitation energy of the x-particle (X) is smaller than the level spacing ω of the environment oscillator (Y), which is assumed to be in its ground state. Starting with a given initial state of X, I will calculate the time evolution of the corresponding density matrix $\hat{\rho}_X$ under the influence of the *vacuum fluctuations* of Y. Excited states of Y contribute only *virtually* here; they cannot become real due to energy conservation.

For illustration, I choose the metastable initial state of X, when classically the particle "rests on top of the hill" ($x = 0$). Quantum mechanically this can be represented by a minimum uncertainty Gaussian wave packet,

$$\psi(x, t = 0) = \pi^{-1/4} w_0^{-1/2} e^{-\frac{1}{2}x^2/w_0^2} , \tag{6}$$

with $w_0 \ll \lambda^{-1/2}$. Also, assume ω to be sufficiently larger than $3/2\lambda$.

First of all, let the system evolve **classically**. Nothing will happen. However, any infinitesimal perturbation of the fine-tuned initial conditions causes X to move "down the hill", left or right (L or R), dragging Y along. There is local chaos in the sense of extreme sensitivity to the initial conditions at $x, y \approx 0$; arbitrarily small uncertainties in the initial conditions lead to a *loss of predictability*. For an *ensemble of initial conditions* X switches with probabilities $p_L(t')$ and $p_R(t') = 1 - p_L(t')$ between L and R, respectively, if at least one trajectory passes $x = 0$ in a certain interval $[t' - \epsilon, t' + \epsilon]$. This corresponds to a loss of information about the actual binary decision "either L or R" and an entropy $S_X = -\sum_{i=L,R} p_i \ln p_i$.[¶] Note that S_X or $p_{L,R}$ are strongly conditional ("fine-grained") quantities. In distinction, the usual classical entropy is calculated after "coarse graining", i.e. by constructing a local probability density $f(t)$ in the phase space of X related to the ensemble average over initial conditions, $S_{c.g.}(t) \equiv \int dx dp \, f(t) \ln f(t) + C$. A chaotic loss of predictability from strongly diverging trajectories in finite regions of phase space causes $S_{c.g.}$ to increase: as time passes, more and more cells of the coarse graining contribute – effectively, the phase space volume occupied by the ensemble grows (without violating Liouville's theorem).

In conclusion, *in a classical system, be it chaotic or not (with or without coarse graining), entropy can only be produced IF there is a physically relevant ensemble of initial conditions.* Thus, one cannot explain altogether classically entropy production or thermalization in a single high-multiplicity event in strong interactions.

Secondly, let the system evolve **quantum mechanically**. To begin with, let there be *no* coupling to the environment ($\omega = 0$). Even with the fine-tuned initial condition, Eq. (6), the amplitude ψ to find X at a particular space-time point begins to flow "down the hill" symmetrically (L and R) due to the quantum spreading of the wave packet. For a free particle $w_0 \rightarrow w(t) = (w_0^2 + w_0^{-2}t^2)^{1/2}$ (for $M = 1$); here one expects an accelerated spreading "downhill", cf. Eq. (12) below. The related probability density $|\psi|^2$ also evolves and stays symmetric; generally, it *cannot* be simulated by the classical evolution starting with an ensemble of initial conditions due to the absence of quantum interference between classical trajectories. In any case, the system remains in a *pure quantum state*. The density matrix is $\hat{\rho}_X(t) = |\psi(t)\rangle\langle\psi(t)|, |\psi(t)\rangle = \exp[-i\hat{H}_0 t]|\psi(0)\rangle$, where $\hat{H}_0$ is the Hamiltonian of X from Eq. (5) with $\omega = 0$. Therefore, $S[\hat{\rho}_X(t)] = -1 \ln 1 = 0$, cf. Eq. (2). Quantum mechanically one knows everything there is to know about a *closed system* (X), given any pure initial state and its Hamiltonian, which consistently yields $S[\hat{\rho}_X] = 0$. Even with an ensemble of initial states, i.e. an impure density matrix $\hat{\rho}_X(0)$, there is *no entropy production,*

[¶] If the ensemble of initial conditions is constrained to preserve the reflection symmetry of the action, Eq. (5), then $p_L(t') = p_R(t') = \frac{1}{2}$ and $S_X(t') = \ln 2$ stay constant.

since $S[\hat{\rho}_X(t)] = S[\exp(-i\hat{H}_0 t)\hat{\rho}_X(0)\exp(+i\hat{H}_0 t)] = S[\hat{\rho}_X(0)]$ stays constant.||

The situation changes completely if the observable subsystem (X) evolves quantum mechanically coupled to the *vacuum fluctuations and virtual excitations of the environment* (Y). According to the above assumptions, the initial density matrix of the total system is:

$$\hat{\rho}(t=0) \equiv \hat{\rho}_X(0) \otimes \hat{\rho}_Y(0) , \qquad (7)$$

with matrix elements $\rho_X(x, x'; 0) = \pi^{-1/2}w_0^{-1}\exp[-\frac{1}{2}(x^2 + x'^2)/w_0^2]$ and $\rho_Y(y, y'; 0) = (\omega/\pi)^{1/2}\exp[-\frac{1}{2}\omega(y^2 + y'^2)]$. The time evolution of the density matrix of the observable subsystem, $\hat{\rho}_X(t) = \mathrm{Tr}_Y \hat{\rho}(t)$, can be calculated with the Feynman-Vernon influence functional technique; I will make use of general results obtained in the first of Refs.[1] The idea is to derive a propagator for $\hat{\rho}_X$, which incorporates the influence of the environment degrees of freedom (Y) exactly. This can be achieved, since Y and its coupling to X are at most quadratic in coordinates and momenta, see Eq. (5).

It should be remarked that the *final state* of the environment is not specified; presently, it may contain virtual excitations of Y.* The relevance of this for a high-multiplicity hadronic (or nuclear) reaction is the following: Even though the QCD vacuum "far away" conforms to the usual one before and after, the additionally produced secondary hadrons all require a dressing of their valence quarks by localized virtual excitations of the vacuum or environment, which obviously makes an essential difference as compared to the initial state.

Presently, the resulting density matrix $\hat{\rho}_X(t)$ is (cf. also the first of Refs.[1]):

$$\begin{aligned}
\rho_X(z_-, z_+, t) = {} & \pi^{-1/2}w^{-1}(t)\, e^{-[z_+ - v(t)t]^2/w^2(t)} \\
& \times e^{-z_-^2\{C + \frac{1}{4}w_0^2 c^2 - d^{-2}[B + \frac{1}{2}w_0^2 bc]^2/w^2(t)\}} \\
& \times e^{iz_-\{(az_+ - 2d^{-1}[B + \frac{1}{2}w_0^2 bc][z_+ - v(t)t]/w^2(t)\}} ,
\end{aligned} \qquad (8)$$

with the effective velocity $v(t) = 0$ for the zero-momentum initial wave packet, the effective width $w(t) \equiv 2\xi|d|^{-1}$, $\xi \equiv (A + \frac{1}{4}w_0^{-2} + \frac{1}{4}w_0^2 b^2)^{1/2}$, and with rather complicated time-dependent coefficients A, B, C, a, b, c, d, to be discussed elsewhere; the coordinates in Eq. (8) are $z_- \equiv x - x'$ and $z_+ \equiv \frac{1}{2}(x + x')$ in terms of ordinary one-dimensional ones. Since we are particularly interested in the decoherence process and entropy production, we consider only the simplest *off-diagonal density matrix elements* here, $\rho_X(x, x' = -x, t) = \rho_X(z_- = 2x, z_+ = 0, t)$. They can be directly related to the **linear entropy** produced in the observable subsystem (X):

$$\begin{aligned}
S^{lin} \equiv {} & \mathrm{Tr}\,[\hat{\rho}_X - \hat{\rho}_X^2] = 1 - \int_{-\infty}^{\infty} dz_- \int_{-\infty}^{\infty} dz_+\, \rho_X(z_-, z_+, t)\, \rho_X(-z_-, z_+, t) \\
= {} & 1 - \frac{1}{2}c_1^{-1/2}w^{-1} ,
\end{aligned} \qquad (9)$$

with $c_1 \equiv C + \frac{1}{4}w_0^2 c^2 - (B + \frac{1}{2}w_0^2 bc)^2/(dw)^2$ and independently of the initial wave packet momentum ($p = 0$ at present). Thus, inserting (9) into (8), one obtains:

$$\rho_X(x, -x, t) = \pi^{-1/2}w^{-1}(t)\,\exp\left\{-x^2 w^{-2}(t)[1 - S^{lin}(t)]^{-2}\right\} , \qquad (10)$$

$$\int_{-\infty}^{\infty} dx\, \rho_X(x, -x, t) = 1 - S^{lin}(t) \geq e^{-S(t)} . \qquad (11)$$

|| The unitary (time evolution) transformation does not change the eigenvalues of $\hat{\rho}_X$. Thus, the statistical entropy, Eq. (2), cannot possibly show a sign of classical chaos in a closed system.

* As a corollary to the *Schmidt decomposition*[1] it is easy to prove that starting with an overall pure state of the complex system, cf. Eq. (7), IF the final state of the environment is a pure state, THEN the observable subsystem ends up in a pure state too (without entropy production).

195

The inequality results from the fact that the linear entropy provides a *lower bound* for the relevant statistical entropy, cf. Eq. (2), as shown in.[1] Note that Eqs. (9)–(11) are completely independent of the time-dependent functions entering there, which are specific for a particular dynamical system. They are based, however, on the Gaussian structure of the subsystem density matrix, Eq. (8).

At this point the attentive reader might wonder what happened to the non-linear interaction $\propto \lambda x^4$ of the double-well potential, see Eq. (5). Of course, it cannot be treated exactly. I employed a mean-field-type approximation, replacing $\frac{1}{4!}\lambda x^4$ by $\frac{1}{2}\lambda\langle x^2\rangle x^2 \equiv \frac{1}{2}\Lambda^2(t)x^2$. As long as one studies only the initial time-evolution over short periods, as compared to the time a classical particle would need to "roll down the hill", one may even set $\Lambda \approx 0$. For the following qualitative considerations, Λ plays the role of an adiabatically changing parameter. However, a more accurate approximation is necessary (and feasible) to follow the truly long-time quasi-periodic motions of the system. Then, one expects periods of increasing decoherence and entropy production, cf. below, followed by periods of *quantum revival* in the observed subsystem. For a more complex environment quantum revivals become more unlikely, since the total system including the environment finds more ways to evolve before a reconstruction of the subsystem initial-state wave function.*

To begin with, it can be checked explicitly that there is no entropy production for a vanishing coupling to the environment, $S^{lin}_{\omega=0}(t) = 0$, cf. Eq. (9). Next, calculating the effective width in the *long-time limit* [†], one finds:

$$w(t) = \left(w_0^2 + w_0^{-2}f_-^{-2} + \omega^3 f_-^{-2}[f_-^2 + \omega^2]^{-1}\right)^{1/2}\frac{f_-^2 + \omega^2}{f_-^2 + f_+^2}\exp t_- , \qquad (12)$$

with $t_- \equiv f_- t$, $f_\pm = [\pm\frac{1}{2}\omega_+^2 + (\frac{1}{4}\omega_+^4 - \omega^2\omega_-^2)^{1/2}]^{1/2}$, and $\omega_\pm^2 \equiv \pm\omega^2 + \Lambda^2 - 1$. Assuming a sufficiently small coupling, $\omega^2 < 1$, note that ω_+^2 is *negative* as long as $\Lambda^2(t) < 1 - \omega^2$. Thus, the *width grows exponentially* with an *effective Lyapunov exponent* f_-. For vanishing coupling to the environment, it reduces to the classical Lyapunov exponent of the inverted oscillator, $f_-^{\omega=0} = (1 - \Lambda^2)^{1/2}$, while the width becomes $w_{\omega=0}(t) = (w_0^2 + [w_0 f_-^{\omega=0}]^{-2})^{1/2}\exp(f_-^{\omega=0}t)$. This suggests quite generally that the time-dependent widths of suitable (Gaussian) wave packets may serve as "quantum indicators" of chaotic behaviour in the corresponding classical system.

It is remarkable how the Lyapunov exponent reflects the dynamics: as the wave packet spreads "downhill", $\Lambda^2(t) \propto \langle x^2\rangle$ increases until f_- reaches zero (becoming purely imaginary afterwards), when $\omega_- = 0$. At this point the behaviour becomes regular in the sense of being governed by harmonic motions close to the minima of the double-well potential with a correspondingly milder time-dependence of the width (cf. the model studied in the first of Refs.[1]).

The second dynamical time scale f_+^{-1} always stays real. It is relevant for certain non-Markovian effects generated by the interaction with the environment ($f_+^{\omega=0} = 0$). These become clearly visible in the entropy evaluated in the same limit as Eq. (12). Using $c_1 = C + O(w_0^2)$, one obtains in leading order:

$$S^{lin}(t) = 1 - w^{-1}(t)\frac{f_-^2 + \omega^2}{\omega^{3/2}f_-} \qquad (13)$$

$$\times \left[[(\frac{f_-[f_+^2 - \omega^2]}{f_+[f_-^2 + f_+^2]} + \frac{f_+[f_-^2 + \omega^2]}{f_-[f_-^2 + f_+^2]})\sin t_+ + \frac{2\omega^2\cos t_+}{f_-^2 + f_+^2}]^2 + \omega^2[\frac{\sin t_+}{f_+} - \frac{\cos t_+}{f_-}]^2\right]^{-1/2} ,$$

---

*Such effects have been experimentally observed in even simpler systems involving a two-state subsystem of one Rydberg atom coupled to a single mode of the electromagnetic field.[5]

[†]Here, hyperbolic functions dominate over trigonometric ones in the time-dependent coefficients in Eq. (8) for times such that a classical particle would still be "rolling down the hill" of the potential; this restriction is presently assumed for simplicity.

196

with $t_+ \equiv f_+ t$. Thus, the linear entropy approaches exponentially its saturation value 1 on the time scale set by the Lyapunov exponent, see Eq. (12), and the von Neumann entropy grows exponentially according to eq. (11), at least as fast. Note that the periodic function multiplying $w^{-1}(t)$ in Eq. (13) is approximately $\propto w^{-1/2}[1 + 2^{1/2} + \sin^2(2^{1/4}\omega t)]^{-1/2}$ for $f_- \approx f_+$ and $\propto w^{-1/2}[1 + \cos^2(2^{1/2}\omega t)]^{-1/2}$ for $f_- \approx 0$; i.e. it persists qualitatively even until the effective Lyapunov exponent becomes imaginary, when the stabilizing effect of the x^4-term in the potential is felt.

CONCLUSIONS

To summarize, in the above specified long-time limit and for the chosen initial conditions, Eq. (7), one obtains the observable subsystem (X) density matrix,

$$
\begin{aligned}
\rho_X(z_-, z_+, t) = \ & \pi^{-1/2} w^{-1}(t) \, e^{-z_+^2 w^{-2}(t)} \\
& \times e^{-\frac{1}{4} z_-^2 w^{-2}(t)[1 - S^{lin}(t)]^{-2}} \, e^{iz_- z_+ f_-} \ .
\end{aligned}
\tag{14}
$$

Even though this density matrix describes the exponential entropy production and its diagonal matrix elements with $z_- = 0$ grow rapidly (apart from the overall normalization factor), the (simplest) off-diagonal matrix elements ($z_+ = 0$) do *not* really decay here as in usual models of quantum decoherence.[1,3] This is no surprise in view of the "poor environment" considered at present, which has only one degree of freedom frozen in its ground state (modulo virtual excitations). Loosely speaking, it is unable to accommodate all the phase information contained in the off-diagonal density matrix elements of the subsystem.

In conclusion, a strong observable entropy production in a quantum system, which shows a chaotic behaviour in the classical limit with exponentially growing modes, requires only a minimal decohering effect due to an environment of vacuum fluctuations coupled to it from a higher energy scale.

In particular, complementary to previous studies,[3,6] one observes here that the environment does not necessarily have to be at any finite temperature for this effect of *partial decoherence* to work. Furthermore, it should be realized that the Schmidt decomposition reveals the remarkable fact that the density submatrices of "subsystem" and "environment" always have identical non-zero eigenvalues.[1] Thus, from the point of view of calculating the entropy, see Eq. (2), their roles can be interchanged and it is a matter of practicability to decide which part of the total Hilbert space is integrated out to find the entropy of the *physically observed subsystem*. The environment-induced quantum decoherence and its relevance for entropy production have presently been illustrated by an elementary example, which, however, points out to interesting consequences for the quantum evolution of classically chaotic non-linear field theories.[7]

Acknowledgments

Special thanks to R. Hagedorn for making it such a difference, if he is around. I thank P. Carruthers, M. Danos, N. E. Mavromatos, B. Müller, and J. Rafelski for stimulating criticism and helpful discussions. This work was supported by the Heisenberg Programme (Deutsche Forschungsgemeinschaft).

REFERENCES

1. H.-Th. Elze, preprint CERN-TH.7131/93 (hep-ph/9404215), submitted to Nucl. Phys. B; CERN-TH.7297/94 (hep-th/9406085), submitted to Phys. Rev. Lett.

2. E. Fermi, Progr. Theor. Phys. 5 (1950) 570; Phys. Rev. 81 (1951) 683;
   L. D. Landau, Izv. Akad. Nauk SSSR, Ser. fiz., 17 (1953) 51;
   S. Z. Belenkij and L. D. Landau, N. Cim. Suppl. 3 (1956) 15;
   E. Stenlund et al., eds., Proc. "Quark Matter '93", Nucl. Phys. A566 (1994).

3. W. H. Zurek, Phys. Today 44, No. 10 (1991) 36;
   R. Omnès, Rev. Mod. Phys. 64 (1992) 339;
   H. D. Zeh, Phys. Lett. A172 (1993) 189.

4. M. Gell-Mann and J. B. Hartle, Phys. Rev. D47 (1993) 3345;
   J. Ellis, N. E. Mavromatos and D. V. Nanopoulos, Phys. Lett. B293 (1992) 37; preprint CERN-TH.7195/94; and references therein.

5. G. Rempe, H. Walther and N. Klein, Phys. Rev. Lett. 58 (1987) 353.

6. W. H. Zurek and J. P. Paz, Phys. Rev. Lett. 72 (1994) 2508.

7. B. Müller, contribution to these Proceedings.

ENTROPY PRODUCTION VIA PARTICLE PRODUCTION

J. Rau

MPI für Kernphysik
Postfach 103980, 69029 Heidelberg, Germany

ABSTRACT

I critically examine the notion of "irreversibility," and discuss in what sense it applies to the spontaneous creation of particles in external fields. The investigation reveals that particle creation in very strong fields can only be described by a non-Markovian transport theory.

WHAT IS THE PROBLEM?

Spontaneous pair creation contributes to thermalisation

A heavy-ion collision appears highly irreversible: while the initial state of the two colliding nuclei is almost pure, particles emerging from the reaction can well be described by a thermal distribution. Some of the mechanisms responsible for this thermalisation are well understood: binary collisions of the microscopic constituents (partons), hadrochemical reactions, or radiation. At very high energies, however, an increasingly important role is played by yet another elementary process: the spontaneous creation of $q\bar{q}$-pairs in strong, coherent chromoelectric fields ('flux tubes').[1] These pairs are produced with a certain momentum distribution, associated with which is a non-zero entropy. It seems therefore that not only particles, but also entropy is produced 'out of the vacuum.' This poses the following question: Is spontaneous pair creation indeed irreversible? And if so, in what sense?

WHAT IS IRREVERSIBILITY?

All investigations of "irreversibility" are based upon a classification of the degrees of freedom

The v.Neumann entropy associated with the full statistical operator of a closed quantum system,

$$S[\rho(t)] := -k \operatorname{tr}(\rho(t) \ln \rho(t)) \quad , \tag{1}$$

is constant in time and hence for our purposes useless. Non-trivial statements about irreversible behavior all refer to a time-dependent *relevant entropy*, which is the entropy associated with the *relevant* degrees of freedom. A distinction between relevant and irrelevant degrees of freedom is thus essential. Often, such a distinction appears in disguise: as a classification of observed vs. unobserved; observable vs. unobservable; subsystem vs. environment; collective vs. noncollective; slow vs. fast; shape vs. randomizing; or extrinsic vs. intrinsic degrees of freedom.

The relevant entropy is obtained by coarse-graining.[2]

The v.Neumann entropy measures the amount of missing information as to the pure state of the system, if one is given the expectation values of *all* observables of the system. In contrast, the relevant entropy measures the amount of missing information as to the pure state of the system, if one is given the expectation values of only the *relevant* observables. Going from the former to the latter involves discarding information about the irrelevant degrees of freedom – a truncation which is reflected in the general inequality

$$S_{\text{rel}}(t) \geq S[\rho(t)] \quad \forall \, t \quad , \tag{2}$$

and often referred to as 'coarse-graining.'

Example: Gibbs vs. Boltzmann entropy

The full state of a classical gas of N indistinguishable particles is given by a symmetric probability density $W(\pi_1 \ldots \pi_N)$ on the $6N$-dimensional N-particle phase space. Associated with this full state is the Gibbs entropy

$$S_G := -k \int d^6\pi_1 \ldots d^6\pi_N \, W(\pi_1 \ldots \pi_N) \ln W(\pi_1 \ldots \pi_N) \tag{3}$$

which, due to Liouville's theorem, stays constant – just like the v.Neumann entropy. For reasons to be discussed shortly, one generally considers only single-particle observables to be relevant, but many-particle correlations to be irrelevant. All information about the relevant degrees of freedom is then contained in the reduced probability density

$$w(\pi) := \int d^6\pi_2 \ldots d^6\pi_N \, W(\pi_1 \ldots \pi_N) \quad , \tag{4}$$

defined on the 6-dimensional single-particle phase space. Associated with this reduced state is the Boltzmann entropy

$$S_B := -kN \int d^6\pi \, w(\pi) \ln w(\pi) \quad . \tag{5}$$

The fact that information about particle correlations has been discarded is reflected in the inequality

$$S_B \geq S_G \quad . \tag{6}$$

It is the coarse-grained Boltzmann entropy to which all non-trivial statements refer, including the famous H-theorem.

The H-theorem relies on a strong separation of time scales

A prediction cannot possibly contain more information than the data on which it is based. Therefore, if the evolution of the relevant degrees of freedom is Markovian, i. e., if their expectation values at time $t + dt$ can be predicted on the basis of their expectation values at time t, then the associated relevant entropy can only increase or stay constant, but never decrease. This is the H-theorem. Clearly, its validity depends crucially on the Markovian property of the evolution. Consider, as an example, the classical gas: If the gas is dilute, and hence the duration of one individual collision much shorter than the average time that elapses between two subsequent collisions, then collisions may be regarded as statistically independent ('Stoßzahlenansatz'); the evolution equation (Boltzmann equation) becomes Markovian; and therefore the H-theorem holds. If, on the other hand, the gas is dense, and hence the duration of a collision approximately equal to the time between two collisions, then there may be memory effects; the evolution is no longer Markovian; and the H-theorem may be temporarily violated.[3]

"Irreversible" is the flow of information from slow to fast degrees of freedom

While the constancy of the v. Neumann entropy shows that complete information about the system is retained in a full microscopic description, the variation of the relevant entropy indicates that the amount of information carried by the relevant degrees of freedom continuously changes. An obvious interpretation is that in the course of the system's evolution, information about the system is being transferred between relevant and irrelevant degrees of freedom. How exactly the relevant entropy behaves, depends crucially on the choice of the 'relevant vs. irrelevant' classification. In general, the flow of information need not have a unique direction; it may occur either from relevant to irrelevant, or from irrelevant to relevant degrees of freedom. As a consequence, the relevant entropy may either increase or decrease. It is only when the relevant degrees of freedom are slow and the irrelevant degrees of freedom are fast that, according to the general H-theorem, the information flow is uniquely directed from relevant to irrelevant degrees of freedom. This 'leaking' of information into fast degrees of freedom is perceived as irreversible. However, the information is not really lost – it only becomes inaccessible to a certain coarse-grained level of description. In the example of a dilute classical gas, information is being transferred from single-particle observables to many-particle correlations. Since in general these correlations will not be measured, part of the information about the system becomes experimentally inaccessible.

The study of irreversible behavior reduces to a time scale analysis

'Irreversibility' refers to the experimenter's ability – or the lack thereof – to prepare, control, or monitor certain degrees of freedom; it thus seems to be a purely 'anthropomorphic' concept. Nevertheless, irreversible features of the dynamics are not entirely subjective. For an observable to be measurable in practice, it is usually necessary that it varies slowly. In this case the experimentally monitored degrees of freedom constitute some subset of the set of all *slow* degrees of freedom. Accordingly, the flow of information from observed to unobserved degrees of freedom is intimately tied to the flow of information from slow to fast degrees of freedom. But the latter is an *objective* property of the dynamics: it is determined by the presence of disparate time scales, and not dependent on any observer. A study of irreversible features of the dynamics should therefore focus on the identification of slow and fast degrees of freedom, and on the analysis of the associated time scales. The choice of the 'relevant vs. irrelevant' classification is then no longer as arbitrary and subjective as it may have seemed at first; rather, the proper choice is determined by objective physical criteria – by the time scales.

IS SPONTANEOUS PAIR CREATION IRREVERSIBLE?

The relevant degrees of freedom are the occupation numbers of momentum states

What is commonly being measured in an experiment, and what enters into most transport equations such as the quantum Boltzmann equation, are the momentum distributions $n_\mp(\vec{p}, t)$ of the produced particles and antiparticles. This strongly suggests choosing the occupation numbers of momentum states as the relevant degrees of freedom. Associated with this choice is the relevant entropy (for spin-1/2 fermions)

$$S_{\text{rel}}(t) := -2k \int_{\vec{p}} \left[\frac{n_-}{2} \ln \frac{n_-}{2} + \left(1 - \frac{n_-}{2} \right) \ln \left(1 - \frac{n_-}{2} \right) + (n_- \leftrightarrow n_+) \right] \tag{7}$$

(where $n_\mp \equiv n_\mp(\vec{p}, t)$). Whether or not the relevant degrees of freedom are sufficiently slow; i. e., whether or not their evolution is Markovian; whether or not the relevant entropy obeys an H-theorem; and hence whether or not in this description pair creation appears irreversible – all this must be the subject of a thorough time scale analysis.

I focus on the pair creation process proper

Irreversible information flows occur at two different stages of the evolution: (i) during the pair creation process proper, and (ii) during the subsequent 'decoherence.' During the pair creation process proper, information leaks from relevant occupation numbers into correlations and relative phases. Under the influence of subsequent collisions, this phase information is then being transferred further to an unobservable 'environment' (or 'heat bath') of high-frequency partons. Such a transfer of phase information to an unobservable environment is often referred to as 'decoherence,' and discussed elsewhere in these Proceedings.[4] Here, I wish to concentrate on stage one of the evolution, the pair creation process proper.

A time scale analysis requires knowledge of the non-Markovian equation of motion

The question to be addressed is the following: Is the initial flow of information from occupation numbers to correlations and relative phases irreversible? In other words, does the relevant entropy (7) obey an H-theorem? According to our previous discussion, the answer to this question hinges upon a careful analysis of time scales. It is necessary to derive a – generally non-Markovian! – equation of motion for the occupation numbers; to identify the memory time of this non-Markovian equation as well as the time scale on which the occupation numbers evolve; to compare these, and thus to find a criterion for the validity of the Markovian approximation. If and only if this criterion is satisfied, the H-theorem holds. If, on the other hand, the criterion is *not* satisfied, then the memory time furnishes the typical time scale on which the relevant entropy may temporarily decrease.

Microscopic model: Schwinger mechanism.[5]

All essential features of the pair creation process are exhibited by the well-known Schwinger mechanism, the spontaneous creation of $e^+ e^-$ pairs in a constant, homogeneous external electric field. It is this simple model which has been considered in much of the literature,[1] and which I now want to subject to further analysis. Let q be the electron charge, $\vec{E}$ the external field, $\vec{p}(t) := \vec{p} + q\vec{E}t$ the time-dependent momentum, m the spin component, and ϕ_{fi} the dynamical phase accumulated between times t_i and t_f. Then in the Heisenberg picture the evolution mixes particle ($a^\dagger$) and antiparticle (b) field operators, with respective amplitudes α_{fi} and β_{fi}:

$$\mathcal{U}(t_2, t_1) \begin{pmatrix} a^\dagger(\vec{p}(t_1), m) \\ b(-\vec{p}(t_1), -m) \end{pmatrix} = \begin{pmatrix} \alpha_{21} & \beta_{21} \\ -\beta_{21}^* & \alpha_{21}^* \end{pmatrix} \begin{pmatrix} e^{i\phi_{21}} & 0 \\ 0 & e^{-i\phi_{21}} \end{pmatrix} \begin{pmatrix} a^\dagger(\vec{p}(t_2), m) \\ b(-\vec{p}(t_2), -m) \end{pmatrix}$$

(8)

Hence on the microscopic level, spontaneous pair creation is described by a time-dependent Bogoliubov transformation.

In the equation of motion, pair creation is accounted for by a non-Markovian source term

The equation of motion for the occupation numbers has the structure of a quantum Boltzmann equation. Aside from the usual acceleration and (possibly) collision terms, it contains an additional source term to account for the spontaneous pair creation. By means of the so-called projection method,[6] this source term may be related to the coefficients of the Bogoliubov transformation:[7]

$$\dot{n}_-^{\text{sou}}(\vec{p}, t) = 4 \, \text{Re} \int_0^{t-t_0} d\tau \; \dot{\beta}^*(-\tau, -\tau) \, e^{-i2\phi(-\tau, 0)} \, \dot{\beta}(0, 0) \cdot S(\vec{p} - q\vec{E}\tau, t - \tau) \qquad (9)$$

Here t_0 denotes the initial time at which the external field is switched on; $\dot{\beta}(t_2, t_1)$ is a shorthand for $\partial\beta(t_2, t_1)/\partial t_2|_{t_2=t_1}$; and the factor $S \equiv [1 - n_-/2][1 - n_+/2] - (1/4)n_- n_+$ accounts for Pauli blocking as well as the possible annihilation of pairs back into the field. Evidently, the source term involves an integration over the entire history of the system and is therefore non-Markovian.

The memory time combines a quantum mechanical and a classical time scale

Careful analysis[7] reveals that significant contributions to the above source term come only from times τ which are smaller than the characteristic memory time

$$\tau_{\text{mem}} \sim \frac{\hbar}{\epsilon_\perp} + \frac{\epsilon_\perp}{qE} \qquad (10)$$

(where $\epsilon_\perp$ denotes the transverse energy or mass). The two terms have very different physical origins. (i) The time $\hbar/\epsilon_\perp$ is quantum mechanical. It corresponds – via the time-energy uncertainty relation – to the time needed to create a virtual particle-antiparticle pair, and may thus be regarded as the 'time between two production attempts.' (ii) The time $\epsilon_\perp/qE$, on the other hand, is classical. It can be interpreted in various ways, depending on the picture employed to visualize the pair creation process. If pair creation is viewed as a tunneling process from the negative to the positive energy continuum, this classical time coincides with the time needed for the wave function to traverse the barrier with the speed of light.

The Markovian approximation is valid only for weak fields

The occupation numbers evolve on a typical time scale set by the inverse production rate $[\dot{n}_-^{\text{sou}}(\vec{p})]^{-1}$. Assuming $p_\parallel = 0$ for simplicity, one obtains as a representative scale

$$\tau_{\text{prod}} \sim \frac{\epsilon_\perp}{qE} \exp\left(\frac{\pi\epsilon_\perp^2}{2\hbar qE}\right) \quad . \qquad (11)$$

This production scale is much larger than the memory time, and hence the evolution is Markovian, only if $E \ll m^2/\hbar q$.

203

First conclusion: The flow of information from occupation numbers to correlations and phases is strictly irreversible only if the field is weak

As long as the field is sufficiently weak, the evolution of occupation numbers is Markovian, the H-theorem holds, and hence the relevant entropy increases monotonically. But as soon as $E \sim m^2/\hbar q$, both memory time and production scale attain the same magnitude $\tau \sim \hbar/m$. As a result, the Markovian approximation breaks down, and there may be temporary violations of the H-theorem (on the same time scale $\tau \sim \hbar/m$). Indeed, oscillations of the relevant entropy have been observed in numerical simulations.[8]

Second conclusion: Pair creation in very strong fields must be described by a non-Markovian transport theory

Whenever the external field is stronger than the critical value $m^2/\hbar q$, there may be sizeable memory effects which cannot be accounted for in a Markovian transport theory. [Particle masses of the order 0.5, 15 or 500 MeV correspond to critical field strengths of the order 10^{-3}, 1 or 10^3 MeV/fm, respectively.] This strong-field domain is in fact the only one which is physically relevant, because only there does spontaneous pair creation occur at an appreciable rate. Hence for all practical purposes pair creation must *always* be described by a *non*-Markovian transport theory.

Acknowledgements

I thank B. Müller for advice and many useful discussions. Financial support by the Studienstiftung des deutschen Volkes, the U.S. Department of Energy (Grant DE-FG05-90ER40592) and the Heidelberger Akademie der Wissenschaften is gratefully acknowledged.

REFERENCES

1. A. Casher, H. Neuberger, and S. Nussinov, *Phys. Rev. D* **20**, 179 (1979); H. Ehtamo, J. Lindfors, and L. McLerran, *Z. Phys. C* **18**, 341 (1981).

2. for a lucid discussion of the entropy concept see: E. T. Jaynes, *Papers on Probability, Statistics and Statistical Physics*, ed. by R. D. Rosenkrantz, Kluwer Academic, Dordrecht, 1989.

3. P. Danielewicz, *Ann. Phys. (N.Y.)* **152**, 239 (1984).

4. H.-Th. Elze, these Proceedings.

5. F. Sauter, *Z. Phys.* **69**, 742 (1931); W. Heisenberg and H. Euler, *Z. Phys.* **98**, 714 (1936); J. Schwinger, *Phys. Rev.* **82**, 664 (1951).

6. see, e. g.: H. Grabert, *Projection Operator Techniques in Nonequilibrium Statistical Mechanics*, Springer, Berlin, Heidelberg, 1982.

7. J. Rau, *Phys. Rev. D* **50** (December 1994); Dissertation, Duke University, 1993.

8. F. Cooper, J. M. Eisenberg, Y. Kluger, E. Mottola, and B. Svetitsky, *Phys. Rev. D* **48**, 190 (1993).

REAL-TIME THERMAL CORRECTIONS
TO PAIR-PRODUCTION PROCESSES
IN HEAVY-ION COLLISIONS

Bo-Sture Skagerstam

Institute of Theoretical Physics
Chalmers University of Technology
and
University of Göteborg
S-412 96 Göteborg, Sweden

INTRODUCTION

It is an honour to address this talk to an audience which celebrates the work done by Rolf Hagedorn on the statistical bootstrap. There is a connection between the physics of the "fire-ball" so beautifully developed by Hagedorn and his many collaborators and what I will describe in my talk: I will focus my attention on some aspects of relativistic thermal quantum field theory in the context of the physics of the quark-gluon plasma. Thermal field theories are in general convenient tools for studying elementary physical processes in the presence of a thermal environment. The physical influence on elementary processes of a thermal environment can sometimes be expressed in terms of effective parameters (see e.g. Refs.1, 2). In QED, and at the one-loop level, the dispersion relation of a charged fermion when a thermal background of *photons* is present can be treated self-consistently and leads to the following effective mass[1]

$$m_{eff} = \frac{m}{2} + \sqrt{\frac{m^2}{4} + \delta m_\beta^2} \ , \tag{1}$$

where $\delta m_\beta = \alpha \pi T^2 / 3m$ and where α is the fine-structure constant. Eq.(1) is valid at least for temperatures below the mass of the particle under consideration. In the soft thermal photon limit it can be shown that the expression Eq.(1) is independent of the spin of the non-thermal particle. The effective mass m_{eff} is also independent of Plancks constant. It is interesting to notice that m_{eff} has actually been measured[3,4] by means of Doppler-free excitations of highly-excited Rydberg levels. Similarly, a thermal environment leads to a modification, $\delta a_e^{\beta,\mu}$, of the anomalous magnetic moment of a charged Dirac fermion. If $m \ll T$, one can e.g. show that[5] for time-like processes

$$\delta \, _e^{\beta,\mu} = -\frac{2\alpha}{3\pi} \int_1^\infty \frac{dx}{x^2} \sqrt{\frac{x+1}{x-1}} f_F^+(xm) \left(1 - 6x^2 + x^3\right) \ , \tag{2}$$

where the chemical potential μ corresponds to charge conservation, β is the inverse temperature, and where $f_F^{\pm}(\omega)$ is the Fermi-Dirac thermodynamical distribution function, i.e.

$$f_F^{\pm}(\omega) = \frac{1}{\exp[\beta(\omega \mp \mu)] + 1} \ . \tag{3}$$

In some models of the hadronization of the quark-gluon plasma, a collimated colour-electric field configuration is supposed to be formed by colliding heavy–ions (for a review see e.g. Ref.6). By the non-perturbative particle-production mechanism of Schwinger,[7] quarks and anti–quarks are supposed to be formed which eventually are dressed into final state hadrons (for some recent reviews see e.g. Refs.8,9). There are at least three important time–scales involved in this complicated physical process: i) the time–scale of pair production t_p, ii) the time–scale of neutralisation of the colour–electric field t_d and, finally, iii) the time–scale(s) t_{th} of thermalization of quarks (or gluons). It can be argued[10] that at RHIC energies $t_p \leq t_{th} \leq t_d$. It may therefore be of interest to study the presence of a colour-electric field in a thermalized quark-gluon plasma. Since we will restrict ourselves to one-loop considerations, it will be sufficient to consider a slowly varying electromagnetic field in QED and a background of electron–positrons at finite temperature and density. Our intent is to make the presentation as self-contained as possible. In the first section below we will present a simple first-quantized picture of thermal real–time fermion propagators. We then outline the derivation of the QED effective action and present some explicit results concerning e.g. the high–temperature limit. We also make some comments on thermal corrections to tunnelling processes in quantum mechanics. The outcome of our considerations is unexpected: there is no real–time thermal correction to the imaginary part of the effective action. To the extent that the imaginary part of the effective action determines the rate for particle-production, we will therefore not find any thermal correction to Schwingers non-perturbative pair-production mechanism in the presence of an external electric field, at least not at the one–loop level.

THERMAL PROPAGATORS – FIRST QUANTIZED PICTURE

For the purpose of calculating the QED effective action we need the fermion external field propagator. This propagator, at finite temperature and density, is easily derived by making use of a first–quantized Feynman-Stückelberg picture. In the vacuum sector, the fermion propagator $iS_F(x'; x|m)$ in the presence of an external, static electromagnetic field is defined by

$$iS_F(x'; x|m) =$$
$$\theta(t' - t) \sum_n \psi_n^{(+)}(\mathbf{x}', t') \overline{\psi}_n^{(+)}(\mathbf{x}, t) \ - \ \theta(t - t') \sum_n \psi_n^{(-)}(\mathbf{x}', t') \overline{\psi}_n^{(-)}(\mathbf{x}, t) \ , \tag{4}$$

where the conjugated spinor $\overline{\psi}_n^{(\pm)}$ is given by $\overline{\psi}_n^{(\pm)} = (\psi_n^{(\pm)})^\dagger \gamma_0$ and n denotes a set of quantum numbers needed in order to completely characterise the physical states. Since the normalised spinor $\psi_n^{(\pm)}(\mathbf{x}, t)$ satisfies the Dirac equation, only the time derivative acting on the step functions in Eq.(4) gives a non- zero contribution, so one finds that

$$(i\not{D} - m)S_F(x'; x|m) = \mathbb{1} \cdot \delta^4(x' - x) \ , \tag{5}$$

where $D_\mu = \partial_\mu + ieA_\mu$ is the conventional external field covariant derivative. The real-time propagator at finite temperature T and chemical potential μ, denoted by $\langle iS_F(x'; x|m) \rangle_{\beta,\mu}$, can now be obtained by the following simple reasoning. Let E_n be the energy of the quantum state under consideration. A particle can then propagate forward in time in a state which is unoccupied by thermal particles, whereas a hole in the occupied states can propagate backwards in time. We can therefore generalize Eq.(4) and write

$$\langle iS_F(x';x|m)\rangle_{\beta,\mu} = \sum_n$$

$$\left[\theta(t'-t)\left([1-f_F^+(E_n)]\psi_n^{(+)}(\mathbf{x}',t')\overline{\psi}_n^{(+)}(\mathbf{x},t) + [1-f_F^+(-E_n)]\psi_n^{(-)}(\mathbf{x}',t')\overline{\psi}_n^{(-)}(\mathbf{x},t)\right)\right.$$
$$\left.-\theta(t-t')\left(f_F^+(-E_n)\psi_n^{(-)}(\mathbf{x}',t')\overline{\psi}_n^{(-)}(\mathbf{x},t) + f_F^+(E_n)\psi_n^{(+)}(\mathbf{x}',t')\overline{\psi}_n^{(+)}(\mathbf{x},t)\right)\right] \;.$$

$$(6)$$

We can now extract the vacuum part Eq.(4) of this propagator and write

$$\langle iS_F(x';x|m)\rangle_{\beta,\mu} = iS_F(x';x|m) + iS_F^{\beta,\mu}(x';x|m) \;, \tag{7}$$

where the thermal part $iS_F^{\beta,\mu}(x';x|m)$ is defined by

$$S_F^{\beta,\mu}(x';x|m) =$$
$$i\sum_n\left(f_F^+(E_n)\psi_n^{(+)}(\mathbf{x}',t')\overline{\psi}_n^{(+)}(\mathbf{x},t) - f_F^-(E_n)\psi_n^{(-)}(\mathbf{x}',t')\overline{\psi}_n^{(-)}(\mathbf{x},t)\right) \;, \tag{8}$$

and where we have defined the distribution $f_F^-(E_n) = 1 - f_F^+(-E_n)$. Notice that there is no time-ordering in $S_F^{\beta,\mu}(x';x|m)$ despite the fact that the time-ordering in Eq.(6) is non-trivial. The thermal propagator Eq.(6) therefore also trivially satisfies Eq.(5). These considerations can, of course, easily be extended to treat particles with Bose-Einstein statistics as well. The result Eq.(8) can also be derived from an explicit calculation using the second-quantized field operators and appropriate thermal averages.[11] It is clear from the derivation presented here that the result prevails for *any* one-particle distribution function $f_F^+(E_n)$, i.e. we do not necessarily have to consider an equilibrium distribution. Even though this seems to be an innocent observation arguments have been presented that the use of an equilibrium distribution is required beyond the one–loop level.[12]

THE QED ONE–LOOP EFFECTIVE ACTION

The generating functional $Z[\bar{\eta},\eta,A_\mu]$ of fermionic Green's functions in an external field is formally defined by

$$Z[\bar{\eta},\eta,A_\mu] =$$
$$\mathrm{Det}\,[(i\,\rlap{/}D - m)]\exp\left[i\int d^4x\left(-\frac{1}{4}F_{\mu\nu}F^{\mu\nu} + \int d^4y\bar{\eta}(x)S_F(x;y|m)\eta(y)\right)\right] \;, \tag{9}$$

which describes second-quantized electrons and positrons interacting with a classical electromagnetic field expressed in terms of the vector potential A_μ. Here $S_F(x;y|m)$ is the external field vacuum propagator as given by Eq.(4). The functional determinant $\mathrm{Det}\,[i\,\rlap{/}D - m]$ gives rise to a contribution to the effective Lagrangian density $\mathcal{L}_{eff}$, i.e.

$$S_{eff} = \int d^4x\mathcal{L}_{eff} = \int d^4x\left[-\frac{1}{4}F_{\mu\nu}F^{\mu\nu}\right] - i\,\mathrm{Tr}\,\log\left[(i\,\rlap{/}D - m)\right] \;. \tag{10}$$

Differentiating Eq.(10) with respect to the fermion mass we can now obtain the one-loop correction to the effective action at finite temperature and density from the expression

$$\frac{\partial\mathcal{L}_{eff}}{\partial m} = i\,\mathrm{tr}\langle iS_F(x';x|m)\rangle_{\beta,\mu} \;, \tag{11}$$

where the trace is now only over spinor indices. Alternatively, we may also write

$$\frac{\partial \mathcal{L}_{eff}}{\partial m} = i \mathrm{tr} \langle x | \frac{1}{\not{p} - m + i\epsilon} - f_F(p_0, A_0) \left(\frac{1}{\not{p} - m + i\epsilon} - \frac{1}{\not{p} - m - i\epsilon} \right) | x \rangle , \quad (12)$$

where $\not{p} = \gamma^\mu(p_\mu - eA_\mu)$. Eq.(12) makes sense only in a time–independent gauge. The thermal distribution function $f_F(p_0, A_0)$ is now written in the following form

$$f_F(p_0, A_0) = \frac{\theta(p_0 - eA_0)}{e^{\beta(p_0 - \mu)} + 1} + \frac{\theta(-p_0 + eA_0)}{e^{\beta(-p_0 + \mu)} + 1} . \quad (13)$$

The distribution function does not have to be chosen to represent an equilibrium distribution. Other choices may be appropriate when the electric field drives the system out of equilibrium. It has been emphasised in the literature (see e.g. Ref.13) that the distribution in Eq.(13) has a non–trivial limit when $T \to 0$. In order to calculate the effective action $\mathcal{L}_{eff}$ from Eq.(12) we make use of Schwingers proper-time formulation, i.e. we write

$$\frac{i}{\mathcal{O} + i\epsilon} = \int_0^\infty ds \exp[i(\mathcal{O} + i\epsilon)s] , \quad (14)$$

where $\mathcal{O}$ is an operator. By calculating the trace in the basis $|p_0, \mathbf{x}\rangle$, we can then derive[14] the following expression for the thermal part, $\mathcal{L}_{eff}^{\beta,\mu}$, of the effective action

$$\frac{\partial \mathcal{L}_{eff}^{\beta,\mu}}{\partial m^2} = \frac{1}{2\pi^{3/2}} \int_{-\infty}^\infty \frac{dp_0}{2\pi} f_F(p_0, A_0) \mathrm{Im} \left\{ \int_0^\infty ds e^2 ab \cot(esa) \coth(esb) \right.$$
$$\left. \times (ih(s) + \epsilon)^{-1/2} \exp \left[-i(m^2 - i\epsilon)s + i\frac{(p_0 - eA_0)^2}{h(s) - i\epsilon} \right] \right\} , \quad (15)$$

where

$$\begin{aligned}
h(s) &= (eF \coth eFs)_{00} , \\
F_{\mu\nu} &= \partial_\mu A_\nu - \partial_\nu A_\mu , \\
a^2 - b^2 &= B^2 - E^2 , \\
ab &= \mathbf{E} \cdot \mathbf{B} , \\
B &= |\mathbf{B}| , \quad E = |\mathbf{E}| .
\end{aligned} \quad (16)$$

We have added $-i\epsilon$ to $h(s)$ in Eq.(15) in order to get the correct branch when $h(s) < 0$, which is determined by Schwinger's formula for the propagator in a constant external field.[7] >From the $\cot(esa)$ factor we find that the s–integral goes through a number of poles on the real axis, $s = k\pi/ea$, which were not apparent in the original expression. In the case of a pure magnetic field it has been shown[15] by making use of the expression Eq.(11) and an explicit form of the corresponding propagator, that the poles actually are absent if we only include a finite number of Landau levels. In order to get the correct result after summing over all Landau levels the s–integration contour has to go slightly below the real axis.

The expression in Eq.(15) can directly be integrated with respect to m^2. The singularity at $s = 0$ should be cured by subtracting the $F_{\mu\nu} = 0$ part or using a ζ–function regularization as in Ref.15. We notice that in the limit $T \to \infty$, and $f_F(p_0, A_0) \to 1/2$, Eq.(15) equals the negative of the real part of the vacuum contribution up to terms which are quadratic in the field. In the high–temperature limit there will therefore be a cancellation between the vacuum one–loop contribution to the effective action and the thermal part of one-loop correction. Such a cancellation does not occur in quantum electrodynamics of charged scalar bosons.[16]

External Magnetic Fields

We shall now find the corresponding correction $S_{eff}^{\beta,\mu} = \int d^4x \mathcal{L}_{eff}^{\beta,\mu}$, to the effective action S_{eff} at finite chemical potential and temperature such that $\mathcal{L}_{eff} = \mathcal{L}_0 + \mathcal{L}_1 + \mathcal{L}_{eff}^{\beta,\mu}$, where $\mathcal{L}_0 = -B^2/2$. In the case of an external magnetic field we e.g. now obtain the well-known result[7] that

$$\mathcal{L}_1 = -\frac{1}{8\pi^2} \int_0^\infty \frac{ds}{s^3} \left[esB \coth(esB) - 1 - \frac{1}{3}(esB)^2 \right] \exp\left[-m^2 s\right] \quad , \tag{17}$$

expressed in terms of renormalized physical parameters. Separating the field independent part we write $\mathcal{L}_{eff}^{\beta,\mu} = \mathcal{L}_0^{\beta,\mu} + \mathcal{L}_1^{\beta,\mu}$, where

$$\mathcal{L}_0^{\beta,\mu} = \frac{1}{3\pi^2} \int_{-\infty}^\infty d\omega \theta(\omega^2 - m^2) f_F(\omega) \left(\omega^2 - m^2\right)^{3/2} \quad , \tag{18}$$

and $f_F(\omega) = \theta(\omega) f_F^+(\omega) + \theta(-\omega) f_F^-(-\omega)$. We therefore conclude that the field independent thermal correction to the Lagrangian density $\mathcal{L}_0^{\beta,\mu}$ can be identified as $\mathcal{L}_0^{\beta,\mu} = \log Z(T,\mu)/\beta V$, where $Z(T,\mu)$ is the partition function for an ideal e^+e^--gas with particle energy $E = \sqrt{k^2 + m^2}$, i.e.

$$\frac{\log Z(T,\mu)}{V} = 2 \int \frac{d^3k}{(2\pi)^3} \left(\log[1 + e^{-\beta(E-\mu)}] + \log[1 + e^{-\beta(E+\mu)}] \right) \tag{19}$$

in a sufficiently large quantization volume V. In general $\mathcal{L}_{eff}^{\beta,\mu}$ is directly related to the partition function $Z(B,T,\mu)$ of the relativistic fermion gas in the presence of an external magnetic field B, i.e. $\mathcal{L}_{eff}^{\beta,\mu} = \log Z(B,T,\mu)/\beta V$. It is possible to write $\mathcal{L}_1^{\beta,\mu}$ in a form similar to the vacuum contribution $\mathcal{L}_1$. Using the identity

$$\frac{\exp[-|x|]}{|x|} = \int_0^\infty \frac{dt}{\sqrt{2\pi t}} \exp\left[-\frac{1}{2}(x^2 t + \frac{1}{t})\right] \quad , \tag{20}$$

the following representation of $\mathcal{L}_1^{\beta,\mu}$, valid for $|\mu| < m$, can actually be derived in a straightforward manner (correcting for a misprint in Ref.15)

$$\mathcal{L}_1^{\beta,\mu} = \frac{1}{4\pi^2} \sum_{l=1}^\infty (-1)^{l+1} \int_0^\infty \frac{ds}{s^3} \exp\left[-\frac{\beta^2 l^2}{4s} - m^2 s\right] \cosh(\beta l \mu)[eBs \coth(eBs) - 1] \quad . \tag{21}$$

In the case $\mu = 0$, Eq.(21) agrees with the result obtained in Refs.17, 18. It is, however, not always obvious, when written in this form, to see how to extract general physical information. In particular it is not obvious how to generalise $\mathcal{L}_{eff}^{\beta,\mu}$ to $|\mu| \geq m$, since then it appears to be divergent. In passing, we observe that the high T behaviour given in Ref.17 is not correct. It is, however, possible to show[15] that Eq.(21) is equal to Eq.(22) given below, which is valid *for all T* and μ. In order to calculate the thermal part $\mathcal{L}_{eff}^{\beta,\mu}$ of the effective action valid for all T and μ in a more useful form, we have made a careful analysis of the convergence and the analytical structure. We get $\mathcal{L}_{eff}^{\beta,\mu} = \mathcal{L}_0^{\beta,\mu} + \mathcal{L}_1^{\beta,\mu}$, where[15]

$$\mathcal{L}_1^{\beta,\mu} =$$
$$\int_{-\infty}^\infty d\omega \theta(\omega^2 - m^2) f_F(\omega) \left[\frac{1}{4\pi^{5/2}} \int_0^\infty \frac{ds}{s^{5/2}} e^{-s(\omega^2 - m^2)} [seB \coth(seB) - 1] \right]$$
$$- \int_{-\infty}^\infty d\omega \theta(\omega^2 - m^2) f_F(\omega) \left[\frac{1}{2\pi^3} \sum_{n=1}^\infty \left(\frac{eB}{n}\right)^{3/2} \sin\left(\frac{\pi}{4} - \frac{\pi n}{eB}(\omega^2 - m^2)\right) \right] .$$

$$\tag{22}$$

The oscillatory term with the sum over n was neglected in Ref.19. It is, however, essential to keep this term in order to get the correct physical result. It leads e.g. to the well-known de Haas - van Alphen oscillations of the magnetic moment.[15] Astrophysical considerations of de Haas - van Alphen oscillations can be found in Ref.15. The thermodynamics of the QED plasma in the presence of an external magnetic field has been considered before in the literature. In Refs.20 one may e.g. find several interesting aspects of the problem. Unfortunately in these considerations the physics is obscured due to the fact that the chemical potential was related to the non-conserved number–operator, which is appropriate in the non–relativistic case but, of course not, in the relativistic domain where one takes charge-conservation into account.

An effective coupling, depending on external parameters like B, T and μ can be defined as follows[21]

$$\frac{1}{\alpha(B, T, \mu)} = -\frac{\partial \mathcal{L}_{eff}}{\partial B} \quad . \tag{23}$$

If $T^2 \gg m^2 \gg eB$ one finds that

$$\mathcal{L}_{eff} =$$
$$-\frac{(eB)^2}{2e^2(\lambda)} \left(1 + \frac{e^2(\lambda)}{12\pi^2} \log\left(\frac{\lambda^2}{m^2}\right)\right) + \frac{(eB)^2}{24\pi^2} \log(\frac{T^2}{m^2}) + \mathcal{L}_0^{\beta,\mu} + \mathcal{O}\left((eB)^4\right) \quad , \tag{24}$$

where λ is a renormalization group scale parameter, i.e.

$$\lambda \frac{de(\lambda)}{d\lambda} = \frac{1}{12\pi^2} e^3(\lambda) + \mathcal{O}\left(e^5(\lambda)\right) \quad . \tag{25}$$

If the arbitrary scale λ is identified with the scale T, we obtain

$$\mathcal{L}_{eff} = -\frac{(eB)^2}{2e^2(T)} + \mathcal{L}_0^{\beta,\mu} \quad , \tag{26}$$

where the running coupling $e(T)$ then is a solution of the renormalization group equation Eq.(25). In more general terms[15] we can write for the running coupling $\alpha(x)$

$$\frac{1}{\alpha(x)} \approx \frac{1}{\alpha(\lambda)} - \frac{2}{3\pi} \log\left(\frac{x}{\lambda}\right) \quad , \tag{27}$$

where $x = \mu, T$ or $\sqrt{eB}$. If λ is identified with any of these scales, we can in each such case write

$$\mathcal{L}_{eff} = -\frac{1}{2} \frac{(eB)^2}{e^2(x)} + \mathcal{L}_0^{\beta,\mu} \quad , \tag{28}$$

when $x \gg m$. Eq.(28) provides for a simple illustration of the role played by the scales $x = \mu, T$ or $\sqrt{eB}$ in perturbative renormalization group improvements.

External Electric Fields

If electric external fields are to be considered it is more difficult to obtain explicit results. For orthogonal electric and magnetic fields and $E = B$, in which case $h(s) = 1/s$ in Eq.(15), one can, however, obtain the following result[14]

$$\mathcal{L}_{eff}^{\beta,\mu}(\mathbf{E} \perp \mathbf{B}, E = B) = \mathcal{L}_{eff}^{\beta,\mu}(F_{\mu\nu} = 0, A_0)$$
$$= \frac{1}{3\pi^2} \int dp_0 f_F(p_0, A_0) \theta((p_0 - eA_0)^2 - m^2)((p_0 - eA_0)^2 - m^2)^{3/2} , \tag{29}$$

a result similar to the absence of quantum corrections in a propagating plane wave at zero temperature.[7] In the high–temperature limit, and $\mu = 0$, we can expand Eq.(29) in powers of

A_0 and extract the Debye screening mass from the coefficient in front of $A_0^2/2$. The result, $m_\gamma^2 = e^2 T^2/3$, agrees with Ref.22 (but not with Ref.18). Similarly, we find the Debye mass at $T = 0$ and large μ to be $m_\gamma^2 = e^2\mu^2/\pi^2$, in agreement with Ref.23. For weak and pure electric fields one can, furthermore, show[14] that in the weak field limit

$$\mathcal{L}_{eff}^{\beta,\mu}(E) - \mathcal{L}_{eff}^{\beta,\mu}(0) \simeq$$

$$\frac{(eE)^2}{24\pi^2} \int_{-\infty}^{\infty} \frac{d\omega}{(\omega^2 - m^2)^{1/2}} \frac{d}{d\omega} \left(f_F(\omega + eA_0) + f_F(-\omega + eA_0)\right) . \tag{30}$$

In the high–T limit we then find[14]

$$\mathcal{L}_{eff}^{\beta,\mu}(E) - \mathcal{L}_{eff}^{\beta,\mu}(0) \rightarrow -\frac{(eE)^2}{24\pi^2} . \tag{31}$$

In analogy with Eq.(23) we define an effective fine structure constant by

$$\frac{1}{\alpha(T)} = \frac{1}{\alpha} + \frac{1}{\alpha E} \left. \frac{\partial \mathcal{L}_{eff}^{\beta,\mu}(E)}{\partial E} \right|_{E \rightarrow 0} . \tag{32}$$

In the high–temperature limit we then have that $\alpha(T) \rightarrow \alpha/(1 - \alpha/3\pi)$, showing a completely different behaviour from the α defined using a magnetic field,[15] which satisfies a zero temperature renormalization group equation as in Eq.(27).

FINAL REMARKS

The rate of particle pair–production in strong external fields is determined by the absorptive part of the effective action $\mathcal{L}_{eff}$. The particle production process can be viewed as a quantum–mechanical tunnelling phenomenon. One expects on general grounds that the rate of particle production should be increased due to the presence of a thermal heat–bath. As an illustrative example of thermal activation of tunnelling processes we mention tunnelling processes in SQUIDs (see e.g. Ref.25). A thermal environment can now, in some sense, be regarded as an external field problem. It is a straightforward calculation to show that the effective, temperature dependent mass of an electron in thermal heat–bath can be obtained from the thermal average of the effective mass of an electron in a plane wave.[5] With this in mind it is interesting to observe that external fields can have a strong effect on tunnelling processes. A monochromatic external source can e.g. almost completely suppress the rate of quantum mechanical tunnelling.[26–28] A large magnetic field parallel to a plane confining electrons, furthermore, leads to an effective potential which makes a metastable state more stable.[29] Returning now to the issue of particle production in an external field, we have by construction according to Eq.(15)

$$\text{Im} \frac{\partial \mathcal{L}_{eff}^{\beta,\mu}}{\partial m} = 0 . \tag{33}$$

Thus we find no real–time thermal correction to Schwingers result for particle production in an external electric field, at least not on the one–loop level. This finding agrees with Ref.24. In the static case (imaginary–time formalism) it was, however, found in Refs.18, 10 that

$$\text{Im}\mathcal{L}_{eff}^{\beta,\mu}(E) \simeq \frac{eET^2}{12} , \tag{34}$$

if $T \gg m$ and $\mu = 0$. The rate of production of heavy quark-antiquarks in ultra–relativistic nuclear collisions by quark-antiquarks and gluon fusion processes has been considered in a pioneering work by Rafelski and Müller.[30] It has been argued that the decay of a thermal gluon, with a non-zero damping constant, dominates the rate of thermal quark-antiquark production if the quark mass m_q is comparable to the temperature scale, i.e. if $m_q/T \simeq \mathcal{O}(1)$. If $\mathcal{R}_{g \to q\bar{q}}$ is the production rate of thermal gluon decay into quark-antiquarks, one finds that[31]

$$\frac{\mathcal{R}_{g \to q\bar{q}}}{T^4} \simeq \mathcal{O}(10^{-2}) \quad . \tag{35}$$

Using Eq.(34) as an estimate of the importance of non-perturbative thermal production we then obtain

$$\frac{\mathrm{Im}\mathcal{L}_{eff}^{\beta,\mu}(E)}{T^4} \simeq \frac{eE}{12T^2}\sqrt{K} \quad . \tag{36}$$

where $K \simeq \mathcal{O}(10)$ is a colour random–walk factor.[8,9] Using $eE \simeq 1\,GeV\,fm^{-1}$ we therefore see that

$$\frac{\mathrm{Im}\mathcal{L}_{eff}^{\beta,\mu}(E)}{T^4} \simeq \left(\frac{0.2}{T_{GeV}}\right)^2 \quad , \tag{37}$$

which can be as important as perturbative production rates. It therefore seems to us that the issue raised in this presentation concerning the calculation of thermal corrections to pair–production processes in strong external electric fields needs further clarification.

Acknowledgements

We are grateful to Per Elmfors, Per Liljenberg as well as David Persson for a fruitful collaboration on the subjects discussed and G. Peressutti for comments on this presentation. It is, furthermore, a pleasure to thank the organizers of this interesting workshop and in particular Jan Rafelski for providing a stimulating atmosphere. Without his enthusiastic and gentle but forceful pressure on me to focus my thinking on the heart of physical problems the present work would have ended up in a different shape. Research supported by the Swedish National Research Council under contract no. 8244-311.

REFERENCES

1. G. Peressutti and B.-S. Skagerstam, "*Finite Temperature Effects in Quantum Field Theory*", Phys. Lett. **110B** (1982) 406.

2. A. E. I. Johansson, G. Peressutti and B.-S. Skagerstam,"*Quantum Field Theory at Finite Temperature*", Nucl. Phys. **B278** (1986) 324.

3. L. Hollberg and J. L. Hall, "*Measurement of the Shift of Rydberg Energy Levels Induced by Blackbody Radiation*", Phys. Rev. Lett. **53** (1984) 230.

4. G. Barton, "*On the Finite-Temperature Quantum Electrodynamics of Free Electrons and Photons*", Ann. Phys. (N.Y.) **200** (1990) 271.

5. B.-S. Skagerstam, "*Thermal Effects in Particle and String Theories*", in *Proceedings of the 1989 Workshop on Superstrings and Particle Theory*, Eds. L. Clavelli and B. Harms (World Scientific, Singapore, 1990); P. Elmfors and B.-S. Skagerstam "*Anomalous Magnetic Moment at Finite Chemical Potential and External Field Effects*", Z. Phys. **C49** (1991) 251 and "*Anomalous QED Magnetic Moment at Finite Chemical Potential*", in *Proceedings of the 2nd Workshop on Thermal Field Theories and Their Applications*, Eds. H. Ezawa, T. Arimitsu and Y. Hashimoto (Elsevier, 1991)

6. H. R. Schmidt and J. Schukraft, "*The Physics of Ultra-Relativistic Heavy-Ion Collisions*", J. Phys. G: Nucl. Part. Phys. **19**(1993) 1705.

7. J. Schwinger, "*On Gauge Invariance and Vacuum Polarization*", Phys. Rev. **82** (1951) 664 and "*Particles, Sources and Fields*", Vol.3 (Addison-Wesley Pub. Co., 1988).

8. K. Sailer, Th. Schönfeld, Zs. Schram, A. Schäfer and W. Greiner, *"Strings and Ropes in Heavy-Ion Collisions: Towards a Semiclassical Unified String-Flux Tube Model"*, J. Phys. G: Nucl. Part. Phys. **17**(1991) 1005.

9. I. D. Flintoft and M. C. Birse, *"Pair Production in Colour-Dielectric Flux Tubes"*, J. Phys. G: Nucl. Part. Phys. **19**(1993) 389.

10. A. K. Ganguly, P. K. Kaw and J. C. Parikh, *"Thermal Tunneling of $q\bar{q}$ pairs in A–A Collisions"*, PRL–TH–93/19.

11. L. Dolan and R. Jackiw, *"Symmetry Behavior at Finite Temperature"*, Phys. Rev. **D9** (1974) 3320.

12. T. Altherr and D. Seibert, *"Problems of Perturbation Series in Non-Equilibrium Quantum Field Theories"*, preprint, CERN-TH.7271/94.

13. B. Müller in *"Proceedings of a NATO Advanced Study Institute on the Vacuum Structure in Intense Fields"*, Cargese 1990, Eds. H. M. Fried and B. Müller (Plenum Press, 1991).

14. P. Elmfors and B.-S. Skagerstam, *"Electromagnetic Fields in a Thermal Background"*, preprint, NORDITA-94/18 P.

15. P. Elmfors, D. Persson and B.-S. Skagerstam, *"The QED Effective Action at Finite Temperature and Density"*, Phys. Rev. Lett. **71** (1993) 480 and *"Real-Time Thermal Propagators an the QED Effective Action for an External Magnetic Field"*, Astroparticle Physics **2** (1994) 299.

16. P. Elmfors, P. Liljenberg, D. Persson and B.-S. Skagerstam, *"Thermal Versus Vacuum Magnetization in QED"*, preprint, Göteborg ITP 94-13.

17. W. Dittrich, *"Effective Lagrangians at Finite Temperature"*, Phys. Rev. **D19** (1979) 2385.

18. M. Loewe and J. C. Rojas, *"Thermal Effects and the Effective Action of Quantum Electrodynamics"*, Phys. Rev. **D46** (1992) 2689.

19. A. Chodos, K. Everding and D. A. Owen, *"QED With a Chemical Potential: The Case of a Constant Magnetic Field"*, Phys. Rev. **D42** (1990) 2881 and A. Chodos in Ref.13.

20. V. Canuto and H.-Y. Chiu, *"Properties of High-Density Matter in Intense Magnetic Fields"*, Phys. Rev. Lett. **21** (1968) 110; *"Quantum Theory of an Electron Gas in Intense Magnetic Fields"*, Phys. Rev. **173** (1968) 1210; *"Thermodynamic Properties of a Magnetized Fermi Gas"*, Phys. Rev. **173** (1968) 1220; *"Magnetic Moment of a Magnetized Fermi Gas"*, Phys. Rev. **173** (1968) 1229; *"Quantum Theory of an Electron Gas with Anomalous Magnetic Moments in Intense Magnetic Fields"*, Phys. Rev. **176** (1968) 1438.

21. A. Chodos, D. A. Owen and C. M. Sommerfield, *"Strong Field Dependence of the Fine Structure Constant"*, Phys. Lett. **B212** (1988) 491.

22. H. A. Weldon, *"Covariant Calculations at Finite Temperture: The Relativistic Plasma"*, Phys. Rev. **D26** (1982) 1394.

23. T. Altherr and U. Kraemmer, *"Gauge Field Theory Methods for Ultra-Degenerate and Ultra-Relativistic Plasmas"*, Astroparticle Physics **1** (1992) (133).

24. P. H. Cox, W. S. Hellman and A. Yildiz, *"Finite Temperature Corrections to Field Theory: Electron Mass and Magnetic Moment, and Vacuum Energy"*, Ann. Phys. (N.Y.), **154** (1984) (211).

25. D. B. Schwartz, B. Sen, C. N. Archie and J. E. Lukens, *"Quantitative Study of the Effect of the Environment on Macroscopic Quantum Tunnelling"*, Phys. Rev. Lett. **55** (1985) 1547.

26. F. Grossmann, T. Dittrich, P. Jung and P. Hänggi, *"Coherent Destruction of Tunneling"*, Phys. Rev. Lett. **67** (1991) 516.

27. R. Bavli and H. Metiu, *"Laser-Induced Localization of an Electron in a Double-Well Quantum Structure"*, Phys. Rev. Lett. **69** (1992) 1986.

28. P. Kamiński, M. Ploszajczak and R. Arvieu, *"Tunnelling Control in the Driven SU(2) N-Body System"*, Europhys. Lett. **26** (1994) 1.

29. P. Ao, *"Magnetic Field Effect on an Electron Tunnelling Out of a Confining Plane"*, Phys. Rev. Lett. **72** (1994) 1898

30. J. Rafelski and B. Müller, *"Strangeness Production in the Quark-Gluon Plasma"*, Phys. Rev. Lett. **48** (1982) 1066 and Phys. Rev. Lett. **56** (1986) 2334(E).

31. T. Altherr and D. Seibert, *"Thermal Quark Production in Ultra-Relativistic Nuclear Collisions"*, preprint, CERN-TH.7038/93.

PIONS, BARYONS AND ENTROPY IN NUCLEAR COLLISIONS

Marek Gaździcki

Institut für Kernphysik, Universität Frankfurt
60486 Frankfurt, August–Euler–Strasse 6, Germany

Collisions of relativistic nuclei allow the systematic study of the properties of highly dense matter. They give a unique possibility to search in the laboratory for the necessary conditions for the creation of the deconfined partonic matter.[1] By changing the masses of the colliding nuclei and the interaction energy we change the volume, the energy density and the baryon density of the system created in the early stage of the interaction.

It is expected that the effective number of degrees of freedom in the deconfined matter is significantly higher than in the hadronic matter. This should cause a high entropy production in the collisions in which the partonic matter was created.[2] This expectation motivated the performed analysis of the entropy production in nuclear collisions as function of the masses of the colliding nuclei and the interaction energy.

The analysis is based on a statistical approach. Description of the high energy collision process by employing statistical methods was introduced by Fermi[3] almost fifty years ago and developed later by Pomeranchuk,[4] Landau,[5] Hagedorn[6] and many others. The model presented in the first part of the paper and used for th data interpretation in the second part is based on Landau's approach.[5]

Let us consider a central collision (impact parameter, $b \approx 0$) of two identical nuclei in the center of mass system. The collision stage in which all incoming matter is excited we call the early stage. In this stage, due to the high energy density, strongly interacting matter is assumed to reach local thermalization and a dominat fraction of the entropy is produced.[5,7] The volume of the interaction region in the early stage, V, is proportional to the Lorentz contracted volumes of the nuclei, V_0:

$$V \sim \frac{V_0}{\gamma} = \frac{2 \cdot m_N \cdot V_0}{\sqrt{s_{NN}}}, \tag{1}$$

where $\sqrt{s_{NN}}$ is the energy in the c. m. system per nucleon pair and m_N is the nucleon mass. Assuming that the inelastic energy released in the collision, E, is proportional to the total available energy we can relate the thermal energy density, ϵ, to the collision energy:

$$\epsilon = \frac{E}{V} \sim (\sqrt{s_{NN}} - 2 \cdot m_N) \cdot \sqrt{s_{NN}}. \tag{2}$$

The early stage is followed by the expansion of the system. The final number of produced particles, observed later in an experiment, is defined at the freeze–out stage when all interactions have stopped. Landau argued that the evolution of relativistic highly dense matter can be approximated by relativistic adiabatic hydrodynamics. Therefore we can assume that the entropy[*] of the system remains constant during the expansion or, at least, that the final state entropy is proportional to the entropy at the early stage. As more than 90% of the produced particles are pions[†] and the entropy per pion is weakly dependent on the freeze–out temperature[5] the mean pion multiplicity should be proportional to the entropy:

$$\langle \pi \rangle \sim S. \tag{3}$$

It can be shown[8] that this relation is not violated by the resonance decays which take place after the freeze–out stage.

In the following we will relate the entropy with the system size and the collision energy, $\sqrt{s_{NN}}$. The entropy, S, can be expressed as:

$$S = V \cdot \sigma, \tag{4}$$

where σ is the entropy density. The entropy density is related to the energy density by an equation of state, which for the matter in the early stage of the collision is assumed to be:[5]

$$p = \frac{1}{3} \cdot \epsilon, \tag{5}$$

where p is the pressure. This leads to:

$$\sigma \sim \epsilon^{3/4}, \tag{6}$$

which together with Eqs. 2 and 4 gives

$$S \sim V \cdot [(\sqrt{s_{NN}} - 2 \cdot m_N) \cdot \sqrt{s_{NN}}]^{3/4}. \tag{7}$$

Due to the fact that even in the central interactions of identical nuclei not all nucleons take part in the collision, the volume of the system in the early stage is proportional to the number of participant nucleons, $\langle N_P \rangle$, and not to the total number of nucleons in the nuclei. Thus taking into account the Lorentz contraction factor we get:

$$\langle \pi \rangle \sim S \sim \langle N_P \rangle \cdot \frac{(\sqrt{s_{NN}} - 2 \cdot m_N)^{3/4}}{\sqrt{s_{NN}}^{1/4}}. \tag{8}$$

The same dependence of the pion multiplicity on the collision energy was obtained by Fermi in his pioneering work on the pion production in high energy interactions.[3] In order to guarantee that in nucleon–nucleon interactions at the threshold energy the pion multiplicity is equal to zero we have to modify the energy factor in Eq. 8 as follows:

$$F_{NN} = \frac{(\sqrt{s_{NN}} - 2 \cdot m_N - m_\pi)^{3/4}}{\sqrt{s_{NN}}^{1/4}}, \tag{9}$$

where m_π is the pion mass. Thus finally we obtain:

$$\langle \pi \rangle \sim S \sim \langle N_P \rangle \cdot F_{NN}. \tag{10}$$

---

[*]In the paper we consider only the entropy related to the 'inelastic' energy released in the early stage of the collision.

[†]The kaon production in the entropy calculation is taken into account in the last part of the paper where the quantative estimations are performed.

We check now the validity of the approach using the data on nucleon–nucleon interactions compiled elsewhere.[9] Due to the assumptions the model is valid only at high collision energies[‡] and therefore the data at $p_{LAB} < 2$ GeV/c are excluded from the following analysis.

The mean multiplicity of pions produced in nucleon–nucleon interactions[§] is presented in Fig. 1 as a function of F_{NN}. The constructed results for nucleon–nucleon interactions, which are defined as a charge symmetric average of p+p, p+n, n+p and n+n interactions,[9,10] are shown for the energies at which the data on nuclear collisions exist.[9] As expected from Eq. 10 for a fixed number of participant nucleons ($\langle N_P \rangle = 2$) the multiplicity increases in proportion to F_{NN}. This justifies a further interpretation of the data on central collisions of identical nuclei[9] in the described approach.

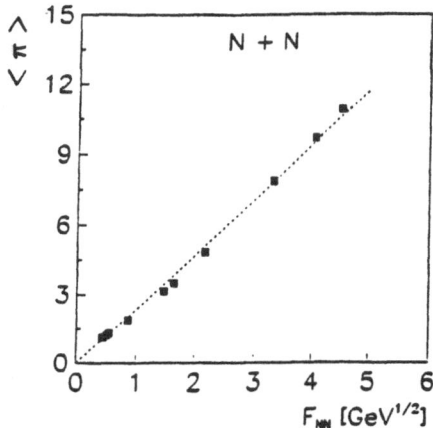

Figure 1. The dependence of the mean multiplicity of pions produced in minimum bias nucleon–nucleon interactions on F_{NN} (see Eq. 9).

At a fixed collision energy per nucleon the pion multiplicity should be proportional to the number of participant nucleons as given by Eq. 10. In order to check this relation we show in Fig. 2 the ratio $\langle \pi \rangle / \langle N_P \rangle$ as a function of $\langle N_P \rangle$ for central nucleus–nucleus collisions at two different collision momenta: ~2.1 GeV/c per nucleon (Ar+KCl and La+La) and 14.6 GeV/c per nucleon (Si+Al and Au+Au). As expected from Eq. 10 the ratio is independent of $\langle N_P \rangle$. Thus in the further study of the energy dependence of the pion multiplicity we remove the $\langle N_P \rangle$ dependence by using the ratio $\langle \pi \rangle / \langle N_P \rangle$.

The ratio $\langle \pi \rangle / \langle N_P \rangle$ is plotted in Fig. 3 as a function of F_{NN} for central nucleus–nucleus collisions and nucleon–nucleon interactions at $p_{LAB} < 15$ GeV/c per nucleon. The multiplicity per participant nucleon for nucleus–nucleus collisions increases linearly with F_{NN} with a slope similar to the slope for nucleon–nucleon interactions. However the intercept of the $\langle \pi \rangle / \langle N_P \rangle$ dependence on F_{NN} is equal about -0.35 instead of being close to zero as observed for nucleon–nucleon interactions.

In order to understand this shift we assume that in central nucleus–nucleus collisions a non–negligible fraction of the 'inelastic' energy and entropy is transfered back to the baryonic sector. This may be due to different distributions of the baryon number in the early stage

---

[‡]This is due to several reasons: a decrease of the entropy per particle at low temperatures, an effect of the 'projection' of the continues entropy onto discrete particle number and invalidity of the used equation of state (Eq. 4) at low collision energies.

[§]The $\langle \pi \rangle$ for charge symmetric collisions[10] was calculated as $\langle \pi \rangle = 3 \cdot \langle h^- \rangle$, where the $\langle h^- \rangle$ was taken from.[9] Thus it includes also a fraction of the kaon multiplicity.

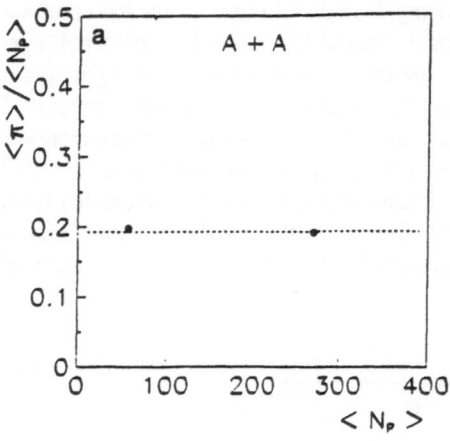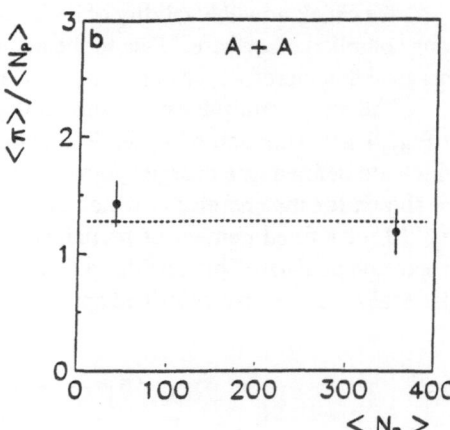

Figure 2. The dependence of the ratio $\langle\pi\rangle/\langle N_P\rangle$ on the mean number of participant nucleons $\langle N_P\rangle$ for $p_{LAB} \approx 2.1$ GeV/c per nucleon (a) and p_{LAB} 14.6 GeV/c per nucleon (b). The Au+Au result at 11.6 GeV/c per nucleon presented in Fig. 2b was extrapolated to 14.6 GeV/c per nucleon using the $\langle\pi\rangle$ dependence on F_{NN} for nucleon–nucleon interactions. The dashed lines show the average values of the data at the corresponding incident momenta per nucleon.

of nucleus–nucleus collisions and nucleon–nucleon interactions as well as a long expansion time in the case of nucleus–nucleus collisions. As the entropy per particle saturates fast with temperature[5] we expect that starting from given 'threshold' values of $\langle N_P\rangle$ and F_{NN} the entropy transferred to baryons should be proportional to the number of baryons ($\langle N_P\rangle$) and independent of the collision energy (F_{NN}). Therefore we can generalized Eq. 10;

$$\langle\pi\rangle + \alpha \cdot \langle N_P\rangle \sim S \sim \langle N_P\rangle \cdot F_{NN}, \tag{11}$$

where the factor α measures the entropy transferred to an average final state baryon. The entropy calculated according to Eq. 11 can be interpreted as the early stage 'inelastic' entropy. It does not contain entropy of the net baryon number carriers in the early stage. It is easy to see that the factor α for central nucleus–nucleus collisions is equal to the constant shift of the pion multiplicity, $\alpha \approx 0.35$. The factor α for nucleon–nucleon interactions, as follows from the previous discussion, is zero. The experimental data on baryon spectra[11] show, in fact, higher 'temperatures' and narrower rapidity spectra of baryons in central nucleus–nucleus collisions than in nucleon–nucleon interactions in qualitative agreement with the above argumentation.

In order to study changes of the entropy in central nucleus–nucleus collisions which are not related to the change of the collision energy (F_{NN}) and the change of the number of participant nucleons ($\langle N_P\rangle$) the difference of the pion multiplicity per participant nucleon for central nucleus–nucleus collisions (AA) and nucleon–nucleon interactions (NN)

$$\Delta\frac{\langle\pi\rangle}{\langle N_P\rangle} = \frac{\langle\pi\rangle_{AA}}{\langle N_P\rangle_{AA}} - \frac{\langle\pi\rangle_{NN}}{\langle N_P\rangle_{NN}} \tag{12}$$

is plotted in Fig. 4 as a function of F_{NN}. The data at low energy ($p_{LAB} < 15$ GeV/c per nucleon) show the previously discussed constant suppression of the pion production in central nucleus–nucleus collisions with respect to nucleon–nucleon interactions due to the entropy transfer to the baryons. The result for central S+S collisions at 200 GeV/c per nucleon obtained by the NA35 Collaboration indicates however an **enhanced** production of pions per

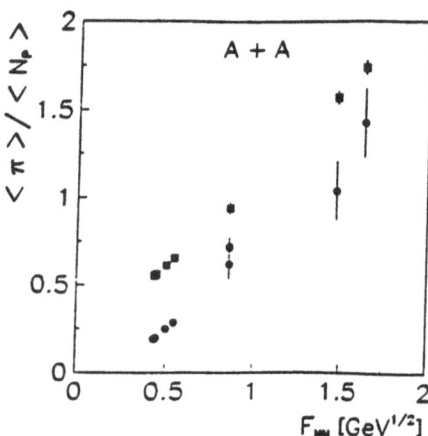

Figure 3. The dependence of the ratio $\langle \pi \rangle / \langle N_P \rangle$ on F_{NN} for data on central collisions of identical nuclei (circles) at F_{NN} below 2 GeV$^{1/2}$ ($p_{LAB} \lesssim 15$ GeV/c per nucleon). The corresponding values for nucleon–nucleon collisions are indicated by open squares.

participant nucleon. This increase of the pion yield (or equivalently increase of the entropy) can be interpreted, in the spirit of our approach, only as due to 'unusual' increase of the entropy density in the early stage. This is because the entropy increase due to the volume and/or the collision energy increase does not affect the difference $\Delta \langle \pi \rangle / \langle N_P \rangle$.

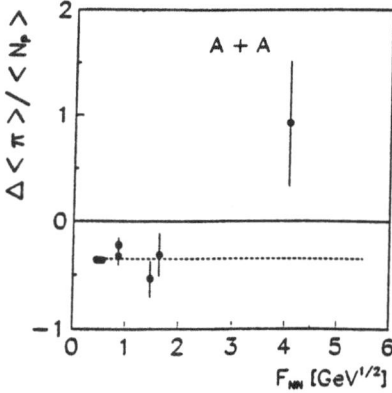

Figure 4. The dependence of the difference $\Delta \langle \pi \rangle / \langle N_P \rangle$ (see Eq. 12) on F_{NN} for the data on central collisions of identical nuclei. The dashed horizonatal line indicates the value -0.35.

As the energy density in the early stage is defined entirely by the collision energy and the geometry of the colliding objects the increase of the entropy density has to be due to the increase of the effective number of degrees of freedom, g, in the early stage of the collision. Such increase of the effective number of degrees of freedom is expected to be a consequence of the creation of the deconfined matter.

The question arises why the enhanced entropy production is not observed in nucleon–nucleon interactions at high energy, in which according to our approach one expects the same energy density as in central nucleus–nucleus collisions at the same energy per nucleon. This may be due to two reasons: higher baryon density in the central part of the collision and significantly larger space–time region in central nucleus–nucleus collisions than in nucleon–nucleon interactions.

An alternative explanation of the entropy enhancement effect, being incompatible with the presented model, can be based on the assumption that in central S+S collisions the energy deposited for particle production is higher than in nucleon–nucleon interactions.[¶] In this case one has to answer the question why this effect is not observed in central nucleus–nucleus collisions at lower energies.

A quantative estimation of the observed increase of the effective number of degrees of freedom is done in the following part of the paper.

From relations

$$\epsilon \sim g \cdot T^4 \tag{13}$$

and

$$\sigma \sim g \cdot T^3, \tag{14}$$

where T is the temperature in the early stage we obtain:

$$\sigma \sim g^{1/4} \tag{15}$$

at a constant energy density. This allows us to introduce the dependence on the effective number of degrees of freedom in Eq. 11:

$$S \sim g^{1/4} \cdot \langle N_P \rangle \cdot F_{NN}. \tag{16}$$

Thus the slope of the increase of the entropy per baryon with F_{NN} is proportional to $g^{1/4}$.

For the final entropy calculation we take into account the production of kaons. Using charge symmetry of the considered systems we can estimate the entropy in the pion entropy units as:

$$S = 3 \cdot (\langle h^- \rangle - \langle K^- \rangle) + \kappa \cdot \langle K/\overline{K} \rangle + \alpha \cdot \langle N_P \rangle, \tag{17}$$

where $\langle h^- \rangle$, $\langle K^- \rangle$ and $\langle K/\overline{K} \rangle$ are the mean multiplicities of negatively charged hadrons, K^-–mesons and kaons and antikaons, respectively. The factor $\kappa \approx 1.4$ is given by the ratio of the entropy per kaon to the entropy per pion at a temperature of 200 MeV. This factor dependends only weakly on the assumed freeze–out temperature. The first component in Eq. 17 estimates the entropy carried by pions, the second component the entropy carried by kaons and the last component estimates the entropy transfered to baryons.

The entropy per baryon calculated according to Eq. 17 is plotted in Fig. 5 as function of F_{NN} for nucleon–nucleon interactions and central nucleus–nucleus collisions. The dashed line follows the dependence obtained for nucleon–nucleon interactions and for central nucleus–nucleus collisions at low energy ($p_{LAB} < 15$ GeV/c per nucleon). The dotted line indicates the dependece of $S/\langle N_P \rangle$ on F_{NN} as defined by the result for central S+S collisions at 200 GeV/c per nucleon. The ratio between the slopes of both straight lines is equal to 1.33±0.13. Therefore the increase of the effective number of degrees of freedom in central S+S collisions at 200 GeV/c per nucleon, g_P, with respect to the effective number of degrees of freedom in nucleon–nucleon and low energy central nucleus–nucleus collisions, g_H, can be estimated to be (see Eq. 16):

$$\frac{g_P}{g_H} = 1.33^4 \approx 3. \tag{18}$$

[¶]The experimental data on baryon production are not precise enough to yield a definite answer concerning this hypothesis.

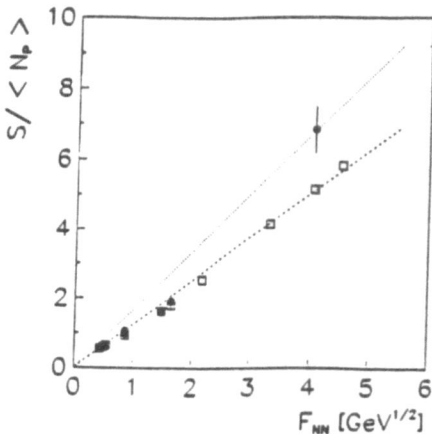

Figure 5. The dependence of the entropy per baryon in the pion entropy units on F_{NN} for nucleon–nucleon interactions (open squares) and central nucleus–nucleus collisions (full circles). The dashed line follows the dependence for nucleon–nucleon interactions and for central nucleus–nucleus collisions at low energy. The dotted line shows the dependence defined by central S+S collisions at 200 GeV/c per nucleon. Note that for the fixed number of degrees of freedom the temperature in the early stage is approximately proportional to F_{NN}.

In summary, the data on the pion multiplicity in nucleon–nucleon interactions and central collisions of identical nuclei were discussed using the statistical approach. The suppression of the pion yield in central nucleus–nucleus collisions at low energy is interpreted as a consequence of the entropy transfer to the baryons. The enhanced entropy production in central S+S collisions at 200 GeV/c per nucleon indicates an increase, by a factor of about 3, of the effective number of degrees of freedom in the early stage of the interaction.

Acknowledgments

I would like to thank to D. Ferenc, St. Mrówczyński, R. Stock and H. Ströbele for numerous discussions and suggestions. I acknowledge also comments of B. L. Friman, J. Knoll and D. Röhrich.

REFERENCES

1. E. V. Shuryak, Phys. Rep. C61:71 (1980) and C115:151 (1984).

2. L. Van Hove, Phys. Lett. B118:138 (1982).
   J. Letessier, A. Tounsi and J. Rafelski, Phys. Lett. B292:417 (1992).

3. E. Fermi, Prog. Theor. Phys. 5:570 (1950).

4. I. Ya. Pomeranchuk, Dokl. Akad. Nauk SSSR 78:884 (1951).

5. L. D. Landau, Izv. Akad. Nauk SSSR, Ser. Fiz. 17:51 (1953).
   S. Z. Belenkij and L. D. Landau, Uspekhi Fiz. Nauk 56:309 (1955).

6. R. Hagedorn, Suppl. Nuovo Cimento 3:147 (1965).

7. J. Letessier, J. Rafelski and A. Tounsi, Paris University Report PAR/LPTHE/94–29 (1994).

8. J. Letessier, J. Rafelski and A. Tounsi, Phys. Lett. B323:393 (1994).

9. M. Gaździcki and D. Röhrich, Frankfurt University Report, IKF–HENPG/3–94 (1994).

10. M. Gaździcki and O. Hansen, Nucl. Phys. A528:754 (1991).

11. J. Bächler et al. Phys. Rev. Lett. 72:1419 (1994).
    M. Gonin et al., Nucl. Phys. A566:601c (1994).
    R. Brockmann et al., Phys. Rev. Lett. 53:2012 (1984).
    Th. Alber et al., Frankfurt University Report IKF–HENPG/1–94 (1994), to be published in
    Z. Phys. C

ENTROPY IN HEAVY ION COLLISIONS

Jean Letessier[1], Johann Rafelski[1,2] and Ahmed Tounsi[1]

[1]Laboratoire de Physique Théorique et Hautes Energies *
Université Paris 7, Tour 24, 5ᵉ étage
2 Place Jussieu, F-75251 Paris Cedex 05, France
[2]Department of Physics, University of Arizona, Tucson, AZ 85721

INTRODUCTION

In relativistic nuclear collisions carried out at CM-energies per nucleon greatly exceeding the rest mass, the final state contains very high particle multiplicity, particularly in the central region of rapidity. This leads to the conclusion that this state comprises a large amount of entropy and this is confirmed by our comprehensive analysis of the experimental data.[1] Here, we address a number of questions arising in this context, in particular:
— How does one measure the entropy produced in the relativistic nuclear collision reaction?
— When and how was this entropy produced?
The underlying premise in this presentation is the generation in the CM of the collision of a space time region of hot, dense hadronic matter with nearly thermal properties, with some fixed energy E, baryonic number B, but zero strangeness $\langle s - \bar{s} \rangle$. We call this region the fireball. Because of the presence of many independent particle-like degrees of freedom, its properties can be well characterized by the usual thermodynamical parameters, the temperature T, the fugacities $\lambda_i = \exp \mu_i / T$ of its constituents and their chemical phase space occupancy factors γ_i.

In the framework of such a largely simplified statistical approach the intriguing question is the possible existence of a color charge deconfined quark and gluon phase referred to as the quark-gluon plasma (QGP). One of the fundamental differences between the QGP and the conventional Hagedorn type hadron gas (HG) structure of the fireball, is the entropy content per baryon (S/B) evaluated at some given (measured) values of statistical parameters. This entropy content can be determined in terms of the final state particle multiplicity.

We have developed a model[2] in particular well applicable in the case that the QGP is formed, allowing us to acquire information about its temperature and entropy evolution. The principal physical provision of our model is the requirement of the balance between the thermal pressure of the QGP and the dynamical compression exercised by the in-flowing

---

*Unité associée au CNRS: D 0280

Hot Hadronic Matter: Theory and Experiment
Edited by J. Letessier *et al.*, Plenum Press, New York, 1995

nuclear matter during the inter-penetration of the projectile and the target. We shall show that within our scheme, akin as we believe to all presently developed more microscopic models of the collision almost 70% of the final state entropy is already produced or indeed it is assumed to be contained in the pre-equilibrium stage of the collision not further explored by the dynamics of the model considered.

We deduce within our model, using improved, perturbative QCD, QGP-equations of state, the evolution in time of the relevant thermodynamical parameters and obtain a remarkable degree of agreement, not only with the final state properties of the S–W/Pb collision system at 200 GeV A, but also surprisingly so, with the Au–Au collisions at 14.6 GeV A. We also apply our method to obtain a full characterization of the forthcoming Pb–Pb collisions. Our detailed results directly determine the spectral shape parameters and abundance of (strange) hadrons and electromagnetic probes (photons, dileptons) produced in the collision, and we explore here as well as in another short contribution[3] some specific experimental consequences.

Our presentation is arranged as follows: First we discuss the different possible descriptions of the final state entropy per baryon based, either directly on the study of the multiplicity or, by means of the equations of state, on the knowledge of the thermal parameters. In part two, we describe the model of the QGP evolution, where the essential parameters are the stopping fractions of energy, momentum and baryonic number. In part three, some consequences of the model on the statistical parameters and comparison with the pertinent experimental results is made.

MEASURE OF THE FINAL STATE ENTROPY

A very useful quantity to explore is the specific entropy per baryon in the fireball S/B, which allows to discriminate between different equations of state of the fireball. The total entropy created in the collision is most easily visible in the final multiplicity of the produced particles. However, in the central rapidity region only few experiments have been able to obtain a full phase space coverage. Because of the need to observe relatively small momentum particles, this is a particularly difficult experimental task and at present the best experimental access to this issue is by means of emulsion techniques. We present here one set of preliminary experimental results which give already a pretty good representation of the specific entropy content in the fireball.

The EMU05 collaboration[4] has studied S–Pb collision at 200 GeV A using a thin Pb-foil placed in front of a emulsion stack. They have concentrated the analysis on central collision events requiring a total final state charged multiplicity to be greater than 300, corresponding to a total central particle multiplicity between 450–1000. They present the multiplicities per interval in rapidity, separately for positive and negative particles. From these data, one can determine the number of protons in the central fireball

$$N_p = \int_{\text{CR}} \frac{d(N^+ - N^-)}{dy} dy \simeq 28.$$

This gives a baryonic number of the fireball B nearly equal to 60 and corresponds to a stopping $\eta_B \simeq 50\%$ for the participant nucleons, the reference value being obtained from the geometric tube-like interaction region model, in this case:

$$N_F = \frac{3}{2} A_P^{2/3} A_T^{1/3} + A_P = \frac{3}{2} 32^{2/3} 207^{1/3} + 32 \simeq 120.$$

If we take for the entropy per particle the thermal value for an isolated system of massless particles $S/N \sim 4$, (in fact at high T for massive hadrons S/N is ≥ 4), we obtain taking the full multiplicity 700 ± 250:

$$S/B \sim 50 \pm 20.$$

A more precise analysis of this experiment can be perform by starting from the ratio

$$D_Q \equiv \frac{\dfrac{dN^+}{dy} - \dfrac{dN^-}{dy}}{\dfrac{dN^+}{dy} + \dfrac{dN^-}{dy}}.$$

which is found in this experiment to be 0.085 ± 0.010 in the central region (see Fig. 1).

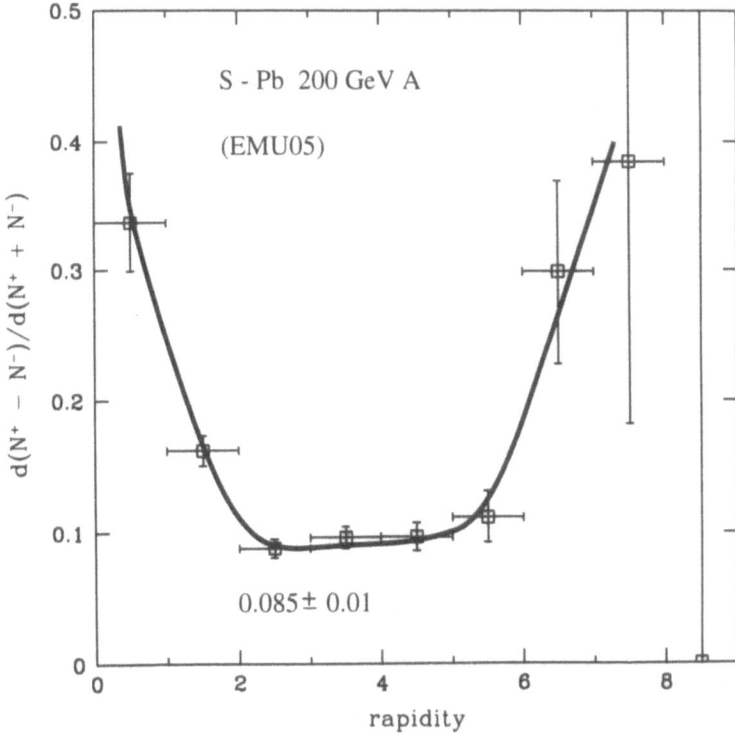

Figure 1. Emulsion data[4] for the charged particle multiplicity from central S–Pb collisions at 200 A GeV as a function of rapidity: the difference of positively and negatively charged particles normalized by the sum of both polarities.

We note that in the numerator of D_Q the charge of particle pairs produced cancels and hence this value is effectively a measure of the baryon number. However, while the denominator is a measure of the total multiplicity, its value is different before or after disintegration of the produced unstable hadronic resonances. Using as input the distribution of final state particles as generated within the hadron gas final state we have shown[1] that $D_Q \cdot S/B$ is nearly independent of the thermal parameters and varies between 4.8, before

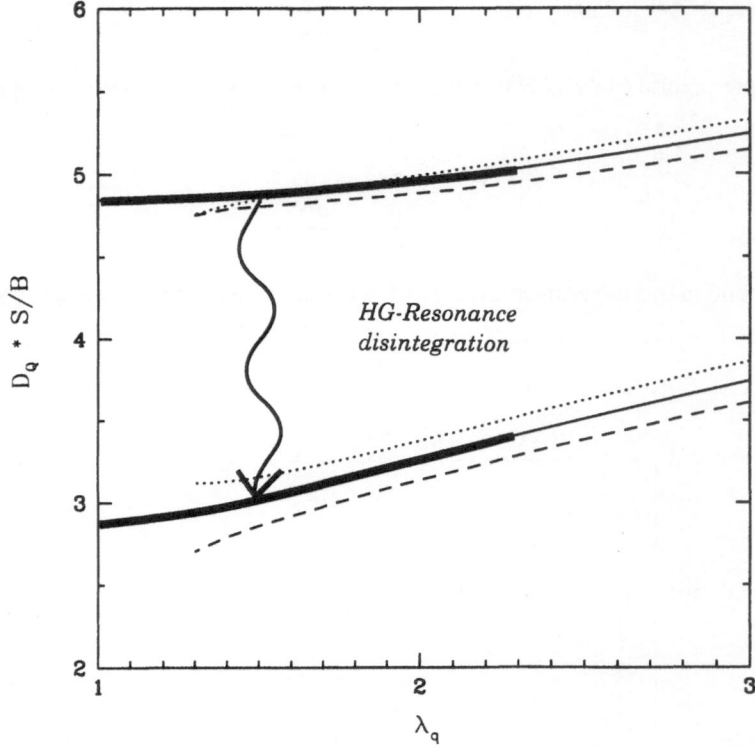

Figure 2. The product $D_Q \cdot (S/B)$ before (upper curves) and after (lower curves) resonance disintegration, as a function of λ_q, for fixed $\lambda_s = 1 \pm 0.05$ and conserved zero strangeness in HG. Note the suppressed zero on the vertical axis.

disintegration of the resonances, to 3 after disintegration (see Fig. 2). It is less than clear that the conservative assumption of the hadronic gas final state is justified, and if we were to assume that e.g. the deconfined state is hadronizing into the final hadronic particles suddenly, production of resonances would be largely suppressed, and thus the greater value 4.8 would apply. Using the above value of D_Q we find for the entropy per baryon of the final state:

$$35 < S/B < 60 \,,$$

with the upper limit applying in the case that few heavy meson resonances are produced.

Another way to determined S/B requires a greater theoretical input and uses the values of the statistical parameters derived within a particular model from strange particle abundances. Consequently the consistency of the model with the data can be assessed. For example we have earlier determined these parameters for the S–W at 200 GeV A (WA85 collaboration[5]), a system similar to the one studied by the EMU05 collaboration. In our study of the ratio of (multi-) strange and anti-strange particles we have determined:[1] $\lambda_s \simeq 1$, $\lambda_q \simeq 1.5$, $\gamma_s \simeq 0.8$ for $T \simeq 230$ MeV. When these values are used taking the partition function of a hadron gas (HG) we obtain

$$S/B \sim 24 \pm 4.$$

But if instead we use a QGP partition function we find, a considerably higher value:

$$S/B \sim 45 \pm 10.$$

So there is a good agreement between the QGP entropy and the observed multiplicity entropy, while at the same time there is strong conflict (a factor 2) of the "measured" multiplicity entropy and the entropy calculated using HG model. We have not found a simple remedy for this discrepancy, in particular we were not able to resolve it by considering changes to the HG equation of state which would be permitted by other data.

The same entropy problem occurs in the analysis of the S–S collisions at 200 GeV A (NA35 collaboration[6]) where a discrepancy of a factor two is also observed between the entropy calculated in a HG model using the statistical parameters determined in a study of the particle ratio[7] and the entropy found using the charged multiplicity.[6] Furthermore, if we were to use the statistical parameters recently extracted using as input other results obtain by the NA36 collaboration for the S–Pb collisions,[8] we would enhance the entropy problem even further. Another view of this problem was also presented here by M. Gaździcki.[9]

MODEL OF QUARK–GLUON PLASMA EVOLUTION

In the above evaluation of the QGP state entropy we needed to use as input the statistical fireball parameters extracted from the analysis of the strange particle multiplicities. The objective of this section is to obtain these parameters in a quantitative way modelling the evolution of the QGP fireball possibly formed in heavy-ion collisions, the existence of this QGP phase being the simplest hypothesis capable to account, in the context of the thermal models, for the experimental data here discussed, including the entropy enhancement.[1] In our model[2] the initial conditions of the QGP-fireball are determined by the fractions of momentum, energy and baryon number which are "stopped" in the central rapidity region.

We use for the partition function an expression of the form

$$\ln Z = \sum_i \frac{g_i V}{2\pi^2} \int \ln \left(1 \pm \gamma_i \lambda_i e^{-\sqrt{m_i^2 + p^2}/T} \right) p^2 \, dp,$$

with $\lambda_{\bar{i}} = \lambda_i^{-1}$ and $\gamma_{\bar{i}} = \gamma_i$ where $i = G, q, \bar{q}, s, \bar{s}$. We take into account the QCD interaction between quarks and gluons by allowing for thermal masses

$$m_i^2 = (m_i^0)^2 + cT^2,$$

and we also correct the quantum degeneracy factors g_i using the perturbative thermal QCD corrections, that we do not explicitly write here.[2] We take the QCD coupling, in the thermal conditions of the experiment, $\alpha_s \simeq 0.6$ and $c \simeq 2$, and for the quark masses

$$m_q^0 = 5 \text{ MeV}, \quad m_s^0 = 160 \text{ MeV}, \quad m_G^0 = 0.$$

In order to describe the temperature and entropy evolution of the fireball, we divide the collision time into four stages:

1) The pre-thermal stage, which is out of the scope of the present description. As we shall show, during this period, nearly 70% of the final particle multiplicity entropy must be produced.

2) The period corresponding to the inter-penetration of the projectile and the target (~ 1 fm/c) nearly corresponding to the time necessary to obtain chemical equilibrium of

non-strange quarks and gluons. The decisive hypothesis we make here is to take the internal pressure of the QGP as constant during this period, and equal to the dynamical compression exercised by the in-flowing matter augmented by the external vacuum pressure:

$$P_{th}(T,\ldots) = P_{dyn} + P_{vac}\,.$$

The pressure due to kinetic motion follows from well-established principles,[10] and can be directly inferred from the energy-momentum tensor:

$$T^{ij}(x) = \int p^i u^j f(x,p)\mathrm{d}^3p\,, \quad i,j = 1,2,3,$$

where $u^j = p^j/E$. We take for the phase-space distribution of incident particles

$$f(x,p) = \rho_0(x)\delta^3(\vec{p} - \vec{p}_{CM}),$$

where $\rho_0(x)$ is the normal nuclear density.

To obtain the pressure exercised by the flow of matter, we consider the pressure component T^{jj}, with j being the direction of $\vec{v}_{CM}$. This gives

$$P_{dyn} = \eta_p \rho_0 \frac{p_{CM}^2}{E_{CM}}\,.$$

Here it is understood that the energy E_{CM} and the momentum P_{CM} are given in the nucleon–nucleon CM frame and η_p is the momentum stopping fraction — only this fraction $0 \leq \eta_p \leq 1$ of the incident CM momentum can be used by a particle incident on the central fireball (the balance remains in the un-stopped longitudinal motion) in order to exercise dynamical pressure. For a target transparent to the incoming flow, there would obviously be no pressure exercised.

The magnitude of stopping has been studied for different reactions. In the S–W/Pb reactions at 200 GeV A, We have seen that $\eta_B \simeq 50\%$. A study of the energy found in the transverse direction to the beam motion[11] suggests that the energy stopping η_E is also nearly 50% in this case. We further shall make the hypothesis that the momentum stopping fraction $\eta_p \simeq \eta_E$. At lower energies (AGS-BNL) we will assume that we move towards full stopping, that is $\eta_E \simeq 1$. When we study collisions of largest nuclei such as Pb–Pb at 160 GeV A, we also anticipate that $\eta_E \simeq 1$.

The initial fireball conditions are now evaluated demanding that the internal thermal pressure P_{th} be balanced by the dynamical and vacuum pressure.

$$P_{th}(T,\ldots) = P_{dyn} + P_{vac} = \eta_p \frac{E_{CM}^2 - m^2}{E_{CM}}\rho_0 + \mathcal{B}\,,$$

where we take $\mathcal{B} = 0.1$ GeV/fm^3 and the left-hand side is determine by using the partition function.

3) During the third time period (2–3 fm/c) the chemical equilibration of the strange quarks takes place. Here we take $\lambda_q = $ Const., corresponding to an entropy of the non-strange quarks and gluons being nearly constant during the accompanying (radial aside of longitudinal) expansion of the fireball volume and the related transfer of thermal energy to collective flow.

4) The last time period corresponds to the final expansion of the fireball accompanied by a decreasing temperature until the freeze-out temperature(s), and production of final state hadron. During this period S/B is nearly constant. We know little about this period which includes hadronization processes, and a crucial hypothesis of this work is that the strange particle multiplicities are already determined by conditions prevailing at the end of period 3).

228

CONSEQUENCES ON THE THERMAL PARAMETERS

During the collision the temperature decrease strongly, essentially due to two mechanism, quark and gluon production and fireball expansion:

T_{th}	temperature where the thermal equilibrium take place
$\downarrow$	*production of q, $\bar{q}$, G*
T_{ch}	temperature of chemical equilibrium for non-strange quarks and gluons
$\downarrow$	*production of s, $\bar{s}$ quarks and fireball expansion*
T_0	temperature of maximal chemical equilibrium
$\downarrow$	*fireball expansion*
T_H	temperature of phase transition, if any (hadronization)
$\downarrow$	*fireball expansion*
T_s	temperature of freeze-out for strange particle, if any
$\downarrow$	*fireball expansion*
T_f	temperature of freeze-out

with $T_{th} > T_{ch} > T_0 > T_H \geq T_s \geq T_f$.

We show in Fig. 3, for the S–W case at $E/B = 8.8$ GeV and stopping parameter $\eta = 0.5$, the qualitative evolutions of T and S/B with time making the hypothesis

$$\gamma_G = \gamma_q = \frac{1}{5}\gamma_s, \qquad \text{when} \quad \gamma_q \leq 1.$$

We see the considerable drop in temperature from the initial stage to the freeze-out point. *A contraire*, the entropy content which determine the final particle multiplicities evolves very little and 70% of the entropy is already present when quarks and gluons are still far from the equilibrium abundance. Practically all the rise is due to the formation of the strange flavor (20%), the remaining 10% arise as consequence of the slight change in the value of λ_q given that we enforce during the collision period the condition of dynamical pressure equilibrium.

In table 1 we display the quantities of interest here, each for a different set of values of the γ_i parameters, for the three qualitatively different collision systems of interest:

— Au–Au   at $E/B = 2.6$ GeV and $\eta = 1$ (complete stopping),
— S–W/Pb   at $E/B = 8.8$ GeV and $\eta = 0.5$,
— Pb–Pb   at $E/B = 8.6$ GeV and $\eta = 1$.

Which are the more remarkable features of these results? Recall that, except for the stopping parameters, the system is fully determined *ab initio* and that the values of the chemical occupancy parameters are fixing in qualitative terms the time period considered. However, within our rough statistical model which does not for example allow for spatial dependence, we obtain for the statistical quantities determined here a very surprising degree of agreement with the values extracted from experiments, assuming QGP formation.

In the following particular cases we compare our results with the "experimental" ones (in parenthesis):

• S–W/Pb	$T_0 = 233$ MeV,	$(232 \pm 5$ MeV$)$,
(8.8 GeV)	$\lambda_q = 1.49$,	(1.48 ± 0.05),
	with a final state $S/B = 49.5$,	(50 ± 15),
• Au–Au	$T_0 = 190$ MeV,	$(190 \pm 20$ MeV$)$,
(2.6 GeV)	$\lambda_q = 4.16$,	$\left(3.68 \, {}^{+0.39}_{-0.20}\right)$,
	with a final state $S/B = 14$,	(13 ± 1).

In S–W collisions, where $\lambda_s \simeq 1$, we have shown that in order to obtained a satisfactory description of the experimental (multi-)strange particle ratios, a low strange particle freeze-out temperature ($T_s \sim 150$ MeV) is needed in a globally hadronizing QGP fireball, with resulting particles escaping without forming an equilibrated HG phase.[12] In the Au–Au case, the observed values of λ_s is near 1.7; if QGP is formed in the initial state, we must have full re-equilibration in the QGP hadronization, before freeze-out take place. This is complicating the argument for or against QGP formation, and implying $T_H > T_s \sim 130$ MeV.[13]

If we now regard closer the differences between the S–Pb and Pb–Pb collisions, we predict a rise in λ_q and we see that the observable temperature rises from $T_0 = 233$ MeV to

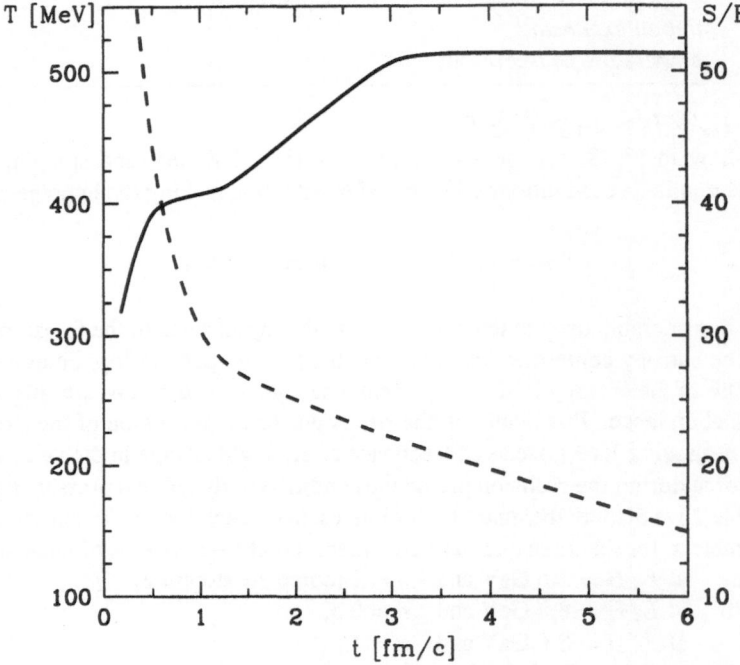

Figure 3. The qualitative evolution of T (dashed line) and S/B (solid line) versus time, for the S–W case at $E/B = 8.8$ GeV and stopping parameter $\eta = 0.5$.

$T_0 = 270$ MeV, which should indeed be soon visible in the high $m_\perp$ particle spectra. Because of the considerable increase in baryon density in the fireball (from $1.8\rho_0$ to $3.2\rho_0$) we see a noticeable drop in specific entropy. This can be understood, since the energy per baryon is similar here to the S–W/Pb collision; the greater energy content per final-state particle at higher temperature thus entails a smaller number of particles per baryon, and hence a smaller entropy content per baryon.

We also observe that the slight rise in λ_q has a considerable impact on the ratio $\overline{\Lambda}/\Lambda$, which is reduced by the factor $(1.49/1.61)^4 = 0.73$ due primarily to a greater number of Λ, given the greater baryon density achieved. The specific yield of strange-particles is also

Table 1. Conditions at different evolution stages in different collision systems: Au–Au at $E/B = 2.6$ GeV, S–W/Pb at $E/B = 8.8$ GeV and Pb–Pb at $E/B = 8.6$ GeV.

Phase-space occupancy	$<s-\bar{s}>=0$ $\lambda_s \equiv 1$	E/B [GeV]		
		2.6	8.8	8.6
		$\eta=1$ Au–Au	$\eta=0.5$ S–W/Pb	$\eta=1$ Pb–Pb
$\gamma_q = 0.2$	T_{th} [GeV]	0.260	0.410	0.471
	λ_q	9.95	1.78	2.00
	n_G/B	0.20	1.55	1.25
	n_q/B	3.00	5.12	3.77
$\gamma_G = 0.2$	$n_{\bar{q}}/B$	0.00	2.12	0.77
	$n_{\bar{s}}/B$	0.02	0.16	0.13
	P_{th} [GeV/fm^3]	0.46	0.79	1.46
$\gamma_s = 0.03$	ρ/ρ_0	3.34	1.70	3.18
	S/B	11.8	40.0	33.4
$\gamma_q = 1$	T_{ch} [GeV]	0.212	0.280	0.324
	λ_q	4.16	1.49	1.61
	n_G/B	0.56	2.50	2.08
	n_q/B	3.11	5.16	4.62
$\gamma_G = 1$	$n_{\bar{q}}/B$	0.11	2.16	1.62
	$n_{\bar{s}}/B$	0.05	0.25	0.21
	P_{ch} [GeV/fm^3]	0.46	0.79	1.46
$\gamma_s = 0.15$	ρ/ρ_0	3.35	1.80	3.19
	S/B	12.3	41.8	34.9
		$\gamma_s = 1$	$\gamma_s = 0.8$	$\gamma_s = 1$
$\gamma_q = 1$	T_0 [GeV]	0.190	0.233	0.270
	λ_q	4.16	1.49	1.61
	n_G/B	0.56	2.50	2.09
$\gamma_G = 1$	n_q/B	3.11	5.12	4.60
	$n_{\bar{q}}/B$	0.11	2.12	1.60
$\gamma_s = 0.8$	$n_{\bar{s}}/B$	0.28	1.27	1.07
or	P_0 [GeV/fm^3]	0.33	0.47	0.84
$\gamma_s = 1$	ρ/ρ_0	2.41	1.05	1.81
	S/B	14.1	49.5	41.7

expected to drop, but it remains relatively high as the number of $\bar{s}$ is greater than the number of $\bar{u}$ or $\bar{d}$ (note that $n_{\bar{q}} = n_{\bar{u}} + n_{\bar{d}}$, and naturally $n_{\bar{s}} = n_s$). Consequently, the ratios of particles such as $\overline{\Xi^-}/\overline{\Lambda}$, which are considered to be more sensitive probes of the QGP phase, will increase by 35–50% as we move to the Pb–Pb system from S–W/Pb.

Interestingly, this ratio will modestly increase while collision energy decreases, as long as the QGP phase is formed, due to an increase in the value of λ_q, which we predict. Note that when E_{CM} drops below the threshold for the formation of QGP, a sudden decrease should follow.

To conclude, we have shown that while we cannot naturally explain the high specific particle yields in terms of the hadronic gas like dense matter, the assumption of the deconfined QGP phase in the fireball not only gives the correct entropy yield, but it allows to develop quantitative models of the collision dynamics, which agree with the presently accessible results. These can be used to extrapolate to the conditions expected in the forthcoming Pb–Pb collisions and we have discussed this here in some detail.

Acknowledgments

J. Rafelski acknowledges partial support by US DOE grant DE-FG02-92ER40733 and thanks his colleagues for their kind hospitality in Paris

REFERENCES

1. J. Letessier, A. Tounsi, U. Heinz, J. Sollfrank and J. Rafelski, *Phys. Rev. Lett.* **70** (1993) 3530. *Strangeness Conservation in hot nuclear fireballs*, Paris preprint LPTHE/92–27R, *submitted to Phys. Rev. D*. U. Heinz, *Nucl. Phys.* **A566** (1994) 205c.
   J. Rafelski, J. Letessier and A. Tounsi, in *Proceedings of the XXVI International Conference on High Energy Physics*, Dallas, Texas, 1992, AIP Conference Proceedings No 272, 983, J.R. Sanford, Editor.

2. J. Letessier, J. Rafelski and A. Tounsi, *Phys. Rev. Lett.* **B333** (1994) 484–493.

3. J. Letessier, J. Rafelski, and A. Tounsi, *Dilepton Spectra in Heavy Ion Collisions,* This Proceedings.

4. Y. Takahashi et al., CERN-EMU05 collaboration, private communication and to be published.

5. S. Abatzis *et al.* (WA85 collaboration), *Phys. Lett.* **B270** (1991) 123; **B259** (1991) 508; **B316** (1993) 615. D. Evans et al. (WA85 collaboration), *Nucl. Phys.* **A566** (1994) 225c.

6. T. Alber *et al*, The NA35 collaboration, *Strange Particle Production in Nuclear Collisions at 200 GeV per Nucleon Z. Phys.* **C** in press, Univ. Frankfurt preprint, IKF-HENPG/1-94, April 1994.

7. J. Sollfrank, M. Gaździcki, U. Heinz and J. Rafelski, *Z. Physik* **C61** (1994) 659.

8. E. Andersen *et al.*, NA36 Collaboration, *Phys. Lett.* **B316** (1993) 603. *Nucl. Phys.* **A566** (1994) 217c; *Phys. Lett.* **B327** (1994) 433.

9. M. Gaździcki, *Pions, Baryons and Entropy in Nuclear Collisions,* This Proceedings.

10. S.R. de Groot, W.A. van Leeuwen, and Ch.G. van Weert, "Relativistic Kinetic Theory", North Holland Pub. Co., Amsterdam (1980).

11. G.W. London, *Result from the CERN Pilot Study of Ultra-Relativistic Nucleus-Nucleus Interaction*, in: *proceedings of the VIII International Conference on Physics in Collision*, Capri, 1988, P. Strolin, ed., Editions Frontières, Paris, 1989, 397.
    I. Otterlund, *Physics of Relativistic Nuclear Collisions*, in: *Particle Production in Highly Excited Matter*, H. H. Gutbrod and J. Rafelski, eds, NATO Physics series Vol. **B 303**, Plenum Press, New York, 1993, 57.
    S.P. Sorensen et al. *Nuclear Stopping Power*, in: *Proc. XXI Int. Symposium on Multiparticle Dynamics 1991*, World Scientific, Singapore, 1992.

12. J. Letessier, J. Rafelski and A. Tounsi, *Phys. Lett.* **B321** (1994) 394.

13. J. Letessier, J. Rafelski and A. Tounsi, *Phys. Lett.* **B328** (1994) 499.

HIGH DENSITY QCD AND ENTROPY
PRODUCTION AT HEAVY ION COLLIDERS

Klaus Geiger

CERN, TH-Division
CH-1211 Geneva 23, Switzerland

INTRODUCTION

In this talk the role of entropy production in the context of probing QCD properties at high densities and finite temperatures in ultra-relativistic collisions of heavy nuclei is inspected. I will argue that the entropy generated in these reactions provides a powerful tool to investigate the space-time evolution and the question whether and how a deconfined plasma of quarks and gluons is formed. I will also address the questions how entropy is produced, and how it is measurable and discuss the uncertainties in predicting the different contributions to the total entropy and particle multiplicities during the course of heavy ion collisions.

I would like to present here a perspective of what we can learn about aspects of QCD at high density with the advent of a new generation of accelerators. Particularly the future heavy ion (AA) colliders, the BNL Relativistic Heavy Ion Collider (RHIC) with maximum available beam energy $\sqrt{s} = 200\,A$ GeV (gold on gold) and the CERN Large Hadron Collider (LHC) with $\sqrt{s} = 5500\,A$ GeV (lead on lead), will provide for the first time the opportunity to study nuclear matter under extreme density compression and very high temperatures, and possibly the formation of a deconfined quark-gluon plasma (QGP). A major goal of the experimental programs at RHIC and LHC is to explore new phenomena associated with the dynamics of quarks and gluons in the hot, ultra-dense environment that is created in these collisions, including the expected (non-) equilibrium QCD phase transition.

THE REGIME OF HIGH DENSITY QCD

Fig. 1 shows the map of QCD as it is understood today. It exhibits three very distinct regions of QCD with quite different physics.[1] On the horizontal axis, r characterizes the space-time distance that can be resolved by probing QCD properties in certain dynamical processes. For instance, when probing a nucleon or nucleus in a deep inelastic scattering event with a momentum transfer Q^2, the photon acts as a microscope with a resolution $r \sim 1/Q$. On the vertical axis, the density ρ_{qg} is the number of quark and gluon quanta with a definite value of rapidity $y = \ln(1/x)$ that is seen by our probe in the transverse plane,

$$\rho_{qg} = \frac{1}{\pi R^2} \frac{dN_{qg}}{dy} \simeq \frac{A\,x f(x, Q^2)}{\pi R^2},$$ (1)

where $f(x, Q^2)$ denotes the sum of quark and gluon structure functions, R is the nucleon radius and A is the number of nucleons.

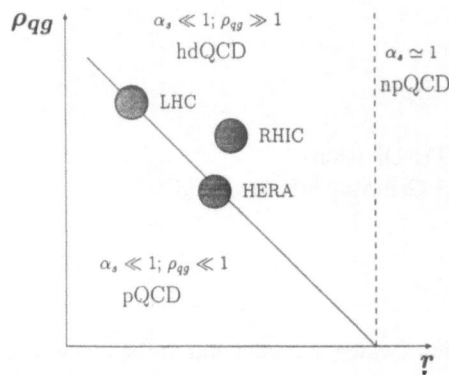

Figure 1. The map of QCD with ρ_{qg} denoting the densities of partons in the transverse plane and r the distance resolved in an experiment.

The three regions of Fig. 1 are:

(i) $r \ll 1$ fm and $\rho \ll R^{-2}$: This is the small distance, low density regime, where the powerful methods of perturbative QCD (pQCD) apply.

(ii) $r \approx 1$ fm: This is the non-perturbative QCD (npQCD) domain, where one has to deal with the complex mechanisms of confinement and relys on lattice calculations and QCD sum rules.

(iii) $r \ll 1$ fm and $\rho \gtrsim R^{-2}$: Here we can probe a high density of partons at short distances (hdQCD). Perturbative techniques can be applied, but in addition, one must tackle important density effects that lead to interference and collective phenomena.

The hdQCD regime is the exciting new field that opens up with the experimental programs at HERA, RHIC and LHC. There are two ways to obtain a system of partons with large density: One way to penetrate hdQCD is deep inelastic ep scattering ($A = 1$) at high energy in the region of very small Bjorken $x \ll 1$. For instance, at HERA, the new experiments measure 30-50(!) gluons in a proton at $x = 10^{-4}$.[2] The other access to the hdQCD region is through collisions of heavy nuclei, in which one can reach high parton densities at not so very high energies or small x, due to the large number of overlapping nucleons ($A \gg 1$). This presumably can be achieved at RHIC ($x \approx 10^{-1}$), but certainly at LHC ($x \approx 10^{-3}$). In particular at LHC both conditions, small x and large A, are combined.

THE ROLE OF ENTROPY IN HEAVY ION COLLISIONS

For the remainder of this talk, I will focus on the hdQCD physics in heavy ion collisions at RHIC and LHC. Qualitatively, the expected space-time evolution of these reactions can be summarized as illustrated in Fig. 2:[6] (1) Immediately after the first nuclear contact, the partons of the colliding nuclei begin to interact frequently with each other, resulting in a vehement materialization of excited quanta. (2) The excited partons can rescatter and emit new particles and thereby evolve through a pre-equilibrium stage towards a thermalized quark gluon plasma state, from which the system evolves further according to the laws of relativistic hydrodynamics. (3) The plasma expands and cools, and eventually a phase transition - perhaps via a mixed parton-hadron phase - into a purely hadronic gas occurs. (4) The freeze-out of this excited hadron matter produces the final hadron yield.

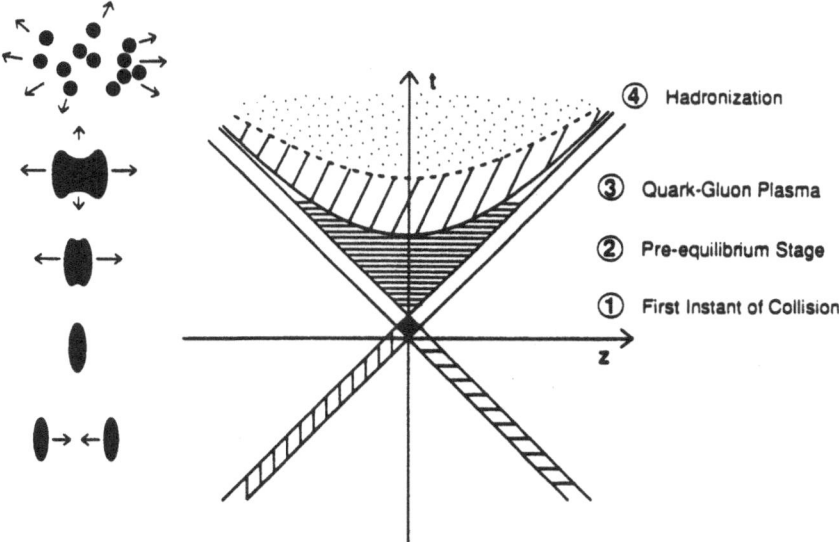

Figure 2. Space-time diagram of the (longitudinal) evolution of a relativistic nucleus-nucleus collision.

One of the most valuable quantities for extracting information about the dynamical evolution is the production of entropy. With each of the above stages (1)-(4) a certain amount of generated entropy is associated. The entropy is generally defined in terms of the density matrix $\hat{\rho}$ of the quantum system, $S = -\text{Tr}\hat{\rho}\ln\hat{\rho}$,[3] or equivalently, in terms of the particle distribution (Wigner) function $W(r,p)$, where $r \equiv r^{\mu} = (t, \vec{r})$, $p \equiv p^{\mu} = (E, \vec{p})$:

$$S(t) = -\int \frac{d^3r\, d^3p}{(2\pi)^3}\, W(r,p)\, \ln W(r,p), \qquad (2)$$

(neglecting Fermi or Bose statistics). Thus S contains valuable information about the time evolution of the system in full phase-space, with the complete history up to time t embodied in the function W. Measuring the entropy as a function of time would therefore allow to trace the space-time evolution of a nuclear collision. Furthermore the rate of entropy production reflects the degree and time scale of equilibration, since $dS/dt \to 0$ as an equilibrium state is reached.[4]

Clearly, it is not easy to extract the time evolution of S from experiment. First of all the entropy is not directly observable, it can be measured only indirectly through the multiplicities of produced particles in a reaction and the amount of transverse energy generated. Secondly, to measure the rate of entropy or particle production, one has to identify signals of characteristic particle emission from different stages of a nuclear collision. Good probes of time rate of change are dileptons, photons, strange and charm production.[5,6] To this end, let me pose the following two questions:

How is entropy produced in heavy ion collisions at RHIC and LHC?

How can we measure entropy in these reactions?

HOW IS ENTROPY PRODUCED?

Most of the entropy and transverse energy is expected to be produced already within the first fm after nuclear contact by very frequent so-called *semihard* parton interactions with only a few GeV momentum transfer.[7,8] This corresponds to the pre-equilibrium regime in Fig. 2. Once the system has reached a thermal equilibrium state, the entropy production vanishes in space-time, and ideally the final transition from the quark-gluon phase to the hadron phase and its freeze out should keep the entropy constant.[9] Let me write the total produced entropy that is produced during the course of a heavy ion collision as

$$\Delta S = \Delta S_{prim}^{(qg)} + \Delta S_{sec}^{(qg)} + \Delta S^{(had)}.$$ (3)

The three contributions arise as follows:

(i) $\Delta S_{prim}^{(qg)}$, *large*: primary parton production due to the materialization of virtual "pre-existing" quanta in the colliding nuclei that are set free by very frequent initial parton scatterings.

(ii) $\Delta S_{sec}^{(qg)}$, *very large*: secondary parton production due to intense gluon bremsstrahlung off partons that have been excited in scatterings, plus production by rescatterings which become increasingly probable as the parton density grows.

(ii) $\Delta S^{(had)}$, *small*: entropy produced through the parton-hadron transition (which ideally should be neglegible) plus additional entropy generated by decay of formed "pre-hadrons" and hadronic resonances into the final stable hadron states.

As an illustrative example, Fig. 3a displays the time development of the specific entropy $(S/N)(t)$ arising from $\Delta S_{prim}^{(qg)}$ and $\Delta S_{sec}^{(qg)}$ as calculated within the parton cascade model[6] for various collider energies. The curves show a rapid build-up of S/N and relax approximately exponential to reach their final values between 3.9 and 4.3. Comparing these values with $(S/N)_{ideal} \simeq 4$ for an ideal gas of non-interacting massless quarks and gluons, one sees that the difference between the resulting entropy of the realistic model calculation and the idealized case amounts less than 10 %. Although the model includes massive quarks and accounts for interactions among the partons, the system of partons behaves effectively like an almost ideal gas. Fig. 3b shows the corresponding relaxation times versus $\sqrt{s}/A$, which are evidently very short and indicate a very rapid thermalization (plasma formation) within less than 0.5 fm.

HOW CAN ENTROPY BE MEASURED?

As mentioned above, the entropy ideally must stay constant once the system has reached an equilibrium (QGP) state. In this context, the only effect of the parton-hadron phase transition is a reorganization of the degrees of freedom from the colored quarks and gluons

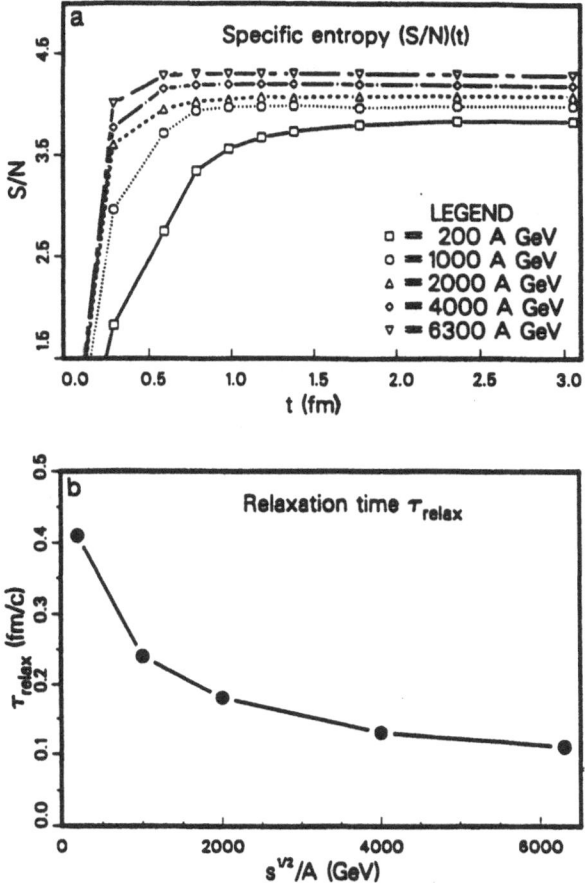

Figure 3. Results of a Monte Carlo simulation of gold on gold collisions at different beam energies. a) Entropy per secondary particle produced as a function of real time. b) Corresponding relaxation times versus beam energy.

to the color singlet hadrons, mostly pions. Now, experiments of deep inelastic ep-scattering, e^+e^--annihilation, Drell Yan, etc., strongly support the hypothesis of local parton-hadron duality.[10] That is, it appears that in high energy processes the mechanism of hadronization is a universal and local phenomenon, independent of the partons prehistory. The measured hadron multiplicity turns out to be simply equal the calculated parton multiplicity times a constant (e.g. $N^{(\pi)} = 1.1 N^{(qg)}$).

On the other hand, if there is a first order QCD phase transition at a temperature T_c, the ratio of the entropy density in the pion plasma, $s^{(\pi)}$, to that of the quark-gluon plasma, $s^{(qg)}$, can be estimated from the effective number of degrees of freedom in the two phases at T_c as $r = s^{(\pi)}/s^{(qg)} \approx 0.7$.[9] Under these assumptions for zero impact parameter collisions the total produced entropy can be measured by

$$\frac{dS}{dy} = c^{(qg)} \left(\frac{dN^{(qg)}}{dy} \right)_{b=0} \approx \frac{c^{(\pi)}}{r} \left(\frac{dN^{(\pi)}}{dy} \right)_{b=0}. \tag{4}$$

where $c^{(qg)} \simeq 4$ and $c^{(\pi)} = 3.6$. Thus, again one has $N^{(\pi)} = c^{(qg)}/c^{(\pi)} N^{(qg)} = 1.1 N^{(qg)}$.

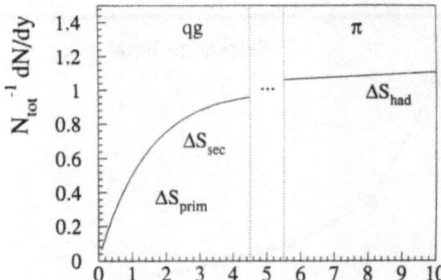

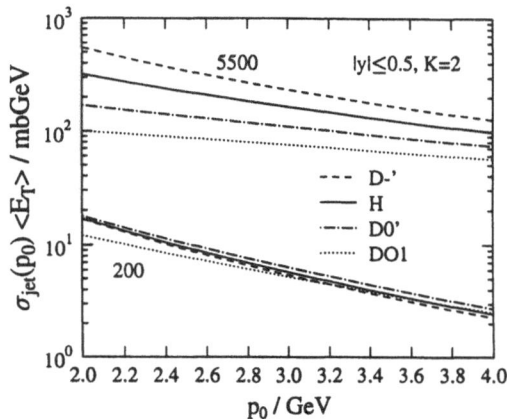

Figure 5. Estimated total transverse energy produced in heavy ion collisions at RHIC and LHC for differents sets of structure function parametrizations [from Ref.[13]].

Fig. 5 shows the impact of structure function dependence on the total transverse energy produced in $Au + Au$ collisions at RHIC and LHC energies.[13] The variation with the choice of structure functions is especially drastic at LHC, where it reaches up to a factor of 5(!) uncertainty.

b) The other source of substantial uncertainty originates from the current lack of better knowledge of the impact of nuclear and dense medium effects which are absent in e.g. hadronic collisions, but become important for heavy ion collisions. The new physics associated with the nuclear and medium effects on the parton level[14] that determine the dynamics of AA collisions drastically evident in Fig. 6. where the various curves a)-e) exhibit the characteristics of different evolution scenarios of simulations of gold on gold collisions,[15] namely, a) 'naive evolution', b) plus nuclear shadowing, c) plus parton fusion and absorption, d) plus Landau-Pomeranchuk-Migdal effect, e) plus soft gluon interference. Shown are the charged particle distributions in pseudorapidity (η), calculated for central $Au + Au$ collisions with $\sqrt{s} = 200$ A GeV and $\sqrt{s} = 6300$ A GeV. Evidently, the particle density in the central rapidity region is reduced successively from the 'naive' calculation (case a) to the default result (case e) by more than a factor of 3.

SUMMARY

Much experimental and theoretical study is required before robust conclusions for experimental observables and open theoretical questions can be drawn - although I believe, already at this point a clear understanding of the fundamental problems on the basis of perturbative QCD and transport theory has been reached.[6] Foremost, we must learn how to utilize medium effects in high density QCD, so that truly quantitative calculations become possible. Almost as importantly, we must develop a microscopic theory for the hadronization of a quark-gluon plasma that is based on QCD, doing away with the presently employed QCD-inspired models that have little predictive power in the new high density regime. If these two goals can be met, quantitative ab-initio calculations of ultrarelativistic heavy ion collisions will become feasible by the time when RHIC and LHC start operation.

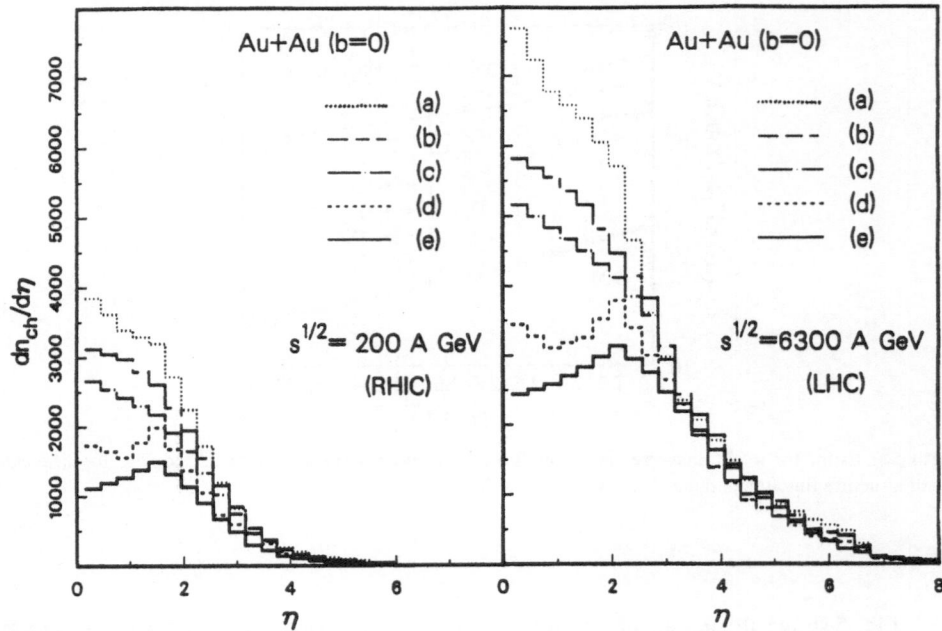

Figure 6. Impact of nuclear and dense matter effects on a) the parton evolution and b) the resulting charged particle spectra in $Au + Au$ collisions at RHIC and LHC energies. The different evolution scenarios are explained in the text.

REFERENCES

1. E. Levin, preprint FERMILAB-CONF-94/068-T (1994).

2. For recent experimental and phenomenological progress, see e.g.: G. Wolf, in the proceedings of the *International Workshop on Deep Inelastic Scattering*, Eilat, Israel, 1994.

3. H.-T. Elze, preprint CERN-TH 7131/93 (1994).

4. S. de Groot, W. A. van Leuwen and C. G. van Weert, *Relativistic Kinetic Theory*, North Holland, Amsterdam 1980.

5. *Quark Matter '93*, Proceedings of the Tenth International Conference on Ultra-relativistic Nucleus-Nucleus Collisions, Borlänge, Sweden, 1993, edited by H. A. Gustafsson *et al.* [Nucl. Phys. **A566**, 1c (1994)].

6. K. Geiger, report CERN-TH 7313/94 (1994).

7. L. V. Gribov, E. M. Levin, and M. G. Ryskin, Phys. Rep. **100**, 1 (1983); E. M. Levin and M. G. Ryskin, Phys. Rep. **189**, 267 (1990).

8. K. Geiger, Nucl. Phys. **A 566**, 257c (1994).

9. J. D. Bjorken, Phys. Rev. **D27**, 140 (1983).

10. Ya. I. Azimov, Yu. L. Dokshitzer, V. A. Khoze and S. I. Troyan, Z. Phys. **C27**, 65 (1985).

11. K. Geiger, Phys. Rev. **D46**, 4986 (1992).

12. A. H. Mueller and J. Qui, Nucl. Phys. **B268**, 427 (1986); K. J. Eskola, J. Qui and X.-N. Wang, Phys. Rev. Lett. **72**, 36 (1994). .

13. K. Eskola, K. Kajantie and P. V. Ruuskanen, preprint HU-TFT-94-6, Helsinki 1994.

14. See e.g.: Proceedings of the *Workshop on Pre-equilibrium Parton Dynamics in Heavy Ion Collisions*, edited by X. N. Wang, LBL-Report 34831, Berkeley 1993.

15. K. Geiger, Phys. Rev. **D47**, 133 (1993).

ABOUT ENTROPY AND THERMALIZATION—
A MINIWORKSHOP PERSPECTIVE

Hans-Thomas Elze[1] and Peter A. Carruthers[2]

[1]CERN-Theory
 CH-1211 Geneva 23, Switzerland
[2]Department of Physics
 University of Arizona, Tucson, AZ 85721, USA

INTRODUCTION

Our intention is to summarize the main ideas brought forth in this miniworkshop on *"Entropy and Thermalization"* in strong interactions at high energy. In particular, some aspects and differing views introduced during the round table discussion, which are not otherwise represented in these Proceedings, will be reported here.

Anticipating our conclusions, there can be no doubt that at present there exists a rich diversity (if not confusion) of concepts related to the measure of entropy to characterize high-multiplicity events in high-energy reactions. This points to the fact that the fundamental problem is far from being solved concerning the characteristic features of these reactions. In terms of a field theory it is still not at all clear, how precisely and most economically to quantify their very complex multiparticle or many degrees of freedom aspects and relate them to experiments. Rather, the general impression is that at best one begins to see the scope of the problem and the first and still quite conventional approaches to describe the essential *disorder* of hot and dense hadronic matter, the *lack of information* on high-order correlations of various kinds, and the *dynamical complexity* of the underlying QCD fields. All of which may be encoded in corresponding measures of entropy. There is a generally shared feeling of the potential richness of collective phenomena hiding in strong interactions at high energy, in particular of heavy nuclei, but mostly attempts to find an adequate formal description to uncover them from experimental findings have been modest.

In the following we try to spread the good news that our present subject forms part of one of the major scientific issues of our time, i.e. the measure and understanding of disorder vs. order, which can be observed in rapid development in several active parts of science and of physics, in particular. We proceed to present some presently rather divergent opinions about the essential features of entropy as well as various first steps in the analysis of entropy production in strong interactions.

CAN WE MAKE SENSE OF DISORDER, LACK OF INFORMATION OR DYNAMICAL COMPLEXITY?*

How should we assess the structures of systems which can exhibit *disordered behaviour* in addition to apparently rather simple coherent and logically structured evolutions? By now there are many ideas about how to approach this problem. Here we consider *entropy* as one useful quantitative measure.

Sometimes apparently coherent motions such as sound waves depend on random microscopic behavior. In other cases a true quantum coherence is essential. Intensity interferometry is rooted in a field ensemble of Gaussian random variables in the most common examples.

Curiously, physicists imagine entropy to be a gross thermodynamic measure, determined by an integration procedure required to convert heat transfer to a perfect differential . Chemists often understand entropy. And computer people have a digital and perhaps better feeling for this concept. Yet they must all be integrated into a single framework covering such diverse aspects as quantum limits of computation and Gödel's theorem.[2] The problem is to define and find the *algorithm* that produces the least computational needs. Yet it must be capable to capture the essential qualities of disordered behaviour and complex structures. Surprisingly, topology nowadays does not (yet?) play any essential role in the field of strong interactions, which is governed by the QCD Lagrangian methodology.

Historically, Boltzmann's genius stands out as the beacon of this subject. A key paradox here is related to *Liouville's theorem*, wherein the many-particle entropy is conserved, contrary to what anybody knows to be the basic issue regarding entropy.

Then, there is *von Neumann's* definition of the *quantum entropy* in terms of the density matrix. This is also a conserved quantity, and seems therefore useless at first sight. However, the natural way out is to consider the von Neumann entropy of intrinsically *open systems* (cf. below), which generally is not a constant of motion. This forms the basis of the work by one of us reported in these Proceedings.[3]

One may also approach the problem in a different way. This is known as *coarse graining*, unfortunately an all-encompassing term, one variant known as the random-phase approximation in the initial conditions. Another attempt to give it a precise meaning in terms of relevant time scales in the context of particle production by an external field is made in Ref.[4] However, there seems to be no general foundation for such procedures, which are introduced studying individual cases. More surprising is that coarse graining does not distinguish the "direction" of time,[5] even though the entropy has to increase in the process of coarse graining. Again, this seems to be open to debate and we only want to mention the idea that string theory may provide a natural coarse graining by having to integrate out unobservable modes and consequently may alter the fundamental quantum mechanical Schrödinger equation in a way which embodies an "arrow of time".[6]

It is well known that Boltzmann's H-theorem was derived from his famous kinetic equation based on a single-particle phase space distribution. By now it is clear that there are many extensions of his approach having to do with higher order correlations and computational complexity among others, i.e. an infinite (quantum) hierarchy of coupled equations and cellular automata, respectively. Very little is known about how the latter or the former BBGKY (and analogous Schwinger-Dyson) hierarchies can be cast into more intuitively comprehensible schemes. This concerns the description, for example, of such striking phenomena as turbulence or multiparticle hadronization processes, which are "understood" to some extent on the basis of "simple" phenomenological equations.

In particle physics we try to calculate the S-matrix. From this we calculate only probabilities of certain events. Usually the most useful formulation is for the so-called

---

*For a related earlier discussion see Ref.[1]

inclusive differential cross sections, for which selected particles in phase space are collected, while all others are averaged over. Then, a sequence of probabilities can be constructed and a hierarchy of entropies follows rigorously, which are in agreement with quantum theory despite their classical appearance. We define, for example, the sequence of *inclusive probability densities*:

$$\rho_1 \;=\; \frac{1}{\sigma}\frac{d\sigma}{d\Gamma_1}\;, \quad \rho_2 \;=\; \frac{1}{\sigma}\frac{d^2\sigma}{d\Gamma_1 d\Gamma_2}\;, \quad \rho_3 \;=\; \frac{1}{\sigma}\frac{d^3\sigma}{d\Gamma_1 d\Gamma_2 d\Gamma_3}\;, \tag{1}$$

etc. and the associated *information entropies*:

$$
\begin{aligned}
S(|B) &= -\sum_A P(A|B)\ln P(A|B)\;, \\
S(|AB) &= -\sum_C P(C|AB)\ln P(C|AB)\;, \\
S(|ABC) &= -\sum_D P(D|ABC)\ln P(D|ABC)\;,
\end{aligned}
\tag{2}
$$

etc. Here $P(D|ABC)$ denotes the conditional probability of finding the value D of an observable keeping values A, B, C of other observables fixed and $\sum_D P(D|ABC) \propto \rho_3$, for example. These entropies can be reformulated in terms of *correlation entropies* analogous to cumulants,[7] which systematically remove irrelevant lower order contributions. These correlations vanish when any variable becomes statistically independent of another. Previous definitions of higher order information entropy miss this point: there, for independent distributions leading to additive entropies, noise potentially obscures the signal of true correlations.

Finally, *all probabilistic entities can be reconstructed from the hierarchy of correlations using generating functional techniques and individual events can be modelled by sampling from the probabilities.*

Next, in order to eliminate one source of confusion, we argue that the above introduced *information entropy is identical to von Neumann's entropy if evaluated for a suitably defined open quantum system.* To see this, we recall Eqs. (1) defining the inclusive densities or, rather, consider the associated probabilities ($\equiv$ density $\times$ flux factor). The same physics of a scattering experiment, for example, can be described in a somewhat unconventional way by calculating a partial trace (with the exclusive variables kept fixed) of the time-evolved density matrix of the total system and integrating over time from $-\infty$ to $+\infty$; here the initial condition has to be specified according to the in-state of the scattering reaction. This defines a time-dependent *density submatrix* which can be diagonalized in the exclusive variables by a unitary transformation (provided these variables correspond to quantum mechanical observables of the system). Applying the formal results from Sect. 2 of Ref.,[8] we conclude that the resulting matrix elements are indeed the *probabilities* to find the corresponding values of observables of the subsystem defined by the exclusive variables. The "inclusive variables" over which one averages or which are integrated out by calculating partial traces, respectively, automatically constitute the *environment* which complements the *open subsystem* in the total closed system. Thus, formally, if we calculate either information entropies according to Eqs. (2) or the corresponding *von Neumann entropies*,

$$
\begin{aligned}
S_{v.N.}(|B) &= -\mathrm{Tr}_A\,\hat{\rho}(A|B)\,\ln\hat{\rho}(A|B)\;, \\
S_{v.N.}(|AB) &= -\mathrm{Tr}_C\,\hat{\rho}(C|AB)\,\ln\hat{\rho}(C|AB)\;,
\end{aligned}
\tag{3}
$$

etc., we obtain the same result. In general, the judicious choice of the exclusive variables is dictated as much by the physical system under consideration as a meaningful separation of subsystem and environment, which is studied for strong interactions in Ref.[8] Both approaches give a quantum mechanically precise meaning to the term *"coarse graining"* by eliminating consistently either inclusive variables or environment degrees of freedom.

Several further points can be made. One is, what if any is the connection with *thermodynamics*? Although not necessary, it is always an interesting limit to consider. The merit of this limit, if justified, is that strongly time-dependent dynamical details become irrelevant in a stationary equilibrium state, concealing our ignorance of the true situation. We remind the reader of the elegance of Landau's application of relativistic fluid mechanics to multiparticle production.

The main point is, what information is obtained and what does it tell us about nature? Despite impressive advances in the precision of experimental data, the conceptual framework for the description of *multihadron production* is still deficient. Can anything of fundamental value come out of the incredibly complicated evolution of hadronic and nuclear collisions being analyzed with theoretical tools which were shaped by experience with few-body final states?

From the moment analysis of multiparticle correlations we can see interesting and strong effects. Recent studies of Bose-Einstein correlations suggest a new and interesting direction. Apart from this, one can imagine that resonance decays account for the main component of the correlation data. This is not very fundamental. At relativistic energies we must sift through enormous data sets in which it is not clear that much of interest has happened. One of the fascinating regularities is the omnipresent negative binomial count distribution, and the associated "linked pair" structure of the cumulant correlations studied by one of us (see Ref.[7] and references therein).

As far as entropy is concerned, the mathematical concept related to probability theory has an intrinsic validity not based on a particular set of variables. However, the variables used to define the probabilities themselves deserve more exploratory thinking in order to reveil some essence of complex dynamical behaviour. The subtleties of the behaviour of systems with many degrees of freedom can defeat the methodology of S-matrix formulations and standard perturbative calculations are doomed to fail if non-linearities are important, such as in semi-classical Yang-Mills fields[9] and hadronizing QCD systems.

To conclude, we are still awaiting specific answers to the question posed in the title above. It appears challenging to study these problems of a rather general nature and of importance beyond the physics of strong interactions.

ASPECTS OF ENTROPY IN STRONG INTERACTIONS

Several members of the round table presented short contributions highlighting their respective views on entropy and related attempts to understand the complex irreversible behaviour in high-energy collisions.

R. Omnès asked the basic question, Given von Neumann's definition of quantum entropy, $S = -\text{Tr} \hat{\rho} \ln \hat{\rho}$, which is a constant of motion for an isolated system, what is the S that increases? He argued that the answer is given by *decoherence theory* in ordinary quantum mechanics and provided an outline thereof. This work is in depth documented, for example, in the review articles.[10]

The underlying reasoning, already implicitly alluded to in the above discussion of an *open subsystem* and its *environment*, is the following: Consider a complex dynamical system with many degrees of freedom (e.g. $N \approx 10^{23}$), such as a piece of solid matter. Select suitable collective observables, such as the center of mass coordinates or momenta etc. Then, split the object ideally into two interacting systems, C and E, defined by the collective ("exclusive") and the complementing environment ("inclusive") degrees of freedom, respectively. The Hamiltonian splits accordingly,

$$H = H_C + H_E + H_{CE} \ , \tag{4}$$

where the last term is responsible for energy exchange between C and E ("*dissipation*"). Starting with an initial state, for example, which is a coherent superposition of states representing the piece of matter being located at x_1 and x_2, respectively, and the environment in its groundstate,

$$|\psi\rangle = a|x_1\rangle_C|0\rangle_E + b|x_2\rangle_C|0\rangle_E , \qquad (5)$$

one can show for suitable model interactions that very rapidly (due to excessively small energy denominators) the environment picks up a little excitation energy through H_{CE}. Most important, however, the relevant collective subsystem density matrix obtained by tracing over the environment degrees of freedom,

$$\begin{aligned}\hat{\rho}_C(t) &\equiv \mathrm{Tr}_E\, \hat{\rho}(t) = \mathrm{Tr}_E\, |\psi(t)\rangle\langle\psi(t)| \\ &\approx |c_1(t)||x_1(t)\rangle\langle x_1(t)| + |c_2(t)||x_2(t)\rangle\langle x_2(t)| , \end{aligned} \qquad (6)$$

where $|c_1| + |c_2| = 1$, becomes essentially diagonal on the same short time scale. This is *environment-induced decoherence*. As far as the collective variables are concerned, a *pure zero-entropy initial state* (chosen only for simplicity) has become a *mixed state with non-zero von Neumann entropy*,

$$S_{v.N.} = -\mathrm{Tr}_C\, \hat{\rho}_C \ln \hat{\rho}_C = -(|c_1|\ln|c_1| + |c_2|\ln|c_2|) . \qquad (7)$$

This effect, mutatis mutandis, is believed to be almost universal, although explicit calculations are restricted to the class of models which eventually can be represented by Gaussian path integrals. Some pertinent questions are: *How complete* is the decoherence effect? *How fast* is it? Under which conditions is the resulting *entropy production irreversible*? Answers from explicit calculations can be given, for example, for the particular quantum mechanical systems studied in Refs.,[3,8] which are constructed with an eye on the relativistic quantum field theory extensions of decoherence theory to be applied to strong interactions.[8,11]

Note that in the above example of a solid piece of matter the microscopic environment (mostly phonons) and the macroscopic collective degrees of freedom are known and rather clearly separated. In general, this will not be the case. For strongly interacting hadronic systems and high-energy collisions, in particular, the main stumbling block preventing a deeper understanding is precisely that we know very little about which are *the* relevant degrees of freedom among an infinity of others. Single-particle observables do not seem to provide a clue.

M. Danos addressed the fundamental question of *irreversibility in quantum physics*, which in a sense is even prior to our discussion of entropy. He presented a provocative point of view, which we quote directly with minor changes:[12]

One of the key points in dissipation in quantum physics is the observation that time-reversal invariant states have probability measure zero. – Generally, the physical states of a system do not exhibit the symmetries of the Hamiltonian. This is so also for the time-reversal symmetry. Since the Hamiltonian itself is time-reversal invariant, time-reversal invariant states must exist, and indeed they do. Only they have "measure zero". – Rather than providing the mathematical derivation of this result,[13] a more physical explanation is given here. Take as the simplest possible example a two-channel system. In a physical state there will be an incoming wave in one channel, say channel 1, and outgoing waves in both channels. The wave function (in the asymptotic region) will be

$$\Psi = e^{-ik_1 x} + ae^{ik_1 x} + be^{ik_2 x} , \qquad (8)$$

where $k_2^2 = k_1^2 - 2mB$ (*), with B the inelasticity of channel 2. The time-reversed wave function is $\Phi = \Psi^*$, which has amplitude- and phase-related incoming waves in both channels, and an outgoing wave in channel 1. To achieve that form the energy matching of Eq. (*) must be fulfilled *exactly*. To actually construct this wave would require an infinite set-up time as a consequence of the time-energy uncertainty relation.

Hence, even though it is easy to write down an expression for the time-reversed state of a physical state for any system, it is principally impossible to *actually construct* such states. Then, the superposition of a state and its time-reversed partner forming a time-reversal invariant state becomes equally impossible. Hence, such states cannot exist in nature.

Unfortunately, we are unable to recall the spirit of the subsequent lively discussion with the audience here, which touched upon doubts about the validity of quantum mechanics, questions about the existence of pure states and the relevance of infinite numbers of degrees of freedom, and several others.

R. Weiner turned the attention to problems closer related to experimental observations. Namely, attempts to understand results of correlation measurements and Hanbury-Brown-Twiss type interferometry with secondary hadronic particles in terms of the space-time and internal structure of their sources. This subject is covered in various ways by his and others' contributions to these Proceedings, in particular those to the miniworkshop "*Multiparticle Dynamics*". We mention some interesting points concerning our present subject.

Consider a system, which is completely characterized by its density matrix $\hat{\rho}$, in terms of *coherent states*, $a_k|\alpha_k\rangle = \alpha_k|\alpha_k\rangle$, where a_k =annihilation operator for mode k. Then, omitting the (non-trivial) sum over modes (due to overcompleteness),

$$\hat{\rho} = \int d^2\alpha \, P(\alpha)|\alpha\rangle\langle\alpha| \ . \tag{9}$$

Quantum statistics allows the weight function to range between the extremes of a coherent distribution and a chaotic one with $P_c(\alpha) = \delta^2(\alpha - \alpha_0)$ and $P_{ch}(\alpha) = \pi^{-1}\bar{n}^{-1}\exp(-|\alpha|^2/\bar{n})$, respectively. It can be shown that a *chaotic distribution is necessary and sufficient to maximize the von Neumann entropy*.[14]

Now, the simplest usually measured *Bose-Einstein correlations* are defined by

$$C_2(k_1, k_2) = \frac{\rho_2(k_1, k_2)}{\rho_1(k_1)\rho_1(k_2)} = \frac{\text{Tr } \hat{\rho} I_1 I_2}{\text{Tr}(\hat{\rho} I_1)\text{Tr}(\hat{\rho} I_2)} \ , \tag{10}$$

where the "intensities" are given by number operators, $I_j = I(k_j) = a_{k_j}^+ a_{k_j}$, and the densities follow from Eqs. (1) with $\Gamma = k$. Thus, in principle, Bose-Einstein correlations measure the density matrix of the system and provide a test for randomization. The very existence of non-trivial correlations, i.e. $C_2(k_1, k_2) > 1$, which is observed in a wide range of experiments, is *evidence for a (partial) randomization and non-vanishing entropy*. However, unfortunately this aspect is usually taken more or less for granted and, so far, empirical parametrizations of these correlations are only employed to derive source geometry and lifetime parameters.

U. Heinz, finally, recalled the apparently strongly disordered outgoing state in a heavy-ion collision. Then, Why is there a $\beta = T^{-1}$ characterizing the exponential slope of major parts of the spectra of secondaries? There are partial answers to this decades old question, but one still feels a lack of understanding. T being a Lagrange multiplier for the variational problem to *maximize S at constant energy*, one asks oneself, Where does the measured large value of this entropy come from? Given that a local thermodynamic equilibrium description of high energy density matter in a collision can be justified, which in turn is still only partly understood,[15] the entropy in terms of particle multiplicities is an important additional piece of information to distinguish various phenomenological equations of state.[16,17]

WHAT DO NEXT?

Instead of providing *the* answer to Lenin's question for our present subject, we discuss the contributions to this miniworkshop, with more details to be found elsewhere in these Proceedings.

The most recent and surprising results of the most conservative approach to entropy in heavy-ion collisions, in the sense of having already a history of four decades, were described by **J. Letessier**.[16] It is based on the hypothesis of the formation of a *fireball*, i.e. a space-time region of hot and dense hadronic matter with approximately thermal properties in the center of mass of these reactions. Its total energy E and baryon number B are assumed to be fixed and their internal properties (particle content, phase space distributions, etc.) to be determined consistently by a *kinetic equilibrium temperature* T and *fugacities* $\lambda_i = \exp \mu_i / T$ of the constituents (as well as additional parameters). The major advantage of this model is its conceptual and technical *simplicity*, in principle, which allows direct comparison with experimental results on multiplicities. Here the measured final state *specific entropy S/B* is employed to discriminate between different equations of state. It is argued that a *hot hadronic gas scenario is unable to fit all available data from the 200 GeV A CERN experiments, whereas incorporating a temporarily existing quark-gluon plasma gives a satisfactory description* ("too good to be true").[16]

The success of this approach crucially depends on the assumptions of (local) thermal equilibrium joined with a simple "macroscopic" collective expansion of the fireball. One further important *consistency* check should be to study two-particle correlations in precisely the same model, as well as single-particle spectra.[18] Finally, it is stated that $\approx 70\%$ *of the measured entropy must be produced already during the pre-thermal phase*, the study of which, thus, becomes crucial also for the understanding of the fireball model parameters.

The above and the following model share the uncertainty ("flexibility") about details of the *hadronization* from the parton phase. According to folklore, there is essentially *no entropy production* during the (non-equilibrium) phase transition. This is important for the interpretation, e.g. in a thermal model, of

$$\frac{dS}{dy} = c_{qg} \frac{dN_{qg}}{dy}\big|_{b=0} \propto \frac{dN_\pi}{dy}\big|_{b=0} , \qquad (11)$$

which is a typical relation between the entropy density calculated in a partonic model, cf. below, and the observed hadron multiplicity.

The study of the early space-time evolution was presented by **K. Geiger**[15] employing a *parton cascade* approach to simulate quark and gluon transport during hadronic or nuclear collisions. This probabilistic scheme is based on state-of-the-art *perturbative QCD* cross sections, which are employed similarly as in simulations of Boltzmann equations including collision terms. Partons are sampled from measured *structure functions* and *propagated classically* in accordance with Altarelli-Parisi type equations. A justification of this procedure for multiple scatterings in extended dense systems, i.e. multiple inter-cascade interactions, within the parton model seems rather difficult, since it goes beyond proven factorization theorems.

Here the total *entropy* arises from three contributions,

$$S = S_{primary} + S_{secondary} + S_{hadronization} . \qquad (12)$$

The *primary contribution*, which amounts to about 40% of the total, stems from the *decoherence process* which sets in once the incoming hadronic wave functions are perturbed by initial state QCD interactions. This is the dynamical origin of the structure functions, which has recently been addressed in Refs.[3,8,11] Another way of stating this is by recalling the definition of structure functions as being related to *inclusive cross sections*; according to our discussion following Eq. (1), there must be an associated entropy. At present, this contribution *cannot* be calculated ab initio or in any quantitative model.

The *secondary contribution*, which accounts for practically all of the rest of the produced entropy, is due to the *production of secondary partons* in elementary bremsstrahlung or scattering processes. This is essentially analogous to what happens in any kind of molecular dynamics simulation, namely a covering of the available classical single-particle phase space via scattering. Finally, the *hadronization contribution* is arguably considered to be small and taken into account by hadronization prescriptions based on universal parton-hadron duality and fitted, for example, to e^+e^--data.

The most interesting result in the present context is the *rapid saturation of entropy production* (together with a thermalization of parton spectra) on a time scale of .5 fm/c or less, with a value of the specific entropy per particle, $S/N \approx 4$, which is more or less the ideal parton gas value.[15]

The *Schwinger mechanism*, i.e. e^+e^--creation in a time-independent homogeneous electric field, was studied by **J. Rau**[4] w.r.t. entropy production and irreversibility. The main result is that the *"relevant" entropy* defined here in terms of the single-particle ($e^\pm$) occupation numbers,

$$S_{rel}(t) = \sum_{all\ modes} \left\{ \tfrac{1}{2}n_- \ln \tfrac{1}{2}n_- + (1 - \tfrac{1}{2}n_-)\ln(1 - \tfrac{1}{2}n_-) + [\, n_- \leftrightarrow n_+ \,] \right\} , \qquad (13)$$

tends to increase. However, there are two essential time scales for the process: the *memory time*, $\tau_{mem} \approx (\hbar/m) + (m/qE)$, and the *production time*, $\tau_{prod} \approx (m/qE)\exp(\pi m^2/2\hbar qE)$. Depending on their relative size the process is essentially Markovian and *irreversible* (weak fields) leading to monotonically increasing S_{rel} or else it shows important *memory effects* (strong fields) leading to oscillations of S_{rel} on the scale of $\tau_{mem} \approx \hbar/m$.[4] These effects are analogous to what happens with the Boltzmann equation depending on the relative size of the time between collisions ("τ_{mem}") and the duration of a single collision ("τ_{prod}").

It seems important to realize how the "relevant" entropy here fits into our above discussion of coarse graining, inclusive variables, and open systems with their environments, respectively. Clearly, the entropy S_{rel} determined by occupation numbers is *relevant* w.r.t. experiments measuring single-particle observables. However, it corresponds to a *chosen cut in the space of observables* of the system. Thus, one deliberately discards information in an inclusive way, e.g. about relative phases of outgoing single-particle waves or higher order correlations (n-point functions with $n > 2$), which amounts to a coarse graining and results in information entropy as before. In distinction, the considerations in Refs.[3,6,8,10] are based on the observation that in some systems or theories (e.g. QCD) there is a *dynamical cut in the space of fundamental modes* of the system, which naturally separates it into an "observable" subsystem and its environment. In this case, the coarse graining is dictated by the complex system itself. Then, the *von Neumann entropy* related to all information available about the subsystem is *not* a constant of motion and is the relevant entropy. An *additional* coarse graining, such as a restriction to inclusive single-particle observables, may still be necessary for practical purposes.

The contribution by **H.-Th. Elze**[3] provides a simple introduction to the mechanism of *environment-induced quantum decoherence* (cf. the above discussion of R. Omnès' presentation) with a view towards strong interactions. In QCD a separation of non-perturbatively interacting almost constant field configurations, which can neither hadronize nor initiate hard scatterings, from the usual high-energy or far off-shell partons seems essential to attack the strong-coupling problem underlying entropy production in multiparticle processes.

We mention two particular results. Employing the *Schmidt decomposition* of the complex system density matrix, see Sect. 2 of Ref.,[8] one finds that the von Neumann entropy for the subsystem always equals the one for its environment. Therefore, one may choose to *eliminate either the environment or the subsystem* degrees of freedom, whichever is simpler. Secondly, in the example of the inverted oscillator,[3] which is partially chaotic in the classical limit, one observes an *exponentially growing entropy production*, which is governed essentially by the classical Lyapunov exponent. Here, the decoherence is induced by the coupling to the *vacuum fluctuations* of only one environment oscillator. Thus, under suitable conditions an extremely simple zero-temperature environment is sufficient to cause entropy production in the subsystem, which might be relevant in the following.

The work on *chaos and entropy production in classical Yang-Mills fields* reported by **B. Müller**[9] addresses the question of entropy production as being connected intimately to the problem of thermalization in strongly interacting systems. One studies the chaotic time evolution of classical Yang-Mills fields employing the lattice gauge theory discretization for the Hamiltonian equations of motion. Thus, it is shown that a random *ensemble of initial field configurations self-thermalizes* rapidly, i.e. the probability distribution of the magnetic plaquette energy evolves into an exponential Boltzmann distribution. Furthermore, the maximal Lyapunov exponent of the time-dependent classical system is demonstrated to yield the *damping rate of coloured collective (plasmon) excitations* $\propto g^2 T$, which is calculated otherwise by finite-temperature QCD perturbation theory ($T \gg T_c$). Its inverse yields a *thermalization time* which rapidly decreases with increasing T from $\tau_S^0 \approx .5$fm/c at $T \approx 200$MeV. Employing the complete Lyapunov spectrum, which presumably corresponds to including other unstable collective excitations at finite T, an even shorter thermalization time can be deduced from the rate of entropy growth, $\tau_S = \bar{S}_{equil}/\partial_t \bar{S}$. Herein, the relevant entropy is the *Kolmogorov-Sinai entropy* $\bar{S}$, which arises by a coarse graining of the classical phase space.[9]

Two related points seem to deserve further study in order to fully understand these remarkable results. Firstly, where does the ensemble of initial field configurations come from? Following Refs.,[3,8] one is led to conjecture that the *environment of high-energy or far off-shell partons* and integrating out these ultraviolet degrees of freedom, respectively, result in the effectively *classical* initial conditions above. Secondly, are this strongly coupled Yang-Mills system under consideration and its evolution stable w.r.t. the ultraviolet quantum fluctuations? Here, *asymptotic freedom* may help to keep such stability, which is necessary in order to relate this approach to actual hadronic or nuclear collisions.

In conclusion, we hope to have raised or rephrased some interesting questions to stimulate further research on entropy and thermalization, particularly in strong interactions. We thank all participants of the miniworkshop for sending copies of their presentations and, especially, J. Rafelski for the intellectual and organizational support *sine qua non*.

REFERENCES

1. P. A. Carruthers, in Proc. NASI "Hot and Dense Nuclear Matter", Bodrum (Turkey) 1993, to be published by Plenum Press.

2. R. Landauer, Phys. Today 45, No. 5 (1991) 23;
   C. J. Chaitan, "Information, Randomness and Incompleteness", World Scientific, Singapore, 1987.

3. H.-Th. Elze, contribution to these Proceedings.

4. J. Rau, contribution to these Proceedings.

5. M. C. Mackey, "Time's Arrow: The Origins of Thermodynamic Behaviour", Springer Verlag, New York, 1992.

6. J. Ellis, N. E. Mavromatos and D. V. Nanopoulos, Phys. Lett. B293 (1992) 37; preprint CERN-TH.7195/94; and references therein;
   M. Gell-Mann and J. B. Hartle, Phys. Rev. D47 (1993) 3345.

7. P. A. Carruthers, Int. J. Mod. Phys. A4 (1989) 5587.

8. H.-Th. Elze, preprint CERN-TH.7131/93 (hep-ph/9404215), to appear in Nucl. Phys. B.

9. B. Müller, contribution to these Proceedings.

10. W. H. Zurek, Phys. Today 44, No. 10 (1991) 36;
    R. Omnès, Rev. Mod. Phys. 64 (1992) 339.

11. H.-Th. Elze, preprint CERN-TH.7297/94 (hep-th/9406085), submitted to Phys. Rev. Lett.

12. M. Danos, private communication.

13. M. Danos, NIST Technical Note 1403 (1993).

14. R. Weiner, private communication.

15. K. Geiger, contribution to these Proceedings.

16. J. Letessier, contribution to these Proceedings.

17. A. Tounsi, contribution to these Proceedings.

18. U. Heinz, contribution to these Proceedings.

NEW DEVELOPMENTS IN CORRELATION STUDIES

B. Buschbeck, P. Lipa and F. Mandl

Institut für Hochenergiephysik der
Österr. Akademie der Wissenschaften
Nikolsdorferg. 18, A-1050 Wien

ABSTRACT

By using the recently developed technique of correlation integrals we measured correlation functions over a wide range of invariant mass 50 GeV $\geq M \geq$ 0.2809 GeV in $\bar{p}p$ reactions at collider energy. A comparison with Monte Carlo models shows that our understanding of the dynamics of multiparticle production is still insufficient. We discuss a possible improvement by including low p_T clustering effects in addition to those of Lund-strings alone.

INTRODUCTION

In principle, the correlation functions both of lower and higher orders provide valuable information on the multiparticle system. Because of their high dimensionality in high energy reactions, they are difficult to measure. However, since Bialas and Peschanski in 1986 have proposed to study the dependence of the factorial moments (FM) on the magnitude of phase space bins,[1] correlation studies have developed rapidly[2] and new tools, the correlation integrals,[3-5] are available now.

The crucial steps leading to the correlation integrals have been:

1. to choose differences of phase space variables between particles (e.g. $\delta y = y_1 - y_2$) and to integrate over all distinct locations in phase space; therefore, new variables like the four momentum transfer $Q_{12}^2 = -(p_1 - p_2)$ and the invariant mass $M = \sqrt{Q_{12}^2 + 4m^2}$ can be analysed,

2. to find the proper normalization* by justifying and quantifying mathematically the event mixing technique,[4]

3. to extend the event mixing technique to the numerator for estimating the cumulant correlation functions.

---

*Q^2 has been used frequently in the past, e.g. when measuring Bose-Einstein correlations, however, the normalization included always some arbitrariness and has been an unsolved problem. An early work of E. Berger et al.[6] should be mentioned here as an exception.

Now we are in the position to apply these tools to the data, to see whether they can help to get more sensitive information on the dynamical origin of multiparticle correlations. This is the aim of the present contribution[†].

DATA, DEFINITIONS AND GENERAL FEATURES

The data sample consists of 160.000 non-single diffractive $\bar{p}p$ reactions at $\sqrt{s} = 630\,\text{GeV}$, measured in the UA1 detector[‡]. In one distinct case, e^+e^- reactions at $\sqrt{s} = 90\,\text{GeV}$ from the DELPHI experiment[§] are presented for comparison. We will investigate in the following the differential density correlation function $r_2(M)$ and the moments $F_i(M)$ (i = 2-5) depending on the invariant mass M over a wide range $50 \geq M \geq 0.2809$.

$$r_2(M) = \frac{\rho_2(M)}{\rho_1 \otimes \rho_1(M)} \tag{1}$$

$$\rho_2(M) = \int_\Omega d^3p_1 \cdot d^3p_2 \cdot \rho_2\left(\vec{p}_1, \vec{p}_2\right) \delta\left(M - m\left(\vec{p}_1, \vec{p}_2\right)\right)$$

$$\rho_1 \otimes \rho_1(M) = \int_\Omega d^3p_1 \cdot d^3p_2 \cdot \rho_1\left(\vec{p}_1\right) \cdot \rho_1\left(\vec{p}_2\right) \delta\left(M - m\left(\vec{p}_1, \vec{p}_2\right)\right)$$

$$\rho_2\left(\vec{p}_1, \vec{p}_2\right) = \frac{1}{\sigma_I} \cdot \frac{d^6\sigma_{\text{incl}}}{d\,\vec{p}_1 \cdot d\,\vec{p}_2}$$

$$\rho_1\left(\vec{p}\right) = \frac{1}{\sigma_I} \frac{d^3\sigma_{\text{incl}}}{d\,\vec{p}}$$

$$m\left(\vec{p}_1, \vec{p}_2\right) = \sqrt{(E_1 + E_2)^2 - (\vec{p}_1 + \vec{p}_2)^2}$$

with $\vec{p}$ being the 3-momenta and E the corresponding energies with pion mass, σ_{incl} and σ_I the inclusive and event cross sections. The integration region Ω in our case is the pseudorapidity interval $-3 \leq \eta \leq 3$ and $p_T > 0.15$ GeV/c[¶].

The numerator and the denominator of eqn.(1) are shown in fig.1a. The ratio is shown in fig.1b. Whereas both quantities in fig.1a show large variations (four orders of magnitude!) over the whole mass range which are due to phase space and the shape of single particle spectrum $\rho_1(\vec{p})$, in the ratio of fig.1b this has cancelled out and the structure of correlations shows up. Beside the differential representation in eqn.(1) we will use also the integral respresentation, the moments

$$F_2(M) = \frac{\int_0^M \rho_2(M')dM'}{\int_0^M \rho_1 \otimes \rho_1(M')dM'} \tag{2}$$

The normalization condition

$$\int_0^{M_{\max}} \rho_2(M)dM = <n(n-1)> \quad \text{in} \quad \Omega \tag{3}$$

$$\int_0^{M_{\max}} \rho_1 \otimes \rho_1(M)dM = <n>^2 \quad \text{in} \quad \Omega$$

[†]Other recent applications of correlation integrals are summarized in[2]
[‡]for details of data selection see[7]
[§]for details of data selection see[8]
[¶]In case of exception, this will be indicated in the text or figure caption.

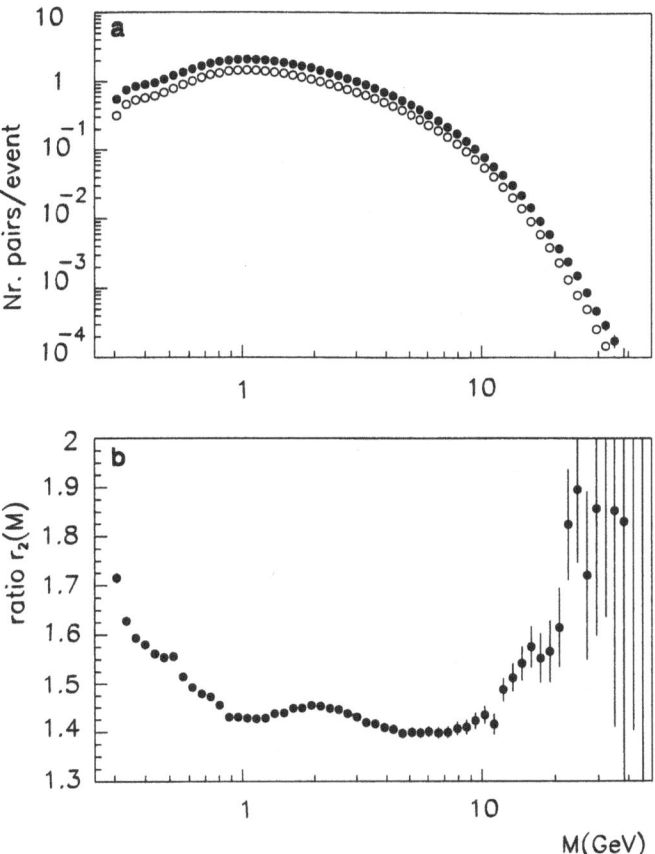

Figure 1. a) The numerator (number of pairs measured, black circles) and denominator (number of uncorrelated pairs, open circles) of eqn.(1). Only the azimuthal angle region of good acceptance is shown here, to avoid problems, which will not occure in the ratio later.
b) The corresponding ratio, eqn. (1).

establishes the connection with the scaled second order factorial moment of the multiplicity distribution in Ω :

$$F_2(M_{\max}) = \frac{<n(n-1)>}{<n>^2} \quad \text{in} \quad \Omega$$

$M_{\max}$ is the highest kinematically possible mass and n is the event multiplicity in Ω.
Higher order moments are defined similarly:

$$
\begin{aligned}
F_i(M) &= \frac{\int_0^M dM \int d^3p_1 \ldots d^3p_i \cdot \rho_i(\vec{p}_1 \ldots \vec{p}_i)\delta(M - m_i(\vec{p}_1 \ldots \vec{p}_i))}{\int_0^M dM \int d^3p_1 \ldots d^3p_i \rho_1(\vec{p}_1) \ldots \rho_1(\vec{p}_i) \cdot \delta(M - m_i(\vec{p}_1 \ldots \vec{p}_i))} \quad (4) \\
F_i(M_{\max}) &= \frac{<n(n-1)\ldots(n-i+1)>}{<n>^i} \quad \text{in} \quad \Omega
\end{aligned}
$$

to define m_i, we used the Grassberger counting convention.[9]

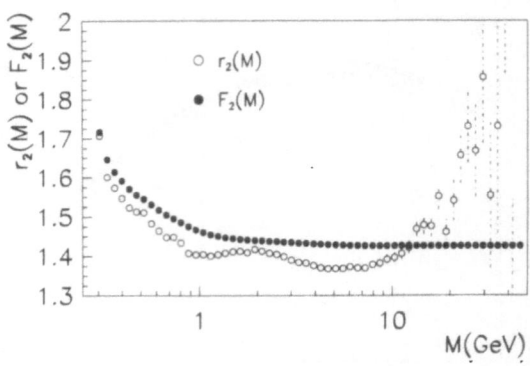

Figure 2. Differential representation (open circles, eqn. (1)) and integral representation (black circles, eqn. (2)) of the second order correlation integrals.

In figs. 1-3 we can read off the whole story of long- and short-range correlations and their respective contributions to the multiplicity of distribution in Ω.

1. The short-range correlations[10] show up as rise of $r_2(M)$ with decreasing M in the region $M < 1\text{GeV}$ (fig.1b). We evaluated from fig.1a that the number of pairs responsible for this rise above the pedestrial value of $r_2 \approx 1.4$ contribute less then 2% to the value of $F_2(M_{\text{max}})$ which is reached already for $M = 50$ GeV in fig.2. Similar considerations hold for the higher moments shown in fig.3.

2. In fig2 $F_2(M)$ decreases for small $M \lesssim 1$ GeV but levels off at larger M because of the presence of long-range correlations. This is seen also in $r_2(M)$, which remains > 1 in the whole range of M ($k_2 = r_2 - 1 > 0$, positive genuine correlations).

3. The values $F_i(M_{\text{max}})$, already reached at $M \approx 50\,\text{GeV}$ in fig.3, represent the multiplicity distribution in Ω. The large values at higher orders i indicate a broad Negative Binomial[||] distribution. From the observations of item 1 we argue that multiplicity distributions in large phase space bins (like Ω) are almost due to long-range correlations only.

In contrast, the F_i of e^+e^- reactions at LEP energy (open circles in fig.3) do show small values at large M which is the signature of an (approximate) Poissonic[**] multiplicity distribution in Ω. Again, the large short-range correlations do contribute only with a small amount to the overall multiplicity distribution and in this case, the long-range correlations are almost absent.

Do we understand these features? Since in e^+e^- reactions they can be reproduced even quantitatively by Monte Carlo models with QCD cascading, we will consider in the following only $\bar{p}p$ reactions.

[||] The factorial moments of the Neg. Bin. distribution have been evaluated already early by A. Giovannini.[11] They provide an equivalent and alternative parametrization of the multiplicity distribution (equivalence theorem[12]).

[**] In case of full phase space, not shown here, the F_i values of e^+e^- are even smaller and indeed near unity.

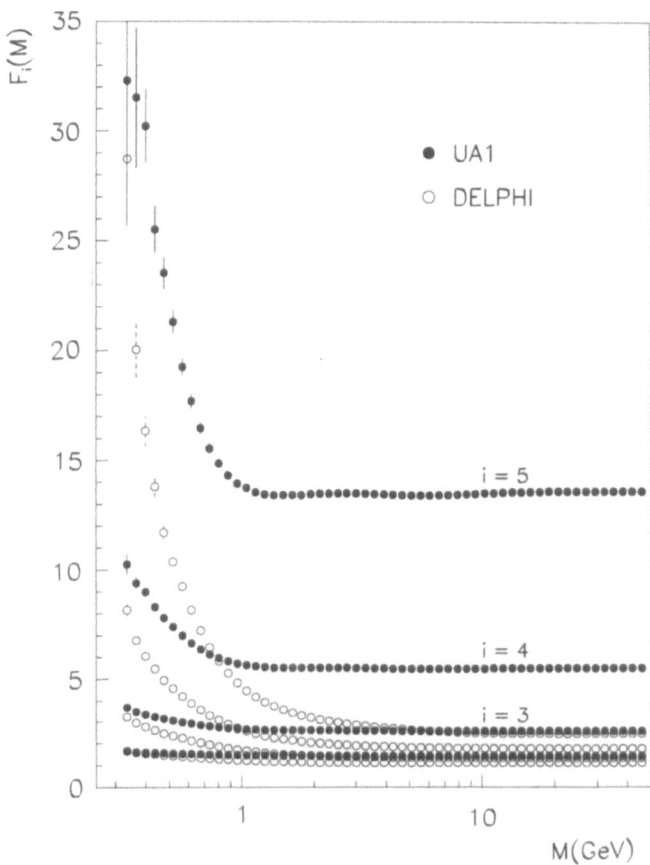

$F_i(M)$

35

30

25

20

15

10

5

0

● UA1
○ DELPHI

i = 5

i = 4

i = 3

1   10

M(GeV)

Figure 3. A comparison of correlation integrals $F_i(M)(i = 2 - 5)$ of $\bar{p}p$ reactions (black circles) with e^+e^- reactions (open circles).

HOW WELL DO WE UNDERSTAND THE DYNAMICAL ORIGIN OF CORRELATIONS IN hh REACTIONS ?

A very detailed physical picture is contained in the Lund Monte Carlo program PYTHIA.[13] A high energy collision is viewed as a collision of two parton beams. Each reaction is initialised by a hard-to-semihard parton-parton scattering with a cross section falling rapidly towards higher partonic transverse momentum p_T according to QCD. The description of inclusive event properties restricts the cut-off parameter $p_{T_{min}}$ to 1.5-2 GeV/c. The broad multiplicity distribution is obtained by varying the number of initial parton scatterings per event from one event to another (varying centrality of the reactions together with Poissonian fluctuations). Both, the mean multiplicity and the shape of the multiplicity distribution depend on $p_{T_{min}}$. Initial and final state radiation is created. Each parton configuration hadronises via Lund-strings spanned between the hard-scattered partons and beam remnants. PYTHIA thus contains jet and resonance production. Bose-Einstein correlations have been added as "final state interactions". Fig.4 shows a comparison of $r_2(M)$ with PYTHIA 5.6 with two selections of $p_{T_{min}}$. We observe 1) an overestimation of correlations with large mass which can be attributed to pairs of high p_T particles, 2) no reproduction of the structure at

2 GeV*, and 3) an underestimation of correlations at small mass. A similar result: overestimation of correlations at high p_T and high multiplicity, and underestimation at low p_T/low multiplicity by PYTHIA has been observed also in the multiplicity and p_T dependence of short-range correlations.[14]

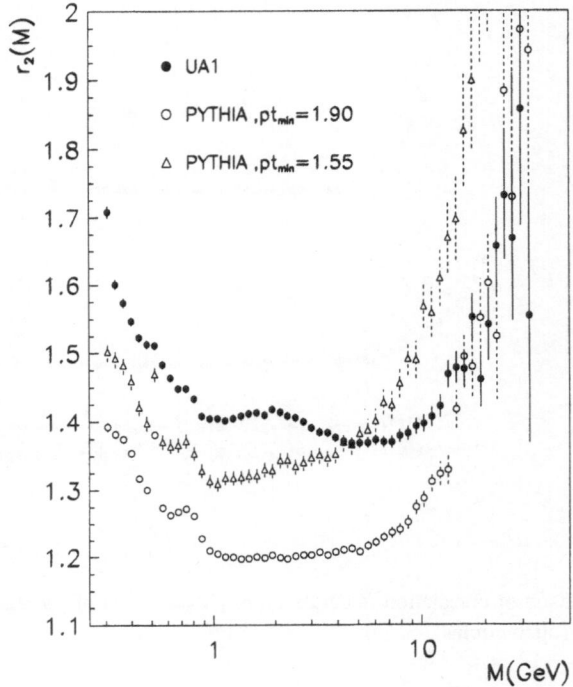

Figure 4. Comparison of $r_2(M)$ in $\bar{p}p$ reactions with PYTHIA 5.6.

How can we obtain a better agreement? Are there additional cluster effects? Convincing evidence for the existence of fireballs has been reviewed by R. Hagedorn at this conference.[15] Does the Lund-string alone really provide an alternative description, or do we have to merge the two pictures: hard scattering processes with subsequent parton showering and string fragmentation on one side and the production of low p_T clusters or fireballs on the other side ?

A first impression, what clustering effects may achieve, can be gained with the UA5 cluster model[†] GENCL.[16] It turns out, that it underestimates high p_T effects as expected and

---

*This structure is likely to be connected with the unknown transition region hard to soft. We verified that it is due to transverse momentum compensation between hadron pairs of relatively low $p_T \approx 0.5$ GeV/c.

†GENCL does by far not contain a detailed physical picture like PYTHIA. Most observations, like the overall multiplicity distribution are put in "by hand" only. However, clustering effects are created carefully and tuned to observed rapidity correlations. A derivation of multiplicity distributions in the framework of the Statistical

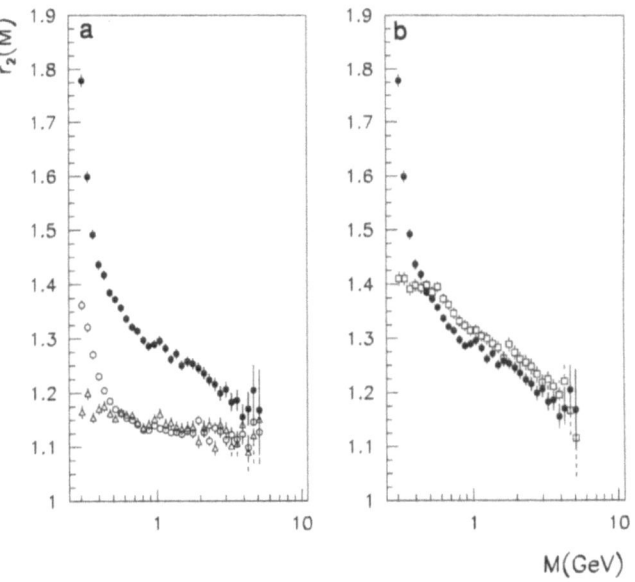

Acknowledgements

We gratefully acknowledge the helpful discussions with T. Sjöstrand.

REFERENCES

1. A. Bialas and R. Peschanski, *Nucl. Phys.* B273 91986) 70; *Nucl. Phys.* B308 (1988) 857.

2. for a review see "Scaling Laws for density correlations and fluctuations in multiparticle dynamics", E.A. De Wolf, L.M. Dremin, W. Kittel, HEN-362 (1993), to appear in *Phys. Rep.*; Some recent results are also given in "Density and Correlation Integrals in Deep Inelastic Muon-Nucleon Scattering at 490 GeV" (Fermilab E665-collab.) MPI-PhE/94-12 and N.M.Agababyan et al. (NA22/EHS-collab.), *Phys. Lett.* B328 (1994) 199, P. Abreu et al. (DELPHI-collab.), *Z. Phys.* C63 (1994) 17.

3. P. Lipa et al., *Phys. Lett.* B285 (1992) 300.

4. H.C. Eggers et al., *Phys. Lett.* B301 (1993) 298.

5. H.C. Eggers et al., *Phys. Rev.* D48 (1993) 2040.

6. E.L. Berger et al., *Phys. Rev.* D15 (1977) 206.

7. N. Neumeister et al. (UA1-collab.), *Z. Phys.* C60 (1993) 633.

8. F. Mandl and B. Buschbeck, "Correlation integral studies in DELPHI and UA1", Proc. 22. Internat. Symp. on Multiparticle Dynamics, Santiago de Compostela, Spain, 1992, Ed. C. Pajares (World Scientific).

9. P. Grassberger, *Nucl. Phys.* B120 (1977) 231.

10. N. Neumeister et al., (UA1-collab.), *Z. Phys.* C60 (1993) 633.

11. A. Giovannini, *Lett. Nuovo Cimento* 6 (1973) 514.

12. Z. Koba, Proc. of 1973 CERN-JNR School of Physics, CERN-Jellow Rep. 73-12; L. Diosi, *Nucl. Instr. Meth.* 138 (1976) 241.

13. T. Sjöstrand, *Comp. Phys. Com.* 82 (1994) 74.

14. C. Albajar et al. (UA1-collab.), *Nucl. Phys.* B345 (1990) 1; Y.F. Wu et al. (UA1-collab.), *Acta Physica Slovaca* V44 (1994) 141.

15. R. Hagedorn, invited lecture, this conference.

16. G.J. Alner et al. (UA5), *Nucl. Phys.* B291 (1987) 445.

17. G.J.H. Burgers, C. Fuglesang, R. Hagedorn and V. Kuvshinov, *Z. Phys.* C46 (1990) 465.

18. P. Aurenche et al., "DTUJET-93. Sampling inelastic pp and $\bar{p}p$ collisions according to the two-component Dual Parton Model", SI-93-3.

HOW TO INVESTIGATE SMALL COLLECTIVE SIGNALS
IN NUCLEUS-NUCLEUS INTERACTIONS

Ivo Derado

Max-Planck-Institut für Physik
80805 München, Germany

The study of correlations amongst secondary particles produced in high-energy collisions has revealed various interesting features of the productions mechanisms which go beyond the information obtained from single particle distributions. In particular, the discovery, in nucleus-nucleus interactions, of rare multiparticle events with large non-statistical fluctuations of the local particle density in phase-space[1] has led to introducing into particle physics the concept of intermittency[2] which provides an additional new method of analysing correlations in such multiparticle events. Here we consider a somewhat broader definition of "intermittency" as any increase of the factorial moments of multiplicity distributions with decreasing of the phase-space region investigated. More recently, considerable improvement has been achieved in the method of analysis by using so-called density integrals and correlation integrals.[3] The method is similar to the method applied in studies of the Bose-Einstein correlations.[4] The ratio of the numbers of all possible ordered q-tuples of particles $(n(n-1)\cdots(n-q+1))$ in some part of the space for each event divided by the number of q-tuples of particles for mixed events. The normalization is such that for full space e.g. $q = 2$ is: $F_2 = \frac{\langle n(n-1)\rangle}{\langle n\rangle^2}$ where $\langle n\rangle$ and $\langle n(n-1)\rangle$ are moments averaged over all events. In the investigation of the fluctuations in O–Au at 200 GeV/nucleon[5] we have shown that this analysis of F_2 moments was sensitive to the addition of 0.07 percent of correlated tracks in the event.

In this work, our interest is to study the possible new coherent phenomena in very high energy nucleus-nucleus interactions. In[5] we have shown that the intermittency analysis shows a very small signal in nucleus-nucleus interactions compared with the signal in deep-inelastic processes.[6] Therefore any strong "intermittency" signal could be an indication for a possible new collective phenomenon.

At present, a theoretical technique for handling coherent effects is far from being developed. On the experimental side there has been very little direct investigation. Some studies on shadowing and cascading were not very informative so far. The total heavy-ion cross section is of the order of many barns so even extremely rare events could be observed if one has a very sensitive selection from the "chaos". The highly relativistic nuclei at RHIC have the advantage of huge multiplicity events, so that we can regard each collision as an experiment, similar to the observation of a stellar object. Cosmic ray experiments[7] have observed some peculiar phenomena: centauro event, clustering of only charged or only γ

particles and low $p_\perp$ ($< 100\,\mathrm{MeV/c}$) chiron events. So far there are no convincing explanations of these effects. Many people strongly believe that this puzzle could be connected with coherent effects in vacuum. The vacuum which, Lorentz invariant, could be a coherent mixture of states of different quantum numbers and so the vacuum expectation value of any quantum number carrying spin-0 field could be different from zero. In high multiplicity events it seems to be natural to ask if some of the observed pions are produced by the coherent decay of a semiclassical pion field with non vanishing pion vacuum expectation value. This is not a normal vacuum and is called in literature[8] "Disoriented Chiral Condensate" (DCC). DCC could be responsible for these cosmic ray observations.

The central problem is: can DCC be produced, and if produced, how can it be detected, and how can we analyse its properties. In order to test our procedure to detect DCC we follow the proposed model "BAKED ALASKA".[9] They speculate that in the interiors of large "fireballs" produced in very high energy collisions, vacuum states of the pions are produced with anomalous chiral order parameters. That leads to anomalously large fluctuations in the charged-to-neutral ratio of produced hadrons. In addition, these coherent pions are emitted from a large volume, implying low $p_\perp$ of pions.

The experimental strategies would be to select the events with charged multiplicities much larger than the average ones and with a somewhat smaller $p_\perp$, and a small average number of gammas. After this selection one should apply the intermittency analysis inside the restricted $p_\perp$, Φ and y (transverse momentum, azimuthal angle around collisions line and rapidity), and select the events which contribute to the increase of F_q in the region of very small phase space. These events are candidates for DCC, and one can then investigate the structure of these events in comparison with "normal events".

In order to test our method we have constructed a two-component Monte Carlo model.

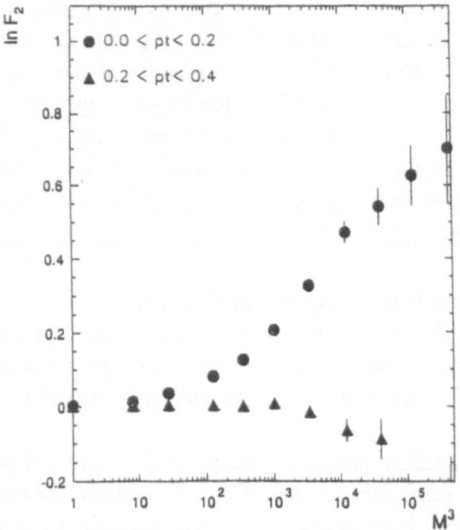

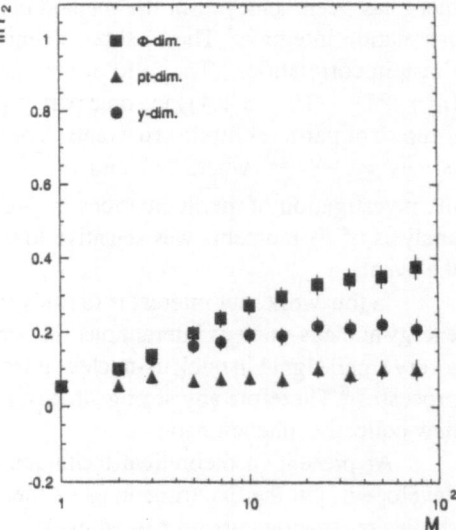

Figure 1. Log-log plot of the normalized factorial moment F_2 vs. the number of divisions (M) in a one-dimensional analysis in Φ, $p_\perp$ and y variables of total phase space (large M is small phase-space region).

Figure 2. Log-log plot of the normalized factorial moment F_2 vs. the M^3 in three dimensions with the cuts $p_\perp < 0.2$ GeV/c and $0.2 < p_\perp < 0.4$ GeV/c.

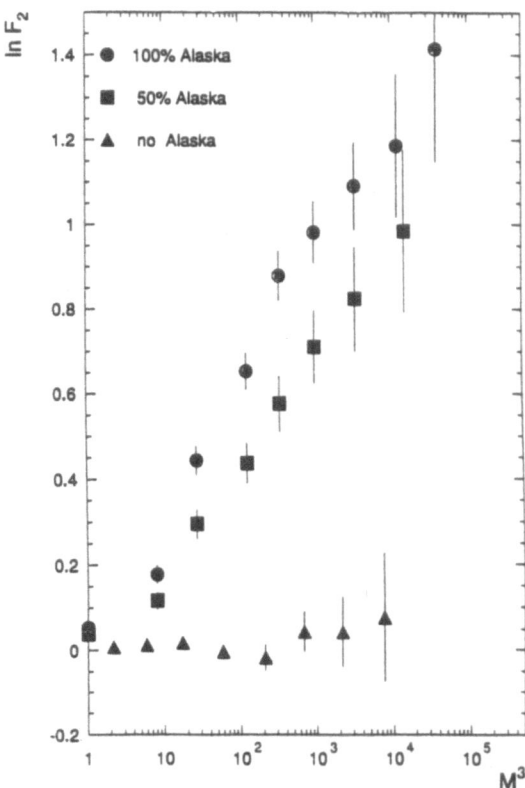

Figure 3. Log-log plot of the normalized factorial moment F_2 vs. the M^3 in three dimensions with the cuts $p_\perp$ 0.2 GeV/c and $y < 0.1$ and different amounts of "Baked Alaska" contributions.

One component is "noncoherent", which is just the standard Fritiof model, the other component is coherent, for which we used the DCC model "Baked Alaska".[9] The noncoherent part consisted of 600 charged particles. The coherent DCC produced in baryon-free region, decays with Poisson distributed pion multiplicities with $\langle n \rangle = 10$, and in its own CMS have the isotropic rapidity distribution and exponential $p_\perp$ distribution with $\langle p_\perp \rangle \sim 0.1$ GeV/c. The DCC was randomly boosted with respect to the cms of the noncoherent part, giving small "microjet structure" in the noncoherent cms.

Fig. 1 shows the F_2 dependence on the number of divisions (M) of total phase space (large M is small phase-space region). In this figure the F_2 is a one-dimensional analysis in Φ, $p_\perp$, y with the requirements that $y < 1$ and $p_\perp < 0.2$ GeV. The increase of F_2 is evident. Fig. 2 shows a three-dimensional analysis with $p_\perp < 0.2$ and for comparison $0.2 < p_\perp < 0.4$. The Baked Alaska effect is very strong. Fig. 3 shows the three-dimensional analysis with $p_\perp$ and y cuts with overall different amounts of Baked Alaska contributions. In the restricted phase-space region the contribution of Baked Alaska is smaller than 0.04 percent. As one can see, it is very easy to select events coming from DCC.

In conclusion, using adequate off-line selection of events and appropriate track cuts in the intermittency analysis, one can observe in nucleus-nucleus interactions any very small collective effects.

ACKNOWLEDGEMENTS

I thank Dr. Krešo Kadija, visitor from the Rudjer Bošković Institute, for adapting the Fritiof Monte Carlo program. I have benefited from the discussion with J.D. Bjorken and E. Seiler about the Baked Alaska Model.

REFERENCES

1. T.H. Burnett et al., Phys. Rev. Lett. **50** (1983) 1062

2. A. Bialas, R. Peschanski, Nucl. Phys. **B273** (1986) 703; **B308** (1988) 857

3. P. Lipa, P. Carruthers, H.C. Eggers, Phys. Lett. **B285** (1992) 300; H.C. Eggers, P. Lipa, Proc. Cracow Workship on Multiparticle Production, April 1993 (Regensburg preprint TPR-93-25)

4. see e.g. M.R. Adams et al., Phys. Lett. **B308** (1993) 418 and references

5. J. Bächler et al., Z. Phys. **C61** (1994) 551

6. I. Derado, G. Jancso, N. Schmitz, Z. Phys. **C56** (1992) 553

7. Mt. Chacaltaya emulsion exposure, BRAZIL-JAPAN Collaboration, Proceedings of the Plovdiv International Conference on Cosmic Ray, Plovdiv, 1977, ed. by B. Betev, Vol. 7, p. 208; JACEE Collaboration, Proceedings of 7th International Symposium on High Energy Cosmic Ray Interactions, Ann Arbor, 1992, ed. L. Jones

8. T.D. Lee and G.C. Wick, Phys. Rev. **D9** (1974) 2291
   J.D. Bjorken and L.D. McLerran, Phys. Rev. **D20** (1979) 2353
   S. Barshay, Phys. Lett. **B227** (1989) 279
   T.D. Lee, Talk given in Niels Bohr Institute 1981, CU-TP-226
   A.A. Anselm and M.G. Ryskin, Phys. Lett. **B266** (1991) 482
   J.P. Blaizot and A. Krzywicki, **D46** (1992) 246
   K. Rajagopal and F. Wilczek, Nucl. Phys. **B404** (1993) 577
   C. Greiner et al., Phys. Lett. **B316** (1993) 226
   R.D. Amado et al., Phys. Rev. Lett. **72** (1994) 960
   S. Gavin et al., Phys. Rev. Lett. (1994) 2143
   Z. Huang et al., Berkeley, LBL-35337, March 1994

9. J.D. Bjorken, K. L. Kowalski, C.C. Taylor, SLAC-PUB-6109 (1993)

NEGATIVE BINOMIAL FITS TO MULTIPLICITY DISTRIBUTIONS FROM CENTRAL COLLISIONS OF ^{16}O+Cu at 14.6A GeV/c AND INTERMITTENCY

M. J. Tannenbaum

Brookhaven National Laboratory
Upton, NY 11973 USA
for the E802/E859 Collaboration
ANL, BNL, UCBerkeley, UCRiverside, Columbia, Hiroshima
INS(Tokyo), Kyoto Kyushu, LLNL, MIT, NYU, Tokyo, Tsukbua

INTRODUCTION

The concept of "Intermittency" was introduced by Bialas and Peschanski[1] to try to explain the 'large' fluctuations of multiplicity in restricted intervals of rapidity or pseudo-rapidity.[2,3] A formalism was proposed[1] to to study non-statistical (more precisely, non-Poisson) fluctuations as a function of the size of rapidity interval, and it was further suggested[1] that the "spikes" in the rapidity fluctuations were evidence of fractal or intermittent behavior, in analogy to turbulence in fluid dynamics which is characterized by self-similar fluctuations at all scales—the absence of a well defined scale of length.[4] Bialas and Peschanski proposed that the data be presented as Normalized Factorial Moments of order q :

$$< F_q(\delta\eta) >= \frac{< n(n-1)\ldots(n-q+1) >}{< n >^q} \quad , \tag{1}$$

where n is the multiplicity in a pseudorapidity interval (bin) of size $\delta\eta$ on a given event and the $< >$ brackets indicate averaging over all events. Intermittency would be indicated by a power-law increase of multiplicity distribution moments over pseudorapidity bins as the bin size is reduced:

$$F_q(\delta\eta) \propto (\delta\eta)^{-\phi_q} \quad . \tag{2}$$

The Normalized Factorial Moment with the clearest interpretation is

$$F_2 = \frac{< n(n-1) >}{< n >^2} = \frac{< n^2 > - < n >}{< n >^2} = \frac{\sigma^2 + < n >^2 - < n >}{< n >^2} = 1 + \frac{\sigma^2}{\mu^2} - \frac{1}{\mu} \tag{3}$$

where $\mu \equiv < n >$ and $\sigma \equiv \sqrt{< n^2 > - < n >^2}$ is the standard deviation. Note that the Normalized Factorial Moments are all equal to unity for a Poisson Distribution.

The formulation of this new concept of "intermittency" in terms of moments was taken by many as an inelegant and confusing development, particularly since the greatest advance in multiplicity distributions in 20 years had recently been made by the UA5 Collaboration[5] who actually determined the functional form of multiplicity distributions. The Negative Binomial Distribution (which had been used sporadically for the total multiplicity[6]) was used by the UA5 collaboration[7] as a "remarkable" description of their measured multiplicity distributions in intervals of rapidity which are not significantly constrained by conservation laws,[8-11] and also for the total multiplicity. Also, a related distribution, the Gamma Distribution, had been used to describe E_T distributions.[12] **One could not help but wonder what "intermittent" behavior would look like in terms of distributions rather than moments—since once the distribution is known, then** *ALL* **the moments are known.**

An "intermittency" analysis of charged particle multiplicity data from the target multiplicity array (TMA) in central collisions of ^{16}O+Cu at 14.6 A·GeV/c has been published by the AGS-E802 collaboration.[13] The centrality cut was made using the Zero degree Calorimeter (ZCAL) and requiring that the forward energy be less than one projectile nucleon (i.e. $T_{ZCAL} < 13.6$ GeV). In agreement with previous measurements,[14] an apparent power-law growth of Normalized Factorial moments with decreasing pseudorapidity interval was observed in the range $1.0 \geq \delta\eta \geq 0.1$. In the present work, multiplicity distributions in individual pseudorapidity bins are presented for the same data. These distributions are excellently fit by Negative Binomial Distributions (NBD) in all $\delta\eta$ bins, allowing, for the first time, a systematic formulation of the subject of "intermittency" in terms of distributions, rather than moments.

NEGATIVE BINOMIAL DISTRIBUTION

The Negative Binomial Distribution of an integer m is defined as

$$P(m) = \frac{(m+k-1)!}{m!(k-1)!} \frac{(\frac{\mu}{k})^m}{(1+\frac{\mu}{k})^{m+k}} \tag{4}$$

where $P(m)$ is normalized for $0 \leq m \leq \infty$, $\mu \equiv <m>$, and some higher moments are:

$$\sigma = \sqrt{\mu(1+\frac{\mu}{k})} \qquad \frac{\sigma^2}{\mu^2} = \frac{1}{\mu} + \frac{1}{k} \qquad F_2 = 1 + \frac{1}{k} \tag{5}$$

The Normalized Factorial Moments and Cumulants $(K_q)^{15,16}$ of the NBD are particularly simple:

$$F_q = F_{(q-1)}(1+\frac{q-1}{k}) \qquad\qquad K_q = \frac{(q-1)!}{k^{q-1}} \tag{6}$$

The NBD, with an additional parameter k compared to a Poisson distribution, becomes Poisson in the limit $k \rightarrow \infty$ and Binomial for k equal to a negative integer (hence the name). The extra parameter has made the NBD useful to Mathematical Statisticians in the Likelihood Ratio Test for whether a distribution is Poisson—more precisely as a "test for independence in rare events."[17] The likelihood ratio test for a Poisson distribution consists of determining whether the NBD parameter $1/k$ is consistent with zero to within its error $s_{\frac{1}{k}}$, which is given[17] as:

$$s_{\frac{1}{k}} = \frac{s_k}{k^2} = \frac{1}{\mu}\sqrt{\frac{2}{N}} \tag{7}$$

where N is the total number of events. For statisticians, the NBD represents the first departure from a Poisson Law. Physicists are more likely to describe the NBD as Bose-Einstein ($k = 1$) or Generalized Bose Einstein $k \neq 1$ distributions.[6]

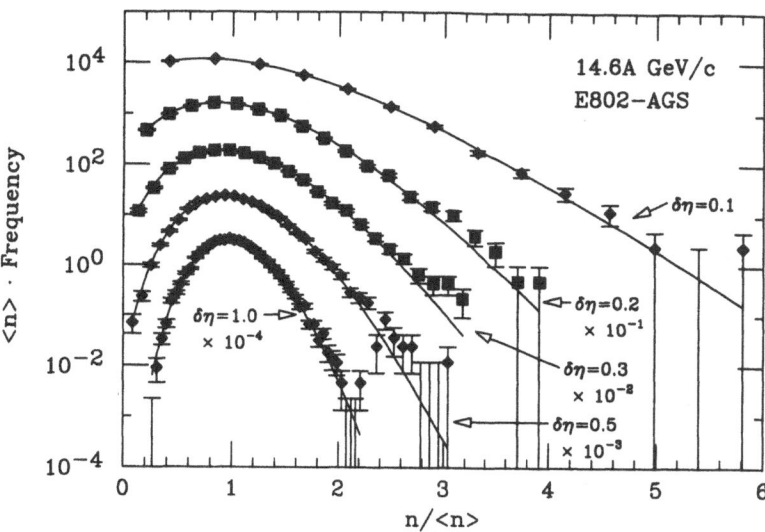

Figure 1. Multiplicity distributions for selected $\delta\eta$ intervals, 0.1,0.2,0.3,0.5,1.0, as indicated. The data for each interval are plotted scaled in multiplicity by $< n >$, the mean multiplicity in the interval. Each successive distribution has been normalized by the factor indicated, for clarity of presentation.

NEGATIVE BINOMIAL FITS TO THE DATA

The AGS-E802 multiplicity distributions for central ^{16}O+Cu in bins of $\delta\eta$= 0.1,0.2,0.3, 0.5,1.0, around a central value of $\eta = 1.7$ in the laboratory, are shown in Fig. 1. Note that $< dn/d\eta >$ is rather constant (2.5% rms variation) for these bins. The distributions for each bin all have approximately the same number of events, 19667. The solid lines on the data are NBD fits. The exact details of the centrality cut are important for Fig. 1, and presumably also for intermittency analyses by moments, since the rising (lower multiplicity) parts of the distributions are determined by the centrality cut. The excellence of the fits of the NBD to the rising as well as the falling parts of all the distributions is attributed to use of the ZCAL centrality cut, which is an indirect cut on multiplicity, rather than the sharp cut on multiplicity which is traditionally used to define centrality. Approximately 3.2% of minimum bias events would be expected to pass the centrality cut, thus the data in Fig. 1 correspond to $\sim$ 0.6 million interactions.

The k parameters of the NBD fits are plotted in Fig. 2 and show a totally unexpected and strikingly steep **linear** dependence on $\delta\eta$—k varies by more than a factor of 3 over 1 unit of $\delta\eta$. This is in sharp contrast to the UA5 results, where k is also linear with $\delta\eta$ but varies by only $\sim$ 10% over the same interval. The linear increase of the NBD parameter $k(\delta\eta)$ with $\delta\eta$ (and thus with $\mu = <n(\delta\eta) >$) indicates that the multiplicity in each $\delta\eta$ bin acts as if it were largely statistically independent of that in the next bin, since the near direct proportionality of the NBD parameter $k(\delta\eta)$ with $<n(\delta\eta) >$ means that the multiplicity distributions convolute as the bin size is extended. The effect of the clear non-zero intercept, $k(0) \neq 0$, is that the ratio $k(\delta\eta)/<n(\delta\eta) >$ does not remain strictly constant with increasing $\delta\eta$, as would be the case for full statistical independence. *The fact that the measured multiplicity distributions are excellently fit by distributions (NBD) with well known properties under convolution enables these observations to be made by inspection.*

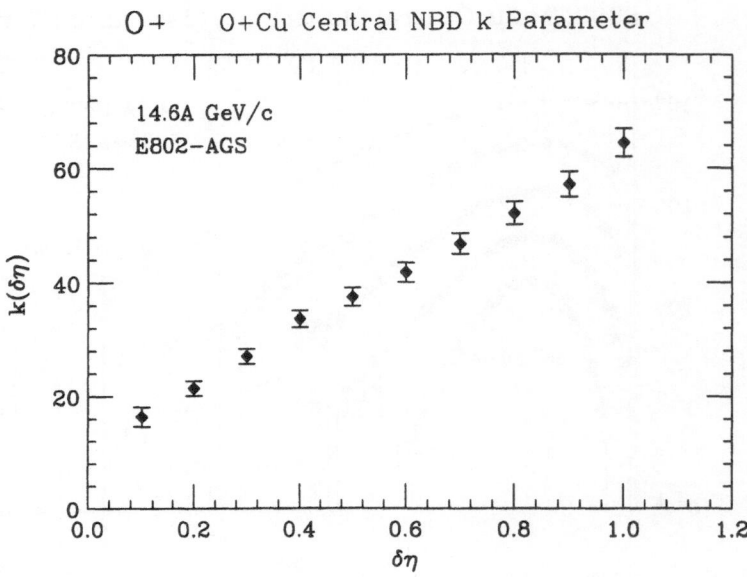

Figure 2. The $k(\delta\eta)$ parameter from NBD fits to the data as a function of the interval $\delta\eta$.

INTERMITTENCY EXPLAINED!

It is now possible to relate the directly measured evolution of the fluctuations of multiplicity with increasing pseudorapidity interval—as described in terms of the Negative Binomial distributions which excellently fit the measurements—to the Normalized Factorial Moment analysis of the same data. The striking linear evolution of the NBD parameter $k(\delta\eta)$ with the width of the interval, explains the observation of fractional power laws based on the "intermittency" formalism in a much more simple, elegant and understandable way. *The apparent power laws with fractional exponent are simply an artifact of using the quantity $F_2 - 1$, which is the inverse of a linear quantity $k(\delta\eta)$.* The "Intermittency" phenomenology, which looks for *self-similar fluctuations at all scales* $\delta\eta$ by a *fractional* power-law increase of bin-averaged Normalized Factorial Moments with decreasing bin size $\delta\eta$, obscures the real physics of multiplicity fluctuations which is given simply and elegantly by the linear evolution of $k(\delta\eta) \equiv 1/(F_2 - 1)$.

Furthermore, for all orders of the Normalized Factorial Moments measured in this experiment[13] the apparent fractional power-law increase with decreasing bin size $\delta\eta$ is entirely given by the Negative Binomial Distribution best fit curves, represented by the single parameter $k(\delta\eta)$—and has nothing to do with the deviations of the measured data points from the best fit curves! The Normalized Factorial moments of all orders can be obtained from the single parameter $k(\delta\eta)$ of the NBD fit (see Eq. 6), and compared point by point with the results of the moment analysis[13] up to 6th order (see Fig. 3). The low order moments agree to well within the statistical errors but there appears to be a small systematic discrepancy between the two methods, which increases slightly with increasing order. Part of the discrepancy may come from the slight difference in the actual data for the two methods and may therefore be real. It is also conceivable[18] that the NBD fits, which give excellent values for the low order moments with the the best statistics, may give smooth values for the high order moments, which miss the fluctuations of the data points at high multiplicity seemingly indicated in

Fig. 1.[19] Two comments are relevant on this possibility: the only visible fluctuations of the data from the curves occur for $n \geq 15$ (and therefore are relevant only for the 15th moment or higher); due to the excellence of the χ^2 of the fits, these fluctuations from the NBD best fit curves are constistent with *statistical fluctuations*. In any case, the slight differences in the results of the two method would not affect the fractal interpretation of either set of data points in Fig. 3 by a "true believer" in the factorial moment formalism of "intermittency".

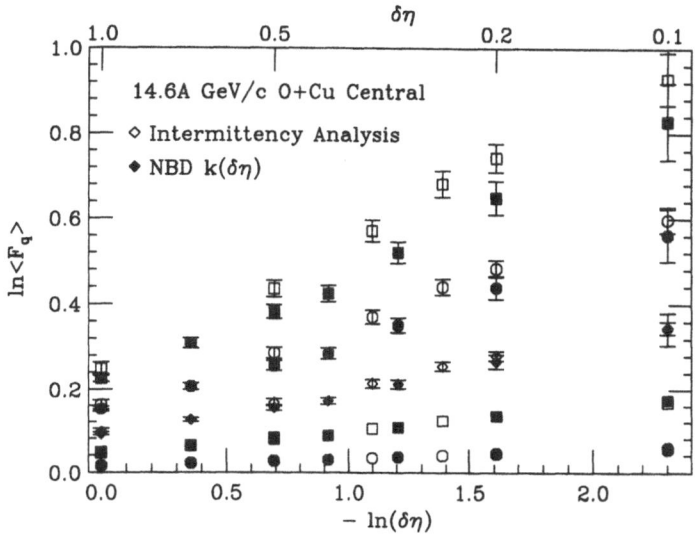

Figure 3. Normalized factorial moments F_q for central collisions of ^{16}O+Cu, for orders q=2,3,4,5,6, from the "intermittency" analysis[13] (open points) compared to the same quantities computed from the NBD parameter $k(\delta\eta)$ of Fig. 2 (solid points).

TWO-PARTICLE CORRELATIONS, THE NBD AND INTERMITTENCY

The importance of two-particle correlations to completely determine the multiplicity distribution was pointed out by Fowler and Weiner,[9] and more recently by Giovannini and Van Hove.[20] The application of two-particle short range correlations to the "intermittency" phenomenology was pioneered by Carruthers, Friedlander, Shih and Weiner,[21] Capella, Fialkowski and Krzywicki,[22] and Carruthers and Sarcevic.[23] The Reduced 2-particle Correlation is parameterized in an exponential form

$$R(y_1, y_2) = \frac{C_2(y_1, y_2)}{\rho_1(y_1)\rho_1(y_2)} = \frac{\rho_2(y_1, y_2)}{\rho_1(y_1)\rho_1(y_2)} - 1 = R(0,0)\, e^{-|y_1 - y_2|/\xi} \tag{8}$$

where $\rho_1(y)$ is the inclusive single particle density (assumed constant), $\rho_2(y_1, y_2)$ is the inclusive two-particle density, $C_2(y_1, y_2)$ is the Mueller[15] 2-particle correlation function, and ξ is the correlation length. Then, the integral can be performed on an interval of full width $\delta\eta$, $0 \leq y_1 \leq \delta\eta$, $0 \leq y_2 \leq \delta\eta$:

$$K_2 \equiv F_2 - 1 = \frac{\int^{\delta\eta} dy_1 dy_2\, C_2(y_1, y_2)}{<n(\delta\eta)>^2} = R(0,0)\,\frac{[1 - \frac{\xi}{\delta\eta}(1 - e^{-\delta\eta/\xi})]}{\delta\eta/2\xi} . \tag{9}$$

For a Negative Binomial Distribution, substitution of the identity $k \equiv 1/(F_2 - 1)$ into Eq. 9 yields the equation for the evolution of the NBD parameter $k(\delta\eta)$:

$$k(\delta\eta) = \frac{1}{F_2 - 1} = \frac{1}{R(0,0)} \frac{\delta\eta/2\xi}{[1 - \frac{\xi}{\delta\eta}(1 - e^{-\delta\eta/\xi})]} \quad . \tag{10}$$

Note that Giovannini and Van Hove[20] were the first give the relationship between the NBD k parameter and the integral of the 2-particle correlation function C_2, and a similar derivation was given by De Wolf.[24] If it is known (eg. from the data) that the multiplicity distribution is Negative Binomial, then the two particle correlation determines the entire distribution. Of course, independently of the distribution, Eq. 10 is valid for the evolution of $1/(F_2 - 1)$ with $\delta\eta$.

This formula, Eq. 10, gives a mathematical explanation of why the linear increase of k with $\delta\eta$ is an indication of the randomness of the multiplicity in adacent $\delta\eta$ bins, while the constancy of k with increasing $\delta\eta$ would be an indication of 100% correlation: in the limit $\delta\eta \ll \xi$, when the $\delta\eta$ interval is well inside the correlation length, $k(\delta\eta) = 1/R(0,0)$, a constant; in the limit $\delta\eta \gg \xi$, k is directly proportional to $\delta\eta$, $k(\delta\eta) \simeq \delta\eta/2\xi$, as expected from convolutions of independent bins.[21] The measured evolution of $k(\delta\eta)$, which appears to be strikingly linear, is equally well described by a fit to Eq. 10 (see Fig. 4) which indicates a weak correlation strength, $R(0,0) = 0.074 \pm 0.005$, and a very short rapidity correlation length, $\xi = 0.12 \pm 0.01$. It is important to note that these results are very sensitive to any short-range two-particle correlation generated by the detector, and in fact, the data of Figs. 1–4 which are uncorrected for instrumental effects, have a known instrumental short range correrelation[13] which constitutes about half the measured effect.

CORRECTION FOR INSTRUMENTAL EFFECTS

A short range correlation was inadvertently built into the Target Multiplicity Array (TMA) used for these measurements, which was constructed of resistive plastic tubes operated in the proportional mode and read out from image signals induced on cathode pads. The detector was composed of individual small panels which were slightly tilted to avoid inefficiency due to the walls of the tubes: the inefficiency was compensated by a small amount of cross-counting on adjacent pads for particles which cross from one wire to another across a tube wall—a built-in short range correlation. The effect of such cross-counting was studied extensively using Monte Carlo (MC) simulations, test beam data, and finally by comparing the measured rate of two-pad clusters on adjacent wires to that predicted by the MC which included all the physical (conversions, decays, multiple scattering) and geometrical effects. The rate on which pads on adjacent wires fired was $(7.45 \pm 0.11)\%$ in the data, compared to $(3.4 \pm 0.3)\%$ in the MC which is composed of a random effect of 2.3%, with only 1.1% from conversions and Dalitz pairs. (The tracks from conversions and Dalitz pairs generally both land on one pad or else the instrumental background from this effect would be much larger.) The difference of 4% was therefore added to the final Monte Carlo used to calculate the instrumental effects. Interestingly, the results of final MC for the instrumental effects, $F_2^{MC} - 1 \equiv K_2^I$, can be rather well represented by Eqs. 8, 9, with parameters $R^I(0,0) = 0.050 \pm 0.010$ and $\xi = 0.072 \pm 0.020$, which is, in fact, a reasonable mathematical description of the built-in short range correlation of the detector.

The NBD analysis is corrected for the instrumental effect by taking the measured two particle correlation $R(y_1, y_2)$ to be the sum of a true effect plus the instrumental effect:

$$R(y_1, y_2) = R^T(y_1, y_2) + R^I(y_1, y_2) \quad , \tag{11}$$

with the further assumption that the instrumental effect has minimal influence on the observed $< n(\delta\eta) >$. It then immediately follows from Eqs. 8–10 that the measured $K_2(\delta\eta) = 1/k(\delta\eta)$

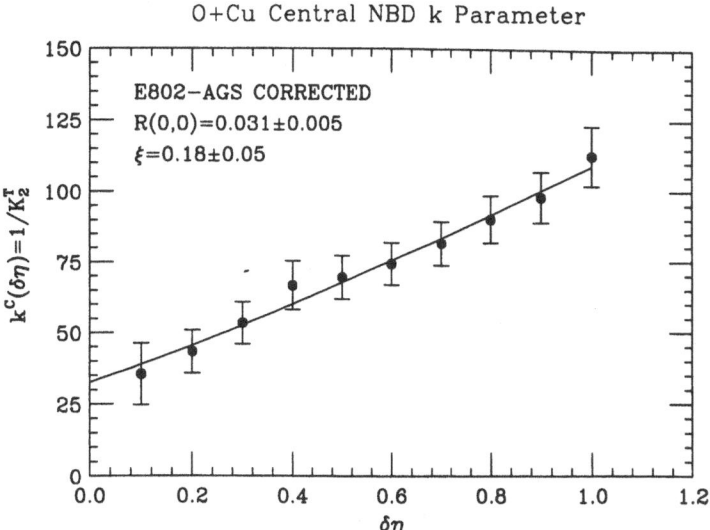

Figure 4. The NBD parameter $k(\delta\eta)$ as a function of the interval $\delta\eta$ (Fig. 2),together with a fit to Eq. 10 with the parameters indicated (solid line).

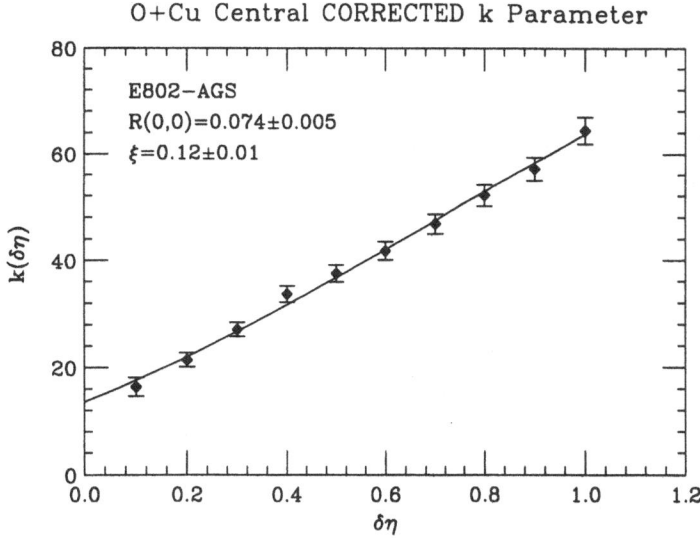

Figure 5. The k parameter from NBD fits to the data, corrected for instrumental effects, as explained in the text, presented as $k^C(\delta\eta) \equiv 1/K_2^T(\delta\eta)$ as a function of the interval $\delta\eta$. The solid line is a fit to Eq. 10 with the parameters indicated.

is just the sum of the integrals of the true plus the instrumental terms, or

$$K_2(\delta\eta) = \frac{1}{k(\delta\eta)} = K_2^T(\delta\eta) + K_2^I(\delta\eta) \quad . \tag{12}$$

The true effect $K_2^T(\delta\eta)$ is then simply

$$K_2^T(\delta\eta) = \frac{1}{k(\delta\eta)} - K_2^I(\delta\eta) \quad . \tag{13}$$

FINAL RESULTS FOR $R(0,0)$ AND ξ

In keeping with the notation based on the NBD, the final results are quoted as $1/K_2^T(\delta\eta)$, denoted $k^C(\delta\eta)$, and are plotted vs $\delta\eta$ in Fig. 5 which clearly illustrates, again, the simple linear evolution and non-zero intercept. The final values of $R(0,0)$ and ξ, corrected for instrumental effects, are derived from a fit of this data to Eq. 10: $R(0,0) = 0.031 \pm 0.005$, $\xi = 0.183^{+0.051}_{-0.042}$ (statistical errors). The systematic error, predominantly from the measured cross-talk uncertainty (4.0% $\pm$ 0.4%), is ± 0.003 for $R(0,0)$ and ± 0.01 for ξ. The hadron correlation length at low energies is known[25] to be roughly $\xi \sim 2$ units of rapidity, with strength $R(0,0) \sim 0.6$. Thus, for the weak correlation strength and small correlation length derived from the E802 data to make sense, it must be that the standard hadronic short range correlation effect is diluted by the random overlap of the multiple collisions in the ^{16}O+Cu reaction. Similar conclusions in the context of the conventional "intermittency" slope parameters were given in references.[21,22,26,27]

This result further demystifies "intermittency". For ^{16}O+Cu central collisions, *"Intermittency" is nothing more than the apparent statistical independence of the multiplicity in small pseudorapidity bins, $\delta\eta \sim 0.2$, due to the surprisingly short two-particle rapidity correlation length!* The 'large' bin-by-bin fluctuations on individual event rapidity distributions from Si+AgBr interactions in cosmic rays[3,28] and the linear evolution of $k(\delta\eta)$ for the present data are both explained by this effect.

It is interesting that exactly the deduced effects from the E802 data—weakened and very short length rapidity correlations in collisions of relativistic heavy ions—were predicted several years ago.[21,22,29,30] In nucleus-nucleus collisions, the conventional short-range correlations should be washed out by the random superposition of correlated sources,[22,26,27] so that eventually only the Quantum-Statistical Bose-Einstein (B-E) correlations should remain.[21,22,31] Other experiments have reported a relationship of "Intermittency" to B-E correlations.[32,33] If B-E correlations were the entire effect, then direct measurements of B-E correlations in the variable $\delta\eta$, instead of the usual variable[34] ($Q_{inv} = p_1 - p_2$), should reproduce the parameters derived from the evolution of $k^C(\delta\eta)$. A preliminary attempt using the E802/E859 spectrometer is shown in Fig. 6. The two pion correlation measurement in the variable Q_{inv}, which is considered to be one of the most demanding in RHI physics, appears to be much easier than the measurement in terms of $\delta\eta$, where the result for ξ is extraordinarily sensitive to the normalization of the correlated and mixed event samples.[35] The results are very preliminary but appear encouraging.

INTERMITTENCY IN TERMS OF DISTRIBUTIONS

Many of the individual components of the present analysis have been noted by previous authors.[9,20–23,31,27,24,32,33,26,29,30,36–38] However, the present data allow for the first time a systematic formulation of the subject of "intermittency" in terms of distributions, rather than moments. Furthermore, the evolution with $\delta\eta$ of the NBD parameters yields a simple, elegant and understandable explanation of the "intermittency" phenomenology. The key to explaining "intermittency", which had not previously been understood, is the dramatic reduction of the two-particle rapidity correlation length for ^{16}O+Cu central collisions from the value in hadron-hadron collisions.[25] Moreover, the correlation length for central ^{16}O+Cu collisions, although smaller than expected, is quite finite and can be measured—which means that a length scale exists in these collisions and therefore there is no intermittency[1,4] in the multiplicity fluctuations.

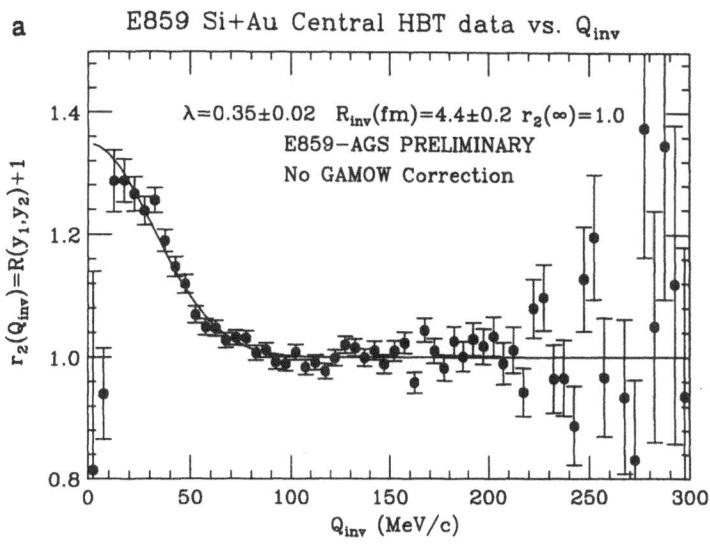

a E859 Si+Au Central HBT data vs. Q_{inv}

$\lambda = 0.35 \pm 0.02$ $R_{inv}(fm) = 4.4 \pm 0.2$ $r_2(\infty) = 1.0$
E859-AGS PRELIMINARY
No GAMOW Correction

$r_2(Q_{inv}) = R(y_1, y_2) + 1$

Q_{inv} (MeV/c)

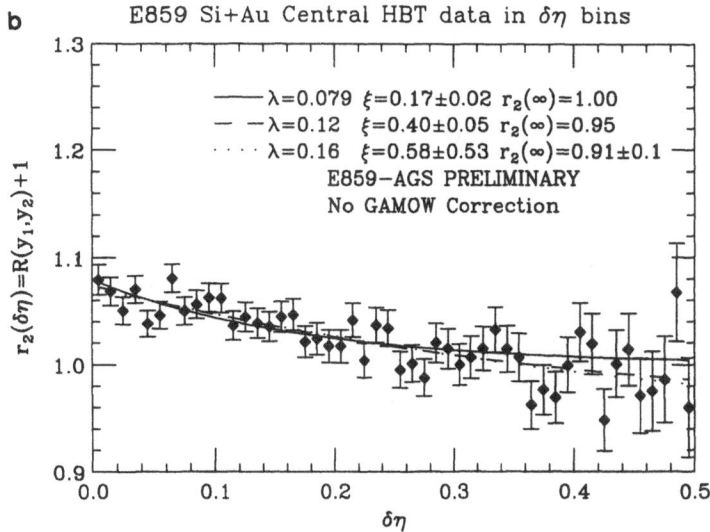

b E859 Si+Au Central HBT data in $\delta\eta$ bins

——— $\lambda = 0.079$ $\xi = 0.17 \pm 0.02$ $r_2(\infty) = 1.00$
— — $\lambda = 0.12$ $\xi = 0.40 \pm 0.05$ $r_2(\infty) = 0.95$
· · · · $\lambda = 0.16$ $\xi = 0.58 \pm 0.53$ $r_2(\infty) = 0.91 \pm 0.1$
E859-AGS PRELIMINARY
No GAMOW Correction

$r_2(\delta\eta) = R(y_1, y_2) + 1$

$\delta\eta$

Figure 6. a) E859 two π^- Bose-Einstein Correlation measurements in the variable Q_{inv} from 231K events in Si+Au central collisions. b) the same data as a function of $\delta\eta$. The curves are fits with different normalization, $r(\infty)$. There is no Gamow Correction and the data are **PRELIMINARY**.

Since the pioneering work of UA5,[5] many other experiments have shown that the NBD provides excellent fits to charged particle multiplicity distributions in restricted $\delta\eta$ intervals for all reactions studied, for example: p+p (NA22[39]), $e^+ + e^-$ (HRS[40]), μ-p DIS (EMC[41]), S+S central (NA35[42]). All these measurements show the same effect as the present data—linear dependences of the NBD parameter $k(\delta\eta)$ with the pseudo-rapidity interval $\delta\eta$, or equivalently with the mean multiplicity in the interval $\mu(\delta\eta)$, with non-zero intercept, $k(0) \neq 0$ (see Fig. 7a). The present data (and also to a certain extent, the other heavy ion data, NA35 S+S) are quantitatively, rather than qualitatively, different from the others in that $k(\delta\eta)$ is much larger, and the dependence on $\delta\eta$ much steeper. True intermittency, with a zero correlation length $\xi \to 0$, would occur if the intercept $k(0) \to 0$, which is not observed in any experiment!

The Clan Model Appears

Amazingly, the parameters of the clan model of Giovannini and Van Hove[20] can be directly read off Fig. 7a, as shown in Fig. 7b: the mean number of clusters, $<N> = k\ln(1.0 + \mu/k)$, and the mean multiplicity in a cluster $<n_c> = \mu/<N> = (\mu/k)/\ln(1.0 + \mu/k)$. For the heavy ion data, $k/\mu \gg 1$ for all cases, so that $<n_c> \simeq 1$ and $<N> \simeq \mu$ to an excellent approximation, with the result that the NA35 and both sets of E802 data are nearly indistinguishable on the line $<n_c> \simeq 1$. Only the UA5 data where $k/\mu < 1$, and to a certain extent NA22, show appreciable multiplicity/per cluster and it will be interesting at RHIC or LHC, which are in the UA5 domain, to see how the parameters evolve from p-p to heavy ion collisions.

The Heavy Ion Data

It is instructive to try to understand the precision obtained for the NBD parameter $k(\delta\eta)$ from the two heavy ion experiments, NA35 and the present experiment, E802. The error estimate, $s_{\frac{1}{k}}$, for the NBD parameter $1/k$ was given above (Eq. 7). Thus, to an excellent approximation, the required number of events N, for a fixed percent error in s_k/k, is

$$N = 2\frac{k^2}{\mu^2}\left(\frac{k}{s_k}\right)^2 . \qquad (14)$$

This explains why the errors for NA35 S+S with 2856 events are so much larger than E802 O+Cu with 19667 events—(k/μ) is 2 to 3 times larger—even though NA35, in distinction to all the other NBD fit experiments, combined the data from all intervals of a given size (as in the normalized factorial moment analysis) to reduce the errors. Interestingly, the fit of Eq. 10 to this data, shown as the dashed line on Fig. 7 with $\xi = 0.6^{+0.4}_{-0.2}$, indicates that k increases with $\delta\eta$ (i.e. $1/\xi \neq 0$) to 99.4% confidence (2.5σ), which is somewhat in disagreement with the conclusions reached by NA35 from this data,[42] and also by EMU01,[43] that "the NBD parameter (1/k) is (within the errors) independent of the width of the rapidity interval." Of course, the present measurement, corrected for instrumental effects, gives a much clearer (3.5σ) effect for the variation of $k(\delta\eta)$.

Higher Order Correlations

A popular misconception is that[44,45] "there are practically no correlations beyond the second order in the heavy ion data, in contrast to the hadron-hadron and e^+e^- collisions." This is evidently incorrect since all the data, heavy ion included, are fit by the NBD, which exhibits finite K_q for all orders (Eq. 6). Of course, since $k \simeq 50 \cdots 100$ for the heavy ion data, $K_3 = 2/k^2 \simeq 0.0008 \cdots 0.0002$, which is roughly two orders of magnitude lower than the sensitivity of previously published direct searches with 'null' results.[46,47]

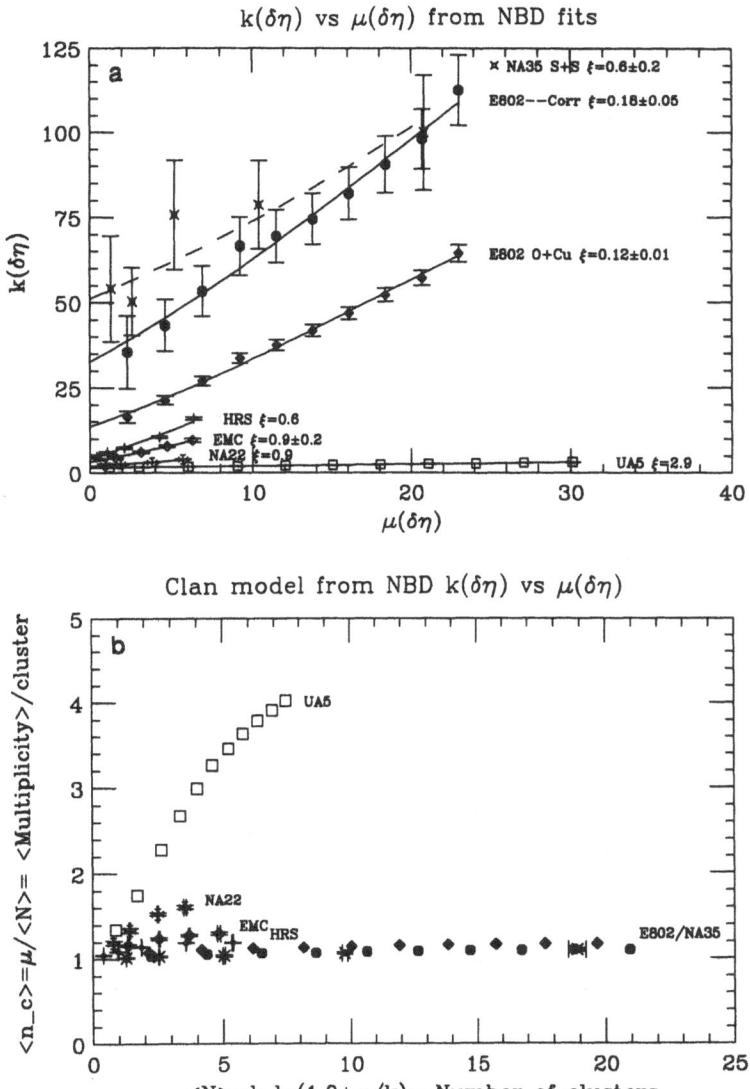

Figure 7. a) The NBD parameter $k(\delta\eta)$ as a function of the mean multiplicity in the pseudo-rapidity interval, $\mu(\delta\eta) = dn/d\eta \times \delta\eta$: UA5 $\bar{p}$+p $\sqrt{s}$ = 540 GeV ($dn/d\eta$=3.01), NA22 p+p $\sqrt{s}$ = 22 GeV (1.90), EMC μ-p DIS $W = 18 - 20$ GeV (1.57), HRS $e^+ + e^-$ 2-Jet $\sqrt{s}$ = 29 GeV (2.12), E802 O+Cu central P_{beam} = 14.6A GeV/c (23.0), E802 corrected $k^C(\delta\eta)$, NA35 S+S central E_{beam} = 200A GeV (10.4). The lines are fits to Eq. 10 with the parameters ξ indicated; b) Parameters of the Clan Model from the same data.

273

SUMMARY AND CONCLUSIONS

E802 ^{16}O+Cu Central(ZCAL) multiplicity distributions in bins of pseudorapidity $\delta\eta = 0.1, 0.2 \ldots 1.0$ show an apparent fractional power-law growth of Normalized Factorial moments with decreasing pseudorapidity interval, in agreement with previous "intermittency" analyses. The same data also exhibit excellent fits to Negative Binomial Distributions. The k parameter of the NBD fits increases steeply and linearly with the $\delta\eta$ interval which is an unexpected and particularly striking result. The linear evolution of the NBD parameter $k(\delta\eta)$ with the width of the interval explains the observation of fractional power laws based on the "intermittency" formalism in a much more simple, elegant and understandable way. *The apparent power laws with fractional exponent of the Normalized Factorial Moments are simply an artifact of using quantities like $F_2 - 1$ which are inversely proportional to a linear quantity*, i.e. $F_2 - 1 = 1/k(\delta\eta)$. Furthermore, the apparent fractional power-law increase of the Normalized Factorial Moments with decreasing bin size $\delta\eta$ for all 6 orders measured in E802 ^{16}O+Cu Central(ZCAL) collisions is entirely given by the Negative Binomial Distribution best fit curves, represented by the single parameter $k(\delta\eta)$—and has nothing to do with the deviations of the measured data points from the best fit curves!

The linear increase of the NBD parameter $k(\delta\eta)$ with $\delta\eta$ can be directly related to the 2-particle short-range rapidity correlation strength and correlation length, conveniently paramterized as an exponential $R(0,0)\, e^{-|y_1 - y_2|/\xi}$, to give the equation

$$ k(\delta\eta) = \frac{1}{F_2 - 1} = \frac{<n(\delta\eta)>^2}{\int^{\delta\eta} dy_1 dy_2\, C_2(y_1, y_2)} = \frac{1}{R(0,0)} \frac{\delta\eta/2\xi}{[1 - \frac{\xi}{\delta\eta}(1 - e^{-\delta\eta/\xi})]} \;, $$

which describes mathematically why the linear increase of k with $\delta\eta$ is an indication of the randomness of the multiplicity in adacent $\delta\eta$ bins ($\delta\eta \gg \xi$), while the constancy of k with increasing $\delta\eta$ would be an indication of 100% correlation ($\delta\eta \ll \xi$). The evolution of $k(\delta\eta)$, which appears to be strikingly linear, is equally well described by a fit to this equation. After correction for instrumental effects, the best fit parameters indicate a weak correlation strength $R(0,0) = 0.031 \pm 0.005$ and a very short rapidity correlation length $\xi = 0.183^{+0.051}_{-0.042}$, e.g. compared to $p - \bar{p}$ collisions where UA5 measured $R(0,0) \sim 2/3$ and $\xi \sim 3$. The weak and very short-range rapidity correlation in nucleus-nucleus collisions had been predicted— since the conventional nucleon-nucleon short-range correlations should be washed out by the random superposition of correlated sources so that eventually only Quantum Statistical (Bose-Einstein) Correlations should remain. The dramatic reduction of the two-particle rapidity correlation length gives a quantitative demystification of "intermittency". For ^{16}O+Cu central collisions, "intermittency" is nothing more than the apparent statistical independence of the multiplicity in small pseudorapidity bins, $\delta\eta \sim 0.2$, *due to the surprisingly short two-particle rapidity correlation length!*

The present data allow for the first time a systematic formulation of the subject of "intermittency" in terms of distributions to complement the normalized factorial moment formalism. In agreement with all previous measurements of NBD fits to multiplicity distributions in hadron and lepton reactions, the k parameter of the NBD fit for central ^{16}O+Cu collisions is found to exhibit an apparently linear increase with the $\delta\eta$ interval, albeit with a much steeper slope than for the other reactions, and a non-zero intercept, $k(0) \neq 0$. True intermittency, $\xi \to 0$, would occur if the intercept $k(0) \to 0$, which is not observed in any experiment. The correlation length for central ^{16}O+Cu collisions, although smaller than expected, is quite finite and can be measured—which means that a length scale exists in these collisions and therefore there is no intermittency in the multiplicity fluctuations. It is clear that the present E802 data have much more in common with the original UA5 observation—an increase in the width of the multiplicity distributions about the average with decreasing $\delta\eta$

interval—than with any of the classical "intermittency" analyses. The difference is quantitave rather than qualitative: the rapidity correlation length is $\xi \sim 3$ in UA5, $\xi \sim 0.2$ in E802.

Acknowledgements

This work has been supported by the U.S. Department of Energy under contracts with ANL (W-31-109-ENG-38), BNL (DE-AC02-76CH00016), Columbia University (DE-FG02-86-ER40281), LLNL (W-7405-ENG-48), MIT (DE-AC02-76ER03069), UC Riverside (DE-FG03-86ER40271), and by NASA (NGR-05-003-513), under contract with the University of California, and by the U.S.-Japan High Energy Physics Collaboration Treaty. The flawless operation of the Tandem-AGS facility at BNL is appreciated. Discussions and/or correspondence with P. Carruthers, G. Ekspong, R. Hwa, P. Lipa, I. Otterlund, D. Seibert, N. Tannenbaum and R. Weiner are acknowledged.

REFERENCES

1. A. Bialas and R. Peschanski, *Nucl. Phys.* B273:703 (1986); B308:847 (1988).

2. J. G. Rushbrooke, contribution 6th High Energy Heavy Study (Berkeley, 1983); P. Carlson, *Proc. 4th Workshop on $\bar{p} - p$ Collider Physics (Bern, 1984)*, CERN Yellow Report 84-09 (1984); UA5 Collaboration, G. J. Alner, *et al.*, Phys. Lett. 138B:304 (1984).

3. JACEE collaboration, T. H. Burnett, *et al.*, *Phys. Rev. Lett.* 50:2062 (1983).

4. B. B. Mandelbrot, *The Fractal Geometry of Nature* (W. H. Freeman, New York, 1983).

5. UA5 collaboration, G. J. Alner, *et al.*, *Phys. Lett.* 160B:193 (1985); see also UA5 collaboration, G. J. Alner, *et al.*, *Physics Reports* 154:247 (1987), and references therein.

6. P. Carruthers and C. C. Shih, *Phys. Lett.* 127B:242 (1983); 165B:209 (1985). See also, G. Ekspong.[7]

7. G. Ekspong, *Festschrift for Leon Van Hove and Proceedings Multiparticle Dynamics*, eds. A. Giovannini and W. Kittel (World Scientific, Singapore 1990), and references therein.

8. W. Thome *et al.*, *Nucl. Phys.* B129:365 (1977); see also K. Eggert, *et al.*, *Nucl. Phys.* B86:201 (1975).

9. G. N. Fowler and R. M. Weiner, *Phys. Lett.* 70B:201 (1977); *Phys. Rev.* D17:3118 (1978); *Nucl. Phys.* A319:349 (1979); *Phys. Rev.* D37:3127 (1988).

10. G. N. Fowler, E. M. Friedlander and R. M. Weiner, *Phys. Lett.* 104B:239 (1981).

11. L. Van Hove, *Physica* 147A:19 (1987); see also L. Van Hove, *Phys. Lett.* B232:509 (1989).

12. See, for example, M. J. Tannenbaum, *Int. J. Mod. Phys.* A4:3377 (1989), and references therein.

13. E802 Collaboration, T. Abbott, *et al.*, "Intermittency in Central Collisions of ^{16}O+A at 14.6 A·GeV/c", BNL-49897, June 15, 1994, *Phys. Lett.*, in press.

14. For an extensive review of this work, see A. Bialas, *Nucl. Phys.* A525:345c (1991).

15. A. H. Mueller, *Phys. Rev.* D4:150 (1971).

16. M. G. Kendall and A. Stuart, *The Advanced Theory of Statistics* (Hafner, New York, 1969).

17. H. Jeffreys, *Theory of Probability* (Clarendon Press, Oxford, 1961).

18. R. C. Hwa and J. C. Pan, *Phys. Rev.* D46:2941 (1992).

19. Interestingly, the data are also described by Gamma distribution fits which give the same conclusions as the NBD fits but which are all higher than the data points at high multiplicity.

20. A. Giovannini and L. Van Hove, *Z. Phys.* C30:391 (1986).

21. P. Carruthers, E. M. Friedlander, C. C. Shih and R. M. Weiner, *Phys. Lett.* B222:487 (1989).

22. A. Capella, K. Fialkowski and A. Krzywicki, *Festschrift for Leon Van Hove and Proceedings Multiparticle Dynamics*, eds. A. Giovannini and W. Kittel (World Scientific, Singapore 1990); *Phys. Lett.* B230:149 (1989).

23. P. Carruthers and Ina Sarcevic, *Phys. Rev. Lett.* 63:1562 (1989); Ina Sarcevic, *Nucl. Phys.* A525:361c (1991); P. Carruthers, H. C. Eggers and Ina Sarcevic, *Phys. Lett.* B254:258 (1991); P. Carruthers, *et al., Int. J. Mod. Phys.* A6:3031 (1991).

24. E. A. De Wolf, *Acta Phys. Pol.* B21:611 (1990).

25. e.g. see L. Foa, *Phys. Rep.* C22:1 (1975); J. Whitmore, *Phys. Rep.* C27:187 (1976); C10:273 (1974); H. Boggild and T. Ferbel, *Ann. Rev. Nucl. Sci.* 24:451 (1974).

26. P. Lipa and B. Buschbeck, *Phys. Lett.* B223:465 (1989).

27. D. Seibert, *Phys. Rev.* D41:3381 (1990).

28. M. J. Tannenbaum, *Mod. Phys. Lett.* A9:89 (1994).

29. K. L. Wieand, S. E. Pratt, A. B. Balantekin, *Phys. Lett.* B274:7 (1992).

30. D. Seibert, *Phys. Rev.* C47:2320 (1994).

31. M. Gyulassy, *Festschrift for Leon Van Hove and Proceedings Multiparticle Dynamics*, eds. A. Giovannini and W. Kittel (World Scientific, Singapore 1990) p. 479.

32. I. Derado, G. Jancso, N. Schmitz, *Z. Phys.* C56:553 (1992).

33. K. Kadija and P. Seyboth, *Phys. Lett.* B287:363 (1992).

34. For recent reviews see B. Lorstad, *Int. J. Mod. Phys.* A4:2861 (1988); W. A. Zajc in *Hadronic Multiparticle Production,* edited by P. Carruthers (World Scientific, Singapore, 1988).

35. E802 Collaboration, Y. Akiba, *et al., Phys. Rev. Lett.* 70:1057 (1993).

36. I. M. Dremin, *Sov. Phys. Uspekhi* 33:647 (1990).

37. S. Barshay, *Z. Phys.* C47:199 (1990).

38. T. Awes, "Koba-Nielsen-Olesen Scaling, Intermittency, and Statistics", ORNL Physics Division Preprint, (January 1990, Unpublished).

39. EHS/NA22 Collaboration, M. Adamus, *et al., Z. Phys.* C37:215 (1988); *Phys. Lett.* B177:239 (1986).

40. HRS Collaboration, M. Derrick, *et al., Phys. Lett.* 168B:299 (1986); *Phys. Rev.* D34:3304 (1986).

41. EMC Collaboration, M. Arnedo, *et al., Z. Phys.* C35:335 (1987).

42. NA35 Collaboration, J. Bächler, *et al., Z. Phys.* C57:541 (1993).

43. EMU01 Collaboration, M. I. Adamovich, *et al., Z. Phys.* C56:509 (1992).

44. A. Bialas, these proceedings.

45. I. Sarcevic, these proceedings; H-T. Elze and I. Sarcevic, *Phys. Rev. Lett.* 68:1988 (1992).

46. EMU01 Collaboration, M. Adamovich, *et al., Phys. Rev. Lett.* 65:412 (1990).

47. R. Holynski, *et al., Phys. Rev. Lett.* 62:733 (1989).

UNIVERSAL PROPERTIES OF ANGULAR CORRELATIONS IN QCD JETS

Wolfgang Ochs

Max-Planck-Institut für Physik
Föhringer Ring 6, D-80805 München, Germany

ABSTRACT

Predictions for angular correlations between an arbitrary number of partons are derived in the high energy limit. The quantities considered depend on angles and primary energy through a single variable ε which implies certain scaling properties and relations between quite different observables. These asymptotic predictions derived in the double log approximation of QCD are checked against Monte Carlo calculations at the parton and hadron level.

INTRODUCTION

Many new results from the detailed study of multiparticle production processes have been obtained in recent years, still there is no fully satisfactory model to describe all these phenomena. One approach is based on perturbative QCD, particularly suited for hard processes involving large momentum transfers, in which an initially scattered parton evolves by gluon Bremsstrahlung into a jet of partons and finally into the observable hadrons, whereby the effect of the color confinement force is modeled. Another approach relies on a statistical and thermodynamic description in which the microscopic degrees of freedom are integrated over and only global quantities are kept. Hadrons are produced from a quark gluon plasma after a phase transition. This approach is applied in particular to nuclear collisions in the search for the quark gluon plasma. As many aspects of particle production change only moderately when going from the more elementary collisions like e^+e^- annihilations to the more complex pp or nuclear collisions it is desirable to develop both the perturbative and the statistical methods and to explore their predictive power.

In any case, for a satisfactory model of multiparticle production we would like to have finally a mathematical model based on a simple principle, there should be only a few parameters, some results should be obtained analytically – even if only approximately – otherwise the structural properties of the theory can hardly be fully explored. A nice example of a model of this type is Hagedorn's bootstrap model: it is based on a single guiding principle, the bootstrap principle, the only parameters being the particle masses and the interaction volume; analytical results on multiparticle observables can be derived from an equation for the generating functional of multiparticle densities.[1] Although it has been attempted to extend the statistical approach to hard processes[2] the most convincing results for such processes are obtained today from the parton cascade model as derived from perturbative QCD.

In this talk I would like to report on results from perturbative QCD on angular correlations inside a parton jet which are obtained in collaboration with Jacek Wosiek.[3,4] The essential parameters of the parton cascade are the scale parameter Λ, which determines the coupling strength and a cutoff Q_o which regulates the infrared and collinear singularities of the gluon Bremsstrahlung. As to the hadronization process we follow here the idea that it is sufficiently soft and the distribution of hadrons follows largely the distribution of partons.[5] In particular for momentum spectra of partons and hadrons such an equality up to a constant has been established, if the parton cascade is evolved down to the hadronic (i.e. pion) mass scale $Q_o \approx m_h$.[6] Such a similarity may also be expected if the observable considered is "infrared save", i.e. does not depend explicitly on the cutoff Q_o. It is one of the motivations of the present study, as to what extent such a hypothesis of soft hadronization is actually correct and supported by the experiment. It would allow for a certain class of observables to be calculated just only in terms of Q_o and Λ parameters, resulting in a scheme of high predictive power.

THEORETICAL SCHEME

Our calculations are based on the double logarithmic approximation (DLA) of QCD. The probability to emit a single gluon of momentum K from a primary parton a of momentum P is given by

$$\mathcal{M}_{P,a}(K)d^3K = c_a a^2(K\Theta_{PK})\frac{dK}{K}\frac{d\Theta_{PK}}{\Theta_{PK}}\frac{d\Phi_{PK}}{2\pi}, \tag{1}$$

where $a^2 \equiv \gamma_0^2 = 6\alpha_s/\pi$ is the QCD anomalous dimension of the multiplicity evolution and $c_g = 1$ or $c_q = 4/9$ for gluon or quark jets respectively. For the running coupling constant we write $a^2(p_T) = \beta^2/(\ln(p_T/Q_0) + \lambda)$ whereby $\lambda = \ln(Q_0/\Lambda)$ and $\beta^2 = 12(\frac{11}{3}N_c - \frac{2}{3}N_f)^{-1} \approx 1.565$ for 5 flavours N_f. In this approximation the recoil effects are neglected (i.e. energy and momentum conservation are violated), integrals are performed by retaining only the leading contributions from collinear and soft divergencies. The interference of the soft gluons is taken into account by the "angular ordering" prescription. In DLA one obtains the leading asymptotic behaviour, but the non-asymptotic corrections of relative order $\sqrt{\alpha_s}$ are potentially large.

The generating functional which includes the "leading" primary parton in the final state is given by the non-linear integral equation[7]

$$Z_{P,a}\{u\} = u(P)\exp\left(\int_{\Gamma_P(K)} \mathcal{M}_{P,a}(K)[u(K)Z_{K,g}\{u\} - 1]d^3K\right). \tag{2}$$

In this form the virtual corrections are included which ensures the proper normalization $Z_P\{u\}\big|_{u=1} = 1$. The density distribution of n partons is then obtained by functional differentiation after the test functions $u(k)$

$$\rho^{(n)}(k_1, ..., k_n) = \delta^n Z\{u\}/\delta u(k_1)...\delta u(k_n) \big|_{u=1} . \tag{3}$$

Similarly, the cumulant (connected) correlation function is derived as in Eq.(3) but with Z replaced by $\ln Z$. Starting with $n = 1$ one obtains from (3) and (2) linear integral equations which can be solved recursively. Here we specialize on angular distributions and the corresponding equations are obtained after integration over the momenta. We have obtained results[3,4] on the general inclusive cumulant correlation function of n particles in their spherical angles. Of special interest are the distribution in the relative polar angle ϑ_{12} between two partons and the multiplicity moments of general order n for particles falling into sidewise angular regions.

It turns out that all these angular observables can be constructed from a generic function $h^{(n)}(\delta, \vartheta, P)$ which fulfils the integral equation

$$h^{(n)}(\delta, \vartheta, P) = d^{(n)}(\delta, \vartheta, P) + \int_{Q_0/\delta}^{P} \frac{dK}{K} \int_{\delta}^{\vartheta} \frac{d\Psi}{\Psi} a^2(p_T) h^{(n)}(\delta, \Psi, K). \tag{4}$$

where δ and ϑ are the small and the large angles of the respective problem and the inhomogeneous term behaves at high energies like $d^{(n)} \sim \exp(2n\beta\sqrt{\ln(P\delta/\Lambda)})$. One finds that the natural variables of the problem, instead of P, ϑ are

$$\varepsilon = \ln(\vartheta/\delta)/\ln(P\vartheta/\Lambda), \qquad \zeta = 1/(\beta\sqrt{\ln(P\vartheta/\Lambda)}). \tag{5}$$

As $\vartheta > \delta \geq Q_0/P > \Lambda/P$, we find $0 \leq \varepsilon \leq 1$. The solution for $\ln h^{(n)}$ can be written as an expansion in $\zeta \sim \sqrt{\alpha_s}$. In the high energy limit (ε fixed, $P \to \infty$) one obtains

$$h^{(n)}(\delta, \vartheta, P) \sim \exp(2\beta\sqrt{\ln(P\vartheta/\Lambda)}\omega(\varepsilon, n)). \tag{6}$$

The scaling function $\omega(\varepsilon, n)$ is known in implicit form.[3] For small ε it has a power expansion $\omega(\varepsilon, n) = n - (n^2 - 1)\varepsilon/2n + \ldots$ Another useful approximation obtains from an expansion in $1/n$ which yields $\omega(\varepsilon, n) \approx n\sqrt{1-\varepsilon}\,(1 - \frac{1}{2n^2}\ln(1-\varepsilon))$ and has a 1% accuracy for $\varepsilon < 0.95$ already for $n = 2$. An asymptotic behaviour of type (6) was also found in the study of azimuthal particle correlations.[8]

POLAR ANGLE CORRELATIONS

First we discuss the correlations $\rho^{(2)}(\vartheta_{12}, P, \Theta) \equiv dn/d\vartheta_{12}$ in the relative polar angle ϑ_{12} of two partons both inside a forward cone around the initial parton of half opening angle Θ. For the study of scaling properties it is more convenient to consider the distribution in the variable $\varepsilon = \ln(\Theta/\vartheta_{12})/\ln(P\Theta/\Lambda)$ given by $\hat{\rho}^{(2)}(\varepsilon) \equiv dn/d\varepsilon = \vartheta_{12}\ln(P\Theta/\Lambda)\rho^{(2)}(\vartheta_{12})$. For the correlation $\hat{r}(\varepsilon) = \hat{\rho}^{(2)}(\varepsilon)/\bar{n}^2$ normalized by the multiplicity in the cone* $\bar{n}$ one obtains in the high energy limit (ε fixed, $P \to \infty$) from Eq. (6) with $h^{(2)}(\vartheta_{12}) \sim \vartheta_{12}\rho^{(2)}(\vartheta_{12})$ for either quark or gluon jet

$$\hat{r}(\varepsilon) = 2\beta\sqrt{\ln(P\Theta/\Lambda)} \exp\left(-2\beta\sqrt{\ln(P\Theta/\Lambda)}\,(2 - \omega(\varepsilon, 2))\right). \tag{7}$$

Differences between quark and gluon jets and the influence of the leading particle show up at finite energies where we obtain

$$\begin{aligned} \hat{r}_a(\varepsilon) &= c_a^{-1} y \exp(-y(2 - \omega(\varepsilon, 2))) - (c_a^{-1} - 1)y \exp(-2y(1 - \sqrt{1-\varepsilon})) \\ &\quad + 2\beta\sqrt{2y}/(c_a f(1-\varepsilon)^{\frac{1}{4}}) \exp(-y(2 - \sqrt{1-\varepsilon})) \end{aligned} \tag{8}$$

with $y = 2\beta\sqrt{\ln(P\Theta/\Lambda)}$, $f = 2\beta K_0(2\beta\sqrt{\lambda})/\sqrt{\pi} \approx 0.145$ for $\lambda = \ln 2$. To explore the scaling properties of $\hat{r}(\varepsilon)$ we consider the quantity $-\ln\hat{r}(\varepsilon)/(2\sqrt{\ln(P\Theta/\Lambda)})$ depending only on ε for any momentum P or cone opening angle Θ. The asymptotic limit and a finite energy result are shown in Fig. 1. As a test of our analytical calculations we also compared

---

*This normalization has better scaling properties[4] than the differential one used earlier[3]

to the Monte Carlo evaluation of the parton cascade.[†] As can be seen from the figure, the predicted asymptotic scaling behaviour is nicely reproduced for small $\varepsilon \leq 0.5$ in a large energy range ($P \geq 20$ GeV) whereas deviations occur for larger ε (smaller relative angles ϑ_{12} approaching the cut-off Q_0/P). It should be noted that the normalization of the above expressions in (7, 8) and in particular the quark and gluon difference at finite energy are of non-leading order in the DLA and therefore less reliable. We have therefore adjusted the overall normalization of the functions in (7, 8) to the Monte Carlo data. We also studied the effect of hadronization provided by the HERWIG MC. Again these effects are negligable for small $\varepsilon \leq 0.5$.[4] Preliminary data from DELPHI[10] have recently given first evidence for approximate ε-scaling in Θ ($\Theta \geq 30^\circ$) and a close proximity of the data to the parton model results.

MULTIPLICITY MOMENTS FOR SIDEWISE ANGULAR CONE OR RING

In a second application we consider particle multiplicities in a sidewise cone $\delta\Omega$ of half opening δ at polar angle ϑ with respect to $\vec{P}$ and in an angular ring of width 2δ symmetrically around the primary parton direction, centered at polar angle ϑ. We will refer to the ring and the cone as the 1D and 2D configurations. We calculate the factorial and cumulant multiplicity moments $f^{(n)}$ and $c^{(n)}$ or normalized by the multiplicities $\bar{n}$ in the respective angular regions $F^{(n)} = f^{(n)}/\bar{n}^n$ and $C^{(n)} = c^{(n)}/\bar{n}^n$ ($f^{(2)} = < n(n-1) >$, $C^{(2)} = F^{(2)} - 1$, etc.). The cumulant moments $c^{(n)}$ can be derived again from the generic equation (4). For the normalized moments we find

$$C^{(n)}(\vartheta, \delta) \sim (\vartheta/\delta)^{D(n-1)} \exp\left(-2\beta\sqrt{\ln(P\vartheta/\Lambda)}(n - \omega(\varepsilon, n))\right) \qquad (9)$$

with $\varepsilon = \ln(\vartheta/\delta)/\ln(P\vartheta/\Lambda)$. The dependence on ε is as in $\hat{r}(\varepsilon)$ discussed above. Similar results have also been found by other groups.[11,12] In the modified LLA including next to leading effects in the exponent[11] the power changes typically by 10%.

The scaling properties of the various moments can again be conveniently investigated by projecting out the ε-dependence of the exponent in Eq.(9), one finds for

$$-\hat{C}^{(n)} \equiv -\frac{\ln[(\delta/\vartheta)^{D(n-1)}C^{(n)}]}{n\sqrt{\ln(P\vartheta/\Lambda)}} = 2\beta(1 - \frac{\omega(\varepsilon, n)}{n}) \qquad (10)$$

in the high energy limit, or $-\hat{C}^{(n)} \simeq 2\beta(1 - \sqrt{1 - \varepsilon})$ for large n independent of n,D.

In Fig. 2a we plot $-\hat{C}^{(2)}$ for the ring ($D = 1$) vs. ε for different primary momenta P for the parton MC. There is a violation of ε-scaling for small ε but the asymptotic prediction is approached for high energies. We have repeated the same calculation as in (10), but for the factorial moments $F^{(n)}$ replacing $C^{(n)}$ in (10), see Fig. 2b. As $F^{(2)} = C^{(2)} + 1$ they approach the same asymptotic limit. Apparently, the nonasymptotic corrections are such that the scaling in ε for small ε sets in already at low energies for the factorial moments. The results for higher moments follow roughly the expectation (10), on the other hand the $D = 2$ moments show a more gradual dependence on ε than predicted.[4]

An interesting aspect of these results is the remarkable universality of the various moments (different n, D) and also of the very different observable $\hat{r}(\varepsilon)$ which all converge against the same limiting function after appropriate rescaling (see Figs. 1,2).

[†]We used the program HERWIG[9] with parameters $\Lambda = 0.15$ GeV, $m_q = m_g = 0.32$ GeV and without non-perturbative gluon splitting for the process $e^+e^- \rightarrow u\bar{u}$.

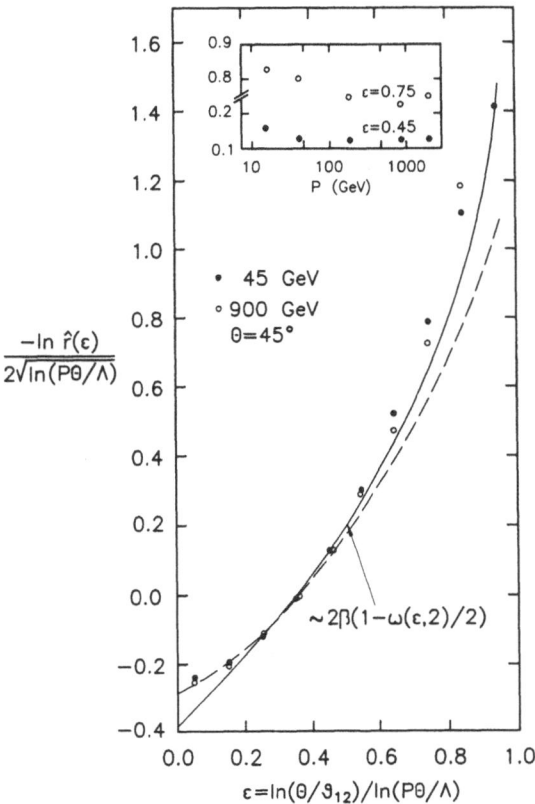

Figure 1. Rescaled polar angle correlation function for the high energy limit, Eq. (7) (full line), and for a quark jet of 45 GeV, Eq. (8) (dashed line), with normalization adjusted to the data as obtained from the HERWIG MC at the parton level. The insert shows the energy dependence of the same quantity for fixed ε.

SUMMARY

The angular observables considered here, after appropriate rescaling, approach a limit in the normalized logarithmic angular variable $\varepsilon = \ln(\vartheta/\delta)/\ln(P\vartheta/\Lambda)$ for ε fixed, $P \to \infty$. The comparison with the parton MC shows that this limit is already approached for 2-particle correlation $\hat{r}(\varepsilon)$ and the factorial moments $F^{(n)}$ at present energies sufficiently far away from the angular cutoff Q_0/P ($\varepsilon \leq 0.5$) whereas the cumulant moments approach the asymptotic limit only at higher energies ($P \sim 1$ TeV). In the region of small ε – not further discussed here – the observables approach a power behaviour in the angular variables and become independent of the cutoff Q_0 ("infrared safe"). It is in this region that also the hadronisation corrections are found small.

The experimental confirmation of the universal high energy behaviour of rather different angular observables and of ε-scaling with two redundant variables P and ϑ can provide further evidence for a soft confinement mechanism and parton hadron duality which allows to calculate hadronic distributions directly from the partonic ones.

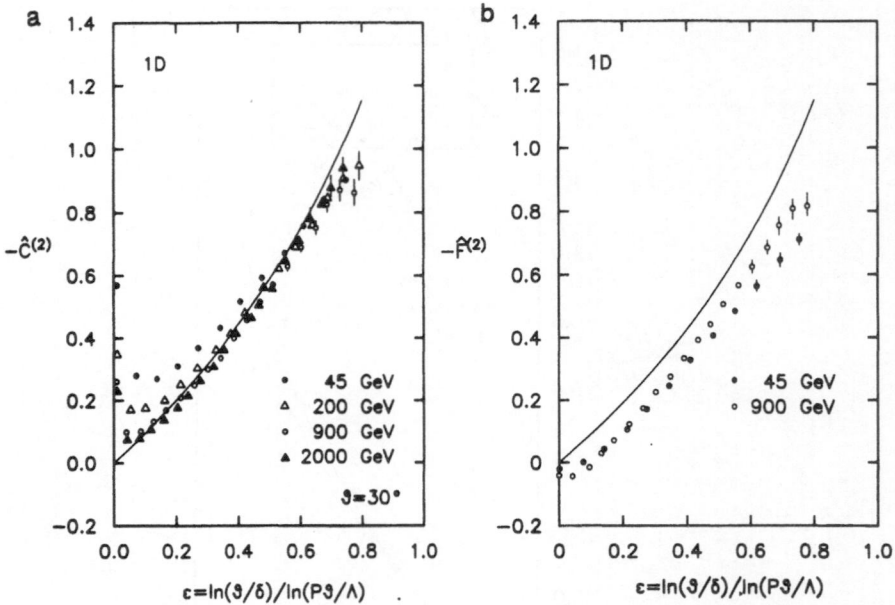

Figure 2. (a) Rescaled cumulant moments for the ring as defined in Eq. (10) for the parton MC for different jet momenta P in comparison with the asymptotic prediction Eq. (10); (b) same as (a) but for factorial moments.

Acknowledgement

I would like to thank R. Hagedorn for the discussions on ref. 2 and J. Wosiek for the collaboration on the topics presented here and many discussions.

REFERENCES

1. R. Hagedorn, I. Montvay, *Nucl. Phys.* B59:45(1973).

2. W. Ochs, *Z. Phys.* C23:131(1984).

3. W. Ochs, J. Wosiek, *Phys. Lett.* B289:159(1992), B304:144(1993).

4. W. Ochs, J. Wosiek, "Angular structure of QCD jets", MPI-preprint to appear.

5. D. Amati, G. Veneziano, *Phys. Lett.* 83:87(1979);
   A. Bassetto, M. Ciafaloni, G. Marchesini, *Phys. Rep.* 100:202(1983);
   Yu.L. Dokshitzer, V.A. Khoze, A.H. Mueller, S.I. Troyan, *Rev. Mod. Phys.* 60:373(1988).

6. Ya.I. Azimov, Yu.L. Dokshitzer, V.A. Khoze, S.I. Troyan, *Z. Phys.* C27:65(1985).

7. V.S. Fadin, *Yad.Fiz.* 37:408(1983); Yu.L. Dokshitzer et al. "Basics of Perturbative QCD", Editions Frontièrs, Gif-sur-Yvette Cedex, France(1991).

8. Yu.L. Dokshitzer, G. Marchesini, G. Oriani, *Nucl. Phys.* B387:675(1992).

9. G. Marchesini, B.R. Webber *Nucl. Phys.* B238:1(1984)1; B310:461(1988).

10. F. Mandl, B. Buschbeck (DELPHI collaboration), to be publ. in Proc. of QCD conference, Montpellier, July 1994; Vienna preprint.

11. Yu.L. Dokshitzer, C.M. Dremin, *Nucl. Phys.* B402:139(1993).

12. Ph. Brax, J.L. Meunier, R. Peschanski, *Z. Phys.* C62:649(1994).

TOWARDS A FIELD THEORETICAL DESCRIPTION
OF MULTIPARTICLE PRODUCTION
IN HIGH ENERGY COLLISIONS

Ina Sarcevic[1] and H. Th. Elze[2]

[1]Department of Physics
 University of Arizona, Tucson, AZ 85721, USA
[2]Theory Division, CERN
 CH-1211 Geneva 23, Switzerland

ABSTRACT

We present an effective field theory of multiparticle correlations based on analogy with Ginzburg-Landau theory of superconductivity. With the assumption that the field represents particle density fluctuations, and in the case of gaussian-type effective action we find that there are no higher-order correlations, in agreement with the recent observations in high energy heavy-ion collisions. We predict that three-dimensional two-particle correlations have Yukawa form. We also present our results for the two-dimensional and one-dimensional two-particle correlations (i.e. cumulants) as projections of our theory to lower dimensions.

INTRODUCTION

Unusually large fluctuations in transverse energy have recently been observed in ultra-relativistic heavy-ion collisions at CERN energies.[1] The independent-collision model which contains the scatterings of the secondary nucleons fails to describe the observed data indicating that perhaps nuclear constituents scatter and produce particle coherently.[2] Similar conclusion has been reached when multiparticle density fluctuations in different phase space regions have been studied via factorial moments, defined as

$$F_q(\delta y) = \frac{1}{M} \sum_{m=1}^{M} \frac{\langle n_m(n_m - 1)\ldots(n_m - i + 1)\rangle}{\langle n_m\rangle^i}, \tag{1}$$

where M is the number of rapidity bins ($\delta y = Y/M$) and n_m is the number of particles in the mth bin.[3] In the past few years, these moments have been measured for different targets and projectiles at energy of 200GeV/nucleon .[4] They were found to increase with decreasing bin size indicating nonstatistical fluctuations and being incompatible with the predictions of the standard classical hadronization models embedded in the existing Monte Carlo models.[4] In addition, the observed effect, sometimes referred to as the "intermittency effect", can not be accounted for by the superposition of independent nucleon-nucleon collisions, even when rescattering and geometrical effects are included.[3]

The possibility of creating the new form of matter, the quark-gluon plasma, in high-energy heavy-ion collisions have inspired intensive theoretical work on identifying the unambiguous QGP signal. Thus it is not surprising that the observation of the unusually large multiparticle density fluctuations has created a new excitement in the field, especially as a possibility of pointing towards the onset of the phase transition from quark-gluon plasma to hadronic matter. Phase transitions in QCD at high temperatures are of general interest – they are directly relevant to cosmology, since such a phase transition occurred throughout the universe during the early moments of the big bang and a first order phase transition could have altered primordial nuclear abundances. Unfortunately, up to now there are no conclusive predictions for detecting the quark matter in heavy-ion collisions and there is no theory to describe the observed "intermittency" phenomenon.[5]

MULTIPARTICLE CORRELATIONS
IN HIGH-ENERGY COLLISIONS

Multiparticle correlations in three "dimensions" are usually measured by subdividing a given total interval $\Omega_{tot} = \Delta Y \, \Delta \phi \, \Delta P$ into M^3 bins of side lengths $(\Delta Y/M, \Delta\phi/M, \Delta P/M)$. With n_{klm} the number of particles in bin (k, l, m) and $n^{[q]} = n!/(n-q)!$ the "vertical" factorial moment is

$$F_q^v(M) \equiv \frac{1}{M^3} \sum_{k,l,m=1}^{M} \frac{\langle n_{klm}^{[q]} \rangle}{\langle n_{klm} \rangle^q} = \frac{1}{M^3} \sum_{k,l,m=1}^{M} \frac{\int_{\Omega_{klm}} \prod_i d^3 x_i \, \rho_q(\vec{x}_1 \ldots \vec{x}_q)}{\left[\int_{\Omega_{klm}} d^3 x \, \rho_1(\vec{x}) \right]^q}. \qquad (2)$$

The second equality illustrates how the factorial moment can be written in terms of integrals of the correlation function ρ_q (Ω_{klm} is the region of integration over bin k, l, m).[6] Because for small bin sizes n_{klm} becomes small and the relative error correspondingly large, an alternative definition is often preferred for three-dimensional analysis, the "horizontal" factorial moment,

$$F_q^h(M) \equiv \frac{1}{M^3} \sum_{k,l,m=1}^{M} \frac{\langle n_{klm}^{[q]} \rangle}{(\langle N \rangle / M^3)^q} = M^{3(q-1)} \sum_{k,l,m=1}^{M} \frac{\int_{\Omega_{klm}} \prod_i d^3 x_i \, \rho_q(\vec{x}_1 \ldots \vec{x}_q)}{\left[\int_{\Omega_{tot}} d^3 x \, \rho_1(\vec{x}) \right]^q}. \qquad (3)$$

This form, while being much more stable, has the drawback that it depends on the shape of the one-particle distribution function ρ_1.

In order to examine the true higher-order correlations, the trivial, combinatoric contributions from two-particle correlations need to be subtracted. The cumulant moments, K_q, which measure the true, dynamical correlations are defined as[7]

$$K_q^v(\Omega_m) = \frac{1}{M^3} \sum_m \int_{\Omega_m} \prod_i d^3 \vec{x}_i \; \frac{k_2(\vec{x}_1, \vec{x}_2 \ldots \vec{x}_q)\rho_1(\vec{x}_1)\rho_1(\vec{x}_2)}{\left[\int_{\Omega_{klm}} d^3 \vec{x} \, \rho_1(\vec{x}) \right]^q}, \qquad (4)$$

where

$$k_2(1,2) = \frac{\rho_2(\vec{x}_1, \vec{x}_2)}{< \rho(\vec{x}_1) >< \rho(\vec{x}_2) >} - 1, \qquad (5)$$

$$k_3(1,2,3) = \frac{\rho_3(\vec{x}_1, \vec{x}_2, \vec{x}_3)}{< \rho(\vec{x}_1) >< \rho(\vec{x}_2) >< \rho(\vec{x}_3) >} - \sum_{perm}^{(3)} \frac{\rho_2(\vec{x}_1, \vec{x}_2)}{< \rho(\vec{x}_1) >< \rho(\vec{x}_2) >} + 2,$$

$$k_4 = \frac{\rho_4(\vec{x}_1, \vec{x}_2, \vec{x}_3, \vec{x}_4))}{<\rho(\vec{x}_1)><\rho(\vec{x}_2)><\rho(\vec{x}_3)><\rho(\vec{x}_4)>} - \sum_{perm}^{(4)} \frac{\rho_3(\vec{x}_1, \vec{x}_2, \vec{x}_3))}{<\rho(\vec{x}_1)><\rho(\vec{x}_2)><\rho(\vec{x}_3)>}$$

$$-\sum_{perm}^{(3)} \frac{\rho_2(\vec{x}_1, \vec{x}_2)\rho_2(\vec{x}_3, \vec{x}_4))}{<\rho(\vec{x}_1)><\rho(\vec{x}_2)><\rho(\vec{x}_3)><\rho(\vec{x}_4)>} + \sum_{perm}^{(12)} \frac{\rho_2(\vec{x}_1, \vec{x}_2)}{<\rho(\vec{x}_1)><\rho(\vec{x}_2)>} - 6. \quad (6)$$

The factorial moments, F_q ($q = 3, 4, 5$), given by Eq. (2) can be expressed in terms of the cumulans in the following way:[8]

$$F_2 = 1 + K_2,$$

$$F_3 = 1 + 3K_2 + K_3,$$

$$F_4 = 1 + 6K_2 + 3(K_2)^2 + 4K_3 + K_4, \quad (7)$$

$$F_5 = 1 + 10K_2 + 15(K_2)^2 + 10K_3K_2 + 10K_3 + 5K_4 + K_5.$$

Clearly, if there are no true, dynamical correlations, the cumulants, K_q vanish and factorial moments approach unity.

It has been found that K_2 decreases from lighter to heavier projectiles, especially in the case of Sulfur.[9] Furthermore, in hadronic collisions K_3 and K_4 are non-negligible (for example, K_3 contributes up to 20% to F_3 at small δy), while in nucleus-nucleus collisions, at the same energy, these cumulants are compatible with zero.[8] This implies that there are no statistically significant correlations of order higher than two for heavy-ion collisions and that the observed increase of the higher-order factorial moments F_q is entirely due to the dynamical two-particle correlations. As an illustration, in Fig. 1 we present the cumulant $K_3(M)$ from the NA35 data for O-Au at 200GeV/nucleon. Recent NA35 two and three dimensional measurements of the factorial moments corroborate our findings.[9]

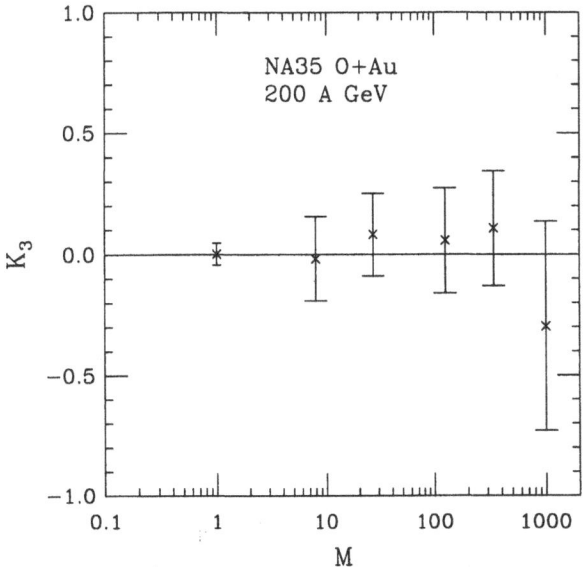

Figure 1. Third order cumulant K_3 as a function of the number of bins M for NA35 OAu data in $(y, \phi, p_\perp)$.[9] Cumulants of higher order are also compatible with zero. This is confirmed in analyses in terms of other variables and different colliding nuclei.[10]

EFFECTIVE FIELD THEORY OF MULTIPARTICLE PRODUCTION

The fact that particles produced in high-energy heavy-ion collisions exhibit only two-particle correlations indicates that perhaps higher-order correlations are washed out by rescattering of the initially correlated particles. Presently, there is no theory that describes this phenomena. Recently, we have proposed a three-dimensional statistical field theory of density fluctuations which has these features.[11,12] This model was formulated in analogy with the Ginzburg-Landau theory of superconductivity. The large number of particles produced in ultrarelativistic heavy-ion collisions justifies the use if a statistical theory of particle production. The formal analogy with the statistical mechanics of a one-dimensional "gas" was first pointed out by Feynman and Wilson and was later further developed by Scalapino and Sugar[13] and many others.[5] The idea is to build a statistical theory of the macroscopic observables by imagining that the microscopic degrees of freedom are integrated out and represented in terms of a few phenomenological parameters and by postulating that this theory will eventually be derived from a more fundamental theory, such as QCD.

While in the G-L theory of superconductivity the field (i.e. the order parameter) represents superconducting pairs, in the particle production problem, the relevant variable is the density fluctuation. The "field" $\Phi(\vec{x})$ is a random variable which depends on the rapidity of the particle and its transverse momentum p_t and it is identified with the density fluctuation.

Even though particles produced in high-energy collisions need not be in thermal equilibrium, one can still introduce a functional of the field Φ, $F[\Phi]$, which plays a role analogous to the free energy in equilibrium statistical mechanics. In principle one should be able to derive this functional from the underlying dynamics.

We define a random field Φ as a function in a three-dimensional space spanned by $(y, \phi, p_\perp)$. Throughout, $p_\perp$ will be implicitly divided by a constant scale $\mathcal{P}$ so that it is dimensionless. Since we are not looking for a phase transition, we omit the quartic term and start with the functional[11]

$$F[\Phi] = \int_0^P dy \int_{-P/2}^{P/2} d^2 p_\perp \left[a^2 \left(\partial \Phi / \partial y \right)^2 + a^2 \left(\nabla_{\vec{p}_\perp} \Phi \right)^2 + \mu^2 \Phi^2 \right] . \tag{8}$$

Taking the appropriate functional derivative, we find for the functional (8) the three-dimensional form of the two-point function

$$\langle \Phi(\vec{x}_1) \Phi(\vec{x}_2) \rangle = \frac{1}{8\pi a^2} \frac{e^{-R/\xi}}{R} , \tag{9}$$

where $\xi = a/\mu$ and $R \equiv [(y_1 - y_2)^2 + p_{\perp 1}^2 + p_{\perp 2}^2 - 2p_{\perp 1} p_{\perp 2} \cos(\phi_1 - \phi_2)]^{1/2}$. Further, we define $\Phi(\vec{x})$ as the fluctuation at the point $\vec{x}$ of the particle density for a given event, $\hat{\rho}_1(\vec{x})$, above/below the mean single particle distribution ρ_1 at that point:

$$\Phi(\vec{x}) \equiv \frac{\hat{\rho}_1(\vec{x})}{\rho_1(\vec{x})} - 1 . \tag{10}$$

Through these definitions, we find that $\langle \Phi(\vec{x}_1) \Phi(\vec{x}_2) \rangle = k_2(\vec{x}_1, \vec{x}_2)$ and that all higher order cumulants become exactly zero, $k_{q \geq 3} = 0$. By means of the specific form of the functional (8) and the definition of Φ as a fluctuation, we take account of the experimental facts in this regard. What is not experimentally certain and is to be tested is whether the second order correlations obey the Yukawa form (9).

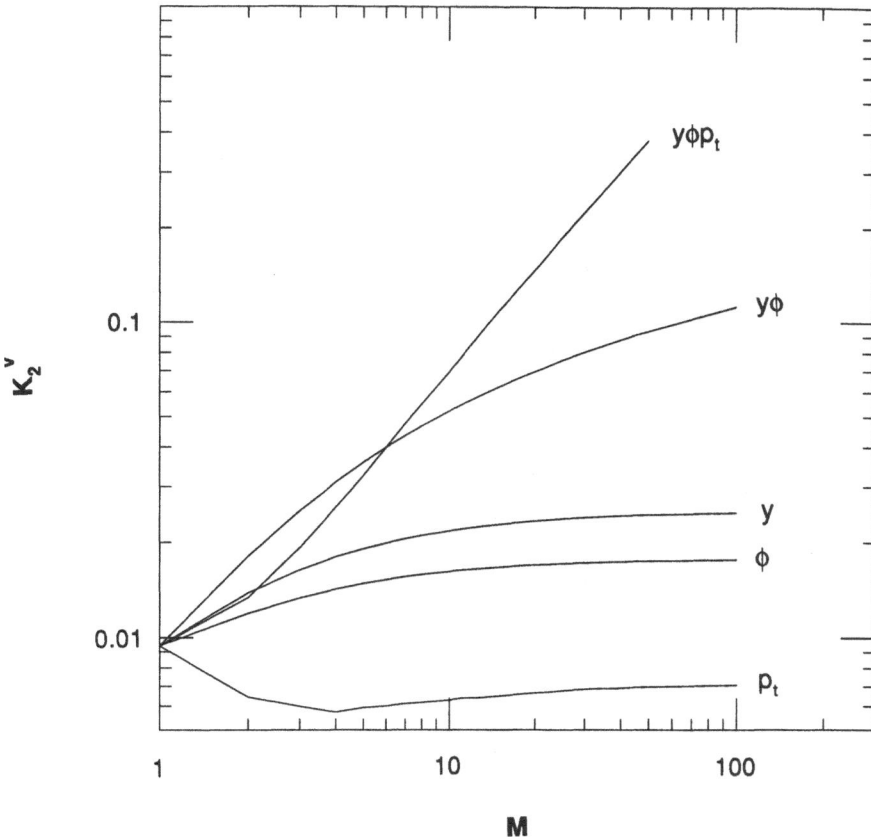

Figure 2. Theoretical predictions for the vertical cumulant moments K_2^v for various dimensions, for fixed parameters $a = 2.0, \xi = 1.0$, incorporating the experimental NA35 rapidity and $p_\perp$ distributions for 200 A GeV O+Au.

PROJECTIONS OF MULTIPARTICLE CORRELATIONS (i.e. CUMULANTS) TO LOWER DIMENSIONS

The second reduced cumulant $k_2 \propto e^{-R/\xi}/R$ can be compared to data only after a suitable integration over its variables. For three dimensions, the vertical integrated cumulant is given by $K_2^v(\delta y, \delta\phi, \delta p) = F_2^v - 1 = M^{-3} \sum_{k,l,m=1}^{M} K_2^v(k, l, m)$, (always taking $\vec{x} \equiv (y, \phi, p_\perp)$), with

$$K_2^v(k, l, m) = \frac{\int_{\Omega_{klm}} d^3\vec{x}_1 \, d^3\vec{x}_2 \, C_2(\vec{x}_1, \vec{x}_2)}{\left[\int_{\Omega_{klm}} d^3\vec{x} \, \rho_1(\vec{x})\right]^2} = \int_{\Omega_{klm}} d^3\vec{x}_1 \, d^3\vec{x}_2 \frac{k_2(\vec{x}_1, \vec{x}_2)\rho_1(\vec{x}_1)\rho_1(\vec{x}_2)}{\left[\int_{\Omega_{klm}} d^3\vec{x} \, \rho_1(\vec{x})\right]^2} , \quad (11)$$

i.e. the integration of k_2 involves a correction due to the shape of the one-particle three-dimensional distribution function $\rho_1(\vec{x})$. Eq. (11) as it stands is exact; horizontal versions have also been derived.[12] A first test of our model would therefore be to see if Eq. (7) or its horizontal equivalent obeys the data in $(y, \phi, p_\perp)$.

The theoretical $k_2(\vec{x}_1, \vec{x}_2)$ is further tested by comparing to factorial cumulant data of lower dimensions. For example, in (y, ϕ), the cumulant is

$$K_2^v(\delta y, \delta \phi) = M^{-2} \sum_{lm} K_2^v(l, m) \qquad (12)$$

with $p_\perp$ integrated over the whole window ΔP,

$$K_2^v(l, m) = \int_{\Omega_m} dy_1 dy_2 \int_{\Omega_l} d\phi_1 d\phi_2 \int_{\Delta P} dp_1 dp_2 \frac{k_2(\vec{x}_1, \vec{x}_2)\, \rho_1(\vec{x}_1)\rho_1(\vec{x}_2)}{\left[\int_{\Omega_m} dy \int_{\Omega_l} d\phi \int_{\Delta P} dp\, \rho_1(\vec{x})\right]^2}. \qquad (13)$$

Cumulants of other variable combinations and lower dimensions are obtained analogously. With these relations it is thus possible, given any three-dimensional theoretical function k_2 to compute factorial cumulants and moments for any combination of its variables. Doing this for different variables serves as a strong test of the theoretical function as the moments probe its different regions.

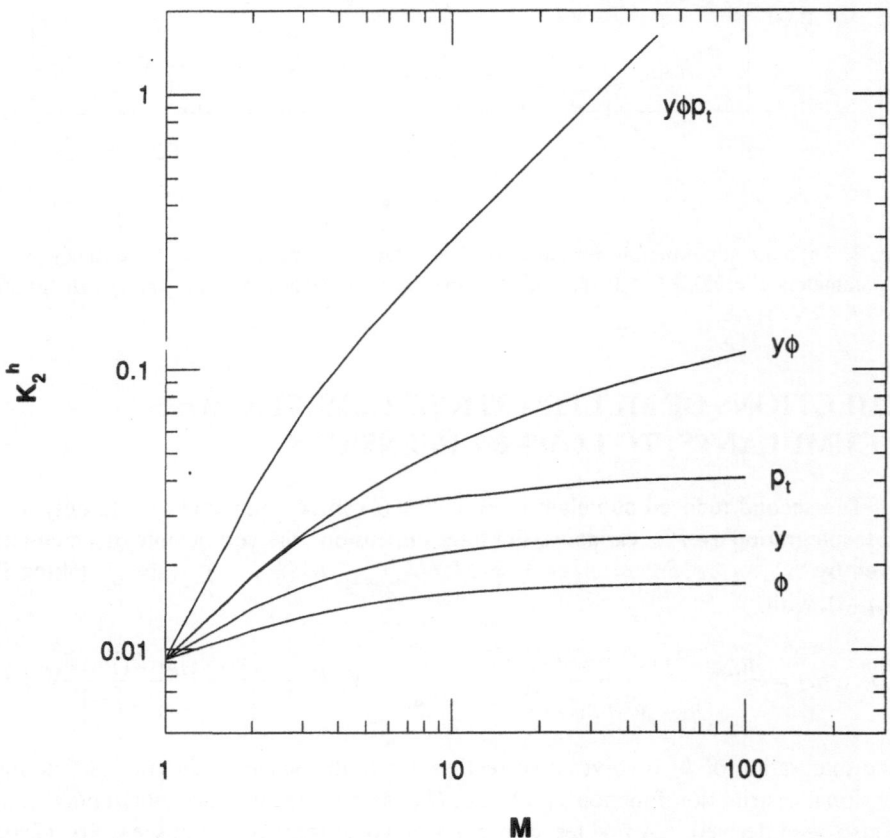

Figure 3. Theoretical predictions for the horizontal cumulant moments K_2^h for the same fixed parameters and NA35 distributions as in Fig. 2. The effect of the $p_\perp$ distribution is apparent. Comparison with NA35 data requires conversion to horizontal factorial moments F_2^h and a fit of a and ξ.

In Figures 2-3 we present our results for the vertical and horizontal factorial moments. In our calculation of the projections, we make the following approximations: We factorize the one-particle distribution into its separate variables: $\rho_1(\vec{x}) = \langle N \rangle_\Omega\, g(y)\, h(\phi)\, f(p_\perp)$, where the three distributions g, h and f are separately normalized over their respective total intervals ΔY, $\Delta\Phi$ and ΔP. The azimuthal distribution is taken as flat, $h(\phi) = 1/\Delta\Phi$. We use the full experimental parametrization for $f(p_\perp)$ provided by NA35. The choice of two parameters a and ξ in Figures 2-3 were given only as an illustration. They need to be determined by comparison with three-dimensional data. Once they are fixed, the two-dimensional and one-dimensional projections are genuine predictions of the theory.

SUMMARY

In summary, we have presented a three-dimensional effective field theory of multi-particle correlations, which gives no higer-order correlations, in agreement with the recent heavy-ion data. In our theory, two-particle correlations have Yukawa form and the corresponding integrated cumulants have singular behavior for small regions of phase space. This prediction seems to be in qualitative agreement with the recent NA35 data.[14] In addition, we have shown that once the parameters a and ξ are determined from comparison with three-dimensional data, our theory gives genuine predictions for the two-dimensional and one-dimensional cumulants. It will be interesting to see whether all our predictions are confirmed with future heavy-ion data.

Acknowledgements

Part of the work presented here was done in collaboration with H. Eggers whom we thank for many interesting discussions. This work was supported in part by the DOE grants DE-FG03-93ER40792 and DE-FG02-85ER40213.

REFERENCES

1. NA34 Collaboration, F. Corriveau *et al.*, *Z. Phys.* **C38**, 15 (1988); NA34 Collaboration, J. Schukraft *et al. Z. Phys.* **C38**, 59 (1988).

2. G. Baym, G. Friedman and I. Sarcevic, *Phys. Lett.* **B219**, 205 (1989).

3. A. Capella, K. Fialkowski and A. Krzywicki, *Phys. Lett.* **B230**, 149 (1989).

4. For a recent experimental review, see W. Kittel, in *Proceedings of the XXIII International Symposium on Multiparticle Dynamics*, eds. M. Block and A. White (World Scientific, Singapore, 1994), pg. 251.

5. For a recent theoretical review, see R. Hwa, in *Proceedings of the XXIII International Symposium on Multiparticle Dynamics*, eds. M. Block and A. White (World Scientific, Singapore, 1994), pg. 239.

6. P. Carruthers and I. Sarcevic, *Phys. Rev. Lett.* **63**, 1562 (1989).

7. P. Carruthers, H.C. Eggers, and I. Sarcevic, *Phys. Lett.* **254B**, 258 (1991).

8. P. Carruthers, H.C. Eggers and I.Sarcevic, *Phys. Rev.* **C44**, 1629 (1991).

9. NA35 Collaboration, I. Derado, in *Proceedings of the Ringberg Workshop on Multiparticle Production*, ed. R.C. Hwa, W. Ochs and N. Schmitz (World Scientific, 1992) pg. 184; NA35 Collaboration, P. Seyboth *et al.* , *Nucl. Phys.* **A544** 293 (1992).

10. H. C. Eggers, Ph. D. Thesis, University of Arizona, 1991.

11. H.-Th. Elze and I. Sarcevic, *Phys. Rev. Lett.* **68**, 1988 (1992).

12. H.C. Eggers, H.T. Elze and I. Sarcevic, *Int. J. Mod. Phys.* **A9**, 3821 (1994).

13. D. J. Scalapino and R. L. Sugar, *Phys. Rev.* **D8**, 2284 (1973); J.C. Botke, D.J. Scalapino and R.L. Sugar, *ibid.*, **D9**, 813 (1974); *ibid.*, **10**, 1604 (1974).

14. NA35 Collaboration, P. Seyboth, in *Proceedings of the XXIII International Symposium on Multiparticle Dynamics*, eds. M. Block and A. White (World Scientific, Singapore, 1994), pg. 325.

INTERMITTENCY AND OTHER SCALING BEHAVIORS IN NUCLEAR COLLISIONS

Rudolph C. Hwa

Institute of Theoretical Science and Department of Physics
University of Oregon, Eugene, OR 97403

GENERAL REMARKS ABOUT INTERMITTENCY

I first make some general remarks about intermittency, particularly for heavy-ion collisions. Then I review scaling behaviors that should be expected in quark-hadron phase transition. First the treatment is for equilibrium systems where the Ginzburg-Landau formalism can be applied. Then, for nonequilibrium systems the approach is changed to using cellular automata, and self-organized criticality is found.

In the Round-Table Discussion on Multiparticle Dynamics the first speaker emphasized the existence of a scale in the correlation function, echoing the theme of this conference in honor of R. Hagedorn, whose discovery of the temperature bearing his name provides a definitive scale in high-energy physics. The remark that I made in that Discussion is to point out that in systems where many-body collective effects are important often the most significant features of their behaviors are without any scale. More specifically, in extended and dense quark-gluon systems there are scaling behaviors in the hadronic observables that characterize the transition from quarks to hadrons. To identify those observables and their scaling behaviors is the subject of this talk.

For several years there have been intensive investigations of the subject of intermittency,[1] which is the scaling behavior of normalized factorial moments F_q as functions of the resolution δ in which the multiplicity fluctuations are observed. In time it was found that the scaling properties are dominated by Bose-Einstein correlation among like-charge particles,[2] and attention was shifted to the study of two-particle correlations. As a consequence the scaling behaviors of F_q, or the cumulants K_q, of the unlike-charge particles have been neglected. Those effects are weaker in comparison to the BE correlation effects, but they are there[3] and should be understood independently. As is often witnessed in many areas of physics, the weaker signal sometimes carries more important information about a system. So my remark here is that the experimental study of F_q and K_q for as large a range of q as possible should be continued. The theoretical significance will be discussed in the following section.

In the nuclear community the interest in the study of intermittency has waned after it was found that the effects are very weak in nuclear collisions.[4] The good fit of the multiplicity distributions by negative binomials has even led to the claim that intermittency is "explained".[5]

However, my opinion is that none of the existing nuclear data have yet probed the dynamical fluctuation that is the heart of intermittency. For that it is important to remember that intermittency is a study of rare events with large multiplicity fluctuations in small bins. Recall the definition of F_q:

$$F_q(\delta) = \langle n(n-1)\cdots(n-q+1)\rangle_\delta / \langle n\rangle_\delta^q \quad . \tag{1}$$

It is nonvanishing only for $n \geq q$ in any bin of width δ, say in rapidity. When δ is small, $\langle n\rangle_\delta$ is also small, so $n/\langle n\rangle_\delta$ becomes large for any fixed q. In NA22 data for hadronic collisions, where $dN/dy \sim 2$, the minimum $n/\langle n\rangle_\delta$ even for $q=2$ varies from 2 to 10 for δ from 0.5 to 0.1. They are therefore from events populating the high tails of the KNO plots of P_n. To achieve the same in the nuclear data where $dN/dy \sim 20$ for O-Cu at $\sqrt{s} = 5 GeV$ and $\langle n\rangle_\delta$ consequently an order of magnitude higher, the minimum n in δ must also be an order of magnitude higher, which is very rare. In other words, in the present nuclear data on F_q for $q \leq 5$ most events contribute, so the dynamical fluctuations are overwhelmed by statistical fluctuations resulting in negligible F_q. It is only when q is large can rare events with large $n/\langle n\rangle_\delta$ contributions exhibit the more interesting dynamics with nonvanishing F_q. In simulation of soft processes by ECCO it has been found that q should be greater than 8 for S-Au collisions before intermittency features of F_q can show up.[6] Thus the study of intermittency in nuclear collisions has barely begun. Further efforts with higher statistics to study high q moments are absolutely necessary before any interesting information about the dynamics of hadronization in dense systems will be manifest.

A corollary to the above remark is that when K_q is found to be small for $2 \leq q \leq 4$, it does not mean that K_q for all higher q are also small. It is therefore very dangerous to formulate an interaction-free theory on the basis of vanishing K_q, as is done in Ref. 7.

Intermittency is the study of the power-law behavior

$$F_q \propto \delta^{-\varphi_q} \quad . \tag{2}$$

It is, however, not the only scaling behavior of interest. Even when (2) fails to be valid over a wide range of δ, other scaling laws may still be valid. Most notable among them is the linear behavior in the so-called Ochs-Wosiek plot,[8] which can be translated to read

$$F_q \propto F_2^{\beta_q} \quad . \tag{3}$$

In this relation the bin width δ is an implicit variable that is varied. It was found empirically that (3) is insensitive to the dimension of the phase space analyzed.[9] There has been no theoretical understanding of that phenomenon in the theory of high energy collisions. In the next section we show how such a behavior arises in a phase transition.

SCALING BEHAVIORS OF EQUILIBRIUM SYSTEMS IN PHASE TRANSITION

It is well known in statistical physics that large fluctuations occur at the critical point. How can we take advantage of that phenomenon in the case of quark-hadron phase transition in heavy-ion collisions and use it as a signature of the critical behavior? To that end the first question is on the choice of the appropriate observable in which one is to examine the fluctuations. The second question is on the choice of a reliable description of phase transition that involves those observables. It turns out that there are natural answers to both questions, not only for second-order phase transition (PT), but also for first order.

But first it is necessary to address the concerns of those who question the usefulness of studying hadronic products as probes of quark matter. There is a strong belief on unproven grounds that there is a hadronic gas phase that would randomize any hadronic signature of phase transition. My view is that if we do not know what the signatures are without the hadronic gas phase, we would never be able to check experimentally whether they are erased in reality, the physics of which is far from being uncontroversial. Besides, if condensed-matter physics is a reliable guide, the physics of critical phenomena is often more interesting than that of the matter on either side of the phase transition.

The formalism chosen to describe PT at a level intermediate between the microscopic theory of lattice QCD and the macroscopic theory of relativistic hydrodynamics is that of Ginzburg-Landau.[10] In that formalism it is possible to calculate the hadronic multiplicity distribution P_n in terms of a few parameters that specify the nature of the PT. For second-order PT the free-energy density is

$$\mathcal{F}[\phi] = a\,|\phi|^2 + b\,|\phi|^4 + c\,|\nabla\phi|^2 \quad , \tag{4}$$

where ϕ is an order parameter related to the particle density ρ by $\rho = |\phi|^2$. The critical point is at where a changes sign, while b and c are positive. For first-order PT, b becomes negative and it is necessary to include a sixth-order term

$$\mathcal{F}[\phi] = a\,|\phi|^2 + b\,|\phi|^4 + k\,|\phi|^6 + c\,|\nabla\phi|^2 \quad . \tag{5}$$

In either case P_n can be obtained by the functional integral[11,12]

$$P_n = Z^{-1} \int \mathcal{D}\phi \, \frac{\langle n \rangle^n}{n!} e^{-\langle n \rangle - F[\phi]} \quad , \tag{6}$$

$$Z = \int \mathcal{D}\phi \, e^{-F[\phi]}, \quad F[\phi] = \int_V d^d r \mathcal{F}[\phi], \quad \langle n \rangle = \int_V d^d r \, |\phi|^2 \quad .$$

Thus the moments F_q can be calculated in terms of the parameters a, b, c, d, k and $V = \delta^d$ for $a < 0$ in the case of second-order PT,[11] and for $b < -2\sqrt{ak}$ in the case of first-order PT.[12]

The homogeneous case where $c = 0$ is easy to treat. The effect of inhomogeneity for $c \neq 0$ has been investigated and found to be insignificant.[13] We summarize here only the results for $c = 0$.

Our calculated results show that in both orders of PT the moments F_q possess the scaling behavior (3). Furthermore, the slope β_q can be well described by the amazingly simple formula

$$\beta_q = (q-1)^\nu \tag{7}$$

Moreover, the exponent ν is independent of $a, b,$ and d in the second order, while in the first order it depends on only one parameter g, where

$$g = -1 - \frac{b}{2\sqrt{ak}} \quad , \tag{8}$$

which characterizes the strength of the first-order PT.

In the case of second-order PT the scaling exponent is found to be

$$\nu = 1.304 \tag{9}$$

It is a number that emerges from the Ginzburg-Landau formalism without any numerical input. For first-order PT ν decreases with g, as shown in Fig. 1. Note that for small g, where the PT is of strong first order, ν is significantly different from (9), but for larger g its value approaches (9), indicating that the PT becomes weak first order.

Ordinarily, first- and second-order PT do not exhibit similar phenomenological features. Here, for particle production we have found that the factorial moments F_q for both cases possess similar scaling behavior (3), with the same formula (7) for β_q. Only the value of ν distinguishes the two cases. Since ν does not depend on a, b, and k separately, it is insensitive to the details of the system undergoing PT; its dependence on g implies that it is sensitive only to the nature of the PT, which is determined by the fundamental properties of QCD, including the number of quarks. Thus we have found a direct link between the observable and the underlying physics with very few intervening factors.

To test the above results would require heavy-ion collisions that can create quark-gluon plasma. So far existing data, when fitted by (3) and (7), all have ν values greater than 1.45.[11] Thus by our criterion, no PT has taken place describable by the Ginzburg-Landau formalism.

There is, however, an unexpected area of physics where our theoretical results can be tested. Since (3), (7), and (9) follow only from (4) and the Ginzburg-Landau description of PT, we need only search for another problem where the same description applies. That was found in quantum optics. A single-mode laser at the threshold of lasing behaves as in a second-order PT. The fluctuation of photon multiplicity is describable by the same formalism as we have used for pion multiplicity above. An experiment to check the scaling behavior (3) and (7) was carried out.[14] Not only is the scaling law verified, the value $\nu = 1.304$ is determined to a high degree of accuracy.

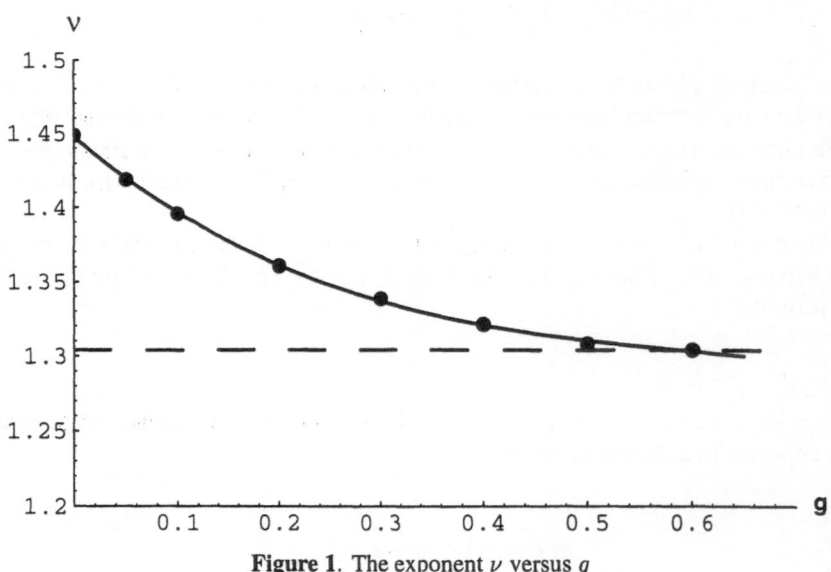

Figure 1. The exponent ν versus g

SELF-ORGANIZED CRITICALITY OF NONEQUILIBRIUM SYSTEMS IN PHASE TRANSITION

If a quark-gluon system is not in thermal equilibrium, phase transition to hadrons must still take place, but the formalism described above would not be applicable. Are there any features about the produced particles that can nevertheless identify the system as having gone through a mixed phase consisting of hadrons and quarks whether or not temperature is a valid notion? We have studied this problem along the lines of some recent development in statistical physics, commonly referred to as self-organized criticality.[15] The aim is to discover some new observables in terms of which the system exhibits scaling behavior without the external tuning of any parameter, such as taking T to the neighborhood of T_c theoretically (even though T is not under experimental control in a heavy-ion collision). What we have found surprisingly is that there is indeed a scaling law, if the observable is the size of a hadronic cluster produced.

Analogous to the avalanches in the sand-pile problem,[15] we measure the clusters that are emitted from the plasma. The clusters are formed either by growth from the initial size of nucleation, or by repeated coalescence between clusters on their way out of the plasma. Because of the complexity of the problem it is is not feasible to study this problem analytically, there being no reliable analytical formulation that can treat the fluctuation of cluster sizes. As with most problems in self-organized criticality, we use cellular automata that are based on simple rules which we adopt as sufficient representation of the essence of the dynamics involved.[16,17]

In brief, we work on a lattice whose initial size $L \times L$ represents a mixed region M in which quarks and hadrons can coexist. Each site is the location of a quark until it is turned into a hadron by nucleation with probability p. All hadrons move to the right with a certain drift velocity. On their way out they take random walks and can collide with one another; upon collision the clusters coalesce and the net size becomes bigger. Growth without coalescence is also possible, when a neighboring site of an existing cluster turns to a hadron with the same probability p as nucleation. When a cluster reaches the boundary, the cluster is emitted and the M region is reduced by the same number of sites, and the boundary is redrawn accordingly. When the right boundary reaches the left boundary, the transition is over. The emitted clusters then contribute to the cluster distribution $P(S)$ where S is the size of a cluster.

After running the cellular automata many times, we found that $P(S)$ exhibits a power-law behavior

$$P(S) \propto S^{-\gamma} \tag{10}$$

where γ is around 1.9. What is significant about this scaling law is that γ is only mildly dependent on the nucleation probability p, and is independent of the initial lattice size L and of the initial hadronic size S_0.[17] This is shown in Fig. 2 where (a) shows the results for a wide range of p and for $L = 16$, (b) for $L = 32$, (c) shows the comparison between $L = 16$ and 32 for $p = 0.2$, and (d) compares $S_0 = 1$ and 2. These universal features make the result insensitive to the experimental conditions of heavy-ion collisions, so long as a plasma of some appreciable size is formed.

Nothing has been adjusted to give rise to the scaling behavior. All of the complications of hadronization in QCD dynamics are summarized in the parameter p. The geometry, the flow of the clusters and the collisions among them all contribute to the origin of the scaling behavior. Due to the assumption of the existence of a mixed region, the simulation describes what is usually considered a first-order phase transition, although we have not explicitly used the temperature variable or the condition of thermal equilibrium.

There is the question of how these clusters are to be observed. While one can go into the intricacies of phenomenology, I have been told by some experimentalists that it is their job to find them, if they are there. What is nice is that it is very hard to think of other competing mechanisms to mimic the effect. Hadrons from jets are different; they are more spread out in phase space (including p_T) than those in these clusters which do not have large relative momenta. Experimental efforts to find the clusters are therefore strongly suggested.

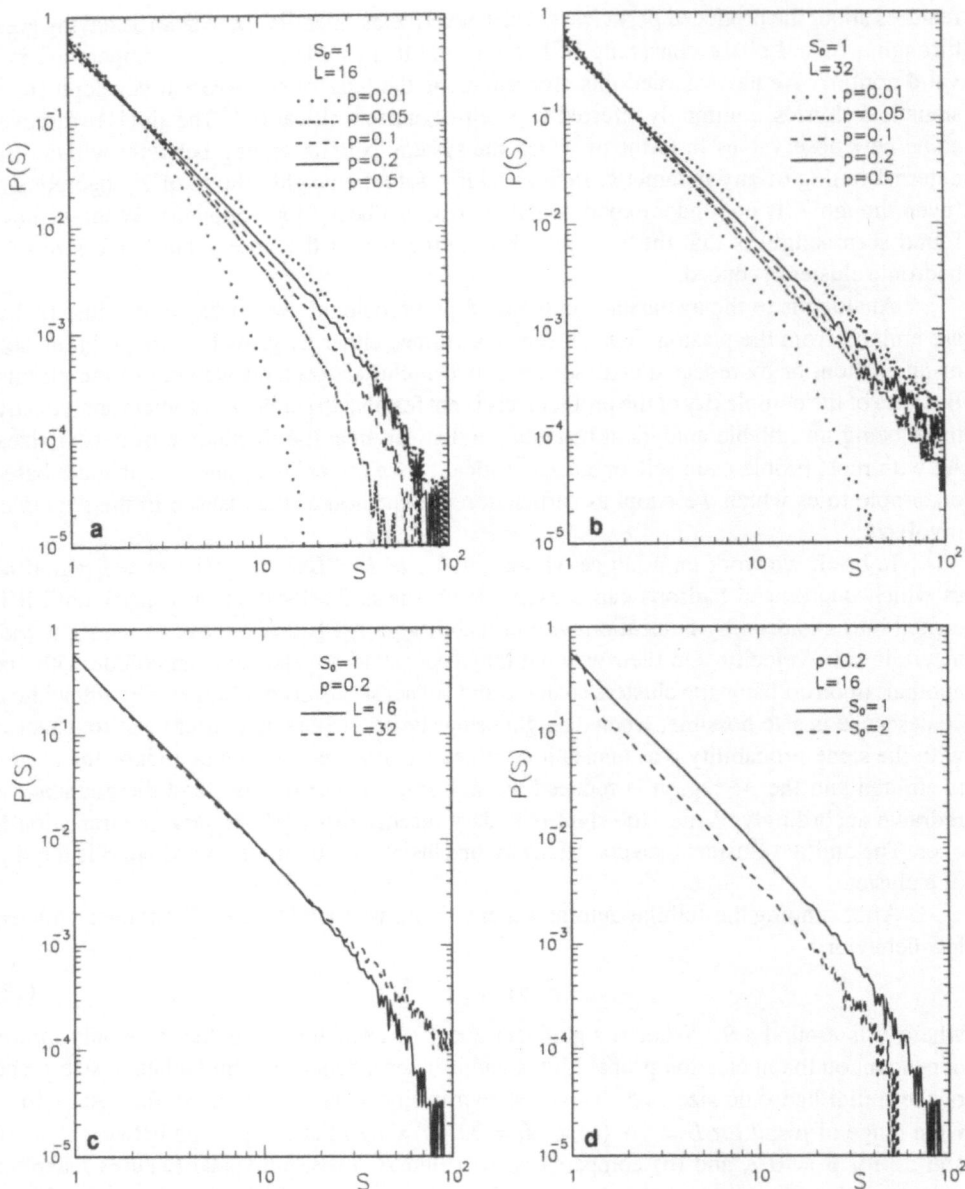

Figure 2. (a) shows the results for a wide range of p and for $L = 16$, (b) for $L = 32$, (c) shows the comparison between $L = 16$ and 32 for $p = 0.2$, and (d) compares $S_0 = 1$ and 2.

Acknowledgments

I am grateful to my collaborators C.S. Lam, M.T. Nazirov, J. Pan and S. Singh, who helped to perform the work reported here. This work was supported, in part, by the U.S. Department of Energy under Grant No. DE-FG06-91ER40637.

REFERENCES

1. A. Białas and R. Peschanski, Nucl. Phys. B273, 703 (1986); 308, 867 (1988).

2. *Soft Physics and Fluctuations*, edited by A. Białas, K. Fiałkowski, K. Zalewski and R.C. Hwa, (World Scientific, Singapore, 1994).

3. N.M. Agababyan *et al.* (NA22), Z. Phys. C59, 405 (1993).

4. P. Seyboth *et al.*, Nucl. Phys. A544, 293 (1992).

5. M.J. Tannenbaum, Proc. of 23rd International Symposium on Multiparticle Dynamics, Aspen, 1993, edited by M.M. Block and A.R. White (World Scientific, Singapore, 1994), p. 261; Mod. Phys. Lett. A9, 89 (1994).

6. R.C. Hwa and J. Pan, Phys. Rev. D 46, 2941 (1992).

7. H.C. Eggers, H.Th. Elze, and I. Sarcevic, Int. J. Mod. Phys. A9, 3821 (1994); I. Sarcevic, Aspen Proceedings;[5] and in these Proceedings.

8. W. Ochs and J. Wosiek, Phys. Lett. B214, 617 (1988).

9. W. Ochs, Phys. Lett. B247, 101 (1990); Z. Phys. C 50, 339 (1991).

10. V.L. Ginzburg and L.D. Landau, Zh. Eksp. Teor. Fiz. 20, 1064 (1950).

11. R.C. Hwa and M.T. Nazirov, Phy. Rev. Lett. 69, 741 (1992).

12. R.C. Hwa, Phys. Rev. C 50, 383 (1994).

13. R.C. Hwa and J. Pan, Phys. Lett. B297, 35 (1992); R.C. Hwa, Phys. Rev. D47, 2773 (1993).

14. M.R. Young, Y. Qu, S. Singh and R.C. Hwa, Optics Comm. 105, 325 (1994).

15. P. Bak, C. Tang, and K. Wiesenfeld, Phys. Rev. A 38, 364 (1988).

16. R.C. Hwa, C.S. Lam and J. Pan, Phys. Rev. Lett. 72, 820 (1994).

17. R.C. Hwa and J. Pan, Phys. Rev. C (to be published), OITS-541.

QCD GENERALISED FACTORIAL MOMENTS

J.-L. Meunier

I.N.L.N., Université de Nice-Sophia Antipolis,
Unité Mixte de Recherche du CNRS, UMR 129
1361 Rt. des Lucioles 06560 Valbonne, France

INTRODUCTION

In this paper we present an improved treatment of the QCD calculations of the Generalised Factorial Moments[1] including energy-momentum conservation and Running Coupling Constant effects in the DLA scheme.

The GFM[1] are a natural generalisation of the standard factorial moments to continuous or fractional orders of a multiplicity distribution. As far as the negative part of its spectrum is concern, the GFM are infrared sensitive quantities which may be of interest in the study of the parton hadronisation.

In the past year, three groups of authors[2-4] have shown that it was possible to define and compute a multi fractal dimension, $\mathcal{D}_q$, for QCD. Technically, this has been possible by computing the positive (and integer) order Factorial Moments of the distribution of particles in a restricted open angle Δ and could be compared with, say, the charged particle distribution in the $\mathcal{Z}_0$ decay at LEP.[5]

$$\mathcal{F}_q(\Delta) \equiv \frac{\langle n(n-1)..(n-q+1)\rangle_\Delta}{\langle n \rangle_\Delta^q} \propto \Delta^{(q-1)(1-\mathcal{D}_q/d)} \tag{1}$$

where d is the dimension of the phase space under consideration ($d = 2$ for the whole angular phase space, and $d = 1$ if one has integrated over, say the azimuthal angle). In the constant coupling case $\mathcal{D}_q$ is well defined and reads:

$$\mathcal{D}_q = \gamma_0 \frac{q+1}{q} \tag{2}$$

where $\gamma_0^2 = 4 C_A \alpha_s / 2\pi$, α_s is the strong interaction coupling constant, C_A is the gluon color factor. Dimensions $\mathcal{D}_q$ are called Fractal because they come from the natural generalisation to discrete variables of the standard moments which are used in the multifractal analysis of a continuous variable.[6] However, in this last field, the index q range is the whole real axis, while in our case it is restricted to positive integers.

The choice of the factorial moments as a specific tool for the study of the scaling behaviour of the high energy multiplicity distributions have been of importance. As a matter of fact, it has been noticed by A.Bialas and R.Peschanski[7] that the use of this observable permits to extract the dynamical signal from the Poisson noise in the Intermittency analysis of the multiplicity signal in high energy reactions. At first sight, the factorial moments $\mathcal{F}_q$ are only defined on integer and positive values of q and does not gives any insight on the negative part of the multifractal spectrum (if any) of the nuclear matter.

This has been noticed some years before by R.Hwa[8] who first proposed a multifractal analysis of the signal by means of the so-called G-moments. However those moments did not have the property of the factorial moments to disentangle the Poisson noise from the dynamical signal, and thus suffer from statistical uncertainties. Further works are in progress in this direction.[9]

From another point of view, one can understand, through their definition, that the standard Factorial Moments of the distribution are sensitive to the occurrence in the distribution of rare events of very high values of n as compared to its mean value n_b. For example, this is why the NA22 event[10] has been so important in the discovery of the intermittent properties of the high energy data.

In contrast, the negative part of the q multifractal spectrum focusses itself on the study of rare events of relatively low values of the studied variable, which corresponds in our case to low multiplicity events (as compared to the mean value of the variable). The moments presented in the following have this property and are a natural and non trivial generalisation of the standard ones.

However, in order to be efficient, one has to work with a multiplicity distribution with relatively high mean value, $n << n_b$. This is why we will not apply this analysis to the intermittency analysis of the data but to the global multiplicity distribution and to its scaling properties with respect to the energy.

Let us recall that the mean particle multiplicity produced by a gluon of energy E disintegrating in a cone of opening angle Θ_0 is given, in the Double Leading-log Approximation (DLA), by:[11]

$$n_b \propto [\frac{E\Theta_0}{\mu}]^{\gamma_0} \qquad (3)$$

were μ is the infrared cut-off of the theory, and that the corresponding global standard Factorial Moments follow by the KNO[12] phenomenon :

$$F_q = \langle n(n-1)..(n-q+1)\rangle = c_q n_b^q \qquad (4)$$

where the c_q are known constants. At first sight, it could be difficult to understand that one could find out some scaling properties from moments which scales with n_b. However, as it will be clear in the following, the Generalised Factorial Moments (GFM) analysis will show up a non trivial behaviour with energy.

On the other hand it is known that the standard QCD factorial moments calculated in the DLA approximation do not fit correctly the experimental data[14][13] and that important corrections are needed in order to describe the experiment reasonably . Some progress has been recently made in this direction.[15] In this paper we first present the asymptotic QCD GFM's and make a careful study of the corrections to be bring in the DLA approximation in order to take into account the Dockshister remark[15] and the effect of the running coupling constant. We show in particular that the introduction of this last effect is rather crucial, specially on the negative part of the GFM's spectrum. In order to gives these results we use a minimal hadronisation MonteCarlo scheme which solves exactly the equations at the partonic level (no colour in a gluonic world) and fully uses the parton hadron duality ideas; one parton disintegrate in n charged hadron (one in mean), accordingly to a poisson distribution. This model is rather crude and, as a consequence, we did not try to make a real fit to the experimental data. The surprise will comes from the fact that the theory is rather close to the experimental one. This is really a surprise because the negative part of the GFM's spectrum is rather sensitive to the infrared structure of the theory and as a consequence to the hadronisation scheme.

The paper is organised as follows : In section 2 we present the Generalised Factorial Moments (GFM) and their behaviour for some useful examples such as Poisson distribution, self-similar structure (KNO) or Negative Binomial distributions. Then, in section 3, the Generalized Factorial Moments of QCD in the Double Leading-log Approximation are presented and we discuss various corrections to the DLA approximation, such as the energy conservation constraints both in the constant coupling case and the running coupling constant case. We conclude in section 4.

THE GENERALISED FACTORIAL MOMENTS

Definition

We first recall the definition and main properties of the GFM. The standard Factorial Moments of a given multiplicity distribution P_n are defined as :

$$
\begin{aligned}
F_q &= \langle n(n-1)(n-2)(\ldots)(n-q+1)\rangle_P \\
&= \sum_0^\infty P_n\, n(n-1)(\ldots)(n-q+1)
\end{aligned}
\tag{5}
$$

which, using the properties of the Γ (Euler) function can be written as :

$$
F_q = \sum_0^\infty P_n \frac{n!}{\Gamma(n-q+1)}
\tag{6}
$$

and under this form can be continued in the complex q plane .

The importance of the Factorial Moments comes mainly from the fact that they can be derived from the generating function $G(z)$; from the theoretical point of view it is in general much easier to handle than the multiplicity itself. One has :

$$
F_q = \frac{\partial^q\, G(z)}{\partial z^q}\Big|_{z=1}, \quad G(z) = \sum_0^\infty z^n\, P_n
\tag{7}
$$

which one has to generalise to continuous or fractional values of q. Let us recall here the properties of the principal value distribution x^{-q-1} defined on $[0,\infty[$ with respect to the convolution of functions defined on the positive real axis (causal functions):[18]

$$
\frac{x^{-q-1} * f(x)}{\Gamma(-q)} = \int_0^x \frac{t^{-q-1}f(x-t)}{\Gamma(-q)} = \partial_q f(x)
\tag{8}
$$

where f is any well behaved function, continuous and indefinitely differentiable at x=0. When the exponent q is positive, the action of x^{-q-1} (principal value) on a test function φ is given by :

$$
(x^{-q-1},\varphi(x)) = \int_0^\infty x^{-q-1}\left(\varphi(x)-\left\{\varphi(0)+x\varphi'(0)+\ldots+\frac{x^{n_q}\varphi^{(n_q)}(0)}{n!}\right\}\right), \quad n_q = \mathrm{Int}(q)
\tag{9}
$$

where $\mathrm{Int}(q)$ is the integer part of the real part of q. When the real part of q is negative, the integral is defined and can be calculated. When q is positive or 0, the principal value ansatz (9) must be used, and, on positives integers the result is just the qth derivative of the function.

This comes from the fact that for q integer and positive the integral diverges together with the Γ function at the denominator of 8. Using this definition of the derivative one can define :

$$F_q = \partial_q G(z)\big|_{z=1} = \frac{1}{\Gamma(-q)} \int_0^1 G(1-t)t^{-q-1}dt \tag{10}$$

In order to be complete, one has to prove that this definition is consistent with that of Eq. (7). It is easy to verify that introducing the definition of the generating function G in Eq. (10) one recovers Eq. (6) thanks to the property of the (B) Euler function of the second kind.

Examples

i) The Poisson case. In this case, P_n is given by:

$$P_n = \frac{n_b^n}{n!} \exp(-n_b), \quad G(z) = \exp(n_b(z-1)) \tag{11}$$

where n_b is the mean value of n, and it is easy to obtain :

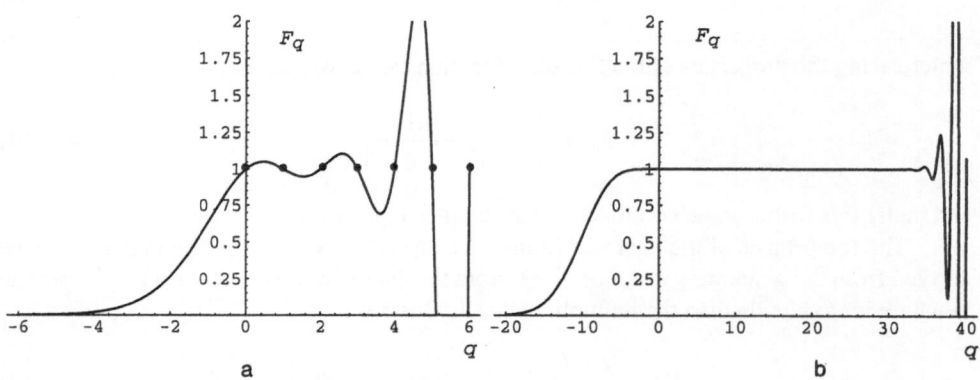

Figure 1. The Poisson GFM; a) : $n_b=1$, b) : $n_b=10$

$$
\begin{aligned}
F_q &= n_b^q \frac{\gamma(-q, n_b)}{\Gamma(-q)} \\
\mathcal{F}_q &= F_q/n_b^q = \gamma^*(-q, n_b)
\end{aligned}
\tag{12}
$$

where $\gamma(-q, n_b)$ is the incomplete γ function and γ^* the analytical incomplete γ function.[22] In any case, the value of γ^* on the integer and positive (or 0) values of q is 1 which is natural since one recovers here the standard factorial moments of the Poisson distribution. But one has to notice that, besides those points, the shape and behaviour of this function depends drastically on the n_b value; This function has in fact two types of behaviour. One for $n_b \simeq 1$ and one for $n_b \gg 1$. This is illustrated in figure 1.

If $n_b = 1$ say, figure (1 a) exhibits a steeply oscillating behaviour for $q \geq 0$ and goes rapidly to 0 when q is negative. This behaviour could prevent us from using those moments for small values of n_b where one cannot wait for a faithful behaviour of the moments and where the numerical formula (7) can be very unstable. This is not a surprise if one considers that those moments are devoted to the study of rare events of low n, ie $n \ll n_b$ For $n_b \gg 1$, say $n_b = 10$, figure (1 b) shows that the γ^* function is practically 1 in a large interval of q : $-n_b \ll q \ll 3n_b$.

From the point of view of the intermittency data, this indicates that the Poisson noise will be disentangled[6] from the dynamical signal only for those q greater than $-n_b$. As in those data the mean value of the number of particle in each bin tends rapidly to 0, one can understand that the fractional part of the moments does not gives any dynamical insight on the basic process.

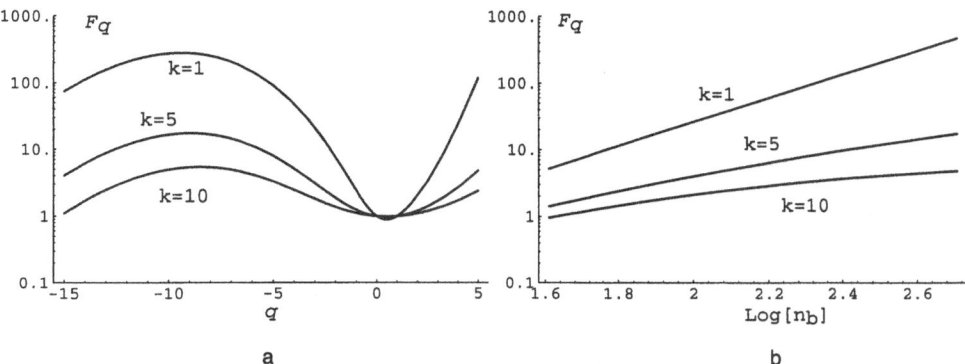

a   b

Figure 2. The GFM of the Negative Binomial Distribution. a) q behaviour for $n_b = 10$; upper curve $k = 1$, intermediate one : $k = 5$, lower one : $k = 10$; b) n_b behaviour at fixed $q = -5$, upper curve $k = 1$, intermediate one : $k = 5$, lower one : $k = 4$

ii) Self similar distributions.   The best way to build an asymptotic self-similar distribution is to construct P_n as a compound Poisson distribution (a particular case of these distributions is the NBD distribution). At the level of the generating function, this gives:

$$G(z) \;=\; H(z-1), \quad H(u) \;=\; \int_0^\infty \varphi(x)\exp(ux)dx, \tag{13}$$

where φ is the usual KNO function :

$$P_n \simeq \lim_{n_b \to \infty} \varphi(n/n_b)/n_b, \tag{14}$$

and one gets:

$$
\begin{aligned}
H(u) &= h(un_b) \\
F_q &= \frac{n_b^q}{\Gamma(-q)} \int_0^{n_b} h(-u)u^{-q-1}du
\end{aligned}
\tag{15}
$$

On this integral, one can notice that if n_b is sufficiently large, $\mathcal{F}_q = F_q/n_b^q$ tends to a constant and we recover the KNO result provides $h(u)u^{-q} \to 0$ when $u \to \infty$. Unless this condition is fulfilled, $\mathcal{F}_q$ will depends on n_b.

If, say, $h(-u) \simeq u^\alpha$, when $u \to \infty$, $\mathcal{F}_q$ will be KNO for $q >> -\alpha$. When $q < -\alpha$, the integral in 15 will be dominated by the high u behaviour of the integrand and

$$\mathcal{F}_q \simeq n_b^{-q+\alpha} \qquad (16)$$

Measuring the generalised moments of the distribution provides a rather nice tool for the study of the high u behaviour of the generating function.

As an example of this situation, we have calculated the moments of the Negative Binomial distribution :

$$h(u) = \frac{1}{(1 - u/k)^k} \qquad (17)$$

which gives :

$$\mathcal{F}_q = \frac{{}_2F_1(k, -q, 1 - q, -n_b/k)}{n_b^q \Gamma(1 - q)} \qquad (18)$$

where ${}_2F_1$ is the hypergeometric function of the second kind.[22]

The general trend of the reduced generalised moments of the distribution is given in figure (2-a) while the power-like behaviour of the moments is shown in fig (2-b).

QCD GENERALISED MOMENTS

QCD Double Leading-log Approximation

The QCD evolution equation for the generating function $G(Q, z)$ of the multiplicity distribution produced by a parton of energy E disintegrating in a cone of opening Θ_0 has been calculated since a while in the DLA approximation[11] and reads :

$$\frac{\partial G(Q, z)}{\partial \log(Q)} = G(Q, z) \int_0^1 \gamma_0^2 \frac{(G(xQ, z) - 1)dx}{x} \qquad (19)$$

where $Q = E\Theta_0$ is the hardness scale. Notice that an implicit infrared cut-off must be understood in this equation, $Qx > Q_0$. This cut-off tends to 0 in the weak coupling regime of the equation and will be of importance in section 3-2.

Let us first fix some notations. As in the preceding section, we define $H(Q, u) = G(Q, 1 - u)$, and the self similar solution (KNO) of the equation 19, $h(v)$, such as $H(Q, u) = h(n_b u)$. Further, for negative u, let us define $\mathbf{h}(y) = h(-\exp(y))$. With these notations the QCD solution obeys an integro-differential equation which does not depends explicitly on the coupling constant γ_0, and reads (using ref.4 with some slight change of notations) :

$$\frac{d^2 \log(\mathbf{h})}{dy^2} = \mathbf{h} - 1 \qquad (20)$$

This equation has been solved in an implicit way (for negative $u = -\exp(y)$[11]) :

$$y - y_+ = \int_{\log(2)}^{X(y)} \frac{du}{\sqrt{2(u - 1 + \exp(-u))}}$$
$$X(y) = \log(1/\mathbf{h}(y)), \quad y_+ = -.251 \qquad (21)$$

Starting from formula (15), the QCD GFM for negative q can be given as an y integral:

$$\mathcal{F}_q = \frac{1}{\Gamma(-q)} \int_{-\infty}^{\log(n_b)} \exp(-X(y) - qy)dy \qquad (22)$$

This expression is well adapted to the steepest descent technique which gives:

$$-X(y) - qy \simeq -(y - y^*)/2\sigma^2 - 1,$$
$$y* = -q - c_0, \quad \sigma^2 = 1 + \exp(-q^2/2 - 1), \quad c_0 = .41 \tag{23}$$

where c_0 has been numerically computed for $y > 2$. This gives asymptotically ($n_b \to \infty$):

$$\log(\mathcal{F}_q^{qcd}) = q^2/2 - (1 - c_0)q - 1 + (q - .5)\log(-q) + \log(\sigma^2)/2,$$
$$q << -2 \tag{24}$$

where we use the same approximation (steepest descent) for the Γ function.

Minimal Hadronisation Scheme

Let us now turn back to formula (19). We have now to take into account the finite infrared cut-off, Q_0 of the theory together with the various kinematical improvement one can bring in the DLA approximation.

In the fixed coupling constant case, the QCD generating function, including energy momentum conservation at the vertex obeys the following integral equation:

$$G(Q, u) = ue^{-S(Q)}$$
$$+ \int_{Q_0/Q}^1 \frac{dx}{x} e^{-(S(Qx)-S(Q))} \int_{Q_0/(Qx)}^1 \gamma_0^2 \frac{dw}{w} G(Qx(1-w), u)G(Qxw, u) \tag{25}$$

where $S(Q)$ is the Sudakov form factor:

$$S(Q) = \int_{Q_0/Q}^1 \frac{dx}{x} \int_{Q_0/(Qx)}^1 \gamma_0^2 \frac{dw}{w} = \gamma_0^2 \frac{\ln Q/Q_0^2}{2} \tag{26}$$

This equation is particularly well adapted to a Monte Carlo treatment.

The hadronisation is introduced via the following "minimal scheme". Each parton is supposed to produce incoherently (Poisson distribution with $\bar{n} = 1$) n hadrons randomly choose in its own available kinematical range (Angular Ordering). This procedure does not exactly fulfill the energy momentum conservation at the hadronic level, and in particular it permits that no hadrons comes out from a parton. This is a procedure close to the realization of a coherent state on a fock space. We verified however that our results does not depends on this particular step of the calculation.

The results are shown in Figure 3 were one has plotted the DLA approximation together with the result of the introduction of the energy conservation at the vertex. As pointed out by Dockshister the modified results brings the Factorial Moments of positive integer closer to the experimental results.However, the negative part of the spectrum is not really modified by introducing such modification in the evolution equation. This is to be understood in the following way: as noticed by Dockshister, the energy momentum conservation constraints are important near the $w = 1$ boundary of the phase space. This induces a strong modification of such quantities which are "infrared safe", namely whose behaviour near $w = 0$ are not pertinent. This is the case of the standard fractional moments.

On the contrary, the negative part of the GFM spectrum is infrared sensible. As a consequence the $w \simeq 1$ region of the phase space in 25 does not matters much.

We now introduce the effect of the running of the coupling constant in the theory. As a first remark, it is well known that as concern the positive part of the GFM spectrum, the DLA approximation of the theory brings to the same structure as that of the fixed coupling regime. This is exemplified in Figure 3 where the DLA approximation of the running coupling

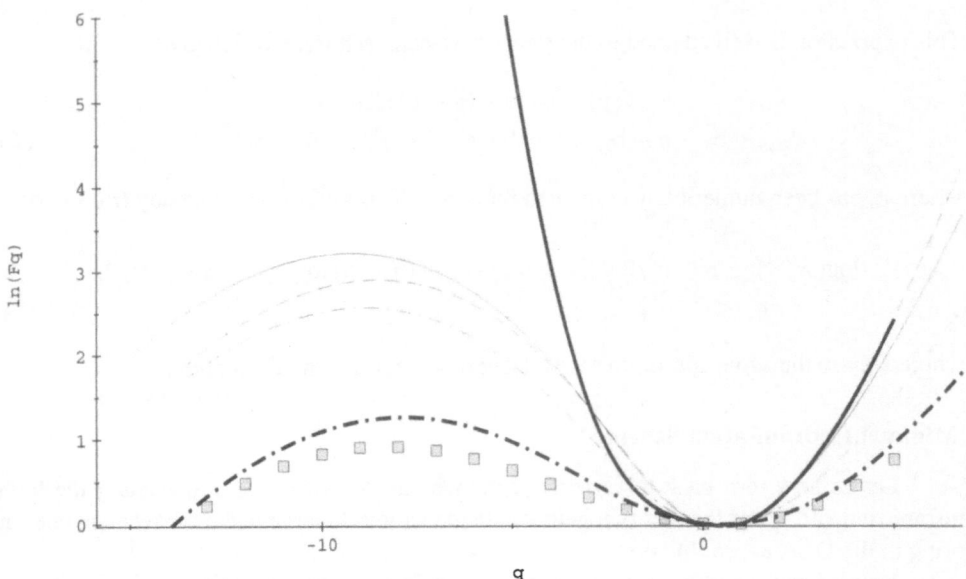

Figure 3. QCD GFM's as a function of q. Thick Continuous line : Asymptotic QCD; Continuous line: DLA ; dotted line : Running Coupling Constant DLA; dotted-dashed line : Improve DLA; Thick dotted-dashed line : Improved Running coupling constant case; squares: OPAL data (one hemisphere multiplicity)

constant version of the theory cannot be distinguished from the fixed coupling constant case. Now, the kinematical corrections to the RCC case brings to the following equation:

$$G(Q,u) = ue^{-S(Q)}$$
$$+ \int_{Q_0/Q}^1 \frac{dx}{x} e^{-(S(Qx)-S(Q))} \int_{Q_0/(Qx)}^1 \gamma^2(k_\perp) \frac{dw_\perp}{w_\perp} G(Qx(1-w),u)G(Qxw,u) \,(27)$$

where $k_\perp = Qxw(1-w)$, $w_\perp = 4k_\perp/Qx$ and $S(Q)$ is now:

$$S(Q) = \int_{Q_0/Q}^1 \frac{dx}{x} \int_{Q_0/(Qx)}^1 \gamma^2(k_\perp) \frac{dw_\perp}{w_\perp} = \gamma_0^2 \ln^2 Q/\Lambda \{ \frac{\gamma_{max}^2}{\gamma_0^2}(\ln(\frac{\gamma_{max}^2}{\gamma_0^2}) - 1) + 1 \} \quad (28)$$

where γ_{max} is the maximal value of the strong interaction coupling constant permitted in the scheme (infrared cut-of of the theory). The results are shown on figure 3 and it is worthwhile to notice that the introduction of the energy corrections in the running coupling constant case greatly improve the results of the theory both for the positive and the negative part of the GFM spectrum. This comes directly from the fact that the increase of the running coupling constant is important both in the $w = 0$ and $w = 1$ limits of the phase space.

The presented results depends on three parameters Λ, Θ_0 the first maximal opening angle, and γ_{max}. We fix Λ to 200 Mev, Θ_0 to $\Pi/4$, and the maximum value of the QCD coupling constant ,γ_{max}, is fixed to 2.2 by requiring that 10.7 particles are produced in one hemisphere (OPAL results[14]). Notice that we have pushed the theory rather far at the end of the cascade. This is to be compared with the experimental results on Intermittency[19] which shows that, at the end of the cascade ($\delta_y \ll 1$), the multifractal dimension of the theory

$(\mathcal{D}_q \simeq \gamma$ is greater than 1) leading to a saturation of the Intermittency pattern. We have also checked that the results do not depend much on the precise value of Θ_0 we used.

CONCLUSIONS

The Generalised Factorial Moments are a natural and comprehensive generalisation of the known factorial moments $(\mathcal{F}_q)$ analysis of a multiplicity distribution. They are defined for all q in the complex plane and, as far as the negative part of its spectrum is concerned $(q < 0)$, are sensitive to the QCD infrared structure .

The QCD calculation of the Generalised Factorial Moments for negative q has been performed in the double leading log accuracy and the role played by the kinematical improvement of the theory has been discussed to within a minimal hadronisation scheme based on parton hadron duality. We have shown that the leading log calculations are rather comparable to experimental data (LEP multiplicity distribution under the $\mathcal{Z}_0$) provided one takes into account both the correct kinematics and the running of the coupling constant.

The surprise comes from the fact that, using a minimal hadronisation scheme, and despite the fact that the negative part of the spectrum is infrared sensitive, the theory gives rather good predictions for this observable.

We will present elsewhere the local results of the theory, namely the intermittent structure of the jet, including all non asymptotic effects. These results can be nicely compared with the asymptotic predictions of.[2]

REFERENCES

1. P. Duclos, J.-L. Meunier, Preprint Nice INLN 94/6, to be published in *Zeit.Phys*. See also J.-L. Meunier in *Soft Physics and Fluctuations* A. Bialas, K. Fialkowski, K. Zalewski and R. Hwa editors, World Scientific 94.

2. Ph. Brax, J.-L. Meunier, R. Peschanski, Preprint Nice INLN 93/1-Saclay Spht/93-011, to be published in *Zeit.Phys*.

3. Y.L. Dokshitzer and I.M. Dremin *Nuclear Physics* **B 402** (1993) 139.

4. W. Ochs and J. Wosiek, *Phys. Lett.* **B289** (1992) 159, and *Phys. Lett.* **B305** (1993) 144.

5. Last LEP results: DELPHI Coll., P. Abreu et al. *Nucl. Phys.***B386** (1992) 471; ALEPH Coll., D. Decamp et al. *Z. Phys.***C53** (1992) 21; See previous references (including OPAL and L3 published results, 1991) therein.

6. For general reviews on the subject: A. Bialas *Nucl. Phys. A***525** (1991) 345c; R. Peschanski, *Int. J. Mod. Phys.***A6** (1991) 3681.

7. A.Bialas and R.Peschanski *Nucl. Phys.* **B273**(1991) 703 ; *Nucl. Phys.* **B309** (1991) 897

8. R.Hwa *Phys. Rev.* **D41**(1991) 1456, C.B.Chiu and R.Hwa, *Phys. Rev.* **D43**(1991) 100, W. Florkowski and R.Hwa *Phys. Rev.* **D43**(1991) 1548.

9. R.Hwa private communication and R.Hwa and J.Pan (to be published)

10. NA22 collaboration: M.Adamous et al., *Phys.Lett.***185B** (1987), 200

11. See for example *Basics of Perturbative QCD* Y.L. Dokshitzer, V.A. Khoze, A.H. Mueller and S.I. Troyan (J. Tran Than Van ed., Editions Frontieres) 1991, and the list of references therein.

12. Z. Koba, H.B. Nielsen and P. Olesen, *Nucl. Phys.* **B40**(1972)317.

13. E.D. Malaza and B.R. Webber, *Phys. Lett.* **B149** (1984) 501.

14. OPAL Collaboration: P.D.Acton et al *Zeit.Phys.***C53** (1992) 539.

15. Y.L. Dokshitzer *Phys. Lett.* **B305** (1993) 295.

16. J.-L. Meunier and R. Peschanski, *Nuclear Physics* **B 374** (1992) 327,

17. Y. Gabellini, J.-L. Meunier and R. Peschanski, *Zeit. Phys.* **C 55** (1992) 455.

18. I.M. Gelfand and G.E. Chilov, *Les Distributions, Dunod* Paris (1965).

19. DELPHI Collaboration *Nuclear Physics* **B 386** (1992) 471,

20. For a recent experimental review, see F. Verbeure, Proceedings of the XXII International Symposium on Multiparticle Dynamics, Santiago de Compostela, 13-17 July 1992 (to be published, World Scientific.)

21. G. Veneziano, Proc. 3rd. Workshop on Current Problems in HEP theory, Florence 1979, eds. Casalbuoni et al; John Hopkins University Press, Baltimore.

22. I.S.Gradshteyn and I.M. Ryzhik, *Table of Integrals, Series and Products* (Academic Press N.Y. 1980).

BOSON SPECTRA AND CORRELATIONS IN SMALL THERMALIZED SYSTEMS

Yuri M. Sinyukov

Bogolyubov Institute for Theoretical Physics
Kiev 252143 Ukraine

INTRODUCTION

The systems formed in ultra-relativistic nucleus-nucleus collisions have a dual nature. On the one hand they are quasi-macroscopic ones producing $10^3 \div 10^5$ hadrons and are eventually thermalized. The classical language of thermodynamics is used often to describe these systems. On the other hand, it follows from current interferometry data that typical effective sizes of the systems are a few *fm* even at the final stage of their evolution. The systems of such kinds have to be considered as quantum objects. They are dense, thermalized and involved in fast collective expansion due to contiguity with vacuum. So, at least locally, the systems can have very small lengths of homogeneity for density, temperature and collective velocities. The spectra and correlations in small thermal relativistic quantum-field systems are not trivial and can be understood on the base of the space-time scale. It includes the total geometrical lengths that the thermalized system occupies, $\overline{R}_i$, the local lengths of homogeneity (hydrodynamic lengths), $\overline{\lambda}_i$, and the wave- length of the quanta, $\lambda_p \propto 1/p^0$.

The statistical mechanics methods for processes of multiparticle production were initially applied by Fermi[1] who used the global equilibrium hadron system in volume V that means actually $\overline{R}_i = \overline{\lambda}_i$. The next step was made by Landau[2] who considered the hydrodynamic one-dimensional expansion of a pion gas that corresponds to the following relations between system lengths: $\overline{R}_T = \overline{\lambda}_T$, $\overline{R}_L >> \overline{\lambda}_L$. The method of calculation of the single particle spectra in a such picture is well-known now [3,4]. The role of the transversal hydrodynamic flows ($\overline{R}_T > \overline{\lambda}_T$, $\overline{R}_L >> \overline{\lambda}_L$) in explanation of the changing slope of the different single particle spectra have been demonstrated recently [5] . The results concerning the particle spectra were based mostly on the calculations using the local Bose-Einstein distribution in expanding hadron or quark-gluon gas. It has been shown[6] that the particle (Wigner) distributions are modified essentially compared with the Bose-Einstein one if $\lambda_p >> \overline{\lambda}_L$ due to specific behavior of relativistic quantum-field systems with small regions of homogeneity. We continue to discuss the effect for wide classes of the systems in this paper.

The smallness of the effective emitting region is the basic condition for the interferometry effect to be realized experimentally. One can expect that for interferometry analysis the hierarchy of the system lengths displays itself even stronger than for spectrometry. Firstly it has been shown[7] that in contrast to small homogeneous systems, $\overline{\mathbf{R}} = \overline{\lambda}$, when the interferometry radius $\mathbf{R}$ coincides with the geometrical radius of the system, $\overline{\mathbf{R}} = \mathbf{R}$, for longitudinally expanding system, $\overline{R}_L \gg \overline{\lambda}_L$, the longitudinal interferometry radius is defined by the length of homogeneity, $R_L \propto \overline{\lambda}_L$. No one correct analytical consideration was present up to now to generalize this statement for the case of three-dimensional expansion: $\overline{R}_T > \overline{\lambda}_T$, $\overline{R}_L \gg \overline{\lambda}_L$. This is also one of the aims of this paper. It has been shown[6] that the structure of the correlation function is modified seriously if the length of homogeneity is smaller than the wavelength of the particles $\lambda_p \gg \overline{\lambda}_L$. This effect appears under the same condition as for single particle spectra and has the same nature. We shall discuss the possibility to reveal this effect experimentally for the system with non-zero chemical potential and chemically non-equilibrium systems.

LENGTH SCALES FOR A LOCALLY THERMALIZED SYSTEM

The peculiarities of the spectra and correlations as we will show are strongly dependent on the ratio between different space-time lengths inherent in the system. The local length of homogeneity $\overline{\lambda}(x_0)$ is defined by the behavior of Wigner function. It means the length within which the deviation of the Wigner function is relatively small and is about the function value[8]

$$\frac{\left| f(p, x_0 + \overline{\lambda}) - f(p, x_0) \right|}{f(p, x_0)} = \qquad (1)$$

So the length of homogeneity will be

$$\left(\overline{\lambda}_i(x, p) \right)^{-1} = \left| \frac{\partial f(p, x)}{\partial x_i} \right| \Big/ f(p, x) \text{ or } \left(\overline{\lambda}_i(x_0) \right)^{-2} = \left| \frac{\partial^2 f(p, x)}{\partial^2 x_i} \right| \Big/ 2 f(p, x) \Big|_{x_0(p)} \qquad (2)$$

The last expression corresponds to the length of homogeneity at the point $x_0(p)$ where the distribution function is maximal, i.e.,

$$\frac{\partial f(p, x)}{\partial x_i} \Big|_{x_0(p)} = 0 \qquad (3)$$

In the framework of phenomenological approach the typical distribution function for bosons forming in ultra-relativistic A+A collisions is the Bose-Einstein distribution in the cylindrical tube with the Gaussian radius $\overline{R}_T$ where the system is considered on the set of "constant proper time" hypersurfaces, $\tau = \sqrt{t^2 - x_L^2}$, has the longitudinally boost-invariant expansion, $v_L = x_L / t$, and also has a transversal expansion, $u_T(r)$. In the standard designations the distribution function at $m_T\beta >>$ ($\beta(r)$ is the inverse of temperature) has the form

$$f(x,p) = \frac{1}{(2\pi)^3} \exp\left[-\beta(m_T \mathrm{ch}(y_L - y)\mathrm{ch}y_T - \mathbf{p}_T \cdot \frac{\mathbf{r}}{r}\mathrm{sh}y_T - \mu)\right] \exp\left[-\frac{r^2}{2\bar{R}_T^2}\right] \quad (4)$$

The first exponent is Bose-Einstein distribution $\exp(\beta p \cdot u(x) - \beta\mu) - 1]^{-1}$ where 4-velocity of the hadron gas u^μ is expressed via longitudinal $y_L(x_L)$ and transversal $y_T(r)$ rapidity; the second exponent is taking into account effectively the finite transversal size of the source. Let us consider the spectral length of homogeneity near the maximum of the distribution function in the longitudinal rest system of the particle, $P_L = 0$ or $y = 0$. Keeping in mind the interferometry analysis, we use longitudinal length of homogeneity $\bar{\lambda}_L(p)$, outward $\bar{\lambda}_O(p)$ along the $\mathbf{p}_T$- direction and $\bar{\lambda}_S(p)$ which is orthogonal to $\mathbf{p}_T$-direction. Let us suppose the gradients of temperature and chemical potential in transversal direction are much smaller than the corresponding gradient of velocity on the freeze-out surface. Then the analysis of the Eq.(3) shows that under the condition

$$\beta p_T \frac{dy_T}{dr}(r_0) << \frac{r_0}{\bar{R}_T^2} \quad (5)$$

the transversal rapidity at point of maximum is $y_T(r_0) << 1$ and the values we are interested in are

$$\bar{\lambda}_L(p) \approx \sqrt{\frac{2T}{m_T}}\tau \ , \quad \bar{\lambda}_S(p) \approx \bar{\lambda}_O(p) \approx \bar{R}_T \quad (6)$$

Under the condition

$$\beta m \frac{dy_T}{dr}(r_0) >> \frac{r_0}{\bar{R}_T^2} \quad (7)$$

th$y_T \approx p_T/m_T$ at the point r_0 and the spectral lengths of homogeneity tend to the thermal ones (in the infinite tube, $\bar{R}_T \rightarrow \infty$) which are

$$\bar{\lambda}_{th,L}(p) = \frac{\sqrt{2Tm}}{m_T}\tau(r_0), \ \bar{\lambda}_{th,S}(p) = \frac{\sqrt{2Tm}}{p_T}r_0, \ \bar{\lambda}_{th,O}(p) = \sqrt{\frac{2T}{m}}\left(\frac{dy_T}{dr}(r_0)\right)^{-1} \quad (8)$$

In the first case the transversal lengths of homogeneity characterized by the thermal Bose-Einstein distribution are larger than the geometrical radius of the system, $\bar{\lambda}_{th,S}\bar{\lambda}_{th,O} >> \bar{R}_T^2$, so the effective lengths of homogeneity approximately coincide with geometrical radius, see Eq.(6). In the second case the thermal lengths of homogeneity are smaller than the geometrical radius. In this case the effective lengths of homogeneity are the thermal ones, see Eq.(8). In these two limits, as we will show, the "interferometry microscope" measures the smaller of a thermal and a geometrical length. In our consideration we suppose that the temperature and chemical potential produce the lengths of homogeneity that are much larger than the collective flows and the geometry produce.

It is important to note that the condition (5) cannot be satisfied properly even at small p_T for specific class of the "soft" transversal flows with logarithmic or more slow dependence of transversal rapidity on the radius. But the qualitative the dependence of the lengths of homogeneity λ at large βm_T is preserved.

Now we consider the average length of homogeneity or hydrodynamic length $\bar{\lambda}(x)$. It is defined as the average of $\bar{\lambda}(x,p)$ over ensemble near point x :

$$\lambda^{-2}(x) = \left\langle \bar{\lambda}^{-2}(x,p) \right\rangle_p \qquad (9)$$

Using the local equilibrium Bose-Einstein function with non-zero chemical potential, one can calculate the length of homogeneity $\bar{\lambda}(x,p)$:

$$\lambda^{-2}(x) = (\nabla_\mu u^\nu)^2 - \frac{3}{T^2}(\nabla T)^2 + \frac{3}{nT}\nabla T \cdot \nabla n - \frac{1}{n^2}(\nabla n)^2 \qquad (10)$$

The last two terms appear only if chemical potential is not constant, $\mu(x) \neq const$. Here

$$\nabla_\mu = \frac{\partial}{\partial x^\mu} - u_\mu u \cdot \frac{\partial}{\partial x}$$

For one-dimensional boost-invariant hydrodynamic expansion the average length of homogeneity coincides with proper time

$$\lambda(x) = \tau \equiv \sqrt{t^2 - x_L^2} \qquad (11)$$

The final length we are introducing is the wave-length of the quanta λ_p. It might be the Compton length for massive field or the de Broglie one for massless field. We will consider

$$\lambda_p = 1/p^0 \qquad (12)$$

The locally equilibrium approach in zero- approximation (without dissipative effects) usually is based on the substitution $\beta = const \to \beta(x)$, $u = const \to u(x)$ in the Wigner function of global equilibrium system. Physically it is a good approach if quanta "feel" only the quasi-homogeneous part of a system. Under this condition the wave-length of the quanta has to be smaller than the length of homogeneity in a thermalized medium, $\lambda_p << \bar{\lambda}$. In the next section we limit ourselves just to this approximation. After that we consider the general case of an arbitrary ratio λ_p to $\bar{\lambda}$.

BOSE-EINSTEIN CORRELATIONS FOR THREE-DIMENSIONAL EXPANDING HADRON SOURCES

The description of the inclusive spectra and correlations for multiparticle production is based on the computation of the following type of averages

$$p^0 \frac{dN}{d\mathbf{p}} = \langle a_{\mathbf{p}}^+ a_{\mathbf{p}} \rangle \;,\; p_1^0 p_2^0 \frac{dN}{d\mathbf{p}_1 d\mathbf{p}_2} = \langle a_{\mathbf{p}_1}^+ a_{\mathbf{p}_2}^+ a_{\mathbf{p}_1} a_{\mathbf{p}_2} \rangle \;,\; \text{etc.} \qquad (13)$$

where $a_{\mathbf{p}}^+$ and $a_{\mathbf{p}}$ are the creation and annihilation operators and brackets $\langle ... \rangle$ mean the average over some density matrix corresponding to the state of the system on some space-time hypersurface. In S-matrix theory the state is $|out\rangle$-state at σ: $t = \infty$. In statistical (thermodynamic) models of multiparticle production the density matrix is chosen to be the statistical operator ρ and σ is the freeze-out hypersurface or it corresponds to the earlier stage of an evolution if one studies the dilepton or photon production from hadron matter and/or quark-gluon plasma.

The double particle spectrum in (13) is usually considered under supposition that the four-operator averages can be written in terms of the irreducible two-operator ones

$$\langle a_{\mathbf{p}_1}^+ a_{\mathbf{p}_2}^+ a_{\mathbf{p}_1} a_{\mathbf{p}_2} \rangle = \langle a_{\mathbf{p}_1}^+ a_{\mathbf{p}_1} \rangle \langle a_{\mathbf{p}_2}^+ a_{\mathbf{p}_2} \rangle + \langle a_{\mathbf{p}_1}^+ a_{\mathbf{p}_2} \rangle \langle a_{\mathbf{p}_2}^+ a_{\mathbf{p}_1} \rangle \qquad (14)$$

In this case the problem of multi-particle spectra is reduced to calculation of the $\langle a_{\mathbf{p}_1}^+ a_{\mathbf{p}_2} \rangle$ averages.

If one considers the identical particles 1 and 2 and uses the on-mass-shell approximation, the two-operator average can be expressed by means of the so-called single-particle Wigner function

$$f(x,p) = (2\pi)^{-3} \int d^4 u \, \delta(p \cdot u) e^{-iu \cdot x} \langle a^+ (p - \frac{u}{2}) a(p + \frac{u}{2}) \rangle \qquad (15)$$

where $p = (p_1 + p_2)/2$. Indeed, if the statistical operator ρ is defined on some hypersurface σ which can be closed by a plane surface $t = $ const, it is possible to use the equation

$$\int d\sigma_\mu \, p^\mu \, e^{ik \cdot x} = (2\pi)^3 \, p^0 \, e^{ik^0 t} \, \delta^3(\mathbf{k}) \quad \text{if} \quad p \cdot k = 0 \qquad (16)$$

that follows from the Gauss theorem. Then

$$\langle a^+(p_1) a(p_2) \rangle_\sigma = \int d^4 u \, p^0 \, e^{i(\Delta p^0 - u^0)t} \, \delta(p \cdot u) \delta^3(\Delta \mathbf{p} - \mathbf{u}) \langle a^+(p - \frac{u}{2}) a(p + \frac{u}{2}) \rangle_\sigma$$

$$= \int d\sigma_\mu p^\mu e^{i\Delta p \cdot x} f(x,p) \qquad (17)$$

If the analysis of the final spectra and correlations is carried out, one has to suppose the Wigner function $f(x,p)_\sigma$ to correspond to free (or weakly interacting) particles on the hypersurface σ. When σ is changed from event to event it is necessary to do the additional averages in σ in all final expressions.

The expressions for the single particle spectrum and the double particle correlation function follow immediately from Eqs. (13)-(15), (17)

$$p^0 \frac{dN}{d\mathbf{p}} = \int d\sigma_\mu p^\mu f(x,p) \qquad (18)$$

$$C(p_1, p_2) = 1 + \left(p_1^0 p_2^0 \frac{dN}{d\mathbf{p}_1} \frac{dN}{d\mathbf{p}_2} \right)^{-} \left| \int d\sigma_\mu P^\mu e^{i\Delta p \cdot x} f(x,P) \right|^2 \qquad (19)$$

where $P = (p_1 + p_2)/2$, $\Delta p = p_1 - p_2$.

As we will show in the next section the Wigner function will be approximately equal to the local Bose-Einstein distribution, $f(x,p) = f_{BE}(x,p)$ if $\lambda_p << \overline{\lambda}$. So, for longitudinal boost-invariant expansion with transversal rapidity distribution $y_T(r)$ and under the conditions: $P^0\tau >> 1$, $M_T\beta >>$ (M_T is half sum of transversal masses) we will use the Wigner function as it is given by the Eq. (4).

The boost-invariant integral measure in Eq. (19) is $d^2x_T dx_L/u^0 = \tau d^2x_T dy_L$. One of the transversal axes, x_o, we direct along the vector $\mathbf{P}_T$. This direction is called the "outward one". The direction orthogonal to it, x_s, is the "sideward one". In this coordinate system the exponent in Eq. (19) can be rewritten as the following:

$$\exp[i\Delta p \cdot x] = \exp\left[-i\left(q_{out}\left(x_o - t\frac{P_T}{P_0}\right) + q_{side}x_s + q_{long}\left(x_L - t\frac{P_L}{P_0}\right)\right)\right] \quad (20)$$

To calculate the integral in the expression (19) for the correlation function we use the saddle point method at the large parameter $m_T\beta >>$ in the Wigner function f(x,p) (4). Then we have for each *long-, out-, side-* projection of the correlation function the structure of the integral we are interested in[9] :

$$F_\alpha(\Delta p) = \int \varphi(x) \exp[\alpha S(x)] e^{i\Delta pg(x)} dx \xrightarrow[\alpha \to \infty]{} \exp[\alpha S(x_0)]\sum_{k=0}^{\infty} c_k \alpha^{-k-\frac{1}{2}} \quad (21)$$

where $\alpha S(x) = \ln f(x)$, $\alpha = \beta m_T$, saddle point x_0 is defined from the equations $S'(x_0) = 0$ that is equivalent to Eqs. (3) and

$$c_k = \frac{\Gamma\left(k + \frac{1}{2}\right)}{(2k)!}\left(\frac{d}{dx}\right)^{2k}\left[\varphi(x)\left(\frac{S(x_0) - S(x)}{(x - x_0)^2}\right)^{-k-\frac{1}{2}} e^{i\Delta pg(x)}\right]\Bigg|_{x=x_0}$$

Leaving the maximal coefficient at each power of Δp in the expansion of $|F_\alpha(\Delta p)|^2$, one can sum series (21) up and find

$$|F_\alpha(\Delta p)| = \left[\frac{2\pi}{\alpha S''(x_0)}\right]^{\frac{1}{2}} \varphi(x_0)\exp[\alpha S(x_0)]\exp\left[\frac{\Delta p^2(g'(x_0))^2}{2\alpha S''(x_0)}\right] \quad (22)$$

Due to $t = f(x_L)$ the correlation function is factorized in the frame of reference. where $P_L = 0$ only:

$$C(\Delta\mathbf{p}, P_L = 0) - 1 = (C(\Delta p_{out}) - 1)(C(\Delta p_{side}) - 1)(C(\Delta p_{long}) - 1) \quad (23)$$

We will use just this reference system. Taking into account that integral (21), (22) corresponds to single particle distribution (18) at $\Delta p = 0$ and the relationship between the Wigner function (4) and αS, $\alpha S(x) = \ln f(x)$, we find for each *i*- projection of the correlation function (19):

$$C(\Delta p_i) = 1 + \exp\left[-\frac{1}{2}(\Delta p_i)^2 R_i^2\right] \quad \text{where } R_i^2 = \frac{2f(x_0)}{\left|\partial^2 f(x_0)/\partial x_i^2\right|}\left(\frac{\partial g_i(x_0)}{\partial x_i}\right)^2 \quad (24)$$

So we have

$$R_i^2(P) = \overline{\lambda}_i^2(x_0(P))(g_i')^2 \tag{25}$$

The form of the functions g_i' is obvious from Eq.(20):

$$\left|g_{long}'\right| = 1, \quad \left|g_{side}'\right| = 1, \quad \left|g_{out}'\right| = 1 - \frac{P_T}{M_T}\frac{d\tau}{dr}(r_0) \tag{26}$$

The result (24) shows that the interferometry radii are actually the spectral lengths of homogeneity $\overline{\lambda}_i(x_0(p))$, see Eq. (2), that are corrected by the factor $g'(x_0)$. For the *side-* and *long- radii* is no correction. As for the *out*-radius, the correction takes into account the duration of radiation $\Delta\tau$ within the length of homogeneity $\overline{\lambda}_{out}(x_0)$. It appears due to the presence of the time-variable in the integral for the correlator (see Eq. (20)). Finally, if $r_0 = x_{out,0} \neq 0$, the interferomery radii in 3-dimentional expanding system in ($P_L = 0$)-frame of reference are:

$$R_L = \overline{\lambda}_L(p) = \tau\sqrt{\frac{2T}{M_T \mathrm{ch} y_T}}, \qquad R_S = \overline{\lambda}_S(p) = r_0\sqrt{\frac{2T}{P_T \mathrm{sh} y_T}}$$

$$R_O = \overline{\lambda}_O(p) + \frac{P_T}{M_T}\Delta\tau = \tag{27}$$

$$\sqrt{\frac{2T}{m}}\left(y_T'' \mathrm{sh}(y_T - \eta_T) + (y_T')^2 \mathrm{ch}(y_T - \eta_T) + T/m\overline{R}_T^2\right)^{-\frac{1}{2}}\left(1 - \frac{P_T}{M_T}\frac{d\tau}{dr}(r_0)\right)$$

where η_T is the average transversal rapidity of a pair, $y_T = y_T(r_0)$, $\tau = \tau(r_0)$.

The correction to the *out*-radius is linear in the duration time $\Delta\tau$. It is typical in general when the radiation process continuous during the matter evolution. In this case there is the dependence between the radius and the freeze-out time $\tau(r)$ because at each time τ most of the particles that are leaving the system are near its surface. For each momentum p the intensity of the radiation is at its greatest near the point $r_0(p)$. In a more artificial picture from the point of view of the hydrodynamics evolution when the process of radiation in time is not correlated with its intensity in configuration space, the correction to the *out*-radius due to duration in time is square-law in $\Delta\tau$ [10]:

$$R_O^2 = \overline{R}_T^2 + \frac{P_T}{M_T}\Delta\tau^2 \tag{28}$$

In our approach we have the result (28) in the case $\tau(r) = const$ after additional averaging in τ with the weight $\sim\exp(-\tau^2/2\Delta\tau^2)$.

If the thermal lengths of homogeneity are much larger than the geometrical sizes, $\overline{\lambda}_{th,S}\overline{\lambda}_{th,O} \gg \overline{R}_T^2$, see Eq.(5), we have the results (6) for spectral lengths of homogeneity or for the interferometry radii (27). In this case *side-* and *out-* radii are approximately a constant, *long-* radius is decreasing as $/\sqrt{M_T}$. For the intensive transversal flow (the large gradient of transversal velocity, see Eq. (7)), the spectral lengths of homogeneity are approximately the thermal ones (8). If the freeze-out proper time $\tau(r_0(p))$ is smooth enough in P_T, the *long*-radius defined by the Eqs.(8),(27) will decrease as $/M_T$. instead of inverse of square root dependence for very soft transversal flows. As to the *side-* and *out-*radii, their behavior strongly depends on $r_0(p_T)$ and $y_T'(r_0(p_T))$ behavior. If these functions are rather smooth, the *side*-radius decreases as $/p_T$ and the *out*-radius is approximately constant. It is easy to illustrate in this important example:

$$y_T(r) = \left(r^2/2\bar{R}_v^2\right)^n \tag{29}$$

Then $r_0^2, y' \propto \ln^n (M_T/m)$ that conforms to the above statement concerning the momentum dependence of the interferometry radii. The important analytically solved case is $n=1$. Then we have for single particle spectra the inverse power behavior at large m_T :

$$p_0 \frac{d^3N}{d^3p} \propto e^{-\beta m_T} \text{ if Eq. (5); } \quad p_0 \frac{d^3N}{d^3p} \propto \left(\frac{2p_T}{m}\right)^{-2-\frac{R_v^2}{\bar{R}_T^2}} \text{ if Eq. (7)} \tag{30}$$

In the last case when conditions (7) is satisfied, i.e., $\beta m \bar{R}_T^2 >> R_v^2$, the single particle spectra can have the power-low decreasing. The P_T -behavior of the correlation functions then changes dramatically in comparision with Eqs.(6), (27):

$$R_L = \tau \frac{\sqrt{2Tm}}{M_T}, \qquad R_S = \frac{2\bar{R}_v \sqrt{Tm} \ln^{\frac{1}{2}}\left(\frac{M_T + P_T}{m}\right)}{P_T}$$

$$R_O = \frac{\bar{R}_v}{\ln^{\frac{1}{2}}\left(\frac{M_T + P_T}{m}\right)} \sqrt{\frac{T}{m}}\left(1 + \frac{P_T}{M_T}\left|\frac{d\tau}{dr}(r_0)\right|\right) \tag{31}$$

All this demonstrates the correlation between the single particle spectrum's behavior and momentum dependence of interferometry radii and how these peculiarities are connected with the length scale in the emitting system.

It is important to mention also the specific class of the transversal flows with logarithmic or more slow dependence of transversal rapidity on the radius, e.g., $v_T = \text{th} y_T = r/\sqrt{\bar{R}_v^2 + r^2}$. Under the conditions (7) when the interferometry radii are equal to the thermal lengths of homogeneity, the *long*-radius is decreasing as $/M_T$, *side*-radius is constant, and *out*-radius grows when P_T increases. It allows one to distinguish the different types of flows.

The general consequence of the intensive transversal flows is that the longitudinal radius decreases stronger then $/\sqrt{M_T}$ when P_T grows, and there are noticeable differences in slopes dR_i/dP_T: $dR_{long}/dP_T \le dR_{side}/dP_T < dR_{out}/dP_T$. The last value is approximately zero or even positive for the systems with developed transversal flows.

METHOD OF LOCALLY-EQUILIBRIUM STATISTICAL OPERATOR

The description of the inclusive spectra and correlations for multiparticle production is based on the computation of the averages (13) over states of a system which consider of some hypersurface σ. The operator ρ for local equilibrium systems is defined from the maximum entropy principle and has the form[11]

$$\rho = e^{-S(\sigma)} = \frac{1}{Z}\exp\left[-\alpha\int d\sigma_\gamma\left(\beta_\nu \,\hat{T}^{\gamma\nu}(x) - \mu\beta\hat{J}^\gamma(x) - \mu_0\beta\hat{J}_0{}^\gamma(x)\right)\right] \equiv \rho(\alpha)|_{\alpha=1} \quad (32)$$

where $\hat{T}^{\gamma\nu}(x)$ is the operator of the energy-momentum tensor, $\hat{J}^\gamma(x)$ is the current density operator, for the pion field it is $\hat{J}^\mu = i(\varphi^*\varphi^{;\mu} - \varphi^{*;\mu}\varphi)$, μ is the charge chemical potential, $\hat{J}_0^\mu$ is the particle flow density operator for chemically non-equilibrium systems, $\hat{J}_0^\mu = i(\varphi^{(-)}\cdot\varphi^{(+);\mu} - \varphi^{(+)}\cdot\varphi^{(-);\mu})$, (+) and (-) corresponds to positive- and negative-frequency field components, μ_0 is a chemical potential for number of the particles, $\beta^\nu = u^\nu/T$ (u^ν is hydrodynamic 4-velocity, T is the temperature), $\beta = \sqrt{\beta_\nu\beta^\nu}$, . The basic model for the consideration is the scalar field.

The main idea of the calculation of the averages (13) is presented in Ref. [6]. Let us introduce the operators for all the field components of the pion triplet:

$$\mathbf{A}(p,\alpha) \equiv \begin{pmatrix} a^+(p,\alpha) \\ a(p,\alpha) \end{pmatrix} = Z^2\rho(\alpha)\mathbf{A}(p)\rho(-\alpha); \quad \mathbf{A}(p) \equiv \begin{pmatrix} a^+(p) \\ a(p) \end{pmatrix} \quad (33)$$

Then

$$\mathbf{A}(p,\alpha=0) = \mathbf{A}(p)$$

$$\langle\mathbf{A}(p)\,a(p')\rangle = \langle a(p')\mathbf{A}(p,\alpha=1)\rangle, \quad \langle\mathbf{A}(p)\,a^+(p')\rangle = \langle a^+(p')\mathbf{A}(p,\alpha=1)\rangle \quad (34)$$

and

$$\frac{\partial\mathbf{A}(p,\alpha)}{\partial\alpha} = [\mathbf{A}(p,\alpha),S(\sigma)] = \int d^3k\,\mathcal{K}(p,k)\mathbf{A}(k,\alpha)\mathcal{K} \quad (35)$$

where for π^0-pions $\mathcal{K} = K$, for π^+: $\mathcal{K} = K + iL$ and for π^-: $\mathcal{K} = K - iL$

$$K(p,k) = \begin{pmatrix} (G+iM_0) & \overline{G} \\ -\overline{G}^* & -(G^*-iM_0^*) \end{pmatrix}, \quad L(p,k) = \begin{pmatrix} M & \overline{M} \\ -\overline{M}^* & -M^*) \end{pmatrix}$$

$$G(p,k) = \frac{-i}{2(2\pi)^3}\int d\sigma\,e^{i(k-p)\cdot x}\frac{\beta(x)}{k^0}\left[k\cdot u\,p\cdot u - (k\cdot p - m^2)/2\right]$$

$$\overline{G}(p,k) = \frac{1}{(2\pi)^3} \int d\sigma \, e^{-i(k+p)\cdot x} \frac{\beta(x)}{k^0} \left[k \cdot u \; p \cdot u - (k \cdot p + m^2)/2 \right]$$

$$M(p,k) = \frac{-i}{2(2\pi)^3} \int d\sigma \, e^{i(k-p)\cdot x} \frac{\beta(x)\mu(x)}{k^0} \left[k \cdot u + p \cdot u \right]$$

$$\overline{M}(p,k) = \frac{-i}{2(2\pi)^3} \int d\sigma \, e^{i(k+p)\cdot x} \frac{\beta(x)\mu(x)}{k^0} \left[p \cdot u - k \cdot u \right]$$

$$M_0 = iM(\mu \to \mu_0)$$

The solution of the integro-differential equation with the Cauchy condition $A(p, \alpha = 0) = A(p)$ is

$$A(p, \alpha = 1) = A(p) + \sum_{n=1}^{\infty} \frac{1}{n!} \int d^3 k \, \mathcal{K}_n(p,k) A(k) \equiv A(p) + \hat{\mathcal{K}}(p,k) A(k) \quad (36)$$

here $\mathcal{K}_n(p,k)$ is the n-th iteration of matrix kernel $\mathcal{K}$.

The integral equations for $\langle a^+(p_1)a(p_2) \rangle$ and $\langle a(p_1)a(p_2) \rangle = \langle a^+(p_1)a^+(p_2) \rangle^*$ then follow from (34) and are

$$\hat{\mathcal{K}}(p,k)\langle a(p') A(k) \rangle = \begin{pmatrix} -p^0 \delta^3(\mathbf{p}-\mathbf{p}') \\ 0 \end{pmatrix} , \; \hat{\mathcal{K}}(p,k)\langle a^+(p') A(k) \rangle = \begin{pmatrix} 0 \\ p^0 \delta^3(\mathbf{p}-\mathbf{p}') \end{pmatrix} (37)$$

The matrix integral equation (37) is the basic one for calculation of the multiparticle spectra (13).

THERMAL WICK'S THEOREM FOR LOCALLY EQUILIBRIUM SYSTEMS

Basing on the Eqs. (34), (35), (37) one can reach the following results (see Ref.[6] for details)

$$\langle A(p_1) A(p_2) ... A(p_n) \rangle = \sum_{\mathcal{P}} \prod_{j' > j} \langle A(p_j) A(p_{j'}) \rangle, \quad (n = 2k) \tag{38}$$

The main peculiarity as compared with the classical result is the presence of additional terms likes $\langle a^+(p_1)a^+(p_2) \rangle$ and $\langle a(p_1)a(p_2) \rangle$ in the expansion. These non-zero terms arise because of the space-time finiteness of the homogeneity regions in locally- equilibrium systems.

BOSON SPECTRA AND CORRELATIONS IN LOCALLY-EQUILIBRIUM SYSTEMS

We consider here a physically important model of one-dimensional boost-invariant hydrodynamic expansion. We shall suppose that the temperature decreases with the radius as $\beta(r) = \beta_0 \, exp\left[r^2/2R_\beta^2\right]$. The solution of the integral equation (37) gives the following results for the Wigner function of the locally thermalized pion field (the method of calculation is given in [6]). In the reference system where particle momentum (or average momentum of pair) is $p = (p_0, p_T, 0, 0)$

$$\frac{dN}{d^3x \, d^3p} = f(x,p)_\sigma = \frac{\langle a^+a \rangle(t=0)}{(2\pi)^3} \tag{39}$$

where $\langle a^+a \rangle(t)$ is the Fourier transformation of $\langle a^+(p_1)a(p_2) \rangle$, and

$$\left\langle a_{\pi^-} a_{\pi^+} \right\rangle = \frac{(2\pi)^3}{2\omega} \overline{G^*} \left[f^+ + f^- \right], \quad \left\langle a_{\pi^-} a_{\pi^+} \right\rangle = \frac{(2\pi)^3}{2\omega} \overline{G^*} \left[f^+ + f^- \right]$$

$$f^+(t,\rho) = \frac{(2\pi)^{-3}}{\exp[\beta(\rho)(\omega(t)-\mu)]-1}, \quad f^- = f^+(\mu \to -\mu) \tag{40}$$

where

$$G(t) = -\frac{ch(\pi t)}{\pi \tau} \int_0^\infty dz \sqrt{(2m_T\tau)^2 + z^2} \; K_{i2t}(z);$$

$$\overline{G}(t) = \frac{i m_T}{2} \int_{-\infty}^\infty dz \frac{e^{i2tz}}{chz} H_1^{(2)}(2m_T\tau chz)$$

$$\omega(t) \equiv \sqrt{(G(t)+\mu_0)^2 - \overline{G}^*(t)\overline{G}(t)}$$

Here $K_\nu(z)$ is a modified Bessel function and $H_1^{(2)}(z)$ is Hankel function of second order. The value τ is the proper time of the expansion, m_T is transversal mass. The chemical potential μ_0 shows the rate of deviation of the pion gas from chemical equilibrium.

The ratio of the Wigner function of bosons in expanding matter to the Bose-Einstein distribution is shown in Fig.1. The main result is that the particle density as well as the momentum spectrum has an enhancement at small p^0 in comparison with the Bose-Einstein distribution at $p^0\tau \leq$. This enhancement becomes stronger when the systems are far from chemical equilibrium ($\mu \to \mu_c$) as it can be showen from (40). The critical value μ_c for Bose condensation reduces as compared with standard one $\mu_c = m$ and according to (40) is $\mu_c = m(m\tau)$ for $m\tau << 1$.

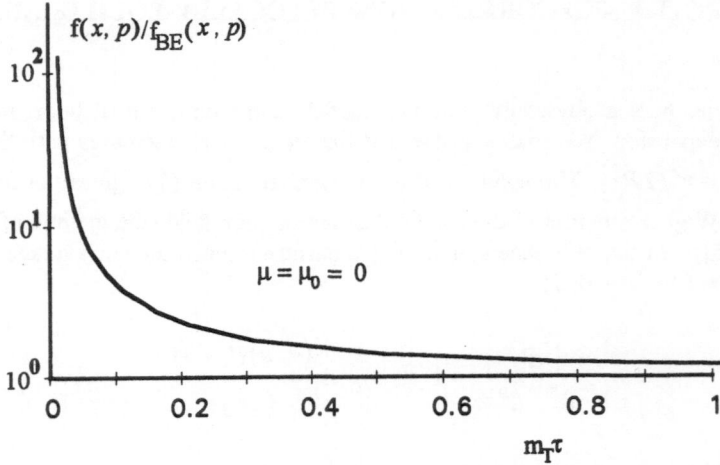

Figure 1. The ratio of the single particle Wigner function of bosons in expanding matter to the Bose-Einstein heat distributuion.

5. TWO- PARTICLE CORRELATIONS

The calculation of two- (many-) particle inclusive spectra is based on the generalized Wick's theorem (38). For true neutral identical particles it gives

$$p_1^0 p_2^0 \frac{dN_{inc}}{d\mathbf{p}_1 d\mathbf{p}_2} = \left\langle a^+(p_1)a(p_1) \right\rangle \left\langle a^+(p_2)a(p_2) \right\rangle + \left\langle a^+(p_1)a(p_2) \right\rangle \left\langle a^+(p_2)a(p_1) \right\rangle + \left\langle a(p_1)a(p_2) \right\rangle \left\langle a^+(p_2)a^+(p_1) \right\rangle$$

The last term is not equal zero due to inhomogeneity of the system. It brings the additional contribution in the correlation function of neutral bosons, e.g. , of $\pi^0\pi^0$ –pairs and due

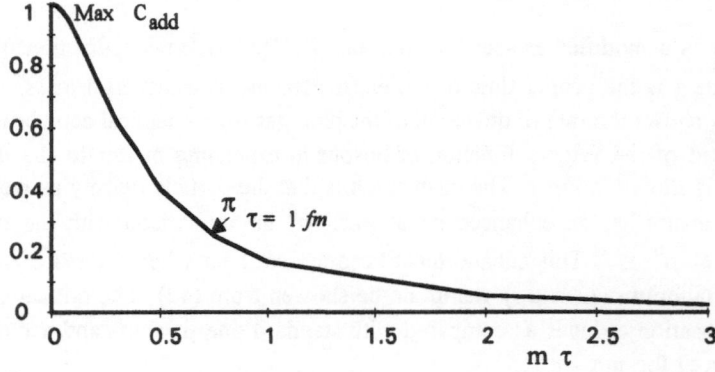

Figure 2. The additional contribution to the correlation peak for neutral and oppositely charged bosons

to isotopic invariance the correlation appears for oppositely charged pions also. The dependence of $\operatorname{Max} C_{add} = C_{add}(\mathbf{p}_1 = \mathbf{p}_2 = 0)$ on $m\tau$ for $\mu_0 = \mu = 0$ is demonstrated in Fig.2. For $\pi^0\pi^0$-pairs the departure from standard maximum value 2 of the correlation function is noticeable for τ around 1 fm and is then 0.25. The maximal value of the correlation function for massless boson will be 3. Notice that parameter $m\tau = \lambda_p/\bar{\lambda}$.

In order to demonstrate the influence of chemical potential on the behaviour of the correlation function at large chemical potentials in "standard" region $M_T\tau >>$ we show its approximation in the reference system where $P = (P_0, P_T, 0, 0)$

$$C(p_1, p_2) = 1 + \alpha_1 e^{-\frac{R_{L1}^2 q_L^2}{2}} e^{-\frac{R_{T1}^2(\mathbf{p}_{1T} - \mathbf{p}_{2T})}{2}} + \alpha_2 e^{-\frac{R_{L2}^2 q_L^2}{2}} e^{-\frac{R_{T2}^2(\mathbf{p}_{1T} + \mathbf{p}_{2T})}{2}} \qquad (41)$$

We will use the designations from the second section. Under the condition $\beta_0(M_T - \mu_0 \pm \mu) >>$ we have

. For $\pi^-\pi^-$-pairs: $\alpha_1 =$, $\alpha_2 = 0$, $R_{L1}^2 = \dfrac{2\tau^2}{\beta_0 M_T} = \bar{\lambda}_L^2$,

$$R_{T1}^2 = \bar{R}_T^2 \text{ if } \bar{R}_T^2 << R_\beta^2 \quad \text{and} \quad R_{T1}^2 = \dfrac{2R_\beta^2}{\beta_0(M_T - \mu_0)} = \bar{\lambda}_T^2 \text{ if } \bar{R}_T^2 >> R_\beta^2$$

. For $\pi^+\pi^-$-pairs: $\alpha_1 = 0$, $\alpha_2 = \left(1 - \dfrac{\mu_0}{M_T}\right)^{-2} \left(1 + \left(\dfrac{2M_T\tau}{\operatorname{ch}(\beta_0\mu)}\right)^2\right)^{-1} \left(1 + \left(\dfrac{2\tau}{\beta_0}\right)^2\right)^{-\frac{1}{2}}$

$$R_{L2}^2 = \dfrac{2\tau^2 \beta_0 M_T}{(\beta_0 M_T)^2 + (2M_T\tau)^2}, \quad R_{T2}^2 = \bar{R}_T^2 \text{ or } R_{T2}^2 = \dfrac{2R_\beta^2}{\beta_0(M_T - \mu_0)} = \bar{\lambda}_T^2 \text{ under the}$$

same conditions as for $\pi^-\pi^-$.

. For $\pi^0\pi^0$-pairs the coefficient α and the corresponding radii R_{T1} and R_{L1} are the same as for $\pi^-\pi^-$-pairs, the coefficient α_2 and the radii R_{T2} and R_{L2} are as for $\pi^+\pi^-$-pairs with $\mu = 0$ in all expressions.

If the chemical potentials are near the critical values, $M_T \approx m$, $\mu + \mu_0 \approx M_T$ we have

. For $\pi^-\pi^-$-pairs: $\alpha_1 = 1$, $\alpha_2 = 0$, $R_{L1}^2 \propto \sqrt{1 - \dfrac{\mu}{M_T - \mu_0}}$, $R_{T1}^2 = \bar{R}_T^2$ or $= 2R_\beta^2$

The transversal interferometry radius here is the smallest of two: $\bar{R}_T$ or $\bar{R}_\beta$

. For $\pi^+\pi^-$-pairs: $\alpha_1 = 0$, $\alpha_2 = 1$, $R_{L2}^2 \propto \sqrt{1 - \dfrac{\mu}{M_T - \mu_0}}$, $R_{T2}^2 = \bar{R}_T^2$ or $= 2R_\beta^2$

. For $\pi^0\pi^0$-pairs the coefficient α and the corresponding radii R_{T1} and R_{L1} are the same as for $\pi^-\pi^-$-pairs, the coefficient α_2 and the radii R_{T2} and R_{L2} are as for $\pi^+\pi^-$-pairs with $\mu = 0$ in all expressions.

As the result we can see that even at large $m\tau$ the additional terms in the correlator of oppositely charged bosons could be experimentally noticeable if the system is far from chemical equilibrium or has the large chemical potential: $|\mu| + \mu_0 \rightarrow m$.

CONCLUSIONS

We show that the experimentally observed particle spectra and correlation are very sensitive to the space-time structure of the emitting matter. The ratios between local lengths of homogeneity of the system, its geometrical sizes and wave-lengths of the particles define the peculiarities of the spectra and correlations. This is reflected in different behavior of the interferometry radii as depending on momentum regions. In particularly, the strong transversal expansion leads to more quick than inverse of $\sqrt{m_T}$ decreasing of the longitudinal interferometry radius. At the same time the side-radius decreases more quickly than the out-radius when transversal momentum increases. Even the structure of the correlation function depends on ratio of the quanta's wave-length to the homogeneity length. The simple analytical approximation for observed values allows to clear up experimentally the space-time structure of the systems formed in ultra-relativistic nucleus-nucleus collisions and the details of its evolution.

ACKNOWLEDGMENT

The research described in this publication was made possible in part by Grant No UC 2000 from the International Science Foundation.

REFERENCES

1. E.Fermi, Progr.Theor.Phys., 5:570 (1950).
2. L.D.Landau, Izvestiya AN SSSR, 17:51 (1953).
3. F.Cooper, G.Frye, Phys.Rev. D10:186 (1974).
4. M.I.Gorenstein, Yu.M.Sinyukov, Phys.Lett. B142:425 (1985).
5. E.Schuedermann, J.Sollfrank, U.Heinz,Phys.Rev. C48: 2462 (1993).
6. Yu.M.Sinyukov, Preprint ITP-93-8E, Kiev (1993), Nucl.Phys. A566:589c (1994).
7. Yu.M.Sinyukov, Nucl.Phys. A498: 151c (1989).
8. R.Balescu. "Equilibrium and Nonequilibrium Statistical Mechanics", Vol.2, A Wiley-Interscience Publication (1976).
9. M.V.Fedoryuk. "Asymptotic, Integrals and Series", Nauka, Moskow (1987).
10. T.Csorgo, S.Pratt, KFKI-1991-28/A (1986).
11. D.N.Zubarev." Nonequilibrium Statistical Thermodynamics" Nauka, Moscow (1971); S.A.Smolyansky, A.D.Panferov." Introduction in Relativistic Statistical Hydrodynanics of Normal Fluid", Saratov University Press, Saratov (1986).

CORRELATIONS AND STRONG INTERACTIONS

R.M. Weiner

Physics Department
University of Marburg, Marburg, Germany

INTRODUCTION

Although Quantum Chromodynamics (QCD) is considered at present to be the best candidate for a theory of strong interactions, it is not directly applicable to the most interesting and important aspects of strong interactions, i.e. to multiparticle production, which is a "soft" process. Instead one uses lattice QCD which predicts in principle the hadronic mass spectrum. The fact that the main content of Hagedorn's statistical bootstrap is also the mass spectrum proves the intuitive power of this approach, developed many years before the advent of QCD.

Why is the mass spectrum so essential for the understanding of strong interactions? In the Hagedorn theory the answer lies in the very derivation of this spectrum. The bootstrap "closes" only for a specific analytical form of this spectrum (the exponential one) and the fact that the bootstrap closes is the reflection of the strong nature of the interaction.

In the following I shall provide a further argument for the importance of the mass spectrum by relating it to another important phenomenological characteristic of strong interactions, namely the multiplicity distribution $P(n)$. I shall then show that the knowledge of this distribution is equivalent with the knowledge of correlations. The remainder of the talk will be devoted to a short review of the most important developments in the field of correlations in the last 5 years.

MULTIPLICITY DISTRIBUTIONS AND CORRELATIONS

In field theory various interactions are characterized by the values of the corresponding coupling constants. Phenomenologically the coupling constant manifests itself in the magnitude of cross sections and in the multiplicities and their distributions. Thus at a given energy of the system the mean multiplicity in electroweak interactions (ew) is much smaller than the corresponding one in strong interactions (s) $\langle n \rangle_{ew} \ll \langle n \rangle_{s}$. For the multiplicity distributions the corresponding inequality reads $P_{ew}(n) \ll P_{s}(n)$ (for $n > 1$). Furthermore we have $P_{ew}(n + 1) \ll P_{ew}(n)$ to be compared with $P_{s}(n + 1) \simeq P_{s}(n)$. In words the last relation* states that in strong interactions the probability to produce $n + 1$ particles is comparable

---

*Here we assume that the energy is high enough so that rest masses do not matter.

Hot Hadronic Matter: Theory and Experiment
Edited by J. Letessier *et al.*, Plenum Press, New York, 1995

with the probability to produce n particles and this can be considered as a phenomenological definition of strong interactions.

Rather than working with multiplicity distributions it is often convenient to consider their moments, to which they are mathematically equivalent. In particular let us consider the second order normalized factorial moment $f_2 = \langle n(n-1) \rangle / \langle n \rangle^2$. Here the averaging is performed with respect to the distribution $P(n)$ i.e. $\langle x^q \rangle = \sum_{x=1}^{\infty} x^q P(x)$. This moment is related to the second order correlation function

$$C_2(k_1, k_2) = \rho_2(k_1, k_2)/[\rho_1(k_1)\rho_1(k_2)] \tag{1}$$

by the equation

$$f_2 = \frac{\int C_2 \rho_1(k_1) \rho_1(k_2) d^3 k_1 d^3 k_2}{\int \rho_1(k_1) \rho_1(k_2) d^3 k_1 d^3 k_2} \tag{2}$$

In Eq.(2) ρ_1 and ρ_2 are the single and double inclusive cross sections. Similar relations hold for the higher order moments and correlations. Here from follows that *multiplicity distributions are determined by the correlations* and from the arguments presented above we conclude that correlations, too, reflect the nature of interactions. As a matter of fact this conclusion could have been reached without going through the multiplicity distributions because lattice calculations provide directly correlation lengths from which the masses of particles are derived.[†] We used the small detour through multiplicity distributions because multiplicity distributions and their moments are important physical quantities in themselves which have been in the center of interest of soft strong interaction phenomenology. They demand different experimental measurements than correlations or mass measurements and are treated usually separately and quite often the link between these two categories is overlooked.

NEW DEVELOPMENTS IN THE FIELD OF CORRELATIONS

Long Range Versus Short Range Correlations

It is customary to distinguish between long range correlations (LRC) and short range correlations (SRC) in momentum (rapidity) space. Although this distinction is not clear cut, in a first approximation one considers $Q = 1 GeV$ as the dividing line between these two types of correlations. Here Q is the invariant four momentum difference between two particles. In the eighties the change with energy of (normalized) $P(n)$ was discovered (KNO scaling violation) and different phenomenological models for this effect were proposed (cf.e.g.[1]), but in the following I shall mention only the quantum optical approach because it applies both to SRC and LRC. In the quantum optical approach one parametrizes the correlation in terms of a coherence length ξ and chaoticity p by writing the correlator of the (pionic) fields π

$$\langle \pi(y_1)\pi(y_2) \rangle = \langle n_{ch} \rangle exp(-|y_2 - y_1|/\xi) \tag{3}$$

$\langle n_{ch} \rangle$ is the mean number of chaotic particles related to the chaoticity p and the mean total multiplicity $\langle n \rangle$ by the equation $p = \langle n_{ch} \rangle / n$. The coherence length ξ which refers to fields is related to the conventional correlation length λ which refers to intensities (i.e. numbers of particles) by the equation $\lambda = \xi/2$. The broadening of the multiplicity distribution with energy was interpreted in this quantum optical approach as an increase of the correlation length and chaoticity with energy.[2] At that time only the UA5 data[3] for multiplicity distributions of charged particles were available and it could not be tested whether this interpretation applies indeed also to identical particles where Bose Einstein correlations (BEC) come into the game

---

[†] So far the correlations obtained from lattice calculations refer only to quark-antiquark states, but in principle more complex correlations between pairs of $q - \bar{q}$ states i.e. meson-meson correlations should be obtainable.

and in which SRC play a major part. With the advent of the newly analyzed UA1[4] and Na22[5] data this situation has changed. It turns out that the asymptotic values of the correlations functions for large Q coincide with the values of the respective moments, which according to eq.(2) proves that: 1.the small Q region, i.e. SRC do not contribute to the moments; 2. *the increase of moments with energy is due to the increase of* LRC.[6]

Property 2. can be due either to a change of the correlation length ξ as suggested previously or to a change of the inelasticity (impact parameter) or number of sources distribution or a combination of these. It appears that the last possibility is realized in nature and that the intensity and range of LRC increase with energy.

It is remarkable that conclusion 2. was reached by studying simultaneously multiplicity distributions *and* BEC which were considered up to that moment as referring almost exclusively to SRC. Further developments in the field of SRC and in particular in BEC will be discussed in the following.

The natural question to ask after this finding was whether this exhausted the energy dependence of correlations or whether there is also an energy dependence of SRC . It turns out that SRC and in particular BEC do also depend on energy.

Bose Einstein Correlations

<u>Correlation Length Versus Radius</u>

Everybody knows that BEC can be used for the determination of radii and lifetimes of sources. What is less known and has been clarified only recently[7] is the fact that actually in BEC at least two distinct length scales enter: a "radius" R and a correlation length L.The length L characterizes the region over which the generating currents J, responsible for the chaotic fields, are correlated. It defines thus in some sense the range of SRC. It is given by the space-time correlator

$$\langle J(x_1)J(x_2)\rangle \equiv C(x-y) = C(0)\exp[-\frac{(x_0-y_0)^2}{2L_0^2} - \frac{(\vec{x}-\vec{y})^2}{2L^2}] \tag{4}$$

for a "static", i.e. velocity independent source and by

$$C \propto exp[-(2(\tau_0^2/L_\eta^2)\sinh^2[(\eta_1-\eta_2)/2]] \tag{5}$$

for an expanding source. Here τ_0 is the proper formation time, η is the space- time rapidity and L_η the corresponding correlation length. For simplicity I have not written out explicitly in the last equation the transverse coordinate dependence and have assumed $\tau_1 = \tau_2 = \tau_0$.

To obtain the correlation function one has still to define the space-time distribution of the source which is characterized by the four-radius R. A typical form for such a distribution reads

$$f(x) \propto exp[-x^2/R^2] \tag{6}$$

With the definitions

$$F(k_1,k_2) = \int \frac{d^4k}{(2\pi)^4} f^*(k_1+k)C(k)f(k+k_2), \quad k_r^0 = E_r, \tag{7}$$

$$d_{12} = \frac{F(k_1,k_2)}{[F(k_1,k_1)\cdot F(k_2,k_2)]^{\frac{1}{2}}} \tag{8}$$

where we used the same symbols for the Fourier transforms of f and C, the second order correlation function reads

$$C_2^{++}(k_1,k_2) = 1 + |d_{12}|^2 \tag{9}$$

The two particle inclusive cross section depends now both on the radius R and on the correlation length L. Applying this formalism to the data at $\sqrt{s} = 22\,GeV$ and $630\,GeV$ one finds[8] that R *increases* with the energy while L *decreases*. It appears therefore that *the range of SRC decreases with s while the range of LRC increases.*

Subsequent to the introduction of the correlation length described above, Sinyukov (cf. ref.[9]) introduced a similar concept in statistical hydrodynamics. He uses the name "length of homogeneity" for it and it is easy to see that it plays in his hydrodynamical approach to BEC the same role as does the correlation length in the classical current approach to BEC.

Intermittency and BEC

This topic has preoccupied the high energy physics community for some time. It started with the suggestion by Bialas and Peschanski[10] that the dependence of factorial moments on the width of the rapidity distribution might reflect an intermittency property and therefore be power like. The fact that experimental data showed such a behaviour was interpreted by some authors as evidence for intermittency. Quite soon, however, it was realized that Bose Einstein correlations would lead to similar effects.[11] This possibility has been definitely confirmed by newer data where it was found that only identical particles showed the "intermittency" property. This is true both for the moments[12] as well as for the correlations of various order.

In order to cope with this observation, Bialas[13] introduced the new idea that the sources have fractal structure and therefore BEC which measure radii of sources show power dependence as a function of, say, Q. It is worth mentioning that the powerlike radius distribution introduced in[13] contains a cut-off length the origin of which is unknown. This cut-off length breaks the scaling behaviour and the comparison with data indicates that it corresponds to a momentum difference of 30-40 MeV, i.e. it represents a length of the order of 6-7 fm. For a hadronic system this is an exceedingly large value (cf. also[7]) and this fact alone may cast doubts about the validity of the entire idea. Nevertheless the concept of a fractal source is a very attractive one and to test this intriguing possibility, new analytical and numerical calculations were performed.[14] We started from a conventional source with a fixed size (eq.6) and calculated the corresponding correlation function at given Q. The resulting $C_2(Q)$ showed as a function of Q power behaviour. The analytical study of $C_2(Q)$ suggests that this happens as a consequence of the cylindrical phase space. Furthermore good agreement with the experimental UA1 data, which were the basis of the considerations of,[13] is obtained. Last but not least all parameters which enter this calculation are well understood physical quantities (e.g. the typical radii and correlation lengths involved are all of the order of 1-fm) and no exotic cut-off is necessary.

We conclude herefrom that conventional BEC with sources of *fixed* i.e. non fractal size, can account for the present experimental observations and that there is no need for intermittency assumptions. This may be a depressing conclusion for many experimentalists and theorists (the number of papers on this subject is a 3 digit figure!) but appears almost unavoidable.

Resonances and Bose-Einstein Correlations

The majority of secondaries produced in high-energy collisions are pions. A large fraction (between 40 % and 80 %) of these pions arise from resonances. Since the resonances have finite lifetimes and momenta, their decay products are created in general outside the production region of the "direct" pions (i.e., pions produced directly from the source) and resonances. As a consequence, the two-particle correlation function of pions reflects not only the geometry of the (primary) source but also the momentum spectra and lifetimes of resonances. Kaons are much less affected by this circumstances; however, correlation experiments with kaons are much more difficult because of the low statistics.

Given the complexity of the problem, there are at present only two main methods to study the influence of resonances on BEC: Monte-Carlo calculations[15] and hydrodynamical calculations.[16] I shall mention in the following some results obtained via hydrodynamics not only because this method is more related to the subject of this meeting but also because hydrodynamics is the only way to obtain information about the equation of state.

The correlation function of two identical particles of momenta p_1 and p_2 can be written as

$$C_2(p_1, p_2) = 1 + \frac{A_{12}A_{21}}{A_{11}A_{22}}, \tag{10}$$

where the matrix elements A_{ij} are given in terms of source functions $g(x, k)$ as

$$A_{ij} = \sqrt{E_i E_j} < a^\dagger(p_i)a(p_j) >= \int d^4x g(x_\mu, k^\mu) e^{iq^\mu x_\mu}. \tag{11}$$

Here, $a^\dagger(p)$ and $a(p)$ are the creation operator and the annihilation operator of a particle of momentum p, and the four-momenta $k^\mu = \frac{1}{2}(p_i^\mu + p_j^\mu)$ and $q^\mu = p_i^\mu - p_j^\mu)$ are the average momentum and the relative momentum of the particle pair, respectively. The interpretation of the source function $g(x_\mu, p^\mu)$ as the quantum analogue of the mean number of particles of momentum p^μ at the space-time point x_μ enables us to decompose g with respect to the origin of the produced hadrons. For instance, if the particles under consideration are two identical pions (e.g., two π^-), one has

$$g(x_\mu, p^\mu) = g_\pi^{dir}(x_\mu, p^\mu) + \sum_{res=p,\omega,\eta,...} g_{res \to \pi}(x_\mu, p^\mu), \tag{12}$$

where the labels dir and $res \to \pi$ refer to direct pions and to pions which are produced through the decay of resonances (such as p, ω, η, etc.), respectively. The contribution from a resonance decay $g_{res \to \pi}(x_\mu, p^\mu)$ can be expressed in terms of the source function of that resonance itself, $g_{res}^{dir}(x^*, p^{*\mu})$, as follows (from here on, quantities related to a resonance will be labeled with a superscript star). Consider a resonance of width Γ, which is created at a space-time point x_μ^* and after a proper time τ decays into $\pi + X$ at $x^\mu = (x^{*\mu} + (\tau/m^*)p^{*\mu}$. As the influences of the decay time τ are described by the probability distribution $\Gamma exp(-\Gamma \tau)$, one obtains

$$g_{res \to \pi}(x_\mu, p\mu) \tag{13}$$
$$= \int \frac{d^3p^*}{E^*} \int d^4x^* \int_0^\infty d\tau \Gamma exp(-\Gamma \tau) \delta^4 \left[x^\mu - \left[x^{*\mu} + \frac{\tau}{m^*} p^{*\mu} \right] \right] \Phi_{res \to \pi} g_{res}^{dir},$$

where $\Phi_{res \to \pi}(p^{*\mu}, p^\mu)$ describes the probability for a resonance of momentum $p^{*\mu}$ to produce a pion of momentum p^μ.

Assuming that the decay is governed by phase space one can determine the functions Φ. The source functions g are then calculated via hydrodynamics assuming a freeze-out at a given temperature T_f. It is worthwhile mentioning that the hydrodynamical code used is a fully 3 dimensional one. The importance of 3 solution resides in the fact that the transverse flow affects drastically not so much the transverse dimension of the system but rather the longitudinal one.[17]

Among the results obtained one should mention a significant difference in the effective longitudinal radii extracted from $\pi^- \pi^-$ correlations and $K^- K^-$ correlations. In the central region one has $R_\parallel^{K^-}/R_\parallel^{\pi^-} \simeq \frac{1}{2}$. This is due to the fact mentioned above that kaons are much less affected by resonances than pions. This effect has been experimentally confirmed by the Na44 collaboration. Another effect predicted by hydrodynamics and which has been apparently confirmed by the Na35 collaboration is the specific dependence of effective radii on rapidity.

New Effects in Bose Einstein Correlations; Breakdown of the Wave Function Approach

We come now to what is one of the most important developments in the field of BEC in the last years. It consists in the realization that the conventional understanding of BEC, namely that it is a correlation restricted to identical particles has to be qualified. In particular there exists a quantum statistical correlation also between particles and antiparticles. Although this new effect is quantitatively small and has not yet been observed experimentally, it is of far reaching significance. As a matter of fact, we are only at the beginning of the understanding of its implications. To judge the impact of this finding it is enough to mention that since the end of 1991 when this surprising effect was found,[18] further three different derivations of the effect were given,[19],[209](cf. also ref.[21] which is based on ref.[9]). The initial motivation of the authors of these three papers was that the results of[18] seemed so surprising that they were "unbelievable".[19] The surprising element is due to the fact that in most conventional approaches to BEC one uses an explicit symmetrization of the wave function which describes the two particle state and such a symmetrization applies of course only for identical particles. That the wave function approach is not always a convenient method for the study of BEC was known for many years because it describes only a state with a fixed number of particles, while experimentally the number of particles is usually not kept fixed (one measures inclusive probabilities rather than exclusive ones). An alternative to the wave function approach is the classical current approach, which is based on field theory i.e. second quantization. Strangely enough, although this alternative was known for many years, it had not been applied until 1991 but to identical particles, although, as will be shown below, it can be applied for particles-antiparticles as well.

Let us start with with some random currents $J_i(x)$ which create pions $\pi_i(x)$. The indices $i = 1, 2, 3$ refer to the isospin components. The simplest form of the Lagrangian for the field-current interaction reads

$$
\begin{aligned}
\mathcal{L}_{int}(x) &= J_1(x)\pi_1(x) + J_2(x)\pi_2(x) + J_3(x)\pi_3(x) \\
&= J_+(x)\pi^-(x) + J_-(x)\pi^+(x) + J_0(x)\pi^0(x)
\end{aligned}
\tag{14}
$$

Charged pions

$$
\pi^\pm = \frac{1}{\sqrt{2}}(\pi_1 \pm i\pi_2)
\tag{15}
$$

are created by complex conjugated currents $J_\mp(x)$, and neutral pions $\pi^0 = \pi_3$ are created by real currents $J_0(x) = J_3(x)$.

We use now the general space-time correlator eq.(4) or eq.(5) and the space-time distribution of the source $f(x)$ and assume a Gaussian form of the density matrix.

In momentum space there are then two types of current correlators

$$
\begin{aligned}
< J_m^*(k_1)J_m(k_2) > &= < J_+^*(k_1)J_+(k_2) > = F(k_1, k_2) \\
< J_m(k_1)J_m(k_2) > &= F(-k_1, k_2) \quad (m = 1, 2, 3)
\end{aligned}
\tag{16}
$$

where F has been defined in eq. (7).

Introducing now the creation and annihilation operators $a_m^\dagger$ and a_m of the pion field one obtains

$$
F(k_1, k_2) = (2\pi)^3\sqrt{4E_1E_2} < a_m^\dagger(k_1)a_m(k_2) >
\tag{17}
$$

$$
F(-k_1, k_2) = -(2\pi)^3\sqrt{4E_1E_2} < a_m(k_1)a_m(k_2) >
\tag{18}
$$

While eq.(17) leads to the usual $\pi^-\pi^-$ or $\pi^+\pi^+$ correlation, Eq.(18) leads to the "new" $\pi^+\pi^-$ correlation. Related to this new correlation is the fact that π^0-π^0 correlations are different from charged pion correlations. Introducing the normalized current correlators

$$d_{rs} = \frac{F(k_r, k_s)}{[F(k_r, k_r) \cdot F(k_s, k_s)]^{\frac{1}{2}}}, \quad \tilde{d}_{rs} = \frac{F(k_r, -k_s)}{[F(k_r, k_r) \cdot F(k_s, k_s)]^{\frac{1}{2}}} \tag{19}$$

one has

$$\begin{aligned}
C_2^{++}(\vec{k_1}, \vec{k_2}) &= 1 + |d_{12}|^2, \\
C_2^{+-}(\vec{k_1}, \vec{k_2}) &= 1 + |\tilde{d}_{12}|^2, \\
C_2^{+0}(\vec{k_1}, \vec{k_2}) &= 1, \\
C_2^{00}(\vec{k_1}, \vec{k_2}) &= 1 + |d_{12}|^2 + |\tilde{d}_{12}|^2
\end{aligned} \tag{20}$$

An estimate of the new terms shows that these are quite small and of the order of $\exp[-(E_1 + E_2)^2\tau^2]$ where $E^2 = k^2 + m^2$ and τ is the lifetime. Their importance however must not be underestimated and it consists, among other things in the fact that according to eq. (18) expectation values of the product of two creation operators do not vanish, which is a manifestation of squeezed states i.e. two particle coherent states, in analogy to usual coherent states for which the expectation value of a single annihilation operator is nonzero. It follows herefrom that squeezed states in particle physics are a quite natural phenomenon.This is at a first look very surprising, because in optics where these states were discovered for the first time, one has to prepare the system in a very special way in order to get them. At a closer look, however, one realizes that the actual observation of these effects in particle physics is not easy at all, because of the constraints imposed by the smallness of momenta of the particles. On the other hand, the fact that the new terms represent an *anti*correlation rather than the conventional corelation may help in their detection.

Another derivation of the new effects this time in the string model, is due to Bowler.[19] Although this derivation is only qualitative, it is interesting because it comes from a different point of view. Also the discussion of the relationship between the classical current approach and the string model is quite instructive.

Last but not least the work of Sinyukov and collaborators is worth mentioning. In ref.[20] it is argued that when using overlapping wave packets in the first quantization the appearance of the new terms is possible albeit in the view of the authors, improbable. On the other hand in second quantization by considering the density matrix appropriate for a system in local (but not global) equilibrium and in particular in boost invariant hydrodynamics a completely different picture emerges (cf.refs.,[21][9]). It turns out that terms of the form eq.(18) must exist and their form agrees with that derived in.[18]‡

An important corollary of[9] is the fact that the single inclusive distribution is also affected by the existence of squeezed states. This follows from the modification of the vacuum by the presence of terms of the form $< aa >$ and $< a^\dagger a^\dagger >$ and should lead to a strong enhancement of the low $p_\perp$ spectrum.

Finally one should mention a general quantum field theoretical derivation of the particle-antiparticle correlation[22] which proves explicitly that this effect is not an artifact of the classical current formalism, is not restricted to dense media and is not specific to any space-time symmetry assumption as the results of ref.[9] might suggest. Furthermore ref.[22] provides an estimate of the magnitude of quantum corrections to the classical current formalism.

---

‡Ref.[21] contains an erroneous statement about the result of ref.[18] referring to the maximum of the correlation function for identical neutral pions.It is stated that in ref.[18] $C_{max}(\pi^0\pi^0) = 3$ "independent of particle mass and source size". This is obviously in contradiction with eqs.(18,22,23) of ref.[18] From these eqs. it is clear that the maximum of the correlation function depends on mass, radius and lifetime and only in very special cases like zero lifetime or vanishing momenta and masses the current approach leads to $C_{max} = 3$.

At the end of this detour into BEC one may ask what are the implications of the recent findings for the starting point of this talk, the Hagedorn mass spectrum. A tentative answer to this question is to say that the parameters of the spectrum, i.e the Hagedorn temperature and the power of the pre-exponential factor must reflect these effects. It is the task of a future theory of strong interactions to show this. We may have to wait until Hagedorn's 80-th anniversary to see this.

Acknowledgments

I am indebted to Michael Plümer for valuable discussions and reading the manuscript.

REFERENCES

1. Correlations and Multiparticle Production, M. Plümer , S. Raha and R.M. Weiner, editors, (CAMP-LESIP IV) World Scientific 1991.

2. G.N. Fowler et al., J.Phys.G16, (1990) 1439.

3. G.J. Alner et al., Phys.Rep.154, (1987) 247.

4. N. Neumeister et al., Phys.Lett.B275, (1992) 186.

5. N. Agagbabyan et al., Z.Phys.C59, (1993) 405.

6. I.V. Andreev et al., Phys.Lett.B321, 1994 277.

7. I.V. Andreev, M. Plümer and R.M. Weiner, Int.J.Mod.Phys.A8, (1993) 4577.

8. I.V. Andreev, M. Plümer, B.R. Schlei and R.M. Weiner, to be published.

9. Yu.M. Sinyukov, Preprint ITP-93-8E, Kiev 1993, to be published in Nucl.Phys.A.

10. A. Bialas and R. Peschanski, Nucl.Phys.B273, (1986) 703.

11. P. Carruthers et al., Phys.Lett.B222, (1989) 487; M. Gyulassy, Multiparticle Dynamics (Festschrift for Leon Van Hove) eds. A. Giovannini and W. Kittel (World Scientific, Singapore, 1990) p.479.

12. I. Derado et al., Z.Phys.C47, (1990) 23.

13. A. Bialas., Acta Phys. Pol.B23, (1992) 561.

14. I.V. Andreev et al., Phys.Lett.B316, (1993) 583; Phys.Rev.D49, (1994) 1217.

15. M. Gyulassy and S.S. Padula, Phys.Rev. C41 (1989) R21.

16. J. Bolz et al., Phys.Rev. D47 (1993) 3860.

17. B. Schlei et al., Phys.Lett. B293 (1992) 275.

18. I.V. Andreev, M. Plümer and R.M. Weiner, Phys.Rev.Lett.67, (1991) 3475.

19. M.Bowler, Phys. Lett. B276 (1992) 237.

20. Yu.M. Sinyukov and A.Yu. Tolstykh, Z.Phys.C61, (1994) 593.

21. Yu.M. Sinyukov and B. Lörstad, Z.Phys.C61, (1994) 587.

22. L. Razumov and R.M. Weiner, to be submitted for publication.

ANALYSIS OF MULTIPARTICLE CORRELATIONS AND THE WAVELET TRANSFORM

Peter Lipa[1,2], Martin Greiner[3] and Peter Carruthers[1]

[1]Department of Physics
University of Arizona, Tucson, AZ–85721, USA
[2]Institut für Hochenergiephysik der Österreichischen Akademie
der Wissenschaften, Nikolsdorfergasse 18, A-1050 Wien, Austria
[3]Institut für Theoretische Physik der Justus-Liebig-Universität
Heinrich-Buff-Ring 16, 35392 Giessen, Germany

INTRODUCTION

The analysis of multiparticle correlations by means of factorial moments[1] and refinements thereof,[2] amounts to counting the abundance of q-tuples of particles in phase space in dependence of the "size" of the tuple. The human brain, however, follows a different strategy to estimate correlations of point patterns: it organizes points/particles in densely populated regions into (hard to quantify) "clumps" or "clusters" and unpopular regions into "voids"; if one looks closer into a particular "clump" it may (or may not) again be organized into "clusters" and "voids", but now with respect to the higher (smooth) background density of the bigger "parent–clump", and so on.

Standard correlation measures (differential q–particle cross sections, associated cumulant functions and various projections/integrals thereof[2]) do not provide a natural description in terms of clusters and voids; for example, one n–particle clump gives $n!/(n - q)!$ different contributions to the differential q–particle cross section $\sigma_q(\boldsymbol{p}_1, \ldots, \boldsymbol{p}_q)$, thereby making it very hard to unfold information on clumps, such as approximate size and shape, from the q-tuple counts.

However, sometimes a choice of a clever basis for the representation of the data can directly unveil information of interest, otherwise scattered in the many dimensions of multiparticle phasespace. We believe that the wavelet transform[5-7] is such a clever change of basis functions for hierarchically organized stochastic processes. Moreover, it provides a formulation of correlations in terms of "clumps and voids at multiple scales" rather than in terms of "q-tuples of particles". We justify our believes by calculation and simulation of correlations in simple, tractable random cascade models.

THE p– AND α–MODELS

We start with a uniform "energy" (or particle) density E in the interval $[0,1]$, the total energy normalized to unity for convenience. For the p-model,[3] this interval is split into two halves with energies $E_1 = pE$ and $E_2 = (1 - p)E$, where E_1 goes randomly, with probability $1/2$, to the left or right subinterval (bin). Subsequently this splitting step is recursively applied to each new subinterval, building up a selfsimilar random histogram with 2^J bins after J cascade steps. For later convenience we set the 'splitting'-parameter equal to $p = p_1 = (1 + \alpha)/2$ and $(1 - p) = p_2 = (1 - \alpha)/2$ with $0 \leq \alpha \leq 1$. Note that in this construction the total energy is conserved at each cascade step.

For the α-model[1] the weighting of the two halves occurs independently (i.e. configurations with $E_1 = p_1 E$ and $E_2 = p_1 E$ may occur*) restricted by the condition that *on average* one has $\langle E_1 + E_2 \rangle = 1$. Thus, in the α-model energy is not conserved at each individual interval splitting. However, the independent weighting of each subinterval makes the α-model easier to generalize to $d = 2, 3, \dots$ dimensions, by building cartesian products of one dimensional α-models.

The "energy density" contained in a certain bin is, as usual, defined as the corresponding energy divided by the binsize; we denote the energy density in the k-th bin ($k = 0, \dots, 2^J - 1$) after J cascade steps as $\epsilon_k^{(J)}$. Correlations between two or more bins are described by the bin-bin correlation densities, the latter being defined as the configuration average $\langle \rangle$ of products of energy densities of the corresponding bins at some scale $j = 0, \dots, J$ (later on we are also interested in correlations at intermediate cascade steps): $\rho_{k_1}^{(j)} = \langle \epsilon_{k_1}^{(j)} \rangle$, $\rho_{k_1, k_2}^{(j)} = \langle \epsilon_{k_1}^{(j)} \epsilon_{k_2}^{(j)} \rangle$, etc..

These can be found from the characteristic function for the correlation densities[†]

$$Z^{(j)}[\vec{\lambda}^{(j)}] = \left\langle \exp\left(i \sum_{k=0}^{2^j-1} \lambda_k^{(j)} \epsilon_k^{(j)} \right) \right\rangle = 1 + i \sum_{k_1} \lambda_{k_1}^{(j)} \rho_{k_1}^{(j)} + \frac{i^2}{2!} \sum_{k_1,k_2} \lambda_{k_1}^{(j)} \lambda_{k_2}^{(j)} \rho_{k_1,k_2}^{(j)} + \dots , \quad (1)$$

by q-th order partial derivatives: $\rho_{k_1,\dots,k_q}^{(j)} = \frac{1}{i^q} \frac{\partial^q Z^{(j)}[\vec{\lambda}^{(j)}]}{\partial \lambda_{k_1}^{(j)} \cdots \partial \lambda_{k_q}^{(j)}} \bigg|_{\vec{\lambda}^{(j)}=0}$.

In Ref.[4] we gave an iterative construction for the characteristic function of the p-model:

$$Z^{(j+1)}[\vec{\lambda}^{(j+1)}] = \frac{1}{2} \left(Z^{(j)}[(1+\alpha)\vec{\lambda}_L^{(j)}] \, Z^{(j)}[(1-\alpha)\vec{\lambda}_R^{(j)}] \right.$$
$$\left. + Z^{(j)}[(1-\alpha)\vec{\lambda}_L^{(j)}] \, Z^{(j)}[(1+\alpha)\vec{\lambda}_R^{(j)}] \right) . \quad (2)$$

This solution can now be used to derive recursion relations for the correlation densities $\rho_{k_1}^{(j)}$, $\rho_{k_1,k_2}^{(j+1)}$, $\dots$. The two-bin correlation density is depicted in Fig. 1: it shows clearly the multiplicative structure of the p-model as the scale becomes finer. The closer two bins are together, the closer is their common ancestor in scale and the stronger they are correlated; a scaling law with decreasing bin-bin distance shows up as a result of the selfsimilarity of the p-model cascade.

WAVELET CORRELATIONS AND MULTISCALE CLUSTERING

We summarize only some very basic concepts of wavelets and the related idea of a multiresolution analysis. For a more profound introduction we refer the reader to some excellent reviews.[5-7]

---

*More general, so called *assymetric*, versions of the α-model are considered in Ref.[1] .

[†]We use the notation $\vec{\lambda}^{(j)} = (\lambda_0, \dots, \lambda_{2^j-1})$.

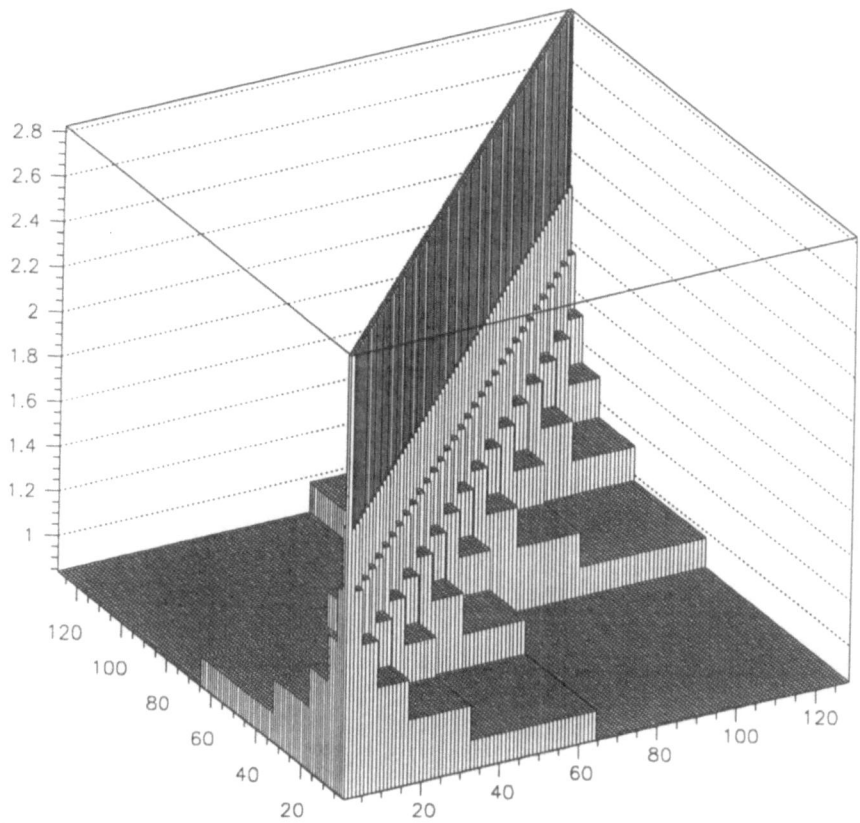

Figure 1. Two-bin correlation density $\rho_{k_1 k_2}$ after $J=7$ p-model cascade steps ($\alpha = 0.4$).

The essence of a multiresolution analysis is a decomposition of a given function $\epsilon(x) \approx \bar{\epsilon}^{(J)}(x)$ at some fine resolution scale J (e.g. the resolution scale of one's detector) in terms of contributions from rougher scales $\bar{\epsilon}^{(j)}(x)$:

$$\bar{\epsilon}^{(J)}(x) = \sum_k \epsilon_k \phi_{Jk}(x) = \bar{\epsilon}^{(0)}(x) + \sum_{j=0}^{J-1} \tilde{\epsilon}^{(j)}(x) = \epsilon_0^{(0)} \phi_{00}(x) + \sum_{j=0}^{J-1} \sum_{k=0}^{2^j-1} \tilde{\epsilon}_{jk} \psi_{jk}(x). \qquad (3)$$

This dual decomposition is exemplified in Fig. 2: the left column represents a sequence of smoothed copies $\bar{\epsilon}^{(j)}(x) = \sum_k \epsilon_k^{(j)} \phi_{jk}(x)$ of the original $\bar{\epsilon}^{(J)}(x)$ at scales $j = 1, \ldots, J$, spanned on a set of basis functions $\phi_{jk}(x) = \phi(2^j x - k)$ with integer $k = 0, 1, \ldots, 2^j - 1$. The latter act as "smoothing filters" when going from a finer scale j to the next coarser scale $j - 1$ and are obtained via dilation (by an integer power of 2) and translation (by an integer k) from *one* judiciously constructed[6] "scaling function[‡]" $\phi(x)$.

Obviously, at each smoothing step $j \rightarrow j - 1$ some detail information on density fluctuations is lost, namely exactly those "clumps" and "voids" whose width is of order $O(2^{-j})$. If we capture this detail information during each of the smoothing steps by taking differences of $\bar{\epsilon}^{(j)}(x)$ at adjacent scales, $\tilde{\epsilon}^{(j)}(x) = \bar{\epsilon}^{(j+1)}(x) - \bar{\epsilon}^{(j)}(x)$ (right column in Fig. 2), we can reconstruct the original function $\bar{\epsilon}^{(J)}(x)$, and any of its smoothed copies at intermediate scales, by adding more and more detail functions $\tilde{\epsilon}^{(j)}(x)$ to the coarsest (nearly

---

[‡]Sometimes also called "father wavelet".

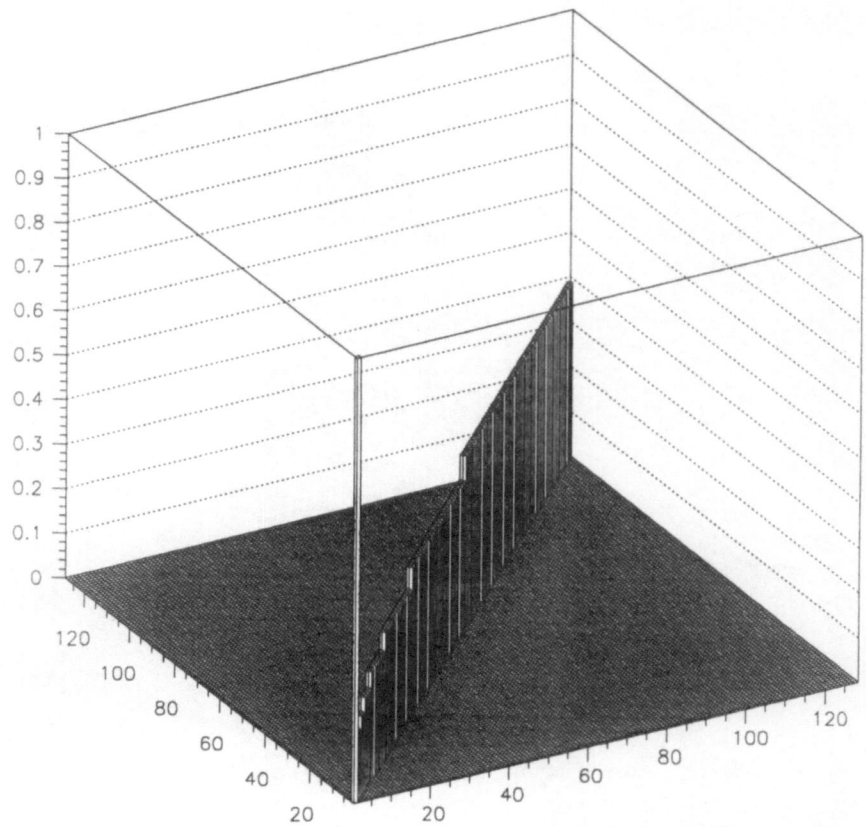

Figure 2. A multiresolution decomposition of a p-model realization (upper left corner) at initial resolution scale $J = 7$ with respect to the Daubechies D12 wavelet basis. The left column shows a sequence of smoothed approximations $\bar{\epsilon}^{(j)}(x)$, $(j = 1, \ldots, J)$, to the original signal, while the right column represents the wavelet transform, or the sequence of mutually orthogonal detail functions, $\tilde{\epsilon}^{(j)}(x) = \bar{\epsilon}^{(j+1)}(x) - \bar{\epsilon}^{(j)}(x)$.

flat) approximation, say $\bar{\epsilon}^{(0)}(x)$. Again, the detail functions are spanned on a family of basis functions (so-called "wavelets") $\psi_{jk}(x) = \psi(2^j x - k)$ obtained by dilations and translations of *one* "mother–wavelet" $\psi(x)$, closely related to the scaling function $\phi(x)$. The coefficents $\tilde{\epsilon}_{jk}$ of the expansion $\tilde{\epsilon}^{(j)}(x) = \sum_{k=0}^{2^j-1} \tilde{\epsilon}_{jk}\psi_{jk}(x)$ are then the wavelet transform of ϵ_k.

A better visual impression of the hierarchy of "clumps" and "voids" is given by density plots of a two-dimensional multiresolution analysis§ in Fig. 3: the top left plot represents a random realization of 500 points obtained by a poisson transform of the two-dimensional α-model; the left column represents the sequence of successive smoothing operations $\bar{\epsilon}^{(j)}$, while the middle column are the detail functions $\tilde{\epsilon}^{(j)} = \bar{\epsilon}^{(j+1)} - \bar{\epsilon}^{(j)}$. For clarity, the detail functions are plotted again in the right column, but this time only with 2 grey-values (black for "voids" and white for "clumps").

One more property of $\phi(x)$ and $\psi(x)$ is crucial in defining a multiresolution analysis¶: one wants the detail functions orthogonal to each other, i.e. $\int \tilde{\epsilon}^{(i)}\tilde{\epsilon}^{(j)} dx \propto \delta_{ij}$, thereby removing redundant information from each detail function. This desire implies (among others) following properties of the basis functions:[6,7] intrascale orthogonality of the ϕ_{jk} (i.e.

---

§Both, Fig. 2 and Fig. 3 were obtained using the 12 coefficients member of the Daubechies' familiy of orthogonal wavelets with the most compact support for a given regularity of the wavelet.[6]

¶Of course, many generalisations and relaxations of these conditions were suggested.

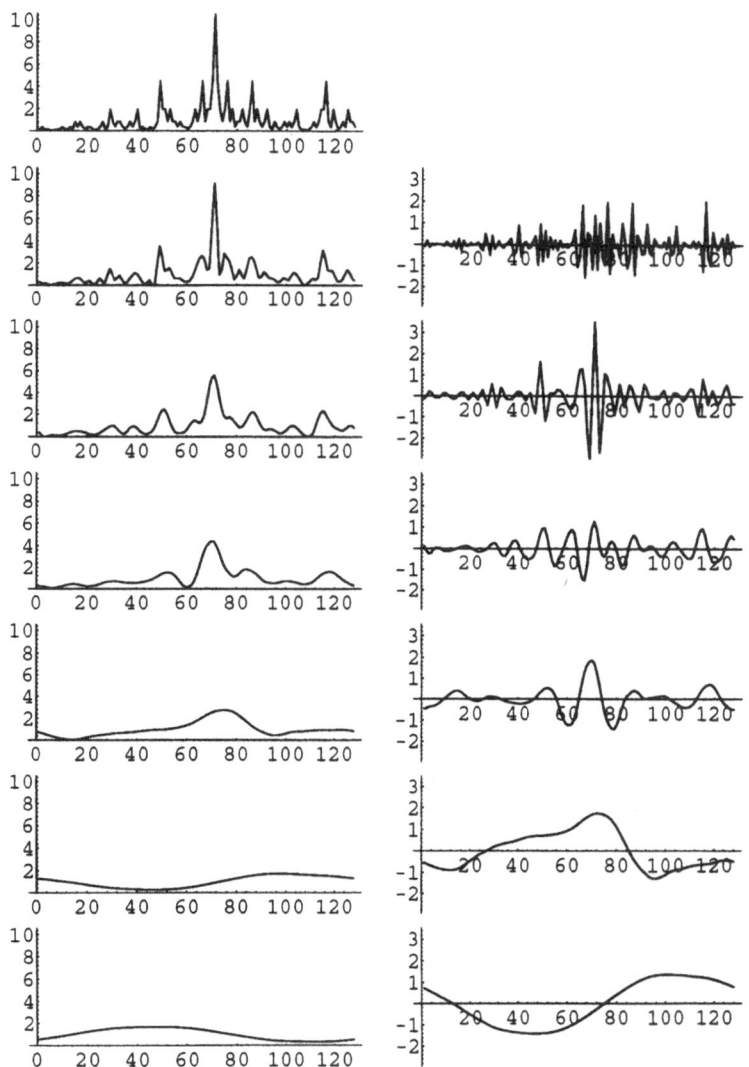

Figure 3. A multiresolution decomposition, with respect to the Daubechies D12 wavelet basis, of a two-dimensional α-model with 500 points, clustered proportional to the local "energy density" $\epsilon^{(J)}$ at initial resolution scale $J = 6$. The left column shows density plots of a sequence of smoothed approximations $\bar{\epsilon}^{(j)}(x)$, $(j = 2, \ldots, J)$, to the original function, while the middle column represents density plots of the wavelet transform, or the sequence of mutually orthogonal detail functions, $\tilde{\epsilon}^{(j)} = \bar{\epsilon}^{(j+1)} - \bar{\epsilon}^{(j)}$. For clarity, the detail functions are exhibited again in the right column, but with only two grey-values: *black* for regions where $\tilde{\epsilon}^{(j)} < 0$, corresponding to "voids" with respect to the scale j, and *white* for regions where $\tilde{\epsilon}^{(j)} > 0$, signalling the appearence of "clumps" of size $O(2^{-j})$ in a smooth background $\bar{\epsilon}^{(j)}$.

$\int \phi_{jk}\phi_{jl}\,dx \propto \delta_{kl}$), inter– and intrascale orthogonality of the ψ_{jk} (i.e. $\int \psi_{ik}\psi_{jl}\,dx \propto \delta_{ij}\delta_{kl}$), and, most notably, the so-called "two–scale equations":

$$\phi(x) = \sum_m h_m\,\phi(2x - m)\,, \qquad \psi(x) = \sum_m g_m\,\phi(2x - m)\,, \qquad (4)$$

where a finite number of nonzero coefficients[||] h_m and $g_m = (-1)^m h_{1-m}$ have to satisfy various conditions stated by Daubechies.[6] Note that $\phi(x)$ is a linear combination of dilated and translated copies of itself (thus the name "scaling function"), and ψ is another linear combination of ϕ's with given coefficients g_m. The simplest (and long known) solutions to (4) is the Haar system, defined by the conjugate filter pair[*] $h_m = \{1/2, 1/2\}$ and $g_m = \{1, -1\}$ leading to a multiresolution decomposition in terms of discontinuous step functions (histograms). Decompositions with arbitrary smoothness can be obtained by constructing filters with more (but still a finite number) of coefficients, thereby rendering ϕ and ψ well localized in both position (x) space and its conjugate Fourier (ω) space.

Once a finite set of admissible h_m is found, the solutions ϕ and ψ can be obtained by (numerical) iteration of eqs. (4), but in practical implementations this is never necessary: Eq. (3) can be viewed as a *linear* transformation of the amplitudes ϵ_k, written compactly as $\vec{\epsilon}$, to the wavelet amplitudes[†] $\tilde{\epsilon}_{jk}$

$$\vec{\tilde{\epsilon}} = (\epsilon_0^{(0)}, \tilde{\epsilon}_{00}, \tilde{\epsilon}_{10}, \tilde{\epsilon}_{11}, \tilde{\epsilon}_{20}, \dots, \tilde{\epsilon}_{23}, \tilde{\epsilon}_{31}, \dots, \tilde{\epsilon}_{J-1,2^{J-1}-1})\,, \qquad (5)$$

so that a wavelet transform amounts to a simple (sparse) matrix operation[‡] $\vec{\tilde{\epsilon}} = \mathbf{W}\vec{\epsilon}$.

If $\epsilon(x) \approx \bar{\epsilon}^{(J)}(x)$ is a random function (such as our p- and α-models) we are interested in correlations between their coefficients $\vec{\epsilon}$. As mentioned in the introduction, the standard bin-bin or particle-particle correlation densities *do not* provide easily accessible information on correlations between "chunks", say of size $O(2^{-j})$. However, this is exactly the kind of information provided by correlations among the *wavelet coefficients* $\tilde{\epsilon}_{jk}$! The latter can be generally obtained by a change-of-basis within the characteristic function (1)

$$Z[\vec{\lambda}] = \left\langle \exp\left(i\vec{\lambda}\cdot\vec{\epsilon}\right)\right\rangle = \left\langle \exp\left(i\vec{\lambda}\mathbf{W}^{-1}\cdot\mathbf{W}\vec{\epsilon}\right)\right\rangle = \left\langle \exp\left(i\vec{\eta}\cdot\vec{\tilde{\epsilon}}\right)\right\rangle = Z[\vec{\eta}] \qquad (6)$$

and taking partial derivatives with respect to the new set of variation parameters $\vec{\eta}$, which are nothing but the wavelet transformed $\vec{\lambda}$'s.

For the second and third order wavelet correlation densities we obtain:

$$\tilde{\rho}_{k_1 k_3} = \sum_{k_2, k_4} (\mathbf{W})_{k_1 k_2}(\mathbf{W})_{k_3 k_4}\rho_{k_2 k_4}\,, \qquad \tilde{\rho}_{k_1 k_3 k_5} = \sum_{k_2, k_4, k_6} (\mathbf{W})_{k_1 k_2}(\mathbf{W})_{k_3 k_4}(\mathbf{W})_{k_5 k_6}\rho_{k_2 k_4 k_6}\,. \qquad (7)$$

This approach is model independent and applicable for any compact wavelet. The results for the second order Haar-wavelet correlation density[4] of the p-model are shown in Fig. 4: this correlation density matrix is diagonal! So to say, the Haar wavelet basis represents the adequate normal coordinates for the p-model. All off-diagonal contributions vanish simply because the average of the product of two differences belonging to different scales is zero; as a consequence there are no second order correlations between different scales (or better,

---

[||]Often called "quadrature mirror filters", where $\{h_m\}$ acts as low–pass filter, removing high frequency contributions, and $\{g_m\}$ as "band–pass" filter, selecting only variations in a specific frequency range. These filters can be easily implemented in hardware, thus making real–time wavelet transforms possible! This opens a wide arena for signal processing and pattern recognition applications.

[*]Similar to Fourier transforms there are various conventions on where to put factors of $\sqrt{2}$. For correlation studies, the "engineer's" assymetric convention is more natural.

[†]Note the customary implicit mapping of double indices (jk) to single indices (i) =(position of $\tilde{\epsilon}_{jk}$ in $\vec{\tilde{\epsilon}}$).

[‡]The transformation matrix $\mathbf{W}$ depends only on the filter pair h_m and g_m of Eqs. (4).

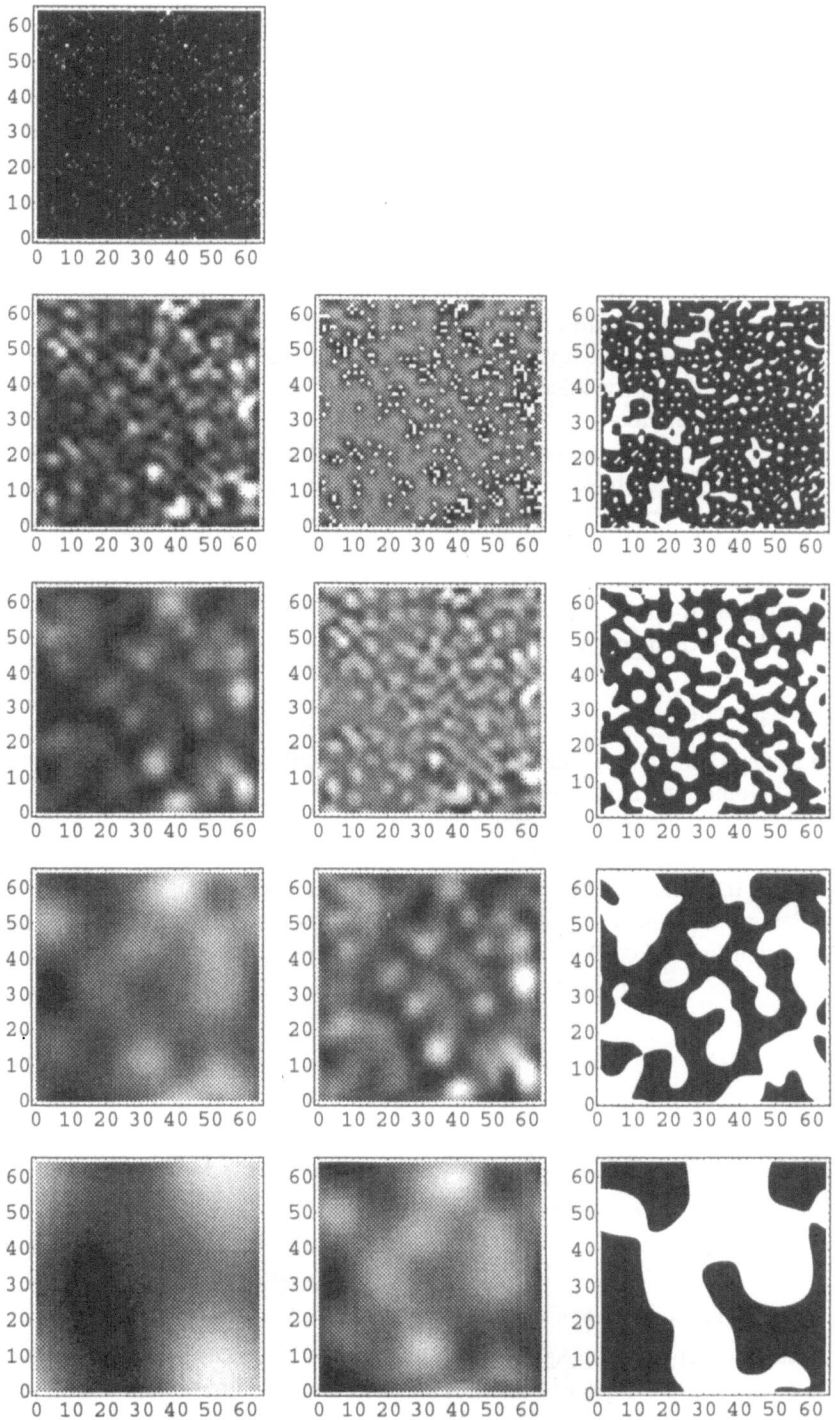

Figure 4. Haar-wavelet transformed two-bin correlation density $\tilde{\rho}_{(j_1 k_1),(j_2 k_2)}$ for the p-model cascade.

"clumps of different sizes") of the p-model cascade! As the resolution j_1 increases, the diagonal contributions reflect a power-law (meaning that "clumps" fluctuate the stronger the smaller they get), which is a clear evidence for the selfsimilarity of the p-model cascade. Compared to the two-bin correlation density, the Haar-wavelet correlation density has a much simpler and clearer structure (compare Fig. 1 with Fig. 4).

For different wavelet bases the dominating contributions are still along the diagonal and also indicate a scaling, which is approximately equal to the scaling occurring in the Haar-wavelet case of Fig. 4. However, off-diagonal contributions arise in bands,[4] which reflect correlations between different scales belonging to the same evolution branch of the p-model cascade tree. Although, as expected, different wavelet bases do not lead to a *complete* diagonalisation of the covariance matrix, they give rise to a "quasi" diagonalisation since off-diagonal contributions are strongly suppressed. This "quasi" diagonalisation of the covariance matrix for hierarchically organized stochastic processes seems to be a general feature of the wavelet basis!

The results of this contribution cannot be more than a very first study; there are still long ways to go. The above observations let us hope, that the successive "smoothing" and "differentiation" operations implied by a wavelet transform will serve as a useful tool to study the selfsimilarity and clustering aspects in complex processes involving multiple scales, ranging from galaxy-clustering to nuclear multifragmentation or QCD parton cascades (jets, mini-jets, mini-mini-jets, . . . , resonances). Of course, these hopes and expectations remain to be verified by further analyses.

Acknowledgements

P.L. would like to thank the organizers of this conference for the opportunity to contribute. He would also like to thank B. Buschbeck for helpful discussions and continuous support. M.G. is grateful to the Alexander von Humboldt Stiftung for its support with a Feodor Lynen Scholarship and to the Deutsche Forschungsgemeinschaft for its successive support with a habilitation scholarship. This work was supported in parts by the US Department of Energy, grant no. DE-FG02-88ER40456, and by the Austrian Fonds zur Förderung der wissenschaftlichen Forschung, project no. P8259–TEC.

REFERENCES

1. A. Białas and R. Peschanski, *Nucl. Phys.* B273 (1986) 70; *Nucl. Phys.* B308 (1988) 857.

2. H.C. Eggers, P. Lipa, P. Carruthers and B. Buschbeck, Phys. Rev. D **48** (1993) 2040.

3. C. Meneveau and K.R. Sreenivasan, Phys. Rev. Lett. **59** (1987) 1424.

4. M. Greiner, P. Lipa and P. Carruthers, "Wavelet Correlations in the p-Model", preprint HEPHY-PUB-587/93 and AZPH-TH/93-23, submitted to *Phys. Rev. E*.

5. S. Mallat, IEEE Trans. Pattern Anal. and Machine Intell. **11** (1989) 674.

6. I. Daubechies, Comm. Pure Appl. Math., 41 (1988) 909; "Ten Lectures on Wavelets", Society for Industrial and Applied Mathematics (SIAM), Philadelphia (1992).

7. Y. Meyer, "Wavelets and Operators", Cambridge University Press, New York (1992); "Wavelets: Algorithms and Applications", Society for Industrial and Applied Mathematics (SIAM), Philadelphia (1993).

MULTIPARTICLE PRODUCTION: SESSION SUMMARY

Andrzej Bialas

Institute of Physics, Jagellonian University
Reymonta 4, 30-059 Cracow, Poland
LPTHE, Universite Paris XI, Orsay, France

LONG–RANGE AND SHORT–RANGE CORRELATIONS

Studying multiparticle dynamics amounts basically to investigation of multiparticle correlations. It took some time before this rather obvious statement found its way into high-energy physics. If I remember correctly, the turning point was the paper by Mueller,[1] who adapted the techniques known from statistical physics and showed explicitly how to relate multiplicity distributions and correlation parameters.

Historically we distinguish long-range and short-range correlations in momentum space. Long-range correlations describe total multiplicity and measure simply the deviation of the data from the Poisson distribution. They are easiest to determine (counting of tracks is sufficient) and, therefore, were studied first. At not too low energies (to avoid constraints from conservation laws) the observed distribution is broader than Poisson. This is interpreted as evidence for a multicomponent nature of high energy collisions. The great success of the negative binomial distribution in description of the data[2] is the best example of this situation. Indeed, the negative binomial spectrum can be written as a (continuous) superposition of Poisson distributions:

$$P(n; \bar{n}, k) = \frac{k^k}{\Gamma(k)} \int z^{k-1} e^{-kz} e^{-\bar{n}z} \frac{(\bar{n}z)^n}{n!} dz \qquad (1)$$

showing explicitly its multicomponent character.

More information is obtained if one measures the momentum dependence of the correlation functions. Numerous studies in the seventies showed that the data can be described in terms of clusters, *i.e.* isotropically decaying objects, rather densely packed in the phase space available for soft particles (it is not clear if the clusters can be reduced to the known resonances but this cannot be excluded). Surprisingly enough, the data in different rapidity intervals can also be described by the negative binomial distribution[2]. This is probably explained by the fact that, as emphasized by Giovannini and Van Hove, the negative binomial distribution can also be written in the form which corresponds to independent emission of clusters[3]. The negative binomial description of the data was discussed extensively during our session — mainly by Tannenbaum[4] who confirmed its excellent agreement with the data in varying intervals of rapidity, and established a simple linear relation between the parameter k of the distribution and the size of the interval in question.

Since few years the main focus of research is on correlations of very short range, much shorter than those related to isotropic cluster decay. This kinematic region — very difficult experimentally — became accessible by increasing statistics and quality of the data, and by the development of new, sensitive tools. It started with factorial moments[5] which played an important role at the beginning. Two new applications of factorial moments were described at this meeting. Derado[6] showed that the technique of factorial moments can be very helpful in tracing the phenomenon of "Baked Alaska"[7,8] even if such events are very rare. Meunier[9] presented generalization of factorial moments to non-integer and to negative ranks. He showed that they can be exceedingly useful in discriminating between different models.

SCALING OF CORRELATION FUNCTIONS

The early observation that the data[10] indicate the presence of very short range correlations led to the suggestion that — perhaps — there exist correlations at all scales[5]. A natural consequence was the idea of scaling, *i.e.* power law dependence of the factorial moments on the size of the phase-space bin (called "intermittency" in ref. 5). Experimental investigations of factorial moments allowed to confirm existence of intermittency in e^+e^- $\bar{p}p$, πp and heavy ion collisions[11].

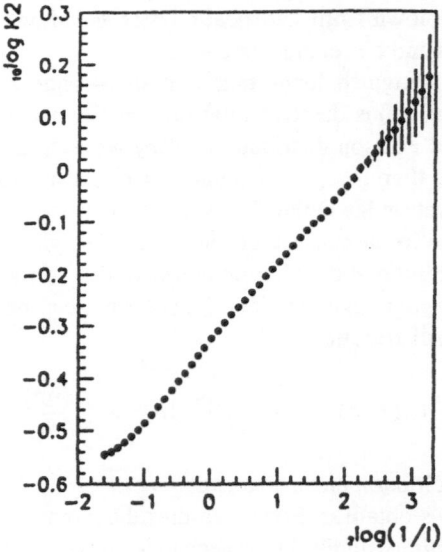

Figure 1. Second cumulant of the charged particles distribution produced in $p\bar{p}$ collisions at 640 GeV/c plotted versus $l = \delta\phi\,\delta y\,\delta(\log\,p_\perp)$. UA1 coll. ref. 13.

Significant progress was possible when a novel technique, called "correlation integrals" was proposed and developed by the Tucson–Vienna collaboration[12]. This turned out to be a great improvement which allowed to measure correlation functions with an unprecedent precision. A typical result is shown in Fig. 1, where the second cumulant of the charged particle distribution in 600 GeV pp collisions[13] is plotted versus the size of the three-dimensional bin in momentum space. One sees that the data follow rather accurately a straight line which, in this log-log plot, represents the power law:

$$C(\delta) \sim \delta^{-f}, \tag{2}$$

i.e. scaling in the 3-dimensional momentum space.

I feel that it is difficult to ignore the data of this quality and therefore it is important to search seriously for a possible physical origin of this phenomenon. Attempts in this direction are described in the next section.

MODELS OF SCALING BEHAVIOUR

It should perhaps be emphasized that observation of scaling does not — and by far — point out uniquely what kind of physics is responsible for it. Therefore, several possibilities were considered.

(a) Cascade Models

Self similar cascade produces scaling behaviour of the factorial moments and therefore served as a guiding example from the very beginning. In particular, the original suggestion[5] of the scaling phenomenon was modelled as a cascade of a "strongly interacting liquid" in analogy to the turbulent flow in hydrodynamics. This analogy emphasizes the relation of the observed phenomenon to the "chaotic behaviour" of the system in question. In view of the recent work by Müller an collaborators[14] who find a clear signal of chaotic behaviour of the gluon fields, it is not unlikely that this may turn out to be more than just a formal analogy.

The first serious self-similar cascade calculation — in the framework of the cluster model — was performed by Ochs and Wosiek[15]. They confirmed the idea of approximate scaling an found a more general regularity which was valid independently of the phase-space dimension

$$C_q(\delta) \sim \{C_2(\delta)\}^{f_q} \tag{3}$$

(in low-dimensional analysis the second moment saturates at small distances, but the eq.(3) is nevertheless valid).

Recently, analytic calculations of the QCD parton cascade became available. Three groups produced results almost simultaneously[16]. These results and very recent progress were reported by Ochs at our session[17]. The major results are (a) the QCD cascade is not actually self-similar because the coupling constant varies along the cascade. Consequently, the scaling is only approximate and the moments tend to saturate at small distances in 3-dimensional phase-space; (b) new observables for testing the predicted behaviour were suggested. They are constructed of angular variables and thus are relatively easy to analyze experimentally.

(b) Phase Transitions

Second order phase transition is well known to imply scaling of the correlation functions[18]. This was realized immediately[19] and the analysis of intermittency in the Ising model was carried out[20]. Recently, Hwa discussed the problem in the Ginzburg-Landau model[21]. He determined the relevant intermittency exponents and found — among other results — that even for the first order transition the Ochs–Wosiek relation (3) is valid (with a rather simple formula for f_q).

It is not clear, of course, if and how the Ising model and/or Ginzburg–Landau theory can actually be applied to particle distributions and what can be the physics of the phase-transition in question.

(c) Self-Organized Criticality

Another attractive mechanism giving scaling of the correlation functions is the self-organized criticality ("theory of avalanches") of Bak et al.[22]. It turns out that many non-linear systems develop automatically the scaling behaviour in a very broad range of the parameters which describe the system. Thus no fine tuning is necessary to obtain scaling (this is in contrast to the phase transition which takes place only at the critical temperature). It was realized already from some time that this provides an interesting possibility of explanation of the scaling behaviour of particle spectra. Only recently, however, Hwa and collaborators found a self-organized critical system which could be applied to multiparticle production. The results were presented during the session[23].

HADRONIZATION

All these explanation of scaling in multiparticle spectra have one common drawback: they are formulated in terms of partons or other non-observable degrees of freedom rather than in terms of the final measured hadrons. This brings immediately the question: why the process of hadronization does not spoil the scaling behaviour. I would like to emphasize that this question is a very serious one: we are talking here about the momentum resolution up to about 40 MeV — significantly below the pion mass — and it is really difficult to understand why the reshuffling of partons into observed hadrons does not affect even such small momentum differences. This puzzled us all for some time an was — I think — a serious obstacle to the progress.

HBT CORRELATIONS

About two years ago the experimental analyzes[24] provided an unambiguous evidence that, as suggested from some time[25], at very small difference of momenta, correlations are dominated by HBT effect, i.e. quantum interference[26]. This is illustrated in Fig. 2, where the data from UA1 and DELPHI collaborations[27] are plotted versus Q, the difference of particle four-momenta squared. One clearly sees that at small Q the correlations between like-sign particles are much stronger than those between unlike-sign ones. In the simplest interpretation this implies that (a) the production of pions in this kinematic range is, at least to a large extent, incoherent (this is the necessary condition for HBT correlations to be present); (b) the correlations observed at low Q are related to the space-time structure of the system rather than to its structure in momentum space.

I would like to emphasize the importance of the conclusion (b): it drastically changes the way of thinking about the problem, as compared to the ideas based on cascade in momentum space described in section 3. What is perhaps even more important, it allows finally to understand why hadronization in not an obstacle for having strong correlations at low Q. Indeed, since the correlations reflect only the space-time structure of the source, they are little sensitive to details of its composition and on momentum distribution of the constituents. That is to say, it does not matter if the source of pions is made of partons, of hadronic fluid, of decaying clusters, or of anything else. Also, the mechanism of formation of the final pions is (almost) irrelevant. What matters is the space-time structure of the source and incoherent character of pion emission.

Thus one annoying problem seems solved. Others remain, however.

The first one : how can one justify the scaling law (2) at small Q ? To underline the importance of this question, the data for like-sign pion correlation function[27] are again shown

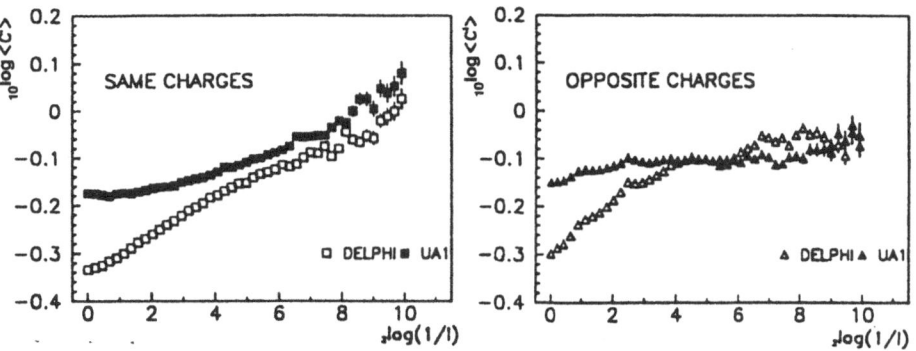

Figure 2. Two-particle correlation integrals for like-sign and unlike-sign particles produced in 640 GeV $p\bar{p}$ collisions and in Z^0 decays. DELPHI and UA1 coll. ref. 27.

in Fig. 3 (in different scale). One sees that they do indeed closely follow the power law at small Q.

There are basically two possible answers: either (a) the observed power law is just a numerical accident which dissimulates a more complicate structure[29] or (b) there is a genuine scaling in the system.

In the case (a) one has to find a reason for the presence of very short range correlations. Two suggestions were made. One is based on the idea of Grassberger[30] which relates the HBT correlation length to the widths of resonances produced in the collision. If a substantial number of relatively narrow resonances is produced, the small correlation length in momentum space is naturally explained[31]. The second suggestion takes advantage of the fact that, as emphasized by Weiner during the session,[29] HBT correlations do not simply reflect the shape of the interaction volume but also correlations inside it. The resulting structure is rich enough to describe the data, as was explicitly shown by Weiner during the morning session. Other detailed features of HBT correlations, related particularly to expansion of the interacting system were discussed during the session by Sinyukov[32].

In the case (b) it is necessary to accept that the region of pion emission does not have a well-defined radius but rather a long tail decreasing as a power of distance (at large distances from the center) [33]. It remains an open question if a fractal structure of the source is also necessary to explain the data. This can be eventually decided when precise data on higher order correlations are available[34].

The second question, which I myself find rather fundamental, can be formulated as follows. The comparison of e^+e^- data and the existing QCD cascade models shows that the region of medium Q (.2 GeV$< Q \lesssim$ few GeV) can possibly be described by the standard quark-gluon cascade followed by hadronization[35]. On the other hand, we have just seen that the very small Q region ($Q \lesssim .1$ GeV) reflects the space-time structure of the system. This — *a priori* — has nothing to do with the anomalous QCD dimensions and the value of the strong coupling constant determining the behaviour of the QCD cascade in the medium Q region. How can one thus explain that the shape of the correlation function (Figs 2, 3) does not visibly change when one passes from one region to another? The problem is most clearly seen when one compares the data for like sign and unlike-sign pions in the data of DELPHI coll. shown in Fig. 2. For unlike pairs there is a clear change of shape at $Q = .2$

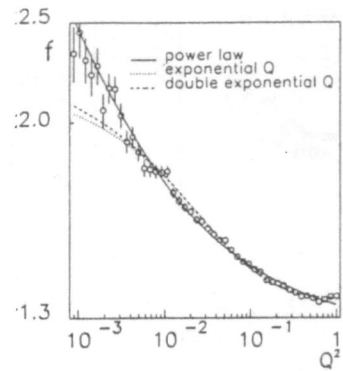

Figure 3. Second factorial moment plotted versus $Q^2 = |(p_1 - p_2)^2|$. UA1 coll. ref. 28.

GeV indicating the end of the scaling region of the quark-gluon cascade. For like-sign pairs, however, the slope continues essentially unchanged until the smallest Q measured. This indicates a surprising connection between the original quark-gluon cascade and the space-time structure of the region of pion emission at the late stage of the collision. It would be of course very interesting to understand better this relation.

UNIVERSALITY

Universality of the correlation functions is a controversial issue which is still far from being resolved. At the present meeting Sarcevic pointed out one marked difference between the heavy ion data and the data from simpler targets[36]. The data show that there are practically no correlations beyond the second order in the heavy ion data, in contrast to the hadron-hadron and e^+e^- collisions. She argued that this observation has an important implication, as it allows to describe particle production in heavy ion collisions by a field theory[36]. Numerous further consequences follow from this approach. It shall be of course very interesting to see if they are confirmed by the future data.

Acknowledgements

I would like to thank all participants of the session for a very friendly cooperation and to J. Rafelski for invitation to the meeting.

REFERENCES

1. A.H. Mueller, *Phys.Rev.* D4:150(1971).

2. UA5 coll., G.J. Alner et al., *Phys. Lett.* B160:193(1985); A. Giovannini, and L. Van Hove, *Z.Phys.* C30:391(1986).

3. A. Giovannini, and L. Van Hove, *Acta Phys.Pol.* B19:495(1988).

4. M. Tannenbaum, *Mod. Phys.Lett.* A9:89(1994) and these proceedings.

5. A. Bialas, and R. Peschanski, *Nucl. Phys.* B273:703(1986); B308:847(1988).

6. I. Derado, these proceedings.

7. J.D. Bjorken, K.L. Kowalski, and C.C. Taylor, Baked Alaska, SLAC-PUB-6109, April 1993; J.D. Bjorken, *Acta Phys. Pol.* B23:561(1992).

8. A. Anselm, *Phys. Lett.* B217:169(1989); J.P. Blaizot, and Krzywicki, *Phys. Rev.* D48:246(1992).

9. P. Duclos, and J.L. Meunier, Nice Preprint INLN 94/6 and J.L. Meunier, these proceedings.

10. JACEE coll., T.H. Burnett et al., *Phys. Rev. Lett.* 50: 2062 (1983).

11. For a recent review, see E.De Wolf, I.Dremin, and W. Kittel, preprint HEN-362 (1993), to be published.

12. P. Lipa, P. Carruthers, H.C. Eggers, and B. Buschbeck, *Phys. Lett.* B285:300(1992); H.C. Eggers, P. Lipa, P. Carruthers, and B. Buschbeck, *Phys. Rev.* D48:2040(1993).

13. UA1 coll., Y.F. Wu et al., Proc. Cracow Workshop on Multiparticle Production (May 1993), World Scientific, Singapore (1994) p. 22.

14. B. Müller, these proceedings and references quoted there.

15. W. Ochs, and J. Wosiek, *Phys. Lett.* B214:617(1988); B232:271 (1989).

16. W. Ochs, and J. Wosiek, *Phys. Lett.* B289:159(1992); B304:144(1993); Ph. Brax, J.L. Meunier, and R. Peschanski, *Z.Phys.* C62:649(1994); Yu.L. Dokshitzer, and I.M. Dremin, *Nucl. Phys.* B402:139(1993).

17. W. Ochs, these proceedings.

18. see e.g. K.G. Wilson, and J. Kogut, *Phys. Rep.* 12:75(1974).

19. J. Wosiek, *Acta Phys. Pol.* B19:863(1988).

20. J. Wosiek, *Nucl. Phys.* B Suppl. 9:640(1989); H. Satz, *Nucl. Phys.* B326:613(1989).

21. R.C. Hwa, and M.T. Nazirov, *Phys. Rev. Lett.* 69:741(1992); R. Hwa, these proceedings.

22. P. Bak, C. Tang, and K. Wiesenfeld, *Phys. Rev. Lett.* 59:381 (1987).

23. R. Hwa et al., *Phys. Rev. Lett.* 72:820(1994); R. Hwa, these proceedings.

24. See ref. 11 for a full list of references.

25. P. Carruthers, E.M. Friedlander, C.C. Shih, and R.M. Weiner, *Phys. Lett.* B222:487(1989); M. Gyulassy, Festschrift L. Van Hove, A. Giovannini, and W. Kittel, eds., World Scientific, Singapore (1990), p. 479.

26. R. Hanbury–Brown, and R.Q. Twiss, *Nature* 177:27(1956).

27. UA1 and DELPHI coll., F. Mandl, and B. Buschbeck, Proc. XXII Int. Symp. on Multiparticle Dynamics, Santiago De Compostela 1992 ed. A. Pajares, World Scientific, Singapore 1993.

28. UA1 coll., N. Neumeister et al., *Z. Phys.* C60:633(1993).

29. I.V. Andreev et al., *Phys. Lett.* B321:277(1994); R. Weiner, these proceedings.

30. P. Grassberger, *Nucl. Phys.* B120:231(1977).

31. A. Capella, A. Krzywicki, and E.M. Levin, *Phys. Rev.* D44:704(1991).

32. Yu. Sinyukov, these proceedings.

33. A. Bialas, *Acta Phys. Pol.* B23:561(1992); *Nucl. Phys.* A545:285c(1992).

34. A. Bialas, and B. Ziaja, *Acta Phys. Pol.* B24:1509(1993).

35. DELPHI coll., P. Abreu et al., *Phys. Lett.* B247:137(1990); A. De Angelis, *Mod. Phys. Lett.* A5:2395(1990).

36. T.H. Elze, and I. Sarcevic, *Phys. Rev. Lett.* 68:1988(1992); I. Sarcevic, these proceedings.

SINGLE PHOTON PRODUCTION IN 200 A·GeV SULPHUR ON GOLD COLLISIONS
WA80 Collaboration

R. Santo[4], R. Albrecht[1], T.C. Awes[5], F. Berger[4], M. Bloomer[2], D. Bock[4], R. Bock[1],
G. Clewing[4], R. Debbe[6], L. Dragon[4], A. Eklund[3], R.L. Ferguson[5],
S. Fokin[7], S. Garpman[3], R. Glasow[4], H.Å. Gustafsson[3], H.H. Gutbrod[1],
O. Hansen[6], G. Hölker[4], J. Idh[3], M. Ippolitov[7], P. Jacobs[2], K.H. Kampert[4],
K. Karadjev[7], B.W. Kolb[1], A. Lebedev[7], H. Löhner[8], I. Lund[8], V. Manko[7],
B. Moskowitz[6], F.E. Obenshain[5], A. Oskarsson[3], I. Otterlund[3], T. Peitzmann[4],
F. Plasil[5], A.M. Poskanzer[2], M. Purschke[1], B. Roters[1], S. Saini[5],
H.R. Schmidt[1], S.P. Sørensen[5,9], K. Steffens[4], P. Steinhaeuser[1], E. Stenlund[3],
D. Stüken[4], A. Vinogradov[7], H. Wegner[6], and G.R. Young[5]

[1]Gesellschaft für Schwerionenforschung, D-6100 Darmstadt, Germany
[1]Lawrence Berkeley Laboratory, Berkeley, California 94720, USA
[2]University of Lund, S-22362 Lund, Sweden
[3]University of Münster, D-4400 Münster, Fed. Rep. of Germany
[4]Oak Ridge National Laboratory, Oak Ridge, Tennessee 37831, USA
[5]Brookhaven National Laboratory, Upton, New York 11973, USA
[6]Kurchatov Institute of Atomic Energy, Moscow 123182, Russia
[7]KVI, University of Groningen, NL-9747 AA Groningen, Netherlands
[8]University of Tennessee, Knoxville, Tennessee 37996, USA

INTRODUCTION

The investigation of highly excited and compressed nuclear matter created in relativistic heavy ion collisions requires probes sensitive to the different stages of the development of the system. Among the characteristic signals listed in table 1, hadrons mainly probe the late stage of the reaction and their final spectra are influenced by their last scatterings. Electromagnetic probes, on the other hand, escape from the reaction zone without rescattering and probe the very early stage of the reaction, where temperatures and densities are highest. Electromagnetic probes are therefore particularly useful to study the possible formation of a Quark-Gluon Plasma, which is expected only at extreme values of temperature and density. The disadvantage of electromagnetic probes is that the signals are usually small and have to be disentangled from a tremendous physical background requiring a very high experimental precision and detailed treatment of the various background sources. In the case of photons this background consists mainly of photons from γ decaying hadrons like π^0 and η. In current experiments the evaluation of this background is performed either by calculating the meson decay photons from an appropriately tuned event generator or by directly measuring

Hot Hadronic Matter: Theory and Experiment
Edited by J. Letessier *et al.*, Plenum Press, New York, 1995

347

Table 1. Characteristic signals from highly excited compressed nuclear matter.

CHARACTERISTIC SIGNALS			
Hadronic		Electroweak	
J/Ψ Suppression ϕ Enhancement	Strangeness- Production	Single Photons (thermal, direct)	Dileptons

those photons in the same experiment. The latter technique has been applied by the WA80 experiment,[1] where π^0 and η mesons are reconstructed from their decay photons by a finely granulated photon spectrometer.

PHOTON PRODUCTION

Single photons, which do not originate from meson decays, are produced by different processes depending on energy, temperature and density of the system. In fig. 1 the basic processes and their characteristic ranges are sketched. 'Direct' photons from hard QCD processes have been identified in hadron-hadron experiments[2] at high p_t and are also expected in heavy ion collisions at appropriately high p_t. In the case of Quark-Gluon Plasma formation, where the same basic graphs are relevant, the photons are expected at much lower p_t and are frequently termed 'thermal'. In a pure hadron gas of appropriate temperature and density

Single Photons from A-A Collisions			
	Hadron Gas	**QGP**	**NN-Collisions**
Elementary processes:	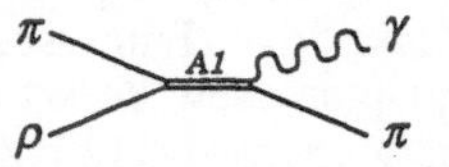 $\pi \dashrightarrow \gamma$ ($A1$) $\rho \dashrightarrow \pi$ $\pi\pi \to \rho\,\gamma$ $\omega \to \pi^0\,\gamma$ $\eta \to \rho^0\,\gamma$, $\eta\gamma$ ... ⋮	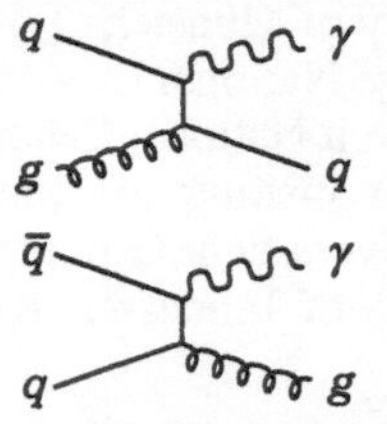 $q \dashrightarrow \gamma$ $g \dashrightarrow q$ $\bar{q} \dashrightarrow \gamma$ $q \dashrightarrow g$	QCD - Compton QCD - Annihilation
Distribution:	Thermal ($\varepsilon, T, S_{HG}, ...$) + radiative decays	Thermal ($\varepsilon, T, S_{QGP}, ...$)	Structure Functions (experiment)
Total Yield:	Integration over Space-Time History		Superposition of NN-Collisions
Relevant Range:	Resonance Decays $p_\perp \leq 2$ GeV/c	"Thermal" Photons $p_\perp \approx 1$–4 GeV/c	"Direct" Photons $p_\perp \geq 4$ GeV/c

Figure 1. Basic processes and relevant p_t ranges for single photon production.

single photons may be produced by interactions of various mesons. While the basic processes above are rather well known, there are large uncertainties in the values of the temperature and energy density and the space time development of the collision. This leads to large uncertainties in the prediction of absolute yields. Following earlier theoretical papers[3-8] a number of calculations are now available[9-16] based on the above graphs and different scenarios of the space time development. Since - at least at CERN energies - the effects are all rather small high statistics high precision data are needed.

EXPERIMENTAL METHOD AND ANALYSIS

The WA80 experiment has concentrated on the measurement of photons by a large acceptance finely granulated leadglass spectrometer consisting of ca. 3800 individual modules covering the pseudotapidity range $2.1 \leq \eta \leq 2.9$. The whole experimental setup is shown in fig. 2. Centrality selection of the events is achieved by setting conditions on the transverse

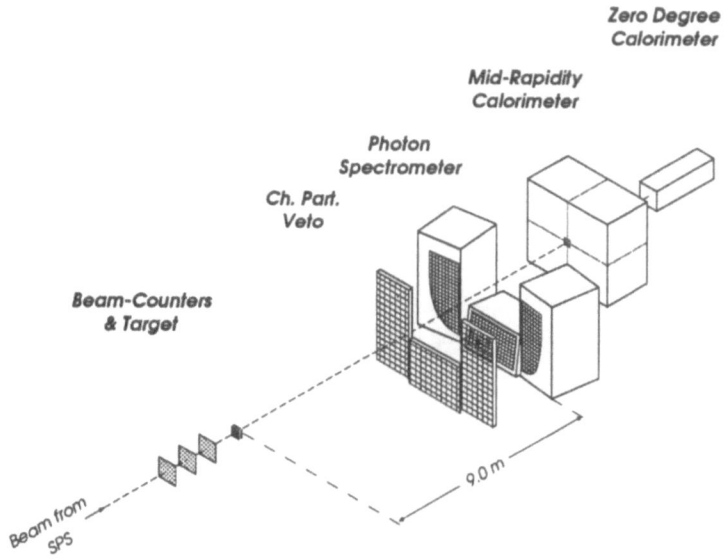

Figure 2. Set up of the WA80 experiment.

energy measured in the mid rapidity calorimeter. π^0 and η are reconstructed by their invariant mass and p_t spectra have been obtained. A summary of π^0 spectra for different centrality conditions are shown in fig. 3. Due to the large p_t range covered it becomes obvious that the spectra are not exponential but display a powerlaw behavior over the whole range measured. In smaller regions of p_t the slopes may, of course, be approximated by exponentials and slope parameters be extracted. In fig. 4, π^0 and η spectra are compared in absolute cross sections vs transverse mass m_t. In this presentation the spectra are found to be remarkably parallel which is in accordance to the empirical recipe of 'm_t-scaling'.[17] It has to be emphasized that for the extraction of single photons to be discussed below the precise knowledge of the π^0 and η yield over a large p_t range is indispensible for a reliable evaluation of the background photons. In fig. 5 the scheme for extracting single photon yields from the inclusive data is displayed. Experimentally, the relevant quantity is the γ/π^0 ratio rather than the cross section of the single photons alone. This is because the experiment has to cope with the large amount of background photons from π^0 decay and the γ/π^0 ratio therefore represents a better measure of the experimental sensitivity. It has to be noted that in our present analysis more than 98% of the backgrond photons are evaluated from meson decays measured in the same experiment and only 2% have to be inferred from extrapolating proton-proton data.

The main problem in the present analysis is, however, not only the determination of the photon background but also the exact determination of the detection and reconstruction

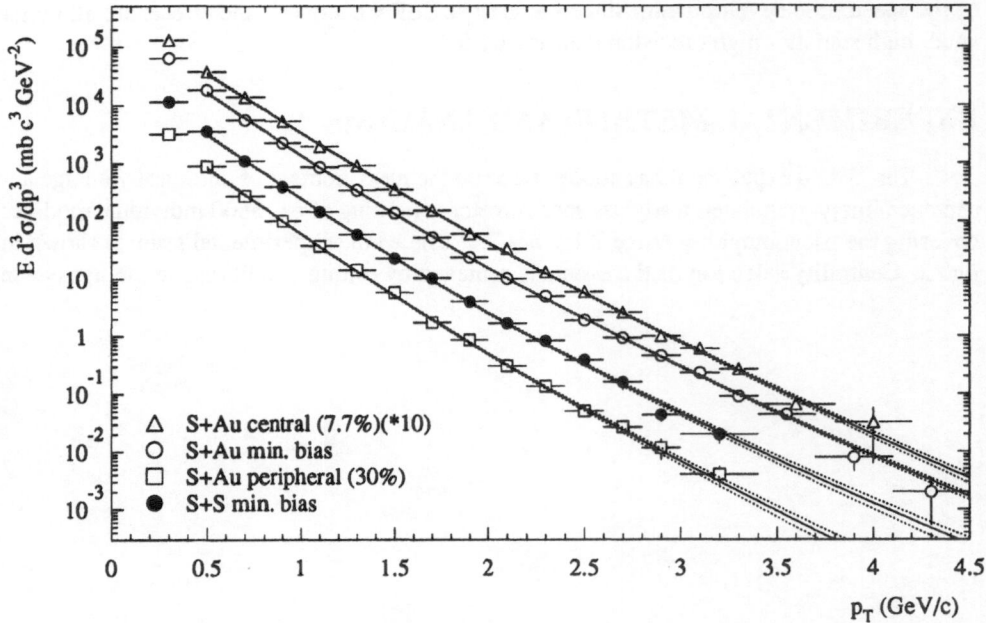

Figure 3. Transverse momentum spectra of identified π^0 for S+S and S+Au collisions at 200 A·GeV. The solid line is a fit to the data according to $Ed^3\sigma/dp^3 = C(\frac{p_0}{p_t+p_0})^n$ suggested by QCD.[19]

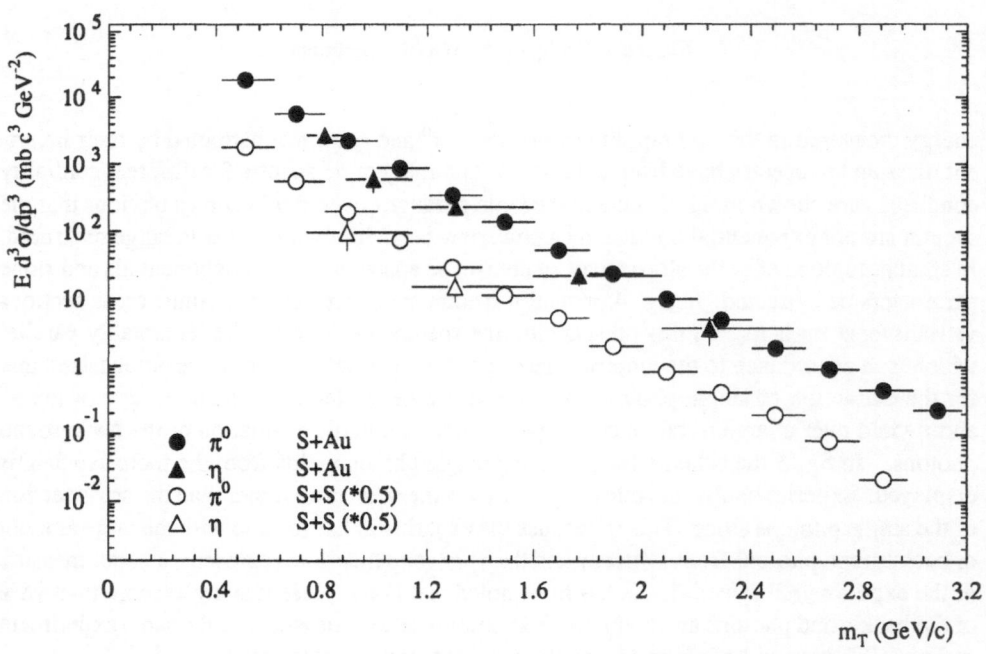

Figure 4. π^0 and η invariant cross sections as function of transverse mass m_t.

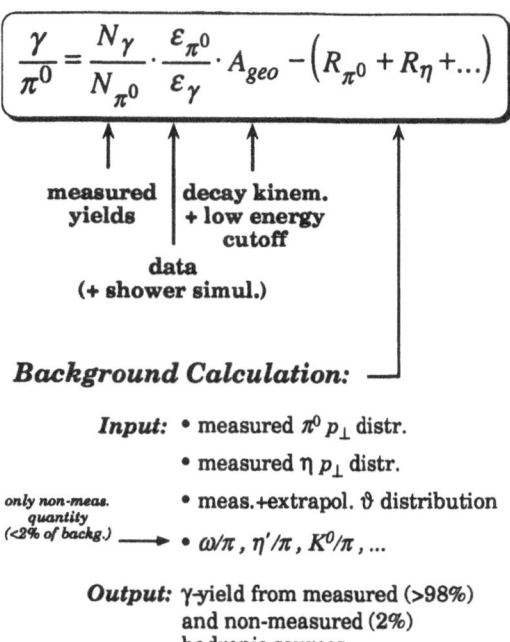

$$\frac{\gamma}{\pi^0} = \frac{N_\gamma}{N_{\pi^0}} \cdot \frac{\varepsilon_{\pi^0}}{\varepsilon_\gamma} \cdot A_{geo} - \left(R_{\pi^0} + R_\eta + ... \right)$$

↑   ↑   ↑   ↑

measured | **decay kinem.**
yields | **+ low energy**
          | **cutoff**

data
(+ shower simul.)

Background Calculation: ──┘

Input: • measured π^0 $p_\perp$ distr.

• measured η $p_\perp$ distr.

only non-meas. • meas.+extrapol. ϑ distribution
quantity
(<2% of backg.) ──→ • ω/π, η'/π, K^0/π, ...

Output: γ-yield from measured (>98%)
and non-measured (2%)
hadronic sources

Figure 5. Sketch of the procedure for extracting single γ/π^o ratios from inclusive photon data.

efficiencies.[18] This problem is specific to heavy ion collisions, where the multiplicities are extremely high, leading to increased overlap between electromagnetic showers or with hadronic showers. A particular method has therefore been developed, in order to determine the photon and π^0 efficiencies, respectively. Their calculation is based on the actual experimental data and proceeds by superimposing single hadronic and electromagnetic showers on a raw-data level to the measured events. The added showers are generated either by a Géant simulation assuming phase space distribution of the particles in accordance to the experimental data, or uses directly experimental data from very peripheral events, where the showers are well separated and identified. The artificial events are then processed with the same reconstruction routines as the real data events. The photon reconstruction efficiency is then defined by the ratio of the reconstructed photon spectrum divided by the known input photon spectrum and is found to depend both on the local particle density and p_t. The described method allows to gradually evaluate efficiencies with high precision up to large multiplicities. The resulting systematic and statistical uncertainty in the reconstruction efficiency is found to be approximately 3%.

RESULTS AND DISCUSSION

Using the procedure outlined above and sketched in fig.5 γ/π^0 ratios have been extracted from the 200 A· GeV data for S on S and S on Au collisions. Preliminary results for single photon yields, obtained after subtracting all contributions from photon decays of hadronic sources are shown in fig. 6 for central S on Au data. Different from peripheral data, where no excess over the hadronic photons is observed, the central data show a posi-

Table 2. Summary of errors for γ/π^0 ratios from different experiments. O+Au data are from Ref. 1

	Data on γ/π^0 from WA80/93			WA98 (expected)	
	O + Au (1986)	S + Au (1990)		Pb + Pb	
No. of events	$2.1 \cdot 10^6$	$10 \cdot 10^6$		$30 \cdot 10^6$	
central collisions	$1.7 \cdot 10^6$	$4 \cdot 10^6$		$20 \cdot 10^6$	
p_T region (GeV/c)	0.6 - 2.2	0.5 - 2.1	2.1 - 3.0	0.5 - 2.0	2.0 - 4.0
statistical error	6.0 %	2.0-3.0 %	4.5-13. %	0.8 %	4.0 %
π^0 accept. + efficiency	3.0 %	3.0 %	2.0 %	3.0 %	2.0 %
nonlinearity	3.0 %	3.0 %	3.0 %	3.0 %	3.0 %
γ reconst. efficiency	4.0-8.0 %	2.0-3.0 %	2.0-3.0 %	2.0 %	2.0 %
η/π^0 induced error	4.0 %	3.5 %	3.5 %	3.0 %	3.0 %
γ's from higher reson.	0.5 %	0.5 %	0.5 %	0.5 %	0.5 %
quadratic sum	9-12 %	6.2-7.0 %	7.0-14.2 %	5.6 %	6.5 %

tive signal with a slight increase towards low p_t. The error bars encompass statistical as well as systematicl errors, which are individually listed in table 2. Assuming gaussian distributions, we arrive at uncertainties of 7-8% and 8-15% for p_t bins below and above 2 GeV/c, respectively. The excess of photons in central data over the hadronic background for p_t <2 GeV/c thus reaches a level of about 2 standard deviations in each point. To compare the data with other experiments one may integrate the data in the relevant p_t range of 0.4< p_t < 2.0 GeV/c to find $(\gamma/\pi^0)_{centr.}$ = 0.091±0.016(stat)±0.058(syst) in central events and (γ/π^0) = 0.013±0.019(stat)±0.058(syst) in peripheral events. NA45 finds[20] for central collisions in the same reaction γ/π^0 = 0.075±0.005(stat)±0.110(syst), while NA34 reported[21] no evidence for a signal within the experimental uncertainty. All results are thus, in the present preliminary state, compatible with each other, but the uncertainties of the two latter experiments are still too large for a sensitive comparison. The most interesting information is probably contained in the p_t spectrum of the single photons, which is shown in fig. 7 in absolute cross sections. It is seen that in this p_t range the data display an exponential behavior with a slope constant of about 180 MeV/c.

Comparing with the rough distinction sketched in fig. 2 the observation of an excess in the range below 2 GeV/c suggests a hadron gas or mixed phase scenario. For a hadron gas the observed slope parameter would translate into a temperature of about 165 MeV/c.

Some theoretical calculations[22-24] have been performed for photon production at the present conditions. Within the scenarios discussed it seems that the experimental yield can only be obtained by assuming long life times of the mixed phase.

CONCLUSION

Transverse momentum spectra of photons from the interaction of 200 A·GeV S ions with S and An targets have been measured with a finely granulated lead glass spectrometer. A preliminary analysis yields an excess of ca. 7% single photons over photons from identified hadronic sources.

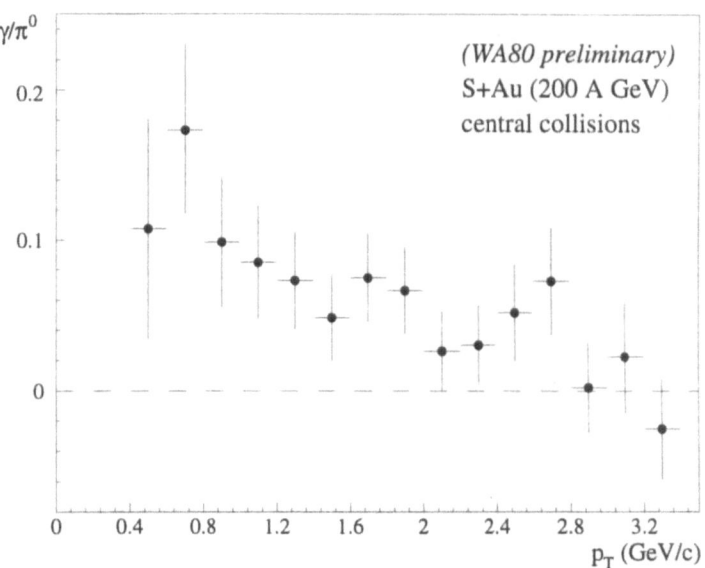

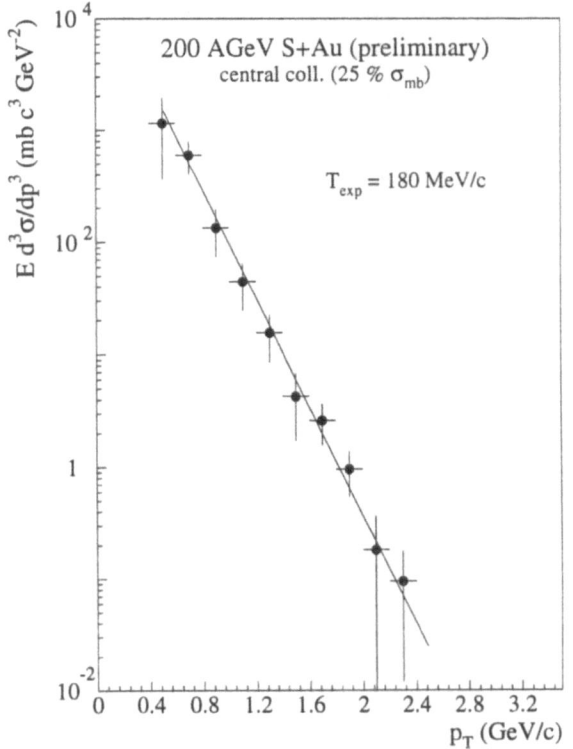

REFERENCES

1. R. Albrecht et al., WA80-Collaboration, Z. Phys. **C51** (1991) 1–10.

2. A. Bonesini et. al., Z. Phys. **C37** (1988) 535.

3. E. L. Feinberg, Nuovo Cimento **34** (1976) 391–412.

4. E. V. Shuryak, Phys. Lett. **78B** (1978) 150–153.

5. K. Kajantie and H. I. Miettinen, Z. Phys. **C9** (1981) 341–345.

6. F. Halzen and H. C. Liu, Phys. Rev. **D25** (1982) 1842–1846.

7. L. D. McLerran and T. Toimela, Phys. Rev. **D31** (1985) 545–563.

8. R. C. Hwa and K. Kajantie, Phys. Rev. **D32** (1985) 1109–1118.

9. M. Neubert, Z. Phys. **C42** (1989) 231–242.

10. P. V. Ruuskanen, Nucl. Phys. **A544** (1992) 169c–182c.

11. D. Bucher, Diploma thesis, University of Münster, 1993.

12. A. Dumitru, et al., Mod. Phys. Lett. **A8** (1993) 1291.

13. J. Kapusta, P. Lichard, and D. Seibert, Phys. Rev. **D44** (1991) 2774–2788.

14. L. Xiong, E. Shuryak, and G. E. Brown, Phys. Rev. **D46** (1992) 3798–3801.

15. J. Kapusta, L. McLerran, and D. K. Srivastava, Phys. Lett. **B283** (1992) 145–150.

16. S. Chakrabarty, et al., Phys. Rev. **D46** (1992) 3802–3806.

17. M. Bourquin and J.M. Gaillard, Nucl. Phys. **B114** (1976) 334.

18. H. Baumeister et.al., Nucl.Instr. and Meth. **A292** (1990) 81.

19. R. Hagedorn, CERN-TM **3684** (1983) preprint.

20. D. Irmscher et.al. (Na45 Call.), Proc. Quark-Matter Conf., Borlänge, Nucl. Phys. **A566** (1994) 347c.

21. T. Åkesson et al., NA34-Collaboration, Z. Phys. **C46** (1990) 369–375.

22. E.V. Shuryak and L. Xiong, Phys. Lett. **B333** (1994) 316.

23. E.V. Srivastava and B. Shinha, preprint VECC Calcutta (1994).

24. J.J. Neumann, D. Seibert and G. Fai, preprint, Kent University, KSVCNR-016-94 (1994) .

LATEST RESULTS
ON DILEPTON PRODUCTION
IN 200 GeV A ION-ION COLLISIONS

Olivier Drapier

Institut de Physique Nucléaire de Lyon
43, B^d du 11 Novembre 1918
F-69622 Villeurbanne CEDEX, France

INTRODUCTION

I present the latest results on dimuon production in the intermediate mass region, as measured by the NA38 and HELIOS-3 collaborations in ion-ion collisions at the CERN SPS. For p-A interactions, signal dimuon mass spectra are well reproduced by a superposition of the known sources (Drell-Yan process, open charm production and resonances). For S-induced reactions, both experiments observe an increase of the dimuon yield, as compared to the one expected from a linear extrapolation of p-A data.

NA38 and HELIOS-3 experiments have measured the dimuon production in collisions induced by 200 GeV/c per nucleon proton and sulphur beams on heavy targets. The dimuon mass spectrum can be divided into three domains :

1. the low mass region, including ρ, ω and ϕ resonances,

2. the intermediate mass region, between ϕ and J/ψ,

3. the high mass region including the J/ψ and ψ' decay dimuons, and muon pairs from the Drell-Yan process.

Several results have already been reported concerning the enhancement of ϕ[1,2] and the suppression of charmonium[3] production. The study of mass continuum dimuons in the intermediate mass range is of great importance, since the production of thermal lepton pairs in this kinematical region has been proposed as a signature of Quark-Gluon Plasma formation[4] . In both experiments, the main contribution to the observed $\mu^+\mu^-$ yield comes from uncorrelated decays of π's and K's. This background is estimated from $\mu^+\mu^+$ and $\mu^-\mu^-$ spectra. The known contributions to the signal spectra (resonance tails, Drell-Yan dimuons and simultaneous semi-leptonic decays of charmed meson pairs) have been simulated by means of Monte-Carlo programs taking into account acceptance and resolution of the detectors.

HELIOS-3 AND NA38 EXPERIMENTS

The Helios-3 and NA38 results reviewed in this paper were first presented at the "Quark Matter '93" conference[5,6]. The Helios-3 layout[6] was designed to detect low transverse mass muon pairs. The muon spectrometer consists of an alumina and iron hadron absorber followed by a dipole magnet. Muon tracks are detected by seven multiwire proportional chambers, and scintillator hodoscopes provide the trigger. The charged particle multiplicity is measured by two silicon ring counters in the pseudorapidity range $3.5 < \eta < 5.2$. Muon pair production was measured in proton and sulphur collisions on a tungsten target, in the kinematical domain defined by :

$$M_\perp \geq \sqrt{4m_\mu^2 + \left(\frac{2P_{min}}{\cosh y_{lab}}\right)^2}$$

$$P_{min} = 7.5 \ GeV/c$$

$$M_\perp \geq 4(7 - 2y_{lab})$$

The NA38 setup[7] is based on a toroidal field muon spectrometer, preceeded by a carbon hadron absorber. Eight hexagonal multiwire proportional chambers detect the muons, and the trigger is performed by four scintillator hodoscopes. The centrality of the collision is estimated from the transverse energy flow of the neutral particles, measured by an electromagnetic calorimeter in the range $1.7 < \eta < 4.1$. NA38 has measured dimuon production in p-W, p-U and S-U interactions, in the following kinematical window :

$$P_\perp \geq 0 \ GeV/c$$

$$3 \leq y_{lab} \leq 4$$

$$-0.5 \leq \cos\theta_{cs} \leq 0.5$$

where θ_{cs} is the polar angle in the Collins-Soper frame. In addition, the analysis reported here is restricted to $m_{\mu\mu} \geq 1.5 \ GeV/c^2$. It is important to notice that most of Helios-3 dimuons lie in a higher rapidity range ($3.5 \leq y_{lab} \leq 5$) than the NA38 ones.

Background Subtraction

The main background source in the "opposite-sign" dimuon ($\mu^+\mu^-$) spectra is the coincidence of two muons originating from the decay of uncorrelated mesons (π and K). Such coincidences lead also to "like-sign" ($\mu^+\mu^+$ and $\mu^-\mu^-$) events, that are used to estimate the number of background opposite-sign events as follows :

$$N_{Bk}^{+-} = 2R\sqrt{N^{++}N^{--}} \tag{1}$$

For NA38, Monte-Carlo studies show that $R = 1$ for S-U data. For proton-nucleus interactions, a simultaneous fit of two data samples acquired with different absorber configurations gives $R = 1.25 \pm 0.04$, in agreement with the Monte-Carlo value. Since the detector acceptance is different for opposite-sign and like-sign events, a cut ("image cut") is applied to the data to avoid any bias coming from the different muon charges. This cut rejects events which would not fulfill the trigger condition if the charge of any of the two muons was changed. In this analysis, the background mass distribution calculated with equation (1) is added to the distribution accounting for the signal events, and a fit is performed on the opposite-sign dimuon spectra. In the Helios-3 analysis, Monte-Carlo simulations lead to $R = 1.09 \pm 0.02$ for S-W data, and $R = 1.57 \pm 0.10$ for protons, due to a higher probability to trigger on opposite-sign than on like-sign events. The precise shape of the combinatorial background spectra is obtained by combining muons from different like-sign events in the same multiplicity class. The background coming from interactions taking place inside the dump was found to be negligible for masses above 1 GeV/c^2.

SIMULATION OF PHYSICAL PROCESSES

Helios-3

The ω cross section[8] was taken as $\sigma(pp \rightarrow \omega + X) = 12.4 \pm 2.4$ mb. The extrapolation of this cross section from p-p to p-W and S-W interactions was done using the multiplicities and the inelastic cross sections calculated with the QGSM[9] model. These predictions were tuned by a comparison with Helios-2 data on O-W interactions, and checked by the DTUNUC 1.02 model[10] . All other contributions were normalized to the ω cross section. The ρ cross section is fixed[8] ($B\sigma_\rho/B\sigma_\omega = 0.6$) without any multiplicity dependence. The Drell-Yan mass spectrum was simulated by Pythia 5.6 using Duke-Owens (1) structure functions with a factor $K = 2.5 \pm 0.5$ ($K = \frac{measured\ cross-section}{leading\ log\ prediction}$). The systematic error due to the choice of the structure functions is estimated to be 20%. From p-W to each multiplicity class of S-W events, the Drell-Yan cross section is scaled according to the corresponding impact parameter range, and corrected for isospin. The semileptonic decay of charmed mesons was simulated using Pythia, with two different values of the inclusive charm production cross-section[8] : $\sigma^{c\bar{c}} = 3.9^{+2.5}_{-1.9}$ μb/nucleon (NA25 measurement), and $\sigma^{c\bar{c}} = 8 \pm 1$ μb/nucleon (NA38, see next section).

NA38

As the events considered in the analysis of NA38 data have a mass $M_{\mu\mu} > 1.5$ GeV/c^2, the contribution of the low mass resonances is neglected[11] . Drell-Yan dimuons were simulated using Pythia 5.6 and GRV LO structure functions. The valence quark composition of the interacting nuclei has been taken into account by generating p-p, p-n, n-p and n-n collisions. In the kinematical domain of interest, the nuclear extrapolations of the cross sections given by Pythia are : $\sigma^{DY}_{pW} = 48.9$ nb, $\sigma^{DY}_{pU} = 62.9$ nb and $\sigma^{DY}_{SU} = 1.79$ μb. The high mass Drell-Yan events are used as normalization, and the K factor can be deduced from a comparison to the experimental spectra (K= 2.3 ± 0.1 for S-U interactions). The charm contribution was simulated by Pythia and GRV LO structure functions and c quark mass $m_c = 1.5$ GeV/c^2. Simulations using other structure function sets and different values of m_c show that the shape of the resulting dimuon mass distribution does not seem to be affected by changing these parameters.

PROTON-NUCLEUS COLLISIONS

In the NA38 analysis,[11] the same following procedure is applied to the four subsamples of p-A data (refered to as p-W 88V, p-W 88D, p-U 87 and p-U 88) : in a first step, a fit is performed to the high mass region in order to obtain the normalization of the Drell-Yan contribution. Since the charm production can be neglected in this mass range, the dimuon mass distribution is assumed to be a superposition of J/ψ, ψ' and Drell-Yan events, the amplitudes of which are free parameters of the fit. The shape of each contribution is fixed and deduced from a simulation taking the full detector response (acceptance and resolution) into account. This fit leads to a value of the Drell-Yan K factor : $K^{DY}_{pW} = 2.3 \pm 0.2$ and $K^{DY}_{pU} = 2.4 \pm 0.4$. The Drell-Yan, J/$\psi$ and ψ' amplitudes are kept from this first fit, and a second fit is performed to the mass region below the J/ψ, taking into account the charm contribution. As can be seen from Figure 1, a good agreement is obtained. The charm cross section in the NA38 kinematical window is estimated to be[11] $B\sigma^{c\bar{c}}_{pW} = 0.405 \pm 0.02$ and $B\sigma^{c\bar{c}}_{pU} = 0.409 \pm 0.02$ nb per nucleon, corresponding to a total charm production cross section : $\sigma^{c\bar{c}}_{pp} = 8 \pm 1$ μb, after correction for phase space and branching ratios (given by Pythia).

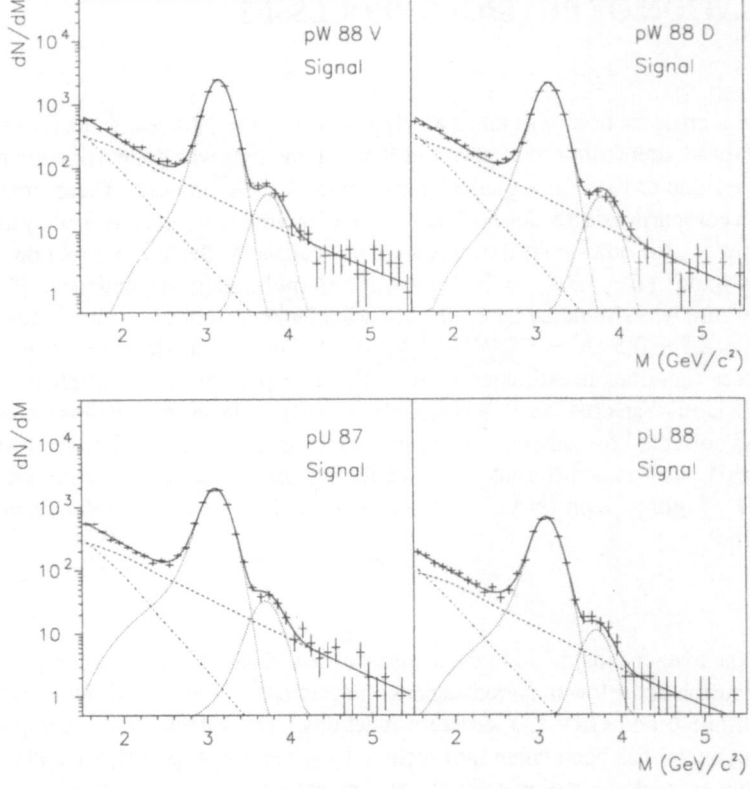

Figure 1. Dimuon mass spectra for p-W and p-U interactions (NA38, background subtracted). These curves are presented in the 'measured' variable (not corrected for acceptance and resolution). The dashed (resp. dash-dotted) curve stands for the Drell-Yan (resp. charm) contribution.

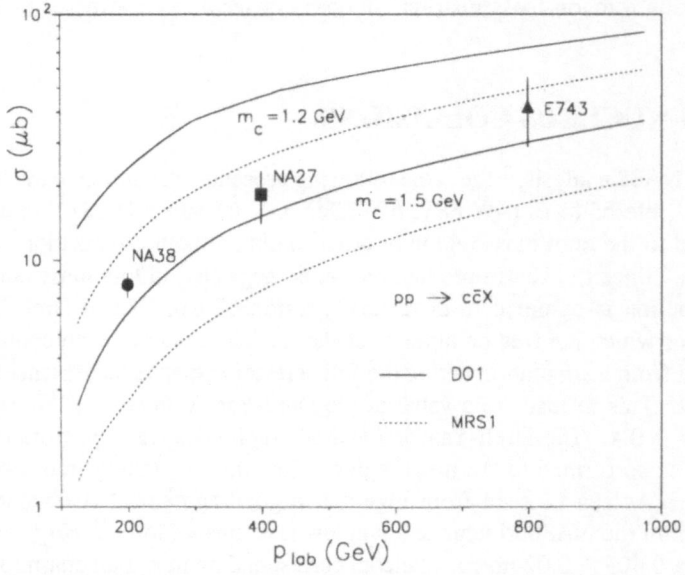

Figure 2. Comparison of the total inclusive charm production cross-section estimated by NA38 with previous measurements at different energies[11,12] (see text for details).

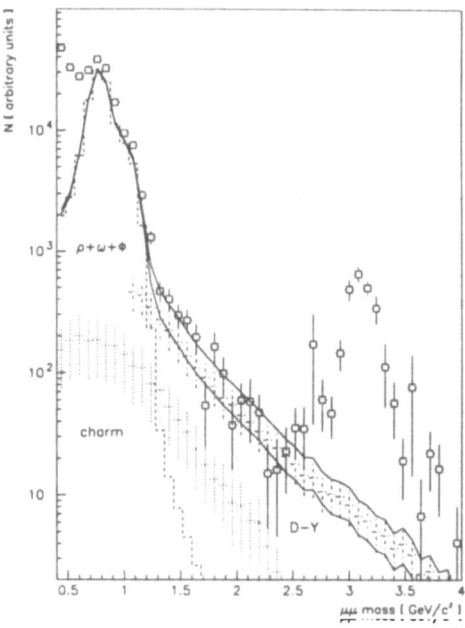

Figure 3. Expected signal (solid curves) as compared to Helios-3 p-W data. This spectrum is corrected for acceptance.

This result is in agreement with NA27 and E743 measurements[12] at different energies, as can be seen in Figure 2. Figure 3 shows the good agreement obtained between Helios-3 p-W data and the expected signal calculated as described in a previous section. In this figure, the "NA25" value was used for the charm production cross section, but another comparison using the "NA38" value does not show any significant difference within the error bars[8, 13]. The amplitudes are obtained via a fit to the number of ω mesons, to which all other contributions are normalized.

NUCLEUS-NUCLEUS COLLISIONS

The expected signal in S-A interactions was calculated, assuming a linear A dependence from protons to sulphur, both for Drell-Yan and charm production. This last assumption is supported by E769 measurements[14] giving $\sigma^{D^0+D^+} \propto A^\alpha, \alpha = 1.00 \pm 0.05$. NA38 results are shown in Figure 4, for the different data subsamples (refered to as S-U 90St, S-U 90Al, S-U 91Phi, and S-U 91Psi) corresponding to slightly different experimental setups. The high mass region is found to be well reproduced, both in shape and in normalisation (the fit to the S-U data gives the same K factor as for p-A data[11] : $K = 2.3 \pm 0.1$). Nevertheless, the linear extrapolation of charm production fails to reproduce the data in the intermediate mass region. The Helios-3 collaboration reaches the same conclusion, as shown in Figure 5 for two different multiplicity intervals. In order to compare the results obtained by the two experiments, the ratio $\frac{D}{S} = \frac{measured\ data}{sum\ of\ the\ expected\ sources}$ in the mass range $(1.5 \leq m_{\mu\mu} \leq 2.5\ \mathrm{GeV/c^2})$ is plotted in Figure 6, as a function of centrality[15] . The Helios-3 multiplicity classes correspond respectively to the following values of impact parameter : $b = (4.3 \pm 1.7), (3.5 \pm 1.5), (2.95 \pm 1.3)$. The NA38 transverse energy intervals correspond to : $b = (7. \pm 0.7), (5.5 \pm 0.7), (4.1 \pm 0.7), (2.4 \pm 0.7)$. For this comparison, the value of the charm cross section corresponding to NA38 data has been taken to calculate the expected signal in the Helios-3 experiment. This value, which is deduced from $\frac{D}{S} = 1$ for NA38 p-A

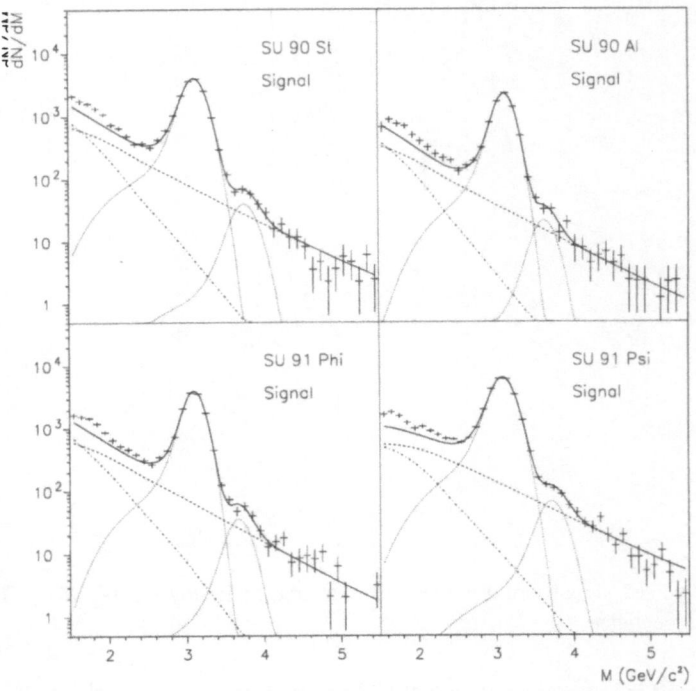

Figure 4. Signal dimuon mass spectra of the four S-U NA38 data subsamples.

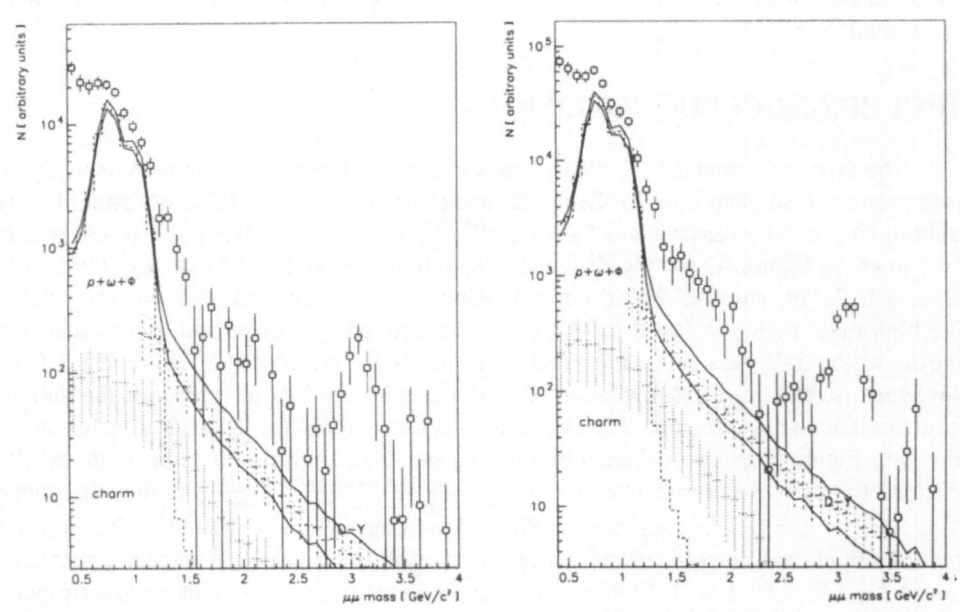

Figure 5. Dimuon mass spectra obtained by Helios-3, for two different multiplicity classes (4 and 5).

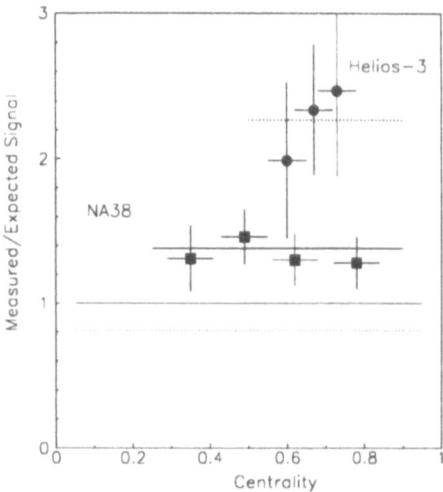

Figure 6. Comparison of NA38 and Helios-3 results, as a function of centrality $(1 - b/b_{max})$.

data (horizontal solid line), leads to $\frac{D}{S} = 0.90 \pm 0.11$ in the Helios-3 analysis (dotted line). Both collaborations measure an "excess", without centrality selection : $\left.\frac{D}{S}\right)_{SU} = 1.39 \pm 0.1$ (NA38), and $\left.\frac{D}{S}\right)_{SW} = 2.27 \pm 0.3$ (Helios-3). The NA38 results do not seem to exhibit any centrality dependence. Given the error bars, one cannot reach any conclusion for the Helios-3 data.

CONCLUDING REMARKS

The Helios-3 and NA38 experiments have measured dimuon production in 200 GeV A proton-nucleus and nucleus-nucleus collisions. Both experiments observe that, in p-A interactions, the intermediate mass region of dimuon spectra can be understood as a superposition of known sources (semi-leptonic decays of charmed particles and muon pairs originating from the Drell-Yan process). A linear extrapolation of these processes fails to reproduce sulphur-nucleus data. Both experiments observe an "excess" of dimuon production with sulphur incident beams. From a theoretical point of view, this result has been compared to the expected amount of thermal dileptons[16] produced in case of quark-gluon plasma formation. Preliminary NA38 analysis indicates that the mass spectrum shape of this unexpected signal is compatible with the distribution of events due to charm production (and represents approximately the same amount of events[5]), and is very different from the combinatorial background spectrum. It is important to notice that both collaborations assumed a linear A-dependence for charm production, supported by E769 measurements[14] with incident pion beams, and that no full simulation of S-A collisions has been performed in order to predict the charm contribution. The excess amounts to 40% of the measured signal in the mass interval $(1.5 \leq m_{\mu\mu} \leq 2.5 \text{ GeV/c}^2)$ for NA38, while it is roughly equal to the signal in Helios-3 data. This discrepancy can be due to the fact the NA38 result is calculated on the reconstructed data, while the Helios-3 $\frac{D}{S}$ ratio is computed after acceptance corrections. Thus, this comparison should be taken with some care. For instance, the NA38 acceptance for Drell-Yan is twice as large as for charm events, due to their quite different angular distributions. Helios-3 collaboration recently showed[13] that the mass distribution of the unexpected signal exhibits a continuous shape from 0.5 to 2.5 GeV/c², and that its slope does not depend on the centrality. This result indicates that it could have the same origin as the dielectron excess observed at lower masses by CERES[17] experiment.

Acknowledgments

It is a great pleasure to thank the organizers, and especially J. Rafelski for this workshop. I also wish to thank G. London (Helios-3 Coll.) and C. Lourenço (NA38 Coll.) for their help and for very usefull discussions.

REFERENCES

1. NA38 Coll., C. Baglin *et al.*, Phys. Lett. **B272** (1991) 449.

2. M.A. Mazzoni, (Helios-3 Coll.), *Quark Matter '91*, Nucl. Phys. **A544** (1992) 623c.

3. NA38 Coll., C. Baglin *et al.*, Phys. Lett. **B255** (1991) 459.
   NA38 Coll., C. Baglin *et al.*, Phys. Lett. **B270** (1991) 105.

4. L.D. McLerran and T. Toimela, Phys. Rev. Lett. **D31** (1985) 545.
   K. Kajantie and P.V. Ruuskanen, Z. Phys. **C44** (1989) 167, and references therein.
   P.V. Ruuskanen, *Quark Matter '91*, Nucl. Phys. **A544** (1992) 169c.

5. C. Lourenço, (NA38 Coll.), *Quark Matter '93*, Nucl. Phys. **A566** (1994) 77c.

6. M.A. Mazzoni, (Helios-3 Coll.), *Quark Matter '93*, Nucl. Phys. **A566** (1994) 95c.
   G. London, *Workshop on dilepton production in relativistic heavy-ion collisions*, GSI, Darmstadt, March 1994.

7. NA38 Coll., C. Baglin *et al.*, Phys. Lett. **B220** (1989) 471.

8. I. Kralik (Helios-3 Coll.), *Helios-3 Note 70*.

9. N.S. Amelin, K.K. Gudima and V.D. Toneev, Sov. J. Nucl. Phys **51** (1990) 1093.

10. H.J. Möhring and J. Ranft, Z. Phys. **C52** (1991) 643.
    I. Kawrakow, H.J. Möhring and J. Ranft, Z. Phys. **C56** (1992) 115.

11. C. Lourenço, (NA38 Coll.), *Workshop on dilepton production in relativistic heavy-ion collisions*, GSI, Darmstadt, March 1994.

12. E.L. Berger, *Advanced Reasearch Workshop on QCD Hard Hadronic Processes*, St. Croix, 1987.

13. M. Masera (Helios-3 Coll.), *27th International Conference on High Energy Physics*, Glasgow, July 1994.

14. E769 Coll., G.A. Alves *et al.*, Phys. Rev. Lett. 70 (1993) 722, and references therein.

15. C. Lourenço, *Fifth Conference on the Intersections of Particle and Nuclear Physics*, St. Petersburg, Florida, June 1994.

16. J. Letessier, J. Rafelski and A. Tounsi, Preprint CERN-TH 7304/94.
    E.V. Shuryak and L. Xiong, Phys. Lett. **B333** (1994) 316.

17. H. Specht (NA45 Coll.), *These proceedings*.
    I. Tserruya (NA45 Coll.), *27th International Conference on High Energy Physics*, Glasgow, July 1994.

DILEPTON SPECTRA IN HEAVY ION COLLISIONS

Jean Letessier,[1] Johann Rafelski,[1,2] and Ahmed Tounsi[1]

[1]Laboratoire de Physique Théorique et Hautes Energies[*]
Université Paris 7, Tour 24, 5e étage
2 Place Jussieu, F-75251 Paris Cedex 05, France
[2]Department of Physics, University of Arizona, Tucson, AZ 85721

DIMUON SPECTRA IN S–W AT 200 GeV A

The theoritical study[1,2] of the deconfined, dense hadronic matter produced in the relativistic collisions has shown that in the present day experiments a relatively high initial value for the light quark–gluon chemical equilibrium temperature is reached, namely $T_{ch} = 280$ MeV in the S–W/Pb case at 200 GeV A. In the forthcoming Pb–Pb experiments at 160 GeV A we expect still a higher value: $T_{ch} = 324$ MeV. At pre-chemical equilibrium even higher temperature values should be present — we recall that the production of the strange quark flavor acts to cool the system to the 'observable' chemical equilibrium temperature values which are about 50 MeV lower. We believe that these relatively high pre-strangeness temperatures can be observed studying the dilepton invariant mass spectra.

The particular merit of studying this electro-magnetic (EM) dilepton signal is that should QGP be produced in relativistic heavy ion collisions, it is expected that EM emissions could be a particularly clean and perhaps characteristic probe of the existence of this interesting transient form of hadronic matter — dileptons and photons can escape relatively easily from the dense phase after emission, and their abundance and spectral represent the properties of the source. Considerable effort has gone in the past decade to establish the methods that allow the precise evaluation of diverse contributing processes to the dilepton radiance and the invariant mass region 1.4–2.6 GeV was identified[3] as the best range for an experimental search of QGP with EM probes. Subsequently, the mixed phase and hadronic gas contributions were studied in realistic calculations,[4] using as input into the dynamics sophisticated models of hadronic gas equations of state. The evolution of the system, including the initial QGP fireball, was described by the relativistic hydrodynamic equations.[5] In this work it was found that the use of a realistic EoS[4,6] for the hadronic gas and mixed state phase and the introduction of transverse expansion effects have strong implications, in particular on phase lifetimes, and consequently on the dilepton yields.

---

[*]Unité associée au CNRS: D 0280

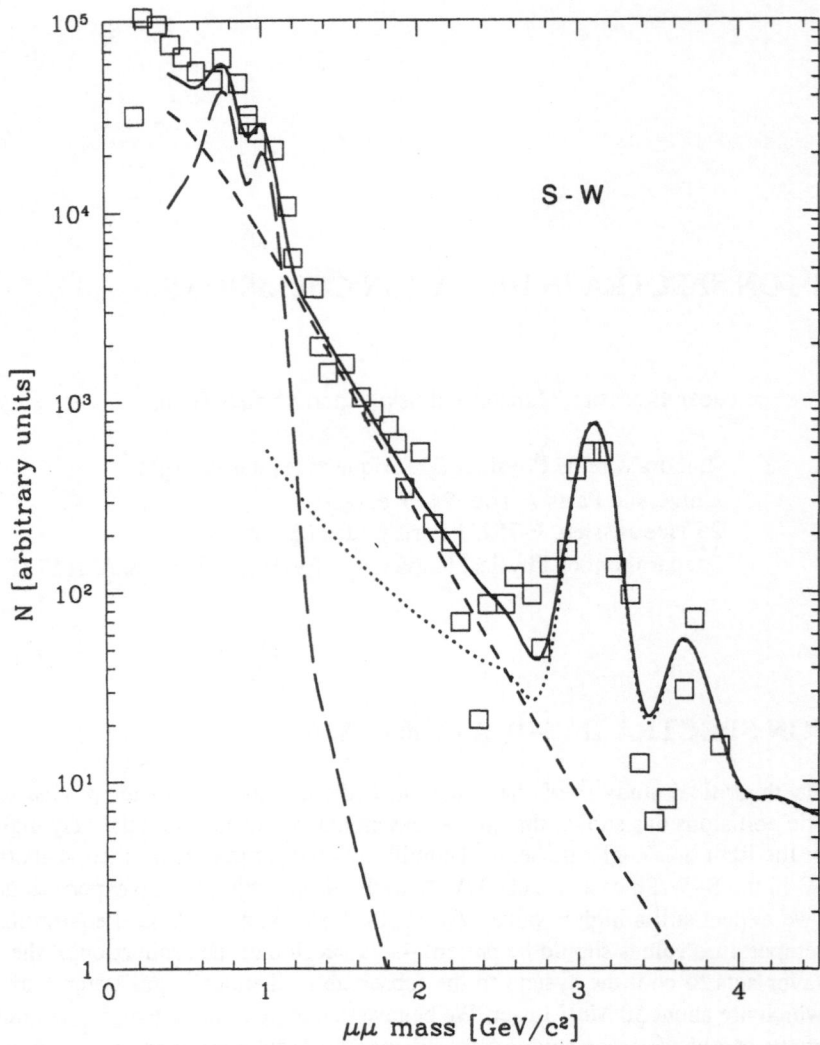

Figure 1. Spectrum of dimuons as a function of the dimuon invariant mass (arbitrary normalization), after Ref. [4] The solid line is the sum of the thermal QGP dimuons (short-dashed contribution), the hadron contribution (long-dashed component) and the Drell-Yan together with normalized J/ψ contributions (dotted line). Experimental results (open squares) are read from Fig. 5, in Ref. [7].

    The high predicted values[2] of the pre-strangeness temperature suggest that the dilepton signal may indeed be visible in the currently explored S–A collisions. Indeed an unexplained dilepton yield was reported recently[7] just where we had predicted the QGP dilepton radiance to be. Therefore, we have interpolated the results of Ref. [4] to the value of the initial temperature applicable in S–W/Pb collision. In Fig. 1 we show, in arbitrary units, the total number of dimuons (solid line) versus the dimuon invariant mass, as the sum of the thermal QGP contribution (short-dashed line), the hadron contribution (long-dashed line) and the Drell–Yan (with $K = 2$) together with renormalized J/ψ contributions (dotted line, chosen to fit the J/ψ peak). The relative yield of the QGP dilepton radiance, in the particularly interesting region between 1.4 and 2.6 GeV of the dimuon invariant mass, arises from these normalizations and is hence not arbitrary.

Let us next comment on other recent work which suggested that these data are difficult to describe: Shuryak and Xiang[8] evaluated dilepton spectra from 200 GeV A collisions and concluded "... even taking parameters at their extreme, one cannot explain the observed photon and dilepton excess in the conventional model which assumes constant velocity of the matter during the expansion...". This assertion could be seen in contradiction with our results. However their estimation of the $\mu^+\mu^-$ yields is based on the hypothesis that at CERN SPS energies the main contribution comes from the mixed phase which "... clearly dominates the space time evolution...". This supposition arises directly from their use of an ideal pion gas EoS for the hadronic gas phase, and their neglect of transverse expansion: in Ref. [4] it was demonstrated that the use of realistic EoS and the transverse expansion results in two main effects:

— the lifetimes of the mixed and hadronic phases are considerably reduced,

— the system spends most of the time in the hadronic phase and the mixed phase plays a lesser role (contrary to the case where we use for the hadronic phase an ideal pion gas without transverse expansion).

DILEPTONS FROM QGP?

While the above result reconfirms that QGP radiance should be visible in these conditions, the visible agreement with the experimental results presented recently (HELIOS3 collaboration[7]) and shown in Fig.1 as open squares is probably just a lucky coincidence. Note that this effect may have also been seen by the NA38 collaboration, despite different experimental conditions prevailing in both experiments.[9] Regarding the comparison of theory an experiment in Fig. 1, two, perhaps compensating, enhancement/reduction effects in the yieldshould be noted: the number of dimuon produced by the source is roughly proportional to the product of the abundance of quarks and antiquarks, which is greatly enhanced should valence quarks be present in the central region. However, the theoretical dilepton work, as mentioned above, did not allow for any stopping of baryon number, and hence this is not accounted her for. On the other hand we have not incorporated any acceptance cuts that are contained in the data and which would reduce the yields.

The rise in temperature, seen as we move from the S–Pb to the Pb–Pb system, suggests a greater visibility of photons and dileptons in the heavier system. However, the situation is here again more complex: the rate of signal to noise of photons and dileptons is proportional to $n_q n_{\bar{q}}/S$, where S is the entropy, which drops by 20% as we move from S–W/Pb to Pb–Pb collisions. Thus while the yield of dileptons in the invariant-mass region of current interest could appear flatter due to the higher temperature, the relative strength of the signal to hadronic background rate will be a bit less. We hope to return in the near future to a more detailed study of the dilepton and photon spectra of the QGP phase using the dynamical evolution determined here in order to assess the sensitivity of this interesting signal to the changes in temperature, acceptance cuts and in particular the presence of baryons in the central rapidity region.

Acknowledgments

J. Rafelski acknowledges partial support by US DOE grant DE-FG02-92ER40733 and thanks his colleagues for their kind hospitality in Paris.

REFERENCES

1. J. Letessier, A. Tounsi, U. Heinz, J. Sollfrank and J. Rafelski, *Phys. Rev. Lett.* **70** (1993) 3530. *Strangeness Conservation in hot nuclear fireballs*, Paris preprint LPTHE/92–27R, *submitted to Phys. Rev. D.*
   U. Heinz, *Nucl. Phys.* **A566** (1994) 205c.
   J. Rafelski, J. Letessier and A. Tounsi, in *Proceedings of the XXVI International Conference on High Energy Physics*, Dallas, Texas, 1992, AIP Conference Proceedings No 272, 983, J.R. Sanford, Editor.

2. J. Letessier, J. Rafelski and A. Tounsi, *Phys. Rev. Lett.* **B333** (1994) 484–493.
   J. Letessier, J. Rafelski, and A. Tounsi, in this volume.

3. J. Letessier and A. Tounsi, *Phys. Rev.* **D40** (1989) 2914.

4. M. Kataja, J. Letessier, P.V. Ruuskanen and A. Tounsi, *Z. Physik* **C55** (1992) 153.

5. L. D. Landau and E. M. Lifshitz, *Fluid Mechanics* Pergamon Press, Oxford, 1959.

6. A. Tounsi, J. Letessier and J. Rafelski, in this volume.

7. M.A. Mazzoni (Helios3 collaboration), *Nucl. Phys.* **A566** (1994) 95c.

8. E.V. Shuryak and L. Xiang, *Where the Excess Photons and Dileptons in SPS Nuclear Collisions Come from?* Stony Brook preprint SUNY-NTG-94-14, March 1994.

9. O. Drapier, in this volume.

PHOTON MULTIPLICITY MEASUREMENT
A NOVEL OBSERVABLE IN HIGH ENERGY
HEAVY ION COLLISION

Subhasis Chattopadhyay

Variable Energy Cyclotron Centre
1/AF, Bidhan Nagar, Calcutta 700 064

For the WA93 Collaboration

ABSTRACT

A high granularity preshower detector has been used in the WA93 experiment for measuring photon multiplicity in the collision of 200 A GeV sulphur on Au and S targets. Preliminary results on N_γ/N_{ch} distributions and $< P_T >$ for photons are described.

INTRODUCTION

The importance of photons in the study of high energy heavy ion collision have been emphasized by several authors for quite some time. The photon multiplicity N_γ measured in conjunction with the charged particle multiplicity N_{ch} in a large region of phase space provides us important observables to study various types of phenomena.

The study of the ratio of N_γ/N_{ch} is important from the viewpoint of studying the possible existence of Disoriented Chiral Condensate (DCC).[1] There have been several recent theoretical studies about the existence of DCC. According to the prediction this type of system leads to the depletion (excess) of N_γ producing CENTAURO (ANTICENTAURO) type of events. Centauro type of events have been found in cosmic ray studies,[2] though they have not been observed in laboratory yet. The distribution of N_γ/N_{ch} are predicted to be an important quantity to look at for such type of studies in the laboratory.

In addition to the study of DCC another important topic is the direct production of real photons as a useful signature of quark-gluon plasma (QGP). It is generally believed that there will be an excess of thermal photons produced in the collision of ultra-relativistic nuclei leading to the creation of a QGP phase.[3] The experimental study for the detection of the produced photons from all possible sources has been carried out in WA80 experiment in detailed manner. The preliminary results suggest the importance of pursuing the study further.

In addition to the photon spectrometry, the photon multiplicity on an event-by-event basis is an elegant and interesting observable.

For non-QGP processes it is expected that $N_{\pi+} \sim N_{\pi-} \sim N_{\pi^0}$ (N_{ch} consists of mostly $\pi^{\pm}$ and N_{γ} consists of mostly $\pi^0 \rightarrow 2\gamma$); and considering the slight proton excess in the incoming channel, we have $N_{\gamma}/N_{ch} \leq 1$. For heavy ion collision, producing large number of secondaries, the fluctuations in N_{γ}/N_{ch} distribution is very small, well within 10% even for S collision at SPS energies. Assuming excess photons are produced in events leading to QGP formation, they should be reflected in the measured N_{γ}/N_{ch} values on event-by-event basis. This excess should alter this distribution which can be statistically analyzed to separate the exotic events of interest. Studies of QGP formation and hadronization[4] suggest large asymmetries in the production of charged and neutral pions.

Measurement of N_{γ} in conjunction with the electromagnetic calorimeter in the same region of phase space can also allow us to derive the average p_T per photon on an event by event basis. In comparison to the charge particle case where the measurement of $< p_T >$ per event requires appropriate tracking of many particles ,for photons,the PMD in conjunction with e.m.calorimeter enable us to sort the events quickly to get $< p_T > /\gamma$.In WA93 setup PMD was sensitive to detect photons coming out of π^0 up p_T as low as 30MeV.

PHOTON MULTIPLICITY DETECTOR

To study these observables, a photon multiplicity detector (PMD) has been incorporated in the WA93 experiment.[5] The layout of the WA93 experiment and the details of the PMD are given elsewhere.[5,6] The PMD is placed in the forward hemisphere in front of the mid-rapidity calorimeter (MIRAC). MIRAC and zero degree calorimeter (ZDC) provide the impact parameter trigger for event selection. A large transverse energy E_T in MIRAC or a low forward energy in ZDC corresponds to "central" collision event.

The Photon Multiplicity Detector is a preshower detector consisting of 3 radiation length of lead converter. 2cm×2cm×3mm scintillator pads placed behind the converter were used as the sensitive medium. The total detector consisted of 100×76 such pads. The light generated inside the scintillator pad was transported via wave-length shifting fibers to the input of an Image Intensifier + CCD camera system. A bundle of 1900 fibers were coupled to one unit of such readout system.[7] The schematic diagram of the detector is given in Fig.1.

About 95 pads were selected throughout the camera which were enlightened with LED separately. The pixels affected by these pads were then used to make the correspondence between all other pads and the pixels which are likely to be affected by them.

Simulation results indicate that $\sim$ 9% of the pads are multiply hit at the preshower level. The detector covers the pseudo-rapidity interval $2.8 \leq \eta \leq 5.2$.

For VENUS γ's the average conversion efficiency is $\sim$ 85% for the $\gamma' s$ falling on the PMD. The hadron contamination probability averaged over the entire detector was estimated to be about 10% of the incident number of hadrons after taking into consideration the effects of the readout resolution.

DATA ANALYSIS AND RESULTS

Method of Analysis for PMD

For PMD the data was taken in two modes (i) pixel mode: where charge deposited in CCD pixels were digitized and stored for each pixel and in (ii) pad mode: in this mode data for each fired pad was stored by adding up all the pixels assigned to each particular pad. The total adc and the number of pixels contributing to each pad was stored.

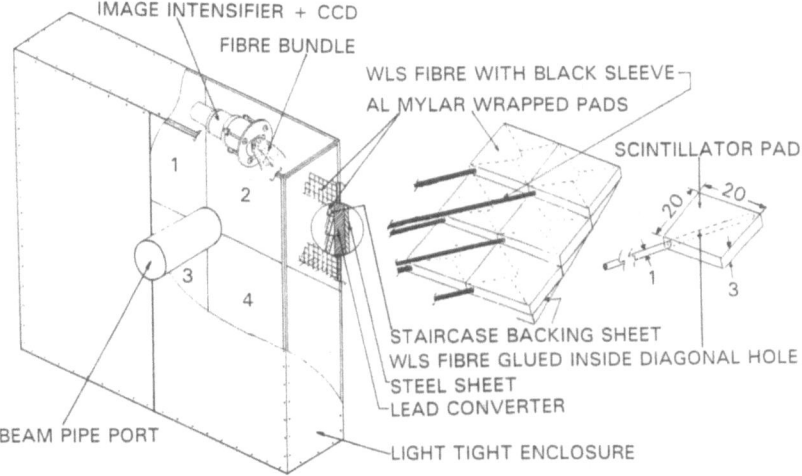

Figure 1. Schematic illustration of the components of PMD having scintillator pads, WLS fibers in diagonal holes and II+CCD readout. Picture on the bottom right shows inside details of the UA2 II+CCD camera system.

Because of the variation in the set of pixels affected by each pad from event to event there is some possibility that the adc values assigned to some particular pad belongs to the neighboring one. A detailed algorithm was thus developed to correct for those leakage. A clustering algorithm was then applied to extract the clusters formed by photon showers. Test data results from electron beam was used to split the big clusters giving proper weight to the affected ones.

Optimization has been performed by using detailed simulation and testdata results in order to obtain an adc threshold above which the clusters were considered as photon clusters.

Because of the Landau tail, some minimum ionizing particles (mips) get mixed even at this threshold and some γ's are lost. Also many of the secondaries from the hadron interactions in lead produce signals greater than the threshold. These introduce the most important systematic error in γ counting. At the present stage it is estimated using VENUS event generator that maximum about 15% of hadrons and their reaction products get mixed up with γ clusters at the working threshold level. For the data shown here, corrections for conversion efficiency have not been applied.

The photon multiplicity obtained by PMD with S-beam have been found to have a nice correspondence with the trigger detectors in WA93 setup. Fig.2(a) shows the anti-correlation between N_γ and the energy measured by the Zero Degree Calorimeter and Fig.2(b) shows the correlation of N_γ measured by PMD with E_T measured by MIRAC.

N_γ and N_{ch} Measurement

Using MIRAC as the trigger detector a sample of events have been analysed for S+Au interactions and the behaviour of N_γ with the charge particle multiplicities for those events have been studied.

N_{ch} was measured with a Si-Drift detector covering an eta region of $2.6 < \eta < 3.5$.[8]

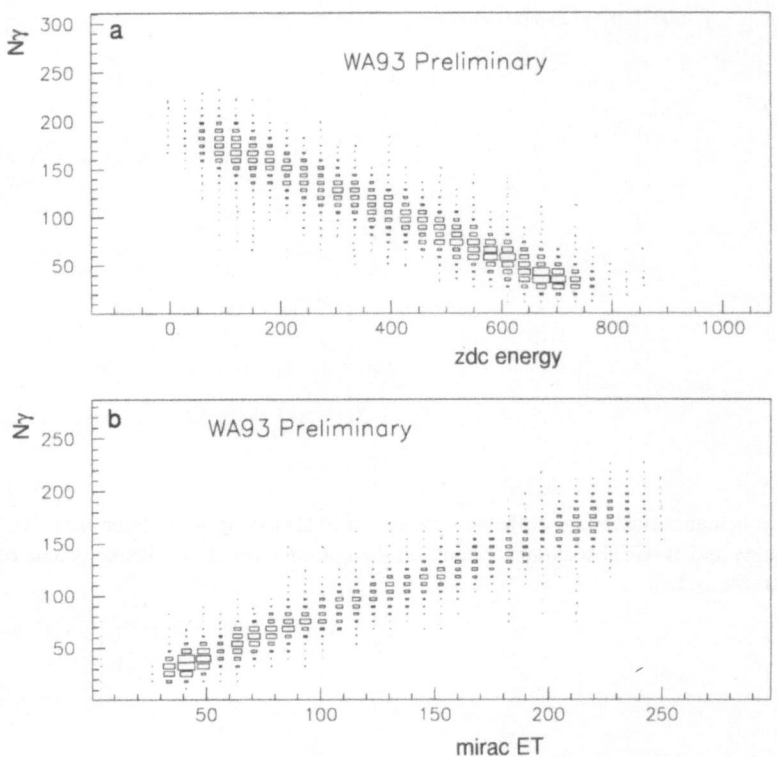

Figure 2. Correlation of PMD with triggers : (a) N_γ vs E_{ZDC}; (b) N_γ vs E_T^{MIRAC}.

The corrections for the systematic error and detection efficiency in gamma counting have not been applied to the present set of results. For N_{ch} only the acceptance correction have been made but the efficiency correction have not been performed.

Fig.3 shows the distribution of $(N_\gamma - N_{ch})/(N_\gamma + N_{ch})$ for central and peripheral events. The distributions in both cases are found to follow nice gaussian behaviour. The distribution is narrow for central events compared to the peripheral ones due to the larger multiplicities in central events.

Mean $< P_T >$ of Photons

As the overlap between PMD and the e.m. calorimeter MIRAC is almost complete in WA93 setup, one can estimate the $< P_T >$ of photons by measuring the ratio of the E_T measured in the e.m.calorimeter and N_γ.

Fig.4 shows the distributions for peripheral and central events. It is seen from the mean of the distributions that the $< P_T >$ of photons observed in central events is $\sim 10\%$ higher than that of photons produced in peripheral events.This result is in accordance with the result obtained by WA80 collaboration using lead glass spectrometer.

We are now studying the systematic and other kind of errors in detail, though the final results are not expected to change drastically after the correction for these effects.

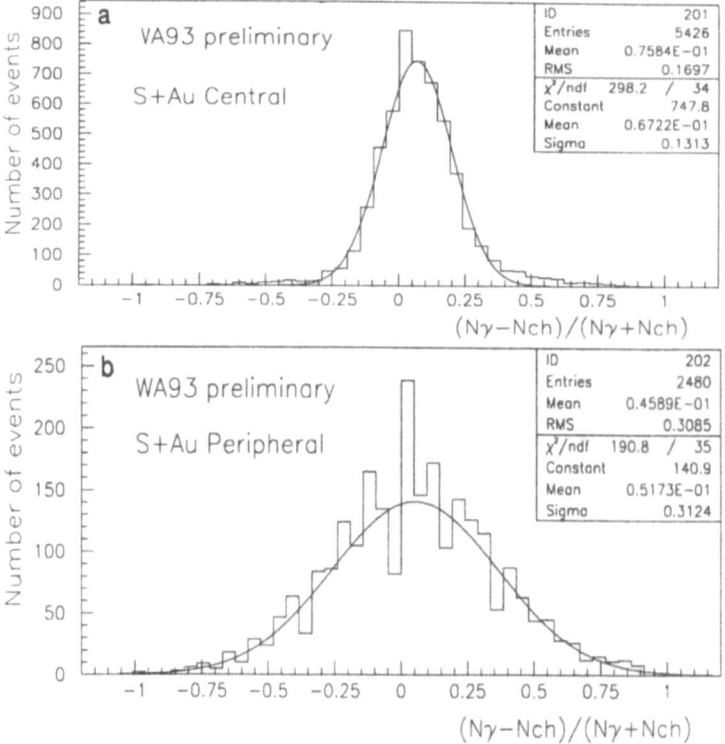

Figure 3. $(N_\gamma - N_{ch})/(N_\gamma + N_{ch})$ distributions for (a) central and (b) peripheral data.

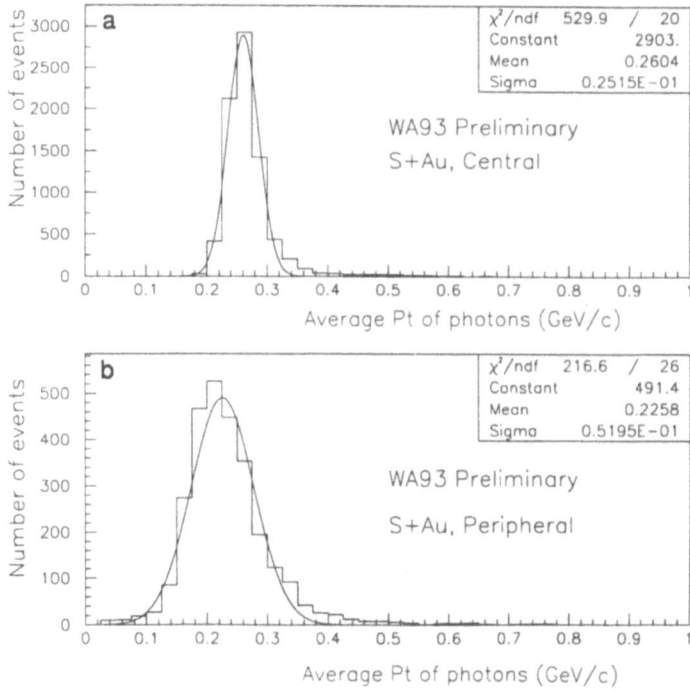

Figure 4. $< P_T >$ distributions of photons for (a) central and (b) peripheral data.

WA98 PHOTON MULTIPLICITY DETECTOR

Encouraged by the preliminary results obtained from PMD in WA93 experiment, we have designed and built a photon multiplicity detector for the lead beam experiment in WA98 experiment.[9] Several modifications have been adopted to increase the light output of this detector in comparison to the WA93 PMD. Test results with pion and electron beam shows considerable improvement. The WA98 setup it designed to have maximum overlap area between PMD and Si-drift, so that we can study the N_γ and N_{ch} ratio over larger region of phase space.

SUMMARY

WA93 Collaboration at CERN has successfully implemented a preshower photon multiplicity detector to study 200 A GeV sulphur interactions. The preliminary results of N_γ/N_{ch} distributions show gaussian behaviour in case of central and peripheral collisions. The average transverse momentum of photons measured in conjunction with MIRAC are found to be consistent with the results obtained in WA80 experiment.

Acknoledgements

We gratefully acknowledge the financial support of the Department of Atomic Energy, Department of Science and Technology and the University Grants Commission of the Government of India, the CERN PPE Division and the Indo-FRG Exchange Programme.

REFERENCES

1. J.D. Bjorken et al., "Baked Alaska", SLAC Preprint SLAC-PUB-6109 (1993); Proc. of Les Rencontres de Physique de la Vallee d'Aoste, ed. M. Greco, Editions Frontiers, p.507 (1993).

2. C.M.G. Lattes, Y. Fujimoto and S. Hasegawa, *Phys. Rep.* 65:151 (1980).

3. S. Raha and B. Sinha, *Int. J. Mod. Phys.* A6:517 (1991). and references therein.

4. S. Pratt, *Phys. Lett.* B301:159 (1993); E. D. Davis et al., Univ. of Arizona preprint AZPH-TH/91-52.

5. WA93 Collaboration, CERN/SPSC/90-14, SPSC/P252 (1990).

6. Y.P. Viyogi, Proc. of the 2nd Int. Conf. on Phys. and Astrophys. of Quark-Gluon Plasma (ICPA-QGP '93), ed. B. Sinha et al., World Sc., Singapore, (in print).

7. R.E. Ansorge et al., *Nucl. Instr. Meth.* A265:33 (1988); J. Aliti et al., *Nucl. Instr. Meth.* A279:364 (1989).

8. P. Steinhaeuser, Ph.D thesis, Univ. of Frankfurt/Main, Germany (1994).

9. WA98 Collaboration, CERN/SPSLC/91-17 SPSLC/P260, May 1991.

DENSITY MODIFICATION OF DILEPTON PRODUCTION IN HOT HADRONIC MATTER

K. Redlich,[1,2] J. Cleymans[3] and V. V. Goloviznin[4]

[1]Fakultät für Physik, Universität Bielefeld
W-4800 Bielefeld 1, Germany
[2]Institute of Theoretical Physics, University of Wroclaw
50-204 Wroclaw, Poland
[3]Department of Physics, University of Cape Town
Cape Town, Rondebosch 7700, South Africa
[4]Kurchatov Institute of Atomic Energy
123182 Moscow, Russia

INTRODUCTION

It is well known that the thermodynamic properties of highly excited high density hadronic matter are qualitatively different from a dilute system. In particular, the results obtained from studies of the thermodynamics of the Statistical Bootstrap Model[1,2] have shown that there should exist a limiting temperature $0.13 < T_H < 0.18$ GeV above which hadronic matter, with hadrons as a fundamental constituents, does not anymore have a meaning. When approaching the limiting temperature the partition function was found to exhibit a singular structure suggesting the possibility of a phase transition. Indeed, recent developments in QCD and in particular in Lattice-Gauge Theory (LGT) have shown that there exists a phase transition from a confined hadron-phase to a deconfined quark-gluon plasma phase. The value of the phase transition temperature T_c found in LGT is compatible with the value of the Hagedorn temperature T_H.[3]

One of the main objectives of ultrarelativistic heavy ion collisions is to study the properties and the nature of high density hadronic matter. It is expected, that for sufficiently large collision energies between heavy ions hadronic matter can be produced in a deconfined quark-gluon plasma phase.[4] The prime objective of the theoretical studies is to find the physical observables sensitive to a phase transition. It has been argued that dilepton production is one of the most promising signals for this.[5]

In this work we will study the structure and the properties of low invariant mass dilepton production. Low mass dileptons can have both thermal and non-thermal origin. Non-thermal dilepton pairs are produced mainly by Dalitz decay of neutral mesons.[6] Thermal dileptons are mainly originating from virtual photon bremsstrahlung by pions and quarks. In a dense medium however, thermal production can be modified by various collective effects like dynamical or Debye screening or by multiple scattering. The last effect is known to be very important when discussing particle bremsstrahlung in a medium.[7-10] It is because it can lead to a substantial suppression of bremsstrahlung spectrum due to the Landau-Pomeranchuk effect (LPE).[8-10] In this work we will discuss the origin and the influence of this effect on the production of dilepton pairs in hot hadronic matter. The cross-sections for real and virtual photon bremsstrahlung in a pion gas and in a quark-gluon plasma will be derived and analyzed. Finally, assuming initial conditions appropriate for U-U collisions at LHC energy, the bremsstrahlung contributions to the overall rate of low mass dilepton pairs will be calculated and compared with the non-thermal background.

MULTIPLE SCATTERING EFFECT AND COHERENCE LENGTH

Many electromagnetic processes, such as e.g. real or virtual photon emission, which occur in the interaction of fast charged particle in a medium, develop in a large region of space along the particle trajectory.[7] The length of this region is called the *coherence length* l_c. The concept of coherence length can be introduced on both the classical and quantum levels. For an illustrative purpose we shall interpret l_c as the length in which the *stripping* of a photon from the radiating charged particle occurs.[11] Suppose that a charged particle colliding with a nucleus radiates a photon of frequency ω. For an ultrarelativistic particle the typical radiation angle θ_B is very small and proportional to the ratio of the particle's mass to its energy E, that is $\theta_B \sim m/E$. Thus, since the radiation is almost parallel to the particle momentum it cannot separate immediately after a collision. The particle and the photon can be considered as independent objects only if they are separated by a distance of the order of the wave length of the radiation $\lambda \sim 1/\omega$. This separation requires some time (radiation formation time) $\Delta t = \lambda/v_r$ with $v_r = 1 - v\cos\theta_B$ being the relative velocity of particle and photon. The coherence length l_c is defined as the distance traveled by the charged particle in time Δt, that is:

$$l_c = v\Delta t \sim \frac{2E^2}{m^2\omega} \tag{1}$$

where we have assumed that $\theta_B < 1$ and also replaced the quantity $(1 - v)$ by $m^2/2E^2$.

The coherence length increases rapidly with increasing particle energy and with decreasing frequency of the radiation. Thus for very energetic particles and soft quanta, l_c can reach macroscopic dimensions. This simple analysis shows that particle bremsstrahlung is not really a point-like interaction, but appears over a large region of space in the longitudinal direction of particles motion.[11] Thus when discussing bremsstrahlung due to the interactions of a charged particle in a medium, one necessarily needs to consider two limiting cases:

(*i*) the coherence length is smaller than the average distance between medium constituents: in this case bremsstrahlung in a medium is undisturbed and equivalent to the radiation in N independent collisions.

(*ii*) the coherence length is large in comparison to the mean free path or to the average distance between the constituents in the medium: here it is necessary to take into account the interaction of the incident charged particle with the medium within the coherence length. Multiple scattering of the incident particle can disturb particle radiation, causing the standard vacuum formula for particle bremsstrahlung to fail. It is mostly due to the interference of the waves radiated in different collisions within the coherence length.

The result for the coherence length introduced in eq.(1) is very general and does not depend on the kind of medium in which the radiation occurs. Thus l_c itself does not determine yet the dynamics of the radiation. In particular, the knowledge of the value of l_c is not sufficient to deduce whether interference within the coherence length leads to enhancement or suppression of particle bremsstrahlung in a medium. This is entirely determined by the medium's properties.

Landau and Pomeranchuk[8] have shown that in amorphous material the interference effect within the coherence length is destructive and consequently leads to a substantial decrease of bremsstrahlung. The above result is known in the literature as the Landau-Pomeranchuk effect. The LPE is not only related with electromagnetic radiation, but it was found to be important in a variety of problems, also those which are entirely non-electromagnetic. Recently LPE was discussed in the context of nuclear shadowing effect,[12] gluon bremsstrahlung[13] and in the parton cascade model in order to account for the time delay of parton emissions.[14] In the following section we will discuss the influence of LPE on real and virtual photon bremsstrahlung in a hot QCD medium.

QUANTITATIVE DESCRIPTION OF BREMSSTRAHLUNG

In order to make quantitative the description of particle bremsstrahlung in a medium and discuss the role of multiple scattering leading to LPE we will start with a semi-classical approach which is valid in the soft photon approximation, i.e. when the frequency of the radiating photon ω is much smaller than the energy of the incident charged particle. In this approach the energy radiated per unit of momentum by the charged particle moving along the trajectory $\vec{r}(t)$ with velocity $\vec{v}(t)$ is given by the following equation:[15]

$$\frac{dI^\gamma}{d^3k} = \frac{\alpha}{4\pi^2} \mid \int dt \, \exp(\, i[\, \omega t \, - \, \vec{k}\vec{r}(t)\,]\,) \, [\vec{n} \times \vec{v}\,] \mid^2 \tag{2}$$

where $\vec{n} \equiv \vec{k}/|\vec{k}|$ is a unit vector in the direction of photon emission.

In the simplest case of scattering of a charged particle by static centers, the particle trajectory is a broken line. The semi-classical formula is known in this case to lead to the same result as the quantum description upon application of the soft photon theorem. In a medium however, the particle trajectory is quite complicated and due to its continuous interactions it is never on-shell. Thus when discussing bremsstrahlung in a medium it is not clear what set of Feynman diagrams corresponds to a given trajectory and what weight should be given to it. The advantage of the semi-classical approach over the quantum treatment is that it can easily be applied to radiation in a thermal medium. For a series of collisions one has to change first the time integral in eq.(2) to a sum of integrals corresponding to all the pieces of the trajectory. Assuming that the particle velocity is constant between two collisions and performing a time integration, the energy radiated by the charged particle undergoing N scatterings is given by:

$$\frac{dI^\gamma}{d^3k} = \frac{\alpha}{4\pi^2} \mid \sum_{j=1}^{N} \frac{\vec{n} \times \vec{v}_j}{\omega - \vec{k}\vec{v}_j} \exp(\, it_{j-1}[\, \omega - \vec{k}\vec{v}_{j-1}\,]\,)$$
$$(1 - \exp(\, i\xi_j[\, \omega - \vec{k}\vec{v}_j\,]\,)) \mid^2 \tag{3}$$

where $\xi_j \equiv t_j - t_{j-1}$ is the time between two collisions .

In a thermal medium many particle trajectories are possible. Thus in order to calculate the total energy radiated by the incident particle one needs to take into account the stochastic nature of the particle interactions and perform an average in the above equation over all possible particle trajectories. This can be done analytically in two limiting cases: *(i)* assuming that after a collision the particle velocities are isotropically distributed (isotropic scattering) and *(ii)* assuming small angle scattering. These two limiting cases correspond to the two distinct phases of QCD matter at finite temperature. The isotropic scattering is approximately valid for pions as the fundamental constituents. Pion-pion elastic scattering is mainly due to s-wave interactions with nearly isotropic final momentum distribution. The small-angle scattering holds in the deconfined phase where the quark interactions are mainly determined by the t-channel Feynman diagrams. This means that small-angle scattering dominates and particle velocities are strongly correlated in successive collisions.

The average over trajectories in a pion gas and in a quark-gluon plasma require separate investigations as the analytical procedure is very different.[16-19]

Bremsstrahlung in a pion gas

As was explained in the previous section the calculation of the cross-section due to particle bremsstrahlung is reduced to finding the average of eq.(3) over all possible particle trajectories in a medium. In a pionic gas the average over trajectories is equivalent to the average over all particle velocities after the collision and over all possible times between two collisions. This is because a given trajectory of a charged pion is parameterized by the scattering angle and the time between collisions. As the pionic gas is assumed to be in thermal equilibrium, many possible velocities can result in between two collisions. Because of this, upon taking the square in eq.(3), all non-diagonal terms will give zero contribution. Indeed, a typical non-diagonal term in eq.(3) has the form:

$$\frac{[\vec{n} \times \vec{v}_j] \cdot [\vec{n} \times \vec{v}_l]}{(\omega - \vec{k} \cdot \vec{v}_j) \cdot (\omega - \vec{k} \cdot \vec{v}_l)} \exp[i(\omega(t_j - t_l) - \vec{k} \cdot (\vec{R}_j - \vec{R}_l))]$$

$$\times (1 - e^{-i(\omega - \vec{k} \cdot \vec{v}_j)\xi_j})(1 - e^{-i(\omega - \vec{k} \cdot \vec{v}_l)\xi_l}) \tag{4}$$

where $t_j, t_l, \vec{R}_j, \vec{R}_l$ are the time and the position of the corresponding scattering event; $\vec{v}_j$ is the velocity of the test particle after j−th collisions and $\vec{n}$ is the unit vector in the direction of the photon emission.

The above expression contains a product of two main factors: the term $[\vec{n} \times \vec{v}_j]$ and the rest. These two terms have a different dependence on $\vec{v}_j$. Indeed, the term $[\vec{n} \times \vec{v}_j]$ depends only on transversal (with respect to the direction of photon emission) component of $\vec{v}_j$ whereas all the rest depends only on its longitudinal component.

To find the final result for the intensity of particle radiation one needs to average the probability of the photon emission over all possible trajectories of the test particle. Our main assumption in the pionic medium is that the scattering keeps no correlation between final and the initial particle velocity. Therefore, one can always find (with equal probability) two trajectories which coincide between points $\vec{R}_l$ and $\vec{R}_j$ but have opposite transversal components of the particles velocity after j-th collision. Thus the off-diagonal terms have different signs for the above trajectories. Consequently, after averaging over velocities these will cancel, leading in general to the vanishing of all non- diagonal terms in eq.(3). We have to point out however that these cancelations are only due to our strict assumption about complete isotropy of particle velocities in each scattering. The deviation from this assumption will naturally imply non-vanishing contributions of some non-diagonal terms in eq.(3). Nevertheless, the pionic medium is very close to the isotropic limit and we assume that only the diagonal terms remain significant.

The average over time between two successive collisions ξ is performed assuming an exponential distribution

$$\frac{dW}{d\xi} = a \exp(-\xi a) \tag{5}$$

with a being the average number of collisions per unit time (inverse mean free time)

$$a(p, T) \equiv g \int \frac{d^3q}{(2\pi)^3} e^{-\sqrt{(p^2 + m_\pi^2)/T}} \sigma_{\pi\pi}(s) v \tag{6}$$

where $\sigma_{\pi\pi}$ is the elastic pion-pion cross-section, $v = \sqrt{s(s - 4m_\pi^2)}/2E_1 E_2$ the relative particle velocity, $s = (p + q)^2$ and g is the degeneracy factor (3 for the pion gas). With the ξ—distribution function (5) the time average in (3) can be done analytically. When relating the intensity of the radiation dI/d^3k with the production cross-section

$$\frac{dI^\gamma}{d^3 k} = N\omega \frac{1}{\sigma_{\pi\pi}} \frac{d\sigma^\gamma}{d^3 k} \tag{7}$$

the final result for the bremsstralung cross-section, after averaging over all possible particle trajectories reads:[15]

$$< \frac{d\sigma^\gamma}{d\omega d\Omega} > = \frac{2\alpha\sigma_{\pi\pi}|\vec{v}|^2}{4\pi^2} \omega \frac{\sin^2\theta}{a^2 + \omega^2(1 - |\vec{v}|\cos\theta)^2} \tag{8}$$

with $d\Omega$ being the unit element of the solid angle.

The medium modification of the bremsstrahlung cross-section due to LPE is in the denominator in the above expression through the term $a(p, T)$. If this term is absent then we are recovering the standard formula for charged particle bremsstralung. How important the modification of the cross-section due to multiple scattering effect is can be seen by considering two limiting solutions for the cross-section in eq.(8). The mean free path λ of incident pions in a thermal medium is given by $\lambda = va^{-1}$. The coherence length according to (1) is determined from $l_c = v[\omega(1 - v\cos(\theta))]^{-1}$. With the above definitions one can find that

$$< \frac{d\sigma^\gamma}{d\omega d\Omega} > = (\frac{d\sigma^\gamma}{d\omega d\Omega})^{(1-\text{sct.})} \times \begin{cases} 1 & \text{if } \lambda >> l_c \\ \\ (\frac{\lambda}{l_c})^2 & \text{if } l_c >> \lambda \end{cases} \tag{9}$$

Thus, when the mean free path exceeds the coherence length the medium acts as the sum of N independent scatterers. In the opposite limit the interference effect within the coherence length starts to be important. The cross-section is modified by the factor $(\lambda/l_c)^2$ leading to the suppression due to LPE. The destructive interference of waves within the coherence length is also seen when analyzing frequency dependence of the bremsstralung cross-section. From (8) one can find that

$$< \frac{d\sigma^\gamma}{d\omega d\Omega} > \sim \begin{cases} \frac{1}{\omega} & \text{if } \lambda >> l_c \\ \\ \omega & \text{if } l_c >> \lambda \end{cases} \tag{10}$$

In the limit of very soft photon, the cross-section decreases linearly with the frequency of the radiation which is in contrast to the standard result which shows $1/\omega$ behaviour. With increasing ω the coherence length is decreasing. Thus, for the value of ω where $\lambda \sim l_c$ the LPE is not important any more and typical $1/\omega$ behaviour for the bremsstrahlung cross-section is again recovered.

Until now we have discussed real photon bremsstrahlung in a pionic medium and derived the cross-section for this electromagnetic process. Assuming that both the momenta and the invariant mass M of virtual photon are small in comparison with the particle energy then under some approximation the formula (8) can be also applied for virtual photon bremsstrahlung by simple replacement of ω by $\sqrt{k^2 + M^2}$.

Bremsstrahlung in a quark-gluon plasma

The averaging over trajectories in the case of quark bremsstrahlung requires a separate investigation. This is mainly because the dynamics in a quark-gluon plasma is different from a pionic medium. The quark cross-section in a plasma is dominated by small angle scattering. Thus, the quark velocities in two subsequent collisions are strongly correlated. In this case non-diagonal terms in (3) cannot be neglected as they will have important contributions to the total cross-section. In order to calculate the energy radiated by a quark moving in a thermal plasma it is convenient to rewrite (2) into the following form:[9, 16]

$$\frac{dI}{d^3k} = \frac{e_q^2 \alpha}{(2\pi)^2} 2\text{Re} \int_0^\infty dt \int_0^\infty d\tau \, \mathbf{K}(t, \tau) \tag{11}$$

where the kernel $\mathbf{K}$ is defined as:

$$\begin{aligned}
\mathbf{K}(t, \tau) \equiv \;& \exp{-i[\omega\tau - \vec{k} \cdot (\vec{r}[t + \tau] - \vec{r}[t])]} \\
& (\vec{v}[t + \tau] \cdot \vec{v}[t] - (\vec{n} \cdot \vec{v}[t + \tau])(\vec{n} \cdot \vec{v}[t]))
\end{aligned} \tag{12}$$

The problem consists now in averaging the above expression over all possible quark trajectories in a plasma. This can be done using functional integral techniques[20] or the method originally proposed by Migdal.[9] In the following we will apply Migdal's prescription to find the indicated averaging. This is done introducing two probability functions: $w_1(\vec{r}, \vec{v}; t)$ - the probability to find at the time t a particle with coordinate $\vec{r}$ and velocity $\vec{v}$ and $w_2(\vec{r}\cdot\vec{v}; t'|\vec{r}, \vec{v}; t)$ - the conditional probability of value $\vec{r}$ and $\vec{v}$, at the time t' under the condition that at the moment of time t these quantities have had the values $\vec{r}$ and $\vec{v}$. Then the averaging over trajectories is equivalent to finding the integral

$$< \mathbf{K} > = \int d\vec{r}\cdot d\vec{v}\cdot d\vec{r}\, d\vec{v} \, w_1 w_2 \mathbf{K} \tag{13}$$

The probability functions w_1, w_2 satisfy the kinetic equation

$$\frac{\partial w}{\partial t} + \vec{v}\frac{\partial w}{\partial \vec{r}} = n(T) \int d\vec{v}\cdot \sigma(\vec{v}\cdot - \vec{v})[w(\vec{r}, \vec{v}\cdot; t) - w(\vec{r}, \vec{v}; t)] \tag{14}$$

where σ is the elastic quark cross-section changing the velocity from $\vec{v}'$ to $\vec{v}$ and $n(T)$ is the density of the medium. The solution of the above differential equation can be found using a Fokker-Planck approximation[9, 17] with the initial conditions

$$w_1(\vec{r}, \vec{v}; 0) = \delta(\vec{r})\delta(\vec{v} - \vec{v}_0) \tag{15}$$

$$w_2(\vec{r}\cdot, \vec{v}\cdot; t|\vec{r}, \vec{v}; t) = \delta(\vec{r}\cdot - \vec{r})\delta(\vec{v}\cdot - \vec{v}) \tag{16}$$

where $\vec{v}_0$ denotes the initial quark velocity.

Using the solutions for the probability function in eqs.(11,13) the integrations can be done analytically with the following final result for the energy radiated by a quark due to virtual photon bremsstrahlung:[17]

$$< \frac{dT^{\gamma^*}}{d\omega dt} > \simeq \frac{8e_q^2 \alpha}{3\pi} \frac{k}{\omega} q(p, T) \left[\left(\frac{m_q^2}{p^2} + \frac{M^2}{\omega^2} \right)^2 + \frac{16q(p, T)}{9\omega} \right]^{-1/2} \tag{17}$$

where $q(p, T) \equiv \frac{1}{4}v \left\langle \theta^2 \right\rangle$ is the characteristic parameter of small angle scattering and

$$< \theta^2 > \equiv n(T)v \int \theta^2 \frac{d\sigma}{d\Omega} d\Omega \tag{18}$$

is the mean square scattering angle per unit path.

The influence of the Landau-Pomeranchuk effect on the virtual photon emission by a moving quark in a medium is contained in the last term in the square bracket in eq.(17). If this term is negligibly small compared to the first one, then the effect is absent. It is easy to see that the above expression remains finite in the limit of massless quarks. Thus, the scattering of particles in a medium regularizes the infrared behaviour of the corresponding amplitudes. For on-shell photons formula (17) coincides with the one previously obtained by Migdal.[9]

The mean square of the angle of multiple scattering for the coherence length l_c is defined as

$$\theta_{MS} \equiv l_c \sqrt{<\theta^2>} \tag{19}$$

whereas a typical emission angle of the bremsstrahlung photon $\theta_B \sim m_q/E$. With these two parameters the energy spectrum of real photon (17) can be written as:

$$< \frac{d\Gamma^\gamma}{d\omega dt} > \simeq \frac{8e_q^2\alpha}{3\pi} (\frac{p}{m_q})^2 q(p,T) \frac{1}{[1 + \frac{2}{9}(\frac{\theta_{MS}}{\theta_B})^2]^{1/2}} \tag{20}$$

Thus, there are two distinct results in the limit when $\theta_{MS} < \theta_B$ and $\theta_{MS} > \theta_B$:

$$< \frac{d\Gamma^\gamma}{d\omega dt} > = [\frac{d\Gamma^\gamma}{d\omega dt}]^{\text{Bethe-Heitler}} \times \begin{cases} 1 & \text{if } \theta_{MS} < \theta_B \\ \\ \sqrt{\frac{9}{2}} \frac{\theta_B}{\theta_{MS}} & \text{if } \theta_{MS} > \theta_B \end{cases} \tag{21}$$

From the above equation one can conclude that if the multiple scattering angle θ_{MS} is smaller than a typical bremsstrahlung angle then there is no medium modification of quark bremsstrahlung cross-section as it is determined by the well known Bethe-Heitler formula. However, the radiation spectrum is suppressed if at the distance l_c the scattering deflects the test quark through an angle θ_{MS} larger than the typical emission angle θ_B.

In the case of small angle scattering the relevant parameter which measures the LPE is the mean square of scattering angle for the coherence length. This is in contrast to the isotropic case where the important parameter was the mean free path of the test particle. Thus, particle bremsstrahlung in a quark-gluon plasma and in a pion gas required different conditions to see Landau-Pomeranchuk suppression effect. From (21) one can also find the following ω dependence of the cross-section

$$< \frac{d\Gamma^\gamma}{d\omega dt} > (\omega) \sim \begin{cases} const. & \text{Bethe} - \text{Heitler} \\ \\ \sqrt{\omega} & \text{Landau} - \text{Pomeranchuk} \end{cases} \tag{22}$$

The photon energy spectrum described by the Bethe-Heitler formula is known to be independent of the frequency of the radiation. The Landau-Pomeranchuk-Migdal effect changes qualitatively this result leading to $\sqrt{\omega}$-behavior. The $\sqrt{\omega}$-behavior has been recently verified experimentally in SLAC experiment E-146[21] where the production of photons due to bremsstrahlung of electron beams traversing uranium target was measured.

As it was explained above the role of the LPE in a quark-gluon plasma is entirely determined by the parameter $\langle \theta^2 \rangle$. To lowest order in α_s the small-angle elastic cross-sections in a quark gluon plasma are described by the t-channel exchange Feynman diagrams which give:

$$\frac{d\sigma_{ab \to ab}}{dt} = \frac{2\pi\alpha_s^2}{t^2} C_{ab} \tag{23}$$

$$C_{ab} = \begin{cases} 4/9 & \text{for } qq \to qq \\ 1 & \text{for } qg \to qg \end{cases} \tag{24}$$

379

When integrating the cross-sections in eq.(18) over $d\Omega$ with a weight factor θ^2, one finds a typical logarithmically divergent integral

$$\int_{\theta_{min}}^{\theta_{max}} \frac{d\theta}{\theta} = \ln[\frac{\theta_{max}}{\theta_{min}}] \tag{25}$$

To estimate this integral one chooses the upper limit θ_{max} to be close to unity: this is the upper boundary for the validity of the small-angle approximation. The lower limit θ_{min} in a Coulomb-like medium is usually cut-off by many-body effects like charge screening. In the plasma a typical screening scale is a Debye mass m_D proportional to gT with g being the coupling constant and T the temperature of the medium.[22] Therefore, $\theta_{max}/\theta_{min} \sim p/gT$, where p is the momentum of the charged particle. In a thermal system, however, temperature gives a natural scale for the particle momentum. Thus,

$$< \theta^2 > \sim \frac{T^3}{p^2} \alpha_s^2 \ln \alpha_s \tag{26}$$

The above parameter is shown to be an increasing function of the temperature. Thus, also the suppression of bremsstrahlung is more effective with increasing temperature as seen in (21). This is to be expected since with increasing temperature the density of the medium is also increasing and consequently multiple scattering is more important.

Multiple scattering and charge screening are not the only collective effects which influence the bremsstrahlung cross-section. Additionally there is the dynamical generation of a thermal mass for fermionic degrees of freedom in the medium. According to[22] fermions in a quark-gluon plasma acquire a thermal mass $m_q^2 \sim 2\pi\alpha_s/3T$ induced by many-body dynamical effects. From eq.(17), however, one can deduce, that when considering the process of real and virtual photon bremsstrahlung in a plasma, the thermal mass generation effect can be small compared with to the multiple scattering one. Indeed, taking the limit $M \to 0$ in (17), one can see, that the second term in the square brackets is of the order of $\alpha_s^2 \ln \alpha_s$, while the first one is only of the order of α_s^2. For this reason the LPE is mainly of importance when discussing the soft electromagnetic signal from a quark-gluon plasma.

SOFT DILEPTON PRODUCTION IN HEAVY ION COLLISIONS

In the last section we have derived the cross-section for real and virtual photon bremsstrahlung in a pion gas and in a quark-gluon plasma under the soft photon approximation. We have discussed qualitatively the modifications of these cross-sections due to the Landau-Pomeranchuk suppression effect. Having in mind phenomenological application of these results in heavy ion collisions we will discuss in this section *dilepton* production due to bremsstrahlung. The cross-section for dilepton production can be related with the one for virtual photon as follows:

$$\frac{d\sigma^{e^+e^-}}{dM} = \frac{2\alpha}{3\pi M} \sqrt{1 - \frac{4m_l^2}{M^2}} \left(1 + \frac{2m_l^2}{M^2}\right) \sigma^{\gamma^\bullet} \tag{27}$$

To obtain $d\sigma^{e^+e^-}/dM$ one needs to integrate the cross-section in a pion gas and in a quark gluon plasma, derived in the previous section, over the virtual photon frequency. The difficulty here, however, is that this procedure is out of the range of the soft photon approximation. We have imposed the upper integration limits for the frequency integral with the help of the energy-momentum conservation laws.[17-19]

In heavy ion collisions one is more interested in the dilepton production rate rather than the cross-section. The expression for the rate is given by kinetic theory[*]

$$\frac{dN^{e^+e^-}}{d^4x} = \int d^3p_1 f(\vec{p}_1) \int d^3p_2 f(\vec{p}_2) |\vec{v}_{12}| \frac{d\sigma^{e^+e^-}}{dM^2} \qquad (28)$$

where the momentum distribution function $f(p)$ of the incident particle and of the particles in the medium are assumed to be thermal and described by a Boltzman distribution function.

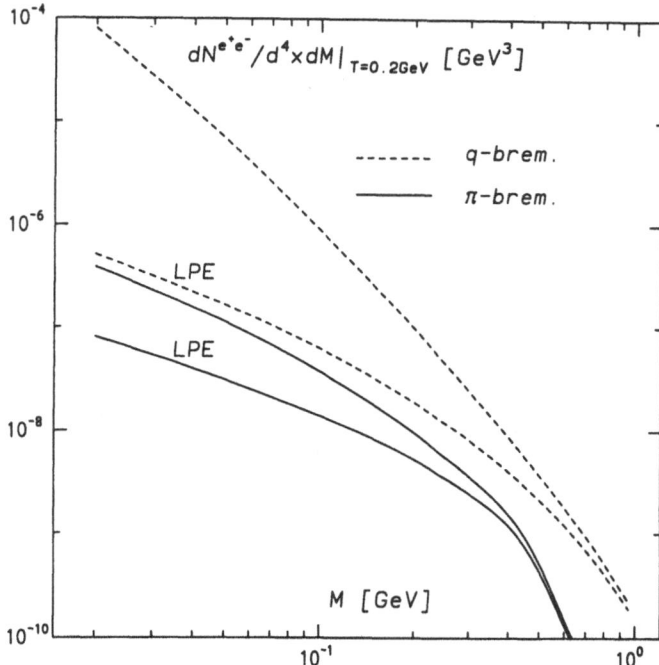

Figure 1. Invariant mass distribution of dilepton pair at $T = 0.2$GeV.

In Fig.1 we show the dilepton production rate in a pion gas (*full line*) and in a quark gluon plasma (*dashed-line*) due to particle bremsstrahlung with and without LPE taken into account. The calculations are performed at fixed temperature T=0.2 GeV. As can be seen in this figure, the Landau-Pomeranchuk effect leads to a significant reduction of dilepton production due to bremsstrahlung of virtual photons from both quark and pion moving in a thermal medium. The reductions of the rates are particularly important in the low invariant mass region. With increasing M the Landau-Pomeranchuk suppression is reduced. This behaviour can be understood intuitively as follows. The time necessary for e^+e^- pair production is proportional to its inverse mass. Increasing invariant mass means also decreasing the production time and

[*]In the quark-gluon plasma case one averages over collisions instead of integrating over the scatterer's momentum distribution.[17]

the coherence length. Thus, the amount of collisions of the test particle with a surrounding medium constituents within l_c is decreasing as well. Consequently, the medium effect is less important. We have to stress however, that when going beyond the soft photon approximation the suppression can be seen for larger invariant mass values. Our approximations should be valid in the part of the spectrum where the invariant mass is less than $M \sim 0.3 - 0.5 GeV$. The results presented in Fig.1, show that in the soft part of dN/d^4xdM spectrum quark bremsstrahlung is of the crucial importance. At fixed temperature the medium suppression of virtual photons radiated from a charged particle is larger in a plasma than in a hadron gas. This behaviour is certainly not only related with differences between the cross-sections as discussed in the previous section. At a fixed temperature the number of degrees of freedom and consequently the density of a plasma is higher than in a hadron gas. This naturally implies higher suppression in a quark-gluon plasma.

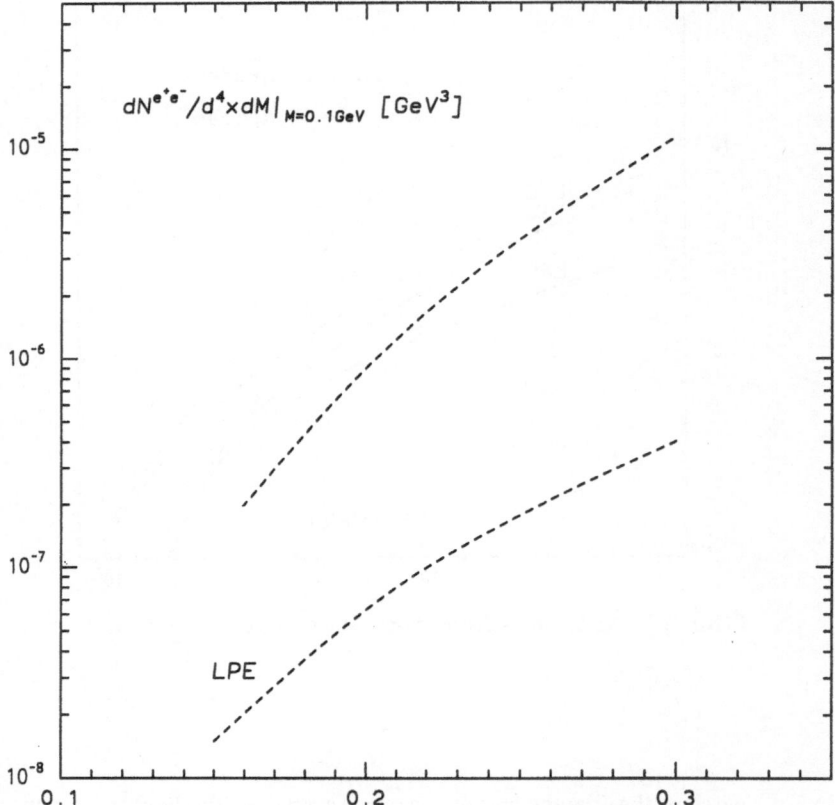

Figure 2. Rate in a plasma at fixed M as a function of T.

The suppression of dilepton production rate due to LPE is not only increasing with decreasing energy of bremsstrahlung photon but also with increasing temperature. This is illustrated in Fig.2 where the rate in a plasma is calculated with and without LPE at fixed invariant mass $M = 0.1$GeV as a function of temperature.

Until now we have studied dilepton production in a medium at fixed temperature. We have to point out, however, that hadronic matter produced in heavy ion collisions is not a static object. It rather undergoes hydrodynamical expansion leading to a time dependence of temperature and of other thermodynamical quantities. The experimentally measured distributions are integrated over the space-time history of heavy ion collisions.

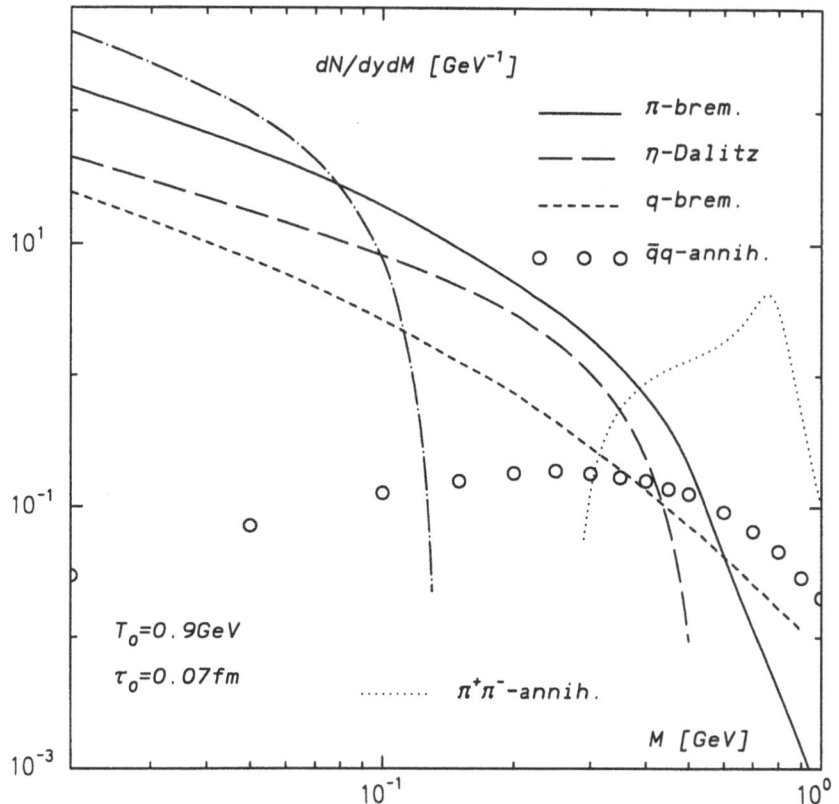

Figure 3. Dilepton rate in the expanding medium.

In Fig.3 we show the results for soft dilepton production rate assuming Bjorken expansion for the space-time evolution of hot hadronic matter produced in a heavy ion collisions.[19] The initial formation time $\tau_0 \sim 0.07$fm and initial temperature $T \sim 0.9$GeV are relevant for U-U collisions at LHC energy.[23] The comparison of quark (*short-dashed line*) and pion (*full line*) bremsstrahlung with non thermal background due to η (*long dashed line*) and π_0 (*dashed-dotted line*) Dalitz decay shows the importance of bremsstrahlung process as a source of soft dilepton production. In the window of invariant mass $0.1 \leq M \leq 0.3$ pion bremsstrahlung acids Dalitz eta decay, quark bremsstrahlung and basic quark annihilation contribution (*circles*). For the invariant mass $M > 0.5$GeV the pion annihilations (*dotted line*) is a dominating process for dilepton productions. Nevertheless, one might expect to observe a quark-gluon plasma signal when measuring double differential spectra, $dN^{e^+e^-}/dydMdp_t$, at high transversal virtual photon momenta, p_t and at low invariant mass. The reason is that a typical momentum distribution is related with the temperature of the emitting system. The plasma phase exists at higher temperatures than the pion gas phase, therefore, the momentum distribution of the pairs emitted from the plasma phase is

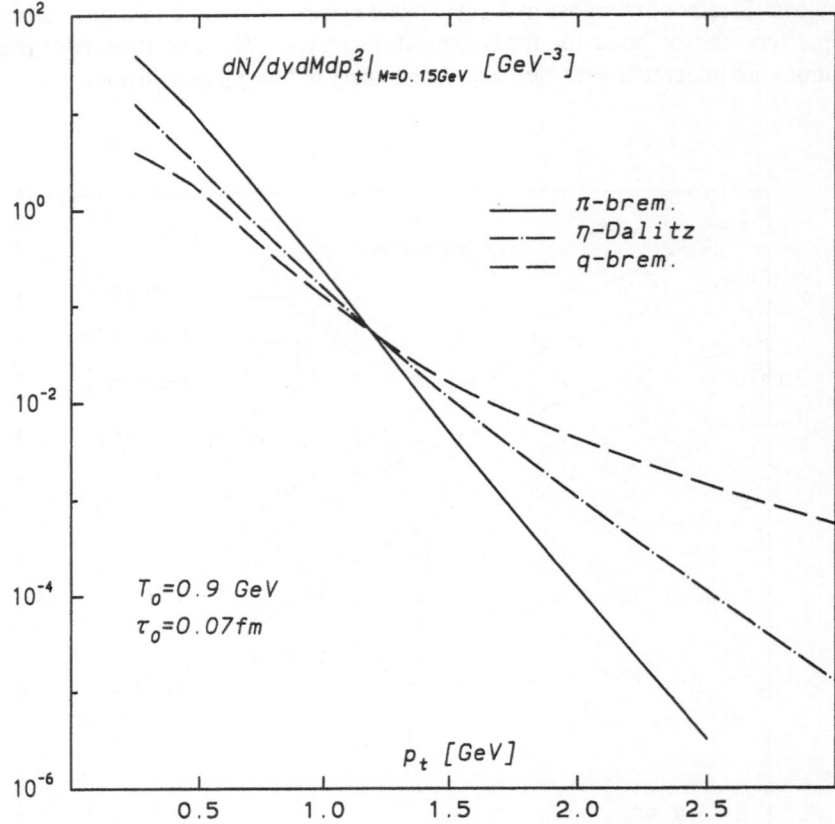

$dN/dy\,dM\,dp_t^2|_{M=0.15GeV}$ $[GeV^{-3}]$

——— π-brem.

—·— η-Dalitz

— — q-brem.

$T_0=0.9$ GeV

$\tau_0=0.07fm$

p_t $[GeV]$

Figure 4. Dilepton rate at fixed value of M as a function of p_t.

expected to be broader than that from the hadronic phase. Hence, at some high enough value of the transversal momentum of the pair, the plasma bremsstrahlung should dominate $dN^{e^+e^-}/dy\,dM\,dp_t$ spectrum. This is illustrated in Fig.4 where the double differential production rate is calculated at fixed invariant mass $M = 0.15\text{GeV}$. At large transverse momentum quark bremsstrahlung (*dashed line*) is indeed dominating, both Dalitz eta decay contribution (*dashed-dotted line*) and pion bremsstrahlung (*full line*). One should, however, keep in mind that the above results are obtained assuming that there is no transverse expansion or that it is very slow. The hydrodynamical model with fast transverse expansion in the late stage of the collision could influence the results in Fig.4. This is because transverse expansion would imply an additional boost of the dilepton pair in the transverse direction, increasing in this way the rate of pion bremsstrahlung for the pair with large transverse momentum. Also the preequilibrium background could still modify the production rate in Fig.4 in the kinematical window of large p_t.

CONCLUSIONS

In this work soft dilepton production due to real and virtual photon bremsstrahlung in hot QCD medium was studied. We have shown that the bremsstrahlung process can be substantially suppressed in a thermal medium due to the Landau-Pomeranchuk effect. This effect is particularly important at high temperature and at the low energy of the bremsstrahlung photons. The analysis of soft dilepton production in an expanding medium shows that particle bremsstrahlung is the main source of thermal pairs.

Assuming optimistic initial conditions for the LHC experiment the thermal bremsstrahlung can exceed the contribution due to the Dalitz eta decay in the kinematical window where the invariant mass $m_\pi < M < 2m_\pi$. These results indicate that thermal dilepton production could be experimentally observed by measuring the soft part of the dilepton production rate at LHC energy.

Acknowledgments

One of us (KR) acknowledges stimulating discussions with R. Baier, E. L. Feinberg, H. Satz and H.J. Specht. This work was partly supported by the Ministry for Research and Technology (BMFT) and the Committee of Scientific Research (KBN).

REFERENCES

1. R. Hagedorn, Suppl. Nuovo Cimento **3** (1965) 147; **6** (1968) 3116; see also *Thermodynamics of Strong Interactions* CERN 71-12 (1971).

2. See Eq. R. Hagedorn and J. Rafelski in *Proceedings Statistical Mechanics of Quarks and Hadrons*, Edited by H. Satz, North - Holland Publishing Company, (1981) 303 - 318.

3. For the compilation of the recent LGT results see Eq. J. Fingberg, U. Heller and F. Karsch, Nucl.Phys. **B392** (1993) 493.

4. H. Satz, in Proceedings of the ECFA Workshop on Large Hadron Collider, Aachen, Germany, 1990, edited by G. Jarlskog and D. Rein (CERN Report No. 90-10, Geneve, 1990) Vol.I,p.188; in Proceedings of QM'91, Nucl. Phys. A (proc. Suppl.), edited by T.C. Awes, F.E. Obenshain, F. Plasil, M.R. Strayer and C.Y. Wong (*to appear*).

5. L.D. McLerran and T. Toimela, Phys. Rev. **D31** (1985) 545; E.V. Shuryak, Phys. Lett. **B78** (1978) 150; K. Kajantie, J. Kapusta, L.D. McLerran and A. Mekjian, Phys. Rev. **D34** (1986) 2746. For the recent reviews see : P.V. Ruuskanen, Proc. Quark Matter '91, Gatlinburg, V. Plasil et al. (eds.); H. Satz, Proc. Int. Lepton-Photon Symp., Geneva, Switzerland (1991); Nucl.Phys. **A525** (1991) 255c.

6. J. Cleymans, K. Redlich and H. Satz, Z. f. Physik **C52** (1991) 517.

7. M. L. Ter-Mikaelyan, Zh. Eksp. Teor. Fiz. **25** (1953) 296.

8. L. Landau and I. Pomeranchuk, Doklady Akad. Nauk S.S.S.R.**92**, No. 3, 535; No.4, (1953) 735.

9. A. B. Migdal, Doklady Akad. Nauk S.S.S.R.**96**, No. 1, (1954) 49; Phys. Rev. **103** (1956) 1811.

10. E. L. Feinberg, Usp. Fiz. Nauk **58** (1956) 193; **132** (1980) 255; E. L. Feinberg and I. Pomeranchuk, Nuovo Cimento Suppl. **3** (1956) 652; N.P. Kalashnikov, V.S. Remizovich and M.I. Ryazanov, in: *Collisions of Fast Charged Particles in Solids*, Gordon and Breach Science Publishers (1985); J. Knoll in Proceedings of the *Workshop on Pre-Equilibrium Parton Dynamics in Heavy Ion Collisions*, LBL, Brekeley (1994) ed. by X. N. Wang.

11. O.R. Frisch and D. N. Olson, Phys.Rev.Lett. **3** (1959) 141; A. I. Akhiezer and N. F. Shul′ga, Sov. Phys. Usp. **30** (1987) 197.

12. D. Kharzeev and H. Satz, Phys. Lett. **B19** (1994) 361; A. Białas and W. Czyż, Phys. Lett. **B328** (1994) 172.

13. M. Gyulassy and X. N. Wang, Preprint CU-TP-598, LBL-32682 (1993). Phys. Rev. **D**.

14. K. Geiger and B. Müller, Nucl.Phys.**B369** (1992) 600.

15. *Classical Electrodynamics*, J. D. Jackson, equation (14.67) J. Wiley & Sons, Inc., New York.

16. J. Cleymans, V.V. Goloviznin and K. Redlich, Phys.Rev.**D47** (1993) 173.

17. J. Cleymans, V.V. Goloviznin and K. Redlich, Phys.Rev.**D47** (1993) 989-997.

18. J. Cleymans, V.V. Goloviznin and K. Redlich, Z. Phys.**C59** (1993) 495.

19. V.V. Goloviznin and K. Redlich, Phys.Lett. **B319** (1993) 520

20. N. V. Laskin, A.S. Mazmanishvili, n.n. Nasonov and N.F. Shul'ga, Zh. Eksp. Teor. Fiz. **89** (1985) 763.

21. S. R. Klein et.al, SLAC PUB-6378 (1993).

22. H.A. Weldon, Phys. Rev. **D26**, (1982) 2738; R.D. Pisarski, Physica **A158**, (1989) 146.

23. J. Kapusta, L.D. McLerran and D.K. Srivastava, Phys. Lett. **B283** (1992) 145.

MINIWORKSHOP ON STRANGENESS

Johann Rafelski

Department of Physics, University of Arizona
Tucson, AZ 85721, USA
and
Laboratoire de Physique Théorique et Hautes Energies*
Université Paris 7, Tour 24, 5e étage
2 Place Jussieu, F-75251 Paris Cedex 05, France

INTRODUCTORY REMARKS

Among the diverse hadronic observables, strange particles and in particular multi-strange antibaryons have emerged in recent years as being very promising probes of the physical properties of the hot hadronic matter formed in relativistic nuclear collisions. A number of experiments has been gathering data at the CERN SPS 200 GeV A beam and the BNL 15 GeV A beams, and there is interesting evidence for new physics emerging from some of these results. This discussion, in an unconventional way, precedes the individual contributions. The intend was to offer a colorful introduction to the uninvolved with little individual bias.

To simplify the transcript the following acronyms are used below:

FA	Federico Antinori (Genua & CERN),	**JC**	Jean Cleymans (Cape Town),
ID	Ivo Derado (München),	**EF**	Evgenyi Feinberg (Moscow),
MG	Marek Gaździcki (IKF-Frankfurt),	**WG**	Walter Geist (Strasbourg),
NG	Norman Glendenning (LBL),	**RH**	Rolf Hagedorn,
UH	Ulrich Heinz (Regensburg),	**BM**	Berndt Müller (Duke),
GO	Grażyna Odyniec (LBL),	**JR**	Johann Rafelski (Arizona),
EQ	Emanuele Quercigh (CERN),	**PS**	Peter Sonderegger (CERN),
JSt	Johanna Stachel (Stony Brook),	**SS**	Stephen Steadman (MIT),
RS	Reinhard Stock (IKF-Frankfurt),	**JS**	John Sullivan (LANL),
AT	Ahmed Tounsi (LPTHE, Paris 7),	**GZ**	Gennady Zinoviev (Kiev).

---

*Unité associée au CNRS: D 0280

The topics/questions put forward were:

1) Why is the strangeness yield interesting?

2) Why are multi-strange antibaryons 'in'?

3) What can be discovered? QGP?

4) How can it be discovered? What conditions are necessary?

5) How can we be sure that we found the deconfined phase?

6) Is there a way to measure the deconfinement temperature as function of baryo-chemical potential, latent heat, color correlation lengths function of temperature, general QCD parameters such as α_s, m_s as function of temperature?

In summary of the discussion, those present felt that:

- Since the initial state contains predominantly the light quark flavors u, d only, strangeness is a tag of newly made hadronic matter. Furthermore, the strange content of the particles is self-analyzing, since the decay of strange particles can be observed in many visual detectors. The strange 'signal' is persistent, that is strangeness produced in the reaction does not disappear before the collision has ended. Finally, strangeness is in normal hadronic interactions suppressed by a factor of magnitude 2 or 3 as compared to production of non-strange matter, and hence should this suppression disappear, we would have indication for presence of new physical phenomena. Among more specific comments we note the possible comparison with $\bar{p}$ yields, the study of chemical equilibration of s, $\bar{s}$ yield. Finally, it was noted that there is a pronounced enhancement effect and thus we have to study it.

- The special role of the strange antibaryons was agreed on, and a number of reasons were cited. If single strange particle yield is enhanced, we should observe the effect 'squared' for doubly strange particles. Furthermore, in conventional single step processes, strange antibaryons are difficult to produce, and thus the conventional physics background is a priori considerably smaller than for other strange particles. A search for anomalous yields has been made and current experiments at CERN SPS indeed obtain anomalous signals, thus these particles are now per se of interest, irrespective of any theoretical argument that can be made. If one imagines that these particles are emerging from a novel form of matter akin to QGP, than their absolute abundance will be sensitive to the space density of strangeness. We have a number of different particles to consider, and thus can explore mechanisms of hadronization and probe the ideas about the statistical (thermal and flavor) equilibration of the source.

- The issue of chemical equilibration and more specifically the determination of the time constants governing the approach to strangeness chemical equilibrium was on top of the agenda. However, the general experimental strategy (acceptances, efficiencies, falsification of models, model-independent results) was more on the mind of the participants than the speculations about specific discoveries.

- General Discussion ensued on the remaining points. It included in particular a discussion of the long term future for the strangeness experimental program. It was argued that one of particularly interesting avenues will be to study the excitation function of certain strange particle observables, which requires variation of the beam energy, rather than only of the projectile and target size.

TRANSCRIPT OF THE DISCUSSION SESSION

The rapporteur edited the transcript — and he bears the responsibility for the accurate reproduction of the physical contents of the discussion. The indication of the name of the person is to be seen as a credit given, all blame rests with the rapporteur.

Why is Strangeness Interesting?

JR: All matter we know is made of u, d flavor, therefore strangeness is attached to newly made matter. Strangeness is self-analyzing, decays and leaves a track of its existence, helps us analyze data. Seeing production of strangeness we see the rate of conversion of kinetic energy to matter, it tells us something about what has happened. If there is a more or less effective production of strangeness, it is in itself an important message.

SS: There is another point: you can look at things such as the antiproton yield — the difference is that strangeness is created in strong interactions but is only destroyed by weak interactions, so it remains a primary you can measure, while things like antiproton can annihilate. So I think in strangeness you have a persistent signal. Strangeness could annihilate if you could have a high density of strangeness, but I think that, too, strangeness has a smaller cross section for strange and anti-strange quarks to annihilate with each other, compared to say u, $\bar{u}$.

BM: Of course you need strangeness annihilation to produce chemical equilibrium, so in that sense annihilation is not negligible . . .

JR: . . . but Steve has a very good point.

GO: I want to support this point considering the present experimental conditions, since we are talking about baryon rich hadronic matter, that may change later when RHIC, LHC come on line.

MG: This is a transparency I have shown at Quark Matter '93 so we know why strangeness is interesting. A kind of review of the situation, a history. Before '88, before we started to work on the experiment, we had a prediction that in the case of QGP creation, we will have a high strangeness production rate. Then we made experiments at CERN '88 and what we discovered is that in the S–S collisions really we have factor two more strangeness production than in the nucleon-nucleon collisions. The conclusion was very simple, the simple models we had at this time, which were based on the superposition of nucleon-nucleon interactions are just discarded immediately. Then we looked more carefully in the data, and we discovered that there is no sign of strangeness enhancement in p–A collisions. If we have built a model which is a little bit more complicated than the superposition of nucleon-nucleon collisions and includes some secondary processes, it would be excluded by the p–A data. Last year we reported strangeness enhancement in S–Ag collisions. What we learned this year is that even very sophisticated models based on hadronic processes are not able to account for the strangeness enhancement, when the programs are tuned to describe the global features of the non-strange particle yields.

JR: . . . this was an operational answer to the question. We have a surprise — that's why strangeness is interesting.

EQ: Strangeness is enhanced for sure — the fact that something is enhanced is interesting, but how 'interesting' is it? That is still a matter of debate.

GO: What's important is that it's not only enhanced but that the rate in S–Ag is enhanced as much as S–S. I think that this is a very interesting puzzle.

Why are Multi-strange Antibaryons In?

SS: I think the multi strangeness anti-baryons is just as interesting at the AGS because we see this fairly large yield of $\overline{\Lambda}$ and we just don't understand what that means. I think it's too early to tell. The trouble is that these computer codes that we run and compare with the data, have to deal here with extremely small cross-sections, and these codes are really designed for looking at the global features of the reaction. I think we really don't have a model yet as to what this yield should be. But it is interesting that we see this fairly large yield of the multi-strange anti-baryons at the AGS, where we are right at the particle production threshold.

RS: Just a little comment about the sensitivity of the signal. I'm just preparing a transparency to show that If you look at single strange probes, the magnitude of an effect of enhancement is always a modest one in contrary of what one would think because between a world, which consists of an equal number of strange and up/down, and a normal baryonic world, where strangeness is suppressed, the difference is of the order of factor 2–2.5. But if you look at multiply strange baryons, then you get to the square of this kind of abundance ratios in the production rates so you magnify your leverage.

WG: I would just like to add that one should probably learn from the ' J/Ψ-suppression'. The lesson for me from that is that one has to extremely well understand the basic processes in p–p collisions: isospin effects, interactions in the target, etc; once all this is understood one can talk about an enhancement or not in a much more precise way. I think also that even for the multi-strange baryons and anti-baryons the margin is only of the order of a factor 2, and so one has to define the enhancement relative to something and this something has to have a very small error, otherwise the effect does not become significant.

UH: Strange anti-baryons are hard to produce, so you try to think of something that is hard to produce — the production process should then go through some exotic state.

GO: You may think of strange anti-baryons as probe of the density of strange quarks which is the issue here. We know that the net strange quark density in QGP and in hadronic gas are different but after hadronization they are almost the same, depending how you deal with the time of evolution of the system. Strange anti-baryons are probing the density of strange quarks and will remove some ambiguity about the source of these particles.

EQ: Forget for a moment the QGP, after all we have still to prove its existence. If you want to create strange anti-baryons, say $\overline{\Omega}$, you need to overcome a CM-threshold of 3.3 GeV, in particular important in secondary collisions. Now, if you want to create this particle in multiple collisions by addition of kaons to anti-nucleons, then you will need to put in three units of strangeness, you will need at least three steps, which is clearly not easy. So tell me how are these particles produced that we see?

JSt: I think the biggest point of interest in multi strange particles would be that if we could understand all the models that have been tuned to explain K production or singly strange hadrons, and if the basic mechanism how they get the enhancement is right, then they should be able to confirm multiply strange objects, while if the basic mechanisms is not right, then we should certainly see discrepancies. It really tests whether the trivial explanations of the enhancement are valid or not.

SS: Right now where we are, we have not actually measured the M_T dependence of these AGS–$\overline{\Lambda}$, but if $\overline{\Lambda}$ have the same M_T dependence of our Λ, then we are seeing yields at the AGS which are ten times what is predicted by ARC and RQMD models, so there would be a serious discrepancy. I say we have to be careful because there could be a different M_T dependence.

EQ: If you have 10,000–20,000 Λ, $\overline{\Lambda}$ you can easily measure it.

BM: In my mind the multi strange baryons and anti baryons actually are important because they sensitively probe the thermodynamic equilibrium that is implicit in Hagedorn's model and everything that is built on it, is complemented in the nuclear reaction also by the flavor equilibrium.

RH: As far as our Statistical Bootstrap Model is concerned, we always deal with chemical and with kinetic equilibrium. There is no difference, we can't distinguish the two. At least we never have done it ...

JR: ... we do now distinguish the two, in other models.

EQ: If one finds chemical equilibrium, but for some reason cannot measure, cannot ascertain the thermal equilibrium, will chemical equilibrium be enough to say that we have reached thermalization?

UH: This is not possible. We cannot prove chemical equilibrium without having thermal equilibrium as a basis to start from, because chemical equilibrium is defined in terms of an abundance relative to the expected abundance at a certain temperature. If you don't know what the temperature is you have no statement.

RS: I will show one transparency about flavor equilibrium. That takes us back to this question how far, how high can the signal be. If you had an ideal plasma, you would have equal abundance of u, d, s, basically. So then this famous Wróblewski ratio λ_W of strange quarks to non strange quarks would be unity. That would be a limit of flavor equilibrium. But then you have a net baryon number in the system which adds u, d, so in such a plasma you could only have a Wróblewski ratio of say $\lambda_W = 0.85$. You have gluons too, which carry half of the energy, and when they convert to quark/anti-quark pairs late in the expansion, they will not have this democratic ratio of $s = u = d$. They freeze out with $\lambda_W \simeq 0.4$ rather than 0.8. What you see after freeze-out is a dilution of the flavor equilibrium into perhaps $\lambda_W \simeq 0.5$. Now, tantalizingly, the current data are at $\lambda_W = 0.35$, so we are not quite there, but we are also above some kind of trivial RQMD limit, which is $\lambda_W = 0.2$–0.25, which would be just reflection of hadronic equilibration. So that comes back to this previous remark I made. How far is the leverage in this entire signal? And we are somehow between $\lambda_W = 0.25$, which is still a boring value, and $\lambda_W = 0.5$, which would be the ultimate fingerprint of a flavor equilibrium in initial QGP.

JR: Factor 2 to the power 3 is eight, remember that!

RS: Yes, Yes.

JC: Yesterday I was reading Hagedorn's lecture, and I was very interested in comment he made on the production of anti-helium. Many years ago we must have dealt with the production rate of anti-helium, I guess it was for p–p scattering at fairly low energies. But as far as I understood it, you had a very good explanation for that, assuming flavor equilibrium. There was nowhere a parameter to take into account that you were away from equilibrium, so in your description you could account for anti-tritium, anti-helium-3, all these very heavy particles at very low energies but with terrific chemical equilibrium.

JR: In other words, what you're saying Jean, is to put into perspective, even in p–p collisions non-strange particles are in flavor equilibrium ...

JC: ... for these very heavy anti-baryons, and if it holds for such low energies, such a small system, why not for a much bigger system?

JR: Yes, this is a very interesting question which reads: Why is non-strange anti-flavor in equilibrium?

UH: I just wanted to expand on that, this surprising co-existence of thermal equilibrium and flavor equilibrium as far as the u and d quarks, or anything that is made up of these particles, is concerned. But then, we also do know that in p–p collisions that is not true for strangeness. The strangeness is not produced at that, same, rate that is expected from that kind of temperature, so there is this famous strangeness suppression that is being observed in p–p and also p–A collisions. That is something that is being partially removed in the nuclear collisions. That is another answer to the question why is strangeness interesting.

WG: I have just a comment on the λ_W, because Reinhard mentioned a kind of an estimated theoretical λ_W which was compared to an experimental measurement. I think it's known already from the e^+e^- collisions that even if you expect a theoretical $\lambda_W \simeq 0.3$, what you observe experimentally is 0.15. In my opinion this arises just from resonance decays, which feed preferentially the light quarks. I would like to add that if you look into the charged particle multiplicities in p–p collisions as a function of the produced particle mass, as you do for single events in e^+e^-, one finds an exponential behavior, with an inverse slope, which is in the order of the standard temperature. To my knowledge this interesting behavior has not been explained in the way of Hagedorn nor was it explained at all.

What Can be Discovered? QGP?

UH: What we certainly can measure directly, or better said decide directly is: first we test whether the concept of thermal equilibrium continues to be true as it was shown for p–p collisions many years ago by Hagedorn and many others. Then based on that, once we have a temperature we can ask whether we also have a chemical equilibrium, and that is a question that we can decide. That's the first thing that can be discovered, whether the time scales in the nuclear collisions are such that we can reach chemical equilibrium, and then we take it from there. Once we have established chemical equilibrium, then the theoreticians have to be asked how could that have happened.

EQ: The situation for an experimentalist is at the frontier which Uli has defined; we can measure particle ratios and see how far from chemical equilibrium we are, coming from the top, or from below, and then it's up to somebody else to decide whether the time scales are as needed, e.g. can in hadronic gas phase production of $\overline{\Omega}$ occur in a reasonable time or if we need plasma or something else to create the needed abundance.

AT: The problem of the time scale is a very important one. If you look to the work of different people, the time scale doesn't seem to be fixed. So some people talk of factor of 100, some talk of factor of 10 between the time scale for the strangeness equilibrium and that of the thermal equilibrium, so when you have a discrepancy like this, this means that probably the kinetic theory has to be redone more precisely, more carefully, this is to me a too big discrepancy to be ignored.

JR: Ahmed reminded us that within a kinetic theory you have as input scattering of particles, and you compute the time to produce other particles. If such big differences remain, either the kinetic theory of particle production or the actual rate calculations need improvement. We have something to discuss here, to decide which of these issues is important ...... *uncontrolled arguments involving several participants follow.*

RH: I think on the theoretical side there are also a lot of uncertainties, and as Tounsi said, time scales are important. Also temperature is important. If you go up into the plasma into very high temperature, then the chemical equilibrium will be more easily restored than at low temperatures, because first of all the masses will decrease. Suppose you have an ideal gas and you create in that ideal gas heavy flavors, then the Boltzmann factor at high temperature is of course larger, at low temperatures it's smaller, and if it is larger you have a larger production rate and a faster approach to chemical equilibrium. So time scale depends very much on the individual collision, whether you have reached a very high temperature or only a very low temperature just above the threshold. Secondly, as it has been pointed out by Danos and Rafelski many years ago, the volume is very important for the attainment of chemical equilibrium. If you have a large volume then it is attained at a different level than in a small volume, because in a small volume, if you create a s–$\bar{s}$ pair, it has a chance to re-annihilate, in a big volume s, $\bar{s}$ fly apart.

RS: (Showing a transparency) May I make another remark about what can go wrong on the experimental side — I am addressing Hagedorn here: we measure particle ratios, as we often time do, in order to find out strange to non-strange ratio. Two things may go fundamentally wrong if you do not have a full phase space acceptance detector. That is illustrated here. If you take this sketch to be a rapidity plot of fireball products, where the fireball sits centrally, and the sensitivity is in the this rapidity y interval between -2 and $+2$ units from 0. And in such a fireball at $T = 200$ MeV, the $\overline{\Lambda}$ would have a narrow rapidity distributions, kaons are shown to be wider, the pions have widest distribution. At fixed temperature the pions spread out much further over phase space, because they are relativistic, whereas the $\overline{\Lambda}$ are still totally non relativistic, so the integral of all these things is made to be equal. In this picture equal total abundance of all particles is shown. Now, if you measure the ratios you can get absolutely anything you want, if you stick to narrow rapidity coverage. Point two, if you measure P_T distributions, or measure in selected P_T or M_T windows, and take yield ratios, then the power in yield is concave as we know, which is this famous low P_T enhancement and the Cronin effect, and what not, and the $\overline{\Lambda}$ yield in P_T would be some kind of shoulder arm structure, and so again you can see that if you take ratios from this, you can get answers, which are missing the target by factors of 5. In other words I plead for as much acceptance in P_T and in y as you can obtain. Otherwise beware of comparing particle yields.

SS: I absolutely agree with Reinhard, it is extremely important to measure these M_T an y dependence and I think whether you use a 4π detector or you use a spectrometer with which you go and scan over the full y and M_T region, you are absolutely right, it is essential to cover the whole phase space.

JR: Since no one dares to disagree, I will have another opinion. I would like to remark that of course, to this comment by Reinhard, the input was a model. The model was a fireball. Therefore I am as well allowed to use as input a fireball model. If I do so, I can correct for using this model for any acceptance that any experiment gives me, and knowing the acceptance, make a model independent analysis as far as I can. Although I will know less, because less information entropy will be observed, when you use small acceptance, I can extract, under circumstances, more precise information.

RS: I would like to disagree, Johann. We know from Hagedorn already for twenty years that in principal one can leave out the longitudinal phase space and look at this thing as if it is boost invariant and put our fireballs everywhere on the axis. But talking about the CERN experiments, we know that we have much more extended longitudinal than transverse phase space. So this single fireball model is just for fun. In reality we have a mixture of fireballs spread over one and a half units. But still, they have their particular spreading to the outside.

EQ: There is the example, which I think, I remember from Professor Hagedorn, of the blind people in front of an elephant. Everybody touching a part of the elephant. You need to see the whole elephant, otherwise you can get *general amusement*.

UH: I would like to make a statement about the question on model dependent or model independent analysis. We already work within the model of thermal equilibrium, whether we have chemical equilibrium or not, but then in order to extract where that comes from, I don't think there is a way of doing that without having an additional model, and that is because as Tounsi pointed out very correctly, there is the uncertainty about the time scales. Unless we can measure the time scale independently, we will not be able to uniquely distinguish one model from another. That brings me to the point that strangeness should be complemented by other signatures, which maybe have more direct experimental information on the time scales.

JS: On the subject of signals, a question, and maybe somebody has an opinion on the subject. I speak as an experimentalist from an experiment where we measure primarily correlation functions of strange and non strange particles including baryons, and one of my questions is if we assume that by the measurement of the radius parameter from one of these correlation functions, I can associate that with the real size of the system, and to what extent would you all think we should see a different size for K, π and baryons, should I see something different in this measurement for K and non-strange particles?

RS: At the recent Helsinki meeting we reached some agreement that if you start for example with a hydrodynamic model and compute the expansion, then you have a freeze-out of π and K synchronously, at the same time. Then you treat resonance contribution properly, they add more to the π-signal than to the K-signal, and in the end such a model, at least qualitatively, describes the observed difference between the effective radial extracted from the correlation function. In other words, inverting the argument, there is not necessarily a proof of the suspicion that the K freeze out earlier and therefore the two observations show you some dynamics of the system. I'm sorry to reflect this kind of trivial explanation, maybe it was not all the truth.

NG: I just wanted to make a remark that I made earlier in the conference, that is that there is nothing that confines K or π to the fireball, which is very hot, and which undergoes a rather long time scale expansion, and so that there can be a spectrum reflecting the thermal history of a fireball before the freeze-out occurs. That can give rise to different temperatures and different thermal populations.

UH: Two comments to both Reinhard and Norman. Number one, it's not true that you cannot implement in a hydrodynamic different freeze-out for K and π — it is just that people have not done it and even without separating that, one can reproduce the difference in the observed radial, because of the excess of resonance decay contributions in the π which is not there in the K. You do not have to assume any different freeze-out at the present stage of the analysis of the data. The other point is that we may have a different concept of what the freeze-out is. But we probably agree that particles that do escape have frozen-out. One should distinguish between freeze-out that happens simultaneously everywhere and a freeze-out that happens continuously from an early time to a very late time. I agree with Norman that in reality we should expect such a picture, in which some particles freeze-out or decouple from the system very early, and other decouple very late, and therefore what is reflected in the data is an integral, or an average over that history.

General Discussion

RH: To the points **3** and **4**. We are looking for phase transition, now I think the most convincing sign of a phase transition is that the temperature reaches some kind of limiting value. Suppose you have a pot where you can't look into, and you wish you had a thermometer outside, and you wish you know the water which is within is going to boil. Well when the thermometer goes to 100^0C it does just stop there, then you say I'm at the phase transition. Now that has been observed since 50 years, in cosmic rays for the first time, the limiting transverse momenta. And I think this is the most convincing sign that there is something. Now, what the details are, that is what we are out to discover.

BM: Things are not quite as simple I think. If you take for example the evolution of the Early Universe, you know that there is a certain point beyond which we cannot look because the matter was in a plasma state, and was not transparent to photons. So it could very well be that there is a certain density, beyond which hadronic matter is so non-transparent that particles basically are not emitted, except from the surface, which typically will be somewhat cooler. On the other hand it does not necessarily imply a phase transition as in the case of an electromagnetic plasma, so there is still the question is there a transition that is smooth or is it a phase transition.

Now, I would like to add one more thing to point **5**. I have tried to make that point whenever I talk about strangeness. I think it is not sufficient to talk about de-confinement, it actually may be that strangeness is a very sensitive probe, a more sensitive probe even of chiral symmetry restoration. And in particular, there is strong evidence from lattice calculations and fundamental considerations in QCD that chiral symmetry breaking is due to instantons. Now instantons act in the direction that is flavor symmetric between u, d, s quarks, and at the moment that we temper with that structure of the QCD vacuum the structure of the QCD vacuum immediately leads to flavor equilibration — this hasn't been explored as far as I know quantitatively.

NG: I guess I 'm going to make a heretical remark concerning the temperature. If you were to consider several different scenarios, one in which there was a limiting temperature due to an exponential rising in the spectrum of particles, or some finite number of particles in which the temperature of the fireball could become very high, in fact what you finally observe in the laboratory are the products of this fireball after it has expanded and cooled, and so I would claim that the common thing that you see and refer to as the ultimate temperature of the body was actually a final temperature of the body, and it would be invariant for any kind of system, essentially given by the density of pions when they can last interact. That happens to be a temperature of 140 MeV, their mass.

JR: . . . some people observe temperature, . . . , inverse slope, of the order of 230 MeV . . .

NG: . . . boosted, as we heard from Heinz, and also as I discussed in '84 in Helsinki by the expansion . . .

RS: . . . I have here still a different opinion, but let me turn to Berndt, you said something about chiral restoration mocking up, or coming in the place of confinement. I have always had another naive question, which is also put forward here to Hagedorn. How, in an entirely hadronic scenario, you could get faster strangeness relaxation constants than we are normally considering 'à la Rafelski'. A view of the relaxation time in the hadronic gas comes from basically considering K–π-baryon gas. Then the relaxation constants for strangeness are very long. At energy density of about 2 GeV fm^{-3}, even if we stay with what ever hadronic scenario would be in, I would also guess that the mass spectrum that we are now looking at is shifted far, far higher, and in such an environment the difference between the strange quark mass as a constituent quark and the ups and downs makes no difference anymore. K* intercommunicates with Δ^* much easier, there are no thresholds, there is not this Boltzmann penalty, the Maxwell factor penalty. Equilibration at this kind of 200–250 MeV initial hadronic temperature with higher masses abundant in the system, could be much faster and it could mock up what we think could only be achieved in a plasma.

JR: The usual hadronic gas situation has been studied carefully, possible strong changes in hadronic gas masses and hence equations of state, which would ensue, have so far not been noticed, and you have always more than just one observable . . . In Helsinki someone said, there's a strong lack of evidence for all this.

SS: From an experimental point of view we see at the AGS that for Au–Au collisions the productions of K$^+$ essentially is just linear with the number of participants. That's somewhat surprising because we had thought that the re-scattering would increase the strangeness production, but it is just linear with the number of participants, so it just becomes an interesting question as we get these Pb-beams at CERN this Fall which behavior we will see (shows a transparency) — we see here this linear behavior and then we see a break, and a much faster rise in terms of the K^+ production. If we saw such a kink like that, such a discontinuity, that it would be very hard to explain that by conventional scenarios. I think this is kind of an opportunity to see a signal. This is future, this is not what we see now at AGS, so far we see linear behavior.

MG: Strangeness enhancement means that you do not see a linear behavior. We already see in S–S collisions your kink, the non-linear behavior. *Arguments are exchanged . . .*

JR: We clearly see there is a difference between AGS and SPS physics, one should record it in memory.

EQ: It's interesting to look at variations of yields of strange baryons and antibaryons as function of the mass number of the projectile. But this discussion reminds me of a letter from Rafelski to SPSC where he stressed that this is only part of the story, the interesting thing would be to look for the same particle ratios as a function of the beam energy, keeping the participants the same. Since the subject is messy enough he wished to simplify it by using also variation in energy. Could somebody comment on this. Rafelski?

JR: No, I will not comment right away, please wait till my lecture in the afternoon, I would like just to make very clear what the question is, because obviously it's an important issue and it should be raised. The point which Emanuele has mentioned is as follows: do we need excitation functions of strange particles and why?

JSt: There is this strangeness suppression factor that Reinhard just showed, λ_W and you see here (shows transparency) the p–p systematic and you see as a function of beam energy that both at the AGS and the SPS strangeness is enhanced significantly. At least as a function of energy, we have these two points at AGS $\sqrt{s_{NN}} \simeq 5.2$ GeV and at the SPS $\sqrt{s_{NN}} \simeq 17.5$ GeV, the result is shown here for the same A: it is S–S for CERN and Si–Al for AGS. We have two points and the situation is not too different. There is a factor 2 enhancement in either case. *Transparency indicates absence of energy dependence . . . Arguments are exchanged.*

EQ: My question was concerning the Pb–Pb at the SPS, which is the new thing, for which data do not yet exist. I just wanted to stress the fact that at issue is not only to study strangeness as function of the mass number of the target or projectile, but also of $\sqrt{s}$ for very heavy systems.

UH: Do we need excitation function? The answer is yes. Why it's yes, that is somewhat a longer answer. I don't think actually that measuring the excitation function will be sufficient, but it's certainly necessary. We see something interesting going on in the strangeness and there is still the question that remains, where does it come from? We have discussed QGP, we have discussed all kinds of exotic modifications to the hadron gas, the thing that we can probably do presently is that we can exclude some models for example naive hadron dynamics. Excitation functions, where do they come in? They come in restricting the theories by a little bit more, eliminating some models.

GO: I think what we really need is a strangeness saturation factor as function of energy. We know that at CERN we are approaching almost 100% saturation. We know with some 'illegal' extrapolation . . . that at AGS it is still relatively low around 30–40%. And now is the question, where does this transition take place? Is it gradual, is it a step function?

SS: I do not agree with the term 'illegal extrapolation', because in the Au–Au experiment at the AGS we measured the K^+ over essentially almost a full rapidity range, so I think we have a very good handle on the yield of strangeness now.

PS: Regarding the experimental measurement of the excitation function — this is an important question and it takes some time of mental and other preparation before we can at CERN for instance think of doing a run at 50 GeV A with Pb or whatever other ion we will have. It implies that the experimentalists get convinced that they can materially do with this, and that we all get convinced that we have to do it because it is a fundamental issue. I think this assembly is important because it is a most favorable meeting place to reach such a decision, so if we want this we should start today to discuss this point further.

Closing Statements

ID: Is it possible from the heavy ion interaction to learn something without use of thermodynamics?

AT: Yes — the thermodynamic approach is one of the approaches, it has given a certain number of results and given a lot of interesting things, but there may be, I am even sure that there are other ways to juggle this problem and to look for a more precise mechanism to describe more accurately these collisions.

RH: Normally you can't measure things of which you have no conception, therefore the thermodynamic concept, which is probably the most popular one for today, is difficult to overcome, so you have to think hard, to think of mechanisms which are not thermodynamic and which give you measurably things.

EF: Experimental results will crucially depend on equations of state of hadronic matter. I will try to show this afternoon that there exists very plausible system beyond hadronic phase actually made of a gas of deconfined constituent quarks, governed by a different equation of state, so maybe there is something else we can discover and maybe we have it already just at the present accelerator energies.

RH: I have a very naive theorist's picture of the future. You might be able to study event by event the interactions of heavy ion collisions, and to trigger not for centrality as you mostly do today in order to get a better chance of finding QGP, but to trigger for medium impact parameters. If you can do that, let's say impact parameter one half, then you would know that in the extreme regions of the rapidity plot, you are dealing with hadron matter, and you have not had a phase transition, even if it was possible. In the inner part you would have better phase transition and there you might distinguish in single experiments what is what.

JS: I just wanted to comment on the analogy of looking at a piece of an elephant. I think if you take many, all events, you run the danger of looking from a far on herd of elephants, that doesn't look so much different than a herd of geese or something like that. You must choose those events carefully.

SS: I agree with Hagedorn. I also think it's extremely important to look at these reactions as a function of impact parameter. I think it is important to look at energy dependence, but as you go up in energy from the AGS value, you go up in thresholds for different particle productions and so forth. It is a little bit more model dependent, so the impact parameter dependence is very important to go from within a given thing that looks like p–p collision to the very central very dense collisions.

WG: I would like just to underline that there is a need for lots of systematic which means measurements of P_T dependence, rapidity dependence, impact parameter, A, and energy dependence in great detail because the margin between an enhancement and the absence of an enhancement may still be just of the order of 2, so the scale has to be known very precisely.

FA: We have a very interesting experimental situation today, the question is if it is all that was there to see?

GO: If you forget for the moment about QGP, strangeness by itself is a tag on a very interesting kind of reactions, so even if we will not end up with the discovery of deconfinement it is worth it.

EQ: It seems clear that we need a comprehensive experimental program to have a handle on the QGP, but this is not new. Quarks and gluons have not been discovered in a single experiment.

SIMILARITIES AND DIFFERENCES IN STRANGENESS PRODUCTION AT BNL AND CERN

Grazyna Odyniec

Lawrence Berkeley Laboratory
University of California
Berkeley, CA 94720

ABSTRACT

Relativistic heavy ion collisions provide, in principle, necessary conditions for the formation of the quark gluon plasma. Strangeness enhancement was amoung the announced specific signals expected from this new, hypothetical state of nuclear matter. Analysis of the first generation of experiments with light and medium sized ions have been completed. Results on strangeness production become precious handles to study in great detail what is actually happening. The current experimental situation is assessed. The emerging picture is still incomplete: however, open questions constrain requirements on future heavy ion experiments.

INTRODUCTION

The search for a quark-gluon plasma has been a long one. The creation of the primordial quark-gluon plasma phase in laboratories and subsequent study of its properties has been an ultimate goal for nuclear scientists since the mid-seventies. Thanks to Quantum Chromodynamics, we know what this new state is likely to be. We know rather precisely the value of the key parameters, such as the temperature and the baryon chemical potential, beyond which nuclear matter can only exist in the form of quarks and gluons in thermal equilibrium. Furthermore, we also know what kind of characteristic signals should mark the formation of a new plasma phase. Strangeness is one of them. Enhancement of strange particle production in the plasma phase as compared to a thermalized hadronic gas, was originally proposed by Koch, Muller and Rafelski[1] almost a decade ago. Indeed, high quark and gluon densities and frequent partonic collisions together with a temperature on the order of 200 MeV (which should a priori correspond to the plasma phase) should enhance the strangeness content of the final state with insufficient time to return to equilibrium in a hadronic phase at lower temperature.

But achieving all the prerequisite conditions experimentally may still not be enough for the formation of a quark-gluon plasma phase in the laboratory. Perhaps there is no simple way to detect it as well. Nevertheless, much can be learned about this new form of highly excited nuclear matter by studying strange particles in general, and multistrange baryons in particular.

In this lecture, the current status of selected experimental results on strangeness production in relativistic ion collisions at CERN and BNL is reviewed. I will try to convey not only my own understanding of the present experimental situation, but also those arising from discussions which took place during this meeting. I would like to apologize in advance to those who may feel that their results are omitted or not fairly represented here. It is clear that due to the time constraints I have had to make choices, emphasizing more the data which are suitable for the comparisons between the two energy regimes.

"FAVORABLE" CONDITIONS AT CERN AND AGS FOR QGP FORMATION

Transition to the hypothetical quark-gluon plasma phase might only take place when the energy density and temperature, reached in the early stages of the collision, exceed their critical values. Creation of a high density requires large energy deposition from the longitudinal motion of the projectile into the interaction volume. In the early eighties, it was realized that there was a chance to reach at least the energy density needed with collisions of heavy ions accelerated in existing machines. Soon after, an exploratory program was successfully launched at the Brookhaven AGS with silicon beams at 14 GeV/c and at the CERN SPS with oxygen and sulphur beams at 200 GeV/c.

The first results were as expected: a significant degree of stopping appeared at both energies. This was the first piece of good news, especially in case of the CERN data, since a rather large degree of transparency was forecast for such a high energy domain, which would, undoubtedly, complicate the interpretation of the data.

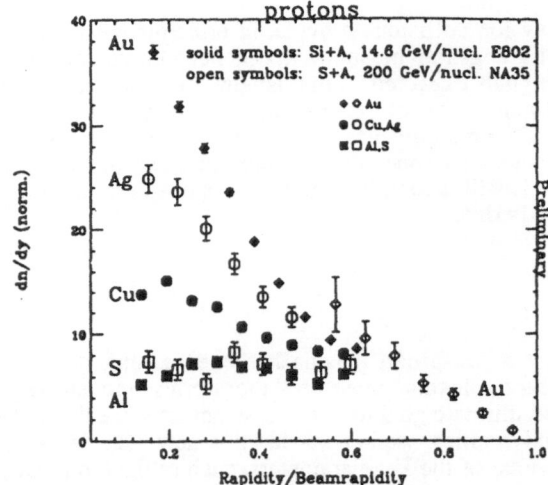

Figure 1. Relative rapidity (y/y_{beam}) distributions of primordial protons in nucleus-nucleus collisions from the AGS E802 experiment (solid symbols) and the CERN NA35 experiment (open symbols).

Fig.1 shows the compilation of the relative rapidity distributions (y/y_{beam}) of net protons for various nucleus-nucleus collisions observed at both energies. Solid symbols represent AGS E802 experiments at 14.6 GeV/c, and an open symbols - CERN NA35 data[a] at 200 GeV/c. First of all, we see that in all reactions there is an appreciable number of baryons at mid-rapidity, and that with increasing target mass more and more projectile and target nucleons are shifted towards mid-rapidity. Momentary stopping of projectiles on target nucleons implies a momentary high baryon energy density. Thus the high energy density seems to be well within reach in experiments at both laboratories.

---

[a] Since the streamer chamber (NA35 CERN experiment) has no identification capability for stable particles the net proton distribution was deduced from the measurement of the charge excess by subtracting negative hadron distributions from positive hadron distributions and applying all the necessary corrections.

400

ENCOURAGING FIRST RESULTS ON STRANGE PARTICLES PRODUCTION

The second piece of good news came in 1988 from the NA35 experiment at CERN and the E802 experiment at BNL.

NA35 observed for the first time a significant enhancement (~factor 2) of Λ, $\bar{\Lambda}$ and K^0 particle production in S+S collisions at 200 GeV[2]. Initial estimates of energy densities in central S+S collisions[3] were of the order of 2 GeV/fm^3 and appeared to be, in principle, adequate for plasma formation. However, this was not likely to happen due to the rather small volume (~ 90 fm^3) which did not seem to be sufficient even for partial thermalization.

Also, the BNL E802 collaboration reported an enhanced K^+/π^+ ratio in Si+Au collisions at 14.6 GeV[4].

This initial excitement stimulated a lot of experimental and theoretical effort. The data, available now, are more mature and better understood; however, theory still does not offer an explanation of the observed phenomena.

RECENT AGS RESULTS ON STRANGENESS

The E859 experiment was designed to improve and expand on single particle inclusive and two-particle correlation data obtained previously by the E802 collaboration. Single particle inclusive measurements have been made for the wealth of particle species, as illustrated in Fig.2. The E859 data analysis provided integrated yields, dN/dy, for both K^+ and K^- in several ranges of rapidities, different centralities and targets. In turn, these results were used to investigate whether the K^+/π^+ enhancement reported earlier by the E802 experiment also persisted in the K^- channel.

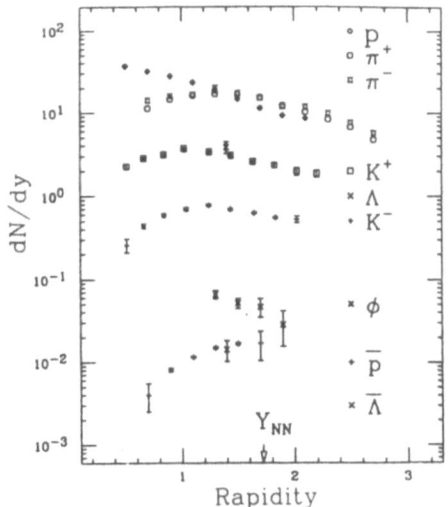

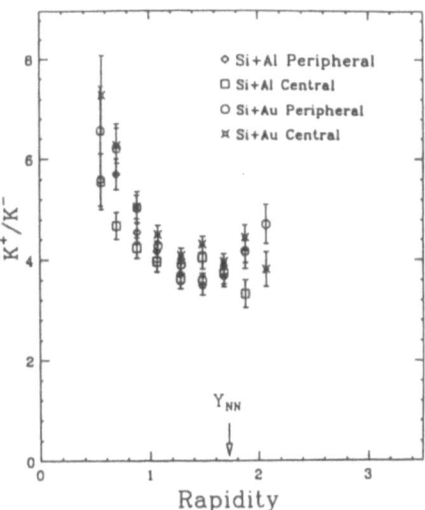

Figure 2. Rapidity distributions for all the particle species measured by the E802 and E859 experiments. Values shown are for central Si+Au collisions at 14.6 GeV/c.

Figure 3. K^+/K^- ratios vs. rapidity for peripheral and central Si+Al and Si+Au collisions from E859 experiment at AGS.

Fig.3 shows the K^+/K^- ratios as a function of rapidity for peripheral and central Si+Al and Si+Au collisions. In both reactions, the K^- and K^+ yields scale almost exactly with each other. Both the magnitude and rapidity dependence remain unchanged. However, the absolute number of kaons in these two reactions differs by more than a factor of 25 and the K/π ratio by a factor of 2. The fact that the enhancement of K^+ and

K⁻ seems to be the same, even though the particles are believed to have very different production mechanisms, will require special attention within the theoretical models.

Furthermore, the K^+/π^+ signal may allow for the assessment of the degree of thermalization. The increase of the K^+/π^+ ratio with the incident ion size is illustrated in Fig.4.

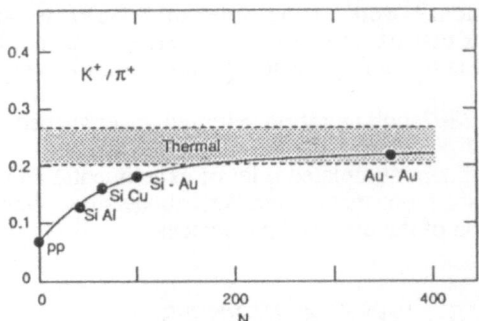

Figure 4. K^+/π^+ ratio versus incident ion size at AGS energies.

It gradually rises by a factor of two from p-p to A-A collisions and then levels off. A thermal system at a temperature of about 200 MeV is expected to favor far more kaons as compared to pions than the typical p-p collision $[m_t(K) \sim m_t(\pi)]$. However, one is rather surprised to see the symptoms of thermalization already at the AGS energies since this energy is believed to be too low for thermalization[5].

Recently the E859 collaboration reported measurements of Λ and $\bar{\Lambda}$. Earlier measurements of Λ production by the E810 experiment were confirmed and expanded, and the first results on $\bar{\Lambda}$ production (one-point, spectrometer integrated yield for central Si+Au collision at mid-rapidity, shown in Fig.2) on a heavy target were obtained. The data with an inverted magnetic field to collect lambdas with the same acceptance as $\bar{\Lambda}$ were used to extract a ratio. At the moment the preliminary value of the $\bar{\Lambda}/\Lambda$ ratio is estimated to be about $(4.0\pm2)\times10^{-3}$ with large systematic uncertainty. Details of the analysis and some very interesting results on antilambda/antiproton ratios are discussed by S.G.Steadman in these proceedings.

The first observation of multistrange baryon production in heavy ion collisions at the AGS was made by E810 collaboration[6]. The lifetime of Ξ^- measured in Si on Pb collisions at 14.6 GeV/c agrees very well with the particle data group value. Fig.5 presents the observed Ξ^- effective mass spectrum (a) and the rapidity distribution (b) after an acceptance correction.

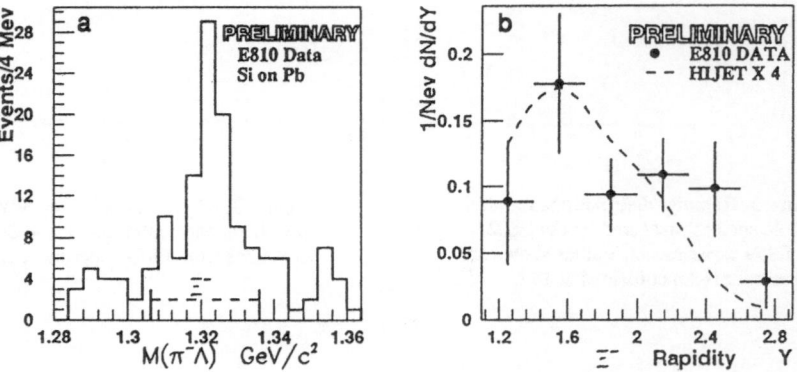

Figure 5. Ξ^- effective mass spectrum (a), and the rapidity distribution (b), in Si on Pb collisions at 14.6 GeV/c (E810 experiment).

HOW CAN WE INTERPRET STRANGENESS DATA FROM THE AGS ?

Some qualitative description of the AGS experimental results was provided by the ARC, a relativistic particle cascade model, without need for any novel mechanism for enhanced strangeness production. The ARC simulations indicate transient presence of a clearly defined high density matter at mid-rapidity. This suggests that the local equilibrium thermal model (fireball) may provide the correct description of the collision. Such an approach was advocated by Rafelski and Danos[7], and with further development by Lettesier, Rafelski and Tounsi[8] to extract chemical potential values from the available experimental data. The basic difference between this method and other models[9] is that instead of testing the validity of a particular model, they developed a method of translating the experimental results into a set of parameters characterizing the collision. Under very few, but rather strong, general assumptions these parameters can be interpreted as the usual thermal parameters (temperature and chemical potentials) of the hadron system at freeze-out[10]. They assume that all central rapidity strange particles are emitted from a thermally equilibrated fireball with u and d flavors in chemical equilibrium, but the strange quark flavor only in relative chemical equilibrium[b]. As a result of this approximation the relative yields of the different particles are described in terms of two fugacity parameters[c]: the strange-quark fugacity λ_s and the light-quark fugacity λ_q. Namely:

$$\bar{\Lambda}/\Lambda = \lambda_s^{-2}\lambda_q^{-4} \qquad \text{and} \qquad K^+/K^- = \lambda_s^{-2}\lambda_q^2 .$$

These permit the determination of fugacity values and chemical potentials from experimental particle yields. The E859 obtained yields with Si+Au at 14.6 GeV/c, even though the data is still preliminary, of :

$$\bar{\Lambda}/\Lambda \sim (4\pm 2)\times 10^{-3} \text{ (in the central rapidity region, see Fig.2), and}$$

$K^+/K^-=4.5\pm 0.5$ (which is constant in the entire central rapidity range $1<y<1.6$, but varies strongly near target and projectile rapidities (see Fig.3))

are presented graphically in the λ_q vs. λ_s plane in Fig.6.

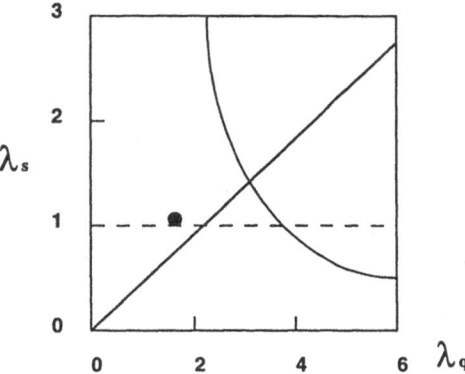

Figure 6. Constraint between the strange-quark fugacity λ_s and the light quark fugacity λ_q, choosing fixed ratios of K^+/K^- and $\bar{\Lambda}/\Lambda$ (from the E859 experiment) in a thermal fireball model. The horizontal dotted line indicates the result expected for QGP. The CERN result is indicated as a solid dot.

---

[b] Strange particle phase space is NOT fully occupied (it is also called "partial saturation" of the strangeness phase-space). When produced, strangeness is distributed among the different particles as governed by the size of the accessible phase space.

[c] fugacity ~ concentration of quarks

The straight line represents $\lambda_s = 0.47\lambda_q$ taken from the kaon ratio and the squared hyperbola[d] curve reflects the parametrization $\lambda_s = 17(1/\lambda_q{}^2)$ deduced from the lambda/antilambda ratio. These lines cross at $\lambda_s \sim 1.5$ and λ_q slightly above 3. The large value of λ_q indicates a relatively high baryon density for a mid-rapidity fireball at the AGS, whereas the strangeness fugacity $\lambda_s \sim 1.5$ implies $\mu_s/T \sim 0.4$ (with $\sim 20\%$ error bar). The latter is considerably different from the CERN result $\lambda_s \sim 1$ (and $\mu_s \sim 0$), as will be discussed later in the paper.

As a consistency check, one would like to independently measure relative yields of p and p̄ and compare p̄/p ratios[e] with the value calculated indirectly from the kaon and lambda ratios[f]. p̄/p (and $\bar{\Lambda}/\Lambda$) ratios are very interesting, but difficult to specify precisely. Since protons are predominantly from participant nuclei and not from produced particles, so their rapidity distribution is very different in shape compare to a p̄. Thus p̄/p has a strong rapidity dependence, even at mid-rapidity[g].

However one can calculate the p̄/p ratio at a fixed rapidity (very small rapidity interval), e.g. at y=1.3 ($\sim$ mid-rapidity) $dN/dy(\bar{p}) \sim 10^{-2}$ and $dN/dy(p)$ is ~ 20, thus p̄/p at y=1.3 amounts to about $(0.5 \pm 0.2) \times 10^{-3}$, which agrees within two standard deviations with the E859 p̄/p value extracted from the $(\bar{\Lambda}/\Lambda)/(K^+/K^-)$ ratio[h].

Our picture is still incomplete. The key signature for the quark-gluon plasma formation is the saturation of the strange quark phase space, due to the shorter time scale expected in the QGP[1]. Assessment of the degree of strangeness saturation would require additional experimental information, e.g. measurements of the $\Lambda/$ p ratio. This is not available presently from existing AGS experiments. We will come back to the more detailed discussion of the strangeness saturation factor after visiting CERN results on strangeness.

CERN RESULTS ON STRANGENESS AND SOME ATTEMPTS AT ITS INTERPRETATION

A vast body of data now exists on the spectra of mesons, baryons and hyperons produced in the collisions initiated by sulphur nuclei at CERN energy. To illustrate the quality of the data and the sophistication of the experiments, I have chosen, as an example, the transverse mass spectra from the NA35 experiment presented in Fig.7. The inverse slope parameters of all hadrons except pions exhibit "thermal" behavior. Their values are 210 ± 20 MeV, which is surprisingly high for a hadronic system. Remarkable consistency between transverse mass distribution of neutral and averaged $(0.5 \times (K^+ + K^-))$ charged kaons produced in S+Ag collisions indicates the quality of the analysis[i].

Equally excellent results emerged from CERN experiments also on strange particle production. The first messenger of the unusual behavior arrived in 1988 (two years after the first heavy ion beams at CERN) with the NA35 report on enhanced Λ, $\bar{\Lambda}$ and K^o yields in S+S collisions at 200 GeV/c[11] These early results have shown that the yields of strange particles relative to the yield of non-strange particles produced in central S+S collisions are higher than the corresponding yields in nucleon-nucleon interactions at the same energies. This observation and the lack of strangeness enhancement in nucleon-nucleus collisions[12] ruled out the interpretations of models based on the superposition of p-p or p-N collisions.

---

[d] here a $(1/x)^2$ dependence

[e] Ratios involving protons should be taken with a grain of salt since experimental values need to be corrected due to contamination by cold projectile and target spectator protons. This task is ussually difficult.

[f] p̄/p ratios can be expressed by the double ratio of $(\bar{\Lambda}/\Lambda)/(K^+/K^-)$.

[g] Note that the slopes of protons and antiprotons are higher than π's and K's. This suggests that at the AGS, they may be freezing in different conditions (the same situation is seen in the CERN data).

[h] In these calculations the resonance decays, which roughly double the yields of both protons and hyperons, are not taken to the account - therefore above relations hold only approximately.

[i] Both data sets were obtained with different detection techniques.

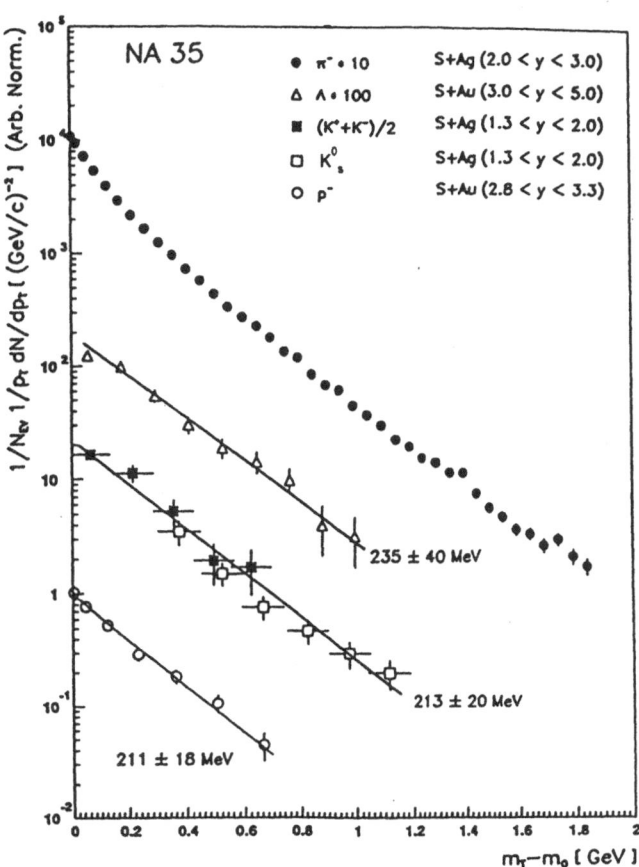

Figure 7. M_t spectra of different particles in S+Au and S+Ag collisions at 200 GeV/c (NA35 experiment).

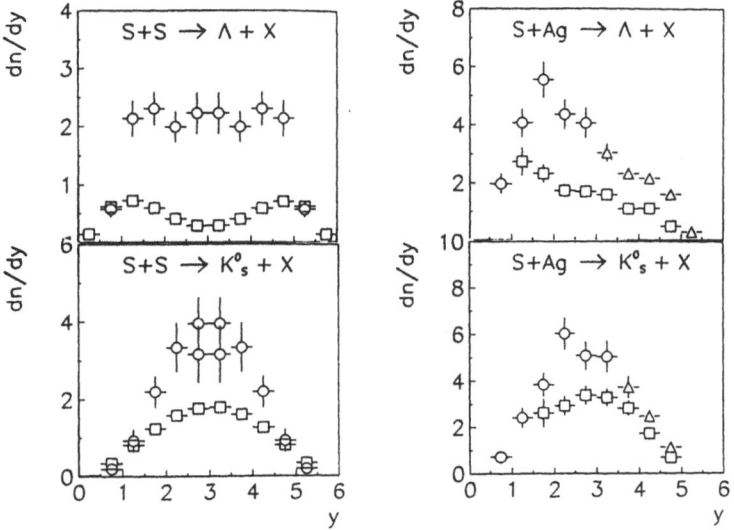

Figure 8. Λ and K^0 rapidity distributions in S+S and S+Ag collisions at 200 GeV/c (NA35).

New results from the NA35 collaboration have shown that the enhanced strangeness production also persists in central S+Ag collisions at 200 GeV/c [13]. Fig.8 shows the Λ and K⁰ rapidity spectra (over full phase space) in collisions of S on Ag (a asymmetric system) and S on S (a symmetric system).

A comparison of the Λ distribution in S+Ag and S+S collisions and a scaled distribution of minimum bias p+S interactions for S+Ag and p+p interactions for S+S (with the scale factor adjusting the multiplicities of negative charged hadrons) shows that the production is not only enhanced, but that this enhancement is most pronounced around midrapidity. The Λ rapidity density at midrapidity and below increases with increasing target mass, whereas for y>4, almost no dependence is observed. The production of K_s^o in S+Ag and S+S collisions is enhanced by factor of 1.7 compared to the scaled yield of p+S (for S+Ag) and p+p (for S+S). The charged kaons show a similar trend. Antilambda production is concentrated at midrapidity and the trend seems to be very much like the one seen in Λ's, although the error bars are still very large [13]. It is quite surprising that the strength of this effect is almost the same in all cases (Λ, Λ̄ and kaons) in both S+S and S+Ag collisions, since one would expect that strangeness enhancement more pronounced in collisions of heavier systems.

Consistently, the strangeness suppression factor which was used to estimate the deviation from flavor symmetry at the quark level in the final state [14] has been found to be similar in S+S and S+Ag collisions and only half as strong as in nucleon-nucleon and nucleon-nucleus collisions.

Other new results from the NA35 experiment which address the systematic study of inverse slope parameters in p+S and S+A interactions at 200 GeV/c are presented in Fig.9.

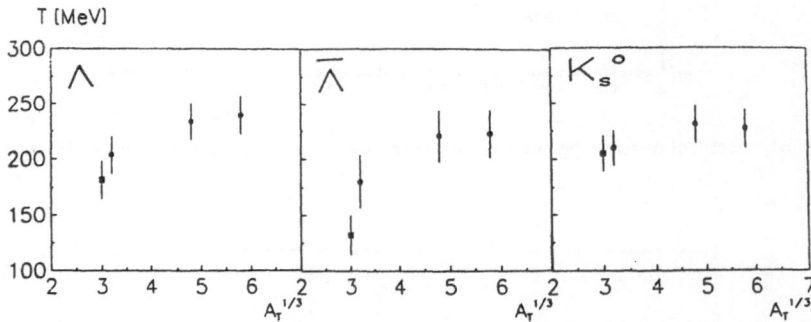

Figure 9. Inverse slope ("temperature" parameter) for Λ, Λ̄ and K⁰ in p+S and S+A collisions at 200 GeV/c (NA35).

The inverse slope for Λ, Λ̄ and K_s^o is higher in S+A than in p+S and further increases with the size of the target nucleus.

The NA35 strange particle ratios, summarized in table 1, were used to extract the quark (baryon number) and strange quark fugacities and the strange quark phase-space occupancy at the freeze-out temperature.

Within the generalized thermal model approach [8,10], discussed in the previous section, the strange quark chemical potential was found to be consistent with zero, and the strange quark abundance was found to be close to the full strangeness saturation as would be expected for the deconfined phase [14]. The same analysis applied to nucleon-nucleon interactions at 200 GeV/c and to nucleus-nucleus collisions at the AGS (see previous paragraph) yields significantly different freeze-out parameters, in particular regarding the degree of strangeness saturation[j] . The freeze-out temperature obtained

---

[j] In p-p only, at AGS the degree of strangeness saturation can not be assessed.

from the particle yields is consistent with the slopes of transverse mass distributions, which support the notion of a common chemical and thermal freeze-out required in explosive disintegration of a high entropy source[14]. An extention of this model avoids the strong assumption of thermal equilibrium (essential at the present stage) in favor of admitting an additional parameter[15]. Consequently, a preliminary estimate suggest that the strangeness phase space saturation, γ, will decrease in value by about 60-70 %.

Table 1. Strange particle yields and ratios from NA35 experiment.

Multiplicity & Ratio	S–S 4π	S–S midrap.	N–N 4π
K^+	12.5 ± 0.4	3.2 ± 0.5	0.24 ± 0.02
K^-	6.9 ± 0.4	2.2 ± 0.5	0.17 ± 0.02
Λ	8.2 ± 0.9	2.05 ± 0.2	0.096 ± 0.015
$\bar{\Lambda}$	1.5 ± 0.4	0.57 ± 0.2	0.013 ± 0.005
$p - \bar{p}$	20 ± 3	3.2 ± 1	0.90 ± 0.09
h^-	98 ± 5	26 ± 1	3.22 ± 0.06
$\dfrac{\bar{\Lambda}}{\Lambda}$	0.18 ± 0.05	0.28 ± 0.1	0.135 ± 0.055
$\dfrac{K^+}{K^-}$	1.8 ± 0.1	1.45 ± 0.4	1.4 ± 0.2
$\dfrac{\Lambda}{p - \bar{p}}$	0.41 ± 0.08	0.64 ± 0.2	0.11 ± 0.02

Reports from NA36, another large CERN experiment designed to study strangeness, support the NA35 results. Fig.10 presents the lambda yields as a function of centrality in S+S collisions measured by the NA35 and NA36 experiments (the acceptance of NA36 was adjusted to match that of NA35).

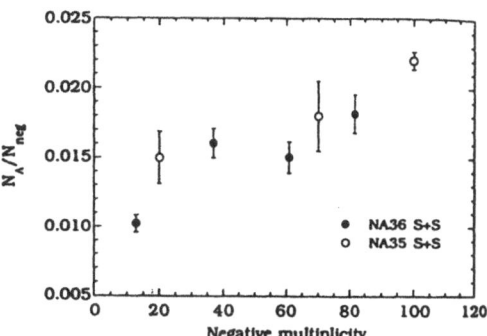

Figure 10. Lambda yields as a function of centrality in S+S collisions at 200 GeV/c (NA35 and NA36).

The comparison shows not only that the production of Λ's in both experiments grow faster with centrality (measured by the multiplicity of negative particles, mostly pions) than the production of negative particles, but also good agreement between these two experiments. The NA36 results, with sulphur collisions on heavier targets, show a monotonically increasing strangeness yield that saturates at high multiplicities, where all the projectile nucleons are involved in the collision, with centrality. Analysis shows that the strangeness production in S+Pb is almost a factor of two higher than in p+Pb[16].

In addition to Λ, $\overline{\Lambda}$ and K^o yields, the NA36 TPC detected of Ξ^- and Ξ^+ hyperon decays. This data is of great importance since the information of singly strange particles alone does not unambiguously distinguish between the hadronic gas and the quark-gluon plasma phase scenarios. The relative abundances of strange (antistrange) baryons with the same or different strangeness content contain a unique characterization of a new form of matter. The particle ratios, after subtraction of the Ξ^-, Ξ^+ contribution to the Λ, $\overline{\Lambda}$ yields, are listed in Table 2. These ratios are also plotted in Fig.11 and compared with a previously published compilation of measurements[17].

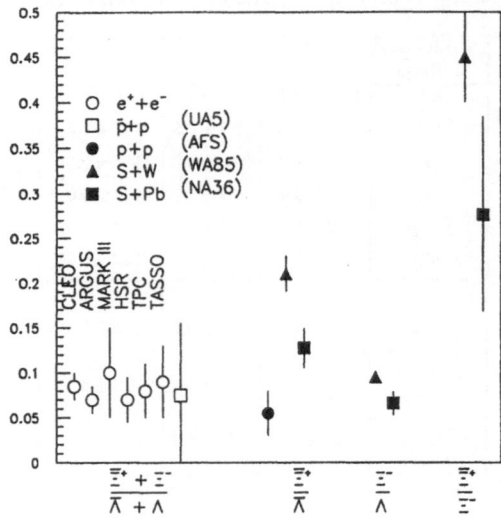

Figure 11. A summary of the world data on various particle ratios involving Λ, $\overline{\Lambda}$, Ξ^- and Ξ^+.

The NA36 results indicate that the enhancement in the $\Xi^+/\overline{\Lambda}$ ratio is at most a factor 2-3 over the p+p results. The particle ratios of Table 2 allow one to extract[18], within the same, already discussed, general thermodynamical scenario, values: $\mu_B/T = 1.73\pm0.15$, $\mu_s/T = 0.03\pm0.06$ and $\gamma_s = 0.38\pm0.04$.

Table 2. Strange particle ratios and their acceptance intervals from NA36 experiment.

	Particle ratio	Acceptance	
$\overline{\Lambda}/\Lambda$	0.117 ± 0.011	$2.0<y<2.5$	$0.6<p_t<1.6\,\mathrm{GeV}/c$
$\overline{\Xi}^+/\Xi^-$	0.276 ± 0.108	$2.0<y<2.5$	$0.8<p_t<1.8\,\mathrm{GeV}/c$
Ξ^-/Λ	0.066 ± 0.013	$1.5<y<2.5$	$0.8<p_t<1.8\,\mathrm{GeV}/c$
$\overline{\Xi}^+/\overline{\Lambda}$	0.127 ± 0.022	$2.0<y<3.0$	$0.6<p_t<1.6\,\mathrm{GeV}/c$

Difference in values of chemical potentials measured by various CERN experiments can be interpreted as a reflection of different degrees of centrality of the collisions. However, μ_s is small and compatible with zero (and consistent with NA35 measurements). This implies that the system formed near central rapidity contains nearly equal number of strange and antistrange quarks. To extract a value for the strange quark saturation factor, γ_s, the ratio of doubly- to singly-strange particle species was used[k]. The calculated γ_s value of 0.38±0.04 differs from that expected for the plasma case ($\gamma_s\sim1$), where the strange quarks would be close to their equilibrium abundance.

---

[k] γ_s^2 is given by product of the Ξ^-/Λ and $\Xi^+/\overline{\Lambda}$ ratio.

An alternative approach, also founded on thermodynamical arguments, was proposed by Cleymans and Satz. They have shown that in a model of an ideal gas composed of hadrons and known hadronic resonances at fixed temperatures and baryon density, which is assumed to be in equilibrium with zero net strange quark density, one can tabulate the production rates of strange particles as a function of T and μ_B. The measured ratios and their experimental uncertainties can then be used to establish an allowed region of the T-μ_B plane, which is necessary to explain the observed yields of Λ, $\bar{\Lambda}$, Ξ^- and Ξ^+. The details of this approach could be found in J.Cleymans contribution to this proceedings.

Slightly different ratios[19] are reported by the CERN WA85 experiment (see the triangles in Fig.11), presumably due to the different phase space coverage[l] and, therefore, different values of μ_B. However the μ_s value remains consistent with zero.

The rare Ω's are of particular interest since they contain three strange quarks; therefore their production by rescattering is practically impossible. The first and only experimental information on omega particles in heavy ion collisions was reported by the WA85 experiment[20]. Amazingly from a data sample of about 60 million triggers, only 7 ± 3.6 Ω^-'s and 4.0 ± 2.0 Ω^-'s survived all cuts applied during the analysis. The estimated ratio $\bar{\Omega}^-/\Omega^-$ for central rapidity ($2.5<y<3$) and high p_t ($p_t>1.6$ GeV/c) is 0.57 ± 0.41. This ratio is not yet corrected for possible differences in acceptance and efficiency for Ω^-'s and $\bar{\Omega}^-$'s.

I will stop the reviewing of experimental results here and conclude that the enhancement of strangeness at CERN has become certain.

But what about theory? Are there any explanations of the observed phenomena within available models?

We have seen that some of the thermodynamical models did manage to describe the measured strange particle ratios fairly well. Let us now examine microscopic models. First of all, there is no ARC type model for CERN energies. All models in which the produced particle system has passed through a phase of thermal equilibrium[21,22] do not predict enhanced population of strange particles observed by the CERN experiments. More successful in describing strangeness enhancement in heavy ion collisions at CERN energies appear to be non-equilibrium models based on a string picture which employ new mechanisms. The formation of "double strings" connected to the same leading quarks to enhance the production of baryons containing strange quarks was introduced in the VENUS model[23]. In the RQMD model, a mechanism of string fusion into "color ropes"[24], which break faster, allowed more frequent production of strange quarks and diquarks resulting in enhanced Λ and p production. Very recent, and still preliminary, NA35 results on p and Λ yields may help shed some light.

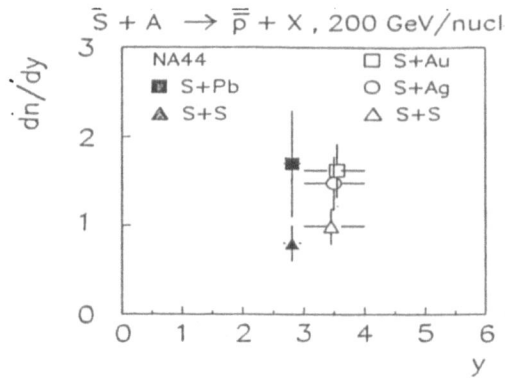

Figure 12. Rapidity density at midrapidity of antiprotons (primordial and from decays) produced in central S+A collisions at 200 GeV/c (NA35 experiment).

---

[l] WA85 experiment acceptance covers only high p_t region of the phase space ($p_t>1$ GeV/c).

The measured ratio $\overline{\Lambda}/p$ near midrapidity is approximately 1 and is therefore significantly larger than the corresponding ratio observed in p+p collisions and minimum bias p+A interactions and shows a slight dependence with the target mass in S+A collisions. Fig.12 summarize the rapidity density of antiprotons at midrapidity observed in central S+A collisions at 200 GeV/c.

However all of these new mechanisms, which allowed models to come closer to experimental measurements, also affect pion yields. At this moment, there are no models which would provide satisfactory simultaneous description of strangeness and pion production at CERN energies.

INSTEAD OF CONCLUSIONS

The local thermal equilibrium model, with allowance for only partial saturation of the strangeness phase space, lets us study the properties of high energy nuclear collisions based on the analysis of the particle multiplicities. Table 3 shows the chemical freeze-out parameters for central collisions at CERN and AGS energies. All three CERN experiments are characterized by vanishing strange quark chemical potential at large baryon density combined with a large degree of strangeness saturation, which differs significantly (by more than 4 σ) from μ_s at the AGS, where baryon chemical potential is more than two times larger than at CERN.

Table 3. Comparison of experimental values of thermal parameters.

	AGS	CERN-WA85	CERN-NA36	CERN-NA35
μ_s/T	0.54±0.11	0.03±0.05	0.03±0.06	0.025
μ_q/T	1.3±0.1	0.39±0.04	0.57±0.05	0.48
γ_s		~ 50 %	~ 40 %	~ 100 %

These numbers invite some speculations: CERN results could be seen as consistent with a scenario of QGP formation followed by a sudden disintegration; $\mu_s=0$ was found despite the expectations that in the hadronization process of QGP the memory about deconfinement will be erased and hadronic particles will reflect conditions of an equilibrated hadronic gas. Interpretation within the hadronic gas scenarios require extremely long fireball lifetimes in order to reach the measured degree of strangeness saturation[m] . In view of the present results, one can not yet exclude the possibility of QGP formation at the AGS. However, the dramatic difference in strangeness chemical potentials suggests that, if it is formed, it must be much slower than at the SPS. Comparison of baryon chemical potentials is complicated by different impact parameters and sizes of the systems measured at both energies.

However, granting the fact that the observed particle ratios for hyperons and antihyperons were reproduced in the scenario of a hot hadronic system freezing out rapidly into the final state, one has to go back and verify the underlying assumptions. This is not going to be easy at all. Therefore, one has to be very careful with any speculative conclusions until more data are available. The future experiments will provide, hopefully, the information on the dependence of "thermal" parameters on the masses of the colliding systems, their impact parameters, and energy[n] . Such a systematic study might allow us to understand the properties of the "macroscopic" high energy density system created in the collisions. In particular, it will be important to determine

---

[m] No signs for such long hadronic lifetime are seen in the data, e.g. HBT analysis.

[n] This calls for Lead-Lead or Gold-Gold collisions at different energies, and NOT only at the highest one.

whether a change of the strangeness saturation factor and chemical fugacity is gradual (Fig.13 a) with $\sqrt{S}$ or sudden (Fig.13 b) as is characteristic of color deconfinement.

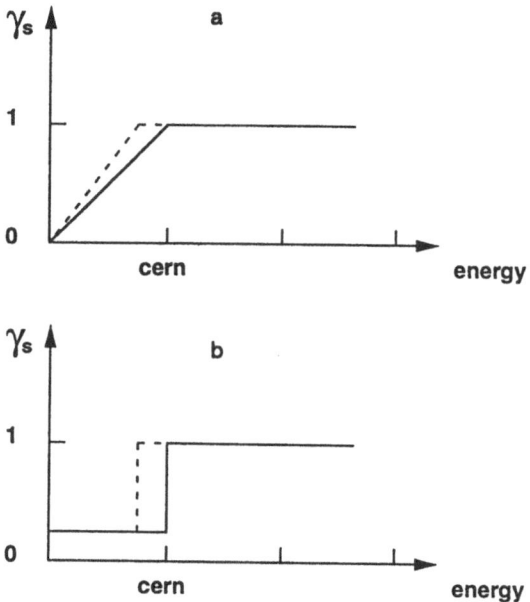

Figure 13. Examples of energy dependence for the saturation factor: a - gradual, b - sudden.

Furthermore, the measurements of other strange and non-strange particle ratios are necessary to constrain the final state better, since presently there are only three independent particle ratio results available[o] .

On the theoretical side, let us hope that the present results provide sufficient motivation for working out a consistent dynamical non-equilibrium hadronization scenario for a quark-gluon plasma.

ACKNOWLEDGMENTS

This work was supported in part by the Director, Office of Energy Research, Division of Nuclear Physics of the Office of High Energy and Nuclear Physics of the U.S. Department of Energy under contract DE-AC03-76SF00098.

REFERENCES

1. P.Koch, B.Muller, J.Rafelski, Phys.Rep.142, 167 (1989)
2. M.Gazdzicki and NA35 Collaboration, Nucl.Phys.A 498 (1989) 375c.
3. A.Bamberger et al.,Phys.Lett.B184(1987)271; J.Bachler et al.,Z.Phys.C 52 (1991) 239.
4. M.Sarabura and E802 Collaboration, Nucl.Phys.A 498 (1989) 409c.
5. M.Jacob, Proceedings of the Fifth International Conference on Nucleus-Nucleus Collisions, Taormina, Italy, 30 May - 4 June, 1994.
6. S.E.Eiseman et al., Phys.Lett.B 325(1994)322

[o] which one tries to reproduce with three parameters.

7. J.Rafelski and M.Danos, Phys.Rev.C - in press.
8. J.Letessier, J.Rafelski and A.Tounsi - Phys.Lett.B 328(1994)499, Phys.Lett.B 333 (1994)484 and preprint CERN-TH 7304/94..
9. e.g. E.Schnedermann, J.Sollfrank and U.Heinz, in: Particle Production from Highly Excited Matter, Plenum Press, New York, 1993, p.175; N.Davidson, H.Miller, R.Quick and J.Cleymans, Phys.Lett.B 255(1991)105; J.Cleymans and H.Satz, Z.Phys.C 57(1993)135; J.Cleymans, K.Redlich, H.Satz and E.Suhonen, Z.Phys.C 58(1993)347
10. J.Letessier, A.Tounsi, U.Heinz, J.Sollfrank and J.Rafelski, preprint PAR/LPTHE/92-27 and TPR-92-28, submitted to Phys.Rev.D.
11. M.Gazdzicki and NA35 Coll., Nucl.Phys.A 498(1989)375c.
12. H.Bialkowska et al., Z.Phys.C 55(1992)491.
13. D.Rohrich and NA35 Coll.,Nucl.Phys.A 566(1994)35c, M.Gazdzicki and NA35 Coll., Nucl.Phys.A 566(1994)503c.
14. J.Sollfrank, M.Gazdzicki, U.Heinz, J.Rafelski, Z.Phys.C 61(1994)659.
15. J.Rafelski - Private communication.
16. E.Andersen et al., Phys.Lett.B 294(1992)127.
17. S.Abatzis et al., Phys.Lett.B 270(1991)123, and references therein.
18. E.Andersen et al., Phys.Lett.B 327(1994)433.
19. D.Evans et al., Nucl.Phys.A 566(1994)225c.
20. Abatzis and WA85 Coll., Nucl.Phys.A 566(1994)491c.
21. H.W.Barz et al., Nucl.Phys.A 525(1991)435c.
22. J.Cleymans, H.Satz, Z.Phys.C 57(1993)135, J.Badier et al., Phys.Lett.B 122(1983)441.
23. K.Werner, J.Aichelin, Phys.Lett.B 308(1993)372.
24. H.Sorge et al., Phys.Lett.B 289(1992)6, H.Sorge et al., Preprint LA-KR-92-1078.

PARTICLE SPECTRA

Ulrich Heinz

Institut für Theoretische Physik, Universität Regensburg
D-93040 Regensburg, Germany

PROLOGUE

The nearly exponential transverse momentum or transverse mass spectra of hadrons produced in high energy collisions were from the beginning one of the driving forces for the development and sophistication of statistical models for multiparticle production. My own first exposure to thermal models of hadronic interactions happened in 1975 when I was working for my diploma thesis on hot nuclear matter and read Rolf Hagedorn's "yellow report" on the Thermodynamics of Strong Interactions.[1] These lecture notes exposed a beautifully selfconsistent theoretical picture which explained the shape of the observed hadron spectra and the apparently limited values of $\langle p_T \rangle$ and T, the inverse slope or "temperature", which in the then accessible energy domain did not increase with the collision energy $\sqrt{s}$. The model gave approximately exponential transverse mass spectra, $dN/dm_T^2 \sim m_T K_1(m_T/T)$, and predicted the excitation of the full (exponential) spectrum of hadronic resonances with thermal weights in high energy collisions. The first feature was already firmly established experimentally at the time.[2] The second one was later nicely verified by experiments[3-6] which identified large fractions of the emitted pions as due to resonance decays after the breakup of the collision fireball and showed[3] that the abundances of these resonances obeyed a Boltzmann distribution with the same temperature T as extracted from the slope of their m_T-spectra (see Figs. 1, 2).

In an attempt to better understand further details of the shape of the m_T-spectra, Hagedorn studied systematically the effect of the decay of thermally distributed unstable resonances on the spectra of the stable daughter particles.[7] It was found that the resulting steeper m_T-spectra of the daughter particles[7] can quantitatively explain[8] the systematic deviation from a purely exponential m_T-dependence at small values of p_T (the so-called "low-p_T anomaly") (see Fig. 2). This success made the model appear even more beautiful.

Little did I know as a young student that exactly at the time of my first exposure to the Hagedorn model its the phenomenological success was shaken severely by the discovery of a power-like tail in the transverse momentum distributions at the higher collider energies (ISR and later S$\bar{p}p$S) and by the rise of $\langle p_T \rangle$ with multiplicity density in high multiplicity events. These new effects were subsequently recognized as hard-scattering phenomena (jets and minijets) which, in contrast to the exponential features of the "soft" phenomena at low p_T, could be described by perturbative scattering between individual partons in the framework

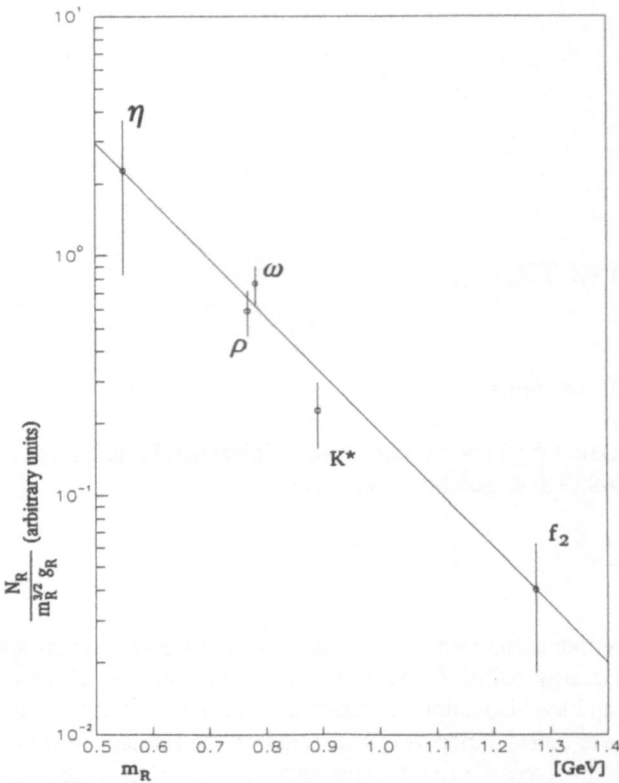

Figure 1. Total multiplicities $N_R/m_R^{3/2} g_R$ of measured resonances at $\sqrt{s} = 53$ GeV in pp collisions,[3] as a function of their mass m_R. (g_R is the spin-isospin degeneracy factor of resonance R.) The solid line represents a Boltzmann fit $\sim \exp(-m_R/T)$ with $T = 180$ MeV.

of QCD. The transition point between perturbative features at high p_T and non-perturbative, statistical features at low p_T is not well-defined. This leads to ambiguities in the extracted "temperature" for the soft, statistical component (in particular if no independent measure of this quantity through resonance to ground state ratios is available, as is the situation in nuclear collisions). This problem continues to haunt the "thermodynamicists" in the high energy community until the present day.

Furthermore, it was never clear which physical mechanism provides the dynamical justification for the assumption of local thermodynamic equilibrium in high energy hadron-hadron collisions. Over the years the Hagedorn model acquired the reputation of a phenomenologically extremely successful model without theoretical justification. Further sophistications like the phenomenological inclusion of a longitudinal flow velocity field[1] (to explain the much broader than thermal rapidity distributions and the so-called central rapidity plateau) and of a finite proper volume of the hadron resonances inside the Hagedorn fireball[11] (which allowed the reinterpretation of the limiting "Hagedorn temperature" as the critical temperature for close-packing in the resonance gas and the transition to quark-gluon matter) added attractive features to the model, but didn't help its somewhat notorious image.

At this meeting two important developments were reported which, I hope, will contribute positively to the image of Rolf Hagedorn's brainchild: Berndt Müller[12] showed that classical color fields created by soft interactions between the scattering hadrons evolve chaotically with a very large Liapunov exponent and thus self-thermalize on a very short time scale.

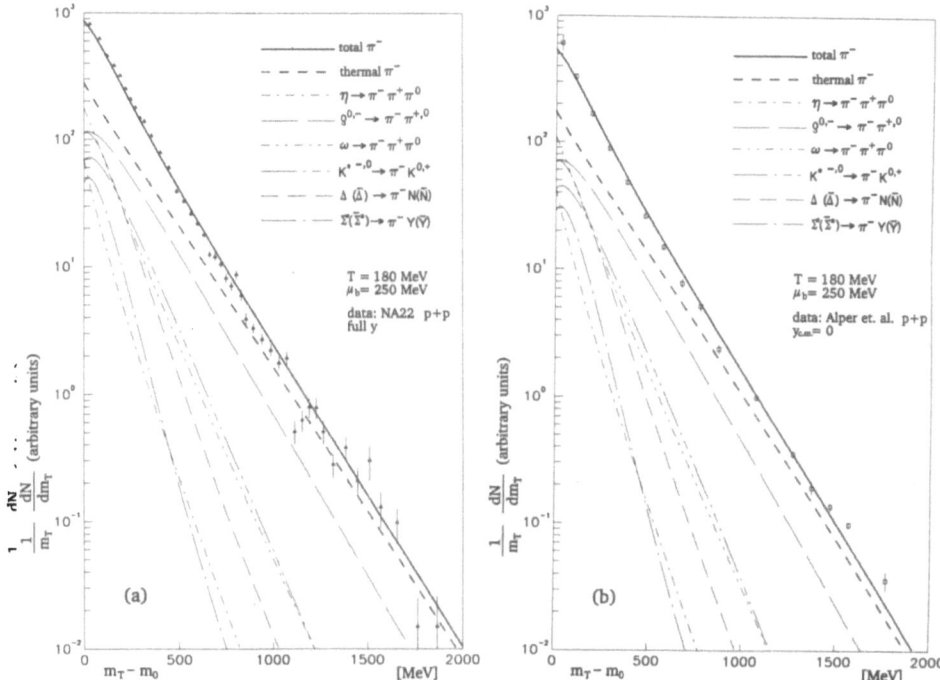

Figure 2. m_T-distribution of negative pions from a resonance gas with temperature $T = 180$ MeV and baryon chemical potential $\mu_b = 250$ MeV, including direct thermal π^- radiation and π^- from various resonance decays.[8] The resulting total spectrum is compared in **a** with NA22 data[9] for pp collisions at 250 GeV/c incident momentum which cover the full rapidity range, and in **b** with midrapidity data from Alper et al.[10] for pp collisions at $\sqrt{s} = 23$ GeV.

This may be relevant even for hadron-hadron interactions and provide the theoretical foundation for the Hagedorn model. In the following I will try to argue that hadron production data from *nuclear* collisions provide further phenomenological support for the model, because they show evidence for another consequence of thermalization which becomes visible when the (transverse) size and lifetime of the collision fireball becomes large enough: transverse hydrodynamic collective flow. These two aspects imbed the static Hagedorn model into a dynamical framework and provide it with additional features of internal consistency.

THERMAL FIREBALL SPECTRA

The Cooper-Frye formula

The invariant momentum spectrum of a locally thermalized emitter is given by the Cooper-Frye formula[13]

$$E\frac{d^3 N_i}{d^3 p} = \frac{dN_i}{\pi \, dy \, dm_T^2} = \frac{1}{(2\pi\hbar)^3} \int_{\Sigma_f} f_i(x, p) \, p^\mu d^3\Sigma_\mu(x) \,. \tag{1}$$

It has two essential ingredients, reflecting two fundamental assumptions:

(*i*) A local equilibrium phase-space distribution function

$$f_i(x,p) = \frac{g_i}{\exp\{[p^\mu u_\mu(x) - b_i\mu_b(x) - s_i\mu_s(x)]/T(x)\} \pm 1}, \tag{2}$$

where g_i is the spin-isospin degeneracy factor of the emitted particle species i, b_i its baryon number, s_i its strangeness, and the ± 1 refers to fermions and bosons, respectively. This distribution depends on 6 local parameters: the temperature field $T(x)$, the space-time distribution of the baryon and strangeness chemical potentials, $\mu_b(x)$ and $\mu_s(x)$, respectively, and the three independent components of a normalized $(u_\mu u^\mu = 1)$ 4-velocity field which describes the local collective flow velocity of a small fluid cell around spacetime point x relative to the frame in which the particle spectrum is measured. Eq. (2) reflects the assumption that before decoupling into free-streaming particles the source has reached a state of local thermodynamic equilibrium, characterized by maximal entropy at given local energy, baryon and strangeness density. These latter quantities develop dynamically, at least at the time of particle freeze-out, according to the laws of ideal hydrodynamics which reflect the conservation of total energy, baryon number and strangeness in the source. (On the time scale of nuclear collisions strangeness is a conserved quantum number.) The hydrodynamic flow velocity field $u_\mu(x)$ describes the conversion of thermal energy into collective flow by the action of pressure gradients, leading to adiabatic cooling of the source via expansion into the surrounding vacuum.

(*ii*) A 3-dimensional freeze-out hypersurface Σ_f, generically parametrized by three (locally orthogonal) coordinates u, v, and w,

$$\Sigma_f(u,v,w) = \left(\Sigma^0(u,v,w), \ldots, \Sigma^3(u,v,w)\right), \tag{3}$$

whose normal vector is given by

$$d^3\Sigma_\mu = -\epsilon_{\mu\nu\lambda\rho}\frac{\partial\Sigma^\nu}{\partial u}\frac{\partial\Sigma^\lambda}{\partial v}\frac{\partial\Sigma^\rho}{\partial w}\,du\,dv\,dw. \tag{4}$$

Eq. (1) thus integrates the normal component of the particle current $p^\mu f_i(x,p)$ over the freeze-out surface. The occurrence of this surface in the Cooper-Frye formula reflects the assumption that the transition from a strongly coupled, local equilibrium system to a decoupled stream of non-interacting particles occurs relatively sudden, on time and distance scales which are small compared to the gradients of the hydrodynamic fields $T(x)$, $\mu_i(x)$, and $u_\mu(x)$. This is an idealization which certainly needs improvement; this requires, however, a more microscopic kinetic treatment which I will not discuss here.

Freeze-out

Local thermodynamic equilibrium and hydrodynamic behaviour can be maintained as long as the collision rate among the particles is much faster than all macroscopic time scales. Freeze-out, that is the breaking of local thermal equilibrium and the decoupling of the particles, can happen by two different mechanisms: the expansion can become so rapid that the system falls out of equilibrium (a well-known example of this "dynamical freeze-out" process is the decoupling of the cosmic background radiation in the early universe – this example also demonstrates that dynamical freeze-out can occur even in spatially infinite systems), or the mean free path of the particles becomes larger than the geometric size of the system ("geometric freeze-out"). For freeze-out to occur, the following criterium should thus be satisfied:

$$\tau_{scatt}^{(i)} \gtrsim \min\left(\tau_{exp}, \tau_{escape}^{(i)}\right). \tag{5}$$

$\tau_{\text{scatt}}^{(i)}$ is the inverse scattering rate of particle species i:

$$\tau_{\text{scatt}}^{(i)} = \frac{\lambda_i}{\langle v_i \rangle} \simeq \frac{1}{\sum_j \langle v_{ij} \sigma_{ij} \rangle \rho_j}. \tag{6}$$

It can be calculated locally from the hydrodynamic densities and the cross sections; the sharp brackets mean an average over the local thermal distribution. It depends on the particle species through the cross sections: strongly interacting particles freeze out later than weakly interacting ones. It also allows to distinguish between *chemical freeze-out, i.e.* the decoupling of particle abundances, by inserting into Eq. (6) the corresponding *inelastic* cross sections for channels which transform different particle species into each other, and *thermal freeze-out*, which describes the decoupling of the momentum distributions and is obtained from Eqs. (5,6) by inserting the *total* cross sections (both elastic and inelastic collisions lead to the equilibration of momenta). Since the particle number changing inelastic processes constitute only a small fraction of the total cross section, chemical freeze-out generally precedes thermal freeze-out.

The expansion time scale

$$\tau_{\text{exp}} = \frac{1}{\partial_\mu u^\mu(x)} \tag{7}$$

can also be computed locally from the hydrodynamic flow profile. The escape time scale

$$\tau_{\text{escape}}^{(i)} = \frac{R}{\langle v_i \rangle} \simeq \frac{R}{c}, \tag{8}$$

through which the finite geometric extension of the collision fireball enters, can be computed as a function of time since the time dependence of the system size is also given by the hydrodynamic equations.

The freeze-out criterium (5) implies that in a realistic collision scenario for heavy-ion collisions the system always begins to decouple at the surface, due to its low densities and large mean free paths there. At this point there is no transverse flow yet, and thus the finite transverse extension plays the dominant role for the freeze-out criterium (possibly helped along by longitudinal expansion). Later on dynamical aspects become more important, and it turns out that for the bulk of the denser interior regions freeze-out is entirely controlled by the development of transverse flow.[14]

Flow spectra

At small impact parameters the collision fireball can be assumed to be cylindrically symmetric. In cylindrical coordinates $(u, v, w) = (r, \phi_s, z)$ the freeze-out surface is parametrized as

$$\Sigma_f^\mu(r, \phi_s, z) = \left(t_f(r, z), r \cos \phi_s, r \sin \phi_s, z \right), \tag{9}$$

and the normal vector reads

$$d^3\Sigma_\mu = \left(1, -\frac{\partial t_f}{\partial r} \cos \phi_s, -\frac{\partial t_f}{\partial r} \sin \phi_s, -\frac{\partial t_f}{\partial z} \right) r \, dr \, d\phi_s \, dz. \tag{10}$$

The 4-momentum is written as

$$p^\mu = (m_T \cosh y, p_T \cos \phi_p, p_T \sin \phi_p, m_T \sinh y), \tag{11}$$

and the flow 4-velocity as

$$u^\mu = \cosh y_T \left(\cosh y_L, \tanh y_T \cos \phi_u, \tanh y_T \sin \phi_u, \sinh y_L \right), \tag{12}$$

where y_L and y_T are the longitudinal and transverse flow rapidities defined through

$$\gamma_{L,T} = \cosh y_{L,T}, \qquad v_{L,T}/c = \tanh y_{L,T}. \tag{13}$$

Orienting the coordinate system such that $\phi_s = \phi_u$ and defining $\phi \equiv \phi_s - \phi_p$, the volume element reads

$$p^\mu d^3\Sigma_\mu = \left[m_T \left(\cosh y - \sinh y \frac{\partial t_f}{\partial z} \right) - p_T \cos\phi \frac{\partial t_f}{\partial r} \right] r\, dr\, d\phi\, dz. \tag{14}$$

The local energy is written as

$$\frac{p \cdot u}{T} = \beta_T \cosh(y_L - y) - \alpha_T \cos\phi, \tag{15}$$

with

$$\alpha_T = \sinh y_T \frac{p_T}{T}, \qquad \beta_T = \cosh y_T \frac{m_T}{T}. \tag{16}$$

A few further steps then lead to the following expression for the invariant momentum spectrum:

$$\begin{aligned} E\frac{d^3N}{d^3p} = & \frac{g\, m_T}{(2\pi)^2} \sum_{n=1}^\infty (\pm)^{n+1} \int_{\Sigma_f} dz\, r dr\, e^{n[\mu/T - \beta_T \cosh(y_L - y)]} \\ & \times \left\{ \left(\cosh y - \sinh y \frac{\partial t_f}{\partial z} \right) I_0(n\alpha_T) - \frac{p_T}{m_T} \frac{\partial t_f}{\partial r} I_1(n\alpha_T) \right\}. \end{aligned} \tag{17}$$

The Bessel functions arise from the ϕ-integration. The shape of the freeze-out surface enters through the derivatives $\frac{\partial t_f}{\partial r}$ and $\frac{\partial t_f}{\partial z}$.

For practical purposes it is often useful to transform from t and z to longitudinally comoving coordinates $\tilde{t}$ and ζ.[15] For the special case of true cylinder symmetry, i.e. a system with a ζ-independent transverse radius $R(\tilde{t})$, this leads to a simplification of the final spectrum because then freeze-out occurs at $\tilde{t}_f = \tilde{t}_f(r)$ independent of ζ. The corresponding expressions for the spectrum[15] are the basis of the spectra shown below.

The transverse mass spectrum is obtained from the above expression by integrating over the rapidity y. If $\partial\tilde{t}_f/\partial r = 0$ (instantaneous transverse freeze-out), one observes that the longitudinal flow y_L completely factorizes from the shape of the transverse mass spectrum! While this factorization is no longer exact if $\partial\tilde{t}_f/\partial r \neq 0$, it is still true approximately, and *the shape of the transverse mass spectrum is therefore primarily dictated by the temperature and the transverse flow rapidity y_T.* The *longitudinal flow velocity determines the rapidity spectrum*, which in turn shows very little sensitivity to T and practically none to the transverse flow.[15,16]

The main effect of transverse flow is a flattening of the transverse mass spectrum. From the expressions above the slope at large m_T can be calculated analytically; one finds

$$\lim_{m_T\to\infty} \frac{d}{dm_T} \ln \frac{dN}{dm_T^2} \approx \frac{1}{T}\sqrt{\frac{1 - v_T/c}{1 + v_T/c}} \equiv \frac{1}{T_{app}}, \tag{18}$$

which can be interpreted in terms of an "apparent" or *blueshifted temperature*, with the blueshift factor given in its familiar form by the transverse expansion velocity. Since in general the transverse velocity is not a constant, but possesses a transverse radial profile, spectra with different blueshift factors are superimposed in the total spectrum. At large m_T the blueshift factor corresponding to the largest expansion velocity across the freeze-out surface dominates and thus dictates the asymptotic exponential fall-off of the m_T-spectrum. In the intermediate m_T region the superposition of spectra with different blueshift factors leads to a clearly visible concave curvature of the m_T-spectra; this will be seen in the figures below. This feature thus is a characteristic consequence of transverse collective expansion.

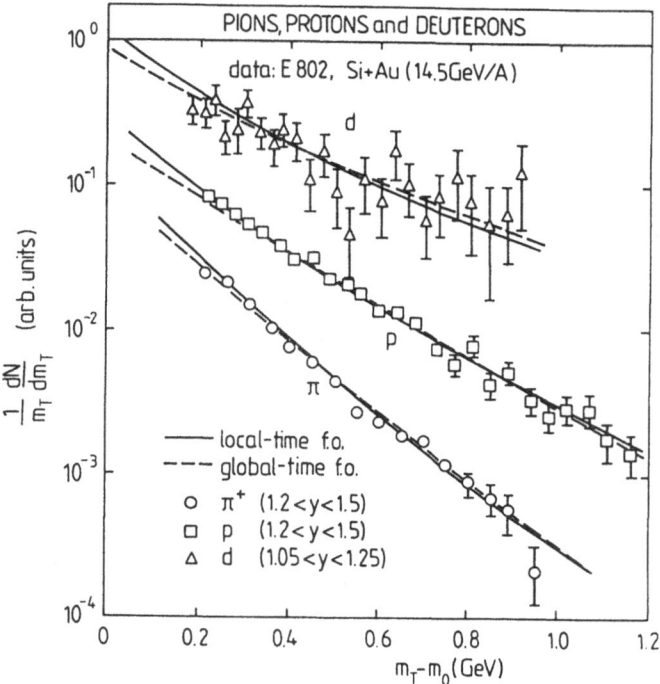

Figure 3. Comparison of the theoretical spectra with the E802 Si+Au data.[18] For details on the two choices of the freeze-out hypersurface see Ref. [19].

Transverse flow and m_T-scaling

This is certainly not the right place to preach about thermal fits to experimental momentum spectra: Rolf Hagedorn has spent a considerable part of his life (and suffered more than his fair share of frustration) advertizing the usefulness of the transverse mass $m_T = \sqrt{m^2 + p_T^2}$ rather than the transverse momentum p_T as the appropriate variable for plotting single particle spectra. Plotted against $m_T - m_0$, all particles from a thermalized emitter should show the same universal exponential behaviour ("m_T scaling"). A particularly clear exposition of this issue was given by Hagedorn in Ref. [7].

However, his arguments are strictly true only for systems without transverse hydrodynamic expansion. Transverse stationarity is a reasonable assumption in hadron-hadron collisions since there, due to the small transverse size of the collision region, transverse flow does not have sufficient time to develop before freeze-out. In nuclear collisions this is different, and a close inspection of Eq. (17) indeed shows that in the presence of transverse flow ($y_T \neq 0$) the shape of the spectrum depends on m_T and p_T independently (through α_T and β_T). At small values of $p_T < m$, *i.e.* in the region of non-relativistic velocities for the emitted particles, this leads to visible distortions of the spectra. They can be qualitatively understood by noting that the collective flow boosts the thermal transverse momentum of the emitted particle according to $p_T = p_T^{\text{thermal}} + m v_T$. This leads to a shift of the peak of the Maxwell distribution towards finite values of p_T and thus to a flattening of the spectrum at low p_T; the effect grows with the mass of the particles, and in the limit of a very large, *r*-independent transverse flow velocity it can even lead to a peak in the spectrum at non-zero values of p_T ("blast wave peak"[17]).

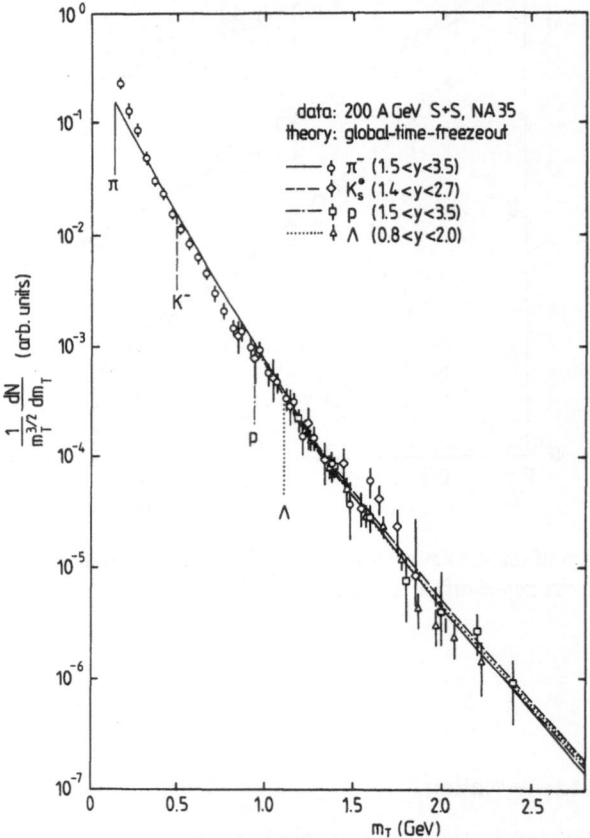

Figure 4. Transverse mass spectra for pions, kaons, protons, and Λ's from central 200 A GeV S+S collisions.[21,23] (Please note that we plot $(1/m_T^{3/2})(dN/dm_T)$!) After arbitrary normalization all slopes are seen to agree with each other, and the spectra fall on a universal theoretical curve[19] for an expanding thermalized fireball. (The discrepancy for pions at low p_T is due to the neglect of resonance decays in this calculation.)

This flattening of the m_T-spectrum at low p_T with increasing particle mass was systematically investigated by Lee et al.[19] We showed there that the effect is seen experimentally in the momentum spectra from 15 A GeV/*c* Si+Au collisions at the Brookhaven AGS measured by the E802 collaboration and argued that this supports the existence of transverse collective flow in these collisions (see Fig. 3). These findings and conclusions have recently been confirmed by the E814 collaboration at the AGS.[20]

At high transverse momenta, $p_T \gg m$, these differences go away, and all particles have the same exponential m_T-slope, independent of their rest mass[19] (see Fig. 4). m_T scaling is recovered in this limit, but in the interpretation of the inverse exponential slope the word "temperature" has to be replaced by "apparent temperature", in which thermal and flow effects are mixed according to Eq. (18). To separate the true temperature T from the flow effects is not an easy task: of course, one may try to exploit the mass dependence of the low-m_T part of the spectrum, but for a quantitative study very accurate data are needed, and even then the resonance decay contributions remain a source of uncertainty. I will describe a complementary approach based on theoretical consistency arguments in the next Section.

Thermal Boltzmann spectra

In order to provide a better basis for the flow analysis of hadron momentum spectra from S-induced heavy-ion collisions at the CERN SPS (200 A GeV/c beam momentum) in the following Section, I will shortly show how in the limit of a stationary fireball (no flow at all, $v_L = v_L = 0$) the above expressions reduce to the well-known thermal spectra long ago advertized by Hagedorn. This also allows me to reiterate some of his remarks on how to properly perform thermal fits to particle production data (hoping to reduce his frustration in the future!). In the stationary case we can choose for the spatial integration hypersurface one of constant time in the observer frame, and $d^3\Sigma_\mu = (d^3x, 0)$ has only a time component. Then the spatial integration $\int_{\Sigma_f} p^\mu d^3\Sigma_\mu \ldots \longrightarrow E \cdot V \cdot \ldots$ results in a simple factor V for the fireball volume in the observer frame.

In a stationary, equilibrated fireball the phase-space distribution function does not depend on x: in Boltzmann approximation we have

$$f(x,p) = \lambda\, e^{-p \cdot u/T}\,, \tag{19}$$

where $\lambda = \exp(\mu/T)$ is the fugacity, and

$$u^\nu = (\cosh y_{FB}, 0, 0, \sinh y_{FB}) \tag{20}$$

is the (constant) 4-velocity of the fireball center-of-mass relative to the observer frame. The spectrum then takes the simple form

$$E\frac{d^3N}{d^3p} = \frac{g\,\lambda\,V}{(2\pi)^3}\, m_T \cosh(y - y_{FB})\, e^{-m_T \cosh(y - y_{FB})/T}\,. \tag{21}$$

From this we see that data which have been collected in a small rapidity window (compared to the total width of the rapidity distribution below, say $\Delta y \leq 0.5$ or so) should be compared with a thermal model by using a fit of the form

$$\frac{dN}{dy\,dm_T^2} \sim m_T e^{-m_T/T_{\text{eff}}} \tag{22}$$

with the effective temperature being related to the true one by

$$T_{\text{eff}} = \frac{T}{\cosh(y - y_{FB})}\,. \tag{23}$$

For the popular fit with a simple exponential in m_T (without the m_T-prefactor) an interpretation of the resulting slope parameter in terms of a temperature is not possible. The factor m_T in front of the exponential in (22) is very important: it causes the theoretical spectrum to become slightly flatter than a pure exponential spectrum at small m_T – in other words, an experimental (dN/dm_T^2)-spectrum at fixed y which appears exactly exponential all the way to values $m_T < T$ already contains excess low-p_T particles relative to the thermal fireball model!

If, on the other hand, data are used which have been integrated over a large rapidity window around y_{FB} (comparable to the total width of the rapidity distribution for a single fireball, say $\Delta y \geq 0.5$ or so), one should fit with the y-integrated form of (21):

$$\frac{dN}{m_T dm_T} = \frac{g \lambda V}{2\pi^2} m_T K_1(m_T/T) \longrightarrow \sim \sqrt{m_T} e^{-m_T/T} , \qquad (24)$$

where the last proportionality holds in the limit $m_T \gg T$. In this case a logarithmic plot of

$$(dN/m_T^{3/2} dm_T) \quad \text{vs.} \quad m_T \qquad (25)$$

(see Fig. 4) should (at least for $m_T > T$) fall on a straight line whose slope reproduces the fireball temperature.

By the way, the form (24) (instead of (22)) should also be used for data collected in only a small rapidity window dy if one suspects that the picture of a single spherically symmetric fireball is unrealistic, and that (in the region of y studied by the experiment) a superposition of many fireballs which move relative to each other with a boost-invariant longitudinal velocity profile is more appropriate (restricted boost-invariance, see below). In this case the summation of the contributions from all these fireballs to the total spectrum is equivalent to integrating Eq. (22) over y, resulting in the expression (24). Since at CERN energies such a picture of restricted longitudinal boost-invariance appears to describe the experimental situation near y_{cm} rather well, *Eq. (24) should be used for all CERN m_T-spectra* (differential in y or not) which contain contributions from the central rapidity region.

The total number of particles in the fireball is obtained by integration over m_T as

$$N_i = \frac{g \lambda_i V T^3}{(2\pi)^2} \left(\frac{m_i}{T}\right)^2 K_2\left(\frac{m_i}{T}\right) , \qquad (26)$$

where K_2 is a modified Bessel function. Once the fugacities λ_i have been determined experimentally through particle ratios and the temperature is known from the slope of the spectrum, this equation can be used to determine the fireball volume.

Integration of Eq. (21) over m_T yields the rapidity spectrum for a stationary spherical fireball (with $y' := y - y_B$):

$$\frac{dN}{dy} = \frac{g \lambda V}{(2\pi)2} m^2 T \, e^{-m \cosh y'/T} \left[1 + 2\frac{T}{m \cosh y'} + 2\left(\frac{T}{m \cosh y'}\right)^2\right] . \qquad (27)$$

By expanding the argument of the exponential which, for $m \geq T$, dominates the shape of this function,

$$\frac{dN}{dy} \sim e^{-m(y-y_{FB})^2/2T} , \qquad (28)$$

we see that the rapidity distribution of a stationary thermal fireball looks like a Gaussian centered at y_{FB}, with a width $\Gamma_{fwhm} = \sqrt{8 \ln 2 T/m} \simeq 2.35\sqrt{T/m}$. Thus the rapidity spectrum has a width which depends on the particle mass and decreases for heavier particles like $1/\sqrt{m}$. For massless particles, the rapidity and pseudorapidity spectra become equal,

$$\frac{dN}{dy} \sim \frac{1}{\cosh^2 y'} , \qquad (29)$$

and look (roughly) like a Gaussian with width $\Gamma_{fwhm} \simeq 1.76$. For small m/T, the deviation from this limit (*i.e.* the influence of the temperature on dN/dy) is very small:

$$\frac{dN}{dy} = \frac{g \lambda V}{(2\pi)^2} T^3 \left(\frac{2}{\cosh^2 y'} + O\left(\frac{m^3}{T^3}\right)\right) . \qquad (30)$$

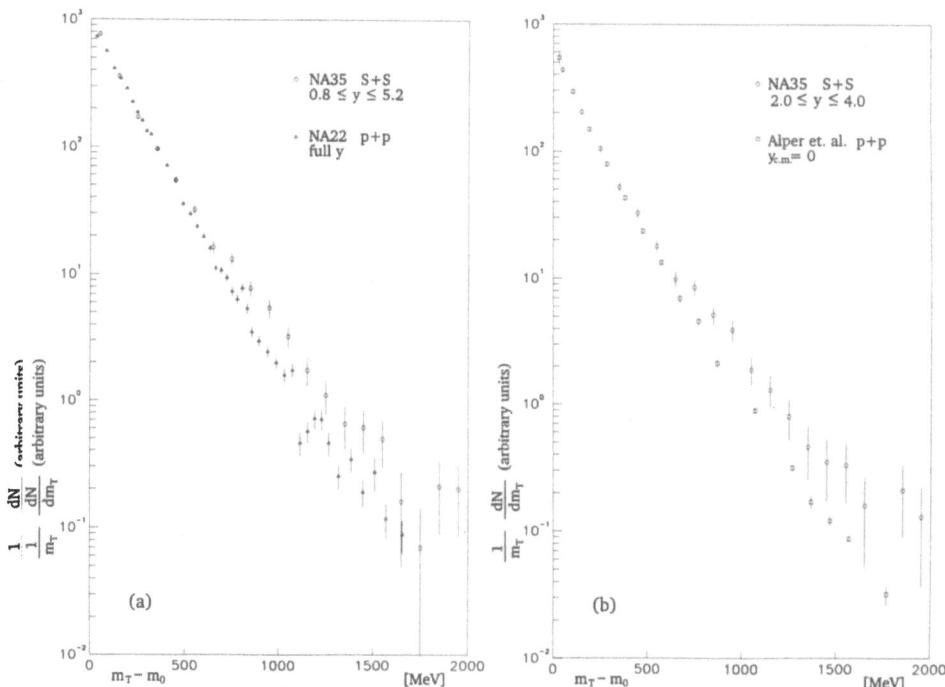

Figure 5. a Comparison of the rapidity integrated pp data[9] of NA22 and the SS data[21] of NA35 $(0.8 \leq y \leq 5.2)$ for negative pions, with arbitrary relative normalization. **b** Comparison of the central rapidity data from pp collisions[10] and from SS collisions.[21]

COLLECTIVE FLOW IN HEAVY-ION COLLISIONS

Before approaching the data with any models, it is instructive to compare the transverse momentum spectra from heavy-ion collisions directly with those from pp collisions at similar beam energies. From Fig. 5 it is clear that the heavy-ion induced pion spectra are flatter, at least in the region $m_T > 500$ MeV. Although space does not allow to show this here, the same is true for all other hadron spectra: thermal fits to the m_T spectra result in apparent temperatures which are consistently somewhat above 200 MeV, compared to the 160–180 MeV slopes from hadron-hadron collisions (see R. Hagedorn's contribution to these proceedings). The situation with pions, shown in Fig. 5, is at first sight even more complicated: the spectrum seems to consist of two components, a flatter one at high m_T and a steeper one at low m_T.

Obviously this flattening at larger m_T cannot be attributed to minijets or similar hard phenomena since the collision energy is too low. However, a similar effect, although less prominent, is already seen in pA collisions at the same energy.[22] The only known way to explain this phenomenon microscopically is by final state rescattering: all event generators agree that a mere superposition of pp collisions including Fermi motion of the nucleons in the initial state gives for nuclear collisions spectra which are indistinguishable from the pp ones, and that only after allowing reinteractions among the produced particles and between produced particles and the target and projectile nucleons the broadening of the m_T-spectra can be reproduced.

Of course, multiple scattering, when it occurs sufficiently often, generates collective behaviour. The transition from multiple scattering (where "multiple" is usually taken to mean "a few times") to collective flow is gradual: in nuclear collisions with their limited total lifetimes it is likely to always remain a subject of debate whether the macroscopic hydrodynamic concepts with their emphasis on collective behaviour are adequate. They are, however, conceptually simple and, if they work even on a qualitative level, useful for a basic understanding of the relevant physics phenomena. This is very much in the original spirit of Hagedorn's model, and I will thus exploit this approach here to its limits.

I will select a specific data set obtained by the NA35 collaboration in 200 A GeV S+S collisions.[21,23] At this energy, small nuclei like sulphur are no longer able to completely stop each other, resulting in a different longitudinal and transverse dynamics; therefore these data are useful to demonstrate the separate extraction of thermal motion and longitudinal and transverse flow. It is also at this energy the most complete set of longitudinal and transverse momentum spectra for a single collision system, with the largest variety of studied particle species. Since the collision system is symmetric, for central collisions contaminations in the spectra from target or projectile spectators, *i.e.* nucleons which do not participate in the formation of the hot and dense reaction region zone near central rapidity, are largely eliminated. This facilitates the theoretical analysis.[24,14–16]

Comparison with a stationary fireball model

We begin by trying to fit the spectra of pions (the most abundantly produced hadrons) with a simple stationary fireball. In Fig. 6a we show the rapidity-integrated π^- m_T-spectrum together with a fit of the form (24). We see that the fit does not describe the data; being constrained by the high-statistics data at low m_T to fall very steeply, with a slope parameter $T = 150$ MeV, the curve fails to reproduce the measured spectrum for $m_T - m_\pi > 600$ MeV.

Similarly, the measured rapidity spectrum (see Fig. 7) is much wider ($\Gamma_{\exp}^{\mathrm{fwhm}} = 3.3 \pm 0.1$)

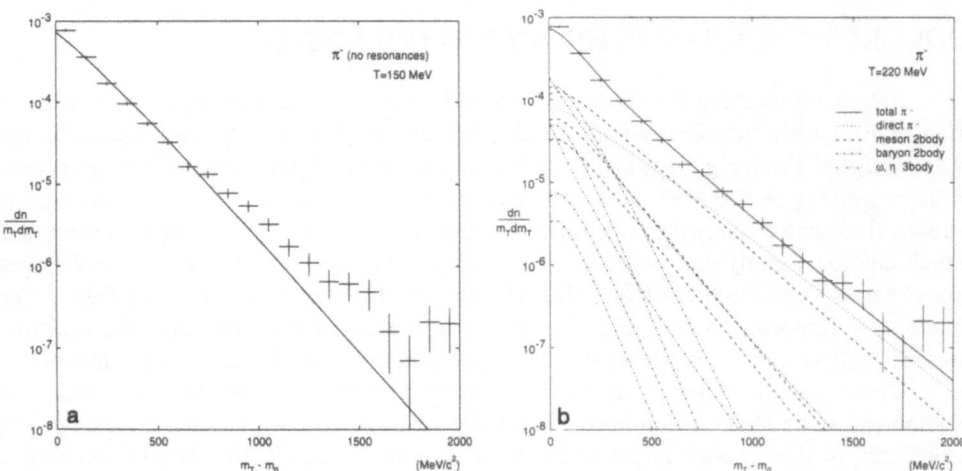

Figure 6. **a** Purely thermal π^- m_T-spectrum in comparison with NA35 data from 200 A GeV/c S+S collisions.[21] **b** Thermal π^- m_T-spectrum including resonance decays from a stationary fireball in comparison with the same data. The temperature T=220 MeV was determined by a thermal fit[8] with $\mu_b = 200$ MeV.

than the prediction from the stationary thermal fireball picture (Eq. (27), dotted line in Fig. 7). This remains true even if one sends $T \to \infty$ ($m/T \to 0$) in the theoretical expression. Thus a purely thermal picture, in spite of its theoretical simplicity, cannot explain the data.

Of course, at $T = 150$ MeV or higher, the thermal fireball picture also predicts the creation of an appreciable number of unstable resonances which, after decoupling, decay and produce additional pions whose spectrum is not thermal (as that of the parent resonances) but concentrated more towards lower momenta. Since the experiment cannot distinguish

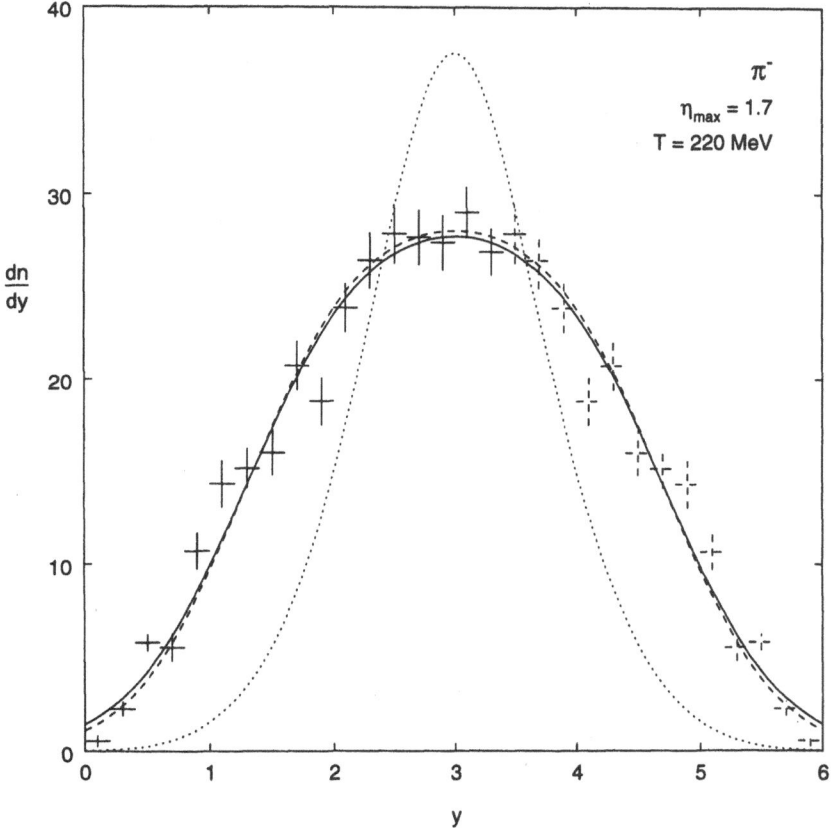

Figure 7. Dotted line: Thermal π^- rapidity-spectrum (including resonance decays) from a stationary fireball. Solid line: π^- rapidity spectrum from a longitudinally expanding source (see text). The effect of resonance decays (dashed line) is marginal. The data are from the same experiment as in Fig. 6.

between directly radiated and decay pions, our fit should take into account such an additional decay contribution, similar to the pp case shown in Fig. 2. The result of a thermal fit to the m_T-spectrum including resonance decays[8] is shown in Fig. 6b: the large and comparatively steep contribution of decay pions at low p_T accounts for the two-component structure of the measured m_T-spectrum, and a perfect fit is achieved. However, this does not help with the rapidity spectrum which still remains very close to the dotted line in Fig. 7.

Longitudinal and transverse flow

The failure of the static fireball picture in the longitudinal direction is even more prominent in hadron-hadron interactions and was mentioned already in the Prologue; it prompted Hagedorn[1] to postulate a superposition of several thermal emitters moving with different rapidities relative to the observer. This prescription can be formalized by generalizing the Hagedorn model to the case of *local* thermal equilibrium and providing it with dynamical aspects according to the laws of hydrodynamics. While Hagedorn discusses this approach in his contribution to this meeting in a short comment on the Fermi-Landau hydrodynamical model, which he rejects for good theoretical reasons, these are not the only possible initial conditions. Based on causality arguments and the inside-outside cascade picture of secondary particle production, Bjorken[25] has developed a hydrodynamical model with boost-invariant longitudinal flow which, when restricted to a finite interval of longitudinal flow rapidities, gives an excellent fit[16] to the measured rapidity spectrum (Fig. 7).

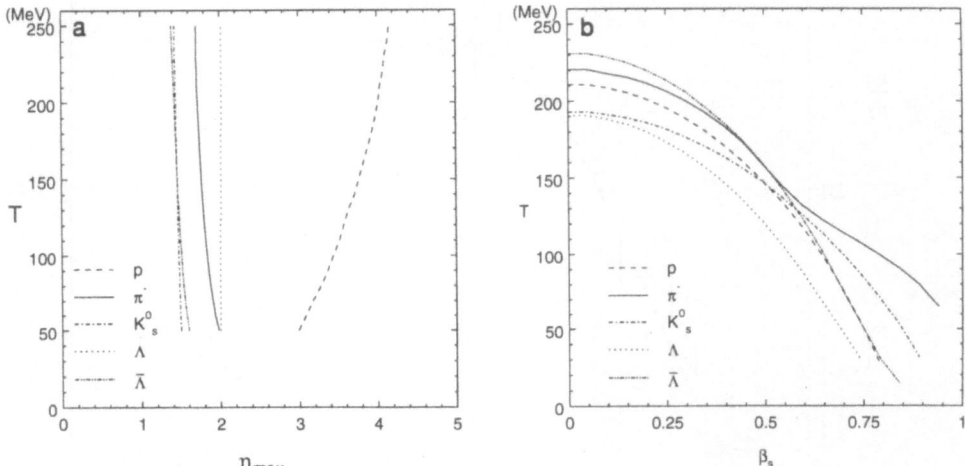

Figure 8. **a** The longitudinal flow extracted from the NA35 rapidity distributions of a variety of particle species. With the exception of the protons, whose rapidity spectrum is strongly contaminated by spectators which have suffered little loss of their beam momentum, the extracted flow rapidities are similar and independent of the choice of the temperature. **b** All fit pairs (T, β_s) of the measured m_T-spectra. Every point on these curves corresponds to a good fit of the computed m_T-spectrum (including decays) to the NA35 measurements.

The maximum flow rapidity $\pm\eta_{max}$ relative to the fireball center-of-mass can be determined independently for various different particle species, by performing independent fits with this model to the respective rapidity distributions.[16] As seen in Fig. 8a, the result is always the same, namely $\eta_{max} = 1.7 \pm 0.1$, and nearly independent of the temperature T with which the tails of the rapidity distribution are smeared out.

One can conclude that the rapidity distributions are quite reasonably described in terms of a sizeable collective longitudinal flow at freeze-out. Why, then, should there not also be some transverse flow? This question is particular relevant since we saw above (see, e.g., Eq. (18)) that the m_T spectra are strongly modified already for moderate values of the transverse flow (e.g. $v_\perp \approx 0.3\,c$). Moreover, as discussed above, the thermal picture implies frequent rescattering of the particles in the fireball. Near the surface multiple scattering acts like a pressure gradient in the system, thus causing collective transverse flow (local equilibrium causes hydrodynamic behaviour). Finally[8] the fit parameters from the thermal fits to the m_T-spectra above are much higher than Hagedorn's limiting temperature resp. the phase transition temperature to the quark-gluon plasma extracted from lattice QCD results, and are inconsistent with the freeze-out condition: at temperatures around 220 MeV the mean free path of pions in a hadron resonance gas is $< 1/3$ fm and thus much smaller than the transverse size of the collision fireball (whose diameter is of the order of 6 fm, corresponding to sulfur projectiles). Under such conditions, it is not clear how pions could freeze-out at all in the absence of transverse collective expansion.

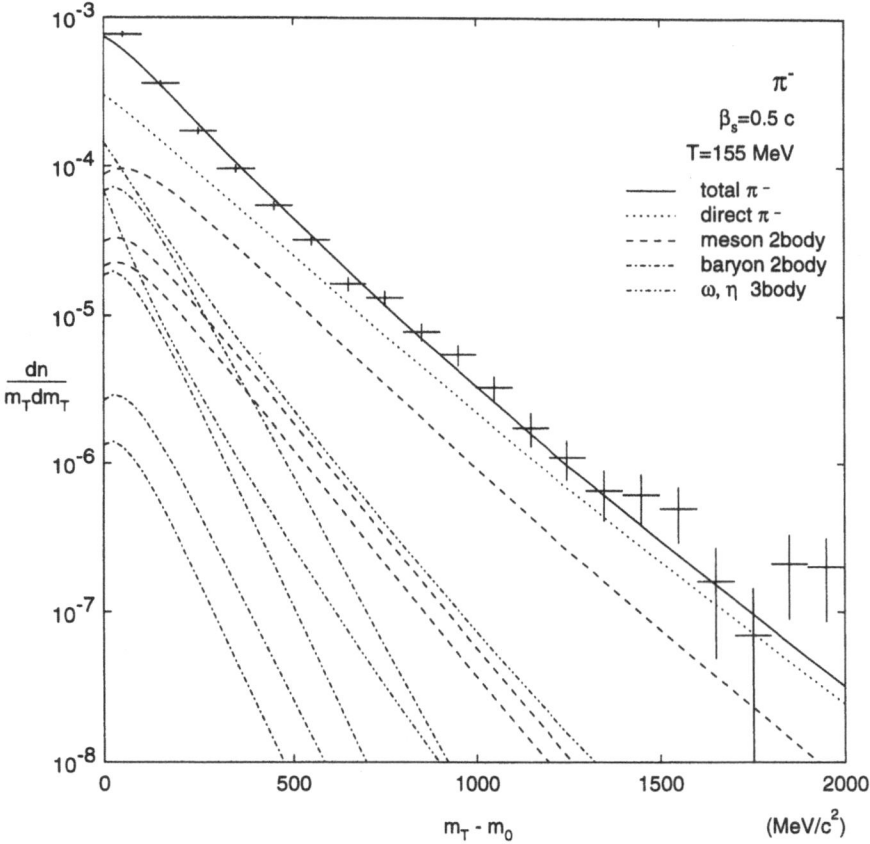

Figure 9. π^- m_T-spectrum including resonance decays and transverse flow in comparison with the same data as in Fig. 6. For the fit T was varied at fixed β_s.

Indeed, the m_T-spectra can also be very nicely fit with lower fireball temperatures by allowing for transverse collective flow. Fig. 9 shows a fit to the pion spectrum with $T = 155$ MeV and $\beta_s = 0.5c$ where β_s is the transverse flow velocity at the fireball surface for a quadratic velocity profile (such that $\langle \beta(r) \rangle = \beta_s/2$). Many more such fits with different values of T can be made by compensating lower temperatures by larger flow velocities according to Eq. (18). For larger transverse flow velocities the resulting curvature in the pion spectrum which was mentioned above even makes up for the loss of resonance decay contributions at the corresponding lower temperatures – a remarkable accident. The same mechanism of trading temperature for flow works in the case of all measured particles (see

Fig. 10), such that in all cases one obtains, within a narrow band, the same set of allowed parameter pairs (T, β_s) which give good fits to the m_T-spectra (see Fig. 8b).

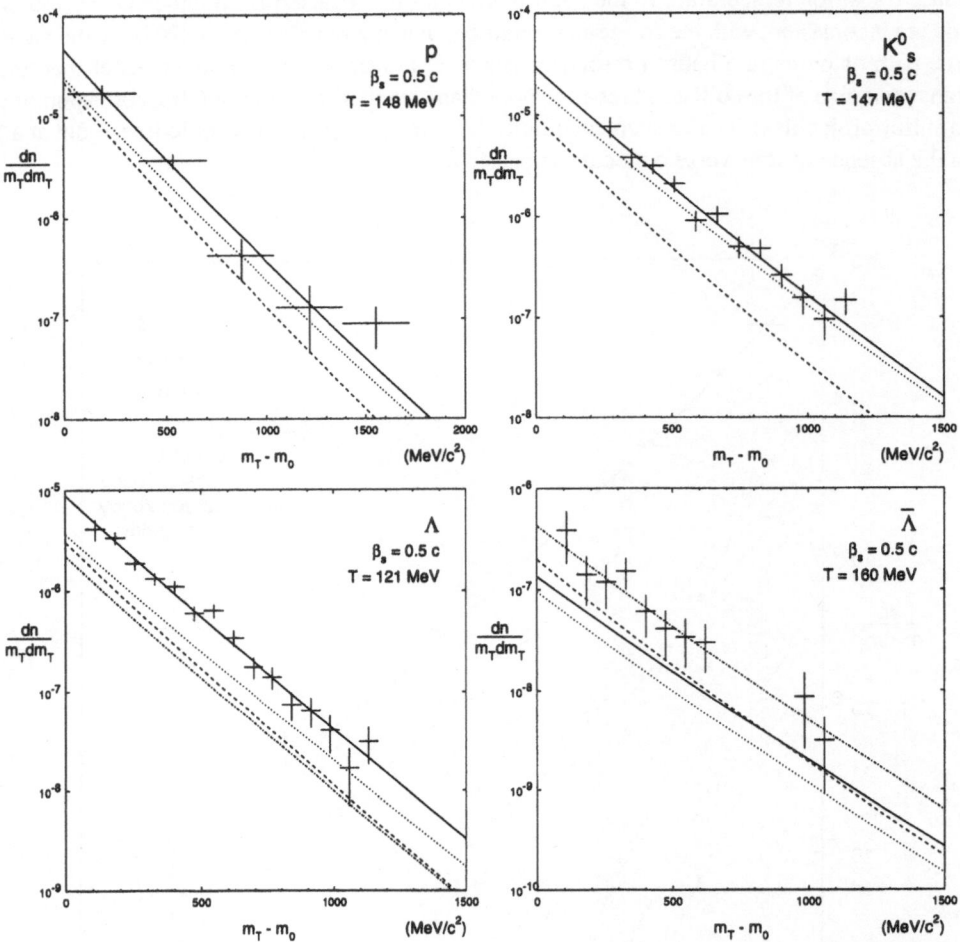

Figure 10. Thermal m_T-spectra of p, K_s^0, Λ and $\overline{\Lambda}$ including resonance decays and transverse flow in comparison with the NA35 data[21,23] from 200 A GeV S+S collisions. In all cases $\beta_s = 0.5$ was fixed and T was adjusted freely.[16]

We have to conclude that the measured m_T-spectra really fix only the apparent temperature $T_{app} = T\sqrt{(1 + v_T)/(1 - v_T)}$, rather than the fireball temperature and its expansion velocity separately. This is a very important lesson. Unless our experimental colleagues can provide us with a direct measurement of the resonance contribution to, say, the pion spectrum, which would fix the temperature, we cannot avoid resorting to theoretical models to correlate some of the data in order to separate the two effects.

A global hydrodynamic model

The crucial missing link in the above phenomenological analysis of the data is the freeze-out criterium (5). It establishes a connection between the temperature (which enters through the densities and averaged cross sections into the scattering time scale) and the expansion flow (which enters the expansion time scale). Rapidly expanding systems can freeze out earlier and at higher temperatures than slowly expanding systems. For this conclusion to be valid freeze-out must be dominated by the dynamics rather than by geometry – this turns out to be true for heavy-ion collisions, but not for hadron-hadron collisions.

We have gone through a systematic hydrodynamic study[24,15,16,14] of S+S collisions at 200 A GeV in order to investigate the following points:

(1) Are the phenomenologically extracted transverse and longitudinal flow rapidities and corresponding freeze-out temperatures mutually consistent within a hydrodynamic model for the expansion? In other words: is it possible to obtain within a 3-dimensional hydrodynamic code simultaneously freeze-out values for all these quantities as they are required by the data? The answer to this question is yes, as can be seen in Fig. 11: the calculated freeze-out values from hydrodynamic calculations with different initial values η_0 for the longitudinal flow lie on a line which intersects the phenomenological band in Fig. 8b of freeze-out values extracted from the slope of the m_T-spectra.

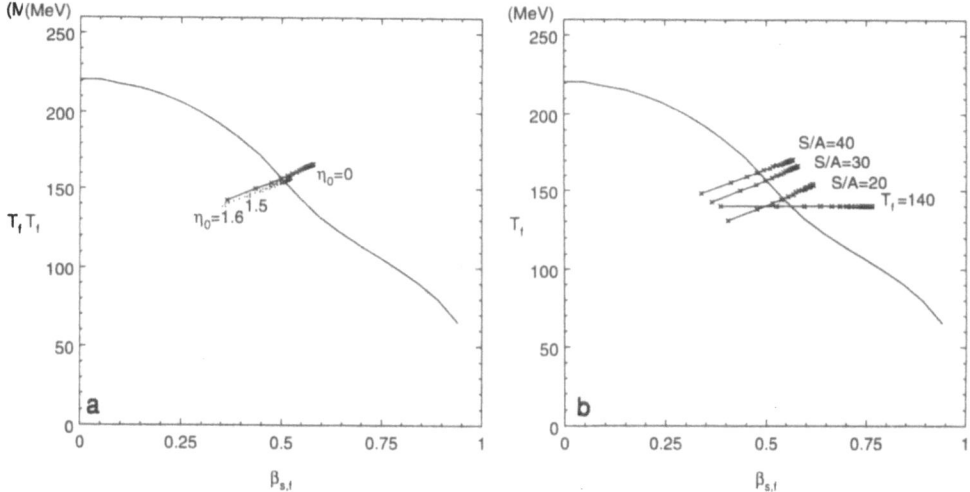

Figure 11. **a** Evidence for transverse flow: the direct correlation between T and β_s provided by the freeze-out criterium following hydrodynamic expansion from states with different initial longitudinal flow η_0 and the anticorrelation provided via Eq. (18) by the fixed m_T-slope (here we show the pion curve taken from Fig. 8b) allow to constrain the freeze-out to the area near the intersection. **b** As in **a**, but for different assumptions on the specific entropy of the expanding fireball and for reasonable variations of the freeze-out criterium (for details see [14]).

(2) Can the existence of transverse flow in the final state be established in a model-independent way? Again Fig. 11 provides a positive answer: the only way to obtain internally consistent freeze-out parameters which also fit the data is by letting the system develop transverse flow. The opposite correlation between T and β_s provided by the freeze-out criterium and by the fixed exponential slope of the measured spectra allows to single out a very narrow region of allowed (T, β_s) pairs near $T = 150$ MeV and (for S+S at 200 A GeV) $\beta_s = 0.5$. The intersection point actually favors a situation with some longitudinal flow already present in the initial state (partial transparency) over Landau-type initial conditions with $\eta_0 = 0$ – this is very much in line with Rolf Hagedorn's conceptual criticism of the latter model to which I already referred. I also find it very striking to observe the consistency of the value of

the freeze-out temperature T_f with the limiting Hagedorn temperature from hadron-hadron collisions and with the QGP transition temperature from lattice QCD!

(3) To what extent does the consistency of the transverse and longitudinal flows at freeze-out with each other and with the data constrain the initial conditions for the hydrodynamic expansion? Unfortunately here the answer is not definitive. A large range of different initial conditions for the longitudinal flow and fireball size, ranging from a very strongly compressed state with little initial flow (close to Landau's full stopping scenario) to complete transparency with rather low initial energy density a la Bjorken, all give very similar freeze-out parameters. The reason for this ambiguity is quite intriguing: Landau initial conditions provide after about 1.3 fm/c a state which is very similar to Bjorken initial conditions. Freeze-out occurs in both cases only after a critical level of *transverse* flow has had time to develop. The total time until freeze-out is quite different in the two extreme cases, but the single particle spectra provide no information on this quantity.

(4) Finally, can the equation of state for the expanding system be constrained? Again the answer is: only within loose limits. It is possible to exclude an ideal gas of massless pions without baryons or resonances, but any realistic, Hagedorn-type resonance gas cannot be distinguished from, e.g., an equation of state with a QGP phase transition. To do so would require direct information from the early stage of the collision which again is not provided by the single particle spectra.

CONCLUDING REMARKS

We have shown that a thermal picture which implements collective longitudinal and transverse flow gives an excellent parametrization of the observed hadron spectra from S+S collisions at CERN. The ambiguity in the transverse spectra between temperature and flow effects was resolved theoretically, by investigating the consistency with the kinetic freeze-out condition and with the hydrodynamical evolution prior to freeze-out. As a result, we found at least circumstantial evidence for a sizeable transverse flow in ultrarelativistic S+S collisions, of the order $\langle v_\perp \rangle \simeq 0.3\,c$. Similar results have been obtained at the lower AGS energies;[20] there an independent measurement of the freeze-out temperature was obtained by reconstructing the $\Delta(1232)$ resonance and determining the Δ/N ratio. It was confirmed that this value of T was well below T_{app} from the m_T-slope, and that the amount of transverse flow needed for compensation is consistent with the dynamical freeze-out criterium. – Confirmation of these tendencies, on a stronger and more quantitative level, in collisions between really heavy ions (Au+Au or Pb+Pb) is desirable.

Single particle momentum spectra cannot distinguish a Landau-like complete stopping scenario from Bjorken-like (partial) transparency; only theoretical consideration of the internal consistency of these two approaches, respectively, or the identification of direct electromagnetic radiation from the hot and dense initial phase can resolve this ambiguity. The spectra also show only a very weak sensitivity to the hadronic equation of state; a phase transition to a QGP phase increases mainly the total fireball lifetime, but finally the freeze-out temperature and the transverse flow which has developed are similar. Perhaps collisions at higher energies, where we dive more deeply into the new QGP phase, yield here a larger lever arm. Hanbury-Brown-Twiss measurements, which in principle can access the total lifetime of the collision region, may also help further.

I have not had time to present our first preliminary results on HBT correlations based on the hydrodynamical approach presented in these lectures. We refer the reader to Refs.[26,27] in which we make a few points of principle as to what and what not one can extract from the HBT correlation function. Again the situation is more complicated than at first thought, and the possibilities to extract information on the early stages are limited. The consistency of our hadronic picture at freeze-out will finally require direct positive confirmation from dilepton and photon spectra.

Acknowledgements

This work was supported in part by BMFT, DFG, and GSI. I also thank my collaborators E. Schnedermann and J. Sollfrank for their invaluable contributions to the work presented here.

REFERENCES

1. R. Hagedorn, "Thermodynamics of Strong Interactions", CERN 71-12, Geneva, 1971.

2. H. Grote, R. Hagedorn, and J. Ranft, "Atlas of Particle Spectra", CERN Report (black!) 1970.

3. G. Jancso et al., *Nucl. Phys.* B124:1 (1977).

4. H. Grässler et al., *Nucl. Phys.* B132:1 (1978).

5. D. Drijard et al., *Z. Phys.* C9:293 (1981).

6. M. Aguilar-Benitez et al., *Z. Phys.* C50:405 (1991).

7. R. Hagedorn, *Riv. Nuovo Cimento*, Vol. 6, Ser. 3, Nr 10 (1983).

8. J. Sollfrank, P. Koch, and U. Heinz, *Z. Phys.* C52:593 (1991).

9. M. Adamus et al., *Z. Phys.* C39:311 (1988).

10. B. Alper et al., *Nucl. Phys.* B100:237 (1975).

11. R. Hagedorn and J. Rafelski, in: "Statistical Mechanics of Quarks and Hadrons", p. 237, H. Satz (ed.), North Holland, Amsterdam, 1981; J. Rafelski and R. Hagedorn, *ibid.*, p. 253.

12. B. Müller, *Colored Chaos*, this volume, and references therein; T. S. Biró, C. Gong, and B. Müller, Duke preprint DUKE-TH-94-73 (1994)

13. F. Cooper and G. Frye, *Phys. Rev.* D10:186 (1974).

14. E. Schnedermann and U. Heinz, *Phys. Rev.* C50:1675 (1994).

15. E. Schnedermann and U. Heinz, *Phys. Rev.* C47:1738 (1993).

16. E. Schnedermann, J. Sollfrank, and U. Heinz, *Phys. Rev.* C48:2462 (1993).

17. P. J. Siemens and J. O. Rasmussen, *Phys. Rev. Lett.* 42:880 (1979).

18. E802 Coll., P. Vincent, M. Sarabura, H. Hamagaki et al., *Nucl. Phys.* A498:67c, 409c, 415c (1989); T. Abbott et al., *Phys. Rev. Lett.* 64:847 (1990)

19. K. S. Lee, U. Heinz, and E. Schnedermann, *Z. Phys.* C48:525 (1990).

20. E814 Coll., J. Barrette et al., *Phys. Lett.* B (1994), in press; P. Braun-Munzinger et al., in: "Hot and Dense Nuclear Matter", H. Stöcker (ed.), NATO ASI Series, Plenum, New York (1994), in press.

21. S. Wenig, Ph.D. thesis, Universität Frankfurt (1990), GSI-Report GSI-90-23.

22. J. W. Cronin et al., *Phys. Rev.* D11:3105 (1975)

23. NA35 Collaboration, J. Bartke et al., *Z. Phys.* C48:191 (1990).

24. E. Schnedermann and U. Heinz, *Phys. Rev. Lett.* 69:2908 (1992).

25. J. D. Bjorken, *Phys. Rev.* D27:140 (1983).

26. U. Mayer, E. Schnedermann, and U. Heinz, *Phys. Lett.* B294:69 (1992).

27. S. Chapman and U. Heinz, TPR-94-20 (1994), *Phys. Lett.* B, in press; S. Chapman, P. Scotto, and U. Heinz, TPR-94-28 (1994), submitted to *Phys. Rev. Lett.*; S. Chapman, P. Scotto, and U. Heinz, TPR-94-29 (1994), *Heavy Ion Physics (Acta Phys. Hungarica, New Series)* 1:1 (1995), in press.

PION AND KAON FREEZEOUT IN NA44

John P. Sullivan
for the NA44 Collaboration

H.Beker[1,a], H. Bøggild[2], J. Boissevain[3], M. Cherney[4], J. Dodd[5],
S. Esumi[6], C.W. Fabjan[1], D. E. Fields[3], A. Franz[1], K.H. Hansen[2],
B. Holzer[1], T. Humanic[12], B.V. Jacak[3], R. Jayanti[7,12], H. Kalechofsky[7],
T. Kobayashi[8,b], R. Kvatadze[1,c], Y.Y. Lee[7], M. Leltchouk[5], B. Lörstad[9],
N. Maeda[6], A. Medvedev[5], Y. Miake[10,e], A. Miyabayashi[9], E. Noteboom[4],
M. Murray[10], S. Nagamiya[5], S. Nishimura[6], S.U. Pandey[12], F. Piuz[1],
V. Polychronakos[11], M. Potekhin[5], G. Poulard[1], A. Sakaguchi[6],
M. Sarabura[3], K. Shigaki[1,e], J. Simon-Gillo[3], W. Sondheim[3], T. Sugitate[6],
J. P. Sullivan[3], Y. Sumi[6], H. van Hecke[3], W.J. Willis[5], and K.Wolf[10]

[1] CERN, CH-1211 Geneva 23, Switzerland.
[2] Niels Bohr Institute, DK-2100 Copenhagen, Denmark.
[3] Los Alamos National Laboratory, Los Alamos, NM 87545, USA.
[4] Creighton University, Omaha, NE, USA.
[5] Columbia University, New York, NY 10027, USA.
[6] Hiroshima University, Higashi-Hiroshima 724, Japan.
[7] University of Pittsburgh, Pittsburgh, PA 15260, USA.
[8] National Laboratory for High Energy Physics, Tsukuba 305, Japan.
[9] University of Lund, S-22362 Lund, Sweden.
[10] Texas A&M University, College Station, TX 77843, USA.
[11] Brookhaven National Laboratory, Upton, NY 11973, USA.
[12] Ohio State University, Columbus,OH 43210, USA
[a] Present address: Rome I Institute, Rome I-00185, Italy
[b] Now at Riken Linac Laboratory, Riken, Saitama 351-01, Japan.
[c] Visitor from Tbilisi State University, Tbilisi, Rep. of Georgia.
[d] Now at Tsukuba University, Tsukuba 305, Japan.
[e] University of Tokyo, Tokyo 113, Japan.

ABSTRACT

Three dimensional source size parameters measured via two-particle interferometry in experiment NA44 for 200 GeV/nucleon S+Pb show a common $1/\sqrt{mt}$ dependence for all three dimensions. This is consistent with a hydrodynamic model for an expanding source. The single particle spectra are also interpreted within a hydrodymanic model.

INTRODUCTION

The NA44 spectrometer is optimized for the study of single and two-particle particle spectra near mid-rapidity for transverse momenta below ≈ 1 GeV/c. A large fraction of all pairs in the spectrometer's acceptance are at low relative momenta, resulting in small statistical uncertainties on the extracted size parameters. In addition, the spectrometer's clean particle identification allows us to measure correlation functions for pions, kaons, and protons. This contribution will concentrate on the source size parameters determined from pion and kaon correlation functions. These size parameters will be compared to calculations from the RQMD event generator[10] and also interpreted in the context of a hydrodynamic model.[3] Finally, the measured single particle spectra will be examined from the viewpoint of hydrodynamics.

EXPERIMENTAL RESULTS

The NA44 Focusing Spectrometer[4-6] uses two dipole magnets and three quadrupoles, covering a momentum range of ±20% around its central momentum setting. For the data shown here, the central momentum settings are 4 GeV/c (lab) for pions and 6 GeV/c for kaons. The two spectrometer angle settings used for π^+ are referred to as low p_T ($< p_T > \approx$ 150 MeV/c) and high p_T ($< p_T > \approx$ 450 MeV/c). The tracking and time-of-flight uses three scintillator hodoscopes whose time resolution is ≈ 100 ps, with a Cherenkov beam counter[7] for the time-of-flight start ($\sigma \approx 35$ ps).

The NA44 data have been analyzed in terms of three components[8,9] of the two-particle momentum difference ($\vec{q} = \vec{p_1} - \vec{p_2}$). The data are analyzed in the frame in which the z-component of the pair momentum ($p_{z1} + p_{z2}$) is zero (the longitudinal center of mass system, or "LCMS"). The momentum difference is resolved into a component (q_{beam}) parallel to the beam direction and a component perpendicular to the beam direction. The perpendicular component is further resolved into q_{out} parallel to the sum of the pair momentum and q_{side}, which is perpendicular to the sum and to the beam. Three corresponding source size parameters (R_{beam}, R_{out}, R_{side}) are determined by fitting eq. 1:

$$C(q_{out}, q_{side}, q_{beam}) = A(1 + \lambda \exp(-q_{out}{}^2 R_{out}^2 - q_{side}^2 R_{side}^2 - q_{beam}^2 R_{beam}^2)) \qquad (1)$$

to measured 3D correlation functions from two different spectrometer settings. The "horizontal" focus spectrometer setting optimizes the acceptance for R_{out}, while the "vertical" focus spectrometer setting is for R_{side}. The resolution in q_{out} and q_{beam} is ≈ 15 MeV/c and in q_{side} is ≈ 30 MeV/c.

DISCUSSION

Fig. 1 compares size parameters from NA44[4-6] to size parameters calculated[10] using the RQMD event generator[1,2] to generate the single particle emission probability distribution at the point of each particle's last interaction. The two-particle correlation function is calculated from the single particle distribution using a Wigner function formalism.[11,12] In fig. 1, the calculations for low p_T π^+ agree with the NA44 results for two of three size components, but the RQMD result is larger (by 3.4σ) than the NA44 result for R_{beam}. The RQMD result also agrees with the NA44 high p_T result for two of three size parameters, put the RQMD value of R_{out} is significantly larger than the NA44 result. For K^+, the NA44 and RQMD results are in good agreement.

Analytical hydrodynamical calculations show that for collective flow with superimposed thermal motion, the radius parameters in different directions become more equal and

Table 1. Results of the m_T dependence on radius parameter

System	$\sqrt{m_T}$	$\sqrt{m_T}R_{to}$	$\sqrt{m_T}R_{ts}$	$\sqrt{m_T}R_l$
High p_T $\pi^+\pi^+$	0.68	2.0 ± 0.1	2.0 ± 0.1	2.1 ± 0.1
Low p_T $\pi^+\pi^+$	0.47	1.9 ± 0.1	2.0 ± 0.1	2.2 ± 0.1
B *Low p_T K^+K^+*	0.74	2.1 ± 0.1	1.9 ± 0.20	2.2 ± 0.2

scale as $1/\sqrt{m_T}$.[3] In table 1 the radius parameters show a $1/\sqrt{m_T}$ dependence. Such behavior is expected if a hydrodynamical expansion takes place, causing correlations between particle positions and momenta in the transverse as well as longitudinal directions.

In order to accept a hydrodynamic model as a reasonable basis for understanding the two-particle correlation data, we must also try to interpret the single particle spectra in the same framework. One such description[13] suggests that for an expanding source in thermal equilibrium, the inverse exponential slope of the m_T spectrum ($dN/dm_T^2 dy$) can be related to the temperature of the source via the expression:

$$\frac{1}{S} = \frac{1}{T}\sqrt{\frac{1 - \beta_T}{1 + \beta_T}} \tag{2}$$

where S is the inverse exponential slope from a fit of $\exp\left(-m_T/S\right)$ to the m_T spectrum, β_T is the transverse expansion velocity, and T is the true temperature of the source. Using this expression, the measured values of S for π^+, K^+, p^+, p^- can be translated into temperatures. A consistent temperature would reinforce the hydrodynamic interpretation of the data.

An expansion velocity, $\beta_T = 0.35$ has been assumed to translate the measured slope parameters to "temperatures" using eq. 2. The values are roughly consistent for K^+, p^+, p^-, but the "temperature" for pions is lower. However, the pion measurement is mainly at low p_T, where the resonance contribution is significant.[10] These resonances contribute a component with a lower apparent temperature and could explain the differences. To examine this possibility, the RQMD event generator was once again used. In the region of the NA44 data, inverse slope of the (RQMD) pion m_T spectrum is about 12% larger than the inverse slope of the (RQMD) spectrum of pions from which pions from η, η', and $\omega(783)$ decay

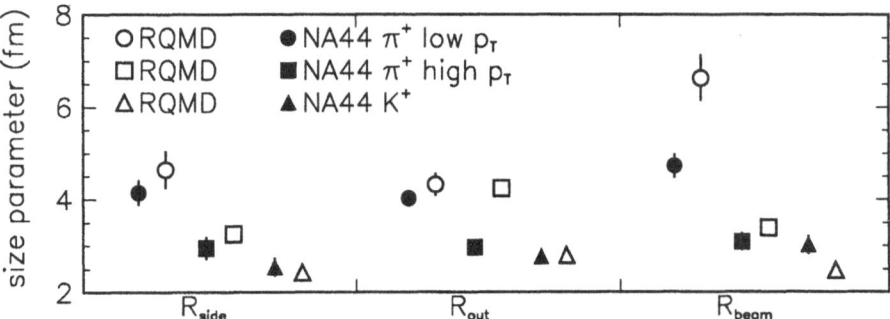

Figure 1. The size parameters, for 200 GeV/nucleon S+Pb, based on fits to 3 dimensional correlation functions, for low p_T π^+ (solid circles), high p_T π^+ (solid squares), and for K^+ pairs (solid triangles) from NA44. The corresponding fit parameters from the correlation functions calculated from RQMD are shown as open symbols.

Table 2. Preliminary NA44 data: For different particle types, the inverse exponential slope parameters (S) from fits to the m_T spectrum over the specified and the "temperatures" calculated assuming $\beta_T = 0.35$ and equation 2.

particle	M_T range fit (GeV)	S (MeV)	T (MeV)
π^+	0 - 0.25	144±16	100±11
K^+	0 - 0.50	206±12	143±8
p^+	0 - 0.90	190±10	132±7
p^-	0 - 0.90	183±10	127±7

have been excluded. This factor can explain 1/2 to 1/3 of the difference between the pion "temperature" in table 2 and those for other particles. However, final data and fits over a more consistent m_T range are needed before drawing final conclusions.

CONCLUSIONS

We have investigated the m_T dependence of boson correlation functions in S+Pb collisions. The transverse radii for pions and kaons show a common $1/\sqrt{m_T}$ dependence. The data indicates the presence of strong position-momentum correlations in the transverse direction, arising from collective flow. This will be further investigated by careful comparison of hydrodynamical model calculations and single particle data, other (e.g. p-p) two particle correlation data, and d/p ratios.

REFERENCES

1. H. Sorge, H. Stöcker and W. Greiner, Nucl Phys. **A498**, 567c (1989); H. Sorge *et al*. Z. Phys. **C47**, 629 (1990).

2. H. Sorge *et al*. Phys. Lett. **B289**, 6 (1992).

3. T. Csörgo and B. Lorstadt, Lund University preprint LUNFD6/(NFFI-7082), submitted to Phys. Rev. Lett. 1994.

4. H. Becker *et al.*, Phys. Lett. **B302**, 510 (1993).

5. H. Bøggild *et al.*, CERN-PPE/(94-75) (1994), in press, Z. Phys.

6. H. Becker *et al.*, submitted to Phys. Rev. Lett. (1994).

7. N. Maeda *et al*, Nucl. Inst. Meth. **A346**, 132 (1994).

8. S. Pratt, Phys. Rev. **D33**, 1314 (1986).

9. G. Bertsch, M. Gong, M. Tohyama, Phys. Rev. **C37**, 1896 (1988).

10. J. P. Sullivan *et al.*, Phys. Rev. Lett. **70**, 3000 (1993).

11. T. Csörgo *et al.*, Phys. Lett. **B241**, 301 (1990).

12. S. Pratt, T. Csörgo, J Zimányi, Phys.Rev **C42**, 2646 (1990).

13. U. Heinz, this conference.

THE DUAL PARTON MODEL
AND HADRON PRODUCTION AT COSMIC RAY ENERGIES

J. Ranft

INFN, Lab. Naz. di Frascati
I–00044 FRASCATI (Roma), Italy

INTRODUCTION

More than 25 years ago I had the privilege to collaborate with R.Hagedorn on the Thermodynamic Model for hadron production in hadron–hadron collisions.[1] This is a model, which is still very relevant for the present and future studies of heavy ion collisions. In this collaboration I learned many things on theoretical particle physics and especially on the subject of multi–hadron production. Hadron production in these years was mainly studied in what we now call the fragmentation region. Many of the topics, I will cover in this contribution like the rise of average transverse momenta with the collision energy or with Feynman–x or the rise of the fraction of heavy hadrons with the collision energy and with the transverse momentum were already studied at these times with the Thermodynamic Model.

A hadron production model to be used at Cosmic Ray energies should take into account all possible information from fixed target experiments and collider experiments at accelerators. There are however important differences: For studying the Cosmic Ray cascade, the main interest is in the forward fragmentation region of hadron–nucleus and nucleus–nucleus collisions. Best studied at accelerators is the central region in hadron–hadron collisions.

In this paper we will discuss hadron production in the framework of the Dual Parton Model with emphasis in the fragmentation region.

the Dual Parton Model (DPM) (a recent review is given in Ref.[2]) has been very successfully describing soft hadronic processes. Observations like rapidity plateaus and average transverse momenta rising with energy, KNO scaling violation, transverse momentum–multiplicity correlations and *minijets* pointed out, that soft and hard processes are closely related. These properties were understood within the two–component Dual Parton Model.[3,4] The hard component is introduced applying lowest order of perturbative hard constituent scattering.[5]

The Dual Parton Model provides a framework not only for the study of hadron–hadron interactions, but also for the description of particle production in hadron–nucleus and nucleus–nucleus collisions at high energies. Within this model the high energy projectile undergoes a multiple scattering as formulated in Glaubers approach; particle production is again realized by the fragmentation of colorless parton–parton chains constructed from the quark content of the interacting hadrons.

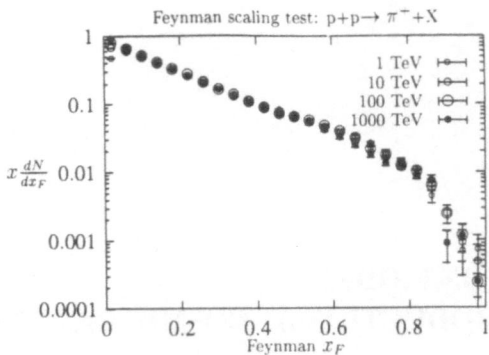

Figure 1. Test of Feynman scaling in the production of π^+ in proton–proton collisions (DPMJET–II).

THE DUAL PARTON MODEL

The Dual Parton Model for hadron–hadron collisions as used here has been fully described in.[3,4] The application to hadron–nucleus and nucleus–nucleus collisions, to Cosmic Ray hadron production and to the Cosmic Ray cascade was described in.[6,7] These papers and further References given in them should be consulted for the details of the models.

In nuclear collisions, the partons at the sea and valence chain ends carry transverse momenta from different sources: (i) The intrinsic parton transverse momentum in the hadron. (ii) A transverse (and longitudinal) momentum resulting from the Fermi motion of the nucleons inside the nucleon. These first two kinds of transverse momentum were implemented into DTUNUC from the beginning. (iii) During the passage of the chain end partons through nuclear matter, they suffer nuclear multiple scattering which changes (usually increases) their transverse momenta.

The multiple scattering of partons is known since a long time to be responsible for the so called *Cronin effect*[8] of particle production at large transverse momentum on nuclear targets. A similar enhancement of particle production in hadron–nucleus and nucleus-nucleus collisions compared to hadron–hadron collisions has been observed in many experiments already at rather modest $p_\perp$.

At large $p_\perp$ this effect can be studied calculating the parton scattering pertubatively. Our rather low $p_\perp$ sea chain ends might be considered as the low $p_\perp$ limit of perturbatively scattered partons. We apply to them and to the hard scattered partons multiple scattering taking into account their path length inside the nuclear matter and adjust the parameters in such a way, that the measured $p_\perp$ ratios at rising transverse momenta are approximately reproduced by the code DTUNUC.

HADRON–HADRON AND HADRON–NUCLEUS COLLISIONS WITH DPMJET–II

DPMJET, version II has been constructed on the basis of the codes DTUNUC–1.04[9] and DTUJET93.[4] Here we report on the study, using DPMJET–II of hadron–hadron, hadron–nucleus and nucleus–nucleus collisions in the Cosmic Ray energy region.

The relevance of an event generator like DPMJET–II based on the Dual Parton Model for hadron production cross sections in the Cosmic Ray energy region can only be claimed, if the model (i) agrees to the best available data in the accelerator energy range and (ii) shows a smooth behaviour in the extrapolation to higher energies.

In order to see, whether data in the accelerator energy range with projectile energies well below 1 TeV are relevant at all, we study first the Feynman scaling behaviour of the model. In Fig. 1 we study the Feynman scaling of the secondary π^+ mesons. In most of the x_F region , say for $0.05 \leq x_F \leq 0.8$, we find Feynman scaling indeed very well satisfied in the Dual Parton Model. The violations of Feynman scaling, which occur around $x_F = 0$ are connected with the well known rise of the rapidity plateau for all kinds of produced particles. This violation is only absent for the leading baryon, where the x_F distribution vanishes at $x_F = 0$. We find also a strong violation of Feynman scaling for secondary nucleons around $x_F = 1$. This is connected with the diffractive component, which clearly violates Feynman scaling.

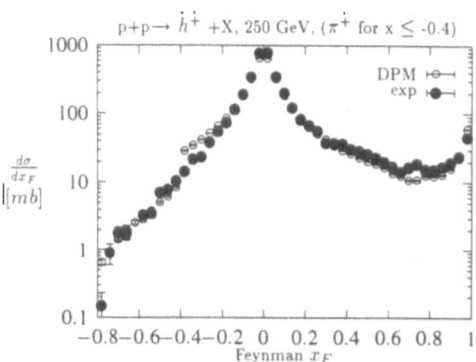

Figure 2. Comparison of Feynman–x distributions. The EHS–NA22 Collaboration[10] and the Dual Parton Model DPMJET–II.

The EHS–NA22 Collaboration[10] has data on $d\sigma/dx_F$ in 250 GeV proton–proton collisions. In Fig. 2 we compare with the production of positively charged hadrons for $x_F \geq$ -0.4 and with π^+ production for $x_F \leq$ -0.4, again the agreement is reasonable. In the projectile fragmentation region at large x_F this distribution is dominated by the leading protons from diffractive and nondiffractive events.

In the fragmentation region the transverse momentum distributions and average transverse momenta are known to depend strongly on Feynman x_F. This effect is known under the name *seagull effect*. In Fig. 3 we compare DPMJET–II with data on the seagull effect measured by the EHS–RCBC Collaboration at 360 GeV and find a reasonable agreement.

In the Cosmic Ray cascade beyond the first generation the interactions of secondary hadrons, mainly Pions and Kaons are as important or even more important than the interaction of nucleons. The Dual Parton Model can be constructed for all hadronic projectiles.

The Glauber model, which is part of DPMJET–II allows to calculate the inelastic hadron–nucleus cross sections. What we need for this calculation is the nuclear geometry and the elementary hadron–nucleus scattering amplitude.

The same information is also needed to construct the inelastic events and indeed usually each run of DPMJET starts with a calculation of the inelastic cross section.

In order to understand the relevance of accelerator data on particle production in hadron–nucleus collisions for the Cosmic Ray cascade, we study again the Feynman scaling behaviour of p+Air→ π^+ +X. As above in proton–proton collisions, we find again, that Feynman scaling

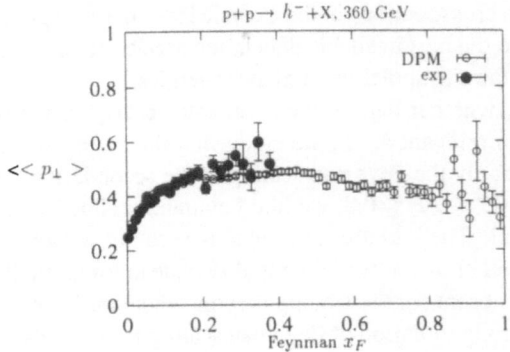

Figure 3. Comparison of the seagull effect. EHS–RCBC Collaboration[11] and the Dual Parton Model DPMJET–II.

is very well satisfied in most of the x_F region. The change of the Feynman x_F distributions with the nuclear target can be described by $\alpha(x_F)$ representing the cross section as

$$x_F \frac{d\sigma}{dx_F}^{p-A} = A^{\alpha(x_F)} x_F \frac{d\sigma}{dx_F}^{p-N} \tag{1}$$

For the transition p–p to p–Air this $A^{\alpha(x_F)}$ behaviour is not relevant, since we know that this kind of extrapolating h–A total cross sections to p–p does not give the correct p–p total cross section. Usually $\alpha(x_F)$ is determined using data for two or more different target nuclei without ever considering p–N collisions.

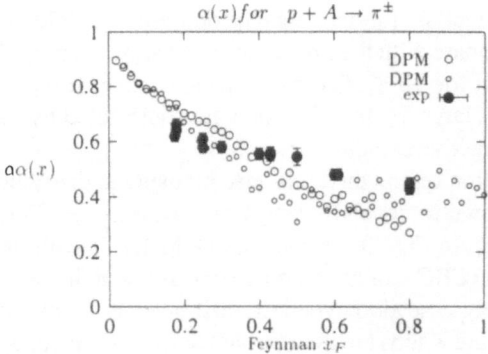

Figure 4. The nuclear dependence of the Feynman–x distribution in hadron–nucleus collisions. Barton et al[12] and the Dual Parton Model DPMJET–II .

The results of double differential cross sections for inclusive hadron production in hadron–nucleus collisions have been represented in the form

$$E \frac{d^3\sigma}{d^3p}^{p-A} = A^{\alpha(x_F, p_\perp)} E \frac{d^3\sigma}{d^3p}^{p-N} \tag{2}$$

With data on two different target nuclei, one can extract $\alpha(x_F, p_\perp)$ without the knowledge of $Ed^3\sigma/d^3p^{p-N}$. The data of Barton et al.[12] at 100 GeV and at a transverse momentum $p_\perp = 0.3$ GeV/c were used to get $\alpha(x_F)$ (in reality : $\alpha(x_F, p_\perp = 0.3 GeV/c)$). In the Monte Carlo calculation it is difficult to get such a good statistics at fixed $p_\perp$, to extract meaningful $\alpha(x_F)$ values. This is just possible for single differential distributions in x_F. In Fig. 4 we compare the $\alpha(x_F)$ as obtained by Barton et al.[12] for pion production at $p_\perp = 0.3$ GeV/c with $\alpha(x_F)$ obtained from DPMJET–II results for all charged hadrons integrated over all $p_\perp$. The agreement in the x_F region of overlap is reasonable. For $x_F \rightarrow 0$ in the Dual Parton Model the limiting $\alpha(x_F)$ value is 1. This is actually also obtained from DPMJET–II. For large values of x_F the limiting $\alpha(x_F \rightarrow 1)$ for the data as well as for the Monte Carlo seems to be around 0.4.

The agreement with these $\alpha(x_F)$ data is the strongest point for the claim, that the Dual Parton Model in the form of the DPMJET–II event generator gives a good description of the nuclear dependence of hadron production in the fragmentation region. We stress however once again, these $\alpha(x_F)$ data are only for fixed $p_\perp$ and it would be highly desirable to obtain better data for the change of hadron production from proton–proton collisions to collisions of protons with light target nuclei.

In order to show the changes in the transverse momentum distributions from p–p to p–A collisions one uses the $\alpha(p_\perp)$ representation

$$E\frac{d^3\sigma^{p-A}}{d^3p} = A^{\alpha(p_\perp)}E\frac{d^3\sigma^{p-N}}{d^3p} . \tag{3}$$

In Fig. 5 we compare DPMJET–II with $\alpha(p_\perp)$ data from Garbutt et al.[13] The data are for identified kinds of secondary hadrons. In the model calculation we present $\alpha(p_\perp)$ only for π^+ and K^+. We find a rather good agreement with the data for π^+, The $\alpha(p_\perp)$ calculated for K^+ are systematic above the values for π^+, but they stay in the region with good statistics below the data for the K^+.

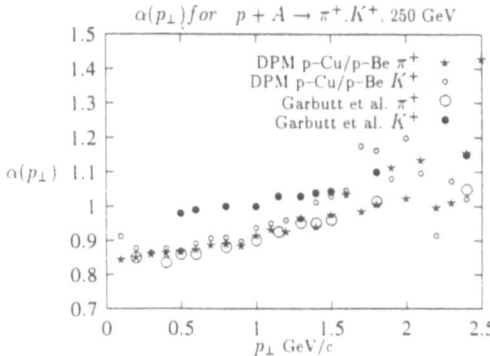

Figure 5. The nuclear dependence ot the $p_\perp$–distributions in proton–nucleus collisions. Garbutt et al.[13] and the Dual Parton Model DPMJET–II.

DPMJET–II AT COSMIC RAY ENERGIES

In[7] we report about the application of DPMJET–II to study the cosmic ray cascade. Here we give only some general properties of DPMJET–II in this energy range.

There is no scientific reason, not to call the two–component Dual Parton Model (the two components are the soft pomeron and the hard pomeron or the minijets) also a minijet model. Minijet models too have a soft and a hard component. The reason not to use the term minijet model for DPMJET is connected with the fact, that the name minijet model so far was only used for models which use a critical pomeron with an intercept of exactly one. In such a minijet scheme, it is then claimed, all the rise of the cross sections with energy is due to the rise of the minijet cross sections. This is not so in our model, therefore we avoid to use the name minijet model.

The supercritical pomeron was used in the two–component DPM from the beginning,[14] while the so called minijet models use the critical pomeron with $\alpha(0) = 1$ from Durand and Pi[15] over Gaisser and Halzen[16], SIBYLL[17] up to HIJING.[18]

There are important differences which result from this different approach:

(i) Both kinds of models determine the free parameters of their model in a fit to total, inelastic and elastic cross sections. Both models obtain acceptable fits, we have reported even about the fits using a critical pomeron elsewhere,[19] but of course, if at the end of the fit we treat the pomeron intercept $\alpha(0)$ as a free parameter instead to fix it to $\alpha(0) = 1$, the fit improves and in all situations (fits using different parton structure functions to calculate the minijet cross sections) we obtain the intercept larger than 1. namely $\alpha(0) \approx 1.07$. These better fits to the data are our main argument for the continuing presence and even rise of soft hadron production at the highest energies.

(ii) Due to these different starting points the chain structure of the models differ: In both models we have a pair of soft valence–valence chains (resulting from cutting one soft pomeron) and in both models we have minijets. Only in the two component Dual Parton Model we have in addition soft sea–sea chains with soft sea quarks at their ends. The number of these chains increases with energy and a substantial part of the rise of the multiplicity and rapidity plateau results from this mechanism.

(iii) The x–distributions of soft sea quarks are determined by the Regge behaviour, which is for soft sea as well as valence quarks like $1/\sqrt{x}$. The minijets are calculated from the deep inelastic structure functions with (depending on the parametrization for the structure functions used) a behaviour like $1/x$ or $1/x^{1.5}$. In the Dual Parton Model the Feynman x_F distributions resulting from fragmenting valence chain ends (which dominate at small energy) and from fragmenting soft sea chain ends do not differ, this is the source of the excellent Feynman scaling and the nearly energy independent spectrum weighted moments. In the minijet models all chains except the single valence chain pair, which dominates at low energy, are minijets with the much softer x–distribution. Therefore in these models Feynman scaling is more strongly violated and the spectrum weighted moments decrease with the collision energy. The rise of the minijet component in the Dual Parton Model leads of course to the same effect, this effect is however smaller, since not all of the rise of particle production is due to the minijets.

In Table 1 we compare the Z_π and Z_K moments calculated with DPMJET–II in p–Air collisions with the ones resulting from HEMAS[20] and SIBYLL.[17] The agreement of the moments, especially the ones from DPMJET–II and from SIBYLL is certainly much better, than expected from the errors of the experimental data used to tune the parameters of the models.

SUMMARY

The excellent Feynman scaling found with DPMJET in large parts of the x_F–region in hadron–hadron and hadron–nucleus collisions gives us the confidence, that accelerator data on Feynman x_F–distributions in the projectile fragmentation region are indeed very relevant for applications in the Cosmic Ray energy region.

Table 1. Comparison of Z_π and Z_K moments in p–Air collisions between DPMJET–II, HEMAS[20] and SIBYLL.[17]

Energy [TeV]	DPMJET Z_π	HEMAS Z_π	SIBYLL Z_π	DPMJET Z_K	HEMAS Z_K	SIBYLL Z_K
1	0.067	0.061	0.072	0.0098	0.0104	0.0073
10	0.069	0.057	0.068	0.0099	0.0113	0.0071
100	0.068	0.056	0.067	0.0102	0.0116	0.0070
1000	0.066	0.056	0.066	0.0101	0.0123	0.0070

The model provides hadron–hadron total, inelastic, elastic and diffractive cross sections consistent with accelerator data. The hadron–Air cross sections derived from this are consistent with hadron–Air cross sections extracted from Cosmic Ray experiments. The model provides also all the necessary cross sections to study nucleus–nucleus collisions in the Cosmic Ray cascade.

As a consequence from the excellent Feynman scaling in the model we find spectrum–weighted moments Z_π and Z_K for hadron–Air collisions, which remain rather constant with increasing collision energy. These moments for h–Air collisions are however smaller than the corresponding moments in hadron–hadron collisions.

The fraction of the primary energy carried by the leading particles in the collision decreases with energy and with the mass of the target nuclei. A large part of this decrease is due to the decreasing fraction of diffractive (single diffractive and double diffractive) events with rising energy and rising target mass.

The model incorporates the Cronin effect and shows a strong seagull effect. Correspondingly the average transverse momenta $< p_\perp >$ rise with the collision energy (mainly due to the rise of the minijet production cross section), with the mass of the nuclear target and projectile and with rising Feynman x_F.

It is important, that the model is able to give a good description of hadron production in nucleus–nucleus collisions. Due to the large fraction of nuclei in primary Cosmic Rays nucleus–Air collisions are of great importance in the Cosmic Ray cascade.

Acknowledgements

The author acknowledges the long and fruitful collaboration with R.Hagedorn on the Thermodynamic Model for hadron production in hadron–hadron collisions. He acknowledges the collaboration with P.Aurenche, F.Bopp, A.Capella, R.Engel, A.Ferrari, H.J.Möhring, C.Pajares, D.Pertermann, S.Roesler, P.Sala and J.Tran Thanh Van on different aspects of the Dual Parton model. The collaboration with G.Battistoni and C.Forti on different aspects of the Cosmic Ray application of the model was essential for the progress of the studies presented. The Author acknowledges the hospitality of Prof. L.Mandelli in Milano and Prof. E.Iarocci and Prof. M.Spinetti in Frascati , where the presented calculations were performed and he thanks INFN for supporting these studies.

REFERENCES

1. R.Hagedorn, J.Ranft, Suppl. Nuv Cim. **VI**, 169 (1966).

2. A. Capella, U. Sukhatme, Chung I Tan and J. Tran Thanh Van, Phys. Rep. **236**, 225 (1994).

3. P. Aurenche, F. W. Bopp, A. Capella, J. Kwiecinski, M. Maire, J. Ranft and J. Tran Thanh Van, Phys. Rev. **D45**, 92 (1992).

4. F. W. Bopp, D. Pertermann, R. Engel and J. Ranft, Phys. Rev. **D 49**, 3236 (1994).

5. B. L. Combridge, J. Kripfganz and J. Ranft, Phys. Lett. **70B**, 234 (1977).

6. J.Ranft, Frascati preprint, subm. to Phys.Rev.D 1994.

7. G.Battistoni, C.Forti and J.Ranft, Frascati preprint, to be published 1994.

8. J. W. Cronin et al., Phys. Rev. **D 11**, 3105 (1975).

9. J. Ranft, A. Capella and J. Tran Thanh Van, Phys. Lett. **B320**, 346 (1994).

10. M.Adamus et al., EHS–NA22 Collaboration, Z. Phys. **C 39**, 311 (1988).

11. J.L.Bailly et al., EHS–RCBC–Collaboration, Z. Phys. **C 35**, 295 (1987).

12. D.S.Barton et al., Phys. Rev. **D 27**, 2580 (1983).

13. D.A.Garbutt et al. , Phys. Lett. **67B**, 355 (1977).

14. A. Capella, J. Tran Thanh Van and J. Kwiecinski, Phys. Rev. Lett. **58**, 2015 (1987).

15. J. Durand and H. Pi, Phys. Rev. Lett. **58**, 2015 (1987).

16. T.K.Gaisser and F.Halzen , Phys. Rev. Lett. **54**, 1754 (1987).

17. R.S.Fletcher, T.K.Gaisser, P.Lipari and T.Stanev, Preprint January 1994 1994.

18. X. N. Wang and M. Gyulassy, Phys. Rev. **D44**, 3501 (1991).

19. R. Engel, F. W. Bopp, D. Pertermann and J. Ranft, Phys. Rev. **D46**, 5192 (1992).

20. C.Forti, H.Biloken, B.d'Ettorre Piazzoli, T.K.Gaisser, L.Satta and T.Stanev, Phys. Rev. **D 42**, 3668 (1990).

PARTICLE PRODUCTION AT THE AGS

S. G. Steadman, P. R. Rothschild, T. W. Sung, and D. Zachary

Physics Department and Laboratory for Nuclear Science
Massachusetts Institute of Technology
Cambridge, MA 02139, USA

OVERVIEW

The systematic behavior of particle production at the AGS has been extensively studied in order to discern if a signal for the formation of the Quark-Gluon Plasma is present in the high baryon density environment created at these energies and to test the adequacy of model descriptions. Beginning with oxygen beams in 1987 and culminating with the commissioning of the Au beam in 1992 there is now a wealth of data available for a wide range of bombarding conditions from pA to Au-Au collisions at AGS energies. Only a few features of the data will be discussed here, especially those features which need a better theoretical understanding.

Rapidity distributions for central Si+Au collisions have been obtained for a wide variety of particle species by the E802 Collaboration, as shown in Fig. 1, with yields spanning four orders of magnitude. Central collisions comprise the upper 7% of the inelastic cross-section. There are also preliminary measurements of Λ and $\bar{\Lambda}$ production at mid-rapidity for a centrality cut of 20%, although these yields require large acceptance corrections for the detector. Furthermore, the $\bar{\Lambda}$ yield is determined by assuming a similar transverse mass distribution to that for Λ production. Recently,[1] the E810 collaboration reported on anti-cascade production, which is also included on this plot. Both the RQMD (version 1.08) and ARC (version 1.9.5) codes generally predict the right magnitudes for the yields for the different particles, although the RQMD results overpredict the pion production by about 30%, a well known feature.

MESONS

By measuring the forward energy in the ZCAL calorimeter (which has an opening of 0.6 deg about the beam direction) the number of spectators (and hence participants) can be inferred, where we have assumed a clean-cut collision geometry in each event, by dividing the forward energy by the incident energy per nucleon. For Au-Au collisions at an incident energy of 11.6 A.GeV/c, the E802 collaboration has measured the fiducial yields[2] for π^+ and K^+ production for a variety of ZCAL energy cuts, as shown in Fig. 2. So far only the limited 1992 data set has been analyzed, consisting of about 100 hours of beam on target. The fiducial yield is the yield measured in the rapidity interval 0.6-1.6 (mid-rapidity) and doubled (from symmetry). This accounts for at least 90% of the total yield. One observes a linear increase for both π^+ and K^+ production as a function of the number of participants. Both the RQMD and ARC codes predict a nonlinear (increasing) rise with the number of participants, which is also the intuitive picture given the likelihood for increased particle production in

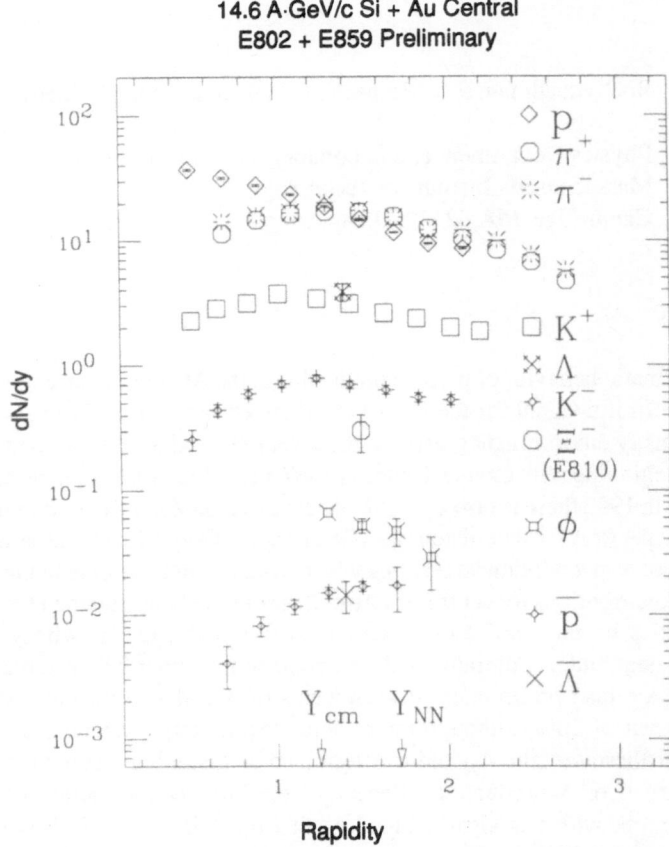

Figure 1. The measured particle rapidity distributions for central Si+Au collisions at 14.6 A GeV/c. All the data shown are from the E802 collaboration, except for the E810 measurement of anti-cascades.

multiple collisions. The saturation in pion production likely results from the large absorption of the produced pions by the nuclear matter they must pass through. Understanding the kaon production systematics is harder, given their lower absorption cross section. Definitive statements await the much better statistics that will be available upon the completed analysis of the 1993 data set. This linear behavior implies a K^+/π^+ yield of 0.21 ± 0.03 in these Au-Au collisions for collision zones larger than ≈ 60 baryons.

A salient feature in the data that has eluded an understanding is the rapidity dependence of K^+/K^- production. The K^+ distributions are broader than the K^- distributions, a feature already seen in p-A collisions. The remarkable aspect of this is the independence of this ratio for a variety of collision conditions, from peripheral Si+Al to central Si+Au and even to central Au-Au collisions (albeit with poor statistics), as shown in Fig. 3. This is surprising,

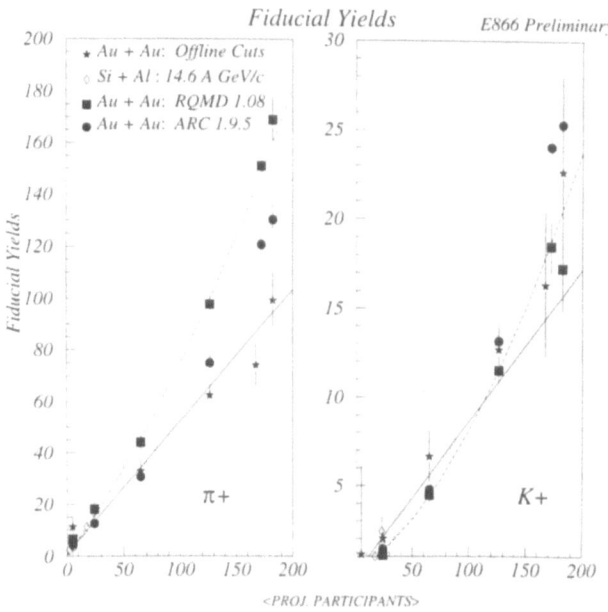

Figure 2. The production of π^+ and K^+ as a function of the number of projectile participants in Si+Al and Au+Au collisions. The number of participants is inferred from the forward energy as measured in ZCAL.

because it is thought that the K^+ and K^- production arise from different mechanisms, with K^+ arising from associated production with Λ and K^- production more by direct K^+K^- pair production. With the vastly larger number of multiple collisions in the large systems it seems surprising that the production ratio should remain a constant.

Another interesting feature of kaon production that is not yet understood is the sharp rise in yield at very low transverse mass, below ≈ 20 MeV in m_t, observed in Si+Pb collisions by the E814 Collaboration. It should be pointed out that a similar sharp rise towards 0 m_t in the same mass range is also seen in the π^- production, as reported by Hemmick, *etal*.[3]

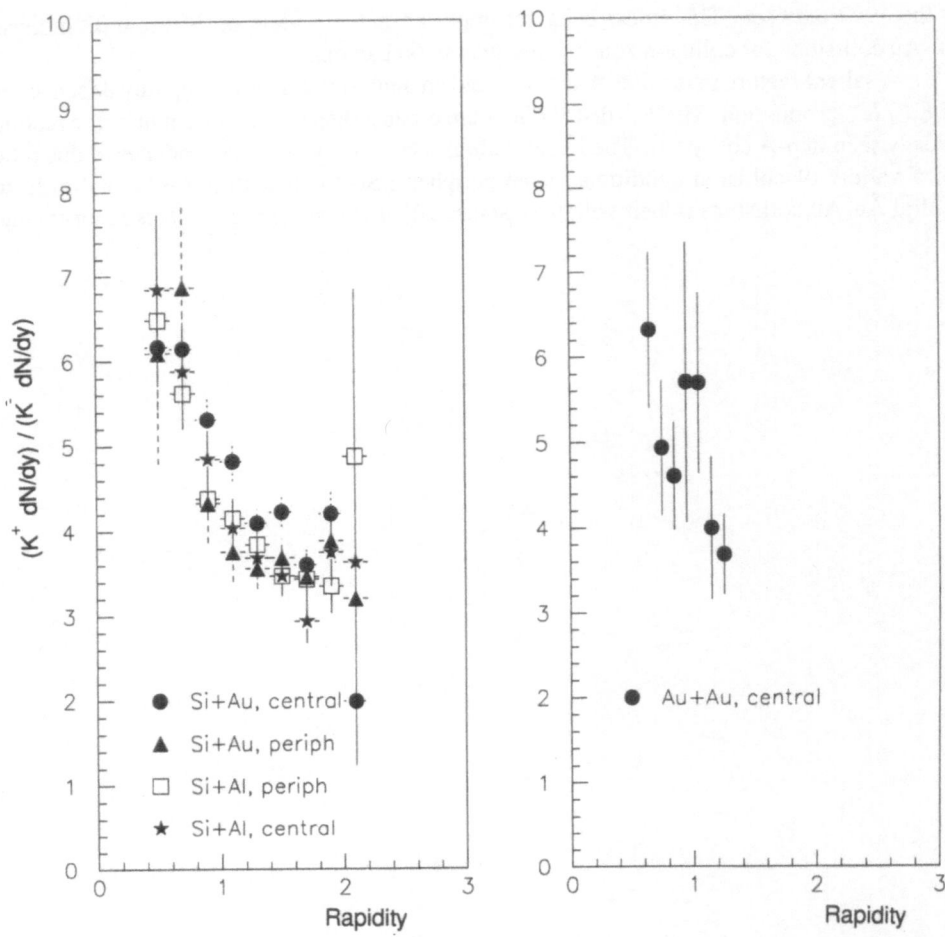

Figure 3. The rapidity distribution for the K^+/K^- production ratio measured by the E802 collaboration.

ANTI-BARYONS

The anti-baryon production is particularly interesting, as all the quarks are produced ones, with none coming from the primordial matter, so that one is unlikely to obtain enhanced production by multiple scattering. It has thus been proposed that strongly enhanced antiproton production would be a good signal for QGP production.[4] However, the anti-protons are also strongly absorbed in the nuclear medium, so the interpretation of their yield is complicated. Using the second-level trigger within E859, which allows on-line tracking and particle ID within 40 μsec, high statistics rapidity distributions are now available for antiproton production for both central and minimum bias Si+Al and Si+Au collisions, as shown in Fig. 4.

For central Si+Au collisions it is noteworthy that the rapidity distribution does not peak at the cm rapidity of 1.25; rather it peaks at a value closer to the nucleon-nucleon rapidity of 1.7. This may arise from the larger absorption of $\bar{p}$ at target rapidities.

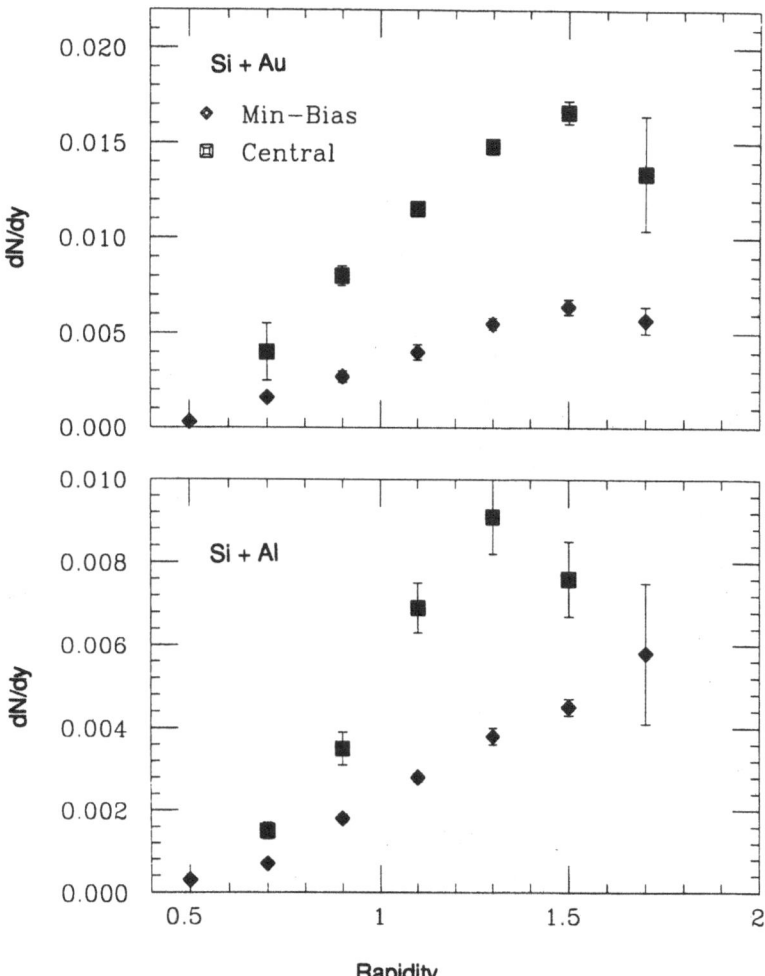

Figure 4. Antiproton production rapidity distributions for central and minimum bias Si+Al and Si+Au collisions at 14.6 GeV/c.

Neither ARC nor RQMD can consistently represent the four data sets. Assuming a formation time of 1.5 fm/c, RQMD is able to account well for the minimum bias yields but underpredicts the yield for central collisions. ARC has a screening mechanism that leads to reduced absorption for central collisions, in agreement with the data, but predicts excess yield for the minimum bias data.

The E802 Collaboration has made a preliminary measurement of $\overline{\Lambda}$ production for central Si+Au collisions. Specifically, they have determined the $\overline{\Lambda}$ to Λ ratio within the spectrometer acceptance. If one assumes a similar p_t distribution for both $\overline{\Lambda}$ and Λ, then one obtains a $\overline{\Lambda}$ yield that implies that about 60% of the $\overline{p}$'s arise from $\overline{\Lambda}$ decay. The absorption for $\overline{\Lambda}$'s in the nuclear medium is thought to be significantly less than for p-bar's. Taking proper account of the $\overline{\Lambda}$ decays may help to "explain" the $\overline{p}$ yields. At present the codes have yet to produce $\overline{\Lambda}$'s with sufficient statistics to make meaningful comparisons with the data.

CONCLUSION

The large amount of available data now places stringent tests on the model calculations. Better statistics data will soon be in hand to confirm the linear dependence of pion and kaon production with the number of participants. The universal rapidity dependence of the K^+/K^- ratio is a simple feature of the data that requires a theoretical understanding. Understanding anti-baryon production has proven to be more difficult, due to the potential feeding contribution of decaying resonances. Both better data and better statistics Monte Carlo calculations are required to make meaningful comparisons. There are extensive amounts of data with the Au beam now being analyzed and definitive results should be available shortly.

Acknowledgments

This work has been supported generously by the U.S. Department of Energy, particularly at MIT under Cooperative Research Agreement DE-FC02-94ER40818. We also thank G.S.F. Stephans for his invaluable help with preparing the manuscript.

REFERENCES

1. S.E. Eiseman, *et al.*, *Phys. Lett.* **B325**:322 (1994).

2. D. Zachary, *Ph. D. thesis*, Mass. Institute of Tech., (1994), unpublished.

3. T. Hemmick, Proceedings of Quark Matter '93, E. Stenlund, et al., eds., Borlange, Sweden.

4. J. Rafelski, *Nucl. Phys.* **A418**:215c (1984).
   J. Rafelski and B. Muller, *Phys. Rev. Lett.* **48**:1066 (1982).

CHEMICAL EQUILIBRIUM AND PARTICLE PRODUCTION IN NUCLEUS–NUCLEUS COLLISIONS AT AGS ENERGY

Peter Braun-Munzinger and Johanna Stachel

Department of Physics
SUNY at Stony Brook
Stony Brook, NY 11794-3800, USA

INTRODUCTION

One of the main goals of the ultra-relativistic heavy-ion program at the BNL AGS and CERN SPS is to study and diagnose highly excited and dense nuclear matter and possibly identify the transition from hadronic matter to deconfined quark-matter with restored chiral symmetry. From the currently running, still comparatively low energy experiments, there is increasing evidence for the creation of matter, perhaps quark-matter, at very high baryon density and moderate temperature ($T < 0.2$ GeV). How to separate such new states of matter from a hot hadron gas scenario remains, however, one of the difficulties in the field.

In this note we expand on a recent study[1] of thermal equilibration and expansion in nucleus-nucleus collisions at AGS energies and focus, in particular, on strangeness production.

Our scenario for a central nucleus-nucleus collision at AGS energy is that a hot and dense system is formed which cools and expands until it decouples after an extended time by disintegrating into a few hundred hadrons. For this scenario we explore quantitatively the predictions of a consistent thermodynamic and hydrodynamic approach to describe the AGS data. Calculations using ideal gas thermodynamics have been reported before[2-4] and compared to particle yield ratios. We use basically the same formalism as[2,3] and, checking some of our numerical results vs. those of,[3] we find good agreement. However, the authors of[2] had a much smaller set of early AGS and SPS data to compare to and our philosophy in fixing the model parameters is different. In contrast to,[4] where the strange quark fugacity is used as a free parameter to reproduce strangeness production, we use chemical equilibrium throughout combined with strangeness conservation.

We determine the baryon chemical potential and freeze-out temperature which appear as parameters in such calculations by simultaneous analysis of nucleon and pion transverse momentum spectra and abundances, and compare the model predictions not only with measured ratios but also with absolute densities. Furthermore, we investigate in more detail the equilibration of strangeness and confront results from our hadron gas scenario with predictions for strangeness production in a quark-gluon plasma. We make use of the compilation[1] of all available results on particle production in Si-nucleus collisions at AGS energies for a comparison with the model.

In order to describe particle distributions and abundances at freeze-out, we need to treat the thermodynamics of the system only after the initally hot and dense fireball has expanded and reached a density at which interactions cease. From this point on the system behaves like a freestreaming ideal gas. In every comoving restframe the system is therefore described by a grand canonical ensemble of fermions and bosons in equilibrium at freeze-out temperature T. Strangeness is also assumed to be in full chemical equilibrium. For an infinite volume the particle number densities are given as integrals over particle momentum p:

$$\rho_i^0 = \frac{g_i}{2\pi^2} \int_0^\infty \frac{p^2 dp}{\exp[(E_i - \mu_b B_i - \mu_s S_i)/T] \pm 1} \tag{1}$$

where g_i is the spin-isospin degeneracy of particle i, E_i, B_i and S_i are its total energy in the local rest frame, baryon number and strangeness, and μ_b and μ_s are the baryon and strangeness chemical potentials ($\hbar$=c=1 unless otherwise noted). For a system of finite size the argument of the integral in equation (1) has to be multiplied[5] by a correction factor. For simplicity, we assume a spherical volume with radius R giving a correction factor

$$f = 1 - \frac{\pi}{4pR} + \frac{3}{16(pR)^2}. \tag{2}$$

In order to account for the volume taken up by baryons and mesons we apply the excluded volume correction and obtain corrected densities

$$\rho_i = \frac{\rho_i^0}{1 + \sum_j V_j \rho_j^0} \tag{3}$$

where V_j is the volume occupied by an individual baryon or meson; we use sharp sphere volumes and radii of 0.8 and 0.6 fm for all baryons and all mesons, respectively. Especially for the higher end of our temperature range, $i.e.$ $T = 0.14$ GeV, the corrections for finite and excluded volumes become substantial, amounting to a factor of 1.8 in particle density.

The temperature range relevant for the present study is 0.10 - 0.15 GeV which sets the scale for the mass range of particles to be considered. We include strange and nonstrange mesons up to a rest mass energy of 1.5 GeV and baryons up to 2 GeV. Limiting this mass range to 1.0 GeV and 1.5 GeV for mesons and baryons, respectively, changes our results by less than 10 %, indicating the insensitivity of the calculations to the precise number of hadron states included.

The range of temperatures to be considered is driven by the experimental observation[6] of the occupation probability of the $\Delta(1232)$ resonance in E814: For a system in thermal equilibrium, temperature values of 0.12 - 0.14 GeV are consistent with the observed abundance. The baryon chemical potential is constrained by the measured pion to nucleon abundance as well as the density of the system at freeze-out (see[7] and below) and we choose a value of μ_b = 0.54 GeV. For a given temperature and baryon chemical potential the strangeness chemical potential is fixed by strangeness conservation. In particular, for $T = 0.120$ and 0.140 GeV one obtains $\mu_s = 0.109$ and 0.138 GeV. Using these input parameters and equations (1) to (3) we find primary particle densities.

To obtain particle densities which can be compared to experimental data one needs to consider decay and feeding. We use all the known branching ratios as given in[8] as well as isospin symmetry and phase space arguments for unknown branching ratios. To illustrate the importance of the feeding and its temperature dependence we note that, at $T = 0.14$ GeV, the primary pion yield is tripled and the nucleon yield is doubled by feeding from higher resonances. At $T = 0.12$ GeV, on the other hand, the pion yield is doubled and the nucleon yield increases by about 60%.

Table 1. Particle ratios calculated in a thermal model for two different temperatures, baryon chemical potential $\mu_b = 0.54$ GeV and strangeness chemical potential μ_s such that overall strangeness is conserved, in comparison to experimental data (with statistical errors in paren theses) for central collisions of 14.6 A GeV/c Si + Au(Pb).

Particles	Thermal Model		Data		
	$T=.120$ GeV	$T=.140$ GeV	exp. ratio	rapidity	ref.
$\pi/(p+n)$	1.29	1.34	1.05(5)	0.6 - 2.8	9,10
$d/(p+n)$	$4.3 \cdot 10^{-2}$	$5.8 \cdot 10^{-2}$	$3.0(3) \cdot 10^{-2}$	0.4 - 1.6	9
$\bar{p}/p$	$1.47 \cdot 10^{-4}$	$5.8 \cdot 10^{-4}$	$4.5(5) \cdot 10^{-4}$	0.8 - 2.2	11
K^+/π^+	0.23	0.27	0.19(2)	0.6 - 2.2	9
K^-/π^-	$5.0 \cdot 10^{-2}$	$6.2 \cdot 10^{-2}$	$3.5(5) \cdot 10^{-2}$	0.6 - 2.3	9
K_s^0/π^+	0.14	0.16	$9.7(15) \cdot 10^{-2}$	2.0 - 3.5	12,9,13
K^+/K^-	4.6	4.3	4.4(4)	0.7 - 2.3	9
$\Lambda/(p+n)$	$9.5 \cdot 10^{-2}$	0.11	$8.0(16) \cdot 10^{-2}$	1.4 - 2.9	12,9,10
$\bar{\Lambda}/\Lambda$	$8.8 \cdot 10^{-4}$	$3.7 \cdot 10^{-3}$	$2.0(8) \cdot 10^{-3}$	1.2 - 1.7	11
$\phi/(K^+ + K^-)$	$2.4 \cdot 10^{-2}$	$3.6 \cdot 10^{-2}$	$1.34(36) \cdot 10^{-2}$	1.2 - 2.0	11
Ξ^-/Λ	$6.4 \cdot 10^{-2}$	$7.2 \cdot 10^{-2}$	0.12(2)	1.4 - 2.9	14
$\bar{d}/\bar{p}$	$1.1 \cdot 10^{-5}$	$4.7 \cdot 10^{-5}$	$1.0(5) \cdot 10^{-5}$	2.0	15

Equation (1) implies random, isotropic emission of all particles. This is not supported by the data (see[1]). Our interpretation, as outlined in,[1] is that the observed distributions are consistent with what one expects for a system which is expanding longitudinally and transversely. Since freeze-out takes place at constant density and temperature, this hydrodynamic flow should not affect angle integrated particle densities, but may severely change particle densities and density ratios at fixed rapidity. We, therefore, compare predictions of the thermal model to experimental quantities integrated over transverse momentum p_t and rapidity y whenever available.

In Table 1 all currently available experimental data on particle ratios measured in central Si + Au(Pb) collisions are compared to the corresponding ratios calculated for two temperatures, 0.12 and 0.14 GeV. As discussed in[1] we compare rapidity integrated yield ratios to thermal model predictions. Comparing the thermal model prediction and experimental particle ratios, overall good agreement is found. In fact, although the abundances vary over six orders of magnitude, all particle ratios are reproduced to better than a factor of two. While our choice for the temperature range considered is driven by the $\Delta(1232)$ abundance and not the yield ratios in Table 1, the range considered nevertheless gives the best overall agreement; the experimental $d/(p+n)$, K/π, $\Lambda/(p+n)$ and ϕ/K ratios favor a slightly lower temperature, the $\bar{p}/p$, $\bar{d}/\bar{p}$ and $\bar{\Lambda}/\Lambda$ ratios are bracketed by the range considered, the K^+/K^- and Ξ^-/Λ ratios favor higher T.

A surprising aspect of our calculation is that the thermal model predictions are in good agreement with composite particle production. Even the $\bar{d}/\bar{p}$ ratio reported recently[15] is well reproduced, without resorting to complicated coalescence model considerations. Furthermore, even absolute densities are in reasonable agreement with our predictions.[1] Maybe the most unanticipated result is that strangeness production is only slightly overpredicted, not providing a clear indication that strangeness is not in chemical equilibrium.

To shed some further light on strangeness production within our hadron gas scenario, we have evaluated the strangeness suppression factor, defined in terms of created quark pairs as

$$\lambda = \frac{2\bar{s}s}{\bar{u}u + \bar{d}d}, \tag{4}$$

by explicit counting of the strange quark content of the hadrons produced in our model. The counting is done prior to resonance decay, in order to compare the result to the recent compilation[16] of data from ultra-relativistic nucleus-nucleus collisions using the Wroblewski procedure.[17] For the same parameters as used in Table 1, i.e. $\mu_b = 0.54$, this leads to $\lambda = 0.54$ and $\lambda = 0.81$ at $T = 0.12$ and $T = 0.14$ GeV, respectively. Since quarks are produced in pairs, this implies an $\bar{s}/\bar{u}$ abundance of 0.54 to 0.81, at $T = 0.12$ and 0.14 GeV. For asymmetric collision systems such as Si+Pb the comparison with experimental results is complicated because data over the full solid angle are not available due to the absence of forward-backward symmetry and because isospin symmetry cannot be invoked. The Wroblewski procedure has, consequently, only been applied to data for the relatively small system Si+Al. Nevertheless, even for this small system, a value of $\lambda = 0.29 \pm 0.03$ was deduced,[16] approximately a factor of two smaller than the predictions of our model. For the system S + S at CERN energies, $\lambda = 0.36 \pm 0.04$, i.e. only a moderate increase.

These λ values should be confronted with what one would expect from a quark-gluon plasma scenario. On the quark level, the chemical potentials are obtained from:

$$\mu_{u,d}^q = \mu_b/3 \qquad \mu_s^q = \mu_b/3 - \mu_s, \tag{5}$$

leading to $\mu_{u,d}^q = 0.18$ GeV, and $\mu_s^q = 0.072(0.045)$ GeV for $T = 0.12(0.14)$ GeV, respectively. Strangeness conservation in a quark-gluon plasma scenario of course implies that $\mu_s^q = 0$. One might be tempted to argue that the relatively small value (compared to, say, the mass of the strange quark) of μ_s^q indicates already that a plasma has been formed. However, for the same μ_s^q, strangeness abundance in a baryon-rich plasma is very different from that in a hadronic gas. On the quark level,

$$\lambda^q = \frac{2\bar{s}s}{\bar{u}u + \bar{d}d} = \frac{\bar{s}}{\bar{u}} = \int_0^\infty \frac{p^2 dp}{\exp[E_s/T] + 1} \; / \; \int_0^\infty \frac{p^2 dp}{\exp[(E_{\bar{u}} + \mu_u^q)/T] + 1}. \tag{6}$$

We have evaluated λ^q assuming negligible $\bar{u}$ mass and $m_{\bar{s}} = 0.18$ GeV. The results for different temperatures and various quark chemical potentials are shown in Fig. 1. Over the range of parameters explored, λ^q is much larger than the hadronic gas values obtained above. To make a proper comparison to the data, one needs to take into account the gluon content and quark production during the hadronization phase. In ref.[18] it is estimated for similar baryo-chemical potential that hadronization leads to an increase of the total number of quarks by a factor of two, while the number of strange quarks increases only by 50%. Keeping in mind that quarks are produced in pairs during hadronization this leads to reduction in λ^q by a factor of 4.8, which would imply $\lambda^q \approx 0.57$ for $\mu_u^q = 0.18$ and $T = 0.14$ GeV. Surprisingly, this is somewhat closer to the data than was estimated in the purely hadronic scenario.

To determine the situation at CERN energies we have repeated the analysis of Sollfrank et al.[19] for the system S+S. The results of our calculations agree well with those of[19] for the K^+/K^- and $\bar{\Lambda}/\Lambda$ ratio if $T = 0.2$ GeV and $\mu_u = 0.077$ GeV are used. Our calculation implies $\lambda = 0.62$, about a factor 1.8 larger than the experimental values determined in.[16] The pure quark-gluon plasma value (see Fig. 1, for $\mu_u = 0.077$ GeV and $T = 0.2$ GeV) is $\lambda^q = 1.3$. Using the hadronization scheme of,[18] we again estimate that after the phase transition $\lambda^q \approx 0.68$. We note in passing that, using the thermal parameters of,[19] one predicts, for $\mu_u = 0.077$ GeV and $T = 0.2$ GeV, a K^+/π^+ ratio of 22%, not consistent with the value of about 13% found experimentally.[20] Furthermore, the pion density at freeze-out is about $0.6/fm^3$ and the nucleon density is close to that of normal nuclear matter, values which seem unrealistically high. A combined analysis using a thermal model including flow, such as done in,[1] would probably yield a more realistic (lower) freeze-out temperature.

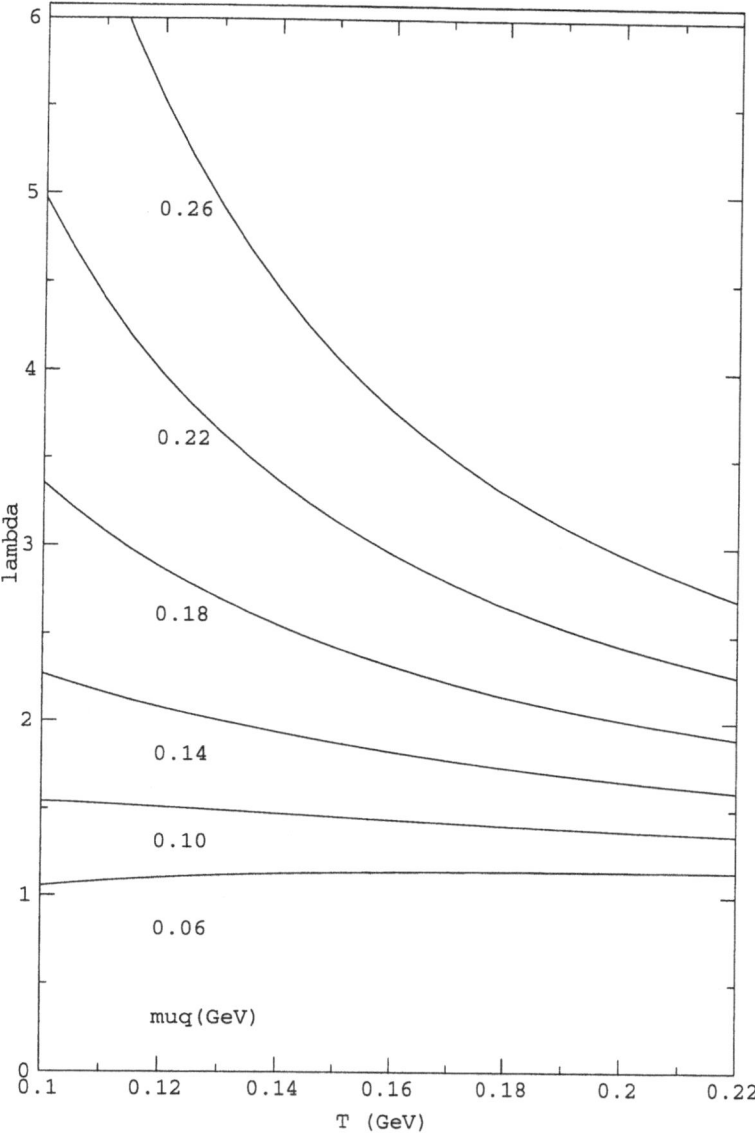

Figure 1. Strangeness suppression factor λ calculated for a plasma of different temperature and u,d-quark chemical potential. The strange quark mass was fixed at 0.18 GeV.

Acknowledgments

This work was supported in part by the National Science Foundation.

REFERENCES

1. P. Braun-Munzinger, J. Stachel, J.P. Wessels, N. Xu, sub. Phys. Lett. **B**, Oct. 1994, preprint SUNY-RHI-94-11;

2. J. Cleymans, H. Satz, E. Suhonen, and D. W. von Oertzen, Phys. Lett. **B242**, 111(1990); N. J. Davidson, H. G. Miller, R. M. Quick, and J. Cleymans, Phys. Lett. **B255**, 105(1991); N. J. Davidson, H. G. Miller, and D. W. von Oertzen, Phys. Lett. **B256**, 554(1991).

3. J. Cleymans and H. Satz, Z. Physik **C57**, 135(1993).

4. J. Letessier, J. Rafelski, A. Tounsi, Phys. Lett. **B328**, 499(1994).

5. H. R. Jaqama, A. Z. Mekjian, and L. Zamick, Phys. Rev. **C29**, 2067(1984).

6. T. K. Hemmick, E814 Coll., Nucl. Phys. **A566**, 435c(1994); J. Barrette *et al.*, E814 Coll., sub. Phys. Lett. **B**, Nov. 1994, preprint SUNY-RHI-94-10;

7. J. Barrette *et al.*, E814 Coll., Phys. Lett. **B333**, 33(1994).

8. Review of Particle Properties, Phys. Rev. **D50**, 1173-1926(1994).

9. T. Abbott *et al.*, E802 Coll., Phys. Rev. **C50**, 1024(1994).

10. J. Barrette *et al.*, E814 Coll., Z. Phys., **C59**, 211(1993); J. Barrette *et al.*, E814 Coll., Phys. Rev. **C50**, in print, preprint SUNY-RHI-94-9.

11. G. S. F. Stephans, E802 Coll., Nucl. Phys. **A566**, 269c(1994).

12. S. E. Eiseman *et al.*, E810 Coll., Phys. Lett. **B297**, 44(1992).

13. T. K. Hemmick, E814 Coll., in Proc. Int. Workshop "Heavy Ion Physics at the AGS '93", G. S. F. Stephans, S. G. Steadman, and W. L. Kehoe eds., MITLNS-2158, p. 204.

14. S. E. Eiseman *et al.*, E810 Coll., Phys. Lett. **B325**, 322(1994).

15. A. Aoki *et al.*, E858 Coll., Phys. Rev. Lett. **69**, 2345(1992).

16. J. Stachel and G.R. Young, Ann. Rev. Nucl. Part. Sci. **42**, 537(1992).

17. A. Wroblewski, Acta Phys. Pol. **B16** 379(1985).

18. H. Barz, B.L. Friman, J. Knoll, and H. Schulz, Nucl. Phys. **A484** 661(1988).

19. J. Sollfrank, M. Gaździcki, U. Heinz, J. Rafelski, Z. Phys. *C61* 659(1994).

20. P. Seyboth *et al.*, Na35 Coll., Nucl. Phys. **A544**, 293c(1992).

MEASUREMENT OF THE Ω/Ξ PRODUCTION RATIO IN CENTRAL S-W INTERACTIONS AT 200 A GEV/c

THE WA85 COLLABORATION

Federico Antinori

CERN, Geneva, Switzerland
and
INFN, Genova, Italy

S. Abatzis[a], A. Andrighetto[e,*], F. Antinori[e], R.P. Barnes[d], A.C. Bayes[d],
M. Benayoun[f], W. Beusch[e], A. Bravar[h], J.N. Carney[d], B. de la Cruz[g],
D. Di Bari[b], J.P. Dufey[e], J.P. Davies[d], D. Elia[b], D. Evans[e], R. Fini[b],
B.R. French[e], B. Ghidini[b], H. Helstrup[c], A.K. Holme[c], A. Jacholkowski[b],
J. Kahane[f], J.B. Kinson[d], A. Kirk[e], K. Knudson[e], J.C. Lassalle[e],
V. Lenti[b], Ph. Leruste[f], V. Manzari[b], F. Navach[b], J.L. Narjoux[f],
A. Penzo[h], E. Quercigh[e], L. Rossi[e,+], K. Šafařík[e], M. Sené[f], R. Sené[f],
G. Vassiliadis[a], O. Villalobos-Baillie[d], A. Volte[f], M.F. Votruba[d]

[a] Athens University, Nuclear Physics Department
GR-15771 Athens, Greece
[b] Dipartimento di Fisica dell' Università
and Sezione INFN, I-70126 Bari, Italy
[c] Universitetet i Bergen, N-5007 Bergen, Norway
[d] University of Birmingham, Birmingham B15 2TT, U.K.
[e] CERN, European Organization for Nuclear Research
CH-1211 Geneva 23, Switzerland
[f] Collège de France and IN2P3, F-75231 Paris, France
[g] CIEMAT, E-28040 Madrid, Spain
[h] INFN, I-34127 Trieste, Italy

[*] Present address: Dipartimento di Fisica dell' Università
I-35131 Padua, Italy
[+] Present address: Dipartimento di Fisica dell' Università
and Sezione INFN, I-16145 Genoa, Italy

ABSTRACT

Strange baryon and in particular multiply strange baryon production is suggested to be a useful probe in the search for Quark-Gluon Plasma formation in heavy-ion collisions. We have measured the $(\Omega^- + \overline{\Omega}^+)/(\Xi^- + \overline{\Xi}^+)$ production ratio at central rapidity and $p_T > 1.6$ GeV/c to be 0.8 ± 0.4.

RESULTS FROM WA85

The WA85 experiment was designed to study strange particle production at central rapidities and medium transverse momenta in sulphur-nucleus collisions using the CERN sulphur ion beam at 200 A GeV/c. Enhancement of K_s^0, Λ, $\overline{\Lambda}$ and $\overline{\Xi}^+$ production in S-W interactions with respect to proton-proton and proton-nucleus interactions have already been reported by our collaboration[1] and have since been used as input by various theoretical models.[2-4] Recently we have reported[5] the observation of 7.0 ± 3.6 Ω^- and 4.0 ± 2.0 $\overline{\Omega}^+$ particles. It becomes now possible to compare, under the same experimental conditions, the relative abundances of particles carrying one, two and three units of strangeness.

In this letter we present the first determination, in S-W collisions, of the relative abundance of the Ω^- with respect to the Ξ^-, the $(\Omega^- + \overline{\Omega}^+)/(\Xi^- + \overline{\Xi}^+)$ production ratio, at central rapidities ($2.5 < y_{LAB} < 3.0$) and medium transverse momenta ($p_T > 1.6$ GeV/c). Ω^-/Ξ^- and $\overline{\Omega}^+/\overline{\Xi}^+$ ratios have not been determined separately because of the small statistics available. The WA85 apparatus and the trigger to identify central collisions have been described elsewhere.[1] We identify Ω^-, Ξ^- and their antiparticles through their two-step cascade decays:

$$\Omega^- \longrightarrow K^-\Lambda(\Lambda \longrightarrow p\pi^-)$$

and

$$\Xi^- \longrightarrow \pi^-\Lambda(\Lambda \longrightarrow p\pi^-).$$

The Ω^-, Ξ^- and Λ decay products are detected by a system of multi-wire proportional chambers. Target, hyperon decay region and chambers are located inside the 1.8 Tesla magnetic field of the CERN-OMEGA Spectrometer. The chambers are modified in such a way as to be only sensitive to central rapidity, medium-high p_T tracks ($p_T > 600$MeV/c for tracks from the target); the apparatus thus records only a small number (≤ 10) of tracks out of the several hundred produced in a central S-W interaction, thus ensuring a good reconstruction efficiency.

The data sample consists of 60 million events of central ($\sim 30\%$ of the total cross-section) S-W collisions at 200 A GeV/c beam momentum. The strategy for the reconstruction of cascade topologies and the criteria for selecting Ω^- and Ξ^- candidates have been described in Ref.[5].

Ω^-, $\overline{\Omega}^+$, Ξ^- and $\overline{\Xi}^+$ candidates were weighted to correct a) for geometrical acceptance losses as a function of their phase-space positions and b) for reconstruction inefficiencies as a function of the track multiplicity of the event, as measured by the total multiplicity detected in the chambers. After subtraction of the residual combinatorial background, the ratio

$$R = \frac{N(\Omega^-) + N(\overline{\Omega}^+)}{N(\Xi^-) + N(\overline{\Xi}^+)}$$

was calculated in two phase-space regions, one defined by a common lower p_T cut and one defined by a common lower m_T cut for Ω^- and Ξ^-. The results are:

$$R = 0.8 \pm 0.4 \text{ for } p_T > 1.6 \text{ GeV}/c$$

and

$$R = 1.7 \pm 0.9 \text{ for } m_T > 2.3 \text{GeV}/c.$$

The quoted errors, because of the low statistics in the Ω^- sample, cannot be interpreted as gaussian. Assuming the number of Ω^- that one would detect in repeated experiments to be described by Poisson statistics, we derive the following lower limits on the values of R:

$$R > 0.39 @ 95\% \text{ c.l. for } p_T > 1.6 \text{GeV}/c$$

and

$$R > 0.79 @ 95\% \text{ c.l. for } m_T > 2.3 \text{GeV}/c.$$

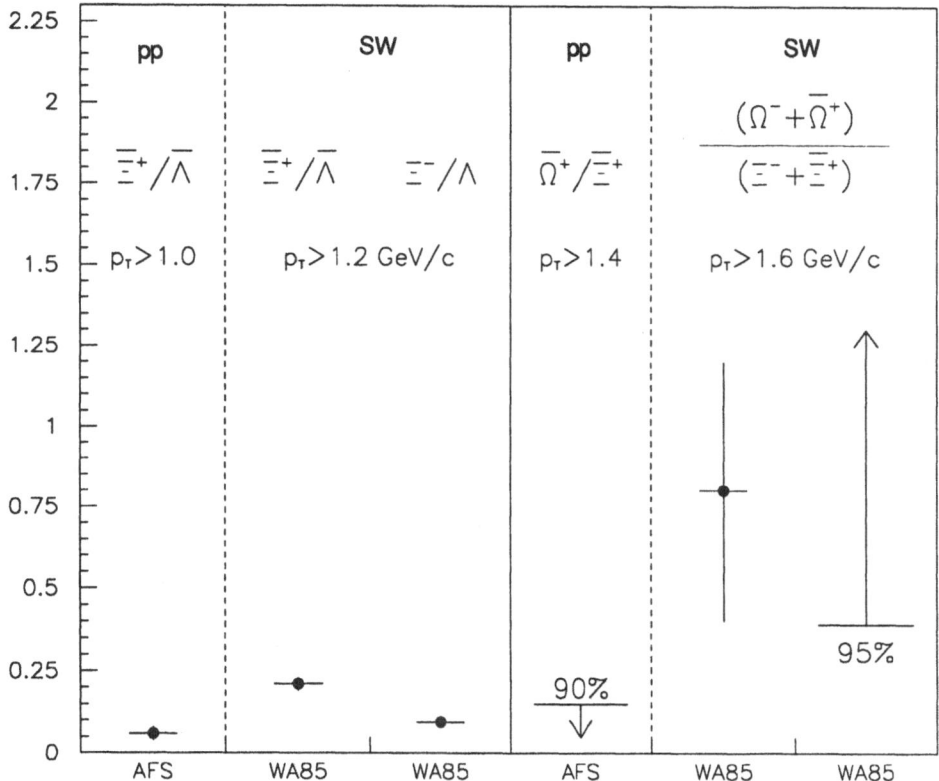

Figure 1. Comparison of Ξ/Λ and Ω/Ξ ratios in proton proton and central S-W collisions.

The only published result on the Ω/Ξ production ratio at central rapidity in proton initiated interactions comes from the AFS collaboration at ISR,[6] who also measured in the same experiment the $\overline{\Xi}^+/\overline{\Lambda}$ ratio. In proton-proton interactions at $\sqrt{s} = 63 \text{ GeV}/c$, they quote an upper limit of 0.15 at 90% c.l. for the $(\overline{\Omega}^+/\overline{\Xi}^+)$ ratio at central rapidity and $p_T > 1.4 \text{GeV}/c$. Published yields for Ω and Ξ production at forward rapidities in proton interactions measured in the analysis of hyperon beam compositions lead to Ω^-/Ξ^- ratios of the order of a few percent (see for example Ref.[7]). The comparison between the WA85 S-W results and the AFS proton-proton results is shown in Fig. 1 both for the Ω/Ξ and for the previously published Ξ/Λ ratios. One notices that the relative rates of $(\Omega^- + \overline{\Omega}^+)/(\Xi^- + \overline{\Xi}^+)$ and $\overline{\Xi}^+/\overline{\Lambda}$ in S-W reactions are significantly higher than similar ratios measured in proton-

proton collisions. Such enhancements are not easy to explain in the standard framework of hadronic interactions. Values for these ratios compatible with our results are instead predicted by models based on the equilibration, at least partial, of strange and nonstrange degrees of freedom,[2,3] as one could expect if a QGP had been formed in the early stages of the interaction.

In conclusion, we have measured the $(\Omega^- + \overline{\Omega}^+)/(\Xi^- + \overline{\Xi}^+)$ production ratio at central rapidity and medium-high p_T in central S-W collisions at 200 A GeV/c laboratory beam momentum. The ratio is of the order of one and it is sensibly higher than the upper limit for the $\overline{\Omega}^+/\overline{\Xi}^+$ ratio obtained in a similar phase-space region in pp collisions.

REFERENCES

1. S. Abatzis et al., Phys. Lett. B244 (1990) 130.
   S. Abatzis et al., Phys. Lett. B259 (1991) 508.
   S. Abatzis et al., Phys. Lett. B270 (1991) 123.
   S. Abatzis et al., Nucl. Phys. A544 (1992) 321c.
   S. Abatzis et al., Nucl. Phys. A566 (1994) 233c.

2. J. Rafelski, Phys. Lett. B262 (1991) 333
   J. Rafelski, Nucl. Phys. A544 (1992) 279c.
   J. Letessier, J. Rafelski, A. Tounsi, Phys. Lett. B321 (1994) 394.
   J. Letessier, J. Rafelski, A. Tounsi, Phys. Lett. B323 (1994) 393.

3. U. Heinz, Nucl. Phys. A566 (1994) 205c.

4. J. Cleymans and H. Satz, Z. Phys. C57 (1993) 135.
   J. Cleymans, K. Redlich, H. Satz, E. Suhonen Z. Phys. C58 (1993) 347.

5. S. Abatzis et al. Phys. Lett. B 316 (1993) 615.

6. T. Akesson et al. Nucl. Phys. B 246 (1984) 1.

7. T. R. Cardello et al. Phys. Rev. D 32 (1985) 1.

NA36
STRANGENESS PRODUCTION
MULTISTRANGE BARYONS

H. Rohringer for NA36 Collaboration

Institut für Hochenergiephysik
der Österr. Akademie der Wissenschaften
Nikolsdorferg. 18, A-1050 Wien

E. Andersen[1], R. Blaes[2], M. Cherney[3], B. de la Cruz[4], C. Fernández[5]
C. Garabatos[5], J.A. Garzón[5], W.M. Geist[2], D.E. Greiner[6], C. Gruhn[6]*
M. Hafidouni[2], J. Hrubec[7], P.G. Jones[6], E.G. Judd[8]†, J.P.M. Kuipers[9]
M. Ladrem[2], P. Ladrón de Guevara[4], G. Løvhøiden[1], J. MacNaughton[7]
J. Mosquera[5], Z. Natkaniec[10], J.M. Nelson[8], G. Neuhofer[7]
C. Pérez de los Heros[4], M. Pló[5], P. Porth[7], B. Powell[9], A, Ramil[5]
H. Rohringer[7], I. Sakrejda[6], T.F. Thorsteinsen[1], J. Traxler[7], C. Voltolini[2]
K. Wozniak[10], A. Yañez[5], and R. Zybert[8]

NA36 mailing address: NA36, c/o Dr. D. E. Greiner, Mailstop 50D
Lawrence Berkeley Laboratory, 1 Cyclotron Road, Berkeley CA 94720, USA

[1]University of Bergen, Dept. of Physics, N-5007 Bergen, Norway
[2]Centre de Recherches Nucléaires, IN2P3-CNRS/Université L. Pasteur
BP 20, F-67037 Strasbourg, France
[3]Creighton University, Dept. of Physics, Omaha, Nebraska 68178, USA
[4]CIEMAT, Div. de Física de Partículas, E-28040 Madrid, Spain
[5]Universidad de Santiago, Dpto. Física de Partículas
E-15706 Santiago de Compostela, Spain
[6]Lawrence Berkeley Laboratory (LBL), Berkeley CA 94720, USA
[7]Institut für Hochenergiephysik (HEPHY), A-1050 Wien, Austria
[8]University of Birmingham, School of Physics and Space Research
Birmingham B15 2TT, UK
[9]European Organization for Nuclear Research (CERN)
CH-1211 Genève 23, Switzerland
[10]Instytut Fizyki Jadrowej, PL-30 055 Krakow 30, Poland

*Present address: CERN, PPE division, Geneva, Switzerland
† Present address: LBL, Berkeley, CA 94720, USA

INTRODUCTION

Experiment NA36 at the CERN North Area was designed to measure, with high statistics, strange particles in Sulfur Lead interactions at 200 GeV/c per nucleon in a wide range of acceptance and to give possible hints to the existence of the Quark Gluon Plasma (QGP). It uses a TPC ($1.0 \times 0.5 \times 0.5\ m^3$) in a 2.7 Tesla field situated approximately 1.5cm above the beam to avoid beam fragments disturbing the chamber. Silicon beam counters , a Čerenkov counter and a forward calorimeter select good beams, trigger on sulfur interactions and determine the centrality of the reactions. The beam hits a lead target of 5% interaction length 60 cm away from the chamber entrance. We use about 1 million events of S+Pb interactions and a smaller sample of p+Pb events for comparison.

Results presented in this paper come from the strange particles acceptance region :

$$1.50 < y < 3.00 \quad 0.5 < p_\perp < 2\,\text{GeV/c} \quad \text{SULPHUR} \quad \text{data}$$
$$2.00 < y < 4.00 \quad 0.5 < p_\perp < 2\,\text{GeV/c} \quad \text{PROTON} \quad \text{data}$$

We reconstruct events with up to 300 charged tracks in the TPC. Tracks reconstructed in the TPC are correlated with the produced negatively charged multiplicity in the reaction via Monte Carlo studies.[1]

Strange Particle Search

Neutral strange particles are found by combining all pairs of oppositely charged tracks of sufficient length. We exclude pairs where both tracks might come from the primary vertex. We also require the line of flight of the reconstructed neutral particle to be compatible with the primary vertex. Additionaly, the decay vertex must be at least 20 cm away from the primary vertex. Less than half of the decays occur outside of the TPC. A geometric fit was used to determine the decay vertex position and the momenta. The cuts used to restrict the backgound were carefully studied using the Monte Carlo simulations. By selecting an appropriate magnet polarity to bend π^+ more copiously into the TPC the $\bar{\Lambda}$ sample is relatively enriched with respect to the Λ. From the Armenteros plot (fig. 1) and the invariant mass plots (fig. 2) one can see clear signals of Λ and $\bar{\Lambda}$ with a mass resolution of about 6 MeV.

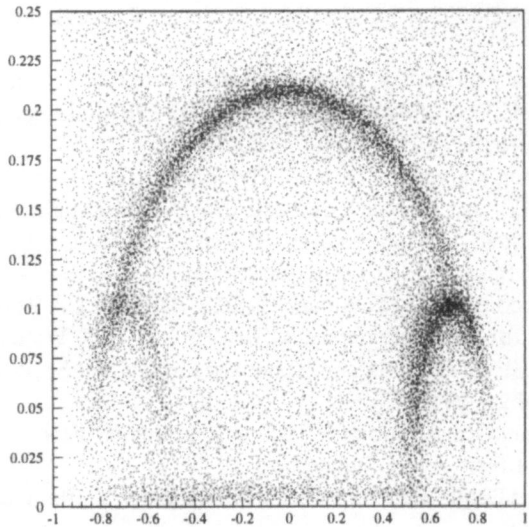

Figure 1. Armenteros - Podolanski plot

Our strange particle reconstruction efficiency is multiplicity dependent and averages about 30%. Depending on y and $p_\perp$ the acceptance for strange particles ranges between

462

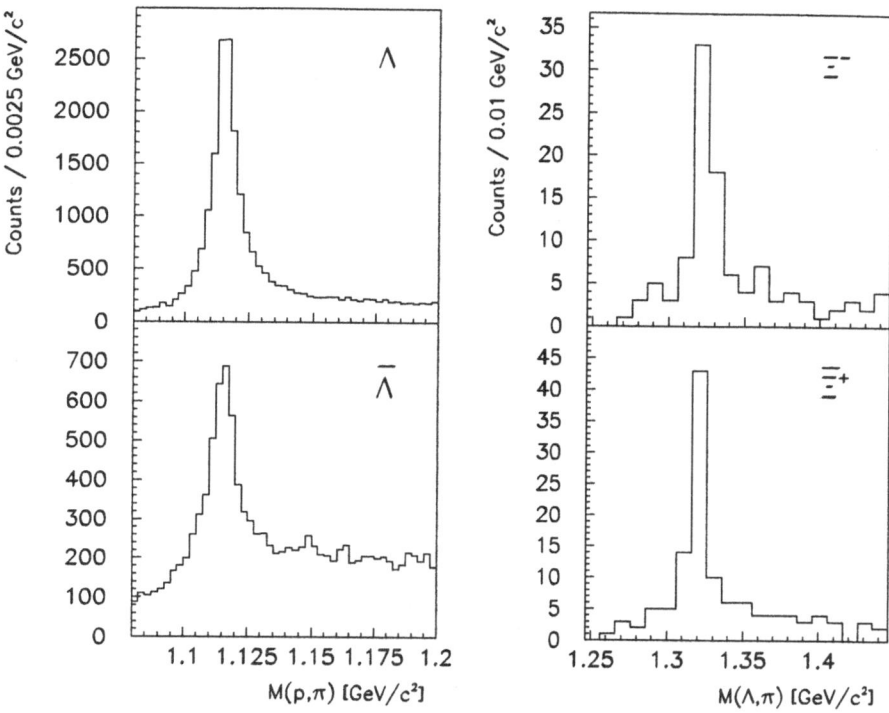

Figure 2. Invariant Mass plots for Λ, $\bar{\Lambda}$, Ξ^- and $\bar{\Xi}^+$

5% and 15%, therefore acceptance and efficiency corrections are needed. A Monte Carlo event generator was used to evaluate these corrections. Currently, the most used Monte Carlo programs for heavy ion collisions are RQMD, VENUS and FRITIOF. They combine particle production with ideas based on experimental facts as to how ION collisions happen. None of them are as precise as the LEP Monte Carlo program. They just reproduce general features of the events and are often tuned to the needs of specific experiments. Ideas for transitions to or from a QGP are not built in. In order to free the efficiency and acceptance corrections from the assumptions built into models an iterative procedure was applied.

DATA ANALYSIS

We looked for the main features of strange baryon production in phasespace to get an insight into the reaction mechanisms.

From the ratios of strange particles one can hopefully extract information on the space - time development of the reactions. A theoretical framework guides this analysis. Advocates of an underlying QGP picture find in such data strong evidence that supports their point of view. In contrast, supporters of a conventional hadron resonance gas picture will find that the experimental data is not incompatible with that model, if some additional features like hard core scattering or rescattering are built into the Monte Carlo program (eg in RQMD). We will try to present both points of view.

Rapidity Distributions

In fig. 3 we compare corrected rapidity distributions for strange particles produced in S+Pb and p+Pb reactions. For ion collisions Λ and $\bar{\Lambda}$ are produced preferably at midrapidity, whereas for the p interactions Λ are concentrated in the target region. So, for ion collisions according to Rafelski[2] this observation could be interpreted as the creation of a hot source (fireball), whereas a leading particle mechanism would be appropriate for strangeness production in the proton induced reactions. In the latter case RQMD[3] is able to reproduce the peak at small y by assuming primary $NN \longrightarrow NYK$ and secondary $KN \longrightarrow NY$ scattering to produce strange particles. For ion collisions a fine effect of a slight shift of the peak value of the rapidity distribution for Λ to lower (if compared with the $\bar{\Lambda}$) rapidity values is visible. It might be explained by a contribution of Λ produced in rescattering of kaons on the target baryons.

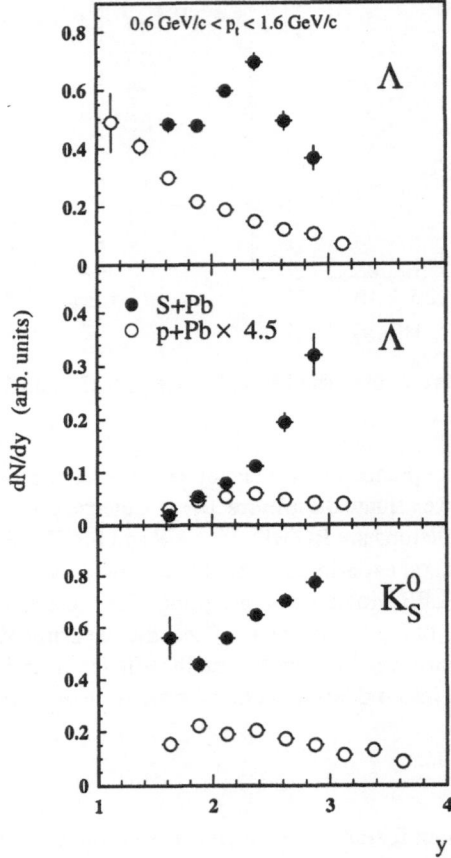

Figure 3. Rapidity distributions for K_s^0, Λ, $\bar{\Lambda}$ in S+Pb and p+Pb reactions

Fireball Picture

Hagedorn has long ago advocated a bootstrap picture of hadronic reactions, where multiple fireballs are created during a collision. Later on they evolve to form other fireballs. This will be reflected in the transverse mass spectra of final state particles, which should demonstrate an exponential behaviour.

$$\frac{1}{p_\perp}\frac{d\sigma}{dp_\perp} \propto \frac{1}{T}\sqrt{m_\perp}\exp\left(-\frac{m_\perp}{T}\right) \tag{1}$$

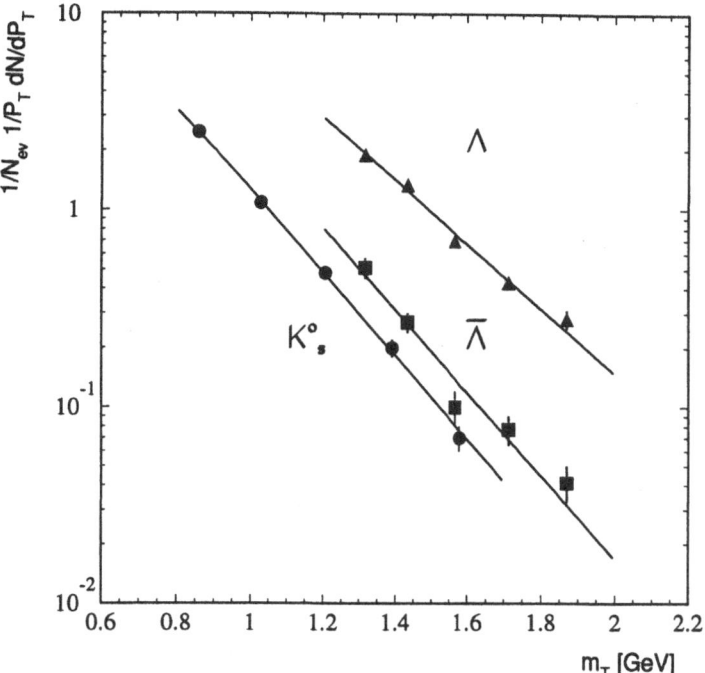

Figure 4. Transverse Mass distributions

In fig. 4 that presents data on a logarithmic scale, we look therefore for a linear behaviour of the transverse mass $m_\perp$ distribution of the produced strange particles. We can then extract a slope parameter T, the so called 'limiting temperature'. T is found to have rather different values in the CERN experiments analyzed so far. Rafelski[2] showed, that the reason are different experimental y - $p_\perp$ acceptance windows. This T_{exp} can be translated regarding these facts to a $T_{freezeout}$, which gives now a common value of 200 MeV independent of experiments. So our experimental data seem to fit a 'Hagedorn picture' with a fireball situated at y = 2.8.[2]

Particle Production

In many experiments a faster growth of strange particle production compared to nonstrange has been observed. In fig. 5 we display therefore the number of strange particles per negative particle as a function of the produced negative multiplicity. In all cases (Λ, $\bar{\Lambda}$ and K) we observe a strong rise of produced strangeness for small multiplicities and then a saturation. This latter effect may point to a saturation in the number of the participating nucleons, if the ratio of strange to nonstrange production is fixed. A similar effect was observed in a Au - Emulsion Experiment at BNL.[4]

According to Rafelski[2] our observation is consistent again with a fireball picture. The fireball size is correlated with the pion yield and the fireball lifetime will determine strangeness production and its possible saturation.

Strangeness Production in Central Collisions

The centrality of a collision can be correlated to the multiplicity up to a certain value. Having observed that the Λ production in ion collisions increases towards midrapidity we

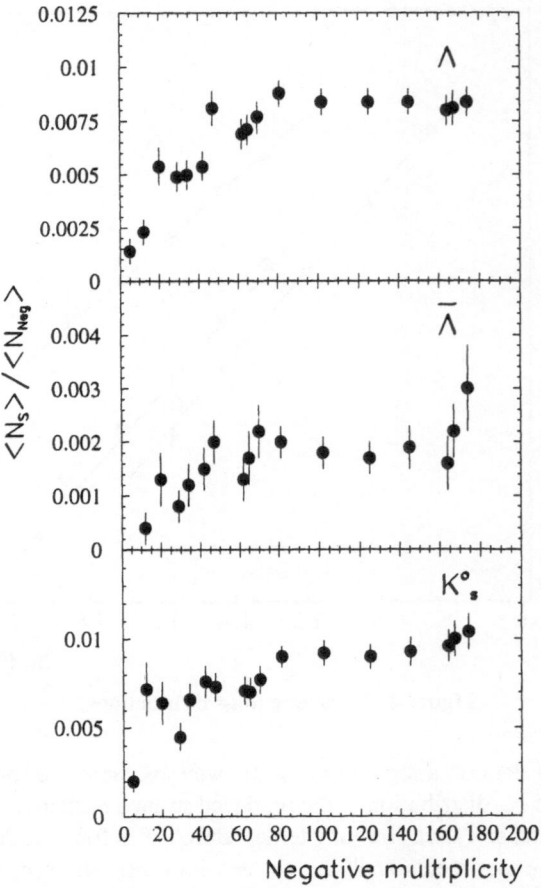

Figure 5. Strange particle production as a function of negative multiplicity

investigate now, whether this enhancement is also dependent on the centrality of the collision. In fig. 6 we show therefore the multiplicity dependence of the relative production rate

$$\frac{\Lambda \text{ in [rapidity centre]}}{\Lambda \text{ at [rapidity edge]}}$$

which clearly grows with increasing multiplicity.

Cascades

Ξ particles have been reconstructed from Λ and $\bar{\Lambda}$ by combining them with a third track fulfilling mainly the criteria to be compatible with a common decay vertex and not to be a primary track. Invariant mass plots for the reconstructed cascades are included in fig. 2.

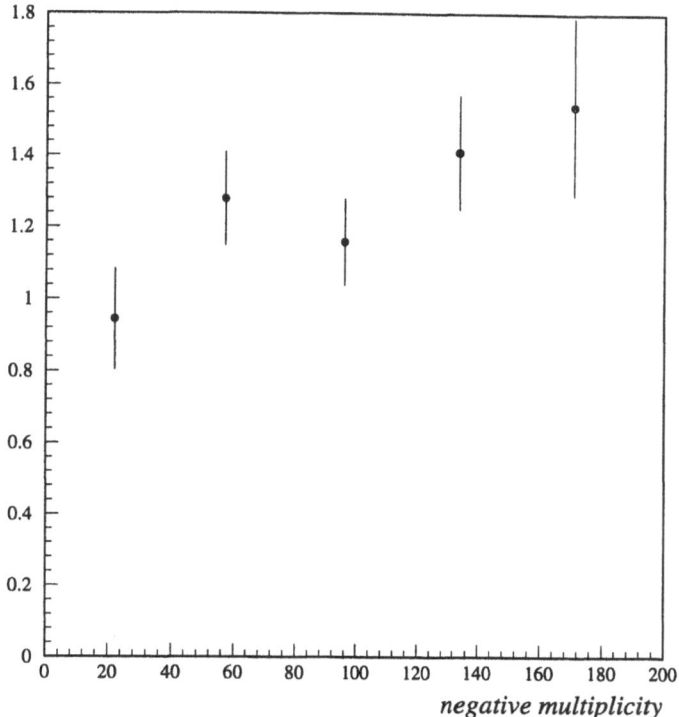

Figure 6. Ratio of cross-sections in midrapidity and target region vs produced negative multiplicity for Λ in S + Pb reactions

Strange Baryon Particle Ratios

Particle ratios will mainly reflect the quark content of the participants and are specifically interesting if strange and antistrange quarks are involved, since these quarktypes have to be created during the interaction. In a thermodynamical approach[5] these ratios depend on the 'Temperature' and chemical potentials of the particles involved. In the case of 'chemical equilibrium' among different particle species these strange particle ratios are fixed and calculable knowing the thermodynamic parameters μ_b, μ_s, T and γ_s, a strangeness saturation factor. In a QGP picture strange quark production will be much faster than in the hadronic gas scenario and therefore strangeness saturation measurements might give insight into the history of strangeness production and its origin.

We now use only data from the small impact parameter region to avoid possible influence from electromagnetic interactions in peripheral collisions. In table 1 our results are given and plotted together with data from other experiments in fig. 7. The data are fully corrected to the acceptance regions stated, for Λ decaying from Ξ and for unseen decay modes.

Extracting Thermodinamic Quantities

If hadrons were composite objects formed by combination of quarks, a probability (P) of producing a particle at an energy (E) and temperature (T) would be:

$$P \propto \Pi g_i, \lambda_i, \gamma_i \exp\left(E_i/\mathsf{T}\right) \qquad i = u, d, s \ldots \tag{2}$$

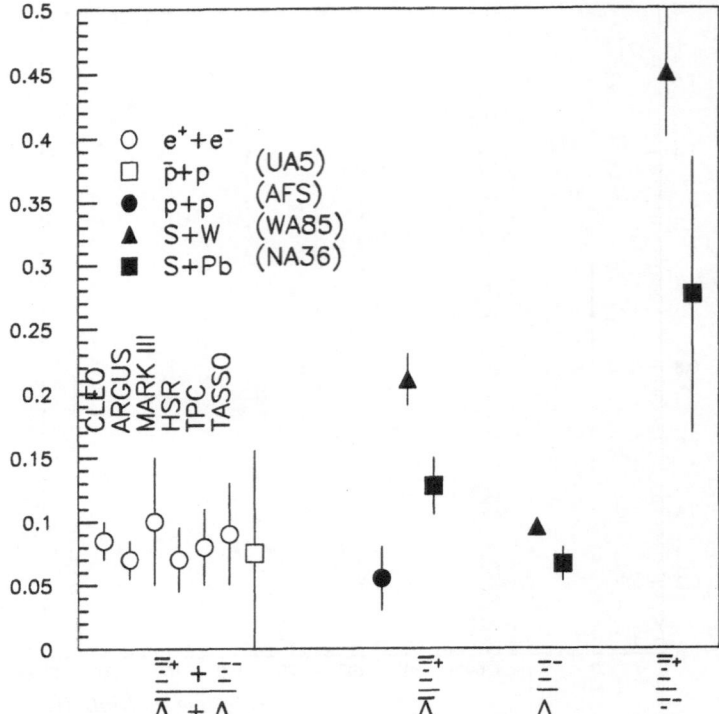

Figure 7. A summary of world data on various Particle ratios

Table 1. Summary of the different particle ratios and their acceptance interval

	Particle ratio	Acceptance	
$\bar{\Lambda}/\Lambda$	0.117 ± 0.011	$2.0 < y < 2.5$	$0.6 < p_\perp < 1.6\,\text{GeV/c}$
$\bar{\Xi}^+/\Xi^-$	0.276 ± 0.108	$2.0 < y < 2.5$	$08 < p_\perp < 1.8\,\text{GeV/c}$
Ξ^-/Λ	0.066 ± 0.013	$1.5 < y < 2.5$	$0.8 < p_\perp < 1.8\,\text{GeV/c}$
$\bar{\Xi}^+/\bar{\Lambda}$	0.127 ± 0.022	$2.0 < y < 3.0$	$0.6 < p_\perp < 1.6\,\text{GeV/c}$

where:

λ_i chemical fugacity . . . particle abundance

g_i statistical factors

γ_i describes amount of quark saturation

From the ratios of the production probabilities
$R(\Lambda) = P(\bar{\Lambda})/P(\Lambda)$ and $R(\Xi) = P(\bar{\Xi}^+)/P(\Xi^-)$ we can show that:

$$R(\Lambda)/R(\Xi)^2 = \exp(6\mu_s/\mathsf{T})$$

$$\frac{\Xi^-}{\Lambda} \times \frac{\bar{\Xi}^+}{\bar{\Lambda}} = \gamma^2$$

$$R(\Lambda)/\sqrt{R(\Xi)} = \exp(-\mu_b/\mathsf{T})$$

These ratios assume that the particle abundances are measured at the same energy. Using fig. 4 Λ and $\bar{\Lambda}$ are extrapolated to the same $m_\perp$ region as the Ξ's.

Then we find :

μ_s / T = 0.03 ± 0.06   **equal number of s and $\bar{s}$ in acceptance region !!**

μ_b / T = 1.73 ± 0.15

γ_s = 0.38 ± 0.04   **strangeness saturation not satified**

In the 'conventional' approach of Satz and Cleymans[6] assuming a hadron resonance gas in ion collisions μ_b and T can be extracted from particle ratios. They provide tables where for fixed T and μ_b different particle ratios are calculated. Due to the experimental error for each particle ratio two lines in the T - μ_b plane can be drawn from these tables, shown in fig. 8. Obviously there is an overlap region where the measured particle ratios have the same T and μ_b. The μ_b/T has about the same value as above, but the extracted T is smaller than what we measure from fig. 4. This discrepancy can be eliminated by assuming a collective transverse flow, which will increase slopes of the transverse mass spectra.[7]

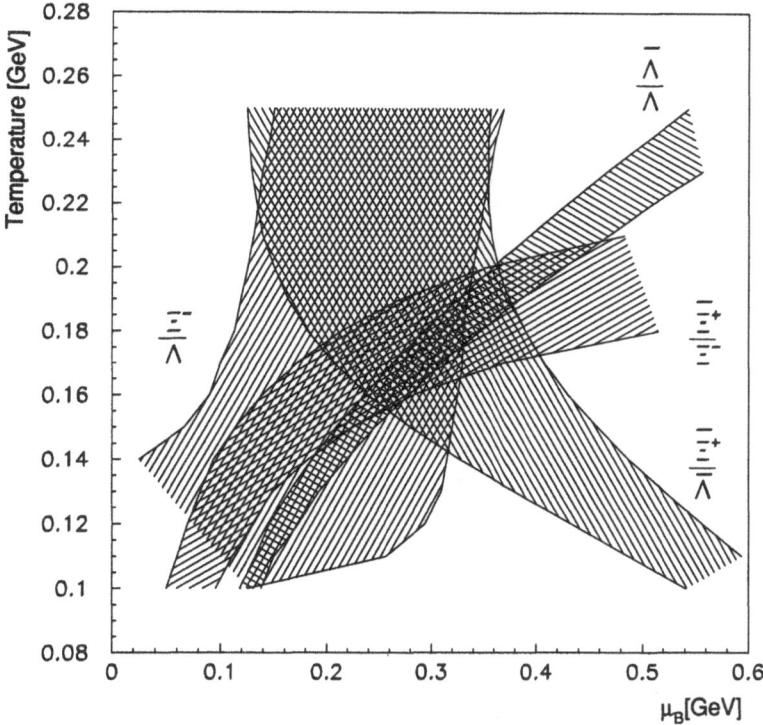

Figure 8. Particle ratios in μ - T plane

CONCLUSIONS

NA36 has shown, that strangeness production occurs predominantly at midrapidity supporting a fireball picture à la Hagedorn. We observe enhanced strangeness production in more central interactions up to a saturation value. From our measurements of strange particle ratios we imply local strangeness conservation ($\mu_s = 0$) at midrapidity . The slope parameter T in the Hagedorn picture is measured to be 200 MeV, which leaves also the possibility for transverse flow. Within a QGP picture the available phasespace for strangeness is not filled up. But using a conventional hadron gas model with acceptable thermodynamic parameters we see no disagreement with our data.

REFERENCES

1. E.Anderson et al. NA36 Coll, Phys.Lett.B 316:603(1993).

2. J.and.H.Rafelski and M.Danos, Phys.Lett.B 294:131(1992).

3. H.Sorge,L.A.Winckelmann,H.Stöcker,W.Greiner, Z.Phys.C 59:85(1993).

4. M.L.Cherry et al. KLMM Coll, Z.Phys.C 62:25(1994).

5. P.Koch,B.Müller and J.Rafelski, Physics Reports 142 no 4(1986).

6. J.Cleymans and H.Satz, Z.Phys.C 57:135(1993).

7. E.Schnedermann and U.Heinz, Phys.Rev.Lett. 69:2908(1992).

ESTIMATES OF THE RATIOS $\bar{\Lambda}/\Lambda$, $\bar{\Xi}/\Xi$ AND $\bar{\Omega}/\Omega$ FROM pp AND pA INTERACTIONS

W.M. Geist and Th. Kachelhoffer

Centre de Recherches Nucléaires
IN2P3–CNRS/Université Louis Pasteur
BP 28, F–67037 Strasbourg Cedex 2, France

INTRODUCTION

The probably large gluonic energy density of a QGP is supposed to lead rapidly to a chemical equilibrium with a high level of (anti-)strange quarks $(\bar{s})s$, especially for a non-vanishing net baryon density.[1] The most direct trace of this enhanced (relative to non-QGP processes) density of s and/or $\bar{s}$ may be the ratio $R(\bar{Y}/Y)$ of production yields of baryons Y $(Y = \Lambda, \Xi, \Omega)$ and their antiparticles. Recent measurements of some of these ratios in S-nucleus collisions at $200\,\text{GeV/c/N}$ show that, for Feynman-$x \approx 0$, $R(\bar{\Lambda}/\Lambda) > [\sigma_{incl}(\bar{\Lambda})/\sigma_{incl}(\Lambda)]_{pp}$ and $R(\bar{\Xi}/\Xi) > R(\bar{\Lambda}/\Lambda)$ (Fig. 1 of Ref.[2]). Unfortunately, there are no corresponding data from pp and/or pA collisions in the same kinematic range. Rather reliable estimates can be, however, be obtained by extrapolation from pA data as discussed below.

PARAMETRIZATION OF DIFFERENTIAL CROSS SECTIONS

For the current analysis of $\overset{(-)}{Y}$ production in pA collisions the invariant differential cross section $E\frac{d\sigma}{dp}(pA \to c) = I(pA \to c)$, $c = \bar{Y}, Y$, is factorized as follows:[3]

$$I(pA \to c) = I(pp \to c)t_c F(A)$$

The function $F(A)$ is parametrized by a power α of A for $A \geq 2$; t_c is an isospin factor. As the measured numerical values of α are compatible with being independent of the type of the secondary hadron c, particle ratios tend to be independent of A:

$$R_{pA}(\bar{Y}/Y) = R_{pp}(\bar{Y}/Y).$$

Only for $Y = \Xi$ one expects t_c to differ from unity at $x < 0$.

The differential cross section $I(pp \to c)$ is factorized as follows:

$$I(pp \to c) = f(x)g(m_c, p_T).$$

The longitudinal part is approximated by $f(x) = (1 - x)^{n_c}$ which reproduces available data well.[3] For the transverse part $g(m_c, p_T)$ a thermal distribution, as function of transverse momentum p_T, with one temperature parameter T ($T = 120\,\text{MeV}$) is used. In this case the ratios $R(\bar{Y}/Y)$ are independent of p_T.

RESULTS AND CONCLUSIONS

Most available data on $I(pA \rightarrow c)$, $c = \bar{p}$, Λ, $\bar{\Lambda}$, Ξ, $\bar{\Xi}$, Ω, $\bar{\Omega}$ are reproduced in the framework of this parametrization.[3] In most cases, however, only the kinematic region $|x| \geq 0.2$ is covered.

It is now easy to extrapolate to $x = 0$ for pA and/or pp collisions. The resulting antiparticle/particle ratios are included in Fig. 1. Obviously, the numerical values of these ratios are of the same order of magnitude as those measured in SA interactions.

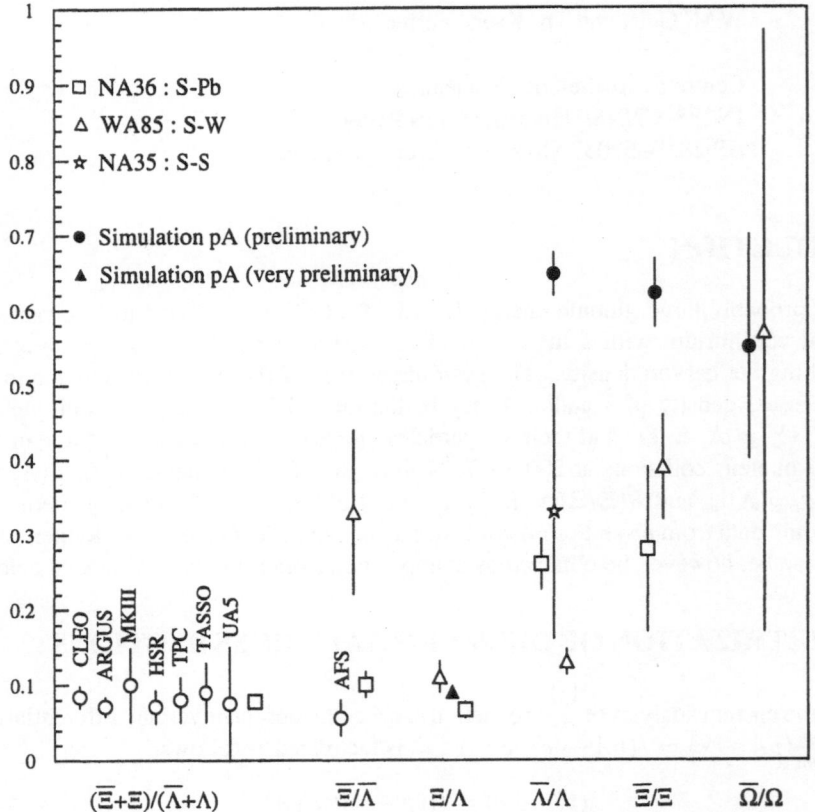

Figure 1. Anti-particules/particules ratios at $x = 0$ for pp and pA collisions from differents experiments together with the results of the simulation.

REFERENCES

1. P. Koch et al., *Phys. Rep.* **C142** (1986) 167

2. E.Andersen et al., Nucl. Phys. **A566** (1994) 217c
   S.Abatzis et al, Phys. Lett. **B316** (1993) 615

3. W. Geist and T. Kachelhoffer, submitted to Zeit. Physik C

THERMALISATION IN HIGH ENERGY HEAVY
ION COLLISIONS AND STRANGE PARTICLE PRODUCTION

J. Cleymans[1] and K. Redlich[2,3]

[1]Department of Physics, University of Cape Town, South Africa
[2]Fakultät für Physik, Universität Bielefeld, Germany
[3]Institute of Theoretical Physics, University of Wroclaw, Poland

ABSTRACT

Two issues related to strange particle production in heavy ion collisions at CERN are discussed : whether or not chemical equilibrium is established and supercooling in the quark-gluon plasma phase takes place.

CHEMICAL EQUILIBRIUM

An enhancement of strange particle production in relativistic ion collisions has now been observed by all experimental groups.[1] This phenomenon is often analysed using the hadronic gas description (for a recent review see Heinz[2]). In such models one assumes that a thermal system is being produced which expands until it reaches freeze-out at which point all the hadronic resonances decay into the lightest stable particles. The observed particle yields thus reflect the properties of the system at freeze-out. The next question to be asked is whether or not the different particle abundancies correspond to a system in chemical equilibrium. Hadronic matter has a certain composition of particle species, characterized by "chemical" potentials. In addition to the baryon number chemical potential μ_B, we have a strangeness chemical potential μ_S. The value of μ_B is fixed by giving the overall baryon number density n_B, and that of μ_S by fixing the overall strangeness to zero. This means that for given values of T and μ_B one tunes the remaining parameter, μ_S, in such a way as to ensure strangeness neutrality. This leads to the values[3] for μ_S shown in figure 1. In a quark-gluon plasma the number of strange quarks balances the number of anti-strange quarks and therefore the relation

$$\mu_S = \mu_B/3$$

holds. In a hadronic gas the relationship is more complicated because particles of opposite strangeness can have different masses (e.g. Λ's and K's).

Note that in figure 1 if the temperature is around 200 MeV it is not possible to distinguish the value of μ_S from the value $\mu_B/3$. This coincidence has led some people to believe that a quark-gluon plasma has been produced since this relationship is natural in a quark-gluon plasma. We do not subscribe to this point of view.

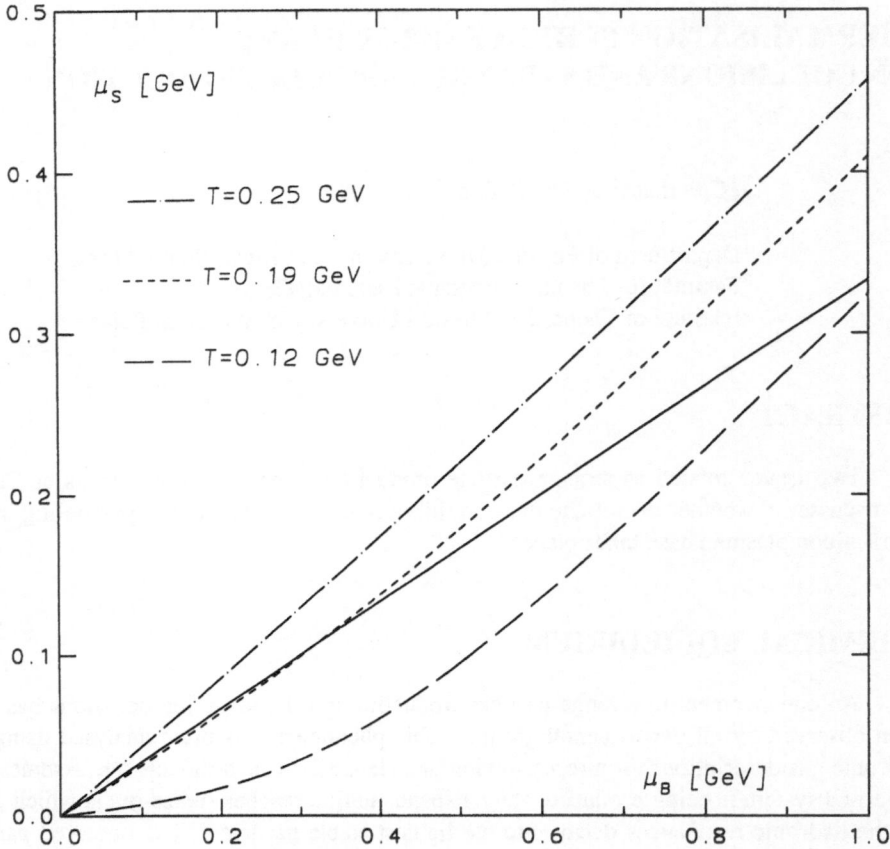

Figure 1. Strangeness chemical potential μ_S as a function of the baryon chemical potential μ_B for different values of the temperature. Also shown is the line $\mu_S = \mu_B/3$ which holds in a quark-gluon plasma.

Preliminary indications from the WA85 collaboration were that chemical equilibrium had been reached, however a more accurate analysis presented at the Quark Matter' 93 meeting[4] indicates that this is not the case.[6]

Each particle ratio leads to a band in the T, μ_B plane, (including the error bar) this gives rise to a region where the hadronic gas reproduces the measured numbers. If all the measured ratios have a common overlap region then one can conclude that the experimental data are consistent with chemical equilibrium. Taking the results of the WA85 collaboratiion[4] as a start it is shown in figure 2 that this is not the case. It has been proposed[5] to parametrize this deviation from chemical equilibrium by a parameter γ_s. This is fully justified in the framework of a quark-gluon plasma but it is not quite correct in a hadronic gas. Here one should work by introducing more chemical potentials. A discussion of different break-up mechanisms has been presented in Ref.[2,8,6].

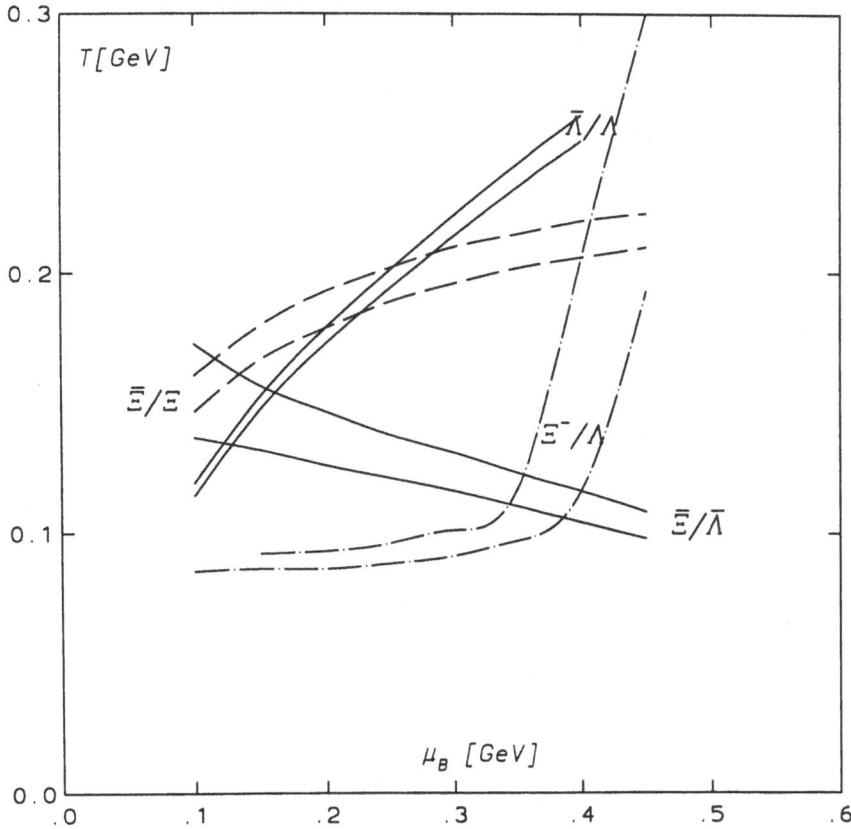

Figure 2. Values of temperature and chemical potential consistent with the experimentally measured particle ratio indicated. As there is no region of common overlap one concludes that there is no chemical equilibrium.

SUPERCOOLED QUARK-GLUON PLASMA

We will here consider the idea that the hadronic gas is produced at a temperature of about 200 MeV and that this temperature is higher than the phase transition temperature.[9,10] Estimates from lattice gauge theories consistently show critical temperatures which are quite low, possibly even below 150 MeV.[7] Due to the uncertain nature of lattice gauge calculations it is probably premature to consider this option very seriously, however it leads to an interesting physical situation. The basic mechanism is depicted in figure 3.

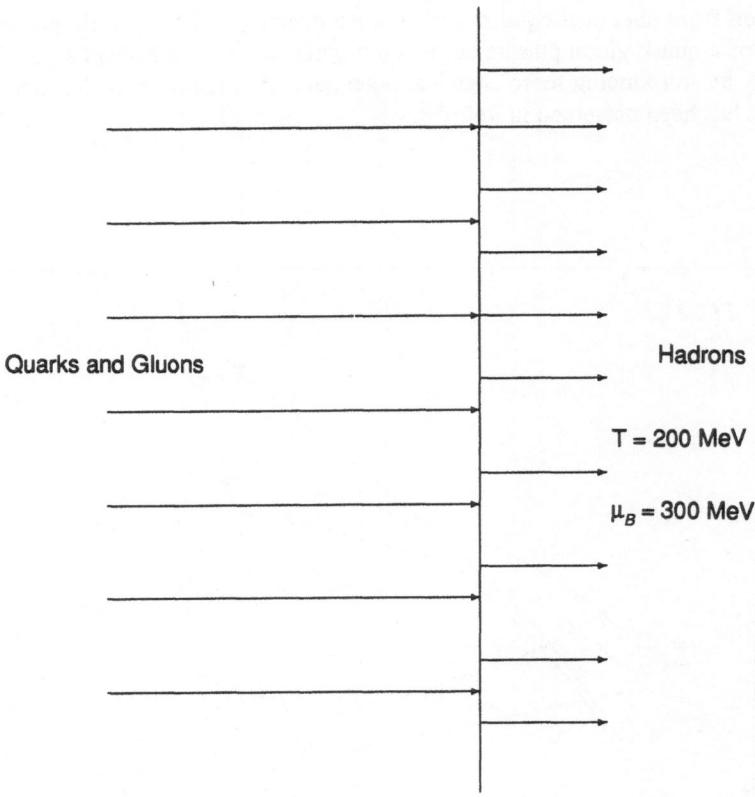

Quarks and Gluons

Hadrons

T = 200 MeV

μ_B = 300 MeV

Figure 3. Schematic presentation of a transition from a quark-gluon plasma phase leading to a hadronic gas with the indicated parameters. The two phases are assumed to be separated by a very sharp front.

On one side we have the final hadronic state which is produced in the superheated state on the other side we have the quark-gluon plasma. Straightforward calculations based on energy-momentum and baryon conservation across the sharp transition front lead to constraints on the state of the quark-gluon plasma shortly before the phase transition took place. A further important constraint is given by the second law of thermodynamics which demands that the entropy increases across the front. It has been shown using fairly realistic descriptions[9,10] for the equation of state that the quark gluon plasma must be in a severely supercooled state. Estimates lead to temperatures as low 60 - 80 MeV. This is shown more quantitatively in figure 4 where we plot the temperature of the supercooled quark-gluon plasma versus the temperature of the superheated hadronic phase. It can be seen that for a temperature in the hadronic gas of about 200 MeV the corresponding temperature in the quark-gluon plasma phas could only have been 70 MeV. Such a severe supercooling will of course strongly encourage the formation of hadronic droplets. These then will modify the equation of state and change the nature of the transition.

It has been shown[10] that if the admixture of hadronic droplets exceeds a certain fraction of the total amount of material, about 20% to be specific then the supercooling is much less severe.

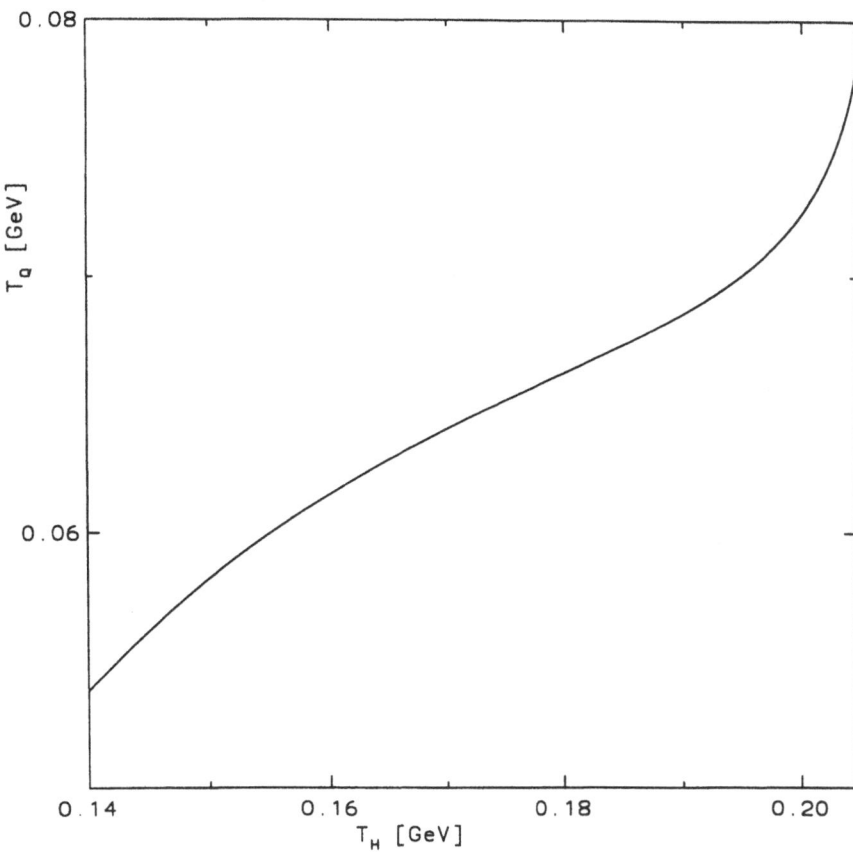

Figure 4. Temperature in the quark-gluon plasma phase as a function of the temperature observed in the hadronic phase.

CONCLUSIONS

Comparing the production rates of different hadrons provides us with a tool for the study of chemical equilibrium in the final state of high energy heavy ion collions. It is also the only tool to obtain the chemical potentials, in particular μ_B. This provides crucial information about the thermodynamic properties of the system at break-up time.

If the observed temperature in the hadronic gas (200 - 230 MeV) is higher than the phase transition temperature (150 MeV?) then one must conclude that the hadronic gas is in a superheated state. This is usually leads to a supercooled quark-gluon plasma phase. An estimate of the degree of supercooling leads to very low temperatures for the quark-gluon plasma phase (below 100 MeV).

Acknowledgment

We gratefully acknowledge useful and stimulating discussions with Helmut Satz, Esko Suhonen and Neven Bilić.

REFERENCES

1. *"Quark Matter '93"*, Proc. of the X'th Int. Conf. on Ultra-Relativistic Nucleus-Nucleus Collisions, Borlänge, 1993, E. Stenlund et al. (editors), Nucl. Phys. **A566** (1994).

2. U. Heinz, Nucl. Phys. **A566**, 205c (1994).

3. J. Cleymans and H. Satz, Z. f. Physik **C57**, 135 (1993).

4. D. Evans et al. (WA85 collaboration), Nucl. Phys. **A566**, 225c (1994).

5. J. Rafelski, Nucl. Phys. **A544**, 279c (1992).

6. K. Redlich, J. Cleymans, H. Satz and E. Suhonen, Nucl. Phys. **A566**, 391c (1994).

7. B. Petersson, Nucl. Phys. **A525**, 237c (1991).

8. J. Cleymans, K. Redlich, H. Satz and E. Suhonen, Z. f. Physik **C58**, 347 (1993).

9. N. Bilić, J. Cleymans, E. Suhonen and D. von Oertzen, Phys. Lett. **B266** (1993) 226.

10. N. Bilić, J. Cleymans, K. Redlich and E. Suhonen, Z. f. Physik **C** (1994) (to be published).

STRANGENESS IN HOT HADRONIC MATTER

Johann Rafelski,[1,2] Jean Letessier,[2] and Ahmed Tounsi[2]

[1]Department of Physics
University of Arizona, Tucson, AZ 85721, USA
[2]Laboratoire de Physique Théorique et Hautes Energies[*]
Université Paris 7, Tour 24, 5è ét.
2 Place Jussieu, F-75251 Paris CEDEX 05, France

THE SEARCH FOR QUARK GLUON PLASMA

After so many distinguished speakers surveyed the role of strange particles in the search for the quark-gluon plasma (QGP), it is hard not to be repetitive, or alternatively, imprecise and vague. We choose therefore to present a somewhat provocative and primarily conceptual summary of the lectures and our views, and more generally, of the experimental and theoretical situation and the future research program. Our point of view is that strangeness has positively provided evidence of new collective physics phenomena occurring in hot and dense hadronic matter formed in highly relativistic heavy ion collisions.

The long aspired substantiation of the deconfined and equilibrated phase we call QGP is now almost within experimental reach. Its discovery will provide a confirmation of the currently established paradigm of strong interactions. Even after that, several important issues will remain. These include in particular the measurement of the parameters and properties of the hadronic phase formed in relativistic nuclear collisions. Here we encounter the fundamental question if a phase transition occurs between the deconfined 'perturbative' and the confining 'hadronic gas' phase, or if a gradual 'cross-over' of the phases is to be expected. What are the properties of the QCD phases, such as the specific heat of the perturbative, deconfining vacuum, and what is the value of the critical temperature? Another fundamental point which will be naturally raised in this context regards the nature of entropy producing processes in the initial instants of the collisions — this poses today a major fundamental challenge to our understanding of the collision dynamics.[1]

Strangeness can, and has already helped in critical ways to address these and other related issues. Practically all strange flavor seen in the different final state particles must be produced in the hadronic interactions occurring in the collision region. Nearly a decade ago it has been recognized that strangeness production processes constitute a bottleneck in

---

[*]Unité associée au CNRS: D 0280

the (chemical) equilibration of strongly interacting matter. In the event of QGP formation, enhanced production of strangeness flavor by glue based processes was predicted to lead to enhanced strange particle abundances, and specifically an enhancement of strange anti-baryons.[2] If the transition from the deconfined state to the final state individual hadrons occurs rapidly, the high density of strangeness in the early phase will manifest itself in the anomalous (high) yields of multiple strange particles. Furthermore, if QGP is not formed, its absence reveals itself by a relative suppression of strange, compared to non-strange particle production, with yields similar to those seen in p–p collisions.

Because of the intrinsic difficulties related to the detection of strange and multi-strange particles, only recently results have been presented which allow a full experimental examination of these ideas. At this meeting several heavy-ion experimental groups[3–5] reminded us of their discovery of the enhancement of strange particle yields in the 200 GeV A collisions of sulphur nuclei with various targets relative to p–p and p–A collisions, enhancement which is increasing with the centrality of the collision. The only presently available comprehensive analysis of these results, including the yields of multiple strange antibaryons is based on a relatively simple, but powerful framework of a locally thermal fireball model.[6] The special virtue of this approach is that the spectra and particle abundances can be described in terms of just a few parameters. At the heart of this approach is the hypothesis that the strong interactions allow to achieve local thermodynamical equilibrium, a fact which is very much in experimental evidence, but which is far from being understood and thus 'confirmed' theoretically.

In many details the thermal model analysis of the experimental results differs fundamentally from comparisons that can be made with results obtained with cascade type models. These contain specific data or extrapolations and assumptions about individual reaction cross sections. However, for the p–p collisions we already know that the attainment of the thermal-type distributions is inexplicable today. Consequently, there is no reason to expect that the 'microscopic' approach which uses multiple series of p–p type interactions leads to better understanding of the thermalization process. If the underlying thermalization processes are different and in particular much faster than those operating in the cascade codes, the whole picture of the collision evolution is ill defined in such models, as we do not implement the microscopic mechanism. In the thermal approach one analyzes the data, assuming that thermalization, though not understood, occurs practically instantaneously in hadronic interactions. The prize one pays is that we loose the ability to describe certain crucial details of the collision evolution.

For example, we have not been able to identify within a thermal model a method to determine the stopping fractions governing the different collisions; generally in the thermal model these need to be extracted from measured data,[7] while the microscopic models can in principle claim to 'derive' e.g. the energy-momentum stopping. On the other hand, we have yet to see which model assumptions and input cross sections do lead us to any of the particular results discussed. It should be noted that at not too high energies the particle production mechanisms coupled with large reaction cross sections show in microscopic model simulations full stopping also seen in the experimental results. In this limit thermal models are as capable to give parameter free predictions, as are microscopic models. We will have to see up to which energy this may occur for the Pb–Pb reactions, where the relative thickness of the nuclear system could lead to full stopping up to initially studied energies near 160 GeV A.

A much discussed question at this workshop was whether current strange particle data are consistent with the fully equilibrated hadronic gas (HG) picture of hot hadronic matter.[8] The rationale for such a study arises since along with our ignorance of the mechanisms of thermalization one could argue for the a priori presence of full equilibrium (thermal and chemical) HG phase. The data is clearly inconsistent with this proposition, and requires, as we shall describe below, non-equilibrium features associated with strange particle chemical equilibration. This implies that a scenario with approach to the chemical equilibrium is developed, which heavily favor in view of the experimental results the alternative including the creation of a transient deconfined (QGP) state.

There is a possible reaction scenario, which could reduce the usefulness of the strange particle results: should the final state hadronization lead to fully equilibrated intermediate hadron gas state, then the memory of the initial QGP state would be lost in single particle observables — initial state anomalies could still remain visible in certain other features, such as specific entropy. Since today's 200 GeV A strange particle ratio data is compatible with the QGP hypothesis, and incompatible with HG, we can conclude that in this kinematic condition, the hadronization of the initial phase must have occurred in a way which left the final hadronic particles no chance to rescatter. It thus appears that we have a favorable situation with the final state observable single particle abundances representing the expected properties of the primordial QGP state. However, at the lower today accessible energy range 10–15 GeV A we have a less clear situation: either there is considerable hadronization re-equilibration, or the QGP phase is not made at all.[6,7]

In the next section we will discuss some important aspects of the data analysis, which were implied, but not fully elucidated by the other speakers, and then we turn our attention to a discussion of possible scenarios of the broader research program we envisage.

PRINCIPLES OF DATA ANALYSIS

Strangeness Flow

The observable 'strangeness' is really not just one quantity which is 'enhanced' by a factor 'two or two and a half only' as several speakers observed. It is important to recognize that there are as many as 10 immediately observable, on hadronic scale, stable strange particles: $K^{\pm}$, K_S, ϕ, Λ, $\overline{\Lambda}$, Ξ, $\overline{\Xi}$, Ω, $\overline{\Omega}$. On the other hand there are also some constraints: since under strong interactions the number of produced strange quarks and antiquarks is equal, and the reaction occurs too fast for WI-decays to set in, we must always have as many s carrying hadrons, as we have carriers of $\bar{s}$. It turns out that this strangeness conservation constraint (see below) is accompanied by considerable differences in the flow pattern of final state particles, as function of the nature of the source (HG versus QGP) and hence this constraint is indeed not reducing the liberty of the system, but helps to clarify the nature of the source. Another possible relation implicit in the abundance of the strange particles is that K_S should closely follow the average of $K^{\pm}$ yields, up to the u–d flavor asymmetry present in heavy nuclei — this provides a useful check of the experimental methods and all three rates are usually studied separately.

It is usual to present the strange particle spectra as function of the Lorentz variables: the rapidity y and transverse mass $m_{\perp}$. To assess the true richness of the subject, let us make a practical and greatly simplifying hypothesis, which reduces the number of observables considerably: we suppose that the shape of these spectra is sufficiently similar for given common range of these quantities, to allow a reduction of all the spectral data to just one primary spectral shape. Now in addition to the spectral parameter usually taken to the the the inverse slope of the $m_{\perp}$ distribution at central rapidity, referred to as temperature, there remain 10 normalization constants. Leaving out an overall normalization constant associated with the reaction volume and another one from among the kaons, there are 8 independent parameters — but we should include in these consideration the closely related yield of antiprotons, the number of interesting yield observables rises to 9. These can be redundantly measured with the help of 36 different strange particle and $\bar{p}$ yield ratios available to us experimentally. It is this set of independent particle yields measured by means of nearly 40 particle ratios that is commonly referred to as *strangeness flow*. Within any sensible theoretical approach there will be just a few undetermined physical parameters available — for example in the thermalized, chemical-off equilibrium model which we have developed there are 4 parameters related to particle abundance (see below). This leaves us with considerable predictive power for the many particle ratios and we are happy to note that the recently measured yields[3,4] at 200 GeV A involving $\bar{p}$, Ω and $\overline{\Omega}$ are in agreement with our prior predictions.[14]

Aside of the yield normalization parameters, there are 11 different spectral shapes in y and $m_\perp$ which we above presumed to be similar — the experimental fact that they are similar and indeed that they are characterized by a common inverse slope parameter is very telling about the underlying physics — i.e. thermalization, and cannot be taken lightly — so far, for 200 GeV A collisions, all measured $m_\perp$ distributions are consistent with the thermal models and suggest strongly that the source of all strange particles is thermalized with a common temperature for all different particles.

The similarity of the spectral shape of particles produced can only be present prior to integration over the yield in the variable $m_\perp$, since the difference in the mass of different particles limits the range of the integration to $m_\perp \geq m_i$ for each particle of mass m_i. Thus in general the $m_\perp$ integrated spectra in rapidity are very dependent on particle mass *even if the distributions in $m_\perp$ for fixed (preferably central y) were initially identical*. For this reason the particle yield comparison has been sometimes misunderstood and the argument has been occasionally made that one has to have a full phase space coverage in order to compare particle yields. We believe that while such full phase space comparison is useful and limits the impact of flow phenomena, comparison between particles can by far more easily be made within narrow windows of rapidity. Should the spectral shapes indeed be the same for different particles, they are also much simpler to interpret prior to integration over y. What the full rapidity coverage allows is to study and understand the underlying reaction picture, e.g. stopping, the formation and properties of the central fireball, matter flow, etc.

Non-Equilibrium Features in the Final State

In order to determine particle abundances, certain non-equilibrium features in the chemical composition of the assumed thermal fireball source and of the hadronization process have to be considered. In particular, we cannot expect, unless very special circumstances prevail, that the strangeness flavor has fully saturated the available phase-space. This necessitates the introduction of a strangeness saturation parameter $0 < \gamma_s \leq 1$ (we imply here that the approach to equilibrium occurs from below) which allows to parametrize the effects of the incomplete chemical equilibration of the strangeness sector.[9] For a QGP state with its fast gluonic strangeness production even the natural short fireball lifetime of only a few fm/c should be nearly sufficient to reach values of γ_s close to one.[2,10,11]

Let us define this off-equilibrium parameter more exactly: since the thermalization (appearance of exponential particle spectra) is seen even if particle yields are not equilibrated, and thus it is presumably occurring at a considerably shorter time scale than the (absolute) chemical equilibration of strangeness, we can characterize the saturation of the strangeness phase space by an average over the momentum distribution:

$$\gamma_s(t) \equiv \frac{\int d^3p\, n_s(\vec{p}, \vec{x}; t)}{\int d^3p\, n_s^\infty(\vec{p}, \vec{x})} \,, \tag{1}$$

Here n_s^∞ is the equilibrium particle density. The factor γ_s enters the phase space momentum distribution as a multiplicative factor, and with the $\vec{x}$ dependence contained solely in the statistical parameters we have:

$$n_s(\vec{p}, \vec{x}; t) \simeq \gamma_s(t) n_s^\infty(\vec{p}; T(\vec{x}, t), \mu_s(\vec{x}, t)) \,, \tag{2}$$

We will show below that it is possible to measure the value of γ_s, and thus this discussion is a very important stepping stone in understanding the behavior of hadronic phases: γ_s can be studied varying a number of parameters of the collision, such as the volume occupied by the fireball (varying size of the colliding nuclei and impact parameter), the trigger condition (e.g. the inelasticity), the energy of colliding nuclei when searching for the threshold energy of abundant strangeness formation.

The theoretical dynamical model to investigate $\gamma_s(t)$ has been developed to considerable detail.[10–12] It arises from a standard population evolution equation. Detailed balance assures that the production and annihilation processes are balancing each other as $\gamma_s \to 1$. At large times the approach to equilibrium takes the form:

$$1 - \gamma_s \to e^{-t/\tau_s} \qquad t \gg \tau_s ,$$

where τ_s is the relaxation time for strangeness equilibration which can be computed[11] using standard QCD methods. This constant is believed[12] to be dominated by the originally proposed[10] glue-fusion processes and is of the magnitude 2–3fm/c for the here important temperature range.

In a purely HG fireball with its expected much longer strangeness saturation time scale, small values of $\gamma_s \sim 0.1$ (like those extracted from N–N collisions) are expected.[13] A measurement of γ_s thus in principle provides important information on the strangeness production time scale and hence a relatively large value of γ_s requires some new physics feature in the structure of the collision fireball, which we like to associate with QGP formation.

If QGP could have been formed, we must allow for the effect of non-equilibrium hadronization. This leads to another principal non-equilibrium feature: we must not take for granted that the relative abundances of mesons and baryons, which are produced by quite different mechanisms, reach their relative chemical equilibrium; yields moreover, even if there is rescattering after their initial formation, the processes which convert mesons into baryon–anti-baryon pairs are relatively slow. We therefore must envision that the abundance of mesons and baryons is different from the equilibrium values, the difference being characterized by concentration factors C_i. On physical grounds one can argue convincingly that there is relative equilibrium within the set of different baryons and, similarly, mesons, since quark exchanges occur frequently in rescattering — processes described by diagrams in which all quark lines are 'flowing' and no quarks are produced or destroyed are most probable according to OZI rule. These quark exchange processes conserve the number of baryons and mesons separately.

We will mostly discuss here abundance of strange particles and thus we have to consider two abundance parameters $C_i^s, i = \mathrm{M}, \mathrm{B}$ for strange mesons and baryons. Given the unknown size of the reaction volume only particle number ratios are considered, and these two parameters enter in the present consideration solely in form of the ratio:

$$R_\mathrm{C}^s \equiv \frac{C_\mathrm{M}^s}{C_\mathrm{B}^s} . \tag{3}$$

We note that the value of R_C for non-strange sector can be greatly different from R_C^s, since the pion yield is largely controlled by the entropy balance in the reaction.

The Fugacities and Chemical Potentials

Once the departure from chemical equilibrium is accounted for by means of the two parameters γ_s and R_C^s we have introduced above, the particle abundances are determined by the usual chemical fugacities λ_i related to the chemical potentials μ_i by:

$$\lambda_i = e^{\mu_i/T} .$$

The fugacity of each hadronic species is simply the product of the valance quark fugacities:

$$\lambda_p = \lambda_u^2 \lambda_d, \qquad \lambda_{K^+} = \lambda_u \lambda_s,$$

etc. Three fugacities are introduced since the flavors u, d, and s are separately conserved on the time scale of hadronic collisions and can only be produced or annihilated in particle-antiparticle pair production processes. However, in many applications it is sufficient to combine the light quarks into one fugacity

$$\lambda_q^2 \equiv \lambda_d \lambda_u, \qquad \mu_q = \frac{\mu_u + \mu_d}{2}.$$

The slight asymmetry in the number of u and d quarks is described by the small quantity $\delta\mu = \mu_d - \mu_u$, which may be estimated by theoretical considerations.

Since the factor γ_s accounts for the deviation from the full phase space occupancy, (which is here assumed for u, d flavors) the chemical potentials for particles and anti-particles are opposite to each other, implying for the respective fugacities

$$\lambda_{\bar{\imath}} = \lambda_i^{-1} \qquad \imath = q, s. \tag{4}$$

Strange Particle Ratios

The abundances of the final state strange particles can be obtained by considering the Laplace transform of the phase space density, which leads to a partition function like expression $\mathcal{Z}_s$. Here the weight of individual components is controlled by the four parameters we have introduced: the non-equilibrium coefficients γ_s, C_i^s and the fugacities λ_q, λ_s:

$$\ln \mathcal{Z}_s = \frac{VT^3}{2\pi^2} \left\{ (\lambda_s \lambda_q^{-1} + \lambda_s^{-1}\lambda_q)\gamma_s C_M^s F_K + (\lambda_s \lambda_q^2 + \lambda_s^{-1}\lambda_q^{-2})\gamma_s C_B^s F_Y \right.$$
$$\left. + (\lambda_s^2 \lambda_q + \lambda_s^{-2}\lambda_q^{-1})\gamma_s^2 C_B^s F_\Xi + (\lambda_s^3 + \lambda_s^{-3})\gamma_s^3 C_B^s F_\Omega \right\}, \tag{5}$$

where the kaon (K), hyperon (Y), cascade (Ξ) and omega (Ω) degrees of freedom are included. In the resonance sums all known strange hadrons should be counted. The phase space factors F_i of the strange particles are explicitly, with g_i describing the statistical degeneracy of each component:

$$\begin{aligned} F_K &= \sum_j g_{K_j} W(m_{K_j}/T), & F_Y &= \sum_j g_{Y_j} W(m_{Y_j}/T), \\ F_\Xi &= \sum_j g_{\Xi_j} W(m_{\Xi_j}/T), & F_\Omega &= \sum_j g_{\Omega_j} W(m_{\Omega_j}/T). \end{aligned} \tag{6}$$

The function $W(x)$ arises from the phase-space integral of the different particle distributions $f(\vec{p})$. For the Boltzmann particle phase space when the integral includes the entire momentum range the well known result is found:

$$W(x_i) \equiv (4\pi)^{-1} \int d^3(p/T_i) f_i(\vec{p}) = x_i^2 K_2(x_i). \tag{7}$$

T is a dimensional parameter, and it is convenient to take the inverse slope for Boltzmann-like distributions, K_2 is the modified Bessel function. It is evident here how to modify this last equality to account for longitudinal flow or limited regions in the phase space covered by experiments.[14]

In the thermal model, also considering the kinematic behavior of the source regarding the longitudinal and collective transverse flow, the shape of the particle spectra for given value of the variable y is a function of $m_\perp$. Thus, studying the suitable ratios of *spectra* of particles with those of their anti-particles within overlapping regions of $m_\perp$ at some fixed rapidity, the spectral shapes and many other statistical factors cancel leading to measured quantities which are directly functions of the four parameters we have defined above: γ_s, R_C^s, λ_q, λ_s which thus can be determined. It was shown[14,16] that this statement remains approximately valid when feed-down by resonance decays is included. One can also determine these parameters fitting the total particle abundances.[15] The lecture by G. Odyniec[6] describes in considerable detail the analysis of the experimental results based on these simple, but very powerful methods.

Strangeness Conservation

The exact balance between s and $\bar{s}$ quarks in hadronic interactions requires non-trivial relations between the parameters characterizing the final state. These are in general difficult to satisfy and the resulting particle distributions are constrained in a way which differs considerably between different scenarios, including in particular the rapidly disintegrating QGP or the equilibrated HG phase. These two alternatives can be distinguished in particular by the value of the strange quark chemical potential μ_s:

1. In a strangeness neutral QGP fireball μ_s is always exactly zero, independent of prevailing temperature and baryon density, since both s and $\bar{s}$ quarks have the same phase-space size.

2. In any state consisting of locally confined hadronic clusters μ_s is generally different from zero at finite baryon density, in order to correct the asymmetry introduced in the phase-space size by a finite baryon content.

At non-zero baryon density, that is for $\mu_B \neq 0$, there is just one (or perhaps at most a few) special value $\mu_B^0(T)$ for which $\langle s \rangle = \langle \bar{s} \rangle$ at $\mu_s^{HG} = 0$, which condition mimics the QGP. For the case of a conventional HG (Hagedorn type) these values have been carefully studied.[14,17] For the final state described by Eq. (5) this condition of strangeness conservation takes a simple analytical form:

$$\mu_B^0 = 3T\cosh^{-1}\left(R_C^s \frac{F_K}{2F_Y} - \gamma_s \frac{F_\Xi}{F_Y}\right), \qquad \mu_s^{HG} = 0. \tag{8}$$

There is at most one real solution, only when the argument on the right hand side in Eq. (8) is greater than unity. We will discuss some important properties of the solutions of Eq. (8) below.

Temperature

The $m_\perp$-spectra of the strange particles used in the multiplicity analysis indicate a common temperature (inverse slope) $T_\perp = 232 \pm 5$ MeV. We emphasize that this temperature is not necessarily the freeze-out temperature of final state particles, but that it is the flow blue-shifted freeze out value: the particles observed in the collision are emanating at some T_f — it cannot be assumed that strange and non-strange particles are freezing out exactly in same condition[21,22] and hence it is convenient to introduce the strange-particle freeze-out temperature T_s and in view of the smaller strange particle cross sections very probably $T_f \leq T_s$. The freeze out could occur in principle well below the temperature $T_\perp$ implied by the transverse particle spectra in which the collective flow velocity v_f, v_s superposes a Doppler blue shift on freeze-out values. We have:

$$T_\perp \leq T_0 \equiv \sqrt{\frac{1 + v_f}{1 - v_f}}T_f, \qquad T_\perp \leq T_0 \equiv \sqrt{\frac{1 + v_s}{1 - v_s}}T_s. \tag{9}$$

As indicated here the initial temperature T_0 is believed to be little higher than the inverse slope spectral temperature $T_\perp$ implied by the high $m_\perp$ particle spectra.

In our opinion it is inaccurate to argue here[23] that the uncertainty in the value of the initial temperature is large, since one has two model dependent extrapolations to perform: from the observed value of the inverse slope the flow component must be separated to arrive at the true freeze-out value of temperature, which then needs to be pursued back in time till the initial time at which the radial flow disappears — we are not at all interested in such a procedure. For the purpose of determining the initial temperature at which strangeness has been produced and radial flow sets in, we do not need to determine the clearly uncertain freeze-out point. Rather, within a model considered, we assume some initial temperature and given the equations of state we can compute accurately the hydrodynamic radial expansion; one finds that at the high $m_\perp \simeq 2$ GeV considered here the inverse slope of the particle distributions is equal or a bit smaller than the initial temperature.[24] Thus we can safely conclude that the observed inverse slope of $\overline{\Lambda}$ is pointing to a primary temperature at which strangeness is present, $T_0 \simeq 235$ MeV.

CURRENT RESULTS

Remarks About the 200 GeV A Data

The experimental results obtained in 200 GeV A S-beam CERN–SPS collisions indicate that the observed strange particle flow is compatible with the source being a rapidly hadronizing QGP state. Here we cite the common results of all experiments:

1) $\mu_s = 0$;

2) the presence of considerable multi-strange particle yields and

3) the anomalous high specific entropy S/B, that is the higher than anticipated yield of pions per participating baryon.[14,18,19]

Furthermore, it was theoretically found[7,20] that the QGP equations of state allow an adequate description of the evolution of the central fireball conditions leading in the freeze-out to the actually observed experimental results.

Can this data be also described by the equilibrated HG model? That this cannot be done we can see taking in Eq. (8) the value $R_C^s = 1$, and solving this equation to determine quantitatively in how far can the strangeness conservation constraint be reconciled with the present data. In Fig. 1 the solid line is just $3\mu_q^0 = \mu_B^0(T; R_C^s = 1, \gamma_s = 0.7)$, see Eq. (8). The dashed and, respectively, dotted curves are the solutions of the strangeness conservation condition for the experimentally permitted range of $\lambda_s = 0.98$ and, respectively, $= 1.08$ (the mean experimental value is at $\lambda_s = 1.03$,[14]). The two straight dissecting curves correspond to the experimental constraint for WA85 data, $\lambda_q = 1.48 \pm 0.05$ (the disagreement of the HG model is even greater for the NA36 results, which, alas, have greater errors). The hatched area is the region of μ_B and T compatible with a hadronic gas source in which meson to baryon abundance equilibrium is reached. Note in the here interesting domain of values of $100 < \mu_B < 400$ MeV there is great sensitivity of the allowed values of μ_B to the value of T. Taking $T_0 \geq 232$ MeV, see Eq. (9), we show in Fig. 1 the experimental cross, 3 s.d. to the right of the range consistent with the experiment.

Before proceeding further we recall that an initial analysis of the first data, the "experimental" point coincided with the hatched area[21,25] in Fig. 1. The increased precision, more recent data, lead in the interim to a shift of the experimental point to the right by about 1 s.d.; the theory was also further developed and includes today all known hadronic resonances. This has moved the solid line ($\lambda_s = 1$) in Fig. 1 by about 1 s.d. to the left, to its current location. Finally, the disagreement between HG-theory and experiment is enhanced by another

1 s.d. since previously we had $\lambda_s > 0.93$ (within 1 s.d.) and now it is $\lambda_s > 0.98$ (dashed line in Fig. 1), which moves the error boundary by one further s.d. to the left, to its current location in Fig. 1. Thus we can infer that a strangeness neutral HG source, even allowing for off-equilibrium strangeness saturation $\gamma_s = 0.7$, is in disagreement with experimental results when the initial temperature is considered.

But considering the incompatibility of the HG phase with the particle yields that Fig. 1 implies, how did the final state particles emerge from say deconfined phase?

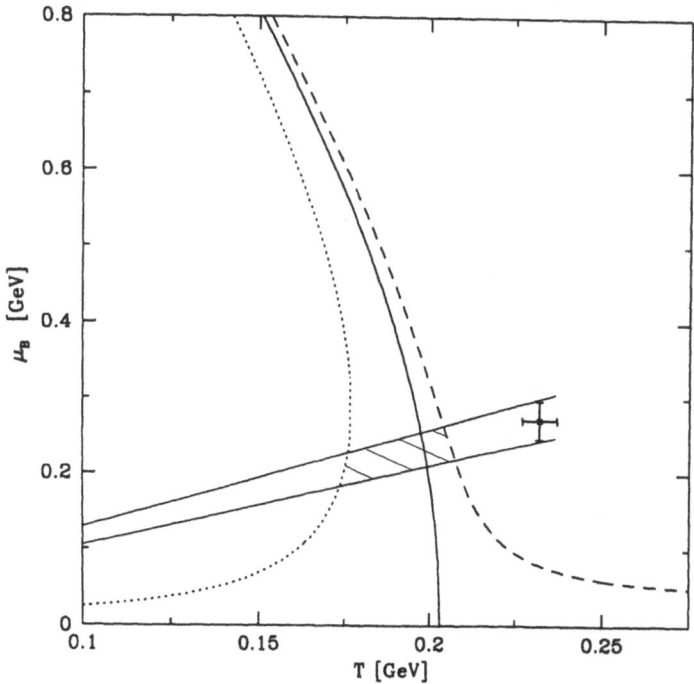

Figure 1. Strangeness conservation constraint in the μ_B–T_f plane. Solid line: $\lambda_s = 1$, dashed curve: $\lambda_s = 0.98$, dotted line: $\lambda_s = 1.08$ (all with $R_C^s = 1$, $\gamma_s = 0.7$). Horizontal intersecting lines correspond to $\lambda_q = 1.48 \pm 0.05$. Hatched area is the region of agreement (within 1 s.d.) of the strangeness conservation condition with the observed values $\lambda_s = 1.03 \pm 0.05$ and $\lambda_q = 1.48 \pm 0.05$. The experimental cross is set at the maximal possible temperature, $T_f = 0.232 \pm 0.005\,\text{GeV}$, consistent with $m_\perp$-spectra.

Case of Sudden disintegration without radial flow:

In order to move the strangeness conservation constraint line to the 'experimental' point, we need to take off-equilibrium value $R_C^s = 1.45$. What this means is that in the primordial source freeze-out, the relative abundance of strange mesons compared to the equilibrium value should be enhanced by a factor 1.45, or that the relative abundance of strange baryons compared to the chemical equilibrium should be suppressed by a factor 0.69, or both at suitable intermediate value.

Case of low temperature freeze-out:

If, however, we are persuaded that the freeze-out temperature is around 150 MeV, which necessitates presence of a freeze-out flow velocity at 0.41c, then it turns out that $R_C^s = 0.39$ and accordingly this implies opposite meson-baryon relative enhancement/suppression trends than just described before.

In either case, it is apparent that the hypothesis only recently entertained of a fireball of hot hadronic gas is nearly (at 3 s.d. level) incompatible with these results. Another serious problem for the HG model source is that it does not reproduce the large measured charged particle multiplicities and thus the entropy content of the final state. Indeed, the final state entropy is largely present in the initial state already, since the fireball is an isolated physical system. From the measured values of temperature and chemical potentials the specific (per baryon) entropy for a given initial state can be evaluated and it turns out to be a characteristic quantity which distinguishes the HG phase from a deconfined state of quarks and gluons.[19,25] The fact that the entropy content in the QGP phase is greater than in HG is due to the excitation of the abundant gluon degrees of freedom.

The question has been raised if we can draw this conclusion that QGP has been formed, considering that there may be some experimental disagreement. As our detailed analysis has shown above, strangeness conservation constraint in hadronization of QGP, as expressed by Eq. (8), is a very sensitive function of the temperature. It is conceivable that different experiments with differing, even though similar triggering and acceptance conditions obtain in their data sample slightly different values of T and hence a quite different value of μ_q^0 (i.e. λ_q^0). As we saw in last subsection, the value of λ_q controls in a sensitive way the particle ratios, such as $\overline{\Lambda}/\Lambda$, and these can easily vary from experiment to experiment, while the *extracted* results regarding the parameters μ_s, γ_s, which are determined by more involved combinations of particle ratios, should agree for different experiments.

It is from this perspective that we view the results of the different experiments presented here (WA85, NA35, NA36) and discern experimental agreement between all the data. The physical quantities we compare are the strange quark chemical potential μ_s and the strangeness occupancy factor γ_s. For example, the greatest difference recorded at the workshop arose between[5] NA36 and WA85 particle ratio results, which yield considerably different values λ_q, the quark fugacity. However, both μ_s and γ_s are consistent between the two experiments.

Remarks about the 10–15 GeV A Results

The experimental results, as well as their theoretical interpretation in terms of diverse cascade models indicate transient presence of a clearly defined region of high density matter at central rapidity, seen in particular in the case of the Au–Au collision simulations. This suggests that a local equilibrium fireball model is fully justified in this range of energies. Several features which are more difficult to account for in a microscopic simulation were studied in the thermal model:[26]

- the quantum nature of hadronic particles, in particular relevant for pions;
- the entropy content of the initial state taken as a parameter, chosen as needed to describe the data;
- consistency of the description with the possible transient presence of a locally deconfined quark-gluon plasma (QGP).

In these calculations state of the art equations of state for both HG and QGP fireballs were used:

- for HG one uses a refined Hagedorn approach — resonances are considered usually to be Boltzmann particles, but since also lower temperatures $T \leq m_\pi$ are explored for freeze-out conditions, pions and all mesons were treated as bosons and nucleons and Δ-resonances as fermions.

- for QGP the analytical form of an ideal quark-gluon gas, amended by[16]
1) inclusion of the effective number of degrees of freedom, reflecting the reduction due to perturbative QCD-interactions, and
2) introduction of non-perturbative thermal mass for quarks and gluons.

What we need to explain and understand is the result[27] which shows that there is profound difference between the 200 GeV A results and the 10–15 GeV case in the freeze-out condition: aside of the expected high value of $\lambda_q \simeq 3.5$ one obtains $\mu_s \neq 0$, specifically $1.5 \leq \lambda_s \leq 1.9$, which implies for the freeze-out temperature $T_s \simeq 130$ MeV that $\mu_s = 68 \pm 15$ MeV. In an equilibrated HG at the above value of λ_q one needs in order to satisfy the strangeness conservation $\mu_s \simeq 50$ MeV. This strongly suggests that we need a freeze out state that is a nearly equilibrated HG. But does this eliminate the option that QGP is formed? Actually not. All we can say is that should QGP be formed, it will hadronize such that in the final state the strange particle abundance is well equilibrated — this conjecture can be made consistent with the need to increase or at least conserve the entropy during the QGP transformation into the HG.[26]

Is it in principle possible to decide that QGP was formed in collisions at 10–15 GeV A? Probably so, should one observe abundant production of multiply strange hadrons, in particular antibaryons. Today, there is already in evidence[28] an unusually high (in comparison to $\bar{p}$) abundance of $\overline{\Lambda}$, and a forward rapidity high yield of $\overline{\Xi}$. A simple probe for QGP in this system is the measurement of the ratios of particles such as $\overline{\Lambda}/\bar{p} \simeq \overline{\Xi^-}/\overline{\Lambda} \propto \gamma_s \lambda_q$, which should thus considerably exceed conventional physics based expectations. Once central rapidity data for these particle ratios become available, we shall return to study the issue again.

TOWARDS THE FUTURE

The Past

We recall that the relativistic heavy ion collision program has been implemented because of the hope and expectation to find a simple, statistical system consisting of nearly freely moving quarks and gluons, deconfined in a limited region of space-time. The goal, when there are many degrees of freedom present, is to find simplicity, simplicity reflected in just a few parameters which can determine the richness of the many different final state particle abundances. In such a simple physical environment, we can in turn expect to learn more about the deconfined state.

Is there a dispute today about the possible formation of QGP? It may be perceived that this is the case since some of us are seeking to prove the existence of HG at temperatures beyond Hagedorn temperature, rather than focusing on the more natural QGP hypothesis, which as we have argued, does work well. The surprise we experience is really elsewhere: the big trouble is that thermal models do work. There is slight disagreement between different groups about the question how well they work, but to a surprising degree, there is seen a stunning degree of thermalization. In some yet not understood way the underlying degrees of freedom have come to some kind of a thermal and even chemical equilibrium, if only for light quarks. One can argue that this means that same physics process occurs in p–p, p–A and A–A interactions. This may be actually true, but we have to realize that in p–p collisions we have small volume and in A–A collisions we have large volume, and it seems that for thermalization this difference does not matter, other than it leads to a lesser strangeness suppression. This is really the puzzle in the field, and until this puzzle is fully resolved, we cannot fully trust our understanding. But that puzzle is the old Hagedorn paradox,[29] just that now, thirty years later, we are finally learning to take Hagedorn very seriously.

The Present

Our current problem is really how to use strangeness to prove or disprove QGP. We believe that there is no single experiment today which could be used to say there is no QGP in the relativistic nuclear collisions, while there is considerable body of evidence in favor of QGP formation in interactions at 200 GeV A. We were looking for asymmetry in the phase space of strange and anti-strange quarks, as expressed by the value of λ_s, the fugacity of strange quarks. At finite baryon density there is a difference because s-quarks like to bind in Λ and $\bar{s}$-quarks in K. Because of this, there is a natural asymmetry in a confined system with $\lambda_s \neq 1$, and a natural symmetry with $\lambda_s \simeq 1$ for the deconfined phase. We found the latter result in all 200 GeV A data, while at 10–15 GeV A the situation is different, but still consistent with possible primordial QGP phase. The mere fact that people come day after day with new results about otherwise rarely produced particles $\bar{\Lambda}$, $\bar{\Xi}$, $\bar{\Omega}$ make us worry. To be able to argue against the formation of QGP we would have liked small yields of strange anti-baryons, because these particles are either not produced at all or they are annihilated in a baryon rich confined and equilibrated environment. Finally, it would seem that to question the formation of QGP we would have to be able to predict the strangeness yield using data from p–p and p–A: as discussed at length,[30] this is certainly not the case at 200 GeV A.

Moreover, we have in hand a full and consistent QGP model of the formation and evolution of QGP[7,20] which results in properties of the freeze-out state as needed to fully account for the production of different strange baryons and anti-baryons measured in recent experiments, and also the total particle multiplicity (entropy). We have demonstrated that for the 200 GeV A collisions a description that makes use of confined hadronic particles miss key properties of the final state. Thus the deconfined QGP alternative lends itself as a natural and indeed expected explanation of the data, but we cannot at this stage exclude the possibility that someone will propose a detailed model based on confined particles or some new exotic state, which will be similarly successful.

The Future

Therefore, the QGP picture of the reaction needs to be confirmed in a convincing, systematic and unique way and we propose that this is done studying the forthcoming Pb–Pb 160 GeV A collisions and in particular recording variation of diverse observables as function of certain easily accessible triggers and eventually, *the beam energy*. There is a great difference between a study of the variation with beam energy, and variation of the impact parameter, characterized normally by different trigger conditions. When we explore the behavior of strangeness observables with the impact parameter, assuming that the picture of a centrally formed fireball applies, we are in fact exploring the dependence of the approach to chemical equilibrium of the strange particle abundance, as we vary the volume. On the other hand some observables in this study are negatively impacted by the presence of considerable amount of spectator matter. In particular the reactions $K+N \leftrightarrow Y + \pi$ on the spectator matter will distort the abundances of easily measurable $\Lambda + \Sigma$ and K. Furthermore, we will depend on kinetic models to establish the variation of the baryon compression (and thus λ_q) as the collision centrality changes. It is nearly certain that different models will give here different answers. We thus believe that alone varying the impact parameter in Pb–Pb collisions we will not be able to identify clearly a change from deconfined to confined hadronic fireball formation.

We are, however, more optimistic regarding the exploration of the energy dependence. As the energy of the collision decreases at this relatively high range of energy,for relatively large volumes available in central Pb–Pb interactions, and if our understanding of the kinetic theory of strangeness production in the QGP phase is correct, the strangeness phase space will remain nearly fully saturated $\gamma_s \simeq 1$. Therefore the ratio

$$R_s \equiv \frac{\overline{\Xi}}{\overline{\Lambda + \Sigma}} \propto \gamma_s \lambda_q$$

(we note here that equivalent definitions of R_s involve the ratios of $\overline{\Lambda}/\bar{p}$ and $\overline{\Omega}/\overline{\Xi}$) should change driven only by the (small) changes of λ_q — it is indeed likely that the baryon compression increases driven by increasing stopping, and thus the initial λ_q could increase. This response to the decrease in the energy of the collision is very counter-intuitive, if we base our thinking on our experience with normal hadronic cross section systematics. For the HG phase in which the abundance of multi strange antibaryons should be far from chemical equilibrium, a decrease in collision energy would decrease the secondary production processes, more so for the heavier and more complex objects. Therefore, once the collision energy is reduced below the minimum needed for the deconfined state we would expect that the ratio R_s rapidly decreases with energy — we expect a rather pronounced change near the transition threshold, from as much as a value of $R_s \simeq 0.5$–0.6 down to 0.1–0.05. Such a threshold behavior coupled with the counter-intuitive response to the energy variation would provide a very convincing confirmation of the QGP formation hypothesis and would determine the transition conditions.

Clearly, we wish that the entropy (multiplicity) production as function of energy is also studied, and expect that disappearance of the QGP phase at threshold energy is accompanied by a relatively sudden reduction in the specific entropy production. Moreover, should in the transition region of energy the formation of QGP occur in a fraction of collisions, triggering on high multiplicity would allow to discriminate against the HG fraction of events, and would enhance abruptness in the transition in the other observables.

Are other, less clear scenarios possible, or are we certain that the above 'smoking gun' signature must occur? Unfortunately, we cannot presently exclude the possibility that QGP formation occurs down to 10 GeV A, and thus (other) changes in strangeness flow encountered as we reduce the energy down from 160 GeV A may be caused by a change in the hadronization dynamics of the primordial state. In order to explore and understand this possibility, a measurement at smallest accessible energy of the Pb–Pb system of the same observables that are being measured at 160 GeV A must be made.

We thus see the following experimental agenda for the Pb–Pb experiments: the Pb–Pb runs should be carried initially out at two not too different energies, in order to establish the trends of the observables in energy. Subsequently, experiments and the accelerators should be tuned to lowest possible energy in a search for qualitatively different behaviors of the observables we have discussed here. The final step would be a search for the transition energy, but the strategy and methods will critically depend on the results of the previous steps.

Acknowledgements

J. R. was partially supported by the US-DOE grant DE-FG02-92ER40733. He thanks his co-authors for their very warm hospitality in Paris.

REFERENCES

1. H.-Th. Elze and P. A. Carruthers, in this volume.

2. J. Rafelski, *Phys. Rep.* **88** (1982) 331;
   J. Rafelski and M. Danos, *Phys. Lett.* **B192** (1987) 432.

3. F. Antinori, in this volume, and
   S. Abatzis *et al.* (WA85 collaboration), *Phys. Lett.* **B270** (1991) 123; **B259** (1991) 508; **B316** (1993) 615.
   D. Evans et al. (WA85 collaboration), *Nucl. Phys.* **A566** (1994) 225c.

4. R. Stock, in this volume, and
   T. Alber *et al*, The NA35 collaboration, *Strange Particle Production in Nuclear Collisions at 200 GeV per Nucleon Z. Phys.* **C** in press, Univ. Frankfurt preprint, IKF-HENPG/1-94, April 1994.

5. H. Rohringer et al. (NA36 collaboration), in this volume;
   E. Andersen *et al.*, NA36 Collaboration, *Phys. Lett.* **B316** (1993) 603; *Nucl. Phys.* **A566** (1994) 217c; *Phys. Lett.* **B327** (1994) 433.

6. G. Odyniec, in this volume.

7. J. Letessier et al, in this volume.

8. J. Cleymans, in this volume.

9. J. Rafelski, *Phys. Lett.* **B262** (1991) 333; *Nucl. Phys.* **A544** (1992) 279c.

10. J. Rafelski and B. Müller, *Phys. Rev. Lett.* **48** (1982) 1066.

11. P. Koch, J. Rafelski and B. Müller, *Phys. Rep* **C142** (1986) 167.

12. N. Bilić, J. Cleymans, I. Dadić and D. Hislop, *Gluon Decay for Strangeness Production in Quark-Gluon Plasma*, University of Cape Town *Preprint August 1995*

13. P. Koch and J. Rafelski, *Nucl. Phys.* **A444** (1985) 678;
    P. Koch, B. Müller and J. Rafelski Z. *Physik* **A324** (1986) 453.

14. J. Letessier, A. Tounsi, U. Heinz, J. Sollfrank and J. Rafelski, *Strangeness Conservation in hot nuclear fireballs*, Paris preprint LPTHE/92–27R, *submitted to Phys. Rev. D.*

15. J. Sollfrank, M. Gaździcki, U. Heinz and J. Rafelski, Z. *Physik* **C61** (1994) 659.

16. J. Letessier, J. Rafelski and A. Tounsi, *Phys. Lett.* **B323** (1994) 393.

17. J. Letessier, J. Rafelski and A. Tounsi, *Phys. Lett.* **B321** (1994) 394.

18. M. Gaździcki, in this volume.

19. J. Letessier, A. Tounsi, U. Heinz, J. Sollfrank and J. Rafelski, *Phys. Rev. Lett.* **70** (1993) 3530.

20. J. Letessier, J. Rafelski and A. Tounsi, *Phys. Lett.* **B333**, (1994) 484.

21. J. Cleymans and H. Satz, Z. *Phys.* **C57** (1993) 135.

22. J. Cleymans, K. Redlich, H. Satz, and E. Suhonen, Z. *Phys.* **C58** (1993) 347.

23. H. Satz, private communication during this meeting.

24. J. Letessier and A. Tounsi, *Phys. Rev.* **D40** (1989) 2914;
    M. Kataja, J. Letessier, P.V. Ruuskanen and A. Tounsi, Z. *Physik* **C55** (1992) 153.

25. J. Letessier, A. Tounsi and J. Rafelski, *Phys. Lett.* **B292** (1992) 417.

26. J. Letessier, J. Rafelski and A. Tounsi, *Phys. Lett.* **B328** (1994) 499.

27. J. Rafelski and M. Danos *Phys. Rev* **C50** (1994) 1684.

28. S. Steadman, in this volume.

29. R. Hagedorn, in this volume.

30. See report on *Miniworkshop on Strangeness* by J. Rafelski, in this volume.

HADRONIC PHYSICS:
THE COSMIC RAY PERSPECTIVE

Lawrence W. Jones*

University of Michigan
Department of Physics
Ann Arbor, Michigan 48109-1120

INTRODUCTION

It is an honor and a great pleasure to join in the recognition of the distinguished career of Dr. Rolf Hagedorn. Before turning to a discussion of cosmic ray hadronic physics, I would like to recall some of my own interactions with Rolf. My first extended visit to CERN was in 1961, following a series of experiments at the Berkeley Bevatron on pion-proton elastic scattering. From these early experiments at momenta below 3 GeV/c, we had seen the diffraction peak and, at larger angles, a small but clear more-or-less continuous component to the elastic scattering which fell steeply with energy.[1] In considering a possible source for this large angle component, I recalled a much earlier Helvitica Physica Acta paper by Weisskopf on "Compound Elastic" scattering in nuclei, wherein a compound nucleus is formed following interaction with an energetic nucleon, and one of its possible decay modes is the original state. As I had read some of Hagedorn's papers on the statistical interpretation of elementary particle interactions, I went to him to discuss the possible applicability of this nuclear concept to the higher-energy pion nucleon problem we had been studying. He was very supportive and interested, and we had many substantial discussions on this subject during my stay at CERN. An early result of these discussions was a short paper on "A Statistical Interpretation of Large-Angle Elastic Scattering".[2] It was always a great pleasure to talk with Rolf; he was a patient teacher and clear expositor as well as an altogether delightful personality.

COSMIC RAY HADRONIC PHYSICS
THE GENERAL PROBLEM

Hadronic interactions of cosmic rays have been studied by many techniques. Of course the most direct is the placement of a detector package in a satellite, or (more commonly) a balloon which can expose the detector to primary cosmic rays above the Earth's atmosphere. The problem with this approach is the very low flux of primary cosmic rays at interesting

---

*Supported in part by the U.S. National Science Fundation

energies. Thus, the flux above 10^{16} eV is only one per m^2 per steradian per year, and even above 10^{14} eV the flux is still only a few per m^2 per steradian per day. Hence the practical upper limit energy explored with even rather ambitious balloon payloads is about 10^{14} eV.

Ground-based electronic detectors have been used to detect muons and the electromagnetic component (the "air showers") of the cascades initiated in the upper atmosphere by primary cosmic rays, and some mountain installations have exploited hadron calorimeters with electonic readout, often in conjunction with muon and shower detectors. From such parameters as the muon-electron ratios vs. energy, information can be obtained on the probable mass spectrum of the primary cosmic rays and on other parameters of the interactions.

Much of the cosmic ray information concerning the nature of high-energy particle interactions comes from "emulsion chambers" exposed at mountain-top elevations (under 500 - 600 g/cm^2 of atmosphere). The emulsion chambers are large areas of stacks of alternating lead plates and nuclear emulsions or X-ray film (sometimes both), built up to 12 or more radiation lengths depth. The chambers have areas of over 30 m^2. In some cases a second, similar chamber is located below the first, with an intervening layer of hydrocarbon in which hadrons may interact. These chambers are exposed for periods of up to a year; the emulsions and X-ray films are then developed, and the film scanned for the clusters of spots formed by electromagnetic cascades.

From the film density at different depths, the energy of the initiating gamma (or electron) can be learned. The threshold energy for gammas is 2 or 4 TeV, typically. Hadron interactions are only seen from the final state energy fraction in gammas (π^0s). The resolution of the emulsions permits substantially parallel cascades to be identified as from the same nuclear interaction in the air above the chambers (or in the hydrocarbon layer of the chamber system itself), and in many cases to extrapolate shower directions back in the atmosphere to the parent nuclear interaction vertex. Such a collection of gamma cascade cores is referred to as a "family". Over the past 25 years a total of greater than 4000 m^2-years of emulsion chambers have been exposed on mountains in Tibet, Japan, Tadjikistan, Kazakstan, and Bolivia, and from these exposures several hundred events have been analyzed.

Where the interaction can be identified, this technique can provide a great deal of information about a particular cosmic ray event; the distribution of transverse momenta, the hadron to gamma (charged to neutral pion) ratio, etc. However the overburden of several interaction mean free paths of air not only reduces the incident flux of energetic nucleons but also dilutes the information on each event with possible products of interactions higher in the atmosphere initiated by the same primary. The situation is further complicated by uncertainty in the primary cosmic ray composition. It is known that primary cosmic rays at energies of 10^{15} to 10^{16} eV contain a mix of nuclei, from protons to iron. Heavier nuclei photo-dissociate in the Coulomb field of air nuclei high in the atmosphere, so that even at high mountain altitudes, cascades initiated by more than one nucleon may be essentially colinear and intertwined.

The cosmic ray emulsion groups typically set a threshold of 100 TeV as the minimum summed visible energy of the cascades in a "gamma family", corresponding to a nucleon of at least two or three times that energy interacting with an air nucleon. Hence these studies have focused on nucleon- nucleon interactions in the range from about 3×10^{14} to over 10^{16} eV. The Fermilab Tevatron has been collecting data from $\bar{p}p$ collisions of 1.8 TeV c.m., equivalent to 1.7×10^{15} eV for a nucleon on a stationary nucleon target, and earlier the CERN $\bar{p}p$ collider had studied reactions at about 1/10 this equivalent cosmic ray energy. Hence it is easy to suppose that all interesting events seen in cosmic rays would have been thoroughly studied at these colliders.

However there is a significant difference in the emphasis between the collider and the cosmic ray studies in this overlapping energy range. At 1.8 TeV c.m., the full kinematic range of rapidity is: $-7.5 < y < +7.5$, corresponding to a lab (cosmic ray) rapidity y from 0 to 15. The collider experiments have focussed specifically on the central region of pseudorapidity (η), mostly the region $\eta < |\ 4\ |$ (here we will consider pseudorapidity, η, and rapidity, y, as equivalent). This is the region where most of the secondary particles are produced and which is certainly the most rich in massive particles produced close to threshold, such as the W, the Z, and the top quark.

Cosmic ray experiments, on the other hand, are sensitive to energy flow, or to d(EN)/dy. Since $E = m_\perp cosh\ y$, for large y and modest, nearly constant average $m_\perp$, the secondary

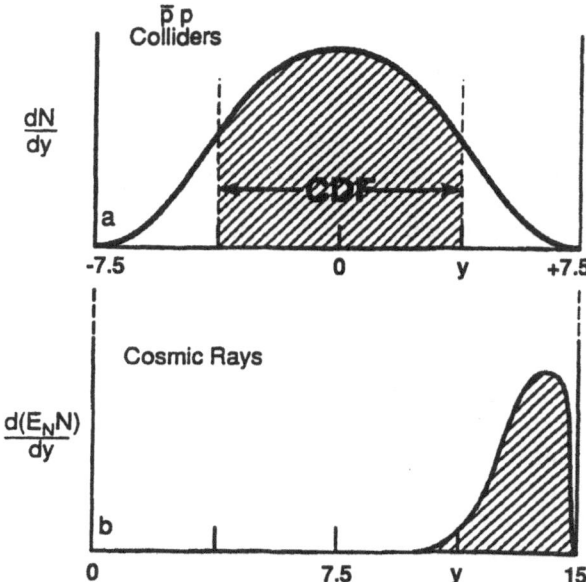

Figure 1. The kinematic range of rapidity, y, for the Tevatron collider (1.8 TeV) and for the equivalent cosmic ray nucleon-nucleon energy (1.7×10^{15} eV). The particle densisty vs. rapidity is shown (approximately) in (a) for the Tevatron collider, with the range of sensitivity of the Fermilab Collider Detector noted. In (b) the same rapidity space for cosmic ray experiments is shown with enegy flow, d(NE)/dy, vs. rapidity plotted. The curves are from T.K. Gaisser.

particle energy is proportional to exp(y), so that

$$d(EN)/dy \propto e^y(dN/dy). \tag{1}$$

These two situations are illustrated in Figure 1, which emphasizes the different kinematic region explored by the two classes of experiment. Thus, it is possible that unusual phenomena may have been observed in cosmic ray experiments and not in accelerator experiments in spite of overlapping energy range due to the different regions of η (y) space studied.

SOME COSMIC RAY RESULTS

From the various techniques discussed above, one conclusion concerning physics in the far-forward region is that "Feynman scaling" does not hold. Feynman scaling would predict that the differential production of secondary particles in a nucleon-nucleon reaction is a function only of $y_{max} - y$, independent of the incident energy. Thus, the cosmic ray data suggest:

$$\frac{1}{\sigma}\frac{d\sigma}{dy} \neq f(y_{max} - y). \tag{2}$$

Specifically, there is evidence that there are fewer fast forward secondaries at higher incident energies than predicted from scaling and the lower energy data.[3]

Another, more controversial conclusion is that the inelasticity decreases with energy. The inelasticity, K, can be defined as:

$$K = <1 - E'/Eo>, \tag{3}$$

where Eo is the incident nucleon energy, and E' is the energy of the most energetic hadron following the collision. At lower energies (100 - 1000 GeV), $K \cong 0.5$. Ding and Ohsawa have argued that K decreases with increasing incident energy, in part based on data from the UA-5 and UA-7 experiments at the CERN collider.[3,4] It should be noted that this conclusion is not universally shared. Slavatinsky, for example, has argued recently that the inelasticity rises to 0.85 at an incident energy of 10^{16} eV.[5]

A third conclusion, from the study of the muon-electron ratio in particular, is that the primary cosmic ray composition becomes heavier (the average value of atomic weight of primary cosmic ray nuclei increases) in the energy interval between 10^{14} and 10^{16} eV. This trend appears to continue a trend seen in the direct, balloon observations of primary cosmic rays below 10^{14} eV, and is consistant with models of magnetic confinement in the galaxy.[6]

Unfortunately, there is one essential problem in the unambiguous interpretation of the indirect (ground-based) cosmic ray data as regards both primary composition and particle physics. The conclusions concerning the character of the forward-production physics are not independant of the assumed mass number spectrum of the incident cosmic ray nuclei. The cosmic ray community sorely needs comprehensive, reliable data on the forward production of all (charged and neutral) particles following a nucleon-nucleon collision at the Tevatron collider energy (1.8 TeV) and below. It is difficult to have a spectrometer that extends through 0 degrees at the Tevatron in view of the vacuum pipe and other obstructions, but I believe that it would be possible to make these measurements with sufficient effort, cleverness, and support. The data from the UA-5 and UA-7 experiments at the CERN collider have been valuable, limited though they may be.

There are many other phenomena reported by and discussed by the cosmic ray emulsion chamber groups. These include the observation of an anomalous number of "aligned" events (wherein the shower cores in a gamma family are aligned rather than more symmetrically distributed), and possible evidence for a "long flying component" in hadron interactions, which gives rise to an attenuation of the cascade in lead more slowly than calculated. Interesting as these and other specific issues may be, I shall chose to focus the remainder of this discussion on a single problem, the Centauro·phenomenon.

CENTAUROS AND ANTI-CENTAUROS

The first group to exploit the emulsion chamber technique (described above) was a Brazil-Japan Collaboration, working at the Mt. Chacaltaya laboratory in Bolivia, at an

496

elevation of 5200 m. They have exposed a series of emulsion chambers over about the past 25 years, totaling hundreds of m²-years of exposure. Of particular interest are data from the "two-storey" chambers, where two emulsion chamber stacks are separated vertically by about 1.8 m, with a 30 cm layer of pitch (hydrocarbon) immediately below the top chamber. Most events (gamma families) observed appear as in Figure 2, wherein there are a number of gamma cascades visible in the top chamber (from neutral pions) and another set of gamma cascades which originate in the lower chamber, probably most of them due to charged pions interacting in the pitch. If the event vertex is not too high above the emulsion chamber, the numbers of charged and neutral pions produced in the parent interaction may be readily deduced; on the average, the ratio of the number of neutral pions to all pions should be one third, and this ratio should show a binomial distribution about this mean value (in the observed region of $\eta - \phi$ space).

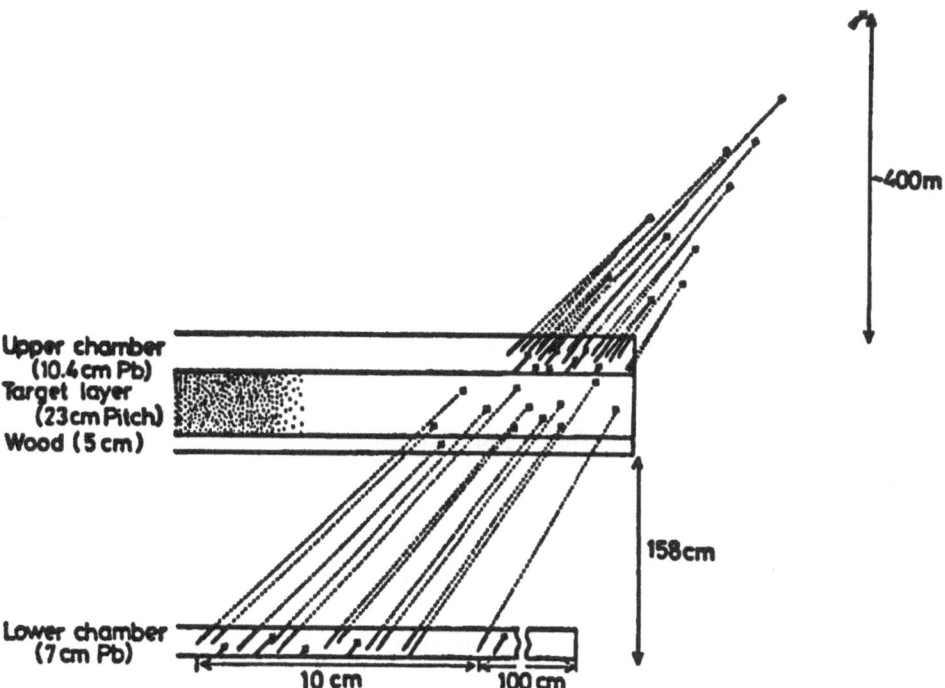

Figure 2. A "typical" event, as recorded in the Brazil-Japan Emulsion Chamber collaboration on Mt. Chacaltaya. Note the different horizontal and vertical scales.

In 1971, the Chacaltaya group reported a most unusual event; a large number of cascades were seen in the lower chamber (suggesting a large number of interacting hadrons in the pitch) and almost no cascades were seen in the upper chamber. Fortunately, triangulation from the cascades demonstrated that the primary interaction occured only 50 m above the apparatus, so the interpretation seemed reasonably clean. As the event as seen in the lower emulsion chamber appeared to come from a totally different "beast" than the event as seen in the upper chamber, it reminded the authors of the mythical Greek creature, the Centaur, whose lower half appeared as a horse and whose upper half appeared as a man.

Subsequent exposures by this group produced more "Centauros"; their characteristics are summarized in a review paper by the group[7] and by papers at various cosmic ray conferences and symposia.[8] As the vertex heights above the detector of these later events are all greater than the 50 m of the first event, they are not as "clean", in that some secondary hadrons may have interacted in the intervening atmosphere and produced neutral pions which gave rise to showers in the upper chamber. Never-the-less, the group reports these as good Centauro candidates. Clearly, they do not fit easily onto a simple binomial distribution of the sort discussed above for charge-independence of the produced secondary pions.

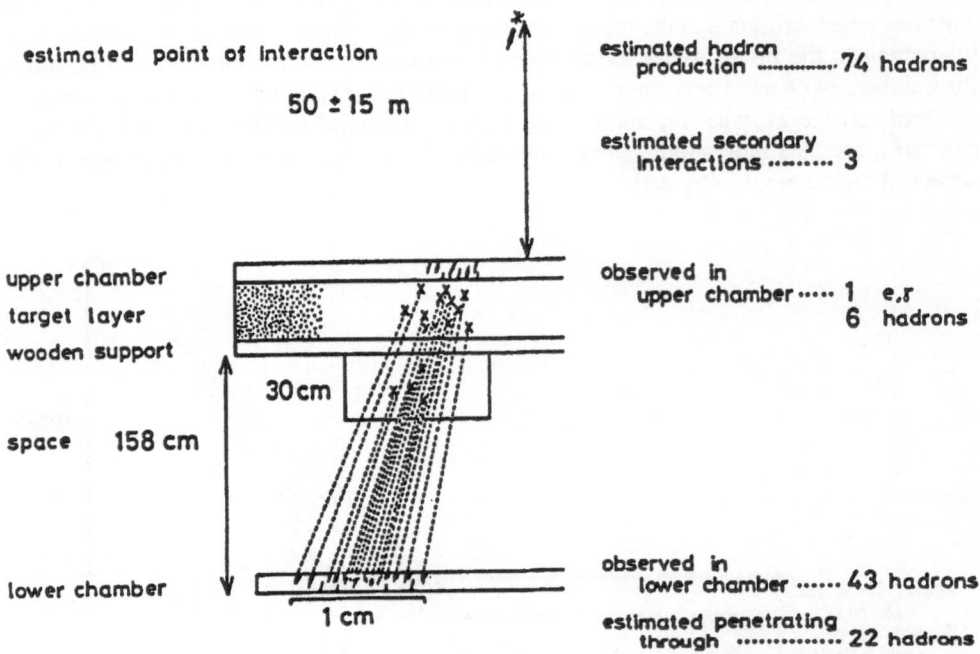

Figure 3. The Centauro I event, from the Brazil-Japan Emulsion Chamber Collaboration.[7]

Table 1. Summary of Properties of Five Centauro Candidates from the Brazil-Japan MT. Chacaltaya Emulsion Chambers

Centauro interaction					
event number	I	II	III	IV	V
height of interaction in meters	50	80	230	500	400
estimated number of A-jets	3	5	13	32	18
hadrons estimated at the interaction					
number	74	71	76	90	63
total energy in TeV	330	370	350	340	350
(e, γ) estimated at the interaction	0	0	0	4	0

A diametrically opposite phenomenon, events containing anomalously few charged hadrons, has been reported by the Japanese American Cooperative Emulsion Experiment from balloon-borne nuclear emulsion stacks. The balloon payload in this series of exposures consisted (from the top down) of nuclear emulsions interleaved with low-Z absorbers as a target layer, a spacer layer to permit cascades to spread laterally, and a calorimeter of alternating lead plates and nuclear emulsions. Flying at the top of the atmosphere (under only 5 or 6 g/cm^2 of air), these balloon packages have studied the primary cosmic ray composition up to about 5×10^{14} eV from the measured ionization in the upper emulsion layers. Primary cosmic rays usually interact in the target layer, and charged secondaries are recorded in lower emulsion layers. In the calorimeter layer electromagnetic cascades develop which identify gammas from the interaction. Gamma energies are, of course, determined; the calibration is verified by reconstructing the invarient mass of $\pi^0 s$.[9]

The observation relevant to this discussion concerns the distribution of charged and neutral (as seen from the gamma cascades) secondaries from energetic interactions. In $\eta - \phi$ space, the distribution of gammas and of charged reaction products is frequently not random but displays clumping outside of expected statistical fluctuations. One particular event is shown in Figure 4 which contains an area in eta-phi space including over 30 gammas and only one charged track. The primary interaction was produced by a proton of about 10^{14} eV. For obvious reasons, such events have been referred to as "Anti-Centauros".

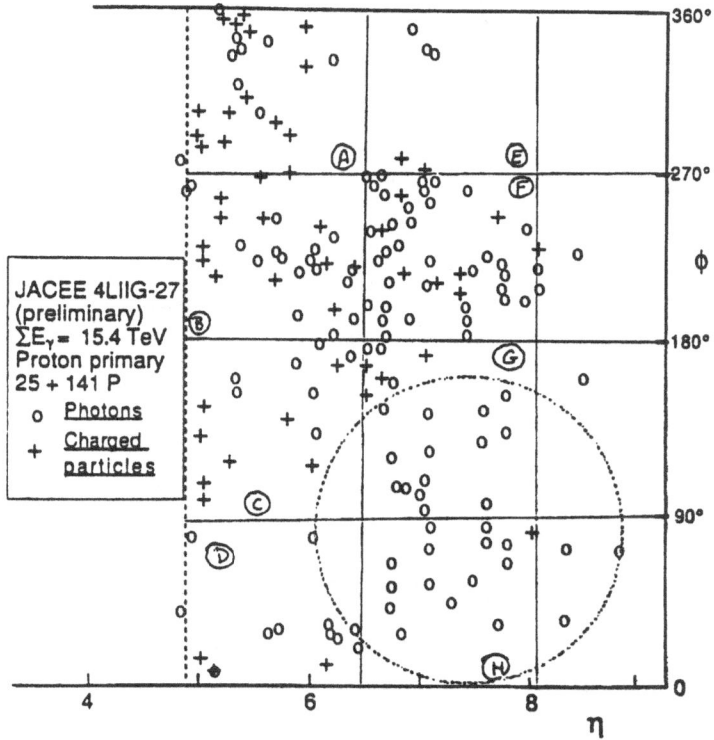

Figure 4. An event from the JACEE collaboration.[9] The region of $\eta - \phi$ space within the dotted circle contains an anomalously high ratio of gammas to charged secondaries.

DISORDERED CHIRAL CONDENSATES AND MINIMAX

J. D. Bjorken and C. Taylor have suggested a mechanism which would explain the Centauro and Anti-Centauro phenomena.[10] They suggest that, following a hard collision between two quarks in a high energy nuclear interaction, there is a highly-excited volume of the vacuum which is expanding rapidly (at the speed of light). Inside this expanding shell, they suggest that the isotopic spin vector may be "frozen", so that when the volume materializes to physical mesons, they may be all of one isospin projection; e.g. all neutral or all charged. More specifically, they define a fraction f, the ratio of the produced $\pi^0 s$ to all pions, and argue that the distribution of f from such "Disordered (or Disoriented) Chiral Condensates" is:

$$N(f) \propto f^{-1/2}. \tag{4}$$

As Dr. Taylor expands on this more completely in his contribution to this Workshop, I will not futher reveal my theoretical naivete.

In an attempt to seek further experimental evidence for this DCC phenomenon, a small experiment has been set up at the Fermilab Tevatron $\bar{p}p$ collider. The experiment, "Minimax", consists of a telescope of multi-wire proportional chambers ranged along side of the accelerator beam pipe to observe secondary charged and neutral particles from beam-beam collisions at an angle of about 50 milliradians. Specifically, the telescope subtends a circle in $\eta - \phi$ space of a radius of about 0.6, centered at $\eta \sim 3.8$. At the downstream end of the chamber array is a matrix of lead-scintillator electro-magnetic calorimeters, and a lead converter can be inserted into the wire chamber plane array to convert gammas among the reaction secondaries. The system has been assembled and tested; secondary particles are cleanly seen from the interaction vertex, and the background problems are understood. A new vacuum chamber is being fabricated and will be installed soon to provide a thin window (rather than a thick flange) between the telescope and the intersection region, and to provide other improvements to reduce background tracks in the chambers. Serious data collection will occur early in 1995. The detector configuration is sketched in Figure 5. A list of the collaborators on the experiment is noted in Table II.

Table 2. Participants in Fermilab Test/Experiment T864

Case Western Reserve University	Fermilab	University of Michigan
K. del Signore	P. Colestock	R. Ball
W. Fickinger	B. Hanna	H.R. Gustafson
T. Jenkins	M. Martens	L. Jones
E. Kangaas		M. Longo
M. Knepley		
K. Kowalski	SLAC	
C. Taylor*	J. Bjorken*	
Duke University	**University of Tennessee**	**Virginia Polytechnic Institute**
S. Oh	A. Weidemann	A. Abashian
W. Walker		D. Haim
		N. Morgan
*Spokesmen		

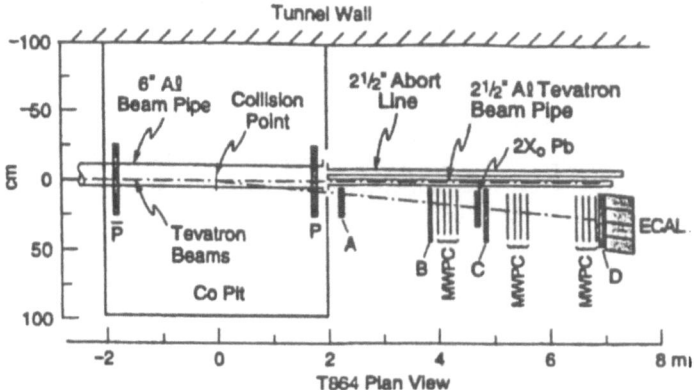

Figure 5. Plan view of the Minimax apparatus installed at the C0 $\bar{p}p$ interaction region at Fermilab.

CONCLUSION

Cosmic ray observations continue to provide hints of unusual phenomena at energies and in kinematic regions unexplored by particle accelerator experiments. While some of these reports may in the end prove to be only manifestations of known physics, experience should teach us to also expect some surprizes. In the case of the Centauro and Anti-Centauro phenomena, an experiment has been set up to explore the charged-to-gamma production ratio at the Fermilab Tevatron in the energy and kinematic region corresponding to the cosmic ray observations. It is our hope that a closer liason between the cosmic ray and accelerator experimental communities will stimulate other opportunities for similar experiments.

REFERENCES

1. K.W. Lai, L.W. Jones, and M.L. Perl, Phys. Rev. Lett. **7**, 125 (1961).

2. G. Fast, R. Hadadorn, and L.W. Jones, Nuovo cimento **27**, 856 (1963).

3. A. Oksawa, p. 466 Proceedings of the VII International Symposium on Very High Energy Cosmic Ray Interactions, Ann Arbor (L.W. Jones, ed) AIP Conf. Proc. 276 (1992), and vol. 4, p. 25 Proceedings of the XXIII International Cosmic Ray Conference, Calgary (1993).

4. L.K. Ding, p. 28 ibid.

5. S. Slavatinsky, Proceedings of the VIII International Symposium on Very High Ener;gy Cosmic Ray Interactions, Tokyo (1994) (to be published).

6. Y. Takahashi, p. 275 Proceedings of the VIII International Symposium on Very High Energy Cosmic Ray Interactions, Ann Arbor, (L.W. Jones, ed) AIP Conf. Proc. 276, (1992).

7. C.M.G. Lattes, Y. Fujimoto, and S. Hasegawa, Phys. Rep. **65**, 151 (1980).

8. S. Hasegawa "Centauro Species in Cosmic-Ray Observation", ICR Report 151-87-5 (1987) (unpublished).

9. J. Lord, and J. Iwai, paper 515 submitted to the International Conference on HEP, (Dallas 1992); J. Iwai (JACEE Collaboration) UWSEA-92-06; T. Burnett, et al, Nucl. Inst. and Methods **A251**, 583 (1986), W.V. Jones, et al. Ann. Rev. Nucl. Sci. **37**, 71 (1987).

10. J.D. Bjorken, Int. J. Mod. Phys **A7**, 4189 (1992). K. Kowalski, and C. Taylor, Preprint CWRU hep-ph/9211282 (1992), and J. Bjorken, K. Kowalski, and C. Taylor, Proc. of the Workshop on Physics at Current Accelerators and Supercolliders (Argonne National Laboratory), J. Hewett, A. White and D. Zeppenfeld, eds) ANL-HEP-CP-93-92 (1992), J.B. Bjorken, International Symposium on Multiparticle Dynamics (Aspen, 1993).

MINIMAX: PROGRESS AND PLANS

Cyrus Taylor

Department of Physics
Case Western Reserve University
Cleveland, OH 44106-7079, USA

INTRODUCTION

It is a great honor to be a part of this conference in recognition of the achievements of Dr. Rolf Hagedorn. His work on the statistical thermodynamics of strongly interacting systems has certainly had a great influence on me. I first carefully studied his work 10 years ago, when I was a beginning post-doctoral fellow excited by the promises of string theory. More recently, his work undoubtedly influenced the way that I (with J. D. Bjorken and K. L Kowalski) tackled the problem of constructing a model for the formation of disoriented chiral condensates. It has been a delight to finally meet and interact with Dr. Hagedorn.

DISORIENTED CHIRAL CONDENSATES

The vacuum is a very complicated place. As understood in quantum field theory, it can carry long-range order parameters, and thus it may be possible to study coherent phenomena associated with the alteration of the properties of the vacuum over long distances.[1] Two questions that immediately arise when approaching this problem are:

1. Can one imagine creating such phenomena in the laboratory?

2. Would you be able to tell?

The second question is much easier to address theoretically.

In the context of spontaneous breaking of chiral symmetry, the disorientation of the chiral condensate within a "bubble" with a radius of several fermis would have a striking signature:[2-8] it will be coherent, and event-by-event, of a given (cartesian) isospin. In events in which the deflection of the vacuum is in the π^0 direction, the produced pions will be neutral; in other events in which the orientation is orthogonal to the π^0 direction, the particles will be charged.

One can make a specific prediction for the distribution of the neutral fraction of such pions fairly easily. A priori, it is equally probable that the condensate will be disoriented towards any of the (cartesian) isospin directions. Remembering that the fields represent probability amplitudes, it follows that

$$P(f)df = \frac{df}{2\sqrt{f}},$$

Hot Hadronic Matter: Theory and Experiment
Edited by J. Letessier *et al.*, Plenum Press, New York, 1995

where f is the fraction of the pions which are neutral.[3] Lest one worry that we are violating conservation of isospin with this argument, we note that it is possible to determine the probability for seeing $2n$ neutral pions out of $2N$ total pions in such a state[6,9,10] from quantum mechanical considerations:

$$P(n, N) = \frac{(N!)^2 2^{2N} (2n)!}{(n!)^2 2^{2n} (2N + 1)!}.$$

Using Stirling's approximation, it follows that $P(n, N) \sim 1/\sqrt{n/N}$ in the limit that bath n and N become large, thus recovering the $1/\sqrt{f}$ found above.

This distribution is significantly different from the distributions normally expected in multiparticle production, even for fairly small values of N. In conventional multiparticle production, the distribution should be be strongly peaked about $f = 1/3$. It is is obvious that a high percentage of dcc's would produce events with anomalously small neutral fractions; by the same token, there would also be a significant number of events with anomalously large neutral fractions.

We have suggested that events with these signatures may have been observed in cosmic-ray events. Specifically, it is tempting to interpret Centauro events as dcc's with an anomalously small neutral fraction.[11-13] (See L. Jones contribution to this volume for a review of Centauro phenomenology.[14])

A bubble of disoriented chiral condensate (dcc) a few fermis in diameter can arguably be described semi-classically, and the evolution of the condensate can be studied using the equations of motion of the sigma model. Much theoretical effort is going into the study of the plausibility of the formation of such 'disoriented chiral condensates' (dcc's). At the time of this workshop, I count 71 papers devoted to the discussion of the formation and/or evolution of dcc's. Of these, some 40 are in the first half of 1994! The field is developing very rapidly, and space does not permit a comprehensive review here. The calculations are hard, involving intrinsically non-perturbative processes and rather severe approximations. At this stage of our understanding of the theory, it appears to be likely that large ($\Delta\eta\Delta\phi >> 1$) domains of dcc are disfavored. Conversely, very small domains of dcc ($\Delta\eta\Delta\phi << 1$) are probably be unobservable.

MINIMAX?

Given both the theoretical uncertainty surrounding dcc's, and the persistent reports of cosmic ray anomalies, it seems appropriate to complement the theoretical work with experimental searches. Heavy ion collisions offer the environment closest to that assumed in most of the theoretical calculations and proton-ion collisions may most closely replicate the environment of the cosmic ray data. Experiments in these environments should be pursued. In the meantime, it may be fruitful to look for dcc's in $p - \bar{p}$ collisions at the Fermilab Tevatron.

To this end, a small test program to initiate the study of this physics (T-864, "MiniMax", J. Bjorken and C. Taylor co-spokesmen[15]) has been proposed, approved, and is underway at the C0 collision area of the Tevatron during the current collider run.

The T-864 Memorandum of Understanding[16] with Fermilab describes the experiment as "T-864 is a simple, staged test program of very modest scope, cost, and impact on the laboratory which responds to the suggestion of the Fermilab Physics Advisory Committee that it "hopes that efforts will continue to develop possible methods for exploring the large rapidity regime." The experiment will initially investigate the background environment in the forward direction at the C0 collision area with a minimal "maximum acceptance" detector (MiniMax).

This will be done initially in noncollider mode in the far forward direction using only MWPC tracking elements and a simple scintillator-based triggering system. Pending successful completion of the initial test program, MiniMax will then carry out studies of charged particles and gamma rays in the fiducial region defined by the MWPC telescope. Physics goals include both generic multiparticle production studies and, of course, a disoriented chiral condensate search. Both studies will proceed in non-collider mode initially, with requests for short collider runs following the successful completion of the initial program." (Under normal operating conditions of the Tevatron, electrostatic separators prevent collisions from occurring in the C0 hall.)

Why go to the forward direction? There are some theoretical arguments which suggest that the signal to noise ratio will be better there than in the central region.[8] Further, if one interprets Centauro events in a dcc framework in a fireball model,[17] one finds that the central rapidity of a Centauro fireball at the Tevatron is $\eta_c \approx 3.3 \pm 0.5$. Thus, in order to definitively search for Centauro events at the Tevatron, one needs to be sensitive to fireballs of rapidity at least as high as $\eta \sim 4$. One also needs good sensitivity to both charged particles and hadrons.

A drawing of the MiniMax detector is included in L. Jones' contribution to these proceedings. The heart of the detector is an MWPC tracking telescope, consisting initially of 12 planes. Each plane has 128 wires and an active area of 32 cm by 32 cm. The chambers operate on 80% Ar/20% CO2, with efficiencies in excess of 95%. The chambers are located 4–6 meters from the collision point, and each chamber has a unique orientation, minimizing problems with 'ghosts' in the tracking algorithms. Provision has been made for placing 1 to 2 radiation lengths of converter after the first four tracking planes in some runs to enable us to detect gammas by their conversion products. This is supplemented by a compact electromagnetic calorimeter at the back of the tracking telescope.

The trigger includes scintillator arrays 2 m upstream and downstream from the collision point, together with scintillator elements in front of, in the middle of, and behind the tracking telescope. The background due to beam-gas interactions is well understood, and an efficient trigger with a signal to noise ratio of about 100 has been developed.

MiniMax was proposed in April of 1993, Approved in May of 1993, and began taking data in November of 1993. It is expected at present to run through the summer of 1995.

PROGRESS AND PLANS

The first, test, phase of the experiment has been completed. Good efficiency at tracking charged particles has been demonstrated, but chamber occupancies are sufficiently high (with a mean of about 10% and a long tail at high occupancies) that we are tracking across the entire acceptance of the detector in only about 60% of collisions. The high occupancies also mean that our efficiency in finding gammas suffers. This is further exacerbated by the fact that the 'window' through which the telescope looks is about 1 radiation length of Al in thickness.

Nevertheless, we regard the first phase of MiniMax as a success. We have a reliable trigger, robust tracking algorithms based on a Hough transform, and a good understanding of the background. The difficulties we have with chamber occupancy were anticipated, and are due to splash from the beampipe, floor, etc. They can now be quantitatively reproduced in GEANT simulations with Pythia input.

There are two strategies for attacking the problems outlined above, so that we can confidently tackle the basic physics goals of MiniMax. First, one can rapidly beat down complications due to beampipe splash by increasing the number of tracking planes. To this end, we doubled the number of tracking planes during the September 1994 shutdown, and are presently commissioning the enlarged detector. Second, we have designed and constructed a new beampipe for the detector. It provides a thin window through which the tracking telescope will look at the collision region, significantly reducing the probability that a particle originating in the collision vertex will undergo interactions before reaching the detector. In addition, the new design provides a stepped-conical beampipe underneath the detector. This should rather significantly reduce the splash seen in the chambers. The pipe should be installed no later than the February 1995 shutdown. T864 is expected to acquire the bulk of its data during the spring of 1995, after the installation of the new pipe.

Acknowledgments

It goes without saying that this work depends crucially on my T-864 collaborators, A. Abashian, R. Ball, P. Colestock, K. Delsignore, W. Fickinger, R. Gustafson, B. Hanna, T. Jenkins, L. Jones, E. Kangas, M. Knepley, K. Kowalski, M. Longo, M. Martens, N. Morgan, S. Oh, J. Streets, W. Walker, and A. Weidemann, and especially my co-spokesperson, J. D. Bjorken. C.-H. Wang of Duke has also provided invaluable support, as has everyone with whom we have interacted at Fermilab. This work was supported in part by the National Science Foundation, grant NSF-PHY-9208651.

REFERENCES

1. Over a decade ago T. D. Lee speculated about the possibility of "vacuum engineering" in *Particle Physics and Introduction to Field Theory* (Harwood Academic Publishers, New York, 1981), p. 826.

2. A. A. Anselm, *Phy. Lett.* **B217**, 169 (1989)

3. A. A. Anselm and M. C. Ryskin, *Phy. Lett.* **B266**, 482 (1991)

4. J-P. Blaizot and A. Krzywicki, *Phys. Rev.* **D 46**, 246 (1992); A. Krzywicki, talk at the *Workshop on QCD Vacuum Structure*, Paris, June 1992, "Collective Pion Emission and QCD Vacuum Structure"

5. J. D. Bjorken, *Int. J. Mod. Phys.* **A7**, 18, 4189 (1992); *Acta. Physica Polonica* **B23**, 561 (1992).

6. K. L. Kowalski and C. C. Taylor, CWRUTH-92-6, hep-ph/9211282 (1992).

7. K. Rajagopal and F. Wilczek, *Nucl. Phys.* **B399**, 395 (1993); *Nucl. Phys.* **B404**, 577 (1993).

8. J. D. Bjorken, K. L. Kowalski, and C. C. Taylor, SLAC-PUB-6109; proceedings of *Les Rencontres de Physique de la Vallee D'Aoste*, La Thuile, March, 1993.

9. V. Karmanov and A. Kudryavtsev, ITEP-88, 1983.

10. D. Horn and R. Silver, *Ann. Phys. (N. Y.)* **66**, 509 (1971).

11. C. M. G. Lattes, Y. Fujimoto, and S. Hasegawa, *Phys. Rep.* **65**, 151 (1980).

12. L. T. Beradzei, et al.(Chacaltaya and Pamir Coll.), *Nucl. Phys.* **B370**, 365 (1992); Chacaltaya and Pamir Coll., Tokyo University preprint ICRR-Report-258-91-27 and references cited therein.

13. J. R. Ren, et al, *Phys. Rev.* **D 38**, 1417 (1988).

14. L. Jones, contribution to this conference.

15. Fermilab Proposal T-864 (J. Bjorken and C. Taylor, spokesmen), "MiniMax", 1 April 1993, unpublished.

16. Memorandum of Understanding, Fermilab Experiment T-864, June 26, 1993.

17. J. D. Bjorken, K. L. Kowalski and C. C. Taylor, Proceedings of the *Workshop on Physics at Current Accelerators and the Supercollider*, p. 637 (1993).

EVENT BY EVENT ANALYSIS OF ULTRARELATIVISTIC NUCLEAR COLLISIONS: A NEW METHOD TO SEARCH FOR CRITICAL FLUCTUATIONS

R. Stock

Fachbereich Physik
Univ. of Frankfurt, Germany

ABSTRACT

The upcoming generation of experiments with ultrarelativistic heavy nuclear projectiles, at the CERN SPS and at RHIC and LHC, will confront us with several thousand identified hadrons per event, suitable detectors provided. An analysis of individual events becomes meaningful concerning a multitude of hadronic signals thought to reveal a transient deconfinement phase transition, or the related critical precursor fluctuations. Transverse momentum spectra, the kaon to pion ratio, and pionic Bose-Einstein correlation are examined, showing how to separate the extreme, probably rare candidate events from the bulk of average events

INTRODUCTION

Event by event analysis of the final state hadrons is, by no means, a novel concept of particle physics: it has been successfully employed in the analysis of two - or three - jet events created in electron-positron collider experiments. We recall the corresponding plots, now shown on textbook covers, exhibiting highly ordered hadron emission patterns, with low background. Owing to the development of suitable collective observables[1,2] such as planarity, thrust, jettiness, triaxiality etc., concerning the relevant multi-hadron momentum space correlations, it was possible to separate the candidate events from millions of irrelevant ones. Here we first encounter the characteristics of all event by event analysis: the average hadronic final state posesses a sufficient number of microscopic degrees of freedom (hadron multiplicity, first of all), in order to support the definition of collective observables characterizing the event as a whole. The signal to noise ratio, governing the collective jettiness observables, is determined by finite particle number statistics, and it was impossible until the advent of high energy colliders to clearly separate the jet correlations from trivial finite number fluctuations. It was possible to isolate a small cross section phenomenon from a predominant soft collision background, the interesting events constituting a minute fraction of the overall collision event sample. This phenomenological pattern, resolving non-trivial, rare candidate events from a predominant background of lesser informational value, will be the subject of the present study.

Event by event analysis in application to relativistic heavy ion collisions, at the lower Bevalac/SIS energies has also demonstrated the existence of collective nuclear matter flow.[3,4] This effect is not seen in a few rare events but in every semi-central collision of heavy nuclei, but its manifestation is in a weak signal only - a small directed momentum component superimposed on an otherwise statistical distribution typical of a Hagedorn fireball emission. A significant analysis requires above hundred charged hadrons per event to be recorded, in order for the collective effect to stand out beyond trivial finite number fluctuations.

The subject of this paper, identification of deconfinement signals in central collisions of Au on Au or Pb on Pb at the CERN SPS and at future colliders, presents a situation in which the interesting events may be rare, and the relevant signal not large. However, an ideal multihadron detector system would now record about 1000 charged particles in an interval of c.m. system rapidity $|y| < 1$, at the SPS Pb beam of 160 GeV per nucleon, this number expected to be even larger, by factors of about four at RHIC and of about 15 at the LHC. Frightening to the detector builder, such high rapidity densities make event by event analysis an extremely powerful, novel possibility of analysis.

At the SPS we may expect the 'ideal' quark gluon continuum state to be formed only very seldom, as the limit of a broad fluctuation in the reaction dynamics trajectory. Event by event analysis may thus be the only viable approach to pin down the relevant hadronic observables.

We are now considering, not some kind of collective ordering in space of the microscopic particle vectors, but properties typical of a thermodynamic description: transverse momentum spectra and 'temperatures', hadron density vs. reaction time, production rates as well as yield ratios of various hadronic species. This is the terminology of the Hagedorn model,[5] however with a different, new aim. That is why our topic lends itself to contribute to the celebration of his 75th birthday.

What is new here: we are one step beyond the traditional fireball model. There, the individual event does not reveal the features that become *visible* in an ensemble of events - temperatures, K/π ratios etc.. Here, each event is a rudimentary ensemble, the whole acting as a heat bath for its parts. A phase transition does, or does not occur in a single, given event. The event itself must be inspected, an ensemble average over many events rendering the signals *invisible* if the relevant events are rare, and if the observable effects represent changes, not by orders of magnitude but by factors of order two or three. This is obviously the case in hadronic signals. For example, an event exhibiting a pion mean transverse momentum of 700 MeV/c, i.e. twice the standard at SPS energy, would obviously be worth further inspection: and a K/π ratio rising, from the standard value of about 0.15, to a value of about 0.5 suggests the sensational assumption of equal flavour population among up, down and strange, i.e. formation of a perfect $n_f=3$ system.

Thus, the margin by which the interesting events differ from the average events is not really dramatic. But if a transient partonic state near equilibrium does, in fact, cause these fluctuations, they must occur *simultaneously* in the candidate events. We need to look for a class of events exhibiting a simultaneous fluctuation in the relevant observables. The more such signals we get proposed from theory, and the more of them can be defined in single events with sufficient significance, the more conclusive is the evidence for an unusual dynamics inferred from the correlation between anomalous features.

Obviously, if all central collision events went through a plasma phase there would be no fundamental need for event by event analysis, with the exception perhaps of jet quenching,[6] deflagration volcanos[7] and other observables consisting in local enhancements in real space. However one intuitively doubts as to whether every event produces the plasma state, at least at the SPS energy: the critical levels of energy density, life-time and reaction volume may be just about reached here,[8] making the equilibrium plasma phase occur only as the limit of precursor situations with local bubbles, i.e. less homogenity, more pre-equilibrium chaos. That such intuitive arguments are applicable in the specific case of a QCD phase transition has recently been shown by Kapusta et al.[9] Furthermore the various parton cascade models now under investigation[10] seem to indicate a partial approach toward chemical equilibrium only,[11] in the average event. The interesting events thus should represent a fluctuation, and relevant observables should be correlated in their fluctuation.

SCHEMATIC EXAMPLES OF EVENT OBSERVABLES

Which of the hadronic observables, regarded to be sensitive to a plasma phase, lend themselves to event by event determination? The answer follows from consideration both of the average multiplicities of hadronic species per event, and of the detection/identification efficiency plus acceptance attained by the detector systems. We will, for the sake of a first survey, leave the second aspect unattended (see next chapter) assuming an 'ideal' detector covering a sufficiently wide interval in longitudinal phase space, $|y| < 1$, or even < 1.5. This is the aim of the experiments NA49, STAR and ALICE at the SPS, RHIC and LHC respectively.[12-14] Clearly, all hadrons with $\langle m \rangle \gg 1$ can be studied event by event in principle but we will not consider Φ, η, K^0 and strange hyperons here because they are recorded by measuring their decay products which is seldom accomplished with high overall efficiency. This leaves $\pi^{+,-,0}$, $K^{+,-}$, p and $\bar{p}$: in central Pb+Pb collisions at 160 GeV/nucleon, the corresponding mid-rapidity densities dN/dy are expected to be about 500 for $\pi^+ + \pi^-$, 65 for $K^+ + K^-$, 80 for p and 10 for $\bar{p}$. We can also define the quantities of total negative and positive hadrons, and h^+ - h^- giving the charge excess distribution, $(h^+ - h^-) \approx 60$ at mid-rapidity. This latter quantity approximates the net baryon density distribution, giving the position of the initial valence quarks in final phase space. An upward fluctuation of dN/dy $(h^+ - h^-)$ at $y=y_{cm}$ indicates events with high stopping of incoming longitudinal quark energy ('hot events').

Transverse Momentum Spectra

As a first example, Fig. 1 shows the transverse momentum distribution of the negative plus positive pions in the domain $2 < y < 4$ in a single central Pb+Pb collision at the SPS energy. It was simulated with the Fritiof event generator.[15] We see a well expressed, typical 'thermal' spectrum up to $p_T \approx 1$ GeV/c, all points being well fitted by an exponential with inverse slope or 'temperature' parameter T=176 MeV. At RHIC and LHC such event spectra will extend up to 1.5 to 2 GeV/c where the influence of hard QCD processes (minijet activity) sets in. At the SPS inspection of the shape differences from an exponential will be of marginal significance but clearly we obtain firmly established values for $\langle p_T \rangle$ and T.

In Fig. 2 I illustrate what is meant by 'firmly' for the temperature T distribution obtained from individual fits of 1000 Fritiof events. The mean value $\langle T \rangle$=183 MeV agrees well with the slope parameter obtained for the p_T distribution of the whole ensemble, which is not surprising in view of the symmetric distribution of single event temperatures. The width of this distribution is small, $\sigma \approx 10$ MeV only! It results, both from the microscopic collision mechanism built into Fritiof which lead to a genuine fluctuation from event to event, and from the degree of accuracy with which the ca. 800 charged pions seen in each event in $2 < y < 4$ fix the inverse slope parameter of each event fit. Thus, any high $\langle p_T \rangle$ and high T event occuring in the real data due to a 'critical opalescence' mechanism, related to the plasma phase transition, would be identified by the event by event fit, even if it occured once only in every hundred events and has a typical T only 20 % higher than the average event. More realistically, such mechanism would produce a shoulder of the distribution and would be selected with reasonably low background by applying a cut at $T > \langle T \rangle + 3\sigma$.

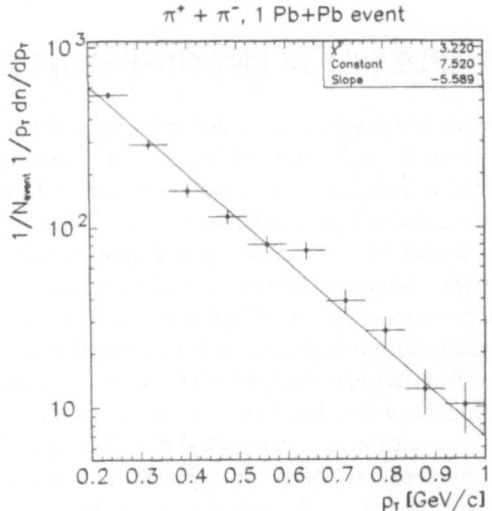

Figure 1. Transverse momentum spectrum of negative and positive pions in a single central Pb+Pb collision at SPS energy; the rapidity domain is $2 < y < 4$; Fritiof event simulation

The Kaon to Pion Ratio

Our second example refers to the K/π ratio which we define here as $(K^+ + K^-)/(\pi^+ + \pi^-)$. From this observable, an estimate can be made[16] of the overall freeze-out ratio of strange to non-strange quarks in the final state. Defining a strangeness suppression factor $\lambda = 2(s + \bar{s})/(u + \bar{u} + d + \bar{d})$ we find a value of λ=0.34± 0.05 in light ion collision at SPS energy, corresponding to a K/π ratio of 0.14 at mid-rapidity. Strangeness enhancement, expected to be a characteristic feature of events with a plasma[17] is, in reality, a reversal of strangeness suppression. Ideally, λ should then approach unity but several factors (presence of u, d valence quarks, finite s-quark mass etc.) let us expect $\lambda \leq 0.65$ in Pb+Pb at the SPS. Thus, the 'plasma candidate' events might exhibit a twofold enhancement of the K/π ratio.

Fig. 3 shows the distribution of the K/π ratios found in the above sample of 1000 Fritiof Pb+Pb events. Again we observe a symmetric, rather narrow distribution. Its relative width, $\sigma/\langle K/\pi \rangle \approx 0.12$ is higher than that of the T distribution due to the modest kaon statistics. Nevertheless, there is zero background at twice the mean, i.e. $K/\pi \approx 0.25$. Thus we encounter the same high selectivity toward fluctuations in this ratio due to the reaction dynamics. We shall see in the next chapter that consideration of the realistic detector dampens our evaluation, the relative width increasing from the ideal 0.12 to about 0.30 in NA49. But even then a twofold increase in the K/π ratio is worth 3σ.

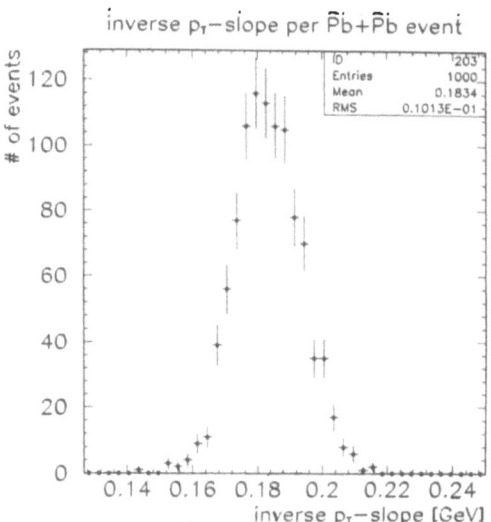

Figure 2. Distribution of the single event inverse slope parameter T from 1000 Fritiof simulated central Pb+Pb events. The width is $\sigma = 10$ MeV, $\langle T \rangle$=183 MeV.

Two Pion Momentum Correlations

Our last example is negative pion Bose-Einstein correlation. I shall only give a very brief sketch of the relevant physics here; the detail may be found in recent state of the art publications,[18] and the idea of event by event pion correlation study has already been explored by Ferenc[19] in a simulation of conditions expected with the LHC detector ALICE.

In a multi-pion final state the symmetry of the total wave function leads to a clumping in momentum space of identical pions as their 4-momentum difference goes to zero. Over a certain distance $\Delta^4 p$ the probability to find pion pairs, triplets etc. is enhanced with respect to the probability expected from the single pion momentum space distribution. Thus, for a pair, $P(p_1, p_2) > P(p_1) \cdot P(p_2)$ over a characteristic distance $|p_1 - p_2| = \Delta p_{1,2}$ which turns out to be related to the average size R in configuration space of the pion emitting source by the relation $\Delta p \cdot R = \hbar$ characteristic of a Fourier transform. More accurately, one defines a correlation function in momentum space, $C(\Delta p_{1,2})$ as the ratio (with proper normalization) of $P(p_1, p_2)/P(p_1) \cdot P(p_2)$. For a pion pair, $C \to 2$ as $\Delta p \to 0$, and $C \to 1$ for large Δp. For a source of average space-time dimensions 10 fm the enhancement domain, where $C > 1$, extends up to about 40 MeV/c in pair momentum difference. We see that the momentum difference of multi-GeV pions must be measured with high accuracy in order to determine the correlation function.

Two important observations are in order here, to understand the physics significance of the correlation effect. Firstly, the total Bose-symmetric wave function of the final multipion state gets established during the time interval Δt in which, on the average, each pion encounters its last strong interaction. This occurs after an expansion of the dense hadronic fireball which takes the average time t_0 in order to arrive at sufficient dilution. The correlation function thus refers to the space-time dimension of the pion source at freeze-out from strong interaction. This is a very late stage of the overall reaction dynamics but the average expansion time t_0 depends critically on the dynamics of the collision. Secondly, closer inspection of the space related components of the correlation function shows[20] that they depend, not only on the

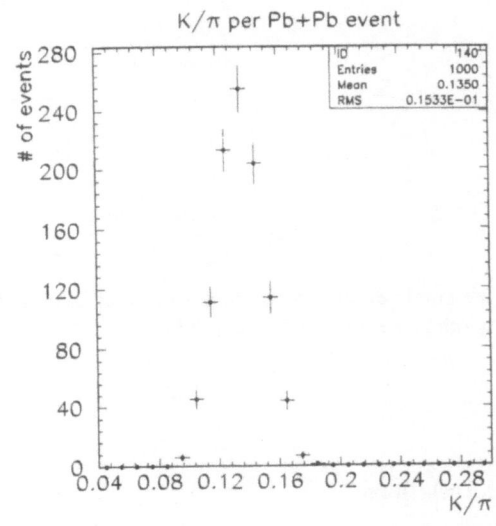

Figure 3. The distribution of the Kaon to Pion ratio from an ensemble of 1000 central Pb+Pb collisions at SPS energy generated with the Fritiof model. The width is σ=0.015, and $\langle K/\pi \rangle$=0.14.

spatial dimensions of the freeze-out state but also on the dynamics through which the system arrived at freeze-out. The dynamics depends on the equation of state of superdense matter, and on the existence and nature of a phase transition from an early plasma to an expanding hadron fireball. This explains why we want to analyze two pion correlation functions at the event by event level.

In practice one analyzes the correlation function in terms of Fourier transforms of plausible Gaussian space-time pion sources. This determines the expansion duration parameter t_0 and an effective transverse dimension of the freeze-out volume, R_{side}. For a definition of this quantity and an evaluation of its significance the reader is referred to ref.[20] The informational value of an event by event study of this quantity will be illustrated below. I refer to the SPS experiment NA49 which will cover the rapidity domain $1.5 < y < 5.5$ for negative pions. Restricting the analysis to the region $1.5 < y_{lab} < 3.5$ (mid-rapidity is at $y_{lab} = 2.8$ in Pb+Pb collisions), NA49 will record about 350 negative pions per event, resulting in a pair statistics of about 60.000 per event. Of these pairs, only a small fraction falls into the relevant domain, of $\Delta p \leq 40$ MeV/c, which contains the relevant information. The significance of fitting that fraction of the two pion correlation function by means of a Gaussian ansatz for $C(\Delta p_{side})$, as a Fourier transform of a spatial Gaussian source distribution $\Phi(R_{side})$, will thus be marginal due to small pair statistics. The expected results are illustrated in Fig. 4. They are taken from a Fritiof-based simulation of Pb+Pb collisions by Ferenc.[21] The Fritiof model ignores the effects of Bose-symmetrization, which are subsequently implemented in a manner analogous to the procedure described by Sullivan et al.[22] This implementation of correlations was carried out for two values of the transverse source dimension, $R_{side} = 7$ and 11 fm. Fig. 4 sketches the resulting distributions of output R_{side} values. A relatively narrow distribution results for the input source size $R_{side} = 7$ fm with $\sigma \approx 4$ fm, but the output distribution for the $R_{side} = 11$ fm source is very broad, with a σ of about 10 fm. This is due to the fact that the region of the correlation function which contains the signal (i.e. $C > 1$) is narrower in relative pair momentum for the source with 11 fm radius, the fit thus based on a smaller fraction of the total pair statistics. Nevertheless, a cut at $R_{side} > 20$ fm would select predominantly $R = 11$ fm events.

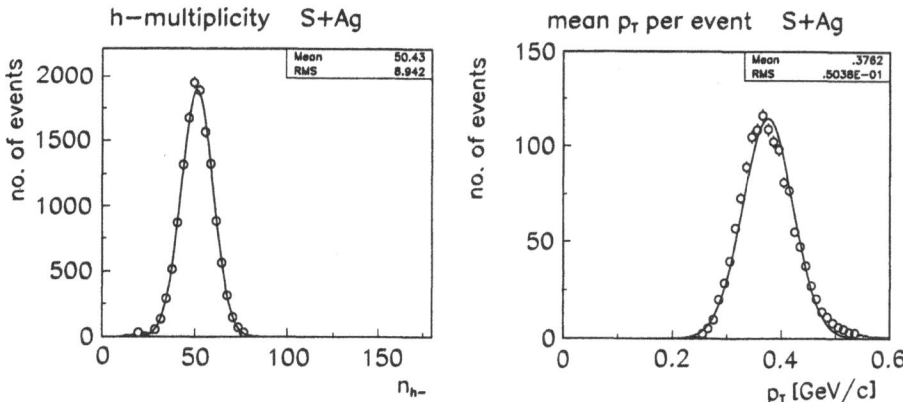

Figure 4. Sketch of the expected outcome of R_{side} determination by two-π^--correlation analysis of single Pb+Pb central collision events. The two distributions correspond to input values R_{side}=7 and 11 fm, the corresponding width are σ=4 and 10 MeV, respectively

We conclude that event by event analysis for the transverse pion source dimension becomes just about meaningfull under SPS Pb+Pb collision conditions. The same holds for an analogous analysis of the total expansion time parameter t_0. The selectivity for non-average events improves at collider energies[19] but not dramatically. We have deliberately chosen the correlation observable to illustrate the most ambitious and complicated end of the range of possible event by event signals.

EVENT BY EVENT ANALYSIS
WITH REALISTIC DETECTOR SYSTEMS

No realistic detector system, of finite cost and dimension, can be constructed with present experiment technique that could uniformly cover a wide rapidity interval providing 'perfect' efficiency for momentum measurement and hadron identification - even if we restrict the task to charged hadrons. The tracking efficiency, determining particle momenta, may well suffice but the track by track hadron identification capabilities remain restricted even if detector layouts combine several signals aiding in the task of identification, such as specific ionisation and time of flight. The fixed target experiments with Pb beams of 160 GeV/A at the SPS are, apparently, confronted with the most complicated task. Mid-rapidity pions have Lab. momenta of about 2-4 GeV/c, the Kaons found at 6-8 GeV/c, and protons/antiprotons at 12-15 GeV/c. Moreover, as a satisfactory rapidity domain needs to be covered, the essential range of hadronic Lab. momenta is of formidable extent, of about 1 to 50 GeV/c, facing experimental technique with an unprecedented task in order to secure a uniform acceptance. Collider experiments are, at first sight, easier because at $|y| < 1.5$ in the c.m. frame, which coincides with the Lab. frame, the momentum spread of the hadrons is more compact, and independent of the accelerator energy, with only the multiplicity density increasing which concerns chiefly the tracking capabilities, the minor difficulty in comparison to hadron identification.

Before discussing the implementation of event by event analysis in realistic detectors, a brief consideration is in order of the non-hadronic observables relevant toward plasma production diagnostics. Among those, direct photon and lepton pair spectroscopy requires an experimental effort commensurable to the task of exclusive hadron spectroscopy. In principle, the ideal solution would be a detector that covers all together. However it turns out that, at least under CERN SPS fixed target conditions, the optimal layouts for exclusive hadron and for lepton/dilepton measurement are contradictory. Moreover, as almost all the lepton/dilepton observables result from subtraction of ensembles the fact that, e.g., a J/Ψ particle was created can not be traced to the event that produced it: this physics is not genuinely an event by event physics, and both approaches must seek separate optimization also at RHIC and LHC.

Proceeding to a consideration of realistic conditions in realizeable experiments, Fig. 5 shows[23] the distributions of negative hadron (essentially negative pion) multiplicity and event mean transverse momentum in 14.000 central S+Ag collisions at 200 GeV/A, as observed in the NA35 streamer chamber. This detector type gets over-exposed by the multiplicity density in this reaction and thus was artificially blindfolded in 2/3 of the total azimuthal acceptance; thus the relatively small observed mean h^- multiplicity of 50 in $1.0 < y < 3.5$. The extrapolated, true $\langle h^- \rangle$ is about 170 but that is irrelevant as far as the quality of event by event analysis is concerned - somehow only the honest, primary data yielded by the detectors are relevant. It is therefore not surprising that the event $\langle p_T \rangle$ distribution is fairly broad, $\sigma / \langle p_T \rangle$ of the entire ensemble being of order 0.15. As $T \approx 0.5 \langle p_T \rangle$ in a thermal system, we would get a $\sigma(T) \approx 25$ MeV in this case. Compare to the plot of Fig. 2: the $\sigma(T) \approx 10$ MeV observed there refer to 800 pions per event. Actually, the width of the $\langle p_T \rangle$ distribution in Fig. 5 shows that we are entirely counting statistics limited in the S+Ag streamer chamber case, $\langle m \rangle \approx 50$ letting us expect about 15 % accuracy. Doing the sampling 14.000 times over, in different events, merely exhibits the full (nearly) Gaussian distribution corresponding to the elementary statistics. The detector momentum tracking accuracy, of order $\Delta p/p = 0.01$, does not surface in this low multiplicity case. If the streamer chamber experiment were designed only for $\langle p_T \rangle$ study its tracking resolution would thus represent an unnecessary level of accuracy. This is not the case in the Pb+Pb simulation of the NA49 design, illustrated in Figs. 1 and 2. Counting statistics alone, with 800 negative plus positive pions per event, would suggest a 3.5 % accuracy of the $\langle p_T \rangle$ and T parameters of each event. The relative width of the T distribution in Fig. 2, however, is larger: $\sigma / \langle T \rangle = 0.055$. At this level of counting statistics the fluctuation at the microscopic level of the single event reaction dynamics becomes visible, and this is precisely the design goal of the experiment. Momentum tracking inaccuracy (not incorporated in the simulations leading to Figs. 1 and

2), at a mean level of $\Delta p/p = 0.01p$, will not significantly change the picture presented by Figs. 1 and 2. However, the outcome of the streamer chamber experiment, Fig. 5, implies a warning; we recognize a non-Gaussian high p_T tail in the data of a real detector. Experience with 4π tracking detectors shows a common trend in developing such tails, perhaps resulting from the asymmetry of the momentum observable which is bounded at $p_T = 0$ MeV/c but unbounded toward high p_T. Poisson statistics analysis may take care of this, to extract the truely non-average events in the high $\langle p_T \rangle$ or T tails.

In concluding this section a brief consideration follows of the real detector implementation of two much more ambitious event signals: the K/π ratio and the pion source dimensions.

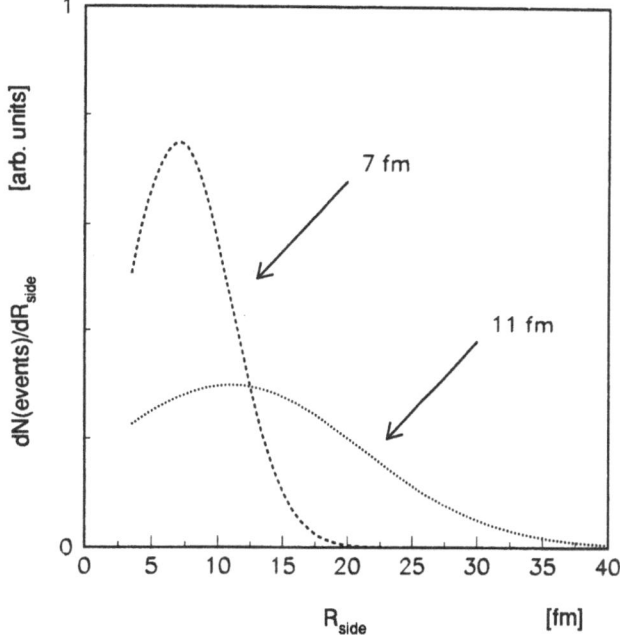

Figure 5. Distributions of multiplicity and mean transverse momentum for negative hadrons in the acceptance of NA35, for single central S+Ag collision events, also showing a Gaussian fit. The $\langle p_T \rangle$ width is $\sigma = 50$ MeV/c.

As we have seen from Fig. 3, elementary counting statistics alone limits the quality of the event K/π ratio signal in central Pb+Pb collisions at the SPS. Furthermore, the simulation resulting in the Fig. 3 data leaves the finite particle identification resolution restrictions unconsidered. This factor, however, dominates the uncertainty of the signal, and it will stay with us in all future collider experiments in spite of the better counting statistics. The difficulty arises from the task of identifying a 15 % fraction of kaons next to the 85 % fraction of pions. A state of the art detector system will achieve a 3 σ separation between kaons and pions. The traditional method of K cross section measurement is based on gathering a huge ensemble of events, then unfolding with high precision the composite particle identification variable (specific ionization, time of flight etc.) with a multi-Gaussian superposition fit for the π, K, p relative composition fractions. Thus, the 3 σ elementary separation may be recovered in a 3

σ quality of relative cross section assessment for the entire ensemble. This method is very unaccurate as applied to the meagre statistics of a single event, the ill-determined tail of the high pion count invading the kaon signal. A new method, based on a track by track maximum likelyhood fit has been developed by Gazdzicki.[24] In its application to a 3 σ average kaon to pion separation detector such as NA49 this model still predicts a considerable broadening of the 'ideal' event K/π distribution (Fig. 3), which thus acquires a relative width of 25-30 %.

Bose-Einstein pion pair correlation analysis of single events, finally, suffers from the principal difficulty to determine the denominator in the correlation function, i.e. the single pion space distribution P(p_i) with sufficient accuracy. In the standard analysis of extended ensembles one uses a sample of pion momenta mixed up from all different events to determine $P(p_i)$ free of pair correlation distortions. One way out may be to employ a primitive event generator, adapted in each event to reproduce the observed pion multiplicity, rapidity distribution, $\langle p_T \rangle$ and T parameters, then generating a high statistics single particle distribution. Alternative methods are now being developed that employ a maximum likelyhood method to extract maximum information from the single event input.[25] From among the properties of a realistic tracking detector, the momentum measurement accuracy and the resolution capability for close tracks play the crucial role here: the most important information stems from pion pairs that are as close as Δp=10 MeV/c if source dimensions in excess of 10 fm are to be analyzed. Thin targets, minimum material in the particles path, high magnetic bending power, low diffusion broadening over the drift-length in TPC-type gas tracking detectors, and high granularity readout systems are the chief characteristics of suitable detector systems. The corresponding experimental technique for collider experiments is still under development.

EVENT BY EVENT ANALYSIS STRATEGY

In this article I have stressed the new methods of event by event analysis, perhaps at the risk of creating an 'ideology' as H. Satz has critically remarked. Let us put it back into perspective at the end:

It is restricted to high cross section hadronic signals, with the exception of jet formation analysis and exotica searches (Centauro-type events etc.).

It is based on the intuive idea that - at least at the low CERN SPS energy - the interesting/relevant events should be rare and need to be recognized one by one. This should be different at the ten times higher c.m. energies at RHIC, and most certainly at the 250 times higher LHC energy.

It is not the end (but only the initial step) of data analysis with detector systems optimized to its specific demands.

In order to illustrate the last remark let us sketch the analysis strategy of such an experiment. Assume that it recorded, say, $5 \cdot 10^6$ central Pb+Pb collision events in one running period. A first round of analysis (actually the most CPU-time consuming step) would determine the event characteristics. From among those we have mentioned only a few here, just enough to develop the ideas. Further event observables are the total particle density and the proton density at mid-rapidity that are related to the total energy density achieved in the event. One may also want to examine the $\bar{p}/p$ and K^-/K^+ ratios at mid-rapidity. This fixes the baryo-chemical and strangeness-chemical potentials, and a complete, approximate, thermo-chemical analysis may ideally be achieved at the event level by determining the temperature, energy density and reaction volume along with these potentials.

One will then define sub-ensembles, say at the 1 % level of $5 \cdot 10^4$ events each, that contain the extreme tail events corresponding to each of the considered event variables. I.e. extremely strange or dense or hot or large (small) or net-baryon rich fluctuations within the overall central event class. If all such signatures are necessary ingredients of a reaction dynamics in the presence of a phase transition the fluctuations have to be correlated. One would then proceed to a final sub-ensemble, say of 10^4 events which are multiply extreme. *It is of prime importance*, at this stage, to be aware of the fact that this ensemble of 'candidate events', stemming from a 4π detector, is still wide enough, containing a total of $10^7 - 10^8$ analyzed hadrons, allowing *not only* the observables mentioned above now to be studied with the traditional ensemble precision, *but also* the detailed shapes of all hadronic p_T distributions, the yield ratios of the neutral strange particles $K^0, \Lambda, \overline{\Lambda}$ and Φ (which do not lend themselves to single event analysis), as well as intermittency, kaon interferometry etc.. If hadronic signals bear any memory at all of the transient deconfinement stage it will be identified here, with little chance of 'trivial scenarios' coming up with a simultaneous explanation of the physics information thus unfolded.

REFERENCES

1. S. L. Wu and G. Zobernig, Z. Phys. C2 (1979) 107;
   S. Brandt and H. Dahmen, Z. Phys. C1 (1979) 61.

2. P. Söding and G. Wolf, Ann. Rev. Nuc. Part. Sci. 31 (1981) 231.

3. H. A. Gustafsson et al., Phys. Rev. Lett. 52 (1984) 1590.

4. P. Danielewicz and G. Odyniec, Phys. Lett. 157B (1985) 146;
   J. J. Molitoris, D. Hahn and H. Stöcker, Nucl. Phys. A447 (1985) 13.

5. R. Hagedorn and J. Ranft, Nuovo Cimento 6 (1968) 300;
   R. Hagedorn, in Cargese Lectures in Physics VI, p. 643,
   Gordon and Breach New York 1973.

6. X. N. Wang and M. Gyulassy, Nucl. Phys. A544 (1992) 559.

7. M. Gyulassy, K. Kajantie and L. McLerran, Nucl. Phys. B237 (1984) 477;
   D. Seibert, Phys. Rev. C47 (1993) 2320.

8. T. Celik, J. Engels, and H. Satz, Nucl. Phys. B205 (1983) 545.

9. J. Kapusta et al., NOC-MINN/94/5T 1994.

10. H. Sorge, H. Stöcker, and W. Greiner, Nucl. Phys. A498 (1989) 567;
    K. Geiger and B. Müller, Nucl. Phys. B369 (1992) 600.

11. K. Geiger, Nucl. Phys. A566 (1994) 257.

12. NA49 proposal CERN/SPS LC 91-31, 1991.

13. J. W. Harris, Nucl. Phys. A566 (1994) 277.

14. ALICE Letter of Intent, CERN/LHC C93-16, 1993.

15. FRITIOF version 1.6 adapted to NA35 data.

16. H. Bialkowska et al., Z. Phys. C55 (1992) 491;
    M. Gazdzicki et al., Nucl. Phys. A566 (1994) 503.

17. J. Rafelski and B. Müller, Phys. Rev. Lett. 48 (1982) 1066.

18. D. Boal, C. K. Gelbke, and B. Jennings, Rev. Mod. Phys. 62 (1990) 553.

19. D. Ferenc, CERN ALICE Note 93-14, 1993.

20. S. Pratt, Phys. Rev. D33 (1986) 33;
    G. Bertsch, M. Gong, and M. Tohyama, Phys. Rev. C37 (1988) 1896.

21. D. Ferenc, private communication.

22. J. P. Sullivan et al., Phys. Rev. Lett. 70 (1993) 3000.

23. D. Brinkmann, to be published, Frankfurt PhD Thesis (1994).

24. M. Gazdzicki, Nucl. Instr. Meth. A345 (1994) 148.

25. J. G. Cramer, D. Ferenc, and M. Gazdzicki, Univ. Frankfurt Report
    IKF-HENPG/92-3, 1992.

OBSERVING STRANGENESS (AND CHARM?) IN HEAVY ION INTERACTIONS

Emanuele Quercigh

CERN
Division PPE
CH - 1211 Geneva 23, Switzerland

STRANGE PARTICLES AND HEAVY ION EXPERIMENTS

As has been pointed out many times in the past and summarized at this Workshop,[1] strange particles and in particular strange baryons are expected to be useful diagnostic tools in the quest for the Quark-Gluon Plasma (QGP), a new state of matter which may be produced during the collision of two heavy ions at ultrarelativistic energies. The onset of a QGP state is in fact expected to enhance the production of these particles with respect to normal hadronic interactions.[2–5]

The experimental challenge posed by the high multiplicity environment of nucleus-nucleus collisions consists in recognizing the strange particles amidst the large background of other tracks present in the region where the detectors need to be placed. As a consequence, up to now, only a few kinds of strange particles have been identified in heavy ion experiments; namely, among mesons the K^{+-} and K^0's and, among the baryons the Λ, Ξ^- and Ω^-. All these particles have a long enough lifetime to decay at a measurable distance from their production point, thus allowing their identification either before their decay, as has been done for the charged kaons, or by kinematic reconstruction of the decay process, as has been done both for charged and neutral kaons and for the hyperons.[6]

As an example of detection in the same experiment of baryons and antibaryons carrying one, two and three units of strangeness, fig. 1a, b, c show the invariant mass spectra for the $p\,\pi^-$, $\Lambda\,\pi^-$ and $\Lambda\,K^-$ systems respectively as measured by the CERN-WA85 experiment in central S-W interactions at 200 • A GeV/c; only track pairs which come from a vertex well separated from the interaction point are entered in the plot. Clear signals appear at the Λ, Ξ^- and Ω^- masses respectively. It is worth noting at this point that the observed strange particles, e.g. the Λ and Ξ^- in this case, can either be directly produced in the collision or be the decay products of some heavier particle; a fact which has to be kept in mind in the physics analysis.

Enhancements of strange particle production in high-energy nucleus-nucleus collisions with respect to proton-proton and proton-nucleus collisions have been observed by several experiments both at BNL and at CERN.[6] At this Workshop enhancements of the Ω/Ξ and $\bar\Lambda/\bar p$ ratios have been reported for the first time by the WA85[7] and NA35[8] Collaborations

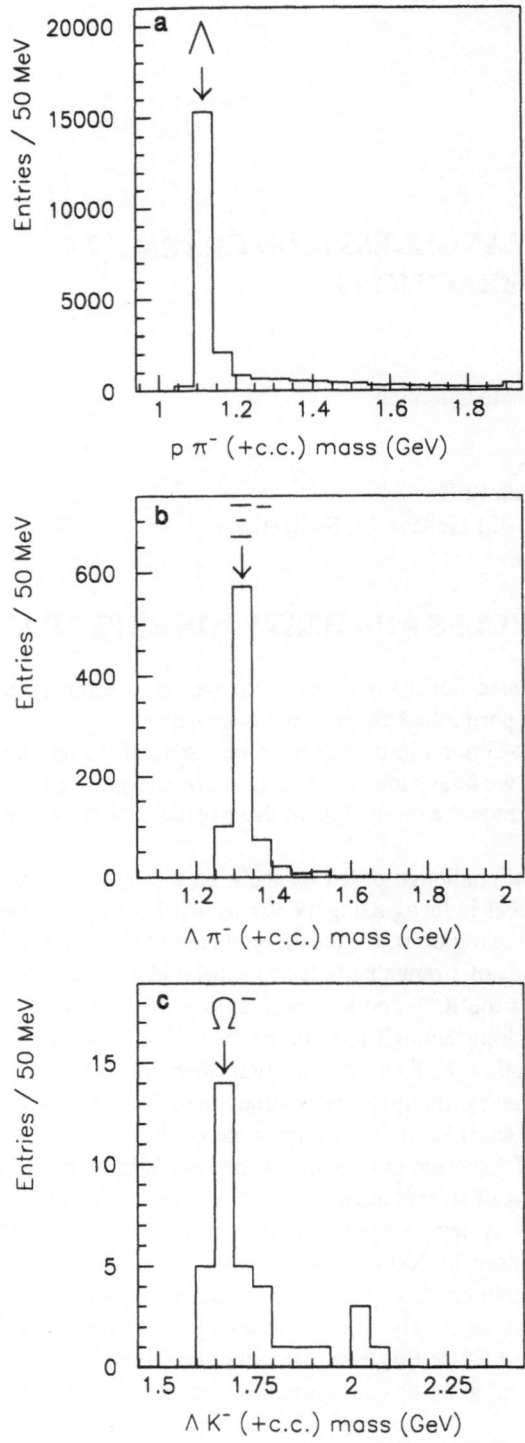

Figure 1. Strange baryon signals (CERN-WA85)

respectively. Such effects were not predicted and are not easy to explain by the conventional models of particle production; they could however be a consequence of the onset of a QGP phase during the collision[4]

The advent of the Au beams at BNL and Pb beams at CERN, gives us the opportunity of searching for a phase transition to QGP in the interaction of truly heavy nuclei. A comprehensive study of strange and multiply strange baryon production at CERN energies has been suggested as a mean to ascertain whether such a phase transition has taken place during the collision. The idea[1] is to determine their "excitation function" by measuring their yields and ratios as a function of the energy density reached during the collision. Such a density can be varied by varying the beam momentum e.g. from 160 to 50 • A GeV/c while keeping the same projectile and target, in order to maintain approximately the same volume and life span of the source. It is however unlikely that the techniques employed up to now would allow the detection of multiply strange baryons e.g. Ξ^- and $\Omega's^-$, from the high multiplicity events expected when the 160 • A GeV/c Pb ion beams become available at CERN. In what follows I shall illustrate the possibility offered for such a study by the silicon microdetectors and in particular by the silicon pixel detectors.

STRANGE PARTICLES AND THE NEW MICRODETECTORS

We shall consider an experiment whose aim is to detect strange baryons produced at central rapidity ($|y| < = \pm 1.5$) in Pb-Pb collisions at 160 • A GeV/c. Strange baryons are reconstructed from the measurement of their decay tracks, which is made possible by placing both target and decay detectors in a constant magnetic field. In order for the apparatus to have enough acceptance to collect, say, one thousand Ω^- decays in a 30 day period, the detectors need to be placed less than one metre from the experimental target, i.e. in a region where the track density over the detector cross section is larger than one track per cm^2. The two track resolution of the detectors must be matched to such a density; however the detector does not need to be very large to obtain a good acceptance. As an example, Fig.2 shows a schematic

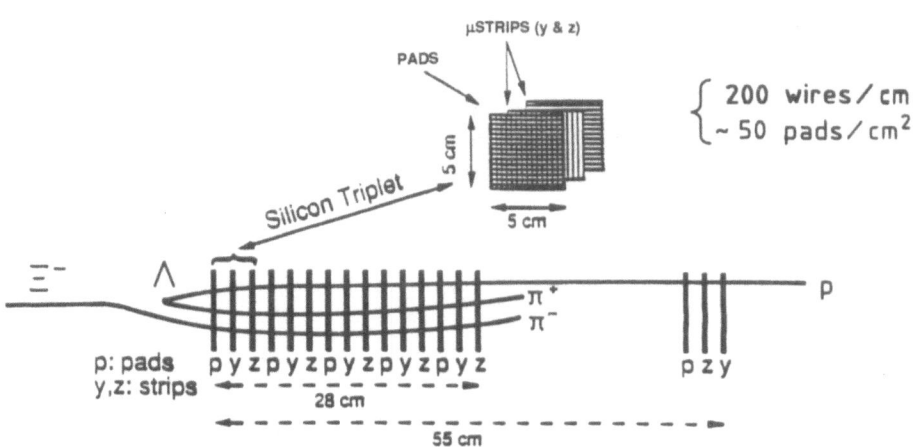

Figure 2. Arrangement of silicon pads and microstrips proposed for the CERN-WA97 telescope. A sketch of a $\Xi^- \rightarrow \Lambda \pi^-$ decay projected on the telescope's bend plane, is also shown.

- **A high-resistivity Silicon matrix of 96 x 64 sensor elements (*pixels*) of size 500 μm x 75 μm ...**

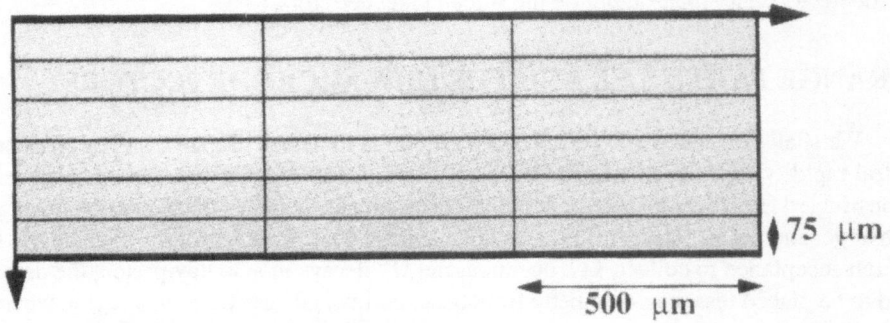

75 μm

500 μm

- **... bump-bonded to 6 read-out chips ("hybrid" technique)**

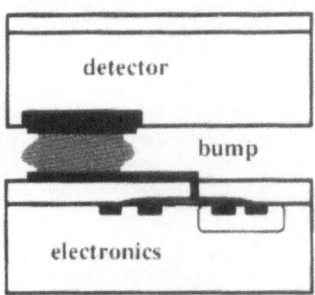

Figure 3. Scheme of the WA97 pixel ladder.

drawing of the hyperon telescope proposed for the CERN-WA97 experiment[9] to study strange baryon production in Pb Pb collisions at 160 • A GeV/c. The telescope is located at about one metre from the target; at such a distance it needs to be only a few centimetres above the beam axis in order to cover the required phase space window. It has a cross section of 5x5 cm^2 and an overall length of 60 cm. Such a relatively modest size is appropriate since, for a good fraction of the interesting events, the decay particles will travel long enough through the detector to allow a precise measurement of their momenta and angles by the silicon microstrip detectors. However, the reconstruction of the decay tracks amidst the several tens of tracks going through the detectors represents a severe task for the pattern recognition program. It requires the use of tracking devices capable to determine the space points on a track directly, i.e. with a two-dimensional readout, thus avoiding the large number of spurious points that would be generated from the intersection of strips.

Silicon pad detectors[10] were, as indicated in Fig.2, the original WA97 choice for such "space point" devices. However, since their granularity can be at most of about 40 pads per cm^2, their usefulness in the pattern recognition is limited to track densities up to 2-3 per cm^2 of detector cross section; moreover, because of the relatively large dimensions of the pads, these detectors have to be used in association with more precise devices, like silicon microstrips. More powerful solutions for high resolution detectors with 2-dimensional readout are offered both by Silicon Drift[11] and Silicon Pixel detectors.[12] In what follows I shall describe the most recent development in Silicon Pixels, namely a detector having 72K sensor elements and covering a total area of 25 cm^2 which has been developed[13] in a collaboration between the WA97 experiment and the CERN-RD19 Project; three such detectors are now being used by WA97 in place of pad detectors.

The basic building block of the WA97 pixel detector is the ladder Fig.3. It consists of a matrix of 96 x 64 rectangular diodes (pixels), ion implanted on a high-resistivity n-type silicon substrate. The physical dimensions of the ladder are 54 x 6 mm^2. The pixel dimensions are 500 x 75 μm^2. Each pixel is connected to a virtual ground via a front-end amplifier on a read out chip by a Pb-Sn solder bump; six chips are bump-bonded to one ladder. Six ladders, spaced by a few mm, are mounted into one array. Two such arrays are mounted face-to-face and staggered in order to cover the 5 x 5 cm^2 area of one detector element using 72000 sensor elements. The average thickness of one detector corresponds to about 1.7 % of a radiation length. Fig.4 is a photograph of the two arrays forming a detector, before assembly; one of the arrays is shown with the associated local driver card and the VME readout module. A new readout chip is being studied which will allow a reduction of the pixel dimensions to 50 x 400 μm^2, thus giving one hundred times more sensor units per cm^2 than were obtainable using the Silicon Pad Detectors initially planned by WA97 for a high statistics study of |S| = 1, 2, 3 baryon production in heavy ion collisions. The availability of such high resolution detectors leads us to consider the possibility of a study of charmed particle production. This is discussed in the next section.

OPEN CHARM DETECTION

The case for studying charmed particle production in heavy ion interactions has been put forward by several authors.[14–16] Charm is considered to be an useful probe for the behaviour of the hadronic matter in the early stages of the interaction since secondary parton scattering during the thermalization phase can produce an enhance ment of the relative abundance of charmed particles in the final state, with respect to proton proton interactions; the magnitude of such an enhancement being related to the length of the thermalization phase.[16] A further enhancement later in the QGP phase could be expected only in the case of the QGP temperatures substantially larger than 300-400 MeV.[17]

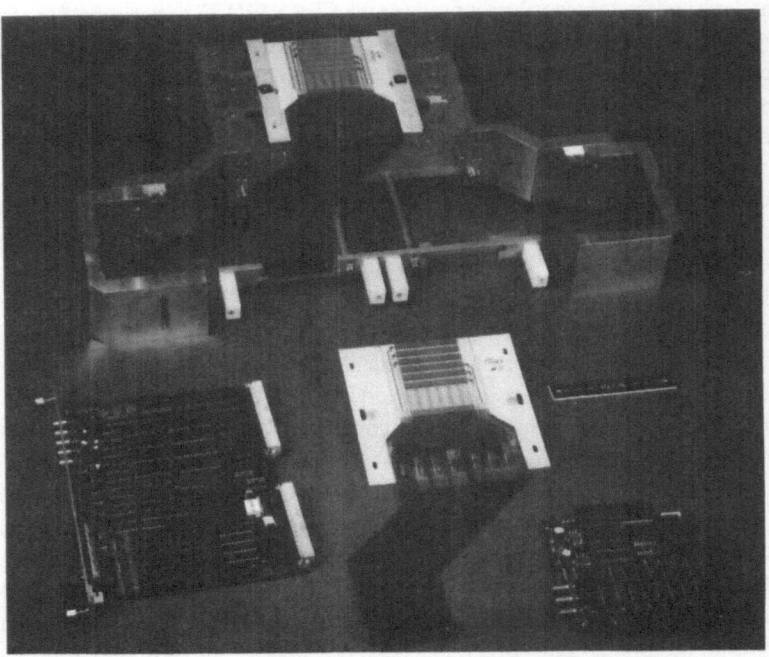

Figure 4. Silicon pixel detector before assembly.

Charmed particles can be identified by the same techniques used for the hyperons, provided that one succeedes in separating the decay vertex from the interaction point and in reconstructing the complete decay process. The main difficulty however, comes from their much shorter lifetime; for example the D^0 and D^+ mesons, which are the first charmed particles observed in hadronic interactions, have $c\tau$ values of 125 and 320 μm respectively i.e less than one hundredth of the corresponding $c\tau$ values for Λ (7.9 cm), Ξ^- (4.91 cm) and Ω^- (2.46 cm). As a consequence, the reconstruction of their decay vertices with enough precision to disentangle them from the interaction point, would require the detector to be located very near to the target. For example if one intended to use 50 μm pitch Si microstrip detectors, as it was done for detecting charm in pion and proton initiated reactions at SPS energies,[18] the first detector will have to be located at about 10 cm from the target i.e. in a region were the density of tracks from a Pb-Pb interaction would exceed 100 per cm^2, thus making the pattern recognition task impossible. It is not surprising therefore that no charmed particle identification has yet been attempted in heavy ion experiments.

Can pixel detectors allow us to meet such a challenge? The detectors of the next generation will have a pixel size of 50 x 400 μm^2, and a 300 μm thickness (0.3 % r.l.) thus allowing a vertex reconstruction error comparable to that reached in the previous charm experiments; their 5000 sensor units per cm^2 however, should allow us to handle a much larger track density. Will this be enough to handle some hundred tracks per cm^2 as as required for a charm experiment with a Pb beam at the SPS? Certainly much work is needed before a definite answer can be given; however it seems now time to start taking seriously a charm experiment for heavy ion collisions.

CONCLUSIONS

A comprehensive programme to study the excitation of strangeness in heavy ion inter-actions at CERN/SPS energies is now starting. The idea to determine an excitation function for $|S| = 1, 2, 3$ baryons, by measuring their yields and ratios as a function of energy density, is attractive and appears to be technically feasible. In this respect, the recent silicon pixel developments are of particular interest. Would it be possible to use this new technique to detect also charmed particles in heavy ion collisions? It certainly seems a difficult task, but it does not look hopeless.

Acknowledgements

The Author wishes to thank F. Antinori, K. Safarik and O. Villalobos Baillie for many useful discussions and suggestions.

REFERENCES

1. J. Rafelski, Contribution at this Workshop

2. P. Koch, B. Muller and J. Rafelski, Phys. Rep. 142 (1986) 167.

3. J. Rafelski, Phys. Lett. B262 (1991) 333;
   J. Rafelski, Nucl. Phys. A544 (1992) 279c;
   H.C. Eggers and J. Rafelski, Int. J. Mod. Phys. A6 (1991) 1067.

4. J. Letessier, J. Rafelski, A. Tounsi, Phys. Lett. B321 (1994) 394;
   J. Letessier, J. Rafelski, A. Tounsi, Phys. Lett. B323 (1994) 393.

5. U. Heinz, Nucl. Phys. A566 (1994) 205c

6. For a summary of recent experimental results see: E. Quercigh, Quark Matter '93 Summary Talk, Nucl.Phys.A 566 (1994) 321c

7. F. Antinori et al. (WA85) , Contribution at this Workshop.

8. S. Kabana et al. (NA35) , Contribution at this Workshop.

9. WA97 Proposal, CERN/SPSLC 91-29, SPSLC/P263 (1991) .

10. R. Beuttenmuller et al., IEEE Trans.Nucl.Sc. NS-33 (1986) 1045;
    R. Beuttenmuller et al., Nucl.Instr.Meth. A252 (1986) 471.

11. P. Rehah and E. Gatti, Nucl.Instr.Meth. A289 (1990) 410;
    W. Chen et al., Nucl.Instr.Meth. A326 (1993) 273.

12. E.H.M. Heijne et al. Nucl.Instr.Meth. A348 (1994) 399 and references therein.

13. E.H.M. Heijne et al. Nucl.Instr.Meth. A349 (1994) 138;
    F. Antinori et al. Nucl.Instr.Meth. A, to be published.

14. E. Shuryak, Sov.J. of Nucl.Phys. 28 (1978) 408;
    E. Shuryak, Phys.Rep. 61 (1980) 72.

15. T. Matsui and H. Satz, Phys.Lett.178B (1986) 416.

16. X.N. Wang and B. Muller, Nucl.Phys.A 566 (1994) 555c.

17. E. Shuryak, Nucl.Phys.A 566 (1994) 559c.

18. see for example M. Adamovich et al. Nucl.Instr.Meth. A309 (1991) 401.

THE PHYSICS AND EXPERIMENTAL PROGRAM
OF THE RELATIVISTIC HEAVY ION COLLIDER (RHIC)

John W. Harris*

Lawrence Berkeley Laboratory
University of California
Berkeley, CA 94720 USA

INTRODUCTION

The primary motivation for studying nucleus-nucleus collisions at relativistic and ultrarelativistic energies is to investigate matter at high energy densities ($\varepsilon \gg 1$ GeV/fm^3). Early speculations of possible exotic states of matter focused on the astrophysical implications of abnormal states of dense nuclear matter.[1] Field theoretical calculations predicted abnormal nuclear states and excitation of the vacuum.[2] This generated an initial interest among particle and nuclear physicists to transform the state of the vacuum by using relativistic nucleus-nucleus collisions.[3] Extremely high temperatures, above the Hagedorn limiting temperature, were expected and a phase transition to a system of deconfined quarks and gluons, the Quark-Gluon Plasma (QGP),[4] was predicted. Such a phase of matter would have implications for both early cosmology and stellar evolution. The understanding of the behavior of high temperature nuclear matter is still in its early stages. However, the dynamics of the initial stages of these collisions, which involve hard parton-parton interactions, can be calculated using perturbative QCD.[5] Various theoretical approaches have resulted in predictions that a high temperature (T ~ 500 MeV) gluon gas will be formed in the first instants (within 0.3 fm/c) of the collision.[6] Furthermore, QCD lattice calculations[7] exhibit a phase transition between a QGP and hadronic matter at a temperature near 250 MeV. Such phases of matter may have existed shortly after the Big Bang and may exist in the cores of dense stars. An important question is whether such states of matter can be created and studied in the laboratory. The Relativistic Heavy Ion Collider (RHIC)[8] and a full complement of detector systems are being constructed at Brookhaven National Laboratory to investigate these new and fundamental properties of matter. [9]

---

* Alexander von Humboldt Foundation U.S. Senior Scientist Award Recipient visiting the Institute für Kernphysik, Universität Frankfurt, Frankfurt, Germany.

THE RELATIVISTIC HEAVY ION COLLIDER (RHIC)

In 1983 the U.S. Nuclear Science Advisory Committee (NSAC) Long Range Plan[10] stated that the construction of RHIC was the highest priority new scientific opportunity in the field. This high priority for RHIC was reaffirmed in subsequent NSAC Meetings and by the National Research Council.[11] The U.S. High Energy Physics Panel (HEPAP) stated[12] in 1985 that "although (RHIC) is in the realm of nuclear physics, it is of interest to the particle physics community. It will be directly testing non-perturbative aspects of the non-Abelian gauge theory of quarks and gluons." RHIC was proposed for construction at Brookhaven National Laboratory and construction was started in January 1991. RHIC is expected to begin operation for physics in March 1999.

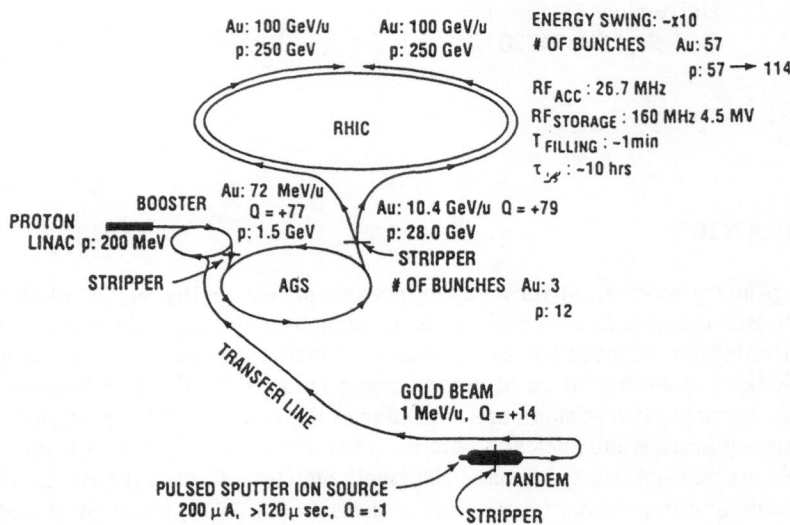

Figure 1. The Relativistic Heavy Ion Collider (RHIC) accelerator complex at Brookhaven National Laboratory. Nuclear beams are accelerated from the tandem Van de Graaff, through the transfer line into the AGS Booster and AGS prior to injection into RHIC. Details of the characteristics of proton and Au beams are also indicated after acceleration in each phase.

A schematic diagram of the RHIC accelerator complex at Brookhaven is displayed in Fig. 1. Nuclear beams are accelerated from the tandem Van de Graaff accelerator through a transfer line into the AGS Booster synchrotron and then into the AGS, which serves as an injector for RHIC. RHIC will accelerate and collide ions from protons up to the heaviest nuclei over a range of energies, up to 250 GeV for protons and 100 GeV/nucleon for Au nuclei. Fig. 2 summarizes the capabilities of the accelerator. In addition to the colliding beams described in Fig. 2, plans are underway to inject and accelerate polarized protons at RHIC in order to study the spin content of the proton.[13]

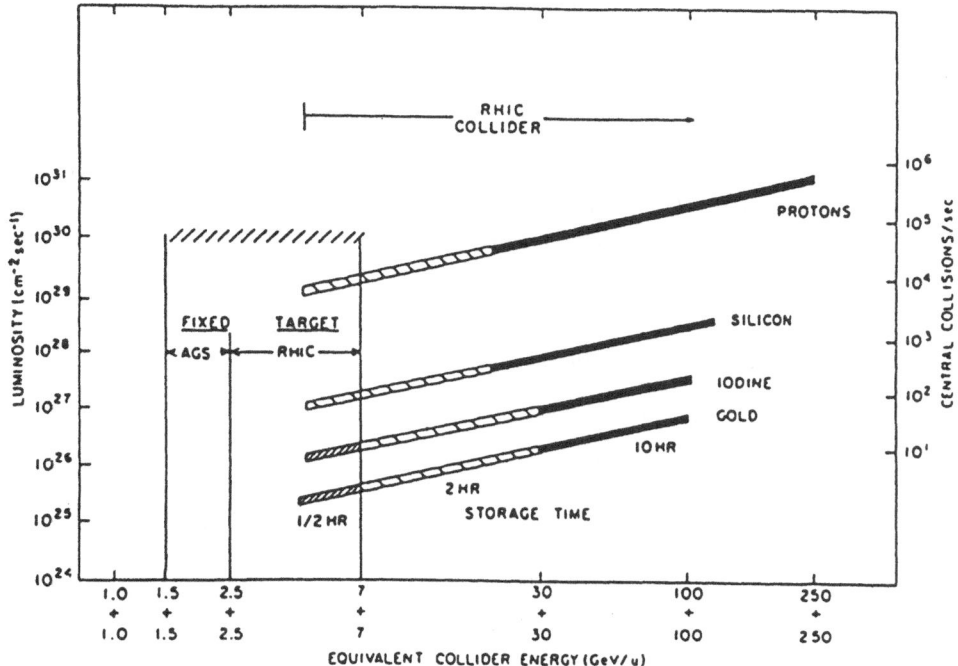

Figure 2. The RHIC luminosity and number of central collisions per second, for impact parameters less than 1 Fermi, are plotted as a function of the colliding beam energies for various projectile systems.

RHIC EXPERIMENTS

Collisions of the heaviest nuclei at impact parameters near zero at RHIC are expected to produce approximately 1000 charged particles per unit pseudorapidity.[14] This presents a formidable environment in which to detect the products of these reactions. The experiments will take various different approaches to search for the QGP. Two large collider detectors have full scientific, technical, cost and schedule approval for construction at RHIC. These are the STAR and PHENIX detector systems, both of which have begun construction and anticipate operation at RHIC start-up. The STAR experiment will concentrate on measurements of hadron production over a large solid angle in order to study global observables on an event-by-event basis. The PHENIX experiment will concentrate on measurements of lepton and photon production and will have the capability of measuring hadrons in a limited range of pseudorapidity. In addition, two smaller experiments have scientific approval and are in the technical development stage. These are a forward and midrapidity hadron spectrometer experiment (BRAHMS) and a compact multiparticle spectrometer (PHOBOS). The collaborations, which intend to construct these detector systems and exploit their capabilities, consist of more than 600 scientists from 80 institutions and 15 countries. A brief description of each experiment and its physics goals will be presented.

The STAR Experiment

The STAR (Solenoidal Tracker At RHIC) experiment[15] will search for signatures of QGP formation and investigate the behavior of strongly interacting matter at high energy density. A flexible detection system will be utilized to simultaneously measure many experimental observables to study signatures of the QGP phase transition as well as the space-time evolution of the collision process. Measurements will be made at midrapidity over a large pseudorapidity range ($|\eta| < 2$) with full azimuthal coverage ($\Delta\phi = 2\pi$) and azimuthal symmetry. The detection system is shown in Fig. 3. It will consist of a silicon vertex tracker (SVT) and time projection chamber (TPC) inside a solenoid magnet with a 0.5 T uniform field for tracking, momentum analysis and particle identification via dE/dx; a time-of-flight system (TOF) surrounding the TPC to extend particle identification to higher momenta; electromagnetic calorimetry (EMC) directly inside the solenoid coil to measure the transverse energy of events and to trigger on and measure high transverse momentum particles and jets; and external time projection chambers (XTPC) to extend the STAR tracking coverage to forward pseudorapidities. The following trigger detectors are also under construction for implementation in STAR (see Fig. 3): a central trigger barrel (CTB), TPC endcap trigger (not shown), vertex position detector (VPD) and the forward veto calorimeters (at large pseudorapidities, not shown).

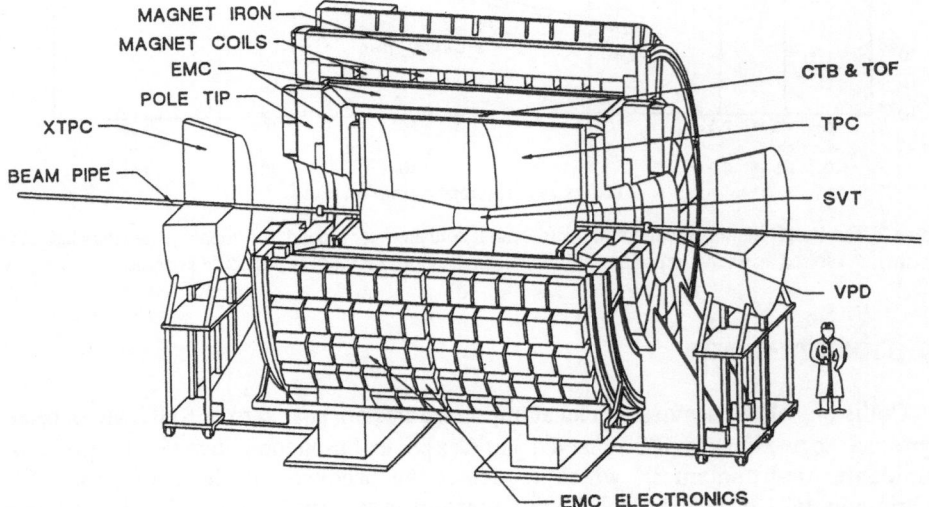

Figure 3. Layout of the STAR experiment at RHIC.

STAR will exploit two aspects of hadron production that are fundamentally new at RHIC. One of these is the measurement of *global observables on an event-by-event basis*. The event-by-event measurements of global observables possible using STAR are:

- Temperature of $\pi^{\pm}$, $<p_t>$ of $\pi^{\pm}$ and $K^{\pm}$
- Flavor content (strangeness using $K^{\pm}$ multiplicity, K/π ratio)
- Charged-particle multiplicity
- Entropy density
- Pion source size
- Reaction dynamics and collision geometry
- Neutral transverse and total transverse energy
- Degree of fluctuations in
    $d^3n/dp_t d\eta d\phi$
    Energy ($d^2E_t/d\eta\, d\phi$)
    Entropy
    Isospin (neutral/charged particle energy)

- Low p_t phenomena
- Presence of high p_t particles, jets
- Stopping & forward baryon distributions
- Particle ratios

The event-by-event measurement of these global observables is possible because of the very high charged particle densities, $dn_{ch}/d\eta \approx 1000$ expected in nucleus-nucleus collisions at RHIC. This will allow novel determination of the thermodynamic properties of single events. For example, the slope of the transverse momentum (p_t) distribution for pions and the $<p_t>$ for pions and kaons can be determined *event-by-event*. Individual events can be characterized by a "temperature" to search for events with extremely high temperature, predicted[16] to result from deflagration of a QGP. Displayed in Fig. 4 are two spectra generated by the Monte Carlo method from Maxwell-Boltzmann distributions with

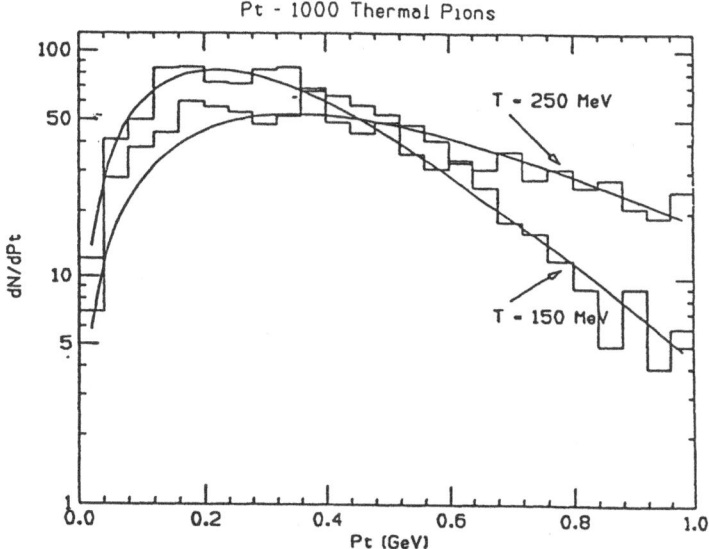

Figure 4. Simulation of the p_t spectrum for one event generated using a Boltzmann distribution of 1000 pions. The histograms correspond to single events generated with T = 150 MeV and 250 MeV. The curves are fits to the histogram using a Maxwell-Boltzmann distribution.

T = 150 and 250 MeV, each containing 1000 pions. The slopes of spectra with T = 150 and 250 MeV derived from fits using a Maxwell-Boltzmann distribution, also shown in Fig. 4, can easily be discriminated at the single event level. The determination of $<p_t>$ for pions can be made very accurately on the single event basis in this experiment, over the expected range of multiplicities in central collisions from Ca + Ca to Au + Au. The $<p_t>$ of kaons can be determined accurately for single events from central Au + Au events.

Correlations between such observables will be made on an event-by-event basis to isolate potentially interesting event types for detailed investigation. Events with special (often extreme) values of many of the above distributions can be available at the trigger level to be able to enhance the sensitivity of STAR, increase statistics for specialized events and increase the likelihood of new discoveries. The full azimuthal coverage, particle identification and continuous tracking of STAR is required to perform these measurements.

In addition to the event-by-event observables, inclusive measurements will be performed in STAR and then correlated on an ensemble basis. From among these inclusive measurements, those of particular interest at RHIC are measurements of φ-mesons, short-lived singly-strange hadrons (Λ, $\overline{\Lambda}$, K^0_s), multiply-strange baryons ($\Xi^{\pm}$, Ω^-), and $K^0_s K^0_s$ particle interferometry. The production cross section, mass and width of φ-mesons can be measured in STAR from the decay $\phi \Rightarrow K^+ + K^-$. The φ production rate, mass and width are expected to be extremely sensitive to changes in the quark masses[17,18,19] due to a chiral phase transition at high energy densities. Enhancements to the strange antibaryon content due to QGP formation have been predicted.[20] Furthermore, multiply-strange baryons ($\Xi^{\pm}$, Ω^-) are expected to be more sensitive to the existence of the QGP than singly-strange hadrons.[21] Examples of the mass spectra expected for Λ and Ξ^- in STAR are shown in Fig.

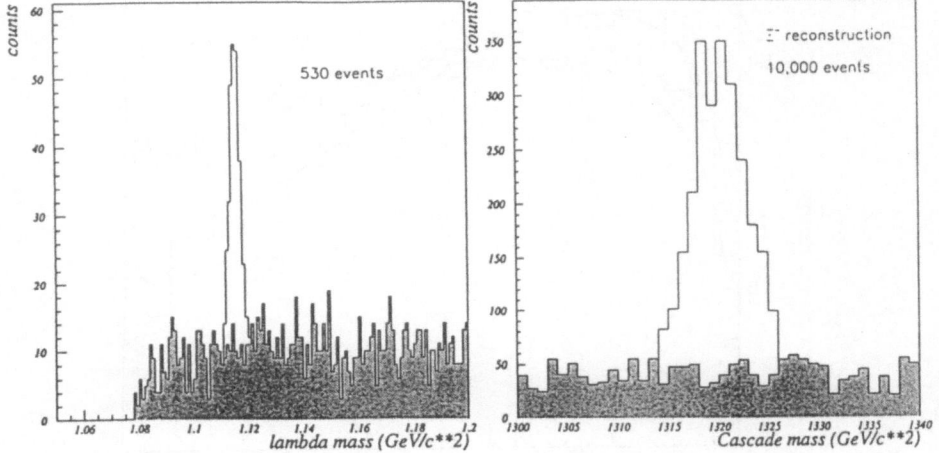

Figure 5. Simulated invariant mass spectra for Λ and Ξ^- measured in STAR using the SVT and TPC.

5. The $K^0_s K^0_s$ correlations will allow two-particle Bose-Einstein measurements without Coulomb distortions.

The other aspect of hadron production that is fundamentally new at RHIC is the use of *hard scattering of partons* as a probe of the properties of high density nuclear matter. Measurable yields of intermediate and high p_t particles and high p_t jets at RHIC will allow investigations of hard QCD processes via both segmented calorimetry and high p_t single particle measurements. Measurements of the remnants of hard-scattered partons will be used as a penetrating probe of the QGP, will provide important new information on gluon distributions in nucleons and gluon shadowing in nuclei, and provide measurements to test perturbative QCD in polarized proton (spin) studies.

The PHENIX Experiment

The physics goals of PHENIX (Pioneering High Energy Nuclear Interaction eXperiment)[22] are to measure as many potential signatures of the QGP as possible as a function of a well-defined common variable such as impact parameter or pseudorapidity density. PHENIX will measure lepton pairs (di-electrons and di-muons), photons and hadrons. The experiment will be sensitive to very small cross section processes such as the production of the J/ψ, ψ', Υ and high p_t spectra. It will also have the capability for high rates with pp and pA collisions.

A diagram of the PHENIX detector system is displayed in Fig. 6. The magnet has an axial field along the beam direction with tracking chambers and detectors for the identification of electrons, muons, photons and hadrons installed outside the field. There are two arms for dielectron measurements, each with 1 steradian acceptance at midrapidity. Each arm is equipped with a ring-imaging Cerenkov detector (RICH), a time-expansion chamber (TEC) for dE/dx measurements, time-of-flight detectors (TOF), and an electromagnetic calorimeter (EM Cal). Photons and hadrons will also be measured at midrapidity with 2 and 0.36 steradian acceptances, respectively. A separate muon spectrometer with 1 steradian acceptance is located at forward rapidities as shown in Fig. 6. The magnetic field in the muon arm is transverse to the beam direction. A silicon multiplicity-vertex detector (MVD) and beam-beam coincidence counters are located near the interaction vertex. The MVD covers $|\eta| < 2.7$ about midrapidity for event selection via charged-particle multiplicity.

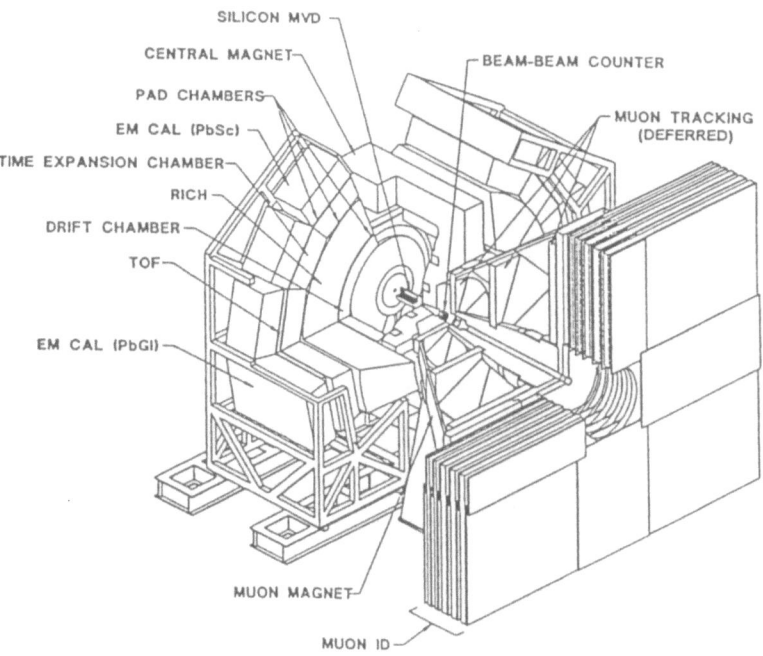

Figure 6. A diagram of the PHENIX experiment at RHIC. The beams collide along the horizontal direction in the center of the detector. The detector components are defined in the text.

PHENIX will investigate the degree of *deconfinement* in RHIC collisions, and thus the possibility of QGP formation, by measuring the yields of J/ψ, ψ', Υ. If the J/ψ radius is larger than the Debye screening length then the QCD potential between the c̄c pairs, which form a J/ψ, is weakened and J/ψ formation suppressed.[23] Since the J/ψ, ψ′, ϒ radii are different, i.e. r(ψ′) > r(J/ψ) > r(ϒ), a study of the relative suppression of each of these is sensitive to the screening in a QGP and will help differentiate between deconfinement and possible dissociation in hot nuclear matter. PHENIX will measure J/ψ via electron-pairs near midrapidity and J/ψ, ψ′, ϒ via muon-pairs at forward angles.

The anticipated raw muon-pair mass spectrum in the forward muon arm is displayed in Fig.7 for one-month of RHIC operation at the design luminosity for central Au + Au collisions. Peaks for the J/ψ, ψ′, ϒ can be observed. With subtraction of like-sign muon-pairs, and application of kinematic cuts the peak-to-background of this spectrum is expected to improve further and the ψ′ peak will become more prominent. For one year of running at the RHIC design luminosity of 2 x 10^{26} cm^{-2}s^{-1}, PHENIX will detect 390K J/ψ, 5.7K ψ′ and 1.2K ϒ for central collisions (top 10 % multiplicity) of Au + Au. The mass resolution is about 100 MeV for the J/ψ and the pion rejection is somewhat better than 10^{-4} at p ≥ 2.5 GeV/c. In addition to these charmonium studies, PHENIX will measure open charm (D$\bar{\text{D}}$ decays) via unlike-sign eμ coincidences using the electron and muon spectrometers. The rate of open charm production is expected to be sensitive to details of the early stages of RHIC collisions and will provide complimentary information to the charmonium studies.

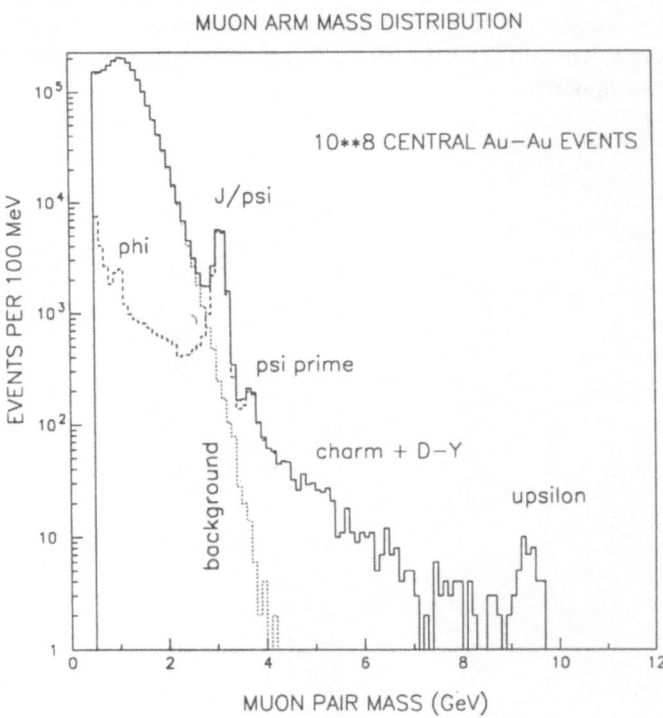

Figure 7. PHENIX muon-pair mass spectrum anticipated in one month of RHIC running for central Au + Au collisions at a luminosity of 2 x 10^{26} cm^{-2}s^{-1}.

PHENIX will investigate possible *chiral symmetry restoration* by performing high resolution measurements of the leptonic and hadronic decays of φ-mesons. In a chirally-restored QGP both the φ-meson and kaon masses may be modified resulting in change in the mass and width of the φ-meson and the branching ratio between leptonic and hadronic channels.[24] The anticipated electron-pair spectrum for 8 days of central trigger Au + Au operation at RHIC is shown in Fig. 8. Extremely sharp peaks for the ω, ϕ and J/ψ are observed, with a 2-to-1 signal-to-noise ratio in the low mass resonance region. The mass resolution is approximately 4 MeV for the $\phi \rightarrow e^+e^-$ (and < 1 MeV for $\phi \rightarrow K^+K^-$ in the hadronic measurement). The pion rejection is better than 10^{-4} for p $\leq$ 4 GeV/c.

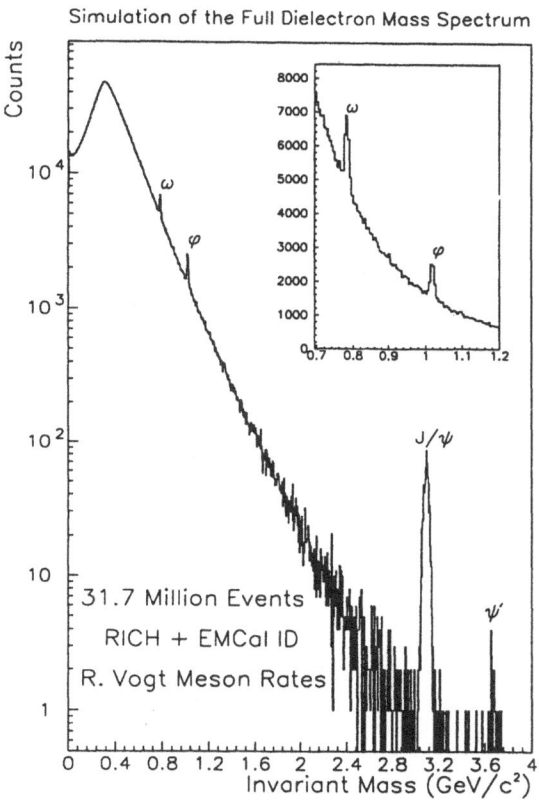

Figure 8. PHENIX electron-pair mass spectrum anticipated in 8 days of RHIC running for central Au + Au collisions at a luminosity of 2×10^{26} cm^{-2}s^{-1}.

There are other measurements of interest that will be undertaken by PHENIX. Measurements will be made using the electromagnetic calorimeter and high resolution photon detector to determine the yield of direct photons (for $p_t \geq$ 0.5 GeV/c) radiated from the nuclear interaction. An enhancement of photons with $p_t \geq$ 2 to 3 GeV/c is expected because of the high gluon concentration in the early high energy-density stage of RHIC collisions.[25] PHENIX will also investigate the order of the QGP phase transition by measuring the $<p_t>$ of identified-charged pions, kaons and protons at mid-rapidity as a function of the energy density. By detecting both hadrons and photons, PHENIX can measure fluctuations in isospin, $\pi^0/(\pi^+ + \pi^-)$, which might occur in a second order phase transition.[26] PHENIX will also study strangeness and charm production for enhancements, jet-quenching, and the space-time evolution of the system as a function of energy density.

The BRAHMS Experiment

The physics goals of the BRAHMS (BRoad RAnge Hadron Magnetic Spectrometers) experiment[27] are to achieve a basic understanding of relativistic heavy ion collisions at RHIC through a systematic study of particle production in AA collisions from the peripheral to the most central in impact parameter. Measurements will be performed using two high resolution magnetic spectrometers at various angles to cover both the baryon-rich fragmentation regions and the high temperature, baryon-depleted midrapidity region. A diagram of the BRAHMS forward and midrapidity spectrometers is shown in Fig. 9. The spectrometers will measure and identify inclusive and semi-inclusive $\pi^{\pm}$, $K^{\pm}$ and $p^{\pm}$, and their momenta with high statistics over a small solid angle and over a wide range of pseudorapidity ($0 \leq \eta \leq 4$) and transverse momentum. Particle identification is performed using various combinations of time-of-flight arrays, and threshold and ring-imaging Cherenkov counters. BRAHMS will measure inclusive and semi-inclusive particle spectra and extract the net baryon densities temperatures from spectral slopes as a function of rapidity to determine whether thermal and chemical equilibrium are reached in these collisions. It will also be able to study both high and low transverse momentum processes. Centrality will be measured using a global multiplicity detector.

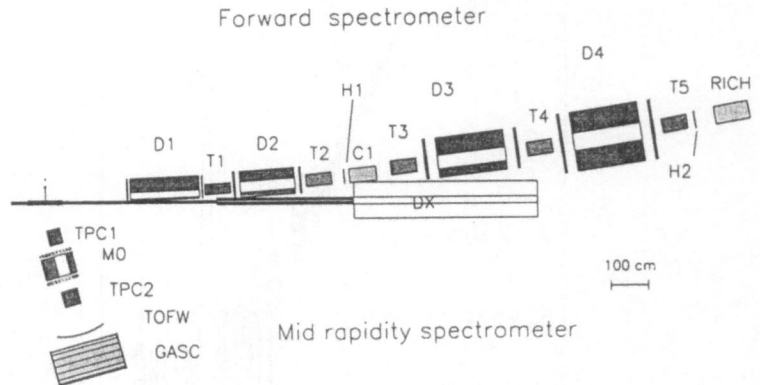

Figure 9. A diagram of the BRAHMS forward and midrapidity spectrometers in the narrow angle hall at RHIC.

The PHOBOS Experiment

The physics goals of the PHOBOS experiment[28] are to measure single particle spectra and correlations between particles with low transverse momenta and to characterize events using a multiplicity detector. Charged particles will be measured and identified in the range $0 \leq y \leq 1.5$ and 15 MeV/c $\leq p_t \leq 600$ MeV/c for pions and 45 MeV/c $\leq p_t \leq 1200$ MeV/c for protons. The range of particles to be studied include γ, $\pi^{\pm}$, $K^{\pm}$, p, $\bar{p}$, ϕ, Λ, $\bar{\Lambda}$, d and $\bar{d}$. Particle ratios, p_t spectra, strangeness production ($K^{\pm}$, ϕ, Λ, $\bar{\Lambda}$) and particle correlations will be studied. An illustration of the experiment is shown in Fig. 10. The top coils and magnet iron are not shown in order to see the other components. The multiplicity detector consists of silicon strip detectors and silicon pad detectors. The two magnetic arms have a field strength of 2 Tesla. Eleven layers of silicon, 5 layers of pads and 6 layers of strips, are installed in each magnetic spectrometer for tracking and momentum measurements.

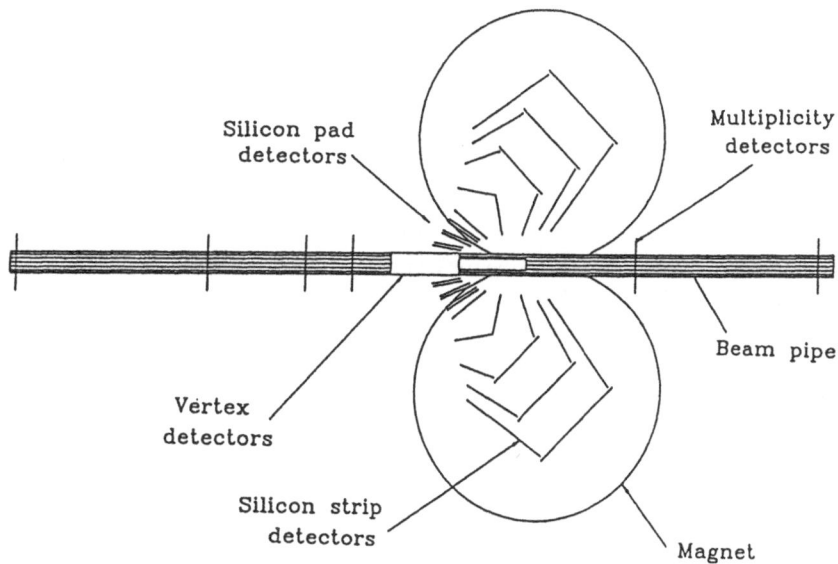

Figure 10. A diagram of the PHOBOS two-arm multiparticle spectrometer. The two arms are located on opposite sides of the beam pipe. Each arm has a 2 Tesla magnet represented by only the lower coils for ease of viewing other parts of the set-up, with silicon detector planes for tracking.

ACKNOWLEDGMENTS

I would like to thank Timothy Hallman for helpful comments and Joy Lofdahl for assistance with the manuscript. I thank my collaborators in STAR and friends in the PHENIX, BRAHMS, AND PHOBOS Collaborations for their work presented in this manuscript. This work was supported in part by the Alexander von Humboldt Foundation and the Director, Office of Energy Research, Division of Nuclear Physics of the Office of High Energy and Nuclear Physics of the U.S. Department of Energy under contract DE-AC03-76SF00098.

REFERENCES

1. E. Feenberg and H. Primakoff, Phys. Rev. 70, 980 (1946); A.R. Bodmer, Phys. Rev. D4, 1601 (1971); G. Baym and S.A. Chin, Phys. Lett. 62B, 241 (1976).
2. T.D. Lee and G.C. Wick, Phys. Rev. D9, 2291 (1974).
3. T.D. Lee, Rev. Mod. Phys. 47, 267 (1975).
4. J.C. Collins and M.J. Perry, Phys. Rev. Lett. 34, 1353 (1975); G. Chapline and M. Nauenberg, Phys. Rev. D16, 450 (1977); L. Susskind, Phys. Rev. D20, 2610 (1979).
5. K. Geiger and B. Mueller, Nucl. Phys. B369, 600 (1992).
6. See presentation of K. Geiger, Proceedings of this Workshop.
7. J. Kapusta, Nucl. Phys. 61, 461 (1980); J. Kuti et al., Phys. Lett. 95B, 75 (1980); H. Satz, Ann. Rev. Nucl. Part. Sci. 35, 245 (1985).
8. Conceptual Design of the Relativistic Heavy Ion Collider, Brookhaven National Laboratory Report BNL 52195 (1989).

9. For future perspectives utilizing heavy nuclear beams in the Large Hadron Collider, please see the presentation of H.Gutbrod, Proceedings of this Workshop.

10. "A Long Range Plan for Nuclear Science," Report of the DOE/NSF Nuclear Science Advisory Committee, December 1983.

11. "Physics through the 1990s," Panel on Nuclear Physics of the Physics Survey Committee and the National Research Council, National Academy Press, Washington, D.C., 1986.

12. U.S. High Energy Physics Panel (HEPAP) Report, 1985.

13. Proposal on Spin Physics Using the RHIC Polarized Collider, RHIC Spin Collaboration (1992).

14. F. Videbaek and T. Throwe, Fourth Workshop on Experiments and Detectors for a Relativistic Heavy Ion Collider, Brookhaven National Laboratory Report BNL-52262 (1990).

15. Conceptual Design Report for the Solenoidal Tracker At RHIC, STAR Collaboration, PUB-5347 (1992).

16. E.V. Shuryak and O.V. Zhirov, Phys. Lett. B89, 253 (1980); E.V. Shuryak and O.V. Zhirov, Phys. Lett. B171, 99 (1986).

17. R. D. Pisarski and F. Wilczek, Phys. Rev. D29, 338 (1984).

18. T. Hatsuda and T. Kunihiro, Phys. Lett. B185, 304 (1987).

19. E.V. Shuryak, Nucl. Phys. A525, 3c (1991).

20. J. Rafelski and A. Schnabel, "Intersections Between Nuclear and Particle Physics," AIP Proceedings No. 176, 1068 (1988), and references therein.

21. J. Rafelski, Phys. Rep. 88, 331 (1982).

22. PHENIX Experiment at RHIC - Conceptual Design Report, PHENIX Collaboration Report (1993).

23. T. Matsui and H. Satz, Phys. Lett. B178 (1986) 416.

24. D. Lissauer and E.V. Shuryak, Phys. Lett. B253 (1991) 15.

25. E. Shuryak and L. Xiong, Phys. Rev. Lett. 70 (1993) 2241; K.J. Eskola and M. Gyulassy, Phys. Rev. C47 (1993) 2239.

26. F. Wilczek, Nuc. Phys. A566 (1994) 123c.

27. Interim Design Report for the BRAHMS Experiment at RHIC, BNL Report, 1994.

28. Conceptual Design Report, PHOBOS Collaboration (1994).

HEAVY ION PHYSICS AT THE LARGE HADRON COLLIDER AT CERN

Hans H. Gutbrod

Gesellschaft für Schwerionenforschung
64220 Darmstadt, Germany
and
CERN, PPE Division
1211 Geneva, Switzerland

For the ALICE Collaboration

ABSTRACT

The Large Hadron Collider (LHC) in the LEP tunnel at CERN is planned also as a heavy ion collider with lead ions colliding at an energy of 2.7 + 2.7 ATeV. This corresponds to collisions of matter with cosmic rays of the utmost energies observed so far. Minor improvements of the newly commissioned lead ion source at the CERN SPS are necessary in order to provide a luminosity of $L = 2 \times 10^{27} \mathrm{cm}^{-2}\mathrm{s}^{-1}$. The detector ALICE has been chosen as the third detector for the LHC and will be dedicated to the physics of these nuclear collisions and also to the large cross section physics in p + p collisions.

INTRODUCTION

In the early seventies, nuclear collisions at laboratory energies below 1 AGeV were investigated at the Bevalac in Berkeley with the aim to learn about the equation of state of nuclear matter at densities where particle production is still small. Nuclei were found to stop each other at these low energies and densities of more than twice nuclear matter densities have been reached at those energies. This corresponds to values as they are of importance in supernova explosions and in neutron stars. Already here, however, some star models require values much larger in order to obtain the supernovae bounce.

The start of the Heavy Ion Program at the AGS at Brookhaven and at the SPS at CERN in 1986 was motivated strongly by the desire to go to much higher densities in order to recreate the first moments of the Early Universe, the Big Bang, right in the laboratory.

The expansion of the quark-gluon dominated Universe ends when the temperature decreases sufficiently to permit hadronization, i.e. formation of the hadronic particles (mesons and baryons). In this phase transition the large scale structure of the luminous matter is established. Laboratory experiments with relativistic nuclei aim to recreate the conditions similar to those encountered before and in the early hadronization period. There are two major differences: the time scale of the Universe expansion is determined by the interplay of the gravitational forces and the Fermi pressure of the hot matter, while in the laboratory work with 'Micro Bangs' there is no gravitation to slow the expansion. Furthermore, the baryon content of the hot hadronic matter is considerable for the present energy scales, given the high baryon number stopping; the early Universe was quite different in this regard.

This lead to the idea to collide heavy nuclei in RHIC presently under construction at Brookhaven and in the proposed LHC in the LEP tunnel at CERN leading to the formation of the highest energy densities presently envisaged to exist in our Universe. Hope is that nuclei are pieces of matter large enough to allow the study of the thermodynamics of hadronic matter in its great simplicity.

LEAD IONS IN THE LHC

After the successful operation of the LEP and its ongoing upgrade program to reach $\sqrt{S} = 200$ GeV the CERN is planning the installation of the LHC in the LEP tunnel to provide p + p collisions at about $\sqrt{S} = 14$ TeV and Pb + Pb collisions at about $\sqrt{S} = 5.5$ ATeV. Two-in-one magnets of very high magnetic fields of 8.5 T are being developed and string tests with full length prototypes are to start in November 1994. A proton luminosity of $2 \times 10^{27} \text{cm}^{-2}\text{s}^{-1}$ is targeted for the two p + p experiments ATLAS and CMS.

The heavy ion option of the LHC requires a further upgrade of the performance of the SPS lead beam intensity by a factor of more than 10. Improving the recently commissioned lead injector and using accumulation schemes in the CERN's low energy rings (LEAR, AAC) or in the PS itself, or a combination of all these measures, allows to expect an increase of even a factor of 30. Having therefore no unsurmountable problems in delivering the necessary lead beam intensities from the injectors there are, however, other limiting factors which cannot be overcome so quickly. They have consequences on the mode of LHC operation with Heavy Ions not yet widely known, but also open up new fields of physics as presented further below.

LUMINOSITY LIMITS

The Intra Beam Scattering (IBS) is the Coulomb Scattering of the ions within the bunch leading to an emittance growth in all the three phase planes of each beam. For a given ion only the number of ions per bunch can reduce this beam blow up, but in order to keep the number of colliding ions the same the number of bunches needs to be increased. With 560 bunches foreseen in the LHC and 9.4×10^7 ions/bunch a 10 hours half-life of the beam due to IBS was calculated. Whilest this effect is the limiting factor for the RHIC at Brookhaven with a $\sqrt{S} = 200$ AGeV, the LHC Heavy Ion beams at several ATeV are further limited in their life time by the extremely large electromagnetic interactions of the ions colliding in the intersection region:

— At the beam intersection, the Electromagnetic Dissociation Weizaecker-Williams (WW) of the ions is a very strong process estimated at more than 200 b in cross-section. Due to the high electromagnetic fields of the lead ions they excite each other, loose one or several nucleons and are therefore not anymore on the correct beam orbit.

— Another process, the Pair Production, limits further the luminosity of the ion beam in the LHC: the pair production leading to an energy loss of the ion followed by an Electron Capture (EC) which leads then to the final loss of the ion. Both processes, WW and EC, are leading to a luminosity decrease by a factor of two after 11.4 hours.

These two effects cannot be influenced by modifying the parameters of the beams, but have to be tolerated and taken into account properly. Furthermore, these effects are the base for a judgement on how low cross section physics can be measured, e.g. with only one intersection region active, i.e. one experiment only for heavy ions at one time, which leads to the highest integrated luminosity, or with two experiments running, which yields only about a quarter of the luminosity per experiment.[1]

PHYSICS EXPECTATIONS

When it is hard to extrapolate from sulfur nucleus data to Pb - Pb collisions at the SPS, then it must be impossible to come close with guesses how the reaction looks like at a 400 times higher $\sqrt{S}$. The data of the TEVATRON at the Fermilab at a $\sqrt{S}$ of 1.8 TeV are arriving just in time to give more guidance. The importance of mini-jet production is being emphasized at this much higher energy compared with RHIC, leading to much shorter thermalisation times and much larger temperature estimates than previously considered. Event simulations are under way[2] using the VENUS, DTUNUNC, the HIJINC, Parton Cascade model, the PYTHIA event generators, or a Hydrodynamical Model. They differ in the maximum particle density by a factor of more than 3, i.e. they predict charged particle density values from 2000 to 8000. Any calculation of the energy density reached is therefore affected by a large uncertainty. One major difference to the nuclear collisions at RHIC is the expectation that at these high energies the low-x gluons are playing the dominant role in the early phase of the collision, forming a 'brick wall' of gluons smashing into each other with a very small thermalisation time of a few tenths of fermi/c, thus creating a gluon plasma followed ca. 1 fm/c later by a quark gluon plasma. Figure 1 shows the energy densities as a function of time as calculated in the Parton Cascade model by T.S. Biró.[3] One observes the very large increase going from RHIC to LHC, attributed to the 'hot-gluon-scenario' in the collisions.

As signals accessible for experimental observations, one expects collective flow of the expanding dense matter produced in the collision to be studied in multi-particle correlations. The thermal photon signal from both the QGP itself and the Mixed and Hadronic Phase should be visible in events with a high particle density at transverse momenta from 1–10 GeV/c. Finally, the design luminosity is high enough to allow a full spectral analysis of $c\bar{c}$ and $b\bar{b}$ bound states, if all efforts are concentrated onto one intersection region only.

Besides the interest for studying matter under extreme conditions and potential phase changes, there are totally new possibilities opening up due to the above mentioned large electromagnetic effects responsible for the limitations in luminosity. Peripheral reactions of ions generates photon-photon collisions at energies up to $\sqrt{S} = 200$ GeV with luminosities way above those of LEP 200.[4] In addition, photon - gluon and photon - quark collisions up to an energy of $\sqrt{S} = 1000$ GeV are predicted[4] and would extend the HERA program substantially. Thus, peripheral collisions could be of large interest to study high mass spectroscopy.

THE UNIVERSAL HEAVY ION DETECTOR ALICE

There was a consensus that one universal detector should be built to study all possible correlations of potent signals, known today or presented tomorrow. It should allow for special runs with specific experimental programs. Earlier on, the two high luminosity detectors CMS and ATLAS of the p-p program were thought to be able to study special channels like, e.g., the di-muons. However, simulations of their performance in a Heavy Ion reaction are not looking convincingly, and strong efforts are now being made to find solutions within one heavy ion detector only. This is in line with the fact that having more than one active intersection region will kill the luminosity faster due to the processes mentioned above.

We are ignorant of the collective phenomena which may happen at this high energy. We are guided in the expectation of the "No-New-Physics" by looking at the data of the Tevatron.

We are surprised that even this knowledge at an energy just a factor of 4 lower allows the event generators to spread by a factor of 4 in their predicted particle densities. The nuclear stopping differs largely in the models. Therefore, a large universal heavy ion detector should measure a large part of the phase space and not just focus into a small part of it.

The detector specifications, like size, length, number of channels etc., are determined by the large multiplicities encountered in the Pb + Pb collisions at a fairly low luminosity of 10^{27} and by the high luminosity of 10^{31} in the p + p mode, i.e. low multiplicity but much higher rate. The detector must measure the rather soft physics below a p_t of $10\,\text{GeV/c}$ not covered by the p + p detectors.

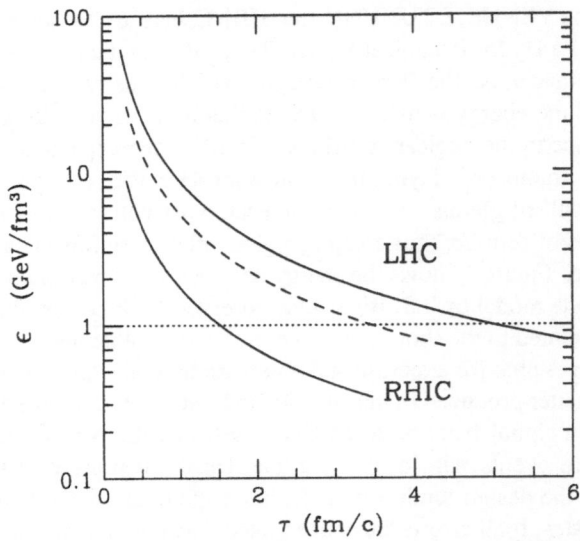

Figure 1. Time dependence of the energy density of the parton plasma at RHIC and LHC energies. The solid lines show the results obtained with the initial values taken from the table given in reverence 3. The dashed line shows the effect of a three times higher initial gluon density at RHIC energy, corresponding to QCD cut-off $p_0 \approx 1.5\,\text{GeV}$ instead of $p_0 = 2.0\,\text{GeV}$. Whereas a well-developed parton plasma is predicted to be formed at the LHC, the prediction for RHIC is uncertain.[3]

EXPERIMENT ALICE

The adopted design[5] shown in Figs 2, 3 and 4 is based on the modification of the existing L3 experiment presently running at LEP at CERN. It contains the largest magnet so far with an inner usable diameter of 11.4 m and a length of 12 m. The field can be as high as 0.6 T but for the major running with heavy ions it would be reduced to 0.2 T. The pole faces have an opening permitting to add forward spectrometers to the central barrel and to cover the rapidity region of $\eta > 2.2$.

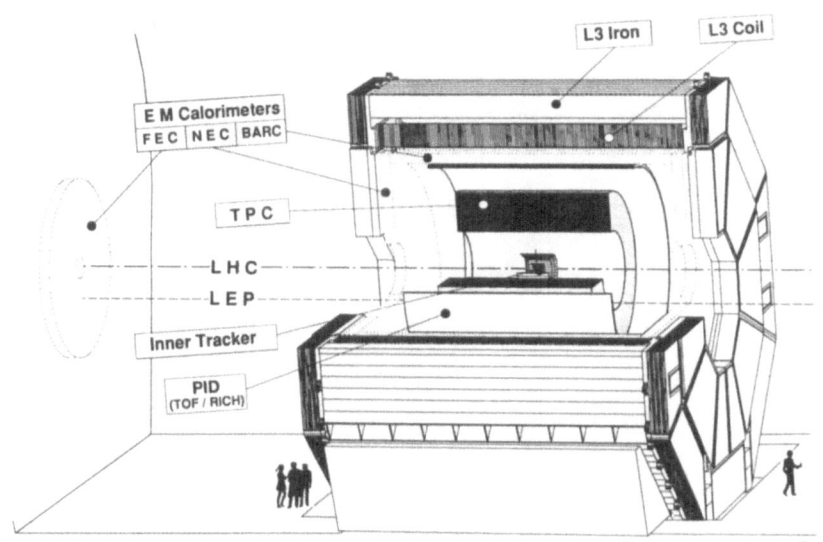

L3 Iron

L3 Coil

E M Calorimeters
FEC | NEC | BARC

T P C

L H C

L E P

Inner Tracker

PID
(TOF / RICH)

jlb

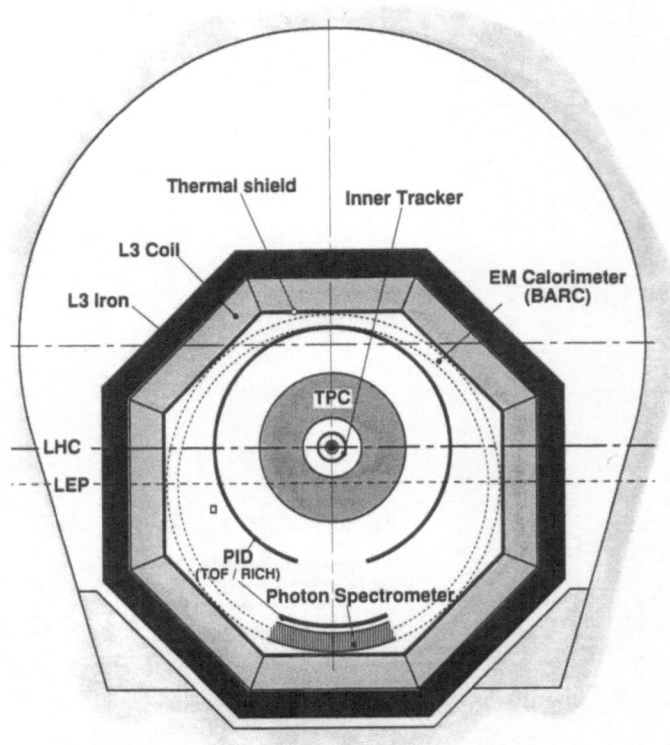

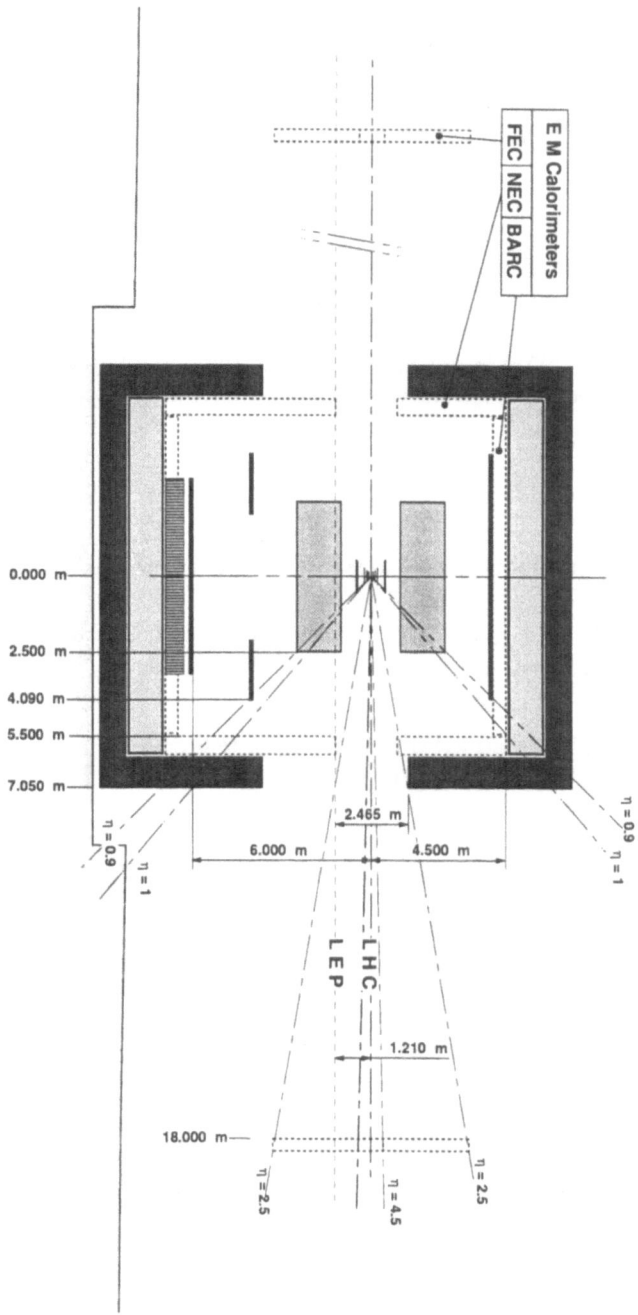

Figure 4. Side view of the ALICE detector

545

large scale energy fluctuations. A liquid Argon calorimeter where the signals are read out with a very economical pad readout presently under development by the University of Lund, Sweden is being studied.

The calorimeter coverage over nearly 8 units of rapidity should allow to study multi jet correlations via high p_t π^0's and photons and uncover features of the possible underlying parton cascade.

The radial dimensions of this detector is determined taking into account the highest particle densities to be expected in view of the recent predictions.

SUMMARY

The future of Relativistic Heavy Ion Physics is very bright if all the planned concepts are realized. Complemented by the RHIC program in the USA this community of scientists will be able to choose doing experiments at $\sqrt{S} = 5$ - 16 AGeV at the SPS, $\sqrt{S} = 60$ - 200 AGeV at RHIC and $\sqrt{S} = 6000$ AGeV at the LHC. Since the opportunities are so many and the challenge so large we in Europe are welcoming interested groups to collaborate in the design and construction of the LHC Universal Heavy Ion Detector ALICE.

REFERENCES

1. D. Brandt, K. Eggert and A. Morsch, SL/Note 94-4 (AP) and LHC Note 263. CERN Internal Notes (1994).

2. P. Giubellino, Proceedings of NATO ASI, Summer School, Il Ciocco (I) on Particle Production in highly excited matter. Plenum Press, London (1993).

3. T.S. Biró et al., LBL-34164 and DUKE-TH-93-46 (1993).

4. S.M. Schneider, W. Greiner and G. Soff, *Phys. Rev.* D46:2930 (1992).

5. N. Antonio et al., ALICE Letter of Intent, CERN / LHCC / 93-16 (1993).

THE HADRONIC FUTURE

Christopher H. Llewellyn Smith

CERN
European Organization For Nuclear Research
CH - 1211 Genève 23

INTRODUCTION

'The Hadronic Future', a title given to me by the organizers, could mean many things: future challenges in understanding QCD that are either theoretical (how to 'solve' QCD, the soft/hard interface, properties of the quark gluon plasma, structure functions at small x...) or experimental (spectroscopy of glueballs and mixed states, small x, heavy ion experiments...), or future experiments involving hadrons (LHC or hadronic physics at future linear colliders). In this talk, I address two of these topics:
1) 'Solving' QCD
and
2) Prospects for constructing larger accelerators and the long-term future of particle physics.

I chose these topics for a number of reasons. First, I have developed some thoughts on both: indeed, the written version of this talk will essentially consist of abstracts of, or introductions to, material written-up elsewhere. Second, I consider that 'solving' QCD is a major challenge for contemporary theoretical particle physics, while the question of prospects for bigger accelerators is of major importance for the future of our subject generally. Finally, I thought that these topics would appeal to Rolf Hagedorn, whose personality and many contributions we have the pleasure of honouring at this meeting.

'SOLVING' QCD

We will never solve QCD in the same sense that we can solve the Schrödinger equation for hydrogen. On the other hand, we cannot solve the Schrödinger equation for helium, let alone — for example — bismuth. Nevertheless, we consider atomic physics 'solved', and have handed it over to the chemists. We have a good qualitative understanding of atomic spectra in terms of simple models and concepts (Thomas Fermi model, LS and jj coupling, etc.), which can be understood from first principles, and we can write computer programmes to calculate to high accuracy if we wish. I believe that we should seek to 'solve' QCD in the same sense.

Twenty years ago, we wished to solve QCD in order to discover whether it is true, but this is now established beyond doubt. However, three powerful motives remain. The first is that scientific curiosity drives us to understand QCD better: we cannot consider our job complete, and hand QCD over to the nuclear physicists, until we have solved QCD in the sense suggested above. The second is that certain properties of QCD which are needed in order to unravel electroweak parameters (e.g. the hadronic parameters which enter B-$\overline{\text{B}}$ mixing) cannot be measured directly and we would therefore like to calculate these properties. The third is that because on the one hand the theory is established, while on the other hand we have an enormous amount of data, QCD provides a wonderful laboratory for developing and testing methods for dealing with strongly coupled field theories, which may be useful in other contexts — studying proposed 'theories of everything', for example.

The question of how to solve QCD is of course much harder to answer than what a 'solution' would mean and why it should be sought. The method described in this talk, which was developed with N.J. Watson, may turn out to run into insurmountable calculational difficulties, but it does illustrate what a 'solution' of QCD might look like.

The underlying idea is simple, even simplistic. It is to attempt to solve the Schrödinger equation for QCD:

$$H\Psi = E\Psi .$$

This proposal runs into three immediate difficulties:

1) It is a functional equation and hence ill-defined.

2) Some means of regulating ultraviolet divergences is required.

3) In order to keep a direct physical interpretation, which is the goal of the whole approach, we wish to avoid ghosts and to maintain rotational invariance. This naturally leads to the gauge choice $A_0 = 0$. In this gauge, however, Gauss's law

$$(\vec{D}.\vec{E} - \rho)\Psi = 0$$

has to be kept as a subsidiary condition. In the case of QED, this equation can be solved explicitly for the longitudinal components of $\vec{E}$ which can then be eliminated from $H(A_0 = 0)$ to give the familiar 'Coulomb gauge' Hamiltonian; in QCD, however, as pointed out by Gribov, the covariant derivative $\vec{D}$ has zero modes and cannot be inverted uniquely.

All these problems can be eliminated by working on a spatial lattice with time continuous, as was first done by Kogut and Susskind. The lattice i) regulates the number of variables, so that solving the Schrödinger equation becomes a well defined many-body problem, ii) non-perturbatively regulates the ultraviolet divergences, and iii) enables Gauss's law, which requires Ψ to be invariant under infinitesimal colour transformations, to be imposed trivially.

Without giving technical details, which may be found in Ref.[1], I indicate here some of the underlying ideas in our attempt to solve the lattice Schrödinger equation, using some concepts that should be familiar from the more usual 'Lagrangian' lattice approach, in which time is also discrete. Throughout, the discussion refers to 'pure' QCD without quarks.

In the 'Hamiltonian' approach (time continuous):

1) The wave function is a function of all possible Wilson loop variables and therefore satisfies Gauss's law.

2) In the 'strong coupling' limit (large lattice spacing), the Schrödinger equation can be solved perturbatively. To lowest order, the wave function is a constant, representing an 'empty lattice'; Wilson loops are then generated in higher orders. The physical limit is, however, at weak coupling.

The approach pioneered by Kogut and Susskind was to calculate to an as-high-as-possible order in strong coupling perturbation theory and then attempt to continue to weak coupling using Padé approximants. However, this approach is not reliable beyond intermediate couplings and furthermore yields only energy eigenvalues, not wave functions or matrix elements.

The approach we espouse, which we call the 'shifted coupled cluster method', has the advantage that not only does it yield wave functions and matrix elements, but it is intrinsically non-perturbative so that it can exhibit non-perturbative phenomena even in the simplest approximation.

It is generally believed that in passing from the strong to the weak coupling limit of QCD there is no phase transition. If so, for all couplings, the wave function of the vacuum may be expressed in terms of sums over all possible sites (n) on the lattice of all possible products of individually independent excitations of the strong coupling vacuum. The multiple excitations all exponentiate, so that we can write

$$\psi_0 = \exp(S) \,,$$

where

$$S = \sum_n S(n) \,,$$

and S(n) describes the excitations centred at n. Similarly, the wave functions of excited (glueball) states may be written

$$\psi_g = G\psi_0 \,,$$

where

$$G = \sum_n G(n)$$

describes excitations, relative to the vacuum, of a given spin and parity. Both S(n) and G(n) may be expanded in terms of any complete set of linearly independent 'linked clusters' of Wilson loops at fixed relative orientation and separation.

In lattice gauge theories, the electric field operator is represented by a vector in colour space E(L) associated with each link L on the lattice, while the magnetic field is represented by a vector $\mathbf{B}(P)$ associated with each plaquette P. The Hamiltonian takes the form

$$H = \sum_L \mathbf{E}(L)^2 + \sum_p \mathbf{B}(P)^2 \,.$$

In Schrödinger representation, E(L) is a first-order partial differential operator with respect to the link variables, out of which the Wilson loops in S and G are constructed. $\mathbf{B}(P)^2$ is — up to a constant which can be dropped — proportional to the sum of the two (directed) Wilson loops formed from the link variables on the sides of P. Thus S and G satisfy the second-order partial differential equations

$$\sum_L (\mathbf{E}^2(L)S + \mathbf{E}(L)S.\mathbf{E}(E)S) + \sum_p B^2(p) = E_0$$

and

$$\sum_L (\mathbf{E}^2(L)G + 2\mathbf{E}(L)S.\mathbf{E}(L)G) = \Delta E_g G \,,$$

where E_0 is the vacuum energy and $\Delta E_g = E_g - E_0$ is the mass gap.

The first step in our attempt to solve these (exact) equations is to choose a complete set of linearly independent linked clusters in which to expand S(n) and G(n). We choose 'shifted linked clusters', consisting of products of Wilson loops each minus their vacuum expectation value, which describe gauge-invariant fluctuations of the fields. It turns out that, to obtain results valid beyond the strong coupling regime, the use of shifted coupled clusters in the expansions is essential. The vacuum expectation values are calculated by a self-consistent iterative procedure using the Feynman–Hellmann theorem and avoiding any group integrals.

The second step is truncate the expansions for S(n) and G(n) according to both the geometrical size and the order of fluctuation of the shifted clusters. The underlying physical idea is that fluctuations on a scale larger than the vacuum correlation length (size of the glueball) should play a negligeable role in $\psi_0 G$. If this is correct, then the results obtained should converge to fixed limiting values as the size of the geometrical cut-off **j** is increased, at fixed lattice spacing.

The imposition of a geometrical cut-off on the size of the clusters of excitations that are retained in S is known in many-body theory as the coupled cluster method. It has been employed successfully in nuclear physics, condensed matter physics and in quantum chemistry (see Ref. [2] for a very clear and readable introduction).

In the case of lattice gauge theories, the advocated procedure is then to verify for fixed lattice spacing (**a**) that a limit is indeed reached as **j** increases, and then compare the limits for decreasing values of **a**. As **a** decreases, the expected scaling behaviour should appear. Once this happens, the wave functions should give a good picture of the physics on scales that are large compared to the 'pixel size' **a**, and the continuum values of masses and matrix elements can be obtained by using the scaling laws to extrapolate to **a** = 0.

We have tried out these ideas for SU(2) gauge theory in two space dimensions. Presumably the physics of SU(2) is similar to that of QCD, which is much more complicated to handle. The physics of systems in two- and three-space dimensions is, of course, different, but the mathematics is similar even though three dimensions are very much harder to handle, and exploratory calculations suggest that the lessons we have learned are relevant also for three dimensions.

Our first discovery was that the coefficients in the power series expansion of S(n) in terms of shifted variables decrease very rapidly with i) the order of the fluctuation, and ii) the geometrical size of the cluster. Furthermore, we can understand this result analytically as a geometrical effect. We therefore believe that this method can yield a good model of the vacuum wave function.

As far as the glueball wave function is concerned, we also find a rapid decrease with the order of the fluctuation but, as expected, the size of the loops that play a significant role — and hence the required magnitude of **j** — increases as the lattice spacing (**a**) decreases, corresponding to the fact that the glueball size, which is constant in physical units, increases in lattice units as **a** decreases. This may well spell the death of this method, as it has of related approaches in the past, since the number of loop variables increases very rapidly with **j**, making it impractical to go to small **a**. However, we find that out of the large number of loops we have retained in our calculations, only a very small number with simple geometries have significant amplitudes. If we could understand this result analytically, it might be possible to justify a scheme in which, for a given **a**, only a limited number of loops are retained as a first approximation.

In conclusion, the shifted coupled cluster method shows some promise. Further progress will depend on physics — in particular, on whether it is true that the glueball wave function may be represented to first approximation in terms of a limited number of loops. If so, our method would be worth pushing further, with more sophisticated computing techniques than used so far to do the algebra, and there is an obvious programme of work beyond studying SU(2) in two dimensions, including SU(3), three dimensions, introduction of Fermions, and extension to other theories such as U(1) — we have already studied Z_2, where the method yields some intriguingly impressive results.[3]

More generally, contemplation of Hamiltonian lattice QCD, with its direct 'physical' interpretation, illustrates what it might mean to 'solve' QCD. The effort invested in studying Hamiltonian lattice QCD has been negligible up to now compared to that which has gone into the Monte Carlo approach: perhaps there are methods other than the coupled cluster method that might be borrowed and adapted from many body theory. However, the success of these techniques will also depend on whether it is possible to construct good approximations in terms of a limited number of variables for small lattice spacings. If not, attempts to 'solve' QCD semi-analytically may be doomed to failure.

LARGER ACCELERATORS AND THE LONG-TERM FUTURE OF PARTICLE PHYSICS

The medium-term prospects for particle physics are bright in Europe[4] (cf: CERN fixed target programme, LEP2, HERA, Gran Sasso) and elsewhere (cf: FNAL, CESR, PEP II, TRISTAN II, SNO, Kamiokande). In the longer term, there are excellent reasons to think that major progress will be made in studying the region opened up by the LHC[4] in the p-p, Pb-Pb and, later, e-p, modes. In the not-too-distant future it should be possible to construct a linear collider that in the region up to 500 GeV would complement the pioneering role of the LHC.

In the longer term, our community will face great technical and fiscal/political challenges. It is now possible to imagine that a 1 TeV linear collider might be built, but a machine of 10 TeV is almost unimaginable (see Ref.[5] for a discussion of limits on future accelerators and references to the specialist literature). Even with all parameters at the limits that be can contemplated — e.g. nm spot sizes — the power needed to run such a machine would probably be prohibitive. With proton machines, the physics 'reach' increases very slowly with energy and luminosity (very roughly like $E^{0.66}L^{0.17}$), and it is very hard to imagine an increase of more than a factor of ten beyond the LHC in energy and luminosity.[5] The problem of dealing with the synchrotron losses sets the most severe limit, but more simply note that even a machine with the radius of the SSC (a strong political and financial upper limit?) and a magnetic field of 30 T (beyond the technical limit?) would have an energy of only 180 TeV, while radiation damage to the detectors puts a limit on possible increases in the luminosity.

These arguments suggest that an increase in reach of more than a factor of three or so beyond the LHC will be very hard to obtain, and such a small step would need a very powerful physics justification. Of course, similar statements turned out to be wrong in the past, so we must be very cautious, but it is possible that the end of accelerator-based particle physics can be glimpsed on the distant horizon beyond the LHC.

The fiscal and political challenges are also great, with budgets for pure science under increasing pressure world-wide. We should not, however, be too gloomy about this. With the facilities listed above, particle physics looks in good shape in the medium-term future. Furthermore, the applied/basic science pendulum will presumably someday swing back. We must help push the pendulum by never missing an opportunity to explain the value of basic science and by doing our best to share the excitement of our work with the general public — an obligation which has how become a necessity.

We must also seek to minimise the cost of our research by maximising the re-use of existing infrastructure and facilities, and to plan and organize the support for big projects on a global scale. The LHC provides an outstanding example of the re-use of existing infrastructure and facilities. If the current momentum for countries in other regions to join the project as partners is maintained, the LHC may become the first global megascience project, thereby setting a wonderful precedent for future projects in particle physics and in other fields.

REFERENCES

1. C.H. Llewellyn Smith and N.J. Watson, *Phys. Lett.* **B 302** (1993) p. 463.

2. R.F. Bishop and H.G. Kümmel, *Physics Today*, March 1987, p. 52.

3. C.H. Llewellyn Smith and N.J. Watson in preparation.

4. C.H. Llewellyn Smith, to be published in Proc. 1994 EPS Conference on Large Facilities in Physics.

5. C.H. Llewellyn Smith, in 'Perkins Conference', ed. R.J. Cashmore and G. Myatt, World Scientific (1995) p. 175.

CONTRIBUTORS

Federico Antinori
CERN-PPE
CH-1211 Geneve 23, Switzerland
E.mail: federico@axomg3.cern.ch

Jürgen Baacke
Inst. f. of Physik, Univ. Dortmund
D-44221 Dortmund, Germany
E.mail:
baacke@het.physik.uni-dortmund.de

Andre Bialas
Institute of Physics
Jagellonian University
ul. Raymonta 4, PL-30 059 Kraków 16
E.mail: bialas@amoco.saclay.cea.fr

Brigitte Buschbeck
Institut für Hochenergiephysik
Österreichische Akademie der Wiss.
Nikolsdorfergasse 18
A-1050 Wien, Austria
E.mail:
v2033dae@helios.edvz.univie.ac.at

Tarik Celik
Department of Physics Engineering
Hacettepe University
Beytepe 06532, Ankara, Turkey
E.mail: fcelik@eti.cc.hun.edu.tr

Masud Chaichian
SEFT University of Helsinki
Res. Inst. for High Energy Physics
Siltavuorenpenger 20 C
SF-00170 Helsinki
E.mail: chaichian@phcu.helsinki.fi

Subhasis Chattopadhyay
VECC/DAE 1/AF, Saltlake, Calcutta
700064 India
E.mail: sub@vecdec@veccal.ernet.in

Jean Cleymans
Theoretical Physics
University of Cape Town
Private Bag, Rondebosch 7700, Cape
R. South Africa
E.mail: cleymans@physci.uct.ac.za

Ivo Derado
Werner-Heisenberg-Institut
Max Planck-Institut für Physik
Postfach 40 12 12
Föhringer Ring 6
D-80805 München, Germany
E.mail: iwd@dmumpiwh.bitnet

Olivier Drapier
IPN, Univ. Claude Bernard Lyon-1
43, bd. du 11 Nov. 1918
F-69622 Villeurbanne Cedex, France
E.mail: drapier@frcpn11.in2p3.fr

John Ellis
CERN-TH
CH-1211 Genèva 23, Switzerland
E.mail: johne@cernvm.cern.ch

Hans-Thomas Elze
CERN-TH
CH-1211 Genèva 23, Switzerland
E.mail: elze@crnvma.cern.ch

Torleif Ericson
CERN-TH
CH-1211 Genèva 23, Switzerland
E.mail: ericson@cernvm.cern.ch

Evgenii L. Feinberg
P.N. Lebedev Physicsl Inst.
Leninsky Prospect 53
RU-117924 Moscow, Russia
E.mail: feinberg@lpi.ac.ru

Steven Frautschi
CALTECH
Ch. Lauritsen Lab.
Pasadena CA 91125, USA

Marek Gaździcki
Inst. für Kernphysik der Uni
August Euler Str. 6
D-60486 Frankfurt, Germany
E.mail:
marek@hvax.ikf.physik.uni-frankfurt.de

Klaus Geiger
CERN-TH
CH-1211 Genèva 23, Switzerland
E.mail: klaus@surya11.cern.ch

Walter Geist
CRN-Strasbourg
Division des Hautes Energies
BP 20 F-67037 Strasbourg cedex
France
E.mail: geist@vxcrna.cern.ch

Norman Glendenning
Nuclear Science Division
LBL, Berkeley, CA 94720, USA
E.mail: nkg@csa5.lbl.gov

Hans H. Gutbrod
CERN-PPE
CH-1211 Genèva 23, Switzerland
E.mail: gutbrod@vxwa80.cern.ch

Rolf Hagedorn
Ch. de la Ramaz
F-01630 Sergy St. Genis Pouilly
France

Remi Hakim
Observatoire de Paris
Section d'Astrophysique
F-92195 Meudon, France
E.mail: hakim@mesiob.obspm.fr

John Harris
Nuclear Science Division
LBL, Berkeley, CA 94720, USA
E.mail: jwharris@lbl.gov

Ulrich Heinz
Inst. Theo. Physik der Uni.
Postfach 397
D-93040 Regensburg, Germany
E.mail:
heinz@vax1.rz.uni-regensburg.d400.de

Rudi Hwa
Department of Physics
University of Oregon
Eugene, OR 97403-5203, USA
E.mail: hwa@oregon.uoregon.edu

Maurice Jacob
CERN-TH
CH-1211 Genèva 23, Switzerland
E.mail: jacob@cernvm.cern.fr

554

Larry Jones
Department of Physics
University of Michigan
Ann Arbor MI 48109-1120, USA
E.mail:
jones@mich.physics.lsa.umich.edu

Burkhard Kämpfer
Institut für Theoretische Physik
Technische Universität Dresden
Mommsenstr. 13, 01062 Dresden,
Germany
E.mail: Kaempfer@fz-rossendorf.de

Jean Letessier
LPTHE, Tour 24-14 5e et
Université Paris 7, 2 place Jussieu
F-75251 Paris Cedex 05, France
E.mail: jletes@lpthe.jussieu.fr

Peter Lipa
Department of Physics,
University of Arizona,
Tucson, AZ–85721, USA
E.mail: lipa@corelli.physics.arizona.edu

Chris Llewellyn-Smith
CERN
CH-1211 Genèva 23, Switzerland

Stefan Mashkevich
Division of High Energy Density Physics
Institute of Theoretical Physics
Ukrainian Academy of Sciences
UA-252143 Kiev 143, Ukraine
E.mail: MASH@phys.unit.no

Jean-Louis Meunier
INLN Univ. Nice, parc Valrose
F-06108 Nice Cedex 2, France
E.mail: meunier@doublon.unice.fr

Berndt Müller
Department of Physics
Duke University
Durhham NC 27706, USA
E.mail: muller@phy.duke.edu

Wolfgang Ochs
Werner-Heisenberg-Institut
Max Planck-Institut für Physik
Postfach 40 12 12, Föhringer Ring 6
D-80805 München, Germany
E.mail: wwo@dmumpiwh.bitnet

Grażyna Odyniec
Nuclear Science Division
LBL, Berkeley, CA 94720, USA
E.mail: odyniec@csa5.lbl.gov

Roland Omnes
Université de Paris XI
F-91405 Orsay cedex, France
E.mail: omnes@stat.th.u-psud.fr

Emanuele Quercigh
CERN
CH-1211 Genèva 23, Switzerland
E.mail: quercigh@cernvm.cern.ch

Johann Rafelski
Department of Physics
University of Arizona
Tucson AZ 85721, USA
E.mail: rafelski@ccit.arizona.edu

Johannes Ranft
Lab. Nationali de Frasati dell'INFN
CP 13, via E. Fermi 140
I-00044 Frascati (Roma), Italy
E.mail: ranft@cernvm.cern.ch

Jochen Rau
MPI-Kernphysik
Postfach 103980
D-69117 Heidelberg, Germany
E.mail: rau@eu10.mpi-hd.mpg.de

Krzysztof Redlich
Fakultät für Physik
Universtität Bielefeld
Postfach 86 40
D-33615 Bielefeld 1, Germany
E.mail: redlich@hrz.uni-bielefeld.de

Juris Reinfelds
Box 30001/Dept CS
New Mexico State University
Las Cruces, NM 88003-0001, USA
E.mail: juris@cs.nmsu.edu

Herbert Rohringer
Östr. Akademische Wissenschaft
Institut für Hochernergiephysik
Nikolsdorfgasse 18, A-1050 Wien
E.mail: HRO@cernvm.cern.ch

R. Santo
Inst. f. Kernphysik
Wilhelm-Klemm-Strasse 9
D-48149 Münster, Germany
E.mail: santo@vsikp0.uni-muenster.de

Ina Sarcevic
Department of Physics
Johns Hopkins Univ.
Baltimore, MD 21218, USA
E.mail: ina@ccit.arizona.edu

Helmut Satz
CERN-TH
CH-1211 Genèva 23, Switzerland
E.mail: satz@crnvma.cern.ch

Luigi Sertorio
Istituto Naxionale di Fisica Nucleare
Via Pietro Giuria, 1
I-10125 Torino, Italy

Yuri Sinyukov
Division of High Energy Density Physics
Institute of Theoretical Physics
Ukrainian Academmy of Sciences
UA-252 143 Kiev 143, Ukraine
E.mail: sinyukov@gluk.apc.org

Bo-Sture Skagerstam
Institute of Theoretical Physics
Chalmers University of Technology
S-412 96 Göteborg, Sweden
E.mail: tfebss@fy.chalmers.se

Johanna Stachel
Department of Physics
State University of New York (SUNY)
Stony Book NY 11794-3800, USA
E.mail:
stachel@nuclear.physics.sunysb.edu

Steve Steadman
MIT, Department of Physics
Cambridge MA 02139, USA
E.mail: sgs@irene.mit.edu

Reinhard Stock
Johann-Wolfgang-Goethe Universität
Institut für Kernphysik
August-Euler-Strasse 6,
D-60 486
Frankfurt/Main, Germany
E.mail: stock@ikf.uni-frankfurt.de

John Sullivan
P2, MS D456
LANL, Los Alamos, NM 87544, USA
E.mail: sullivan@p2hp4.lanl.gov

Mike Tannenbaum
Brookhaven National Laboratory
Department of Physics
Upton NY 11973, USA
E.mail: sapin@bnldag.ags.bnl.gov

C. Taylor
Department of Physics
Case Western Reserve University
Cleveland, OH 44106-7079, USA
E.mail: cct@po.cwru.edu

Ahmed Tounsi
LPTHE Tour 24-14, 5e et
Université Paris 7, 2 place Jussieu
F-75251 Paris Cedex 05, France
E.mail: tounsi@lpthe.jussieu.fr

Ludwig Turko
Institute of Theoretical Physics
University of Wroclaw
Plac Maxa Borna 9
PL-50 204 Wroclaw, Poland
E.mail: lturko@proton.ift.uni.wroc.pl

Gabriele Veneziano
CERN-TH
CH-1211 Genèva 23, Switzerland
E.mail: Veneziano@cernvm.cern.ch

Richard Weiner
Fachbereich Physik
Philipps-Universität Marburg
Mainzer Gasse 33, D-35037
Marburg an der Lahn, Germany
E.mail: weiner@vax.hrz.uni-marburg.de

Gena Zinoviev
Division of High Energy Density Physics
Institute of Theoretical Physics
Ukrainian Academmy of Sciences
UA-252 143 Kiev 143, Ukraine
E.mail: gezin@glas.apc.org

INDEX